深入浅出 MySQL
数据库开发、优化与管理维护
（第 3 版）

翟振兴　张恒岩　崔春华　黄荣　董骐铭　著

人民邮电出版社

北京

图书在版编目（CIP）数据

深入浅出MySQL：数据库开发、优化与管理维护：第3版 / 翟振兴等著. -- 北京：人民邮电出版社，2019.9（2023.2重印）
 ISBN 978-7-115-51539-1

Ⅰ．①深… Ⅱ．①翟… Ⅲ．①SQL语言－程序设计 Ⅳ．①TP311.132.3

中国版本图书馆CIP数据核字(2019)第126910号

内 容 提 要

本书源自网易公司多位资深数据库专家数年的经验总结和MySQL数据库的使用心得，在之前版本的基础之上，基于MySQL 5.7版本进行了内容升级，同时也对MySQL 8.0的重要功能进行了介绍。除了对原有内容的更新之外，本书还新增了作者在高可用架构、数据库自动化运维，以及数据库中间件方面的实践和积累。

本书分为"基础篇""开发篇""优化篇""管理维护篇"和"架构篇"5个部分，共32章。基础篇面向MySQL的初学者，介绍了MySQL的安装与配置、SQL基础、MySQL支持的数据类型、MySQL中的运算符、常用函数等内容。开发篇面向的是MySQL设计和开发人员，内容涵盖了表类型（存储引擎）的选择、选择合适的数据类型、字符集、索引的设计和使用、开发常用数据库对象、事务控制和锁定语句、SQL中的安全问题、SQL Mode及相关问题、MySQL分区等。优化篇针对的是开发人员和数据库管理人员，内容包括SQL优化、锁问题、优化MySQL Server、磁盘I/O问题、应用优化、PS/SYS数据库、故障诊断等内容。管理维护篇适合数据库管理员阅读，介绍了MySQL高级安装和升级、MySQL中的常用工具、MySQL日志、备份与恢复、MySQL权限与安全、MySQL监控、MySQL常见问题和应用技巧、自动化运维系统的开发等内容。架构篇主要面向高级数据库管理人员和数据库架构设计师，内容包括MySQL复制、高可用架构、MySQL中间件等内容。

本书内容实用，覆盖广泛，讲解由浅入深，还提供了大量来自一线的工作实例，进一步提升了本书的实战性和可操作性。本书适合数据库管理人员、数据库开发人员、系统维护人员、数据库初学者及其他数据库从业人员阅读，也适合用作大中专院校相关专业师生的参考用书和相关培训机构的培训教材。

◆ 著　　翟振兴 张恒岩 崔春华 黄荣 董骐铭
　责任编辑　傅道坤
　责任印制　焦志炜

◆ 人民邮电出版社出版发行　北京市丰台区成寿寺路11号
　邮编　100164　电子邮件　315@ptpress.com.cn
　网址　http://www.ptpress.com.cn
　北京七彩京通数码快印有限公司印刷

◆ 开本：787×1092　1/16
　印张：43.75　　　　　　　2019年9月第1版
　字数：1 070千字　　　　 2023年2月北京第11次印刷

定价：158.00元

读者服务热线：(010)81055410　印装质量热线：(010)81055316
反盗版热线：(010)81055315
广告经营许可证：京东市监广登字20170147号

作者简介

翟振兴，网易技术专家，毕业于清华大学软件学院，2005年入职网易，经历了网易多个核心系统的数据库设计和运维工作，对高并发下的数据库架构变迁有着深刻的理解，目前主要对自动化运维、大数据、NewSQL等新技术有着较多的兴趣和研究。

张恒岩，网易技术经理，2010年毕业于北京科技大学，在网易DBA组工作9年，负责过网易内部多个核心数据库的运维以及数据库自动化运维系统的设计和开发。在数据库架构设计、性能优化、故障诊断以及自动化运维等方面有丰富的经验。

崔春华，网易资深DBA，有10多年的数据库技术领域从业经验，深刻理解数据库原理并具有丰富的实战经验，拥有Oracle 9i OCP证书，曾先后从业于石化、电信行业，加入网易后，负责网易多个核心业务的数据库设计和维护工作，擅长数据库优化、故障诊断、架构设计，主要参与Cetus开源项目和私有云RDS底层设计开发。

黄荣，网易资深数据库工程师，毕业于北京邮电大学，曾参与网易多个核心业务的数据库设计和开发。至今有12年数据库相关工作经验，尤其擅长高可用架构设计、数据库优化及故障诊断。现专注于研究MySQL内部原理、探索MySQL新技术和开发自动化运维平台。

董骐铭，网易高级数据库工程师，毕业于北京航空航天大学，负责网易数据库运维平台的研发工作，参与了Cetus及其管理工具项目的开发，专注于自动化与智能运维相关技术的探索与实现。

前言

　　MySQL 是由 David Axmark、Allan Larsson 和 Michael Widenius 3 名瑞典人于 20 世纪 90 年代开发的一个关系型数据库。最初,他们的目的是用自己的 ISAM(Indexed Sequential Access Method,索引顺序存取方法)和 mSQL(mini SQL,一种轻量级 SQL 数据库引擎技术)来连接访问表格,但后来发现 mSQL 的速度和灵活性不能满足需求,于是他们开发了几乎与 mSQL API 接口相同的数据库引擎,并用创始人之一 Michael Widenius 女儿 My 的名字命名,这就是 MySQL 的来由。

　　说到 MySQL 就不得不提到开源软件。MySQL 在设计之初就考虑了以后引入第三方代码的方便性,并于 2000 年开始采用 GNU GPL(General Public License)许可协议,使自己成了开源软件的一分子。开源战略对 MySQL 的发展和广泛应用,可以说起到了至关重要的作用。从 MySQL 的历史就可以看出,它最早起源于开源软件 mSQL,并从中借鉴了许多东西。不仅开发 MySQL 用到了许多开源工具,而且 MySQL 的许多重要组件都直接来自其他第三方的贡献,如 BDB 存储引擎来自 Berkeley DB,其具有里程碑意义的 InnoDB 数据库存储引擎也是来自芬兰 Innobase OY 公司的贡献。

　　进入 21 世纪,MySQL 的发展步入了快车道。MySQL 自 2001 年开始引入 InnoDB 存储引擎,并于 2002 年正式宣布 MySQL 全面支持事务,满足事务 ACID 属性(Atomicity,原子性;Consistent,一致性;Isolation,隔离性;Durable,持久性),并支持外键约束,使 MySQL 具备了支持关键应用的最基本特性。2003 年,MySQL 4.0 发布,开始支持集合操作 UNION。2004 年,MySQL 4.1 发布,增加了对子查询的支持。2005 年,MySQL 5.0 发布,增加了对视图(View)、数据库存储过程(Stored Procedure)、触发器(Trigger)、服务器端游标(Cursor),以及分布式事务协议 XA 等高级特性的支持。再加上从 3.23.15 就开始支持的复制特性,至此,MySQL 从功能上已经具备了支持企业级应用的主要特性。在实际应用方面,LAMP(Linux + Apache + MySQL + Perl/PHP/Python)也逐渐成了 IT 业广泛使用的 Web 应用架构。

　　大家知道 ANSI/ISO SQL 是公认的关系数据库标准。从 SQL 标准的符合性来说,MySQL 不仅无法跟成熟的商业数据库相比,在开源数据库中也远不是最好的,比如 PostgreSQL 就是业界公认的 ANSI/ISO SQL 标准符合性最好的开源数据库,MySQL 直到 5.0 版本才支持的一些特性,PostgreSQL 早都实现了。既然如此,为什么 MySQL 却在开源数据库中独占鳌头呢? 根本的原因就在于它的性能!

有专门机构的调查研究显示，许多数据库提供的功能特性，只有 40%会被用户经常使用，而一些复杂的高级功能特性不仅会增加系统的复杂性，而且往往还会引起性能问题。PostgreSQL 是美国加州大学伯克利分校以教学为目的开发的数据库系统，以追求功能实现的"完美"为首要目标，虽然在标准的遵从性上比 MySQL 领先，但性能一直是其短板，很难支持较大的应用。而 MySQL 的开发者在性能与标准的取舍上，一直坚持性能优先的原则，从不为追求标准的符合性而牺牲性能。SQL 标准符合性差是 MySQL 的弱点，但通过上述策略保证了 MySQL 在性能方面的优势。这就是 MySQL 在互联网行业非常流行的另一个重要原因，因为 Web 应用往往需要支持大量的数据和并发请求，性能常常是首要因素。

随着 MySQL 功能的不断完善，其性能不断提高，可靠性不断增强，从 2005 年开始，我们陆续将一些重要数据库迁移到 MySQL。虽然相对于 Oracle 来说，MySQL 比较简单，管理维护相对容易（这也是 MySQL 的另外一个优势），但在迁移及其后的管理维护过程中，我们也经常遇到一些问题，例如，MySQL 提供了许多存储引擎，这些存储引擎各有特点，在实际应用中应该怎样来选择？MySQL 出现了性能问题，应该如何来诊断和优化？在数据安全方面，究竟需要注意些什么？MySQL 的锁机制有什么特点，如何减少锁冲突，提高并发度？

遇到诸如此类的问题，自然会想到查阅 MySQL 文档、上网搜索、到论坛找类似问题的答案或寻求帮助等。通过上述途径当然可以解决许多问题，但却需要花费大量的精力和时间，效率很低。因为我们发现 MySQL 的文档很"精练"，也很零碎，远没有 Oracle 的文档那么详细与系统；网上一搜，结果可能数以万计，面对浩如烟海的网页，要找出真正有用的信息决非易事（搜索引擎还有许多改进的余地）；至于论坛上的答案，又往往是五花八门，让人无所适从。我们作为专职 DBA（数据库管理员）尚且如此，其他开发、维护人员可能就感到更困难了。而且，不同的 DBA 或开发人员遇到同一个问题，可能还要再次去寻求解决方案，造成不必要的重复劳动。

为改变这种状况，我们决定将 DBA 平时使用 MySQL 积累的经验、解决问题的方法和思路，以及我们对 MySQL 的认识等整理出来，编写一本《MySQL 实用手册》，供 DBA 组及公司其他同事参考。

在编写过程中，我们根据自己的经验列出了 MySQL 开发、管理过程中可能遇到的一些问题，收集了以前解决 MySQL 问题的方法，形成了实用手册的基本内容框架。在此基础上，我们又研究了 MySQL 官方手册，筛选了比较重要、实用但 MySQL 手册讲得不够详细或内容过于零散的部分作为补充。为力求准确，我们还专门做了许多测试，比如有关 MySQL 锁的测试、事务完整性的测试等。基于实用为主的原则，《MySQL 实用手册》的第一稿篇幅不大，包括"开发篇""优化篇"和"管理篇"3 个部分。"开发篇"主要介绍了与 MySQL 数据库开发设计相关的一些问题，包括存储引擎选择原则、如何选择合适的数据类型、不同字符集的特点及设置、索引的设计原则、SQL 注入的类型，以及程序设计实现中如何防范 SQL 注入类安全漏洞等。"优化篇"首先介绍了 MySQL 数据库优化调整的一般步骤和方法，随后分别就索引问题、SQL 优化、数据库对象调整、锁问题、MySQL 关键性能参数设置、I/O 优化、应用程序优化等做了讨论，并介绍了两个简单实用的优化命令，特别对锁问题做了比较系统、详细的介绍和讨论。"管理篇"除备份恢复等基本内容外，还介绍了 MySQL 安全配置管理应注意的各个方面，以及管理维护中一些常用的命令和小技巧。

《MySQL 实用手册》第一稿出来后，我们发放给了公司的一些同事，大家反馈内容很实用，对 MySQL 的开发管理很有帮助。得到这个评价，我们心里已经很是欣慰了。当有同事建议我们将其出版时，更是超出了我们的预期，开始只是随口答应了一下，并没敢当真。后来，热心的同事替我们与出版社取得了联系，出版社看过内容介绍和提纲后，觉得内容不错，做一些补充和修改就可以出版。

得到这个回复，我们非常高兴。但高兴之余，心里也很忐忑。在我们的概念中，出书都是作家、专家，及各类名人的"专利"，我们这些无名之辈，有这个资格吗？此时，上级领导的支持、同事的不断鼓励，给了我们信心。虽然我们不是什么专家，写不出多么高深的内容，但作为工作在数据库开发管理第一线的工程师，遇到的实际问题可能更多、更具体，写出来的东西也许更实用。MySQL 是一个开源数据库，开源的精神就是分享和交流，基于这一点，我们的顾虑就少了，也更加坦然了。在《MySQL 实用手册》第一稿的基础上，根据同事和出版社的建议，对内容做了一些补充和修订，增加了"基础篇"，以利于初次接触 MySQL 的读者阅读；在"管理篇"中补充了有关复制、日志管理和 MySQL 集群的介绍，使本书的内容更加丰富、完善。在内容编排上，我们基本遵循由易到难、循序渐进的原则，最后就形成了《深入浅出 MySQL：数据库开发、优化与管理维护》一书。

本书的第 1 版于 2008 年出版后得到了不少读者的支持和肯定，这给了我们极大的鼓舞。2013 年，我们又基于 MySQL 5.5 进行了大量修订，推出了第 2 版。这些年 MySQL 5.7、MySQL 8.0 相继发布，书中的一些内容已然过时，同时，我们近几年在解决 MySQL 高并发、高可用、可扩展和自动化维护方面积累了不少经验，也希望能分享给读者。于是我们又聚在一起修订出第 3 版，以期回馈读者长久以来对我们的喜爱和支持。

除了补充版本升级相关内容之外，本次修订的重点是分享我们在高可用架构、数据库自动化运维，以及数据库中间件方面的一些实践和积累，具体如下。

- 本书基于 MySQL 5.7 版本进行内容升级，同时也对 MySQL 8.0 的重要功能进行了介绍，因此本书适合使用 MySQL 5.7/8.0 的读者（本书中 MySQL 5.7 和 8.0 版本的测试环境版本分别为 5.7.22 和 8.0.11）。
- 数据库服务云化、容器化，以及管理维护自动化，是数据库运维的必然趋势。本书增加了第 29 章（自动化运维系统的开发），介绍了作者在数据库自动化运维开发方面的实践，为大家打造自己的自动化运维平台提供了一个参考。
- 架构篇是这次修订的重点，作者全面改写了第 30 章（MySQL 复制）和第 31 章（高可用架构）两章，并增加了第 32 章（MySQL 中间件），介绍了作者在数据库中间件研发方面的经验。

本书组织结构

本书第 3 版分为基础篇、开发篇、优化篇、管理维护篇、架构篇 5 个部分，共 32 章。本书的具体章节安排如下。

- 第一部分，基础篇（第 1~5 章）：主要面向 MySQL 的初学者，包括 MySQL 的安装与配置、SQL 基础、MySQL 支持的数据类型、MySQL 中的运算符、常用函数等内容。通过这部分内容的学习，读者可以熟悉 MySQL 基本的安装和相关使用方法。
- 第二部分，开发篇（第 6~14 章）：主要面向 MySQL 的设计和开发人员，包括表类型（存储引擎）的选择、选择合适的数据类型、字符集、索引的设计和使用、开发常用数据库对象、事务控制和锁定语句、SQL 中的安全问题、SQL Mode 及相关问题、MySQL 分区等内容。通过这部分内容的学习，读者可以了解 MySQL 设计和开发中需要关注的问题。

- 第三部分，优化篇（第 15~21 章）：主要面向开发人员和数据库管理员，包括 SQL 优化、锁问题、优化 MySQL Server、磁盘 I/O 问题、应用优化、PS/SYS 数据库、故障诊断等内容。通过这部分内容的学习，读者可以了解 MySQL 中需要优化的对象和常用的优化方法。
- 第四部分，管理维护篇（第 22~29 章）：主要面向数据库管理员，包括 MySQL 高级安装和升级、MySQL 中的常用工具、MySQL 日志、备份与恢复、MySQL 权限与安全、MySQL 监控、MySQL 常见问题和应用技巧、自动化运维系统的开发等内容。通过这部分内容的学习，读者可以了解在 MySQL 中常用的管理维护方法。
- 第五部分，架构篇（第 30~32 章）：主要面向高级数据库管理人员和数据库架构设计师，包括 MySQL 复制、高可用架构、MySQL 中间件等内容。通过这部分内容的学习，读者可以了解一些 MySQL 的高级应用。

致谢

非常感谢网易公司管理层对本书的支持，尤其感谢本书前两版的第一作者、技术部技术总监唐汉明先生的大力支持。

回首来时路，十年踪迹十年心。编写和修订本书的过程，既是我们对 MySQL 再学习、再认识的过程，也是我们见证 MySQL 应用在各行各业高速发展的阶段。MySQL 有许多优势，当然也还存在许多不足，但现在的 MySQL 应该说是值得信赖的，伴随着开源软件的发展，相信 MySQL 的未来也是值得期待的。这也正是我们把 MySQL 介绍给大家的原因，希望能对大家有所帮助。

由于我们水平有限，书中难免会有谬误，欢迎大家通过微信公众号（lede_dba）和我们进行沟通并给予指正。

资源与支持

本书由异步社区出品，社区（https://www.epubit.com/）为您提供相关资源和后续服务。

提交勘误

作者和编辑尽最大努力来确保书中内容的准确性，但难免会存在疏漏。欢迎您将发现的问题反馈给我们，帮助我们提升图书的质量。

当您发现错误时，请登录异步社区，按书名搜索，进入本书页面，点击"提交勘误"，输入勘误信息，点击"提交"按钮即可。本书的作者和编辑会对您提交的勘误进行审核，确认并接受后，您将获赠异步社区的 100 积分。积分可用于在异步社区兑换优惠券、样书或奖品。

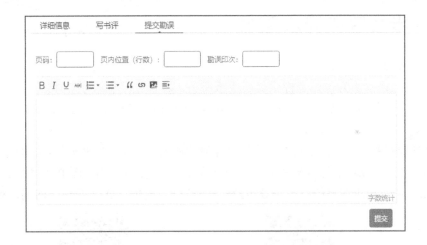

扫码关注本书

扫描下方二维码，您将会在异步社区微信服务号中看到本书信息及相关的服务提示。

与我们联系

我们的联系邮箱是 contact@epubit.com.cn。

如果您对本书有任何疑问或建议，请您发邮件给我们，并请在邮件标题中注明本书书名，以便我们更高效地做出反馈。

如果您有兴趣出版图书、录制教学视频，或者参与图书翻译、技术审校等工作，可以发邮件给我们；有意出版图书的作者也可以到异步社区在线提交投稿（直接访问www.epubit.com/selfpublish/submission 即可）。

如果您是学校、培训机构或企业，想批量购买本书或异步社区出版的其他图书，也可以发邮件给我们。

如果您在网上发现有针对异步社区出品图书的各种形式的盗版行为，包括对图书全部或部分内容的非授权传播，请您将怀疑有侵权行为的链接发邮件给我们。您的这一举动是对作者权益的保护，也是我们持续为您提供有价值的内容的动力之源。

关于异步社区和异步图书

"异步社区"是人民邮电出版社旗下IT专业图书社区，致力于出版精品IT技术图书和相关学习产品，为作译者提供优质出版服务。异步社区创办于2015年8月，提供大量精品IT技术图书和电子书，以及高品质技术文章和视频课程。更多详情请访问异步社区官网 https://www.epubit.com。

"异步图书"是由异步社区编辑团队策划出版的精品IT专业图书的品牌，依托于人民邮电出版社近30年的计算机图书出版积累和专业编辑团队，相关图书在封面上印有异步图书的LOGO。异步图书的出版领域包括软件开发、大数据、AI、测试、前端、网络技术等。

异步社区

微信服务号

目录

第一部分 基础篇

第1章 MySQL的安装与配置 2
1.1 MySQL的下载 2
   1.1.1 在Windows平台下下载MySQL 3
   1.1.2 在Linux平台下下载MySQL 3
1.2 MySQL的安装 5
   1.2.1 在Windows平台下安装MySQL 5
   1.2.2 在Linux平台下安装MySQL 8
1.3 MySQL的配置 12
   1.3.1 Windows平台下配置MySQL 12
   1.3.2 Linux平台下配置MySQL 13
1.4 启动和关闭MySQL服务 13
   1.4.1 在Windows平台下启动和关闭MySQL服务 13
   1.4.2 在Linux平台下启动和关闭MySQL服务 13
1.5 小结 14

第2章 SQL基础 15
2.1 SQL简介 15
2.2 （My）SQL使用入门 15
   2.2.1 SQL分类 15
   2.2.2 DDL语句 16
   2.2.3 DML语句 23
   2.2.4 DCL语句 33
2.3 帮助的使用 34
   2.3.1 按照层次看帮助 34
   2.3.2 快速查阅帮助 35
2.4 查询元数据信息 36
2.5 小结 37

第3章 MySQL支持的数据类型 38
3.1 数值类型 38
3.2 日期时间类型 43
3.3 字符串类型 49
   3.3.1 CHAR和VARCHAR类型 50
   3.3.2 BINARY和VARBINARY类型 51
   3.3.3 ENUM类型 51
   3.3.4 SET类型 52
3.4 JSON类型 52
3.5 小结 54

第4章 MySQL中的运算符 55
4.1 算术运算符 55
4.2 比较运算符 56
4.3 逻辑运算符 59
4.4 位运算符 60
4.5 运算符的优先级 61
4.6 小结 62

第5章 常用函数 63
5.1 字符串函数 63
5.2 数值函数 66
5.3 日期和时间函数 68

5.4 流程函数 71
5.5 JSON 函数 73
    5.5.1 创建 JSON 函数 74
    5.5.2 查询 JSON 函数 75
    5.5.3 修改 JSON 的函数 79
    5.5.4 查询 JSON 元数据函数 81
    5.5.5 JSON 工具函数 83
5.6 窗口函数 86
    5.6.1 ROW_NUMBER() 87
    5.6.2 RANK()/DENSE_RANK() 89
    5.6.3 PERCENT_RANK()/CUME_DIST() 89
    5.6.4 NTILE(N) 90
    5.6.5 NTH_VALUE(expr,N) 91
    5.6.6 LAG(expr,N)/LEAD(expr,N) 91
    5.6.7 FIRST_VALUE（expr）/LAST_VALUE(expr) 92
    5.6.8 聚合函数作为窗口函数 93
5.7 其他常用函数 93
5.8 小结 95

第二部分 开发篇

第 6 章 表类型（存储引擎）的选择 98
6.1 MySQL 存储引擎概述 98
6.2 各种存储引擎的特性 100
    6.2.1 MyISAM 101
    6.2.2 InnoDB 102
    6.2.3 MEMORY 108
    6.2.4 MERGE 109
    6.2.5 TokuDB 111
6.3 如何选择合适的存储引擎 112
6.4 小结 113

第 7 章 选择合适的数据类型 114
7.1 CHAR 与 VARCHAR 114
7.2 TEXT 与 BLOB 115
7.3 浮点数与定点数 118
7.4 日期类型选择 120
7.5 小结 120

第 8 章 字符集 121
8.1 字符集概述 121
8.2 Unicode 简述 121
8.3 汉字及一些常见字符集 123
8.4 怎样选择合适的字符集 124
8.5 MySQL 支持的字符集简介 125
8.6 MySQL 字符集的设置 126
    8.6.1 服务器字符集和排序规则 126
    8.6.2 数据库字符集和排序规则 127
    8.6.3 表字符集和排序规则 127
    8.6.4 列字符集和排序规则 128
    8.6.5 连接字符集和排序规则 128
8.7 字符集的修改步骤 129
8.8 小结 129

第 9 章 索引的设计和使用 130
9.1 索引概述 130
9.2 设计索引的原则 131
9.3 索引设计的误区 132
9.4 索引设计的一般步骤 132
9.5 BTREE 索引与 HASH 索引 133
9.6 索引在 MySQL 8.0 中的改进 134
    9.6.1 不可见索引 134
    9.6.2 倒序索引 135
9.7 小结 136

第 10 章 开发常用数据库对象 137
10.1 视图 137

10.1.1 什么是视图　137
10.1.2 视图操作　137
10.1.3 创建或者修改视图　137
10.1.4 删除视图　139
10.1.5 查看视图　139
10.2 存储过程和函数　140
10.2.1 什么是存储过程和函数　141
10.2.2 存储过程和函数的相关操作　141
10.2.3 创建、修改存储过程或者函数　141
10.2.4 删除存储过程或者函数　144
10.2.5 查看存储过程或者函数　144
10.2.6 变量的使用　146
10.2.7 定义条件和处理　146
10.2.8 光标的使用　148
10.2.9 流程控制　149
10.2.10 事件调度器　152
10.3 触发器　155
10.3.1 创建触发器　155
10.3.2 删除触发器　157
10.3.3 查看触发器　157
10.3.4 触发器的使用　158
10.4 小结　159

第 11 章 事务控制和锁定语句　160

11.1 LOCK TABLES 和 UNLOCK TABLES　160
11.2 事务控制　161
11.3 分布式事务的使用　166
11.3.1 分布式事务的原理　166
11.3.2 分布式事务的语法　166
11.3.3 存在的问题　168
11.4 小结　171

第 12 章 SQL 中的安全问题　172

12.1 SQL 注入简介　172

12.2 应用开发中可以采取的应对措施　173
12.2.1 PrepareStatement+Bind-Variable　173
12.2.2 使用应用程序提供的转换函数　174
12.2.3 自己定义函数进行校验　174
12.3 小结　175

第 13 章 SQL Mode 及相关问题　176

13.1 MySQL SQL Mode 简介　176
13.2 SQL Mode 的常见功能　178
13.3 常用的 SQL Mode　180
13.4 SQL Mode 在迁移中如何使用　182
13.5 小结　183

第 14 章 MySQL 分区　184

14.1 分区概述　184
14.2 分区类型　185
14.2.1 RANGE 分区　187
14.2.2 LIST 分区　188
14.2.3 COLUMNS 分区　189
14.2.4 HASH 分区　192
14.2.5 KEY 分区　195
14.2.6 子分区　196
14.2.7 MySQL 分区处理 NULL 值的方式　197
14.3 分区管理　199
14.3.1 RANGE 与 LIST 分区管理　199
14.3.2 HASH 与 KEY 分区管理　205
14.3.3 交换分区　206
14.4 小结　208

第三部分　优化篇

第 15 章 SQL 优化　210

15.1 优化 SQL 语句的一般步骤　210

15.1.1 通过 show status 命令了解各种 SQL 的执行频率　210
15.1.2 定位执行效率较低的 SQL 语句　211

15.1.3 通过 EXPLAIN 分析低效 SQL 的执行计划 211
15.1.4 通过 show profile 分析 SQL 216
15.1.5 通过 trace 分析优化器如何选择执行计划 219
15.1.6 确定问题并采取相应的优化措施 220

15.2 索引问题 220
15.2.1 索引的存储分类 220
15.2.2 MySQL 如何使用索引 222
15.2.3 查看索引使用情况 231

15.3 两个简单实用的优化方法 231
15.3.1 定期分析表和检查表 232
15.3.2 定期优化表 233

15.4 常用 SQL 的优化 233
15.4.1 大批量插入数据 234
15.4.2 优化 INSERT 语句 235
15.4.3 优化 ORDER BY 语句 235
15.4.4 优化 GROUP BY 语句 239
15.4.5 优化 JOIN 操作 239
15.4.6 优化嵌套查询 243
15.4.7 MySQL 如何优化 OR 条件 245
15.4.8 优化分页查询 246
15.4.9 使用 SQL 提示 248

15.5 直方图 250
15.5.1 什么是直方图 250
15.5.2 直方图的分类 251
15.5.3 直方图实例应用 252
15.5.4 直方图小结 256

15.6 使用查询重写 256

15.7 常用 SQL 技巧 259
15.7.1 正则表达式的使用 259
15.7.2 巧用 RAND() 提取随机行 261
15.7.3 利用 GROUP BY 的 WITH ROLLUP 子句 262
15.7.4 用 BIT GROUP FUNCTIONS 做统计 263
15.7.5 数据库名、表名大小写问题 265
15.7.6 使用外键需要注意的问题 265

15.8 小结 266

第 16 章 锁问题 267

16.1 MySQL 锁概述 267

16.2 MyISAM 表锁 268
16.2.1 查询表级锁争用情况 268
16.2.2 MySQL 表级锁的锁模式 268
16.2.3 如何加表锁 269
16.2.4 并发插入（Concurrent Inserts）271
16.2.5 MyISAM 的锁调度 272

16.3 InnoDB 锁问题 273
16.3.1 背景知识 273
16.3.2 获取 InnoDB 行锁争用情况 275
16.3.3 InnoDB 的行锁模式及加锁方法 276
16.3.4 InnoDB 行锁实现方式 279
16.3.5 Next-Key 锁 283
16.3.6 恢复和复制的需要，对 InnoDB 锁机制的影响 284
16.3.7 InnoDB 在不同隔离级别下的一致性读及锁的差异 287
16.3.8 什么时候使用表锁 288
16.3.9 关于死锁 289

16.4 小结 294

第 17 章 优化 MySQL Server 296

17.1 MySQL 体系结构概览 296

17.2 MySQL 内存管理及优化 298
17.2.1 内存优化原则 298
17.2.2 MyISAM 内存优化 298
17.2.3 InnoDB 内存优化 301
17.2.4 调整用户服务线程排序缓存区 305

17.3 InnoDB log 机制及优化 305
17.3.1 InnoDB 重做日志 305
17.3.2 innodb_flush_log_at_trx_commit 的设置 306
17.3.3 设置 innodb_log_file_size，控制检查点 307
17.3.4 调整 innodb_log_buffer_size 308

17.4 调整 MySQL 并发相关的参数 308
17.4.1 调整 max_connections，提高并发连接 308

17.4.2 调整 back_log 309
17.4.3 调整 table_open_cache 309
17.4.4 调整 thread_cache_size 309
17.4.5 innodb_lock_wait_timeout 的设置 309
17.5 持久化全局变量 309
17.6 使用资源组 310
17.7 小结 312

第 18 章 磁盘 I/O 问题 313
18.1 使用固态硬盘 313
18.2 使用磁盘阵列 314
　18.2.1 常见 RAID 级别及其特性 314
　18.2.2 如何选择 RAID 级别 315
18.3 虚拟文件卷或软 RAID 315
18.4 使用 Symbolic Links 分布 I/O 315
18.5 禁止操作系统更新文件的 atime 属性 316
18.6 调整 I/O 调度算法 316
18.7 RAID 卡电池充放电问题 318
　18.7.1 什么是 RAID 卡电池充放电 318
　18.7.2 RAID 卡缓存策略 319
　18.7.3 如何应对 RAID 卡电池充放电带来的 I/O 性能波动 321
18.8 NUMA 架构优化 322
18.9 小结 325

第 19 章 应用优化 326
19.1 优化数据表的设计 326
　19.1.1 优化表的数据类型 326
　19.1.2 通过拆分提高表的访问效率 328
　19.1.3 逆规范化 329
19.2 数据库应用优化 330
　19.2.1 使用连接池 330
　19.2.2 减少对 MySQL 的访问 330
　19.2.3 负载均衡 331
19.3 小结 331

第 20 章 PS/SYS 数据库 332
20.1 Performance Schema 库 332
　20.1.1 如何开启 PS 库 332
　20.1.2 PS 库的表 333
20.2 SYS 库 335
　20.2.1 SYS 库的对象 335
　20.2.2 SYS 对象的实际应用 336
20.3 小结 340

第 21 章 故障诊断 341
21.1 故障诊断和处理的原则 341
21.2 故障处理一般流程 343
　21.2.1 故障发现 343
　21.2.2 故障定位 345
　21.2.3 故障解决 346
21.3 典型故障案例 349
　21.3.1 案例 1 349
　21.3.2 案例 2 353
21.4 小结 356

第四部分 管理维护篇

第 22 章 MySQL 高级安装和升级 358
22.1 Linux/UNIX 平台下的安装 358
　22.1.1 安装包比较 358
　22.1.2 安装二进制包 359
　22.1.3 安装源码包 359
　22.1.4 参数设置方法 360
22.2 升级 MySQL 361
22.3 小结 363

第 23 章 MySQL 中的常用工具 364
23.1 MySQL 官方工具 364
　23.1.1 mysql（客户端连接工具） 364
　23.1.2 mysqladmin（MySQL 管理工具） 371
　23.1.3 mysqlbinlog（日志管理工具） 371

23.1.4　mysqlcheck（表维护工具）　378
23.1.5　mysqldump（数据导出工具）　380
23.1.6　mysqlpump（并行的数据导出工具）　384
23.1.7　mysqlimport（数据导入工具）　385
23.1.8　mysqlshow（数据库对象查看工具）　385
23.1.9　perror（错误代码查看工具）　387
23.1.10　MySQL Shell　387

23.2　Percona 工具包　390
23.2.1　pt-archiver（数据归档工具）　391
23.2.2　pt-config-diff（参数对比工具）　393
23.2.3　pt-duplicate-key-checker（检查冗余索引工具）　394
23.2.4　pt-find（查找工具）　395
23.2.5　pt-heartbeat（监控主从延迟工具）　395
23.2.6　pt-kill（杀死会话工具）　397
23.2.7　pt-online-schema-change（在线修改表结构工具）　397
23.2.8　pt-query-digest（SQL 分析工具）　399
23.2.9　pt-table-checksum（数据检验工具）　401
23.2.10　pt-table-sync（数据同步工具）　402

23.3　小结　403

第 24 章　MySQL 日志　404

24.1　错误日志　404
24.2　二进制日志　405
24.2.1　日志的位置和格式　405
24.2.2　日志的读取　406
24.2.3　日志的删除　407
24.2.4　日志的事件　411
24.2.5　日志闪回　412
24.3　查询日志　415
24.3.1　日志的位置和格式　415
24.3.2　日志的读取　416
24.4　慢查询日志　416
24.4.1　文件位置和格式　416
24.4.2　日志的读取　417
24.4.3　Anemometer 简介　419
24.5　小结　421

第 25 章　备份与恢复　422

25.1　备份/恢复策略　422
25.2　逻辑备份和恢复　422
25.2.1　备份　423
25.2.2　完全恢复　425
25.2.3　基于时间点恢复　427
25.2.4　基于位置恢复　427
25.2.5　并行恢复　427
25.3　物理备份和恢复　428
25.3.1　冷备份和热备份　428
25.3.2　MyISAM 存储引擎的热备份　429
25.3.3　InnoDB 存储引擎的热备份　429
25.4　表的导入和导出　438
25.4.1　导出　438
25.4.2　导入　442
25.5　小结　445

第 26 章　MySQL 权限与安全　446

26.1　MySQL 权限管理　446
26.1.1　权限系统的工作原理　446
26.1.2　权限表的存取　446
26.1.3　账号管理　449
26.2　MySQL 安全问题　461
26.2.1　操作系统相关的安全问题　461
26.2.2　数据库相关的安全问题　463
26.3　其他安全设置选项　471
26.3.1　密码插件　471
26.3.2　safe-user-create　472
26.3.3　表空间加密　473
26.3.4　skip-grant-tables　474
26.3.5　skip-networking　474
26.3.6　skip-show-database　475
26.4　小结　475

第 27 章　MySQL 监控　476

27.1　如何选择一个监控方案　476
27.1.1　选择何种监控方式　476
27.1.2　如何选择适合自己的监控工具　477
27.2　常用的网络监控工具　477
27.2.1　Open-Falcon 简介　477
27.2.2　Nagios 简介　479
27.2.3　Zabbix 简介　480

27.2.4 几种常见开源软件比较　481
27.3 Zabbix 部署　481
　27.3.1 Zabbix Server 软件安装　482
　27.3.2 Zabbix Server 配置与启动　482
　27.3.3 配置 Zabbix Web 服务端　483
　27.3.4 Zabbix Agent 安装和配置　485
　27.3.5 PMP 插件介绍和部署　486
　27.3.6 Zabbix Web 端操作　489
27.4 性能医生 orzdba　491
　27.4.1 orzdba 安装　491
　27.4.2 orzdba 使用　492
27.5 小结　492

第 28 章　MySQL 常见问题和应用技巧　493

28.1 忘记 MySQL 的 root 密码　493
28.2 数据目录磁盘空间不足的问题　494
28.3 mysql.sock 丢失后如何连接数据库　495
28.4 从 mysqldump 文件抽取需要恢复的表　496
28.5 使用 innobackupex 备份恢复单表　497
28.6 分析 BINLOG，找出写的热点表　498
28.7 在线 DDL　499
28.8 小结　502

第 29 章　自动化运维系统的开发　503

29.1 MySQL 自动化运维背景　503
29.2 CMDB 系统搭建　504
　29.2.1 CMDB 数据库　504
　29.2.2 批量管理系统　505
　29.2.3 后台 API　511
29.3 任务调度系统　515
　29.3.1 Celery 安装　515
　29.3.2 Celery 任务部署　516
　29.3.3 Flower 监控　518
29.4 客户端搭建　519
　29.4.1 Vue.js 简介　519
　29.4.2 Vue 项目搭建　520
29.5 自动化运维平台实战　525
　29.5.1 搭建 CMDB　525
　29.5.2 搭建任务调度平台　528
　29.5.3 搭建客户端　529
　29.5.4 项目演示　532
29.6 小结　532

第五部分　架构篇

第 30 章　MySQL 复制　534

30.1 复制概述　534
　30.1.1 复制中的各类文件　536
　30.1.2 3 种复制方式　537
　30.1.3 复制的 4 种常见架构　540
30.2 复制搭建　543
　30.2.1 异步复制　543
　30.2.2 多线程复制　547
　30.2.3 增强半同步复制　553
30.3 GTID（Global Transaction Identifier）　558
　30.3.1 格式与存储　558
　30.3.2 gtid_purged　561
　30.3.3 复制搭建　565
　30.3.4 主从切换　571
　30.3.5 常见问题　572
30.4 主要复制启动选项　576
　30.4.1 log-slave-updates　576
　30.4.2 read-only/super_read_only　576
　30.4.3 指定复制的数据库或者表　577
　30.4.4 slave-skip-errors　579
30.5 日常管理维护　579
　30.5.1 查看从库复制状态和进度　579
　30.5.2 主从复制问题集锦　580
　30.5.3 多主复制时的自增长变量冲突问题　582
　30.5.4 如何提高复制的性能　584
30.6 小结　588

第31章 高可用架构 589

31.1 MHA 架构 589
- 31.1.1 安装部署 MHA 591
- 31.1.2 应用连接配置 598
- 31.1.3 自动 failover 605
- 31.1.4 网络问题触发的 failover 操作 614
- 31.1.5 手动 failover 614
- 31.1.6 在线进行切换 615
- 31.1.7 修复宕掉的 Master 617

31.2 MGR 架构 617
- 31.2.1 安装部署 MGR 618
- 31.2.2 监控 625
- 31.2.3 primary 成员切换 626
- 31.2.4 重要特性 628
- 31.2.5 常见问题 629

31.3 InnoDB Cluster 636
- 31.3.1 安装部署 637
- 31.3.2 初始化 MySQL Router 642
- 31.3.3 集群 Metadata 643
- 31.3.4 集群成员角色切换 644
- 31.3.5 集群删除/增加节点 646
- 31.3.6 重新加入节点 648

31.4 小结 649

第32章 MySQL 中间件 650

32.1 MySQL Router 650
- 32.1.1 MySQL Router 的安装 651
- 32.1.2 MySQL Router 的初始化 653
- 32.1.3 MySQL Router 策略验证 656

32.2 Cetus 架构 658
- 32.2.1 Cetus 的安装配置 659
- 32.2.2 Cetus 的使用 670
- 32.2.3 Cetus 日志文件 674
- 32.2.4 Cetus 的后端管理 675
- 32.2.5 Cetus 的路由策略 678
- 32.2.6 常见问题 681

32.3 小结 682

第一部分
基础篇

第 1 章 MySQL 的安装与配置

近几年，开源数据库逐渐流行起来。由于具有免费使用、配置简单、稳定性好、性能优良等优点，开源数据库在中低端应用中占据了很大的市场份额，而 MySQL 正是开源数据库中的杰出代表。

MySQL 数据库隶属于 MySQL AB 公司，总部位于瑞典。公司名中的"AB"是瑞典语"aktiebolag"或"股份公司"的首字母缩写。MySQL 支持几乎所有的操作系统，并且支持容量很大的表（MyISAM 存储引擎支持的最大表大小为 65536TB，InnoDB 为 64TB）。这些特性使得 MySQL 的发展非常迅猛，目前已经广泛应用在各个行业中。

1.1 MySQL 的下载

用户通常可以到 MySQL 官方网站下载最新版本的 MySQL 数据库。按照用户群的分类，MySQL 数据库目前分为社区版（Community Server）和企业版（Enterprise）。它们最重要的区别在于：社区版是自由下载而且完全免费，但是官方不提供任何技术支持，适用于大多数普通用户；企业版是收费的，下载的软件有试用期，但相应地提供了更多的功能和更完备的技术支持，更适合于对数据库的功能和可靠性要求较高的企业客户。

MySQL 的版本更新很快，目前最新的 GA 版本为 8.0。这些不同版本之间的主要区别如表 1-1 所示。

表 1-1　　　　　　　　　　MySQL 不同版本的重要改进

版　本	重　要　改　进
4.1	增加了子查询的支持；字符集中增加了对 UTF-8 的支持
5.0	增加了视图、过程、触发器的支持，增加了 information_schema 系统数据库
5.1	增加了表分区的支持；支持基于行的复制（row-based replication）
5.5	InnoDB 成为默认存储引擎；支持半同步复制；引入 performance_schema 动态性能视图
5.6	支持部分 Online DDL 操作；支持 ICP/BKA/MRR 等优化器改进；引入 GTID；支持多库并行复制
5.7	密码安全性提高；支持多线程并行复制、多源复制；支持 JSON；引入 sys 系统库；引入 MGR
8.0	在线持久化全局参数；大幅提高数据字典性能；引入窗口函数、ROLE、直方图、降序索引、不可见索引；修复自增列重启 BUG

对于不同的操作系统平台，MySQL 提供了相应的版本。本章以 Windows 平台下的图形化安装包以及 Linux 平台下的 RPM 包为例，说明 MySQL 的下载、安装、配置、启动和关闭过程。

本章的测试环境分别是 64 位的 Windows 10 和 x86-64 平台上的 RedHat Linux 6，MySQL 版本为最新的 8.0.11。

1.1.1 在 Windows 平台下下载 MySQL

Windows 平台提供了两种类型的安装包，如表 1-2 所示。

表 1-2　　　　　　　　　　　　Windows 平台下的安装包

类　　型	特　　点
noinstall 压缩包	安装简单，解压即可使用；灵活性差，无法自主选择组件
MySQL Installer	安装简单，用向导式一步步提示安装，可以灵活选择安装、删除、更改 MySQL 提供的所有组件

其中，MySQL Installer 为官方推荐安装方式，它的安装包又分为 mysql-installer-web-community 和 mysql-installer-community 两种类型，主要区别是前者安装包很小，安装时需要连接互联网，组件需要在线下载最新版本；而后者安装时则不需要连网，安装包中包含了完整的组件，安装包较前者大很多，下载过程慢，但安装过程快。以 8.0.11 版本为例，mysql-installer-web-community 的安装文件为 15.8MB，而 mysql-installer-community 为 230MB。本节以 mysql-installer-community 的安装包为例来演示下载过程。

在 MySQL 官网的 DOWNLOADS 页面中，单击左侧的 MySQL on Windows，然后在打开的页面组件列表中单击 MySQL Installer，进入如图 1-1 所示的安装包列表页面。

单击 "mysql-installer-community-8.0.11.0.msi" 一栏对应的 Download 按钮进入下一页，如图 1-2 所示。

图 1-1　Windows 平台安装包列表

图 1-2　安装包下载页面

若忽略以上登录界面，直接单击图 1-2 中框选标识的链接，可直接下载文件到浏览器的默认下载目录。

1.1.2 在 Linux 平台下下载 MySQL

在 Linux 平台下，MySQL 官方也提供了多种安装方式。不同的安装方式需要下载的安装包也不同，具体如表 1-3 所示。

表 1-3　　　　　　　　　　　　Linux 平台下的安装包

类　　型	特　　点
Yum/APT/SUSE Repository	安装仓库包极小，版本安装简单灵活，升级方便；其中 Yum Repository 适合 Red Hat Enterprise Linux/Oracle Linux/Fedora 平台；APT Repository 适合 Debian/Ubuntu 平台；SUSE 适合 SUSE Linux

续表

类型	特点
RPM 包	安装简单；灵活性差，无法灵活选择版本、升级
通用二进制包	安装较复杂，灵活性高，平台通用性好
源码包	安装最复杂，时间长，参数设置灵活，性能好

要下载 MySQL RPM 包，可以采用以下两种方法。

1. 通过网页直接下载

（1）在 MySQL 官网的 DOWNLOADS 页面中，单击左侧的 MySQL Community Server，然后在页面下方的 Generally Available(GA)Releases 区域中，选择合适的操作系统和对应版本。本例中分别选择 Red Hat Enterprise Linux /Oracle Linux 和 Red Hat Enterprise Linux 6/Oracle Linux 6（x86，64-bit），下面的软件列表会自动进行过滤，只显示符合条件的 RPM 包，如图 1-3 所示。在这些包中，"RPM Bundle"是全部安装包的集合，其余则是各个独立的子安装包。通常选择"RPM Package, MySQL Server"和"RPM Package, Client Utilities"及其依赖包即可满足基本使用。从图 1-3 中的框选提示也可以看出，MySQL 官方给出了 Yum Repository 的下载链接地址，推荐大家使用这种更简便的方式进行安装。

（2）以下载 Server 包为例，单击"RPM Package, MySQL Server"一栏对应的 Download 按钮，进入如图 1-4 所示的页面。

忽略上面的登录界面，只需要单击下面框选标识的链接，即可直接下载文件到浏览器的默认下载目录中。

（3）将下载后的文件用 FTP 等工具传送到 Linux 服务器上。

图 1-3 RHEL6 Linux 平台 RPM 列表

图 1-4 RPM 下载界面

2. 通过命令行方式下载

（1）得到下载地址的 URL（用鼠标右键单击图 1-4 中的"No thanks, just start my download"链接，在弹出的菜单中单击"复制链接地址"）。

（2）使用 wget 命令以及复制的链接地址在 Linux 服务器上直接下载 Server 和 Client 及其依赖软件包。

1.2 MySQL 的安装

MySQL 的安装步骤在不同平台上有所区别。下面以 Windows 10 平台和 RedHat Linux 6 平台为例，介绍 MySQL 在不同操作系统平台上的安装方法。

1.2.1 在 Windows 平台下安装 MySQL

Windows 平台下的安装包有两种，一种是 noinstall 包，顾名思义，不需要安装就可以直接使用；另一种是通过 MySQL Installer 进行图形化安装，这种方式更加简单，可以灵活地安装和配置 MySQL 提供的所有组件，也是官方所推荐的安装方式，本节将主要介绍这种安装方式。

（1）双击安装文件 mysql-installer-community-8.0.11.0.msi，进入 MySQL 安装界面，如图 1-5 所示。

（2）选中条款左下角 "I accept the license terms"，接受条款，单击 Next 按钮，进入 Choosing a Setup Type 界面，选择 MySQL 安装类型，如图 1-6 所示。

左边以单选按钮的形式列出了 5 种安装类型，右上部的文本框描述了选定类型所对应的安装组件，右下部分的单选按钮可以选择安装所有组件还是只安装已经正式发行（GA）的组件，这里保留照默认选项，即满足开发者需求的所有组件。

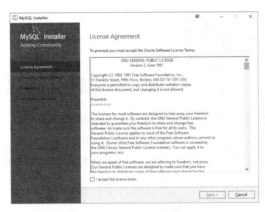

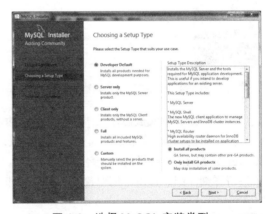

图 1-5 License 条款　　　　　　　　图 1-6 选择 MySQL 安装类型

（3）单击 Next 按钮，进入 Installation 界面，如图 1-7 所示。

此界面列出了 MySQL 即将安装或者升级的组件列表，如果安装包中的组件是最新版本，则执行安装；否则将先升级到最新版本后再安装。

（4）单击 Execute 按钮，开始安装升级过程，界面显示安装进度，全部安装成功后，显示如图 1-8 所示的界面。

（5）单击 Next 按钮，进入配置 MySQL 组件界面，如图 1-9 所示。

（6）单击 Next 按钮，开始配置 MySQL Server 8.0.11，由于配置内容较多，这里不一一演示，配置内容包括组复制、网络配置、授权方式、账户和角色、服务的自动启停等多个选项，按照默认选项全部单击 Next 按钮，最后到达 Apply Configuration 界面后单击 Execute 按钮完成配置，配置完成后界面如图 1-10 所示。

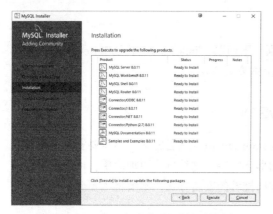

图 1-7　MySQL 安装或升级的组件列表

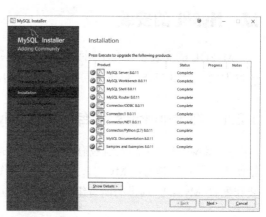

图 1-8　MySQL 安装完成

图 1-9　MySQL 组件配置

图 1-10　MySQL Server 配置完成

（7）单击 Finish 按钮，返回如图 1-11 所示的主配置界面。

（8）单击 Next 按钮，进入 MySQL Router Configuration 界面，如图 1-12 所示，MySQL Router 是 MySQL 提供的一个轻量级中间件，可以实现读写分离等路由功能，具体会在第 32 章进行详细介绍。这里不用配置，直接跳过即可。

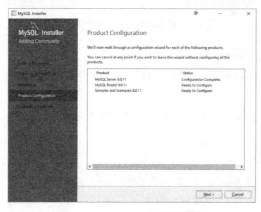

图 1-11　MySQL 主配置界面

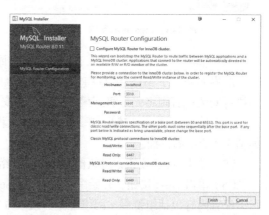

图 1-12　MySQL Router Configuration 界面

1.2 MySQL 的安装

（9）单击 Finish 按钮，回到主配置界面，单击 Next 按钮，进行最后一项样例数据库的配置。在 8.0 版本中，会创建两个样例数据库，分别是 world 和 sakila，其中 world 数据模型较简单，只有 3 张表；sakila 数据模型较复杂，有 22 张表，并且有视图触发器等对象，方便做各种测试。要创建样例库，首先需要和 MySQL 实例进行连接，图 1-13 所示为测试数据库连接的界面，root 下面输入实例配置时的密码，单击 Check 按钮，绿色的状态栏显示测试成功。

（10）单击 Next 按钮，显示样例库安装脚本执行界面；单击 Execute 按钮，完成所有配置过程。再次单击 Next 按钮，显示安装完成，如图 1-14 所示，界面中可以选择启动 Workbench 或者是 MySQL Shell 来连接实例。Workbench 是一个官方提供的免费图形化管理工具，使用很简单，本书不做详细介绍；MySQL Shell 是 8.0 版本新提供的一个客户端工具，后面的章节会详细介绍。

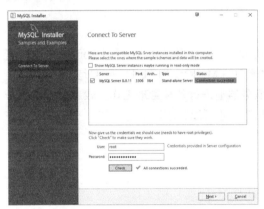

图 1-13　MySQL 连接测试　　　　　　　　图 1-14　MySQL 配置完毕

（11）这里使用传统的 MySQL 客户端尝试连接，客户端工具在"开始"菜单中可以找到，如图 1-15 所示。

单击图 1-15 中的 MySQL 8.0 Command Line Client 菜单，启动 MySQL 客户端工具，然后输入 root 密码，如图 1-16 所示。

输入之前设置的密码后，登录成功。

图 1-15　MySQL 客户端　　　　　　　　图 1-16　登录到 MySQL

1.2.2 在 Linux 平台下安装 MySQL

在 Linux 平台下安装 MySQL 和在 Windows 平台下安装有所不同，在 Linux 平台下不用图形化的方式来安装，并且 Linux 支持 RPM 包、通用二进制包、源码包 3 种安装方式。下面以 RPM 包为例来介绍如何在 Linux 平台下安装 MySQL，其他安装方式会在第 22 章详细介绍。

RPM 是 Redhat Package Manage 的缩写。通过 RPM 的管理，用户可以把源代码包装成一种以 rpm 为后缀的文件形式，更易于安装。对于 RPM 包的下载和安装，MySQL 官方提供了两种方式：一种是直接在页面上按需下载对应的包；另一种是使用版本仓库（Repository）的方式进行安装，这种方式最大的特点是安装简单方便，对于不同版本的软件安装和升级尤为方便，本节将分别介绍这两种安装方式。

1. 直接安装 RPM 包

MySQL 的 RPM 包包括很多套件，老一点的版本一般只安装 Server 和 Client 就可以。其中 Server 包是 MySQL 服务端套件，为用户提供核心的 MySQL 服务；Client 包是连接 MySQL 服务的客户端工具，方便管理员和开发人员在服务器上进行各种管理工作。较新的版本由于包之间存在更多的依赖关系，通常要下载以下几个包才可以完成标准的安装。

- mysql-community-server。
- mysql-community-client。
- mysql-community-libs。
- mysql-community-common。
- mysql-community-libs-compat。

在本例中，安装 RPM 包的具体操作步骤如下。

（1）切换到 root 下（只有 root 才可以执行 RPM 包）。

（2）按照以下顺序安装 MySQL 相关包（顺序不对可能提示包依赖）。

```
[root@hz_10_120_240_251 ~]# rpm -ivh mysql-community-common-8.0.11-1.el6.x86_64.rpm
warning: mysql-community-common-8.0.11-1.el6.x86_64.rpm: Header V3 DSA/SHA1 Signature, key ID 5072e1f5: NOKEY
Preparing...                ########################################### [100%]
   1:mysql-community-common ########################################### [100%]
[root@hz_10_120_240_251 ~]# rpm -ivh mysql-community-libs-8.0.11-1.el6.x86_64.rpm
warning: mysql-community-libs-8.0.11-1.el6.x86_64.rpm: Header V3 DSA/SHA1 Signature, key ID 5072e1f5: NOKEY
Preparing...                ########################################### [100%]
   1:mysql-community-libs   ########################################### [100%]
[root@hz_10_120_240_251 ~]# rpm -ivh mysql-community-libs-compat-8.0.11-1.el6.x86_64.rpm
warning: mysql-community-libs-compat-8.0.11-1.el6.x86_64.rpm: Header V3 DSA/SHA1 Signature, key ID 5072e1f5: NOKEY
Preparing...                ########################################### [100%]
   1:mysql-community-libs-co########################################### [100%]
[root@hz_10_120_240_251 ~]# rpm -ivh mysql-community-client-8.0.11-1.el6.x86_64.rpm
warning: mysql-community-client-8.0.11-1.el6.x86_64.rpm: Header V3 DSA/SHA1 Signature, key ID 5072e1f5: NOKEY
Preparing...                ########################################### [100%]
   1:mysql-community-client ########################################### [100%]
[root@hz_10_120_240_251 ~]# rpm -ivh mysql-community-server-8.0.11-1.el6.x86_64.rpm
warning: mysql-community-server-8.0.11-1.el6.x86_64.rpm: Header V3 DSA/SHA1 Signature, key ID 5072e1f5: NOKEY
Preparing...                ########################################### [100%]
   1:mysql-community-server ########################################### [100%]
```

1.2 MySQL 的安装

（3）启动 MySQL。

```
[root@hz_10_120_240_251 lib]# service mysqld start
Initializing MySQL database:                               [  OK  ]
Starting mysqld:                                           [  OK  ]
```

如果 OS 是 redhadt7，则命令为 systemctl start mysqld.service。通过下面的 chkconfig 命令，可以让操作系统重启时自动启动 MySQL。

```
chkconfig --level 2345 mysqld on
```

执行下面的命令，可以检查自启动状态，确认生效。

```
[root@hz_10_120_240_251 ~]# chkconfig --list|grep mysql
mysqld          0:off   1:off   2:on    3:on    4:on    5:on    6:off
```

（4）登录 MySQL。

```
[root@hz_10_120_240_251 log]# mysql -uroot
ERROR 1045 (28000): Access denied for user 'root'@'localhost' (using password: NO)
```

MySQL 5.7 之后的版本在默认安装时去掉了 root 用户的空密码，初次启动 MySQL 时系统会生成一个临时 root 密码，可以通过查看 /var/log/mysqld.log 来查看。在本例中，临时密码以粗体代码显示：

```
[root@hz_10_120_240_251 log]# more mysqld.log
2018-07-11T06:42:22.233799Z 0 [System] [MY-013169] [Server] /usr/sbin/mysqld (mysqld 8.0.11)
initializing of server in progress as process 28218
2018-07-11T06:42:26.645905Z 5 [Note] [MY-010454] [Server] A temporary password is generated
for root@localhost: N&Rg1lDdrw16
2018-07-11T06:42:28.555800Z 0 [System] [MY-013170] [Server] /usr/sbin/mysqld (mysqld 8.0.11)
initializing of server has completed
```

用临时密码登录：

```
[root@hz_10_120_240_251 log]# mysql -uroot -p
Enter password:
Welcome to the MySQL monitor.  Commands end with ; or \g.
Your MySQL connection id is 9
Server version: 8.0.11

Copyright (c) 2000, 2018, Oracle and/or its affiliates. All rights reserved.

Oracle is a registered trademark of Oracle Corporation and/or its
affiliates. Other names may be trademarks of their respective
owners.

Type 'help;' or '\h' for help. Type '\c' to clear the current input statement.

mysql>
```

但临时密码登录无法进行大多数操作，比如想查看当前数据库列表，会提示如下错误：

```
mysql> show databases;
ERROR 1820 (HY000): You must reset your password using ALTER USER statement before executing
this statement.
```

提示必须要更改初始密码，才可以执行此命令，可以用以下命令进行更改：

```
mysql> alter user 'root'@'localhost' identified by "Test@123";
Query OK, 0 rows affected (0.18 sec)
```

新密码必须要满足密码强度规则，默认规则如下：

```
mysql> SHOW VARIABLES LIKE 'validate_password%';
```

Variable_name	Value
validate_password.check_user_name	ON
validate_password.dictionary_file	

```
| validate_password.length            | 8      |
| validate_password.mixed_case_count  | 1      |
| validate_password.number_count      | 1      |
| validate_password.policy            | MEDIUM |
| validate_password.special_char_count| 1      |
+-------------------------------------+--------+
```

简单解释一下,新密码必须至少为 8 位(validate_password.length),其中至少要包含一个数字(validate_password.number_count)、一个特殊字符(validate_password.special_char_count),以及大小写字符至少各包含一个(validate_password.mixed_case_count)。如果不满足这些条件,则提示如下错误:

```
mysql> alter user 'root'@'localhost' identified by "test@123";
ERROR 1819 (HY000): Your password does not satisfy the current policy requirements
```

至此,MySQL 安装完毕。

2. 通过 Yum Repository 安装

通过上面的介绍,我们发现 RPM 包安装方式虽然简单,但也有一些缺陷,比如需要下载的包较多,且包之间安装有先后依赖关系,最重要的是升级不方便。如果有新版本,则需要重新下载所有包进行替换。

为了解决这些不便之处,MySQL 官方提供了一种新的安装方式——Yum Repository。Yum(全称为 Yellow dog Updater, Modified)是一个在 Fedora 和 RedHat 以及 CentOS 中的 Shell 前端软件包管理器。基于 RPM 包管理,能够从指定的服务器自动下载 RPM 包并且安装,可以自动处理依赖性关系,并且一次安装所有依赖的软件包,无须烦琐地一次次下载和安装。

Yum Repository 的安装包非常小,8.0 版本只有 25KB,下载方式和其他 RPM 包类似,这里不再赘述。下面详细介绍 Yum Repository 的安装和使用方法。

(1)安装 Yum Repository。

```
[root@hz_10_120_240_251 ~]# rpm -ivh mysql80-community-release-el6-1.noarch.rpm
warning: mysql80-community-release-el6-1.noarch.rpm: Header V3 DSA/SHA1 Signature, key ID 5072e1f5: NOKEY
Preparing...                          ########################################### [100%]
   1:mysql80-community-relea########################################### [100%]
```

安装完毕后,在 /etc/yum.repos.d 下多了 mysql-community.repo 和 mysql-community-source.repo 这两个文件,它们分别是 MySQL 社区版 RPM 包和源码包的 Yum 源文件,里面记录了支持的软件版本和下载相关的一些参数。

(2)使用 Yum Repository 来安装 MySQL 8.0。

用 cat 命令截取 mysql-community.repo 的部分内容,如下所示:

```
……
# Enable to use MySQL 5.7
[mysql57-community]
name=MySQL 5.7 Community Server
baseurl=http://repo.mysql.com/yum/mysql-5.7-community/el/6/$basearch/
enabled=0
gpgcheck=1
gpgkey=file:///etc/pki/rpm-gpg/RPM-GPG-KEY-mysql

[mysql80-community]
name=MySQL 8.0 Community Server
baseurl=http://repo.mysql.com/yum/mysql-8.0-community/el/6/$basearch/
enabled=1
gpgcheck=1
gpgkey=file:///etc/pki/rpm-gpg/RPM-GPG-KEY-mysql
……
```

可以看出，最新 GA 版本 8.0 的 enabled=1，其他版本均为 0。如果安装最新 GA 版本的 MySQL，则不用做任何设置，直接执行如下命令：

```
[root@hz-10-120-13-227 yum.repos.d]# yum install mysql-community-server
……
Resolving Dependencies
--> Running transaction check
---> Package mysql-community-embedded.x86_64 0:5.7.17-1.el6 will be obsoleted
---> Package mysql-community-embedded-devel.x86_64 0:5.7.17-1.el6 will be obsoleted
---> Package mysql-community-server.x86_64 0:5.7.17-1.el6 will be updated
---> Package mysql-community-server.x86_64 0:8.0.11-1.el6 will be obsoleting
--> Processing Dependency: mysql-community-common(x86-64) = 8.0.11-1.el6 for package: mysql-community-server-8.0.11-1.el6.x86_64
--> Processing Dependency: mysql-community-client(x86-64) >= 8.0.0 for package: mysql-community-server-8.0.11-1.el6.x86_64
--> Running transaction check
---> Package mysql-community-client.x86_64 0:5.7.17-1.el6 will be updated
---> Package mysql-community-client.x86_64 0:8.0.11-1.el6 will be an update
--> Processing Dependency: mysql-community-libs(x86-64) >= 8.0.0 for package: mysql-community-client-8.0.11-1.el6.x86_64
---> Package mysql-community-common.x86_64 0:5.7.17-1.el6 will be updated
---> Package mysql-community-common.x86_64 0:8.0.11-1.el6 will be an update
--> Running transaction check
---> Package mysql-community-libs.x86_64 0:5.7.17-1.el6 will be updated
--> Processing Dependency: libmysqlclient.so.20()(64bit) for package: mysql-community-devel-5.7.17-1.el6.x86_64
---> Package mysql-community-libs.x86_64 0:8.0.11-1.el6 will be an update
--> Running transaction check
---> Package mysql-community-devel.x86_64 0:5.7.17-1.el6 will be updated
---> Package mysql-community-devel.x86_64 0:8.0.11-1.el6 will be an update
--> Finished Dependency Resolution

Dependencies Resolved

================================================================================
 Package                    Arch          Version            Repository    Size
================================================================================
Installing:
 mysql-community-server     x86_64        8.0.11-1.el6       mysql80-community
                                                                          378 M
     replacing  mysql-community-embedded.x86_64 5.7.17-1.el6
     replacing  mysql-community-embedded-devel.x86_64 5.7.17-1.el6
Updating for dependencies:
 mysql-community-client     x86_64        8.0.11-1.el6       mysql80-community
                                                                           28 M
 mysql-community-common     x86_64        8.0.11-1.el6       mysql80-community
                                                                          656 k
 mysql-community-devel      x86_64        8.0.11-1.el6       mysql80-community
                                                                          4.2 M
 mysql-community-libs       x86_64        8.0.11-1.el6       mysql80-community
                                                                          2.5 M

Transaction Summary
================================================================================
Install       1 Package(s)
Upgrade       4 Package(s)

Total download size: 413 M
Is this ok [y/N]: y
```

从安装过程可以看出，一些包被废弃，一些被更新，包之间的依赖被自动处理，整个过程几乎不需要人工介入，非常方便。

如果特殊需求，要安装其他低版本的 MySQL，比如要安装 5.7 版本，那么可以将源文件中[mysql80-community]下的配置项 enabled 改为 0，同时将[mysql57-community]下对应参数改

为 1 即可。或者使用如下命令更改更方便:

```
yum-config-manager --disable mysql80-community
yum-config-manager --enable mysql57-community
```

1.3 MySQL 的配置

MySQL 安装完毕后,大多数情况下都可以直接启动 MySQL 服务,而不需要设置参数。因为系统中所有的参数都有一个默认值。如果要修改默认值,则必须要配置参数文件。下面就 Windows 和 Linux 两种平台下的配置方法进行介绍。

1.3.1 Windows 平台下配置 MySQL

对于图形化的安装方式,MySQL 提供了一个图形化的实例配置向导,可以引导用户逐步进行实例参数的设置,具体操作步骤如下。

(1)单击"开始"→MySQL Installer Community 菜单,进入安装界面,如图 1-17 所示。

(2)选择产品下面的 MySQL Server,单击图 1-17 中的 Reconfigure,进入选择配置类型界面。可以发现,与安装 MySQL 过程中配置产品的界面和内容一致,这里根据自己需求来设置即可,我们不再赘述。

(3)通过 MySQL Installer 提供的图形化功能,可以对一些重要实例参数进行设置,但对于更详细的参数则无能为力。此时,常用的方法是通过修改配置文件进行设置。Windows 中配置文件命名为 my.ini,通常位于 MySQL 安装目录下,本例中位于 C:\ProgramData\MySQL\MySQL Server 8.0\my.ini。下面是过滤掉部分注释后的内容。

图 1-17 MySQL 实例配置界面

```
[client]
port=3306

[mysql]
no-beep

[mysqld]

port=3306
datadir=C:/ProgramData/MySQL/MySQL Server 8.0/Data

# The default authentication plugin to be used when connecting to the server
default_authentication_plugin=caching_sha2_password

# The default storage engine that will be used when create new tables when
default-storage-engine=INNODB

# Set the SQL mode to strict
sql-mode="STRICT_TRANS_TABLES,NO_ENGINE_SUBSTITUTION"

# General and Slow logging.
log-output=FILE
general-log=0
general_log_file="BIH-L-2673.log"
slow-query-log=1
slow_query_log_file="BIH-L-2673-slow.log"
long_query_time=10
```

1.4 启动和关闭 MySQL 服务

上面示例中的粗体代码代表了不同模块的参数，通常配置最多的参数模块是[mysqld]，也就是 MySQL 实例相关参数，新增或修改参数只需要增加或者修改此模块下的条目即可。

1.3.2 Linux 平台下配置 MySQL

在 Linux 下配置 MySQL 和在 Windows 下通过编辑参数文件配置非常类似，区别在于参数文件的位置和文件名不同。在 Linux 下也可以在多个位置部署配置文件，通过 RPM 包安装的放在/etc 下，文件名称只能是 my.cnf（在 Windows 下文件名默认是 my.ini）。

在 MySQL 5.6 之前，MySQL 提供了多个参数模板，用来适应不同环境的需求，它们的名称类似于 my-***.cnf，其中***代表了环境对于资源大小的需求，比如 my-huge.cnf、my-large.cnf、my-medium.cnf、my-small.cnf 等，在 8.0 版本中已经取消了这些模板，需要直接编辑 my.cnf 或 my.ini。

1.4 启动和关闭 MySQL 服务

安装配置完毕 MySQL 后，接下来就该启动 MySQL 服务了。这里强调一下，MySQL 服务和 MySQL 数据库不同，MySQL 服务是一系列后台进程，而 MySQL 数据库则是一系列的数据目录和数据文件；MySQL 数据库必须在 MySQL 服务启动之后才可以进行访问。下面就针对 Windows 和 Linux 两种平台，介绍 MySQL 服务的启动和关闭方法。

1.4.1 在 Windows 平台下启动和关闭 MySQL 服务

对于采用图形化方式安装的 MySQL，可以直接通过 Windows 的"开始"菜单（单击"开始"→"控制面板"→"管理工具"→"服务"菜单）中找到 MySQL80 服务，双击后进入如图 1-18 所示的界面，单击"启动"或"停止"按钮来启动或关闭 MySQL 服务。

用户也可以在命令行中手工启动和关闭 MySQL 服务，如下所示。

（1）启动服务：

```
c:\>net start mysql80
MySQL80 服务正在启动.
MySQL80 服务已经启动成功.
```

（2）关闭服务：

```
c:\>net stop mysql80
MySQL80 服务正在停止.
MySQL80 服务已成功停止.
```

图 1-18　服务列表中启动和关闭 MySQL80

1.4.2 在 Linux 平台下启动和关闭 MySQL 服务

在 Linux 平台下，可以采用如下命令查看 MySQL 服务的状态：

```
[root@localhost bin]# netstat -nlp
Active Internet connections (only servers)
Proto Recv-Q Send-Q Local Address      Foreign Address     State       PID/Program name
tcp        0      0 0.0.0.0:3306       0.0.0.0:*           LISTEN      3168/mysqld
```

```
tcp        0      0   :::9922              :::*        LISTEN      1864/sshd
Active UNIX domain sockets (only servers)
Proto RefCnt Flags    Type     State      I-Node   PID/Program name Path
unix  2      [ ACC ]  STREAM   LISTENING  16537243 3168/mysqld      /var/lib/mysql/ mysql.sock
unix  2      [ ACC ]  STREAM   LISTENING  4875     1915/xfs         /tmp/.font-unix/fs7100
```

其中 3306 端口是 MySQL 服务器默认监听端口。

与在 Windows 平台上类似，在 Linux 平台上启动和关闭 MySQL 也有两种方法，一种是通过命令行方式启动和关闭，另一种是通过服务的方式启动和关闭（适用于 RPM 包安装方式）。下面分别对这两种方法进行介绍。

1. 命令行方式

在命令行方式下，启动和关闭 MySQL 服务命令如下所示。

（1）启动服务：

```
[root@localhost bin]# cd /usr/bin
[root@localhost bin]# ./mysqld_safe &
[1] 23013
[root@localhost bin]# Starting mysqld daemon with databases from /var/lib/mysql
```

（2）关闭服务：

```
[root@localhost bin]# mysqladmin -uroot shutdown
STOPPING server from pid file /var/lib/mysql/localhost.localdomain.pid
070820 04:36:30  mysqld ended

[1]+  Done                    ./mysqld_safe
```

2. 服务的方式

如果 MySQL 是用 RPM 包安装的，则启动和关闭 MySQL 服务过程如下所示。

（1）启动服务：

```
[root@localhost]# service mysqld start
Starting mysqld:                                           [  OK  ]
```

如果在启动状态，需要重启服务，可以用以下命令直接重启，而不需要先关闭再启动：

```
[root@localhost]# service mysqld restart
Stopping mysqld:                                           [  OK  ]
Starting mysqld:                                           [  OK  ]
```

（2）关闭服务：

```
[root@localhost]# service mysqld stop
Stopping mysqld:                                           [  OK  ]
```

> 注意：在命令行启动 MySQL 时，如果不加 "--console"，启动关闭信息将不会在界面中显示，而是记录在安装目录下的 data 目录里面，文件名字一般是 hostname.err，可以通过此文件查看 MySQL 的控制台信息。

1.5 小结

本章以 Windows 平台和 Linux 平台为例，讲述了 MySQL 8.0 在不同操作系统平台下的下载、安装、配置、启动和关闭的过程。其中在 Windows 平台下介绍了 MySQL Installer 图形化安装包；在 Linux 平台下只介绍了 RPM 包的两种安装方法，而没有介绍二进制包和源码包。之所以选择这几种包进行介绍，主要是因为它们比较简单，适合初学者快速入门。第 22 章将会对 Linux 下的二进制包和源码包进行详细的介绍。

第 2 章 SQL 基础

本章将通过丰富的实例对 SQL 语言的基础进行详细介绍,读者不但能够学习到标准 SQL 的使用方法,还能够学习到 MySQL 中一些扩展 SQL 的使用方法。

2.1 SQL 简介

当面对一个陌生的数据库时,通常需要一种方式与它进行交互,以完成用户所需要的各种工作。这个时候,就要用到 SQL 语言了。

SQL 是 Structure Query Language(结构化查询语言)的缩写,它是使用关系模型的数据库应用语言,由 IBM 在 20 世纪 70 年代开发出来,作为 IBM 关系数据库原型 System R 的原型关系语言,实现了关系数据库中的信息检索。

20 世纪 80 年代初,美国国家标准局(ANSI)开始着手制定 SQL 标准,最早的 ANSI 标准于 1986 年完成,就被叫做 SQL-86。标准的出台使 SQL 作为标准关系数据库语言的地位得到了加强。SQL 标准目前已几经修改,更趋完善。

正是由于 SQL 语言的标准化,所以大多数关系型数据库系统都支持 SQL 语言,它已经发展成为多种平台进行交互操作的底层会话语言。

2.2 (My)SQL 使用入门

这里用了(My)SQL 这样的标题,目的是在介绍标准 SQL 的同时,也将一些 MySQL 在标准 SQL 上的扩展一同介绍给读者。希望读者看完本节后,能够对标准 SQL 的基本语法和 MySQL 的部分扩展语法有所了解。

2.2.1 SQL 分类

SQL 语句主要可以划分为以下 3 个类别。

- DDL(Data Definition Language)语句:数据定义语言,这些语句定义了不同的数据段、数据库、表、列、索引等数据库对象。常用的语句关键字主要包括 create、drop、alter 等。
- DML(Data Manipulation Language)语句:数据操纵语句,用于添加、删除、更新和查询数据库记录,并检查数据完整性。常用的语句关键字主要包括 insert、delete、update

和 select 等。

- DCL（Data Control Language）语句：数据控制语句，用于控制不同数据段直接的许可和访问级别的语句。这些语句定义了数据库、表、字段、用户的访问权限和安全级别。常用的语句关键字主要包括 grant、revoke 等。

2.2.2 DDL 语句

简单来说，DDL 就是对数据库内部的对象进行创建、删除、修改等操作的语言。它和 DML 语句的最大区别是 DML 只是操作表内部的数据，而不涉及表的定义、结构的修改，更不会涉及其他对象。DDL 语句更多地是由数据库管理员（DBA）使用，开发人员一般很少使用。

下面通过一些例子来介绍 MySQL 中常用 DDL 语句的使用方法。

1. 创建数据库

启动 MySQL 服务之后，输入以下命令连接到 MySQL 服务器：

```
[root~]# mysql -uroot -p
Enter password:
Welcome to the MySQL monitor.  Commands end with ; or \g.
Your MySQL connection id is 31
Server version: 8.0.11 MySQL Community Server - GPL

Copyright (c) 2000, 2018, Oracle and/or its affiliates. All rights reserved.

Oracle is a registered trademark of Oracle Corporation and/or its
affiliates. Other names may be trademarks of their respective
owners.

Type 'help;' or '\h' for help. Type '\c' to clear the current input statement.

mysql>
```

在以上命令行中，mysql 代表客户端命令，"-u"后面跟连接的数据库用户，"-p"表示需要输入密码。

如果数据库设置正常，并输入了正确的密码，将看到上面一段欢迎界面和一个"mysql>"提示符。在欢迎界面中说明了以下几部分内容。

- 命令的结束符，用";"或者"\g"结束。
- 客户端的连接 ID，这个数字记录了 MySQL 服务到目前为止的连接次数；每一个新连接都会自动加 1，本例中是 31。
- MySQL 服务器的版本和类型，本例中是"8.0.11 MySQL Community Server - GPL"，说明是 8.0.11 的社区发行版。
- 通过"help;"或者"\h"命令来显示帮助内容，通过"\c"命令来清除命令行 buffer。

在 mysql>提示符后面输入所要执行的 SQL 语句，每个 SQL 语句以分号（;）或者"\g"结束，按回车键执行。

因为所有的数据都存储在数据库中，因此需要学习的第一个命令是创建数据库，语法如下所示：

```
CREATE DATABASE dbname
```

例如，创建数据库 test1，命令执行如下：

```
mysql> create database test1;
Query OK, 1 row affected (0.00 sec)
```

可以发现，执行完创建命令后，下面有一行提示"Query OK, 1 row affected (0.00 sec)"，这段提示可以分为 3 个部分。"Query OK"表示上面的命令执行成功。读者可能会觉得奇怪，又不是执行查询操作，为什么显示查询成功？其实这是 MySQL 的一个特点，所有的 DDL 和 DML（不包括 SELECT）操作执行成功后都显示"Query OK"，这里理解为执行成功就可以了。"1 row affected"表示操作只影响了数据库中一行的记录，"0.00 sec"则记录了操作执行的时间。

如果已经存在这个数据库，系统会提示：

```
mysql> create database test1;
ERROR 1007 (HY000): Can't create database 'test1'; database exists
```

这时，如果需要知道系统中都存在哪些数据库，可以用以下命令来查看：

```
mysql> show databases;
+--------------------+
| Database           |
+--------------------+
| information_schema |
| mysql              |
| performance_schema |
| sys                |
| test1              |
+--------------------+
5 rows in set (0.00 sec)
```

可以发现，在上面的列表中除了刚刚创建的 test1 外，还有另外 4 个数据库，它们都是安装 MySQL 时系统自动创建的，其功能分别如下。

- information_schema：主要存储系统中的一些数据库对象信息，比如用户表信息、列信息、权限信息、字符集信息、分区信息等。每个用户都可以查看这个数据库，但根据权限的不同看到的内容不同。
- performance_schema：MySQL 5.5 引入的系统库，用于存储系统性能相关的动态参数表。
- sys：MySQL 5.7 引入的系统库，本身不记录系统数据，基于 information_schema 和 performance_schema 之上，封装了一层更加易于调优和诊断的系统视图。
- mysql：存储系统的用户权限信息。

在查看系统中已有的数据库后，可以用如下命令选择要操作的数据库：

USE dbname

例如，选择数据库 test1：

```
mysql> use test1
Database changed
```

然后再用以下命令来查看 test1 数据库中创建的所有数据表：

```
mysql> show tables;
Empty set (0.00 sec)
```

由于 test1 是刚创建的数据库，还没有表，所以显示为空。命令行下面的"Empty set"表示操作的结果集为空。如果查看一下 mysql 数据库里面的表，则可以得到以下信息：

```
mysql> use mysql

mysql> show tables;
+---------------------------+
| Tables_in_mysql           |
+---------------------------+
| columns_priv              |
```

```
| component                |
| db                       |
| default_roles            |
| engine_cost              |
| func                     |
| general_log              |
| global_grants            |
| gtid_executed            |
| help_category            |
| help_keyword             |
| help_relation            |
| help_topic               |
| innodb_index_stats       |
| innodb_table_stats       |
| password_history         |
| plugin                   |
| procs_priv               |
| proxies_priv             |
| role_edges               |
| server_cost              |
| servers                  |
| slave_master_info        |
| slave_relay_log_info     |
| slave_worker_info        |
| slow_log                 |
| tables_priv              |
| time_zone                |
| time_zone_leap_second    |
| time_zone_name           |
| time_zone_transition     |
| time_zone_transition_type|
| user                     |
+--------------------------+
33 rows in set (0.00 sec)
```

2. 删除数据库

删除数据库的语法很简单，如下所示：

```
drop database dbname;
```

例如，要删除 test1 数据库可以使用以下语句：

```
mysql> drop database test1;
Query OK, 0 rows affected (0.00 sec)
```

可以发现，提示操作成功后，后面却显示 "0 rows affected"，这个提示的含义是前一次 MySQL 操作所影响的记录行数，通常只对增删改操作生效，drop 等 DDL 操作通常显示 "0 rows affected"。

> **注意**：数据库删除后，下面的所有表数据都会全部删除，所以删除前一定要仔细检查并做好相应备份。

3. 创建表

在数据库中创建一张表的基本语法如下：

```
CREATE TABLE tablename (
column_name_1 column_type_1 constraints,
column_name_2 column_type_2 constraints,
…
column_name_n column_type_n constraints)
```

因为 MySQL 的表名是以文件的形式保存在磁盘上的，所以表名的字符可以用任何文件名允许的字符。column_name 是列的名字；column_type 是列的数据类型；constraints 是这个列的约束条件，在后面的章节中会详细介绍。

例如,创建一个名称为 emp 的表。表中包括 ename(姓名)、hiredate(雇用日期)和 sal(薪水)3 个字段,字段类型分别为 varchar(10)、date、int(2)(关于字段类型将会在第 3 章中介绍):

```
mysql> create table emp(ename varchar(10),hiredate date,sal decimal(10,2),deptno int(2));
Query OK, 0 rows affected (0.02 sec)
```

表创建完毕后,如果需要查看一下表的定义,可以使用如下命令:

DESC tablename

例如,查看 emp 表,将输出以下信息:

```
mysql> desc emp;
+---------+---------------+------+-----+---------+-------+
| Field   | Type          | Null | Key | Default | Extra |
+---------+---------------+------+-----+---------+-------+
| ename   | varchar(10)   | YES  |     |         |       |
| hiredate| date          | YES  |     |         |       |
| sal     | decimal(10,2) | YES  |     |         |       |
| deptno  | int(2)        | YES  |     |         |       |
+---------+---------------+------+-----+---------+-------+
4 rows in set (0.00 sec)
```

虽然 desc 命令可以查看表定义,但是其输出的信息还是不够全面。为了得到更全面的表定义信息,有时就需要查看创建表的 SQL 语句,可以使用如下命令查看:

```
mysql> show create table emp \G;
*************************** 1. row ***************************
       Table: emp
Create Table: CREATE TABLE 'emp' (
  'ename' varchar(10) DEFAULT NULL,
  'hiredate' date DEFAULT NULL,
  'sal' decimal(10,2) DEFAULT NULL,
  'deptno' int(2) DEFAULT NULL,
  KEY 'idx_emp_ename' ('ename')
) ENGINE=InnoDB DEFAULT CHARSET=gbk
1 row in set (0.02 sec)

ERROR:
No query specified

mysql>
```

从上面创建表的 SQL 语句中,除了可以看到表定义以外,还可以看到表的 engine(存储引擎)和 charset(字符集)等信息。"\G" 选项的含义是使得记录能够按照字段竖向排列,以便更好地显示内容较长的记录。

4. 删除表

表的删除命令如下:

DROP TABLE tablename

例如,要删除数据库 emp 可以使用以下命令:

```
mysql> drop table emp;
Query OK, 0 rows affected (0.00 sec)
```

5. 修改表

对于已经创建好的表,尤其是已经有大量数据的表,如果需要做一些结构上的改变,可以先将表删除(drop),然后再按照新的表定义重建。这样做没有问题,但是必然要做一些额外的工作,比如数据的重新加载。而且,如果有服务在访问表,也会对服务产生影响。

因此,在大多数情况下,表结构的更改都使用 alter table 语句,以下是一些常用的命令。

(1) 修改表类型，语法如下：

```
ALTER TABLE tablename MODIFY [COLUMN] column_definition [FIRST | AFTER col_name]
```

例如，修改表 emp 的 ename 字段定义，将 varchar(10)改为 varchar(20)：

```
mysql> desc emp;
+---------+---------------+------+-----+---------+-------+
| Field   | Type          | Null | Key | Default | Extra |
+---------+---------------+------+-----+---------+-------+
| ename   | varchar(10)   | YES  |     |         |       |
| hiredate| date          | YES  |     |         |       |
| sal     | decimal(10,2) | YES  |     |         |       |
| deptno  | int(2)        | YES  |     |         |       |
+---------+---------------+------+-----+---------+-------+
4 rows in set (0.00 sec)
mysql> alter table emp modify ename varchar(20);
Query OK, 0 rows affected (0.03 sec)
Records: 0  Duplicates: 0  Warnings: 0
mysql> desc emp;
+---------+---------------+------+-----+---------+-------+
| Field   | Type          | Null | Key | Default | Extra |
+---------+---------------+------+-----+---------+-------+
| ename   | varchar(20)   | YES  |     |         |       |
| hiredate| date          | YES  |     |         |       |
| sal     | decimal(10,2) | YES  |     |         |       |
| deptno  | int(2)        | YES  |     |         |       |
+---------+---------------+------+-----+---------+-------+
4 rows in set (0.00 sec)
```

(2) 增加表字段，语法如下：

```
ALTER TABLE tablename ADD [COLUMN] column_definition [FIRST | AFTER col_name]
```

例如，在表 emp 中新增加字段 age，类型为 int(3)：

```
mysql> desc emp;
+---------+---------------+------+-----+---------+-------+
| Field   | Type          | Null | Key | Default | Extra |
+---------+---------------+------+-----+---------+-------+
| ename   | varchar(20)   | YES  |     |         |       |
| hiredate| date          | YES  |     |         |       |
| sal     | decimal(10,2) | YES  |     |         |       |
| deptno  | int(2)        | YES  |     |         |       |
+---------+---------------+------+-----+---------+-------+
4 rows in set (0.00 sec)

mysql> alter table emp add column age int(3);
Query OK, 0 rows affected (0.03 sec)
Records: 0  Duplicates: 0  Warnings: 0

mysql> desc emp;
+---------+---------------+------+-----+---------+-------+
| Field   | Type          | Null | Key | Default | Extra |
+---------+---------------+------+-----+---------+-------+
| ename   | varchar(20)   | YES  |     |         |       |
| hiredate| date          | YES  |     |         |       |
| sal     | decimal(10,2) | YES  |     |         |       |
| deptno  | int(2)        | YES  |     |         |       |
| age     | int(3)        | YES  |     |         |       |
+---------+---------------+------+-----+---------+-------+
5 rows in set (0.00 sec)
```

(3) 删除表字段，语法如下：

```
ALTER TABLE tablename DROP [COLUMN] col_name
```

例如，将字段 age 删除掉：

```
mysql> desc emp;
+---------+---------------+------+-----+---------+-------+
| Field   | Type          | Null | Key | Default | Extra |
```

```
+----------+---------------+------+-----+---------+-------+
| ename    | varchar(20)   | YES  |     |         |       |
| hiredate | date          | YES  |     |         |       |
| sal      | decimal(10,2) | YES  |     |         |       |
| deptno   | int(2)        | YES  |     |         |       |
| age      | int(3)        | YES  |     |         |       |
+----------+---------------+------+-----+---------+-------+
5 rows in set (0.00 sec)
mysql> alter table emp drop column age;
Query OK, 0 rows affected (0.04 sec)
Records: 0  Duplicates: 0  Warnings: 0
mysql> desc emp;
+----------+---------------+------+-----+---------+-------+
| Field    | Type          | Null | Key | Default | Extra |
+----------+---------------+------+-----+---------+-------+
| ename    | varchar(20)   | YES  |     |         |       |
| hiredate | date          | YES  |     |         |       |
| sal      | decimal(10,2) | YES  |     |         |       |
| deptno   | int(2)        | YES  |     |         |       |
+----------+---------------+------+-----+---------+-------+
4 rows in set (0.00 sec)
```

（4）字段改名，语法如下：

```
ALTER TABLE tablename CHANGE [COLUMN] old_col_name column_definition
[FIRST|AFTER col_name]
```

例如，将 age 改名为 age1，同时修改字段类型为 int(4)：

```
mysql> desc emp;
+----------+---------------+------+-----+---------+-------+
| Field    | Type          | Null | Key | Default | Extra |
+----------+---------------+------+-----+---------+-------+
| ename    | varchar(20)   | YES  |     |         |       |
| hiredate | date          | YES  |     |         |       |
| sal      | decimal(10,2) | YES  |     |         |       |
| deptno   | int(2)        | YES  |     |         |       |
| age      | int(3)        | YES  |     |         |       |
+----------+---------------+------+-----+---------+-------+
mysql> alter table emp change age age1 int(4) ;
Query OK, 0 rows affected (0.02 sec)
Records: 0  Duplicates: 0  Warnings: 0

mysql> desc emp
    -> ;
+----------+---------------+------+-----+---------+-------+
| Field    | Type          | Null | Key | Default | Extra |
+----------+---------------+------+-----+---------+-------+
| ename    | varchar(20)   | YES  |     |         |       |
| hiredate | date          | YES  |     |         |       |
| sal      | decimal(10,2) | YES  |     |         |       |
| deptno   | int(2)        | YES  |     |         |       |
| age1     | int(4)        | YES  |     |         |       |
+----------+---------------+------+-----+---------+-------+
5 rows in set (0.00 sec)
```

注意：change 和 modify 都可以修改表的定义，不同的是 change 后面需要写两次列名，不方便。但是 change 的优点是可以修改列名称，modify 则不能。

（5）修改字段排列顺序。

前面介绍的字段增加和修改语法（ADD/CHANGE/MODIFY）中，都有一个可选项 first|after column_name，这个选项可以用来修改字段在表中的位置，ADD 增加的新字段默认是加在表的最后位置，而 CHANGE/MODIFY 默认都不会改变字段的位置。

例如，将新增的字段 birth date 加在 ename 之后：

```
mysql> desc emp;
```

```
+---------+---------------+------+-----+---------+-------+
| Field   | Type          | Null | Key | Default | Extra |
+---------+---------------+------+-----+---------+-------+
| ename   | varchar(20)   | YES  |     |         |       |
| hiredate| date          | YES  |     |         |       |
| sal     | decimal(10,2) | YES  |     |         |       |
| deptno  | int(2)        | YES  |     |         |       |
| age     | int(3)        | YES  |     |         |       |
+---------+---------------+------+-----+---------+-------+
5 rows in set (0.00 sec)

mysql> alter table emp add birth date after ename;
Query OK, 0 rows affected (0.03 sec)
Records: 0  Duplicates: 0  Warnings: 0

mysql> desc emp;
+---------+---------------+------+-----+---------+-------+
| Field   | Type          | Null | Key | Default | Extra |
+---------+---------------+------+-----+---------+-------+
| ename   | varchar(20)   | YES  |     |         |       |
| birth   | date          | YES  |     |         |       |
| hiredate| date          | YES  |     |         |       |
| sal     | decimal(10,2) | YES  |     |         |       |
| deptno  | int(2)        | YES  |     |         |       |
| age     | int(3)        | YES  |     |         |       |
+---------+---------------+------+-----+---------+-------+
6 rows in set (0.00 sec)
```

修改字段 age，将它放在最前面：

```
mysql> alter table emp modify age int(3) first;
Query OK, 0 rows affected (0.03 sec)
Records: 0  Duplicates: 0  Warnings: 0

mysql> desc emp;
+---------+---------------+------+-----+---------+-------+
| Field   | Type          | Null | Key | Default | Extra |
+---------+---------------+------+-----+---------+-------+
| age     | int(3)        | YES  |     |         |       |
| ename   | varchar(20)   | YES  |     |         |       |
| birth   | date          | YES  |     |         |       |
| hiredate| date          | YES  |     |         |       |
| sal     | decimal(10,2) | YES  |     |         |       |
| deptno  | int(2)        | YES  |     |         |       |
+---------+---------------+------+-----+---------+-------+
6 rows in set (0.00 sec)
```

注意：CHANGE/FIRST|AFTER COLUMN 这些关键字都属于 MySQL 在标准 SQL 上的扩展，在其他数据库上不一定适用。

（6）更改表名，语法如下：

```
ALTER TABLE  tablename RENAME  [TO] new_tablename
```

例如，将表 emp 改名为 emp1，命令如下：

```
mysql> alter table emp rename emp1;
Query OK, 0 rows affected (0.00 sec)

mysql> desc emp;
ERROR 1146 (42S02): Table 'sakila.emp' doesn't exist
mysql> desc emp1;
+---------+---------------+------+-----+---------+-------+
| Field   | Type          | Null | Key | Default | Extra |
+---------+---------------+------+-----+---------+-------+
| age     | int(3)        | YES  |     |         |       |
| ename   | varchar(20)   | YES  |     |         |       |
| birth   | date          | YES  |     |         |       |
| hiredate| date          | YES  |     |         |       |
| sal     | decimal(10,2) | YES  |     |         |       |
| deptno  | int(2)        | YES  |     |         |       |
+---------+---------------+------+-----+---------+-------+
6 rows in set (0.00 sec)
```

2.2.3 DML 语句

DML 操作是指对数据库中表记录的操作，主要包括表记录的插入（insert）、更新（update）、删除（delete）和查询（select），是开发人员日常频繁使用的操作。下面将依次对它们进行介绍。

1. 插入记录

表创建好后，就可以往里插入记录了。插入记录的基本语法如下：

```
INSERT INTO tablename (field1,field2,…,fieldn) VALUES(value1,value2,…,valuen);
```

例如，向表 emp 中插入以下记录，即 ename 为 zzx1，hiredate 为 2000-01-01，sal 为 2000，deptno 为 1，命令执行如下：

```
mysql> insert into emp (ename,hiredate,sal,deptno) values('zzx1','2000-01- 01','2000',1);
Query OK, 1 row affected (0.00 sec)
```

也可以不用指定字段名称，但是 values 后面的顺序应该和字段的排列顺序一致：

```
mysql> insert into emp  values('lisa','2003-02-01','3000',2);
Query OK, 1 row affected (0.00 sec)
```

含可空的字段、非空但是含有默认值的字段以及自增字段，可以不在 insert 后的字段列表里面出现，values 后面只写对应字段名称的值。这些没写的字段自动设置为 NULL、默认值、自增的下一个数字，这样在某些情况下可以大大缩短 SQL 语句的复杂性。

例如，只对表中的 ename 和 sal 字段显式插入值：

```
mysql> insert into emp  (ename,sal) values('dony',1000);
Query OK, 1 row affected (0.00 sec)
```

来查看一下实际插入值：

```
mysql> select * from emp;
+--------+------------+---------+---------+
| ename  | hiredate   | sal     | deptno  |
+--------+------------+---------+---------+
| zzx    | 2000-01-01 | 100.00  |       1 |
| lisa   | 2003-02-01 | 400.00  |       2 |
| bjguan | 2004-04-02 | 100.00  |       1 |
| dony   | NULL       | 1000.00 |    NULL |
+--------+------------+---------+---------+
```

果然，设置为可空的两个字段都显示为 NULL。在 MySQL 中，insert 语句还有一个很好的特性，可以一次性插入多条记录，语法如下：

```
INSERT INTO tablename (field1, field2, …, fieldn)
VALUES
(record1_value1, record1_value2, …, record1_valuesn),
(record2_value1, record2_value2, …, record2_valuesn),
…
(recordn_value1, recordn_value2, …, recordn_valuesn)
;
```

可以看出，每条记录之间都用逗号进行了分隔。下面的例子中，对表 dept 一次插入两条记录：

```
mysql> insert into dept values(5,'dept5'),(6,'dept6');
Query OK, 2 rows affected (0.04 sec)
Records: 2  Duplicates: 0  Warnings: 0

mysql> select * from dept;
+--------+----------+
| deptno | deptname |
+--------+----------+
| 1      | tech     |
| 2      | sale     |
```

```
| 5      | fin       |
| 5      | dept5     |
| 6      | dept6     |
+--------+-----------+
5 rows in set (0.00 sec)
```

这个特性可以使得 MySQL 在插入大量记录时，节省很多的网络开销，大大提高插入效率。

2. 更新记录

表里的记录值可以通过 update 命令进行更改，语法如下：

`UPDATE tablename SET field1=value1,field2.=value2,…,fieldn=valuen [WHERE CONDITION]`

例如，将表 emp 中 ename 为 "lisa" 的薪水（sal）从 3000 更改为 4000：

```
mysql> update emp set sal=4000 where ename='lisa';
Query OK, 1 row affected (0.00 sec)
Rows matched: 1  Changed: 1  Warnings: 0
```

在 MySQL 中，update 命令可以同时更新多个表中数据，语法如下：

`UPDATE t1,t2,…,tn set t1.field1=expr1,tn.fieldn=exprn [WHERE CONDITION]`

在下例中，同时更新表 emp 中的字段 sal 和表 dept 中的字段 deptname：

```
mysql> select * from emp;
+--------+------------+---------+--------+
| ename  | hiredate   | sal     | deptno |
+--------+------------+---------+--------+
| zzx    | 2000-01-01 | 100.00  | 1      |
| lisa   | 2003-02-01 | 200.00  | 2      |
| bjguan | 2004-04-02 | 100.00  | 1      |
| dony   | 2005-02-05 | 2000.00 | 4      |
+--------+------------+---------+--------+
4 rows in set (0.00 sec)

mysql> select * from dept;
+--------+----------+
| deptno | deptname |
+--------+----------+
| 1      | tech     |
| 2      | sale     |
| 5      | fin      |
+--------+----------+
3 rows in set (0.00 sec)

mysql> update emp a,dept b set a.sal=a.sal*b.deptno,b.deptname=a.ename where a.deptno=b.deptno;
Query OK, 3 rows affected (0.04 sec)
Rows matched: 5  Changed: 3  Warnings: 0

mysql> select * from emp;
+--------+------------+---------+--------+
| ename  | hiredate   | sal     | deptno |
+--------+------------+---------+--------+
| zzx    | 2000-01-01 | 100.00  | 1      |
| lisa   | 2003-02-01 | 400.00  | 2      |
| bjguan | 2004-04-02 | 100.00  | 1      |
| dony   | 2005-02-05 | 2000.00 | 4      |
+--------+------------+---------+--------+
4 rows in set (0.01 sec)

mysql> select * from dept;
+--------+----------+
| deptno | deptname |
+--------+----------+
| 1      | zzx      |
| 2      | lisa     |
| 5      | fin      |
+--------+----------+
3 rows in set (0.00 sec)
```

至此，两个表的数据同时进行了更新。

注意：多表更新的语法更多地用于根据一个表的字段来动态地更新另外一个表的字段。

3. 删除记录

如果记录不再需要，则可以用 delete 命令进行删除，语法如下：

```
DELETE FROM tablename [WHERE CONDITION]
```

例如，在 emp 中将 ename 为 "dony" 的记录全部删除，命令如下：

```
mysql> delete from emp where ename='dony';
Query OK, 1 row affected (0.00 sec)
```

在 MySQL 中可以一次删除多个表的数据，语法如下：

```
DELETE t1,t2,…,tn FROM t1,t2,…,tn [WHERE CONDITION]
```

如果 from 后面的表名用别名，则 delete 后面也要用相应的别名，否则会提示语法错误。

在下例中，同时删除表 emp 和 dept 中 deptno 为 3 的记录：

```
mysql> select * from emp;
+--------+------------+---------+--------+
| ename  | hiredate   | sal     | deptno |
+--------+------------+---------+--------+
| zzx    | 2000-01-01 | 100.00  | 1      |
| lisa   | 2003-02-01 | 200.00  | 2      |
| bjguan | 2004-04-02 | 100.00  | 1      |
| bzshen | 2005-04-01 | 300.00  | 3      |
| dony   | 2005-02-05 | 2000.00 | 4      |
+--------+------------+---------+--------+
5 rows in set (0.00 sec)

mysql> select * from dept;
+--------+----------+
| deptno | deptname |
+--------+----------+
| 1      | tech     |
| 2      | sale     |
| 3      | hr       |
| 5      | fin      |
+--------+----------+
4 rows in set (0.00 sec)
mysql> delete a,b from emp a,dept b where a.deptno=b.deptno and a.deptno=3;
Query OK, 2 rows affected (0.04 sec)

mysql>
mysql>
mysql> select * from emp;
+--------+------------+---------+--------+
| ename  | hiredate   | sal     | deptno |
+--------+------------+---------+--------+
| zzx    | 2000-01-01 | 100.00  | 1      |
| lisa   | 2003-02-01 | 200.00  | 2      |
| bjguan | 2004-04-02 | 100.00  | 1      |
| dony   | 2005-02-05 | 2000.00 | 4      |
+--------+------------+---------+--------+
4 rows in set (0.00 sec)

mysql> select * from dept;
+--------+----------+
| deptno | deptname |
+--------+----------+
| 1      | tech     |
| 2      | sale     |
| 5      | fin      |
+--------+----------+
3 rows in set (0.00 sec)
```

注意：不管是单表还是多表，不加 where 条件将会把表的所有记录删除，所以操作时一定要小心。

4．查询记录

数据插入到数据库中后，就可以用 SELECT 命令进行各种各样的查询，使得输出的结果符合用户的要求。SELECT 的语法很复杂，这里只介绍最基本的语法：

```
SELECT * FROM tablename [WHERE CONDITION]
```

查询最简单的方式是将记录全部选出。在下面的例子中，将表 emp 中的记录全部查询出来：

```
mysql> select * from emp;
+--------+------------+---------+--------+
| ename  | hiredate   | sal     | deptno |
+--------+------------+---------+--------+
| zzx    | 2000-01-01 | 2000.00 | 1      |
| lisa   | 2003-02-01 | 4000.00 | 2      |
| bjguan | 2004-04-02 | 5000.00 | 3      |
+--------+------------+---------+--------+
3 rows in set (0.00 sec)
```

其中"*"表示要将所有的记录都选出来，也可以用逗号分割所有字段来代替，例如，以下两个查询是等价的：

```
mysql> select * from emp;
+--------+------------+---------+--------+
| ename  | hiredate   | sal     | deptno |
+--------+------------+---------+--------+
| zzx    | 2000-01-01 | 2000.00 | 1      |
| lisa   | 2003-02-01 | 4000.00 | 2      |
| bjguan | 2004-04-02 | 5000.00 | 3      |
+--------+------------+---------+--------+
3 rows in set (0.00 sec)

mysql> select ename,hiredate,sal,deptno from emp;
+--------+------------+---------+--------+
| ename  | hiredate   | sal     | deptno |
+--------+------------+---------+--------+
| zzx    | 2000-01-01 | 2000.00 | 1      |
| lisa   | 2003-02-01 | 4000.00 | 2      |
| bjguan | 2004-04-02 | 5000.00 | 3      |
+--------+------------+---------+--------+
3 rows in set (0.00 sec)
```

"*"的好处是当需要查询所有字段信息时，查询语句很简单，但是只查询部分字段的时候，必须要将字段一个一个列出来。

上例中已经介绍了查询全部记录的语法，但是在实际应用中，用户还会遇到各种各样的查询要求，下面将分别进行介绍。

（1）查询不重复的记录。

有时需要将表中的记录去掉重复后显示出来，可以用 distinct 关键字来实现：

```
mysql> select ename,hiredate,sal,deptno from emp;
+--------+------------+---------+--------+
| ename  | hiredate   | sal     | deptno |
+--------+------------+---------+--------+
| zzx    | 2000-01-01 | 2000.00 | 1      |
| lisa   | 2003-02-01 | 4000.00 | 2      |
| bjguan | 2004-04-02 | 5000.00 | 1      |
+--------+------------+---------+--------+
3 rows in set (0.00 sec)

mysql> select distinct deptno from emp;
+--------+
| deptno |
```

```
+--------+
| 1      |
| 2      |
+--------+
2 rows in set (0.00 sec)
```

（2）条件查询。

在很多情况下，用户并不需要查询所有的记录，而只是需要根据限定条件来查询一部分数据，用 where 关键字可以实现这样的操作。

例如，需要查询所有 deptno 为 1 的记录：

```
mysql> select * from emp;
+--------+------------+---------+--------+
| ename  | hiredate   | sal     | deptno |
+--------+------------+---------+--------+
| zzx    | 2000-01-01 | 2000.00 | 1      |
| lisa   | 2003-02-01 | 4000.00 | 2      |
| bjguan | 2004-04-02 | 5000.00 | 1      |
+--------+------------+---------+--------+
3 rows in set (0.00 sec)

mysql> select * from emp where deptno=1;
+--------+------------+---------+--------+
| ename  | hiredate   | sal     | deptno |
+--------+------------+---------+--------+
| zzx    | 2000-01-01 | 2000.00 | 1      |
| bjguan | 2004-04-02 | 5000.00 | 1      |
+--------+------------+---------+--------+
2 rows in set (0.00 sec)
```

结果集中将符合条件的记录列出来。上面的例子中，where 后面的条件是一个字段的=比较，除了=外，还可以使用>、<、>=、<=、!=等比较运算符；多个条件之间还可以使用 or、and 等逻辑运算符进行多条件联合查询，运算符会在以后的章节中详细讲解。

以下是一个使用多字段条件查询的例子：

```
mysql> select * from emp where deptno=1 and sal<3000;
+--------+------------+---------+--------+
| ename  | hiredate   | sal     | deptno |
+--------+------------+---------+--------+
| zzx    | 2000-01-01 | 2000.00 | 1      |
+--------+------------+---------+--------+
1 row in set (0.00 sec)
```

（3）排序和限制。

我们经常会有这样的需求，取出按照某个字段进行排序后的记录结果集，这就用到了数据库的排序操作，用关键字 ORDER BY 来实现，语法如下：

*SELECT * FROM tablename [WHERE CONDITION] [ORDER BY field1 [DESC|ASC], field2 [DESC|ASC],…, fieldn [DESC|ASC]]*

其中，DESC 和 ASC 是排序顺序关键字，DESC 表示按照字段进行降序排列，ASC 则表示升序排列，如果不写此关键字默认是升序排列。ORDER BY 后面可以跟多个不同的排序字段，并且每个排序字段可以有不同的排序顺序。

例如，把 emp 表中的记录按照工资高低进行显示：

```
mysql> select * from emp order by sal;
+--------+------------+---------+--------+
| ename  | hiredate   | sal     | deptno |
+--------+------------+---------+--------+
| zzx    | 2000-01-01 | 2000.00 | 1      |
| bzshen | 2005-04-01 | 3000.00 | 3      |
| lisa   | 2003-02-01 | 4000.00 | 2      |
| bjguan | 2004-04-02 | 5000.00 | 1      |
```

```
+--------+------------+---------+--------+
4 rows in set (0.00 sec)
```

如果排序字段的值一样,则值相同的字段按照第二个排序字段进行排序,依次类推。如果只有一个排序字段,则这些字段相同的记录将会无序排列。

例如,把 emp 表中的记录按照部门编号 deptno 字段排序:

```
mysql> select * from emp order by deptno;
+--------+------------+---------+--------+
| ename  | hiredate   | sal     | deptno |
+--------+------------+---------+--------+
| zzx    | 2000-01-01 | 2000.00 | 1      |
| bjguan | 2004-04-02 | 5000.00 | 1      |
| lisa   | 2003-02-01 | 4000.00 | 2      |
| bzshen | 2005-04-01 | 4000.00 | 3      |
+--------+------------+---------+--------+
4 rows in set (0.00 sec)
```

对于 deptno 相同的前两条记录,如果要按照工资由高到低排序,可以使用以下命令:

```
mysql> select * from emp order by deptno,sal desc;
+--------+------------+---------+--------+
| ename  | hiredate   | sal     | deptno |
+--------+------------+---------+--------+
| bjguan | 2004-04-02 | 5000.00 | 1      |
| zzx    | 2000-01-01 | 2000.00 | 1      |
| lisa   | 2003-02-01 | 4000.00 | 2      |
| bzshen | 2005-04-01 | 4000.00 | 3      |
+--------+------------+---------+--------+
4 rows in set (0.00 sec)
```

对于排序后的记录,如果希望只显示一部分,而不是全部,则可以使用 LIMIT 关键字来实现。LIMIT 的语法如下:

```
SELECT …[LIMIT offset_start,row_count]
```

其中 *offset_start* 表示记录的起始偏移量,*row_count* 表示显示的行数。

在默认情况下,起始偏移量为 0,只需要写记录的行数就可以,这时,实际显示的就是前 n 条记录。例如,显示 emp 表中按照 sal 排序后的前 3 条记录:

```
mysql> select * from emp order by sal limit 3;
+--------+------------+---------+--------+
| ename  | hiredate   | sal     | deptno |
+--------+------------+---------+--------+
| zzx    | 2000-01-01 | 2000.00 | 1      |
| lisa   | 2003-02-01 | 4000.00 | 2      |
| bzshen | 2005-04-01 | 4000.00 | 3      |
+--------+------------+---------+--------+
3 rows in set (0.00 sec)
```

如果要显示 emp 表中按照 sal 排序后从第二条记录开始的 3 条记录,可以使用以下命令:

```
mysql> select * from emp order by sal limit 1,3;
+--------+------------+---------+--------+
| ename  | hiredate   | sal     | deptno |
+--------+------------+---------+--------+
| lisa   | 2003-02-01 | 4000.00 | 2      |
| bzshen | 2005-04-01 | 4000.00 | 3      |
| bjguan | 2004-04-02 | 5000.00 | 1      |
+--------+------------+---------+--------+
3 rows in set (0.00 sec)
```

limit 经常和 order by 一起配合使用来进行记录的分页显示。

注意:limit 属于 MySQL 扩展 SQL92 后的语法,在其他数据库上不能通用。

（4）聚合。

很多情况下，用户都需要进行一些汇总操作，比如统计整个公司的人数或者统计每个部门的人数，这时就要用到 SQL 的聚合操作。

聚合操作的语法如下：

```
SELECT [field1,field2,…,fieldn] fun_name
FROM tablename
[WHERE where_contition]
[GROUP BY field1,field2,…,fieldn
[WITH ROLLUP]]
[HAVING where_contition]
```

其参数说明如下。

- fun_name 表示要做的聚合操作，也就是聚合函数，常用的有 sum（求和）、count(*)（记录数）、avg（平均值）、max（最大值）、min（最小值）。
- GROUP BY 关键字表示要进行分类聚合的字段，比如要按照部门分类统计员工数量，部门就应该写在 GROUP BY 后面。
- WITH ROLLUP 是可选参数，表明是否对分类聚合后的结果进行再汇总。
- HAVING 关键字表示对分类后的结果再进行条件的过滤。

注意：HAVING 和 WHERE 的区别在于，HAVING 是对聚合后的结果进行条件的过滤，而 WHERE 是在聚合前就对记录进行过滤。如果逻辑允许，我们尽可能用 WHERE 先过滤记录，因为这样结果集减小，聚合的效率将大大提高，最后再根据逻辑看是否用 HAVING 进行再过滤。

例如，要在 emp 表中统计公司的总人数：

```
mysql> select count(1) from emp;
+----------+
| count(1) |
+----------+
|        4 |
+----------+
1 row in set (0.00 sec)
```

在此基础上，要统计各个部门的人数：

```
mysql> select deptno,count(1) from emp group by deptno;
+--------+----------+
| deptno | count(1) |
+--------+----------+
|      1 |        2 |
|      2 |        1 |
|      4 |        1 |
+--------+----------+
3 rows in set (0.00 sec)
```

更细一些，既要统计各部门人数，又要统计总人数：

```
mysql> select deptno,count(1) from emp group by deptno with rollup;
+--------+----------+
| deptno | count(1) |
+--------+----------+
|      1 |        2 |
|      2 |        1 |
|      4 |        1 |
|   NULL |        4 |
+--------+----------+
4 rows in set (0.00 sec)
```

统计人数大于 1 人的部门：

```
mysql> select deptno,count(1) from emp group by deptno having count(1)>1;
+--------+----------+
```

```
| deptno | count(1) |
+--------+----------+
|   1    |    2     |
+--------+----------+
1 row in set (0.00 sec)
```

最后统计公司所有员工的薪水总额、最高和最低薪水：

```
mysql> select * from emp;
+--------+------------+---------+--------+
| ename  | hiredate   | sal     | deptno |
+--------+------------+---------+--------+
| zzx    | 2000-01-01 |  100.00 |   1    |
| lisa   | 2003-02-01 |  400.00 |   2    |
| bjguan | 2004-04-02 |  100.00 |   1    |
| dony   | 2005-02-05 | 2000.00 |   4    |
+--------+------------+---------+--------+
4 rows in set (0.00 sec)

mysql> select sum(sal),max(sal),min(sal) from emp;
+----------+----------+----------+
| sum(sal) | max(sal) | min(sal) |
+----------+----------+----------+
| 2600.00  | 2000.00  |  100.00  |
+----------+----------+----------+
1 row in set (0.00 sec)
```

（5）表连接。

当需要同时显示多个表中的字段时，可以用表连接来实现。从大类上分，表连接分为内连接和外连接，它们之间的最主要区别是，内连接仅选出两张表中互相匹配的记录，而外连接会选出其他不匹配的记录。我们最常用的是内连接。

例如，查询出所有雇员的名字和所在部门名称，因为雇员名称和部门分别存放在表 emp 和 dept 中，因此，需要使用表连接来进行查询：

```
mysql> select * from emp;
+--------+------------+---------+--------+
| ename  | hiredate   | sal     | deptno |
+--------+------------+---------+--------+
| zzx    | 2000-01-01 | 2000.00 |   1    |
| lisa   | 2003-02-01 | 4000.00 |   2    |
| bjguan | 2004-04-02 | 5000.00 |   1    |
| bzshen | 2005-04-01 | 4000.00 |   3    |
+--------+------------+---------+--------+
4 rows in set (0.00 sec)

mysql> select * from dept;
+--------+----------+
| deptno | deptname |
+--------+----------+
|   1    | tech     |
|   2    | sale     |
|   3    | hr       |
+--------+----------+
3 rows in set (0.00 sec)

mysql> select ename,deptname from emp,dept where emp.deptno=dept.deptno;
+--------+----------+
| ename  | deptname |
+--------+----------+
| zzx    | tech     |
| lisa   | sale     |
| bjguan | tech     |
| bzshen | hr       |
+--------+----------+
4 rows in set (0.00 sec)
```

外连接又分为左连接和右连接，具体定义如下。

- 左连接：包含所有的左边表中的记录甚至是右边表中没有和它匹配的记录。
- 右连接：包含所有的右边表中的记录甚至是左边表中没有和它匹配的记录。

例如，查询 emp 中所有用户名和所在部门名称：

```
mysql> select * from emp;
+--------+------------+---------+--------+
| ename  | hiredate   | sal     | deptno |
+--------+------------+---------+--------+
| zzx    | 2000-01-01 | 2000.00 | 1      |
| lisa   | 2003-02-01 | 4000.00 | 2      |
| bjguan | 2004-04-02 | 5000.00 | 1      |
| bzshen | 2005-04-01 | 4000.00 | 3      |
| dony   | 2005-02-05 | 2000.00 | 4      |
+--------+------------+---------+--------+
5 rows in set (0.00 sec)
mysql> select * from dept;
+--------+----------+
| deptno | deptname |
+--------+----------+
| 1      | tech     |
| 2      | sale     |
| 3      | hr       |
+--------+----------+
3 rows in set (0.00 sec)
mysql> select ename,deptname from emp left join dept on emp.deptno=dept.deptno;
+--------+----------+
| ename  | deptname |
+--------+----------+
| zzx    | tech     |
| lisa   | sale     |
| bjguan | tech     |
| bzshen | hr       |
| dony   |          |
+--------+----------+
5 rows in set (0.00 sec)
```

比较这个查询和上例中的查询，都是查询用户名和部门名，两者的区别在于本例中列出了所有的用户名，即使有的用户名（dony）并不存在合法的部门名称（部门号为 4，在 dept 中没有这个部门）；而上例中仅仅列出了存在合法部门的用户名和部门名称。

右连接和左连接类似，两者之间可以互相转化，例如，上面的例子可以改写为如下的右连接：

```
mysql> select ename,deptname from dept right join emp on dept.deptno=emp.deptno;
+--------+----------+
| ename  | deptname |
+--------+----------+
| zzx    | tech     |
| lisa   | sale     |
| bjguan | tech     |
| bzshen | hr       |
| dony   |          |
+--------+----------+
5 rows in set (0.00 sec)
```

（6）子查询。

某些情况下，当进行查询的时候，需要的条件是另外一个 select 语句的结果，这个时候，就要用到子查询。用于子查询的关键字主要包括 in、not in、=、!=、exists、not exists 等。

例如，从 emp 表中查询出所有部门在 dept 表中的所有记录：

```
mysql> select * from emp;
+--------+------------+---------+--------+
| ename  | hiredate   | sal     | deptno |
+--------+------------+---------+--------+
| zzx    | 2000-01-01 | 2000.00 | 1      |
| lisa   | 2003-02-01 | 4000.00 | 2      |
| bjguan | 2004-04-02 | 5000.00 | 1      |
```

```
| bzshen | 2005-04-01 | 4000.00 | 3 |
| dony   | 2005-02-05 | 2000.00 | 4 |
+--------+------------+---------+--------+
5 rows in set (0.00 sec)

mysql> select * from dept;
+--------+----------+
| deptno | deptname |
+--------+----------+
| 1      | tech     |
| 2      | sale     |
| 3      | hr       |
| 5      | fin      |
+--------+----------+
4 rows in set (0.00 sec)

mysql> select * from emp where deptno in(select deptno from dept);
+--------+------------+---------+--------+
| ename  | hiredate   | sal     | deptno |
+--------+------------+---------+--------+
| zzx    | 2000-01-01 | 2000.00 | 1      |
| lisa   | 2003-02-01 | 4000.00 | 2      |
| bjguan | 2004-04-02 | 5000.00 | 1      |
| bzshen | 2005-04-01 | 4000.00 | 3      |
+--------+------------+---------+--------+
4 rows in set (0.00 sec)
```

如果子查询记录数唯一,还可以用=代替 in:

```
mysql> select * from emp where deptno = (select deptno from dept);
ERROR 1242 (21000): Subquery returns more than 1 row
mysql> select * from emp where deptno = (select deptno from dept limit 1);
+--------+------------+---------+--------+
| ename  | hiredate   | sal     | deptno |
+--------+------------+---------+--------+
| zzx    | 2000-01-01 | 2000.00 | 1      |
| bjguan | 2004-04-02 | 5000.00 | 1      |
+--------+------------+---------+--------+
2 rows in set (0.00 sec)
```

某些情况下,子查询可以转化为表连接,例如:

```
mysql> select * from emp where deptno in(select deptno from dept);
+--------+------------+---------+--------+
| ename  | hiredate   | sal     | deptno |
+--------+------------+---------+--------+
| zzx    | 2000-01-01 | 2000.00 | 1      |
| lisa   | 2003-02-01 | 4000.00 | 2      |
| bjguan | 2004-04-02 | 5000.00 | 1      |
| bzshen | 2005-04-01 | 4000.00 | 3      |
+--------+------------+---------+--------+
4 rows in set (0.00 sec)
```

转换为表连接后:

```
mysql> select emp.* from emp ,dept where emp.deptno=dept.deptno;
+--------+------------+---------+--------+
| ename  | hiredate   | sal     | deptno |
+--------+------------+---------+--------+
| zzx    | 2000-01-01 | 2000.00 | 1      |
| lisa   | 2003-02-01 | 4000.00 | 2      |
| bjguan | 2004-04-02 | 5000.00 | 1      |
| bzshen | 2005-04-01 | 4000.00 | 3      |
+--------+------------+---------+--------+
4 rows in set (0.00 sec)
```

注意:表连接在很多情况下用于优化子查询。

(7) 记录联合。

我们经常会碰到这样的应用,将两个表的数据按照一定的查询条件查询出来后,将结果合并到一起显示出来，这个时候，就需要用 union 和 union all 关键字来实现这样的功能，具体语法如下：

```
SELECT * FROM t1
UNION|UNION ALL
SELECT * FROM t2
...
UNION|UNION ALL
SELECT * FROM tn;
```

UNION 和 UNION ALL 的主要区别是 UNION ALL 是把结果集直接合并在一起，而 UNION 是将 UNION ALL 后的结果进行一次 DISTINCT，去除重复记录后的结果。

来看下面的例子，将 emp 和 dept 表中的部门编号的集合显示出来：

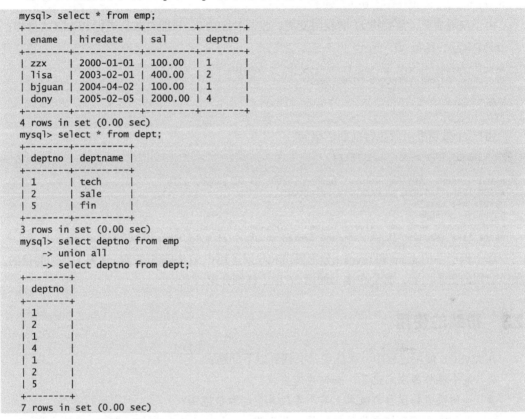

将结果去掉重复记录后显示如下：

```
mysql> select deptno from emp
    -> union
    -> select deptno from dept;
+--------+
| deptno |
+--------+
|   1    |
|   2    |
|   4    |
|   5    |
+--------+
4 rows in set (0.00 sec)
```

2.2.4 DCL 语句

DCL 语句主要是 DBA 用来管理系统中的对象权限时使用，一般开发人员很少使用。下面通过一个例子简单说明一下。

创建一个数据库用户 z1，具有对 sakila 数据库中所有表的 SELECT/INSERT 权限：

```
mysql> grant select,insert on sakila.* to 'z1'@'localhost' identified by '123';
Query OK, 0 rows affected (0.00 sec)

mysql> exit
Bye
[mysql@db3 ~]$ mysql -uz1 -p123

mysql> use sakila
Database changed
mysql> insert into emp values('bzshen','2005-04-01',3000,'3');
Query OK, 1 row affected (0.04 sec)
```

由于权限变更，需要将 z1 的权限变更，收回 INSERT，只能对数据进行 SELECT 操作：

```
[mysql@db3 ~]$ mysql -uroot

mysql> revoke insert on sakila.* from 'z1'@'localhost';
Query OK, 0 rows affected (0.00 sec)

mysql> exit
Bye
```

用户 z1 重新登录后执行前面的语句：

```
[mysql@db3 ~]$ mysql -uz1 -p123

mysql> insert into emp values('bzshen','2005-04-01',3000,'3');
ERROR 1046 (3D000): No database selected
mysql> use sakila
Database changed
mysql> insert into emp values('bzshen','2005-04-01',3000,'3');
ERROR 1142 (42000): INSERT command denied to user 'z1'@'localhost' for table 'emp'
```

以上例子中的 grant 和 revoke 分别授出和收回了用户 z1 的部分权限，达到了我们的目的。关于权限的更多内容，将会在本书的第 26 章中详细介绍。

2.3 帮助的使用

在 MySQL 使用过程中，可能经常会遇到以下问题：
- 某个操作语法忘记了，如何快速查找？
- 如何快速知道当前版本上某个字段类型的取值范围？
- 当前版本都支持哪些函数？请举例说明。
- 当前版本是否支持某个功能？

对于上面列出的各种问题，我们可能想到的办法是查找 MySQL 的文档。不错，这些问题在 MySQL 官方文档中都可以很清楚地查到，但是却要耗费大量的时间和精力。

所以对于以上问题，最好的解决办法是使用 MySQL 安装后自带的帮助文档，这样当遇到问题时就可以方便快捷地进行查询。

2.3.1 按照层次看帮助

如果不知道帮助能够提供些什么，那么就可以用 "? contents" 命令来显示所有可供查询的分类，如下例所示：

```
mysql> ? contents
You asked for help about help category: "Contents"
```

```
For more information, type 'help <item>', where <item> is one of the following
categories:
   Account Management
   Administration
   Compound Statements
   Data Definition
   Data Manipulation
   Data Types
   Functions
   Functions and Modifiers for Use with GROUP BY
   Geographic Features
   Help Metadata
   Language Structure
   Plugins
   Procedures
   Storage Engines
   Table Maintenance
   Transactions
   User-Defined Functions
   Utility
```

对于列出的分类,可以使用"? 类别名称"的方式针对用户感兴趣的内容做进一步的查看。例如,想看看 MySQL 中都支持哪些数据类型,可以执行"? data types"命令:

```
mysql> ? data types
You asked for help about help category: "Data Types"
For more information, type 'help <item>', where <item> is one of the following
topics:
   AUTO_INCREMENT
   BIGINT
   BINARY
   BIT
   BLOB
   BLOB DATA TYPE
   BOOLEAN
   ……
```

上面列出了此版本支持的所有数据类型,如果想知道 int 类型的具体介绍,也可以利用上面的方法进一步查看:

```
mysql> ? int
Name: 'INT'
Description:
INT[(M)] [UNSIGNED] [ZEROFILL]

A normal-size integer. The signed range is -2147483648 to 2147483647.
The unsigned range is 0 to 4294967295.
```

帮助文档中显示了 int 类型的详细描述。通过这种"? 类别名称"的方式,就可以一层层地往下查找用户所关心的主题内容。

2.3.2 快速查阅帮助

在实际应用当中,如果需要快速查阅某项语法时,可以使用关键字进行快速查询。例如,想知道 show 命令都能看些什么,可以用如下命令:

```
mysql> ? show
Name: 'SHOW'
Description:
SHOW has many forms that provide information about databases, tables,
columns, or status information about the server. This section describes
those following:

SHOW AUTHORS
SHOW CHARACTER SET [LIKE 'pattern']
SHOW COLLATION [LIKE 'pattern']
```

```
SHOW [FULL] COLUMNS FROM tbl_name [FROM db_name] [LIKE 'pattern']
SHOW CONTRIBUTORS
SHOW CREATE DATABASE db_name
SHOW CREATE EVENT event_name
SHOW CREATE FUNCTION funcname
...
```

例如，如果想参看 CREATE TABLE 的语法，可以使用以下命令：

```
mysql> ? create table
Name: 'CREATE TABLE'
Description:
Syntax:
CREATE [TEMPORARY] TABLE [IF NOT EXISTS] tbl_name
    (create_definition,...)
    [table_option ...]
    [partition_options]

CREATE [TEMPORARY] TABLE [IF NOT EXISTS] tbl_name
    [(create_definition,...)]
    [table_options]
    [partition_options]
    [IGNORE | REPLACE]
    [AS] query_expression

CREATE [TEMPORARY] TABLE [IF NOT EXISTS] tbl_name
    { LIKE old_tbl_name | (LIKE old_tbl_name) }
......
```

2.4 查询元数据信息

在日常工作中，我们经常会遇到类似下面的应用场景：

- 删除数据库 test1 下所有前缀为 tmp 的表；
- 将数据库 test1 下所有存储引擎为 myisam 的表改为 innodb。

对于这类需求，在 MySQL 5.0 之前只能通过 show tables、show create table 或者 show table status 等命令来得到指定数据库下的表名和存储引擎，但通过这些命令显示的内容有限且不适合进行字符串的批量编辑。如果表很多，则操作起来非常低效。

MySQL 5.0 之后提供了一个新的数据库 information_schema，用来记录 MySQL 中的元数据信息。元数据指的是数据的数据，比如表名、列名、列类型、索引名等表的各种属性名称。这个库比较特殊，它是一个虚拟数据库，物理上并不存在相关的目录和文件；库里 show tables 显示的各种"表"也并不是实际存在的物理表，而全部是视图。对于上面的两个需求，可以简单地通过两个命令得到需要的 SQL 语句：

```
select concat('drop table test1.',table_name,';') from tables where table_schema='test1' and
table_name like 'tmp%';
```

```
select concat('alter  table test1.',table_name,' engine=innodb;') from tables where table_
schema='test1' and engine='MyISAM';
```

下面列出一些比较常用的视图。

- SCHEMATA：该表提供了当前 MySQL 实例中所有数据库的信息，show databases 的结果取之此表。
- TABLES：该表提供了关于数据库中的表的信息（包括视图），详细表述了某个表属于哪个 schema、表类型、表引擎、创建时间等信息。show tables from schemaname 的结果取之此表。
- COLUMNS：该表提供了表中的列信息，详细表述了某张表的所有列以及每个列的信息。show columns from schemaname.tablename 的结果取之此表。

- STATISTICS: 该表提供了关于表索引的信息。show index from schemaname.tablename 的结果取之此表。

2.5 小结

本章简单地介绍了 SQL 语句的基本分类 DML/DDL/DCL，并对每一种分类下的常用 SQL 的用法进行了举例说明。MySQL 在标准 SQL 的基础上进行了很多扩展，本章对常用的一些语法做了简单介绍，对于更详细的说明，读者可以参考 MySQL 的帮助或者官方文档。在本章的最后，还介绍了用户应如何使用 MySQL 中的帮助文档，以便快速查找各种语法定义。

第 3 章 MySQL 支持的数据类型

每一个常量、变量和参数都有数据类型,它用来指定一定的存储格式、约束和有效范围。MySQL 提供了多种数据类型,主要包括数值型、字符串类型、日期和时间类型。不同的 MySQL 版本支持的数据类型可能会稍有不同,用户可以通过查询相应版本的帮助文件来获得具体信息。本章将以 MySQL 5.7 为例,详细介绍 MySQL 中的各种数据类型。

3.1 数值类型

MySQL 支持所有标准 SQL 中的数值类型,其中包括严格数值类型(INTEGER、SMALLINT、DECIMAL 和 NUMERIC),近似数值数据类型(FLOAT、REAL 和 DOUBLE PRECISION),并在此基础上做了扩展。扩展后增加了 TINYINT、MEDIUMINT 和 BIGINT 这 3 种长度不同的整型,并增加了 BIT 类型,用来存放位数据。表 3-1 中列出了 MySQL 5.7 中支持的所有数值类型,其中 INT 是 INTEGER 的同名词,DEC 是 DECIMAL 的同名词。

表 3-1   MySQL 中的数值类型

整数类型	字节	最小值	最大值
TINYINT	1	有符号 -128 无符号 0	有符号 127 无符号 255
SMALLINT	2	有符号 -32768 无符号 0	有符号 32767 无符号 65535
MEDIUMINT	3	有符号 -8388608 无符号 0	有符号 8388607 无符号 16777215
INT、INTEGER	4	有符号 -2147483648 无符号 0	有符号 2147483647 无符号 4294967295
BIGINT	8	有符号 -9223372036854775808 无符号 0	有符号 9223372036854775807 无符号 18446744073709551615
浮点数类型	字节	最小值	最大值
FLOAT	4	±1.175494351E-38	±3.402823466E+38
DOUBLE	8	±2.2250738585072014E-308	±1.7976931348623157E+308
定点数类型	字节	描述	
DEC(M,D), DECIMAL(M,D)	M+2	最大取值范围与 DOUBLE 相同,给定 DECIMAL 的有效取值范围由 M 和 D 决定	

续表

位 类 型	字 节	最 小 值	最 大 值
BIT(M)	1~8	BIT(1)	BIT(64)

在整数类型中，按照取值范围和存储方式的不同，分为 tinyint、smallint、mediumint、int 和 bigint 这 5 个类型。如果超出类型范围的操作，会发生"Out of range"错误提示。为了避免此类问题发生，在选择数据类型时要根据应用的实际情况确定其取值范围，最后根据确定的结果慎重选择数据类型。

对于整型数据，MySQL 还支持在类型名称后面的小括号内指定显示宽度。例如，int(5) 表示当数值宽度小于 5 位时在数字前面填满宽度，如果不显示指定宽度，则默认为 int(11)。一般配合 zerofill 使用，顾名思义，zerofill 就是用"0"填充的意思，也就是在数字位数不够的空间用字符"0"填满。以下几个例子分别描述了填充前后的区别。

（1）创建表 t1，有 id1 和 id2 两个字段，指定其数值宽度分别为 int 和 int(5)：

```
mysql> create table t1 (id1 int,id2 int(5));
Query OK, 0 rows affected (0.03 sec)
mysql> desc t1;
+-------+---------+------+-----+---------+-------+
| Field | Type    | Null | Key | Default | Extra |
+-------+---------+------+-----+---------+-------+
| id1   | int(11) | YES  |     | NULL    |       |
| id2   | int(5)  | YES  |     | NULL    |       |
+-------+---------+------+-----+---------+-------+
2 rows in set (0.00 sec)
```

（2）在 id1 和 id2 中都插入数值 1，可以发现格式没有异常：

```
mysql> insert into t1 values(1,1);
Query OK, 1 row affected (0.00 sec)

mysql> select * from t1;
+------+------+
| id1  | id2  |
+------+------+
|    1 |    1 |
+------+------+
1 row in set (0.00 sec)
```

（3）分别修改 id1 和 id2 的字段类型，加入 zerofill 参数：

```
mysql> alter table t1 modify id1 int zerofill;
Query OK, 1 row affected (0.04 sec)
Records: 1  Duplicates: 0  Warnings: 0

mysql> alter table t1 modify id2 int(5) zerofill;
Query OK, 1 row affected (0.03 sec)
Records: 1  Duplicates: 0  Warnings: 0

mysql> select * from t1;
+------------+-------+
| id1        | id2   |
+------------+-------+
| 0000000001 | 00001 |
+------------+-------+
1 row in set (0.00 sec)
```

可以发现，在数值前面用字符"0"填充了剩余的宽度。大家可能会有所疑问，设置了宽度限制后，如果插入大于宽度限制的值，会不会截断或者插不进去报错？答案是肯定的：不会对插入的数据有任何影响，还是按照类型的实际精度进行保存，这时，宽度格式实际已经没有意义，左边不会再填充任何的"0"字符。下面在表 t1 的字段 id1 中插入数值 1，id2 中插入数值 1111111，位数为 7，大于 id2 的显示位数 5，再观察一下测试结果：

```
mysql> insert into t1 values(1,1111111);
Query OK, 1 row affected (0.00 sec)

mysql> select * from t1;
+------------+---------+
| id1        | id2     |
+------------+---------+
| 0000000001 |   00001 |
| 0000000001 | 1111111 |
+------------+---------+
2 rows in set (0.00 sec)
```

很显然，如上面所说，id2 中显示了正确的数值，并没有受宽度限制影响。

所有的整数类型都有一个可选属性 UNSIGNED（无符号），如果需要在字段里面保存非负数或者需要较大的上限值时，可以用此选项，它的取值范围是正常值的下限取 0，上限取原值的 2 倍，例如，tinyint 有符号范围是 –128～+127，而无符号范围是 0～255。如果一个列指定为 zerofill，则 MySQL 自动为该列添加 UNSIGNED 属性。

另外，整数类型还有一个属性：AUTO_INCREMENT。在需要产生唯一标识符或顺序值时，可利用此属性，这个属性只用于整数类型。AUTO_INCREMENT 值一般从 1 开始，每行增加 1。在插入 NULL 到一个 AUTO_INCREMENT 列时，MySQL 插入一个比该列中当前最大值大 1 的值。一个表中最多只能有一个 AUTO_INCREMENT 列。对于任何想要使用 AUTO_INCREMENT 的列，应该定义为 NOT NULL，并定义为 PRIMARY KEY 或定义为 UNIQUE 键。例如，可按下列任何一种方式定义 AUTO_INCREMENT 列：

```
CREATE TABLE AI (ID INT AUTO_INCREMENT NOT NULL PRIMARY KEY);
CREATE TABLE AI(ID INT AUTO_INCREMENT NOT NULL ,PRIMARY KEY(ID));
CREATE TABLE AI (ID INT AUTO_INCREMENT NOT NULL ,UNIQUE(ID));
```

对于小数的表示，MySQL 分为两种方式：浮点数和定点数。浮点数包括 float（单精度）和 double（双精度），而定点数则只有 decimal 一种表示。定点数在 MySQL 内部以字符串形式存放，比浮点数更精确，适合用来表示货币等精度高的数据。

浮点数和定点数都可以用类型名称后加 "(M,D)" 的方式来进行表示，"(M,D)" 表示该值一共显示 M 位数字（整数位+小数位），其中 D 位位于小数点后面，M 和 D 又称为精度和标度。例如，定义为 float(7,4) 的一个列可以显示为 -999.9999。MySQL 保存值时进行四舍五入，因此如果在 float(7,4) 列内插入 999.00009，近似结果是 999.0001。值得注意的是，浮点数后面跟 "(M,D)" 的用法是非标准用法，如果要用于数据库的迁移，则最好不要这么使用。float 和 double 在不指定精度时，默认会按照实际的精度（由实际的硬件和操作系统决定）来显示，而 decimal 在不指定精度时，默认的整数位为 10，默认的小数位为 0。

下面通过一个例子来比较 float、double 和 decimal 三者之间的不同。

（1）创建测试表，分别将 id1、id2、id3 字段设置为 float(5,2)、double(5,2)、decimal(5,2)：

```
CREATE TABLE 't1' (
  'id1' float(5,2) default NULL,
  'id2' double(5,2) default NULL,
  'id3' decimal(5,2) default NULL
)
```

（2）往 id1、id2 和 id3 这 3 个字段中插入数据 1.23：

```
mysql> insert into t1 values(1.23,1.23,1.23);
Query OK, 1 row affected (0.00 sec)

mysql>
```

```
mysql> select * from t1;
+------+------+------+
| id1  | id2  | id3  |
+------+------+------+
| 1.23 | 1.23 | 1.23 |
+------+------+------+
1 row in set (0.00 sec)
```

可以发现，数据都正确地插入了表 t1。

（3）再向 id1 和 id2 字段中插入数据 1.234，而 id3 字段中仍然插入 1.23：

```
mysql> insert into t1 values(1.234,1.234,1.23);
Query OK, 1 row affected (0.00 sec)

mysql> select * from t1;
+------+------+------+
| id1  | id2  | id3  |
+------+------+------+
| 1.23 | 1.23 | 1.23 |
| 1.23 | 1.23 | 1.23 |
+------+------+------+
2 rows in set (0.00 sec)
```

可以发现，id1、id2、id3 都插入了表 t1，但是 id1 和 id2 由于标度的限制，舍去了最后一位，数据变为了 1.23。

（4）同时向 id1、id2、id3 字段中都插入数据 1.234：

```
mysql> insert into t1 values(1.234,1.234,1.234);
Query OK, 1 row affected, 1 warning (0.00 sec)

mysql> show warnings;
+-------+------+-------------------------------------------+
| Level | Code | Message                                   |
+-------+------+-------------------------------------------+
| Note  | 1265 | Data truncated for column 'id3' at row 1  |
+-------+------+-------------------------------------------+
1 row in set (0.00 sec)

mysql> select * from t1;
+------+------+------+
| id1  | id2  | id3  |
+------+------+------+
| 1.23 | 1.23 | 1.23 |
| 1.23 | 1.23 | 1.23 |
| 1.23 | 1.23 | 1.23 |
+------+------+------+
3 rows in set (0.00 sec)
```

此时发现，虽然数据都插入进去，但是系统出现了一个 Warning，报告 id3 被截断。如果是在传统的 SQLMode（第 13 章会详细介绍 SQL Mode）下，这条记录是无法插入的。

（5）将 id1、id2、id3 字段的精度和标度全部去掉，再次插入数据 1.23：

```
mysql> alter table t1 modify id1 float;
Query OK, 3 rows affected (0.03 sec)
Records: 3  Duplicates: 0  Warnings: 0

mysql> alter table t1 modify id2 double;
Query OK, 3 rows affected (0.04 sec)
Records: 3  Duplicates: 0  Warnings: 0

mysql> alter table t1 modify id3 decimal;
Query OK, 3 rows affected, 3 warnings (0.02 sec)
Records: 3  Duplicates: 0  Warnings: 0
mysql> desc t1;
+-------+---------------+------+-----+---------+-------+
| Field | Type          | Null | Key | Default | Extra |
+-------+---------------+------+-----+---------+-------+
```

```
| id1   | float         | YES |     | NULL    |       |
| id2   | double        | YES |     | NULL    |       |
| id3   | decimal(10,0) | YES |     | NULL    |       |
+-------+---------------+-----+-----+---------+-------+
3 rows in set (0.00 sec)

mysql> insert into t1 values(1.234,1.234,1.234);
Query OK, 1 row affected, 1 warning (0.00 sec)

mysql> show warnings;
+-------+------+-----------------------------------------+
| Level | Code | Message                                 |
+-------+------+-----------------------------------------+
| Note  | 1265 | Data truncated for column 'id3' at row 1|
+-------+------+-----------------------------------------+
1 row in set (0.00 sec)

mysql> select * from t1;
+-------+-------+------+
| id1   | id2   | id3  |
+-------+-------+------+
| 1.234 | 1.234 |    1 |
+-------+-------+------+
1 row in set (0.00 sec)
```

这个时候，可以发现 id1、id2 字段中可以正常插入数据，而 id3 字段的小数位被截断。

上面这个例子验证了上面提到的浮点数如果不写精度和标度，则会按照实际精度值显示，如果有精度和标度，则会自动将四舍五入后的结果插入，系统不会报错；定点数如果不写精度和标度，则按照默认值 decimal(10,0)来进行操作，并且如果数据超越了精度和标度值，系统则会报错。

对于 BIT（位）类型，用于存放位字段值，BIT(M)可以用来存放多位二进制数，M 范围为 1~64，如果不写，则默认为 1 位。对于位字段，直接使用 SELECT 命令将不会看到结果，可以用 bin()（显示为二进制格式）或者 hex()（显示为十六进制格式）函数进行读取。

下面的例子中，对测试表 t2 中的 bit 类型字段 id 做 insert 和 select 操作，这里重点观察一下 select 的结果：

```
mysql> desc t2;
+-------+--------+------+-----+---------+-------+
| Field | Type   | Null | Key | Default | Extra |
+-------+--------+------+-----+---------+-------+
| id    | bit(1) | YES  |     | NULL    |       |
+-------+--------+------+-----+---------+-------+
1 row in set (0.00 sec)

mysql> insert into t2 values(1);
Query OK, 1 row affected (0.00 sec)
mysql> select * from t2;
+------+
| id   |
+------+
|      |
+------+
1 row in set (0.00 sec)
```

可以发现，直接 select * 的结果为 NULL。下面改用 bin()和 hex()函数再试试：

```
mysql> select bin(id),hex(id) from t2;
+---------+---------+
| bin(id) | hex(id) |
+---------+---------+
| 1       | 1       |
+---------+---------+
1 row in set (0.00 sec)
```

结果可以正常显示为二进制数字和十六进制数字。

数据插入 bit 类型字段时，首先转换为二进制，如果位数允许，将成功插入；如果位数小于实际定义的位数，则插入失败。下面的例子中，在 t2 表插入数字 2，因为它的二进制码是 "10"，而 id 的定义是 bit(1)，将无法进行插入：

```
mysql> insert into t2 values(2);
Query OK, 1 row affected, 1 warning (0.00 sec)

mysql> show warnings;
+---------+------+--------------------------------------------------------+
| Level   | Code | Message                                                |
+---------+------+--------------------------------------------------------+
| Warning | 1264 | Out of range value adjusted for column 'id' at row 1   |
+---------+------+--------------------------------------------------------+
1 row in set (0.01 sec)
```

将 ID 定义修改为 bit(2) 后，重新插入，插入成功：

```
mysql> alter table t2 modify id bit(2);
Query OK, 1 row affected (0.02 sec)
Records: 1  Duplicates: 0  Warnings: 0

mysql> insert into t2 values(2);
Query OK, 1 row affected (0.00 sec)

mysql> select bin(id),hex(id) from t2;
+---------+---------+
| bin(id) | hex(id) |
+---------+---------+
| 1       | 1       |
| 10      | 2       |
+---------+---------+
2 rows in set (0.00 sec)
```

3.2 日期时间类型

MySQL 中有多种数据类型可以用于日期和时间的表示，不同的版本可能有所差异，表 3-2 列出了 MySQL 5.7 中所支持的日期和时间类型。

表 3-2　　　　　　　　　　MySQL 中的日期和时间类型

日期和时间类型	字　节	最　小　值	最　大　值
DATE	4	1000-01-01	9999-12-31
DATETIME	8	1000-01-01 00:00:00	9999-12-31 23:59:59
TIMESTAMP	4	19700101080001	2038 年的某个时刻
TIME	3	−838:59:59	838:59:59
YEAR	1	1901	2155

这些数据类型的主要区别如下。

- 如果要用来表示年月日，通常用 DATE 来表示。
- 如果要用来表示年月日时分秒，通常用 DATETIME 或者 TIMESTAMP 表示。两者的主要区别在本章的后面会详述。
- 如果只用来表示时分秒，通常用 TIME 来表示。
- 如果只是表示年份，可以用 YEAR 来表示，它比 DATE 占用更少的空间。YEAR 有 2 位或 4 位格式的年。默认是 4 位格式。在 4 位格式中，允许的值是 1901～2155 和 0000。在 2 位格式中，允许的值是 70～69，表示从 1970～2069 年。MySQL 以 YYYY 格式显示 YEAR

值（从 5.5.27 开始，2 位格式的 year 已经不被支持）。

从表 3-2 中可以看出，每种日期时间类型都有一个有效值范围，如果超出这个范围，在默认的 SQLMode 下，系统会进行错误提示，并将以零值来进行存储。不同日期类型零值的表示如表 3-3 所示。

表 3-3　　　　　　　　　　MySQL 中日期和时间类型的零值表示

数 据 类 型	零 值 表 示
DATETIME	0000-00-00 00:00:00
DATE	0000-00-00
TIMESTAMP	00000000000000
TIME	00:00:00
YEAR	0000

DATE、TIME 和 DATETIME 是经常使用的 3 种日期类型。以下例子在 3 种类型字段插入了相同的日期值，来看看它们的显示结果。

创建表 t，字段分别为 date、time、datetime 这 3 种日期类型：

```
mysql> create table t (d date,t time,dt datetime);
Query OK, 0 rows affected (0.01 sec)

mysql> desc t;
+-------+----------+------+-----+---------+-------+
| Field | Type     | Null | Key | Default | Extra |
+-------+----------+------+-----+---------+-------+
| d     | date     | YES  |     | NULL    |       |
| t     | time     | YES  |     | NULL    |       |
| dt    | datetime | YES  |     | NULL    |       |
+-------+----------+------+-----+---------+-------+
3 rows in set (0.01 sec)
```

用 now() 函数插入当前日期：

```
mysql> insert into t values(now(),now(),now());
Query OK, 1 row affected (0.00 sec)
```

查看显示结果：

```
mysql> select * from t;
+------------+----------+---------------------+
| d          | t        | dt                  |
+------------+----------+---------------------+
| 2007-07-19 | 17:41:13 | 2007-07-19 17:41:13 |
+------------+----------+---------------------+
1 row in set (0.00 sec)
```

显而易见，DATETIME 是 DATE 和 TIME 的组合。用户可以根据不同的需要，来选择不同的日期或时间类型以满足不同的应用。

下面对 TIMESTAMP 类型的特性进行一些测试。

首先看一下 explicit_defaults_for_timestamp（5.6 版本后引入）参数的默认值：

```
mysql> show variables like 'explicit%';
+---------------------------------+-------+
| Variable_name                   | Value |
+---------------------------------+-------+
| explicit_defaults_for_timestamp | OFF   |
+---------------------------------+-------+
1 row in set (0.00 sec)
```

创建测试表 t，字段 id1 为 TIMESTAMP 类型：

3.2 日期时间类型

```
mysql> create table t (id1 timestamp);
Query OK, 0 rows affected (0.03 sec)
mysql> desc t;
+-------+-----------+------+-----+-------------------+-----------------------------+
| Field | Type      | Null | Key | Default           | Extra                       |
+-------+-----------+------+-----+-------------------+-----------------------------+
| id1   | timestamp | NO   |     | CURRENT_TIMESTAMP | on update CURRENT_TIMESTAMP |
+-------+-----------+------+-----+-------------------+-----------------------------+
1 row in set (0.00 sec)
```

可以发现，系统给 tm 自动创建了默认值 CURRENT_TIMESTAMP（系统日期），并且设置了 not null 和 on update CURRENT_TIMESTAMP 属性。插入一个 NULL 值试试：

```
mysql> insert into t values(null);
Query OK, 1 row affected (0.01 sec)

mysql> select * from t;
+---------------------+
| id1                 |
+---------------------+
| 2018-08-01 15:46:38 |
+---------------------+
1 row in set (0.00 sec)
```

果然，t 中自动插入了系统日期。注意，MySQL 只给表中的第一个 TIMESTAMP 字段设置默认值为系统日期，如果有第二个 TIMESTAMP 类型，则默认值设置为 0 值，测试如下：

```
mysql> alter table t add id2 timestamp;
Query OK, 0 rows affected (0.08 sec)
Records: 0  Duplicates: 0  Warnings: 0

mysql> desc t;
+-------+-----------+------+-----+---------------------+-----------------------------+
| Field | Type      | Null | Key | Default             | Extra                       |
+-------+-----------+------+-----+---------------------+-----------------------------+
| id1   | timestamp | NO   |     | CURRENT_TIMESTAMP   | on update CURRENT_TIMESTAMP |
| id2   | timestamp | NO   |     | 0000-00-00 00:00:00 |                             |
+-------+-----------+------+-----+---------------------+-----------------------------+
2 rows in set (0.00 sec)
```

在 MySQL 5.6 之前，可以修改 id2 的默认值为其他常量日期，但是不能再修改为 current_timestamp，因为 MySQL 规定 TIMESTAMP 类型字段只能有一列的默认值为 current_timestamp，如果强制修改，系统会报如下错误提示：

```
mysql> alter table t modify id2 timestamp default current_timestamp;
ERROR 1293 (HY000): Incorrect table definition; there can be only one TIMESTAMP column with CURRENT_TIMESTAMP in DEFAULT or ON UPDATE clause
```

MySQL 5.6 版本之后，这个限制已经去掉，可以随意修改，如下例所示：

```
mysql> alter table t modify id2 timestamp default current_timestamp on update CURRENT_TIMESTAMP;
Query OK, 0 rows affected (0.00 sec)
Records: 0  Duplicates: 0  Warnings: 0

mysql> desc t;
+-------+-----------+------+-----+-------------------+-----------------------------+
| Field | Type      | Null | Key | Default           | Extra                       |
+-------+-----------+------+-----+-------------------+-----------------------------+
| id1   | timestamp | NO   |     | CURRENT_TIMESTAMP | on update CURRENT_TIMESTAMP |
| id2   | timestamp | NO   |     | CURRENT_TIMESTAMP | on update CURRENT_TIMESTAMP |
+-------+-----------+------+-----+-------------------+-----------------------------+
2 rows in set (0.00 sec)
```

如果将 explicit_defaults_for_timestamp 设置为 on，则默认值、not null 和 on update CURRENT_TIMESTAMP 属性都不会自动设置，需要手工操作，具体如下：

```
mysql> set explicit_defaults_for_timestamp=on;
Query OK, 0 rows affected (0.02 sec)
```

```
mysql> create table t (id timestamp);
Query OK, 0 rows affected (0.01 sec)

mysql> desc t;
+-------+-----------+------+-----+---------+-------+
| Field | Type      | Null | Key | Default | Extra |
+-------+-----------+------+-----+---------+-------+
| id    | timestamp | YES  |     | NULL    |       |
+-------+-----------+------+-----+---------+-------+
1 row in set (0.00 sec)
```

TIMESTAMP 还有一个重要特点，就是和时区相关。当插入日期时，会先转换为本地时区后存放；而从数据库里面取出时，也同样需要将日期转换为本地时区后显示。这样，两个不同时区的用户看到的同一个日期可能是不一样的，下面的例子演示了这个差别。

（1）创建表 t8，包含字段 id1（TIMESTAMP）和 id2（DATETIME），设置 id2 的目的是为了和 id1 做对比：

```
CREATE TABLE 't8' (
  'id1' timestamp NOT NULL default CURRENT_TIMESTAMP,
  'id2' datetime default NULL
)
Query OK, 0 rows affected (0.03 sec)
```

（2）查看当前时区：

```
mysql> show variables like 'time_zone';
+---------------+--------+
| Variable_name | Value  |
+---------------+--------+
| time_zone     | SYSTEM |
+---------------+--------+
1 row in set (0.00 sec)
```

可以发现，时区的值为"SYSTEM"，这个值默认是和主机的时区值一致的，因为我们在中国，这里的"SYSTEM"实际是东八区（+8:00）。

（3）用 now() 函数插入当前日期：

```
mysql> select * from t8;
+---------------------+---------------------+
| id1                 | id2                 |
+---------------------+---------------------+
| 2018-09-25 17:26:50 | 2018-09-25 17:26:50 |
+---------------------+---------------------+
1 row in set (0.01 sec)
```

结果显示 id1 和 id2 的值完全相同。

（4）修改时区为东九区，再次查看表中日期：

```
mysql> set time_zone='+9:00';
Query OK, 0 rows affected (0.00 sec)

mysql> select * from t8;
+---------------------+---------------------+
| id1                 | id2                 |
+---------------------+---------------------+
| 2018-09-25 18:26:50 | 2018-09-25 17:26:50 |
+---------------------+---------------------+
1 row in set (0.00 sec)
```

在结果中可以发现，id1 的值比 id2 的值快了 1 个小时，也就是说，东九区的人看到的"2018-09-25 18:26:50"是当地时区的实际日期，也就是东八区的"2018-09-25 17:26:50"，如果还是以"2018-09-25 17:26:50"理解时间必然导致误差。

TIMESTAMP 的取值范围为 19700101080001 到 2038 年的某一天，因此它不适合存放比较久远的日期。下面简单测试一下这个范围：

```
mysql> insert into t values (19700101080001);
Query OK, 1 row affected (0.00 sec)

mysql> select * from t;
+---------------------+
| t                   |
+---------------------+
| 1970-01-01 08:00:01 |
+---------------------+
1 row in set (0.00 sec)

mysql> insert into t values (19700101080000);
Query OK, 1 row affected, 1 warning (0.00 sec)
```

其中 19700101080000 超出了 tm 的下限，系统出现警告提示。查询一下，发现插入值变成了 0 值：

```
mysql> select * from t;
+---------------------+
| t                   |
+---------------------+
| 1970-01-01 08:00:01 |
| 0000-00-00 00:00:00 |
+---------------------+
2 rows in set (0.00 sec)
```

再来测试一下 TIMESTAMP 的上限值：

```
mysql> insert into t values('2038-01-19 11:14:07');
Query OK, 1 row affected (0.00 sec)

mysql> select * from t;
+---------------------+
| t                   |
+---------------------+
| 2038-01-19 11:14:07 |
+---------------------+
1 row in set (0.00 sec)

mysql> insert into t values('2038-01-19 11:14:08');
Query OK, 1 row affected, 1 warning (0.00 sec)

mysql> select * from t;
+---------------------+
| t                   |
+---------------------+
| 2038-01-19 11:14:07 |
| 0000-00-00 00:00:00 |
+---------------------+
2 rows in set (0.00 sec)
```

从上面的例子可以看出，TIMESTAMP 和 DATETIME 的表示方法非常类似，主要有以下几个区别。

- TIMESTAMP 支持的时间范围较小，其取值范围从 19700101080001 到 2038 年的某个时间，而 DATETIME 是从 1000-01-01 00:00:00 到 9999-12-31 23:59:59，范围更大。两者都可以设置默认值和 ON UPDATE CURRENT_TIMESTAMP 属性，使得日期列可以随其他列的更新而自动更新为最新时间。

- TIMESTAMP 在 MySQL 5.6.6 版本之后增加了控制参数 explicit_defaults_for_timestamp，如果设置为 on，则 TIMESTAMP 需要显式指定默认值和 ON UPDATE CURRENT_TIMESTAMP 属性；如果设置为 off，则会自动设置默认值为 CURRENT_TIMESTAMP（系统时间）和 ON UPDATE CURRENT_TIMESTAMP 属性，并且自动设置为 not null。MySQL 8.0.2

后此参数默认为 on；之前版本默认为 off。

- 当 explicit_defaults_for_timestamp 设置为 off 时，表中的第一个 TIMESTAMP 列自动设置为系统时间。如果在一个 TIMESTAMP 列中插入 NULL，则该列值将自动设置为当前的日期和时间。在插入或更新一行但不明确给 TIMESTAMP 列赋值时也会自动设置该列的值为当前的日期和时间，当插入的值超出取值范围时，MySQL 认为该值溢出，使用 "0000-00-00 00:00:00" 进行填补。

- TIMESTAMP 的插入和查询都受当地时区的影响，更能反映出实际的日期。而 DATETIME 则只能反映出插入时当地的时区，其他时区的人查看数据必然会有误差的。

- TIMESTAMP 的属性受 MySQL 版本和服务器 SQLMode 的影响很大，本章都是以 MySQL 5.7 为例进行介绍的，对于不同的版本可以参考相应的 MySQL 帮助文档。

YEAR 类型主要用来表示年份，当应用只需要记录年份时，用 YEAR 比 DATE 将更节省空间。下面的例子在表 t 中定义了一个 YEAR 类型字段，并插入一条记录：

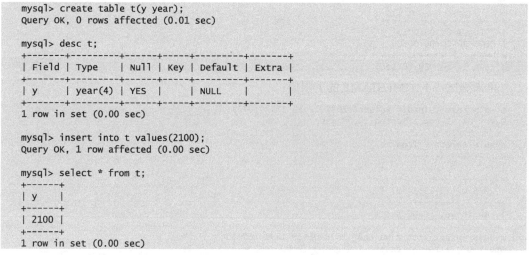

MySQL 以 YYYY 格式检索和显示 YEAR 值，范围是 1901～2155。当使用两位字符串表示年份时，其范围为 "00" 到 "99"。

- "00" 到 "69" 范围的值被转换为 2000～2069 范围的 YEAR 值。
- "70" 到 "99" 范围的值被转换为 1970～1999 范围的 YEAR 值。

细心的读者可能发现，在上面的例子中，日期类型的插入格式有很多，包括整数（如 2100）、字符串（如 2038-01-19 11:14:08）、函数（如 NOW()）等，读者可能会感到疑惑，到底什么样的格式才能够正确地插入到对应的日期字段中呢？

下面以 DATETIME 为例进行说明。

- YYYY-MM-DD HH:MM:SS 或 YY-MM-DD HH:MM:SS 格式的字符串。允许 "不严格" 语法，即任何标点符都可以用做日期部分或时间部分之间的间隔符。例如，"98-12-31 11:30:45" "98.12.31 11+30+45" "98/12/31 11*30*45" 和 "98@12@31 11^30^45" 是等价的。对于包括日期部分间隔符的字符串值，如果日和月的值小于 10，不需要指定两位数。"1979-6-9" 与 "1979-06-09" 是相同的。同样，对于包括时间部分间隔符的字符串值，如果时、分和秒的值小于 10，则不需要指定两位数。"1979-10-30 1:2:3" 与 "1979-10-30 01:02:03" 相同。

- YYYYMMDDHHMMSS 或 YYMMDDHHMMSS 格式的没有间隔符的字符串，假定

字符串对于日期类型是有意义的。例如，"19970523091528"和"970523091528"被解释为"1997-05-23 09:15:28"，但"971122129015"是不合法的（它有一个没有意义的分钟部分），将变为"0000-00-00 00:00:00"。

- YYYYMMDDHHMMSS 或 YYMMDDHHMMSS 格式的数字，假定数字对于日期类型是有意义的。例如，19830905132800 和 830905132800 被解释为"1983-09-05 13:28:00"。数字值应为 6、8、12 或者 14 位长。如果一个数值是 8 位或 14 位长，则假定为 YYYYMMDD 或 YYYYMMDD HHMMSS 格式，前 4 位数表示年。如果数字是 6 位或 12 位长，则假定为 YYMMDD 或 YYMMDDHHMMSS 格式，前两位数表示年。其他数字被解释为仿佛用零填充到了最近的长度。

- 函数返回的结果，其值适合 DATETIME、DATE 或 TIMESTAMP 上下文，例如 NOW() 或 CURRENT_DATE。

对于其他数据类型，其使用原则与上面的内容类似，限于篇幅，这里就不再赘述。

最后通过一个例子，说明如何采用不同的格式将日期"2007-9-3 12:10:10"插入到 DATETIME 列中。

```
mysql> create table t6(dt datetime);
Query OK, 0 rows affected (0.03 sec)
mysql> insert into t6 values('2007-9-3 12:10:10');
Query OK, 1 row affected (0.00 sec)

mysql> insert into t6 values('2007/9/3 12+10+10');
Query OK, 1 row affected (0.00 sec)

mysql> insert into t6 values('20070903121010');
Query OK, 1 row affected (0.01 sec)

mysql> insert into t6 values(20070903121010);
Query OK, 1 row affected (0.00 sec)

mysql> select * from t6;
+---------------------+
| dt                  |
+---------------------+
| 2007-09-03 12:10:10 |
| 2007-09-03 12:10:10 |
| 2007-09-03 12:10:10 |
| 2007-09-03 12:10:10 |
+---------------------+
4 rows in set (0.00 sec)
```

3.3 字符串类型

MySQL 中提供了多种对字符数据的存储类型，不同的版本可能有所差异。以 5.7 版本为例，MySQL 包括了 CHAR、VARCHAR、BINARY、VARBINARY、BLOB、TEXT、ENUM 和 SET 等多种字符串类型。表 3-4 详细列出了这些字符类型的比较。

表 3-4　　　　　　　　　　　　　　MySQL 中的字符类型

字符串类型	字节/字符	描述及存储需求
CHAR(M)	M 个字符	M 为 0~255 之间的整数
VARCHAR(M)	0~M 个字符	M 为 0~65535 之间的整数，值的长度+1 个字节
BINARY(M)	M 个字节	M 为 0~255 之间的整数

字符串类型	字节	描述及存储需求
VARBINARY(M)	0～M 个字节	M 为 0～65535 之间的整数,值的长度+1 个字节
TINYBLOB		允许长度 0～255 字节,值的长度+1 个字节
BLOB		允许长度 0～65535 字节,值的长度+2 个字节
MEDIUMBLOB		允许长度 0～167772150 字节,值的长度+3 个字节
LONGBLOB		允许长度 0～4294967295 字节,值的长度+4 个字节
TINYTEXT		允许长度 0～255 字节,值的长度+1 个字节
TEXT		允许长度 0～65535 字节,值的长度+2 个字节
MEDIUMTEXT		允许长度 0～167772150 字节,值的长度+3 个字节
LONGTEXT		允许长度 0～4294967295 字节,值的长度+4 个字节

下面将分别对这些字符串类型做详细的介绍。

3.3.1 CHAR 和 VARCHAR 类型

CHAR 和 VARCHAR 很类似,都用来保存 MySQL 中较短的字符串。两者的主要区别在于存储方式的不同:CHAR 列的长度固定为创建表时声明的长度,长度可以为从 0～255 的任何值;而 VARCHAR 列中的值为可变长字符串,长度可以指定为 0～65535 之间的值。在检索的时候,CHAR 列删除了尾部的空格,而 VARCHAR 则保留这些空格。下面的例子中通过给表 vc 中的 VARCHAR(4)和 char(4)字段插入相同的字符串来描述这个区别。

(1)创建测试表 vc,并定义两个字段 "v VARCHAR(4)" 和 "c CHAR(4)":

```
mysql> CREATE TABLE vc (v VARCHAR(4), c CHAR(4));
Query OK, 0 rows affected (0.16 sec)
```

(2)v 列和 c 列中同时插入字符串 "ab   ":

```
mysql> INSERT INTO vc VALUES ('ab  ', 'ab  ');
Query OK, 1 row affected (0.05 sec)
```

(3)显示查询结果:

```
mysql> select length(v),length(c) from vc;
+-----------+-----------+
| length(v) | length(c) |
+-----------+-----------+
|         4 |         2 |
+-----------+-----------+
1 row in set (0.01 sec)
```

可以发现,c 字段的 length 只有 2。给两个字段分别追加一个 "+" 字符,结果可以看得更清楚:

```
mysql> SELECT CONCAT(v, '+'), CONCAT(c, '+') FROM vc;
+----------------+----------------+
| CONCAT(v, '+') | CONCAT(c, '+') |
+----------------+----------------+
| ab  +          | ab+            |
+----------------+----------------+
1 row in set (0.00 sec)
```

显然,CHAR 列最后的空格在做操作时都已经被删除,而 VARCHAR 依然保留空格。

3.3.2　BINARY 和 VARBINARY 类型

BINARY 和 VARBINARY 类似于 CHAR 和 VARCHAR，不同的是它们包含二进制字符串而不包含非二进制字符串。在下面的例子中，对表 t 中的 binary 字段 c 插入一个字符，研究一下这个字符到底是如何存储的。

（1）创建测试表 t，字段为 c BINARY(3)：

```
mysql> CREATE TABLE t (c BINARY(3));
Query OK, 0 rows affected (0.14 sec)
```

（2）往 c 字段中插入字符 "a"：

```
mysql> INSERT INTO t SET  c='a';
Query OK, 1 row affected (0.06 sec)
```

（3）分别用以下几种模式来查看 c 列的内容：

```
mysql> select *,hex(c),c='a',c='a\0',c='a\0\0'  from t;
+------+--------+-------+---------+-----------+
| c    | hex(c) | c='a' | c='a\0' | c='a\0\0' |
+------+--------+-------+---------+-----------+
| a    | 610000 |     0 |       0 |         1 |
+------+--------+-------+---------+-----------+
1 row in set (0.00 sec)
```

可以发现，当保存 BINARY 值时，在值的最后通过填充 "0x00"（零字节）以达到指定的字段定义长度。从上例中看出，对于一个 BINARY(3) 列，当插入时 "a" 变为 "a\0\0"。

3.3.3　ENUM 类型

ENUM 中文名称叫枚举类型，它的值范围需要在创建表时通过枚举方式显式指定，对 1～255 个成员的枚举需要 1 个字节存储；对于 255～65535 个成员，需要 2 个字节存储。最多允许有 65535 个成员。下面往测试表 t 中插入几条记录来看看 ENUM 的使用方法。

（1）创建测试表 t，定义 gender 字段为枚举类型，成员为 "M" 和 "F"：

```
mysql> create table t (gender enum('M','F'));
Query OK, 0 rows affected (0.14 sec)
```

（2）插入 4 条不同的记录：

```
mysql> INSERT INTO t  VALUES('M'),('1'),('f'),(NULL);
Query OK, 4 rows affected (0.00 sec)
Records: 4  Duplicates: 0  Warnings: 0

mysql> select * from t;
+--------+
| gender |
+--------+
| M      |
| M      |
| F      |
| NULL   |
+--------+
4 rows in set (0.01 sec)
```

从上面的例子中，可以看出 ENUM 类型是忽略大小写的，在存储 'M'、'f' 时将它们都转成了大写；还可以插入数字来替代对应的值，上例中的字符 '1' 会隐式转换为数字 1，实际代表了枚举中的第一个元素 'M'，所以并没有报错；本例中如果插入 'M' / 'F'（忽略大小写）的其他字符或者 '1' / '2' 之外的其他数字，都将报 "ERROR 1265 (01000): Data truncated for column 'gender' at row 1" 的错误。读者可以自行验证。

另外，ENUM 类型只允许从值集合中选取单个值，而不能一次取多个值。

3.3.4 SET 类型

SET 和 ENUM 类型非常类似，也是一个字符串对象，里面可以包含 0~64 个成员。根据成员的不同，存储上也有所不同。

- 1~8 成员的集合，占 1 个字节。
- 9~16 成员的集合，占 2 个字节。
- 17~24 成员的集合，占 3 个字节。
- 25~32 成员的集合，占 4 个字节。
- 33~64 成员的集合，占 8 个字节。

SET 和 ENUM 除了存储之外，最主要的区别在于 SET 类型一次可以选取多个成员，而 ENUM 则只能选一个。下面的例子在表 t 中插入了多组不同的成员：

```
Create table  t (col  set ('a','b','c','d');
insert into t  values('a,b'),('a,d,a'),('a,b'),('a,c'),('a');
mysql> select * from t;
+------+
| col  |
+------+
| a,b  |
| a,d  |
| a,b  |
| a,c  |
| a    |
+------+
5 rows in set (0.00 sec)
```

SET 类型可以从允许值集合中选择任意 1 个或多个元素进行组合，所以对于输入的值只要是在允许值的组合范围内，都可以正确地注入到 SET 类型的列中。对于超出允许值范围的值例如（'a,d,f '）将不允许注入到上面例子中设置的 SET 类型列中，而对于（'a,d,a'）这样包含重复成员的集合将只取一次，写入后的结果为"a,d"，这一点请注意。

3.4 JSON 类型

JSON 是 JavaScript Object Notation 的缩写，它是一种数据交换格式。JSON 出现之前，数据交换大多使用 XML 来传递数据，但随着 XML 的规范越来越复杂，给开发人员的使用带来越来越多的困难。为了解决这个问题，雅虎工程师 Douglas Crockford 于 2002 年发明了 JSON 这种超级轻量级的数据交换格式，随后风靡 Web 世界。

自 5.7.8 版本起，MySQL 开始支持 JSON 类型，在此之前，通常使用 VARCHAR 或 TEXT 来保存 JSON 格式数据。JSON 类型比字符类型有如下优点：

- JSON 数据类型会自动校验数据是否为 JSON 格式，如果不是 JSON 格式数据，则会报错。
- MySQL 提供了一组操作 JSON 数据的内置函数，可以方便地提取各类数据，可以修改特定的键值。
- 优化的存储格式，存储在 JSON 列中的 JSON 数据被转换成内部的存储格式，允许快速读取。

3.4 JSON 类型

简单地说，JSON 实际就是 JavaScript 的一个子集，支持的数据类型包括 NUMBER、STRING、BOOLEAN、NULL、ARRAY、OBJECT 共 6 种，一个 JSON 中的元素可以是这 6 种类型元素的任意组合，其中 BOOLEAN 使用 true/false 的字面值文本表示；null 使用 null 的文本表示；字符串和日期类型都用双引号引起来表示；ARRAY 要用中括号引起来，OBJECT 保存的 KV 对要用大括号引起来，其中的 K 也要用双引号引起。下面是几个格式正确的例子：

```
["abc", 10, null, true, false]
{"k1": "value", "k2": 10}
["12:18:29.000000", "2015-07-29", "2015-07-29 12:18:29.000000"]
```

ARRAY 和 OBJECT 也可以嵌套引用，比如：

```
[99, {"id": "HK500", "cost": 75.99}, ["hot", "cold"]]
{"k1": "value", "k2": [10, 20]}
```

下面我们举例看一下 JSON 在 MySQL 中的使用。

首先，创建表 t1，列 id1 为 JSON 类型。

```
mysql> create table t1(id1 json);
Query OK, 0 rows affected (0.00 sec)
```

往表 t1 中插入如下 JSON 格式数据：

```
mysql> insert into t1 values('{"age":20, "time":"2018-07-14 10:52:00"}');
Query OK, 1 row affected (0.00 sec)
```

如果插入有语法错误的 JSON 数据，则直接报错：

```
mysql> INSERT INTO t1 VALUES('[1, 2,');
ERROR 3140 (22032): Invalid JSON text: "Invalid value." at position 6 in value for column 't1.id1'.
```

通过 JSON_TYPE 函数可以看到插入的 JSON 数据是哪种类型：

```
mysql> select json_type('"abc"') js1,json_type('[1,2,"abc"]') js2,json_type('{"k1": "value"}') js3;
+--------+-------+--------+
| js1    | js2   | js3    |
+--------+-------+--------+
| STRING | ARRAY | OBJECT |
+--------+-------+--------+
1 row in set (0.00 sec)
```

JSON 数据类型对于大小写是敏感的，'x'和'X'是不同的两个 JSON 数据，常见的 null、true、false 必须是小写才合法，通过 JSON_VALID 函数可以判断一个 JSON 数据是否合法，如下例所示：

```
mysql> SELECT JSON_VALID('null') n1, JSON_VALID('NULL') n2,JSON_VALID('false') f1, JSON_VALID('FALSE') f2;
+------+------+------+------+
| n1   | n2   | f1   | f2   |
+------+------+------+------+
|    1 |    0 |    1 |    0 |
+------+------+------+------+
1 row in set (0.00 sec)
```

可以看出，'null'和'false'语法合法，而'NULL'和'FALSE'则语法不合法。对大小写敏感的原因是 JSON 的默认排序规则是 utf8mb4_bin（具体见字符集一章）所致，这里不再详述。

还有一种特殊情况，如果 JSON 数据的 value 中字符串 value 中包括双引号或者单引号，则插入时需要加反斜线进行转义，如下例插入 value 为 ab"c 的一个 OBJECT 类型的 JSON 数据。

- 显式插入：

```
mysql> insert into t1 values(json_object("name","ab\"c"));
Query OK, 1 row affected (0.00 sec)
```

- 隐式插入：

```
mysql> insert into t1 values('{"name":"ab\"c"}');
ERROR 3140 (22032): Invalid JSON text: "Missing a comma or '}' after an object member." at position 13 in value for column 't1.id1'.
mysql> insert into t1 values('{"name":"ab\\"c"}');
Query OK, 1 row affected (0.01 sec)
```

可以发现，隐式插入时，要多加一个反斜线才可以正常识别。

MySQL 对 JSON 的存储做了一些限制，JSON 列不可有默认值，且文本的最大长度取决于系统常量：max_allowed_packet。该值仅在服务器进行存储的时候进行限制，在内存中进行计算的时候是允许超过该值的。

对于 JSON 数据的常用操作，MySQL 提供了很多函数，比如上面提到的 JSON_VALID/JSON_OBJECT 等，更多的函数在第 5 章中详细讲解。

3.5 小结

本章主要介绍了 MySQL 支持的各种数据类型，并通过多个实例对它们的使用方法做了详细的说明。学完本章后，读者可以对每种数据类型的用途、物理存储、表示范围等有一个概要的了解。这样在面对具体应用时，就可以根据相应的特点来选择合适的数据类型，使得我们能够争取在满足应用的基础上，用较小的存储代价换来较高的数据库性能。

第 4 章 MySQL 中的运算符

MySQL 支持多种类型的运算符,这些运算符可以用来连接表达式的项。运算符的类型主要包括算术运算符、比较运算符、逻辑运算符和位运算符。本章将通过实例对 MySQL 5.7 支持的这几种运算符进行详细地介绍。

4.1 算术运算符

MySQL 支持的算术运算符包括加、减、乘、除和模运算。它们是最常使用、最简单的一类运算符。表 4-1 列出了这些运算符及其作用。

表 4-1　　　　　　　　　　MySQL 支持的算术运算符

运 算 符	作　　用
+	加法
–	减法
*	乘法
/, DIV	除法,返回商
%, MOD	除法,返回余数

下例中简单地描述了这几种运算符的使用方法。

```
mysql> select 0.1+ 0.3333 ,0.1-0.3333, 0.1*0.3333, 1/2,1%2;
+------------+------------+------------+--------+------+
| 0.1+ 0.3333 | 0.1-0.3333 | 0.1*0.3333 | 1/2    | 1%2  |
+------------+------------+------------+--------+------+
|     0.4333 |    -0.2333 |    0.03333 | 0.5000 |    1 |
+------------+------------+------------+--------+------+
1 row in set (0.00 sec)
```

- "+" 运算符用于获得一个或多个值的和。
- "–" 运算符用于从一个值中减去另一个值。
- "*" 运算符使数字相乘,得到两个或多个值的乘积。
- "/" 运算符用一个值除以另一个值得到商。
- "%" 运算符用一个值除以另外一个值得到余数。

在除法运算和模运算中,如果除数为 0,将是非法除数,返回结果为 NULL,如下例所示:

```
mysql> select 1/0, 100%0 ;
+------+-------+
| 1/0  | 100%0 |
```

```
+------+------+
| NULL | NULL |
+------+------+
1 row in set (0.02 sec)
```

对于模运算，还有另一种表达方式，使用 MOD(a,b)函数与 a%b 效果一样：

```
mysql> select 3%2,mod(3,2);
+-----+----------+
| 3%2 | mod(3,2) |
+-----+----------+
|   1 |        1 |
+-----+----------+
1 row in set (0.01 sec)
```

4.2 比较运算符

熟悉了最简单的算术运算符，再来看一下比较运算符。当使用 SELECT 语句进行查询时，MySQL 允许用户对表达式的左边操作数和右边操作数进行比较，比较结果为真，则返回 1，为假则返回 0，比较结果不确定则返回 NULL。表 4-2 列出了 MySQL 5.7 支持的各种比较运算符。

表 4-2　　　　　　　　　　MySQL 支持的比较运算符

运算符	作用
=	等于
<>或!=	不等于
<=>	NULL 安全的等于（NULL-safe）
<	小于
<=	小于等于
>	大于
>=	大于等于
BETWEEN	存在于指定范围
IN	存在于指定集合
IS NULL	为 NULL
IS NOT NULL	不为 NULL
LIKE	通配符匹配
REGEXP 或 RLIKE	正则表达式匹配

比较运算符可以用于比较数字、字符串和表达式。数字作为浮点数比较，而字符串以不区分大小写的方式进行比较。下面通过实例来学习各种比较运算符的使用。

● "="运算符，用于比较运算符两侧的操作数是否相等，如果两侧操作数相等，则返回值为 1，否则为 0。注意 NULL 不能用于"="比较。

```
mysql> select 1=0,1=1,NULL=NULL;
+-----+-----+-----------+
| 1=0 | 1=1 | NULL=NULL |
+-----+-----+-----------+
|   0 |   1 |      NULL |
+-----+-----+-----------+
1 row in set (0.00 sec)
```

4.2 比较运算符

- "<>" 运算符，和 "=" 相反，如果两侧操作数不等，则值为 1，否则为 0。NULL 不能用于 "<>" 比较。

```
mysql> select 1<>0,1<>1,null<>null;
+------+------+-----------+
| 1<>0 | 1<>1 | null<>null |
+------+------+-----------+
|    1 |    0 |      NULL |
+------+------+-----------+
1 row in set (0.00 sec)
```

- "<=>" 运算符，和 "=" 类似，在操作数相等时值为 1，不同之处在于即使操作的值为 NULL 也可以正确比较。

```
mysql> select 1<=>1,2<=>0 ,0<=>0, NULL <=>NULL;
+-------+-------+-------+-------------+
| 1<=>1 | 2<=>0 | 0<=>0 | NULL <=>NULL |
+-------+-------+-------+-------------+
|     1 |     0 |     1 |           1 |
+-------+-------+-------+-------------+
1 row in set (0.17 sec)
```

- "<" 运算符，当左侧操作数小于右侧操作数时，其返回值为 1，否则其值为 0。

```
mysql> select 'a'<'b' ,'a'<'a' ,'a'<'c',1<2;
+---------+---------+---------+-----+
| 'a'<'b' | 'a'<'a' | 'a'<'c' | 1<2 |
+---------+---------+---------+-----+
|       1 |       0 |       1 |   1 |
+---------+---------+---------+-----+
1 row in set (0.03 sec)
```

- "<=" 运算符，当左侧操作数小于等于右侧操作数时，其返回值为 1，否则返回值为 0。

```
mysql> select 'bdf'<='b','b'<='b' ,0<1;
+-----------+----------+-----+
| 'bdf'<='b' | 'b'<='b' | 0<1 |
+-----------+----------+-----+
|         0 |        1 |   1 |
+-----------+----------+-----+
1 row in set (0.00 sec)
```

- ">" 运算符，当左侧操作数大于右侧操作数时，其返回值为 1，否则返回值为 0。

```
mysql> select 'a'>'b','abc'>'a' ,1>0;
+---------+-----------+-----+
| 'a'>'b' | 'abc'>'a' | 1>0 |
+---------+-----------+-----+
|       0 |         1 |   1 |
+---------+-----------+-----+
1 row in set (0.03 sec)
```

- ">=" 运算符，当左侧操作数大于等于右侧操作数时，其返回值为 1，否则返回值为 0。

```
mysql> select 'a'>='b','abc'>='a' ,1>=0 ,1>=1;
+----------+------------+------+------+
| 'a'>='b' | 'abc'>='a' | 1>=0 | 1>=1 |
+----------+------------+------+------+
|        0 |          1 |    1 |    1 |
+----------+------------+------+------+
1 row in set (0.00 sec)
```

- "BETWEEN" 运算符的使用格式为 "a BETWEEN min AND max"，当 a 大于等于 min 并且小于等于 max，则返回值为 1，否则返回 0；当操作数 a、min、max 类型相同时，此表达式等价于（a>=min and a<=max），当操作数类型不同时，比较时会遵循类型转换原则进行转换后，再进行比较运算。下例中描述了 BETWEEN 的用法：

```
mysql> select 10 between 10 and 20, 9 between 10 and 20;
+----------------------+---------------------+
| 10 between 10 and 20 | 9 between 10 and 20 |
+----------------------+---------------------+
|                    1 |                   0 |
+----------------------+---------------------+
1 row in set (0.00 sec)
```

- "IN"运算符的使用格式为"a IN (value1,value2,…)",当 a 的值存在于列表中时,则整个比较表达式返回的值为 1,否则返回 0。

```
mysql> select 1 in (1,2,3) , 't' in ('t','a','b','l','e'),0 in (1,2);
+--------------+-----------------------------+------------+
| 1 in (1,2,3) | 't' in ('t','a','b','l','e') | 0 in (1,2) |
+--------------+-----------------------------+------------+
|            1 |                           1 |          0 |
+--------------+-----------------------------+------------+
1 row in set (0.00 sec)
```

- "IS NULL"运算符的使用格式为"a IS NULL",当 a 的值为 NULL,则返回值为 1,否则返回 0。

```
mysql> select 0 is null, null is null;
+-----------+--------------+
| 0 is null | null is null |
+-----------+--------------+
|         0 |            1 |
+-----------+--------------+
1 row in set (0.02 sec)
```

- "IS NOT NULL"运算符的使用格式为"a IS NOT NULL"。和"IS NULL"相反,当 a 的值不为 NULL,则返回值为 1,否则返回 0。

```
mysql> select 0 is not  null, null is not  null;
+---------------+------------------+
| 0 is not null | null is not null |
+---------------+------------------+
|             1 |                0 |
+---------------+------------------+
1 row in set (0.00 sec)
```

- "LIKE"运算符的使用格式为"a LIKE %123%",当 a 中含有字符串"123"时,则返回值为 1,否则返回 0。

```
mysql> select 123456 like '123%',123456 like '%123%',123456 like '%321%';
+--------------------+---------------------+---------------------+
| 123456 like '123%' | 123456 like '%123%' | 123456 like '%321%' |
+--------------------+---------------------+---------------------+
|                  1 |                   1 |                   0 |
+--------------------+---------------------+---------------------+
1 row in set (0.00 sec)
```

- "REGEXP"运算符的使用格式为"str   REGEXP str_pat",当 str 字符串中含有 str_pat 相匹配的字符串时,则返回值为 1,否则返回 0。REGEXP 运算符的使用方法将会在第 15 章中详细介绍。

```
mysql> select 'abcdef' regexp 'ab' ,'abcdefg' regexp 'k';
+----------------------+----------------------+
| 'abcdef' regexp 'ab' | 'abcdefg' regexp 'k' |
+----------------------+----------------------+
|                    1 |                    0 |
+----------------------+----------------------+
1 row in set (0.00 sec)
```

4.3 逻辑运算符

逻辑运算符又称为布尔运算符，用来确认表达式的真和假。MySQL 支持 4 种逻辑运算符，如表 4-3 所示。

表 4-3　　　　　　　　　　　　MySQL 中的逻辑运算符

运算符	作用
NOT 或 !	逻辑非
AND 或 &&	逻辑与
OR 或 ||	逻辑或
XOR	逻辑异或

- "NOT" 或 "!" 表示逻辑非。返回和操作数相反的结果：当操作数为 0（假），则返回值为 1，否则值为 0。但是有一点除外，那就是 NOT NULL 的返回值为 NULL，如下例所示。

```
mysql> select not 0, not 1, not null ;
+-------+-------+----------+
| not 0 | not 1 | not null |
+-------+-------+----------+
|     1 |     0 |     NULL |
+-------+-------+----------+
1 row in set (0.00 sec)
```

- "AND" 或 "&&" 表示逻辑与运算。当所有操作数均为非零值并且不为 NULL 时，计算所得结果为 1；当一个或多个操作数为 0 时，所得结果为 0；操作数中有任何一个为 NULL，则返回值为 NULL，如下例所示。

```
mysql> select (1 and 1),(0 and 1) ,(3 and 1 ) ,(1 and null);
+---------+---------+----------+------------+
| (1 and 1) | (0 and 1) | (3 and 1 ) | (1 and null) |
+---------+---------+----------+------------+
|       1 |       0 |        1 |       NULL |
+---------+---------+----------+------------+
1 row in set (0.00 sec)
```

- "OR" 或 "||" 表示逻辑或运算。当两个操作数均为非 NULL 值时，如有任意一个操作数为非零值，则结果为 1，否则结果为 0；当有一个操作数为 NULL 时，如另一个操作数为非零值，则结果为 1，否则结果为 NULL；假如两个操作数均为 NULL，则所得结果为 NULL，如下例所示。

```
mysql> select (1 or 0) ,(0 or 0),(1 or null) ,(1 or 1),(null or null);
+--------+--------+-----------+--------+-------------+
| (1 or 0) | (0 or 0) | (1 or null) | (1 or 1) | (null or null) |
+--------+--------+-----------+--------+-------------+
|      1 |      0 |         1 |      1 |        NULL |
+--------+--------+-----------+--------+-------------+
1 row in set (0.00 sec)
```

- "XOR" 表示逻辑异或。当任意一个操作数为 NULL 时，返回值为 NULL；对于非 NULL 的操作数，如果两个的逻辑真假值相异，则返回结果 1，否则返回 0，如下例所示。

```
mysql> select 1 xor 1 ,0 xor 0,1 xor 0,0 xor 1,null xor 1;
+---------+---------+---------+---------+------------+
| 1 xor 1 | 0 xor 0 | 1 xor 0 | 0 xor 1 | null xor 1 |
```

```
|       0 |       0 |       1 |       1 |    NULL |
+---------+---------+---------+---------+---------+
1 row in set (0.00 sec)
```

4.4 位运算符

位运算是将给定的操作数转化为二进制后,对各个操作数每一位都进行指定的逻辑运算,得到的二进制结果转换为十进制数后就是位运算的结果。MySQL 5.7 支持 6 种位运算符,如表 4-4 所示。

表 4-4　　　　　　　　　　MySQL 支持的位运算符

运算符	作用
&	位与（位 AND）
|	位或（位 OR）
^	位异或（位 XOR）
~	位取反
>>	位右移
<<	位左移

可以发现,位运算符中的位与 "&"、位或 "|" 和前面介绍的逻辑与和逻辑或非常类似。其他操作符和逻辑操作有所不同,下面将分别举例介绍。

● "位与" 对多个操作数的二进制位做逻辑与操作,例如 2&3,因为 2 的二进制数是 10,3 是 11,所以 10&11 的结果是 10,十进制数字还是 2,来看实际结果:

```
mysql> select 2&3;
+-----+
| 2&3 |
+-----+
|   2 |
+-----+
1 row in set (0.00 sec)
```

可以对两个以上操作数做 "或" 操作,测试一下 2&3&4,因为 4 的二进制是 100,和上面的 10 做与操作 100&010 后,结果应该是 000,可以看实际结果:

```
mysql> select 2&3&4;
+-------+
| 2&3&4 |
+-------+
|     0 |
+-------+
1 row in set (0.00 sec)
```

● "位或" 对多个操作数的二进制位做逻辑或操作,还是上面的例子,2|3 的结果应该是 10|11,结果还是 11,应该是 3,实际结果如下:

```
mysql> select 2|3;
+-----+
| 2|3 |
+-----+
|   3 |
+-----+
1 row in set (0.00 sec)
```

● "位异或" 对操作数的二进制位做异或操作,10^11 的结果是 01,结果应该是 1,可以看实际结果为:

```
mysql> select 2^3;
+-----+
| 2^3 |
+-----+
|  1  |
+-----+
1 row in set (0.00 sec)
```

- "位取反"对操作数的二进制位做 NOT 操作，这里的操作数只能是一位。下面看一个经典的取反例子，对 1 做位取反，具体如下所示：

```
mysql> select ~1 ,~ 18446744073709551614;
+----------------------+---------------------------+
| ~1                   | ~ 18446744073709551614    |
+----------------------+---------------------------+
| 18446744073709551614 |            1              |
+----------------------+---------------------------+
1 row in set (0.00 sec)
```

结果可能会令人感到疑惑，1 的位取反怎么会是这么大的数字？来研究一下，在 MySQL 中，常量数字默认会以 8 个字节来表示，8 个字节就是 64 位，常量 1 的二进制表示为 63 个 "0" 加 1 个 "1"，位取反后就是 63 个 "1" 加一个 "0"，转换为二进制后就是 18446744073709551614，实际结果如下：

```
mysql> select bin(18446744073709551614);
+------------------------------------------------------------------+
| bin(18446744073709551614)                                        |
+------------------------------------------------------------------+
| 1111111111111111111111111111111111111111111111111111111111111110 |
+------------------------------------------------------------------+
1 row in set (0.00 sec)
```

- "位右移"对左操作数向右移动右操作数指定的位数。例如 100>>3，就是对 100 的二进制数 0001100100 右移 3 位，左边补 0，结果是 0000001100，转换为二进制数是 12，实际结果如下：

```
mysql> select 100>>3;
+--------+
| 100>>3 |
+--------+
|   12   |
+--------+
1 row in set (0.00 sec)
```

- "位左移"对左操作数向左移动右操作数指定的位数。例如 100<<3，就是对 100 的二进制数 0001100100 左移 3 位，右边补 0，结果是 1100100000，转换为二进制数是 800，实际结果如下：

```
mysql> select 100<<3;
+--------+
| 100<<3 |
+--------+
|  800   |
+--------+
1 row in set (0.00 sec)
```

4.5 运算符的优先级

前面介绍了 MySQL 支持的各种运算符的使用方法。在实际应用中，很可能将这些运算符进行混合运算，那么应该先进行哪些运算符的操作呢？表 4-5 中列出了所有的运算符，优先级由低到高排列，同一行中的运算符具有相同的优先级。

表 4-5　　　　　　　　　　　MySQL 中的运算符优先级

优先级顺序	运 算 符
1	:=
2	||、OR、XOR
3	&&、AND
4	NOT
5	BETWEEN、CASE、WHEN、THEN 和 ELSE
6	=、<=>、>=、>、<=、<、<>、!=、IS、LIKE、REGEXP 和 IN
7	|
8	&
9	<<和>>
10	-和+
11	*、/、DIV、%和 MOD
12	^
13	-（一元减号）和~（一元比特反转）
14	!

在实际运行的时候，可以参考表 4-5 中的优先级。实际上，很少有人能将这些优先级熟练记忆，很多情况下我们都是用"( )"将需要优先的操作括起来，这样既起到了优先的作用，又使得其他用户看起来更易于理解。

4.6　小结

本章主要介绍了 MySQL 中支持的各种运算符。这些运算符可以帮助用户完成算术、比较、逻辑和位逻辑操作，读者在使用时要注意运算符的优先级。另外，在使用比较运算符时要保证比较的操作数类型是一致的，这样可以避免由于操作数类型的不一致而得出错误的数据。

第 5 章 常用函数

经常编写程序的朋友一定体会得到函数的重要性，丰富的函数往往能使用户的工作事半功倍。函数能帮助用户做很多事情，比如说字符串的处理、数值的运算、日期的运算等，在这方面 MySQL 提供了多种内建函数帮助开发人员编写简单快捷的 SQL 语句，其中常用的函数有字符串函数、日期函数和数值函数。

在 MySQL 数据库中，函数可以用在 SELECT 语句及其子句（例如 WHERE、ORDER BY、HAVING 等）中，也可以用在 UPDATE、DELETE 语句及其子句中。本章将配合一些实例对这些常用函数进行详细的介绍。

5.1 字符串函数

字符串函数是最常用的一种函数，如果读者编写过程序，不妨回过头去看看自己使用过的函数，可能会惊讶地发现字符串处理的相关函数占已使用过的函数很大一部分。在 MySQL 中，字符串函数同样是最丰富的一类函数。表 5-1 列出了这些常用字符串函数，以供参考。

表 5-1　　　　　　　　　MySQL 中的常用字符串函数

函　　数	功　　能
CONCAT(s1,s2,...,sn)	连接 s1,s2,...,sn 为一个字符串
INSERT(str,x,y,instr)	将字符串 str 从第 x 位置开始，y 个字符长的子串替换为字符串 instr
LOWER(str)	将字符串 str 中所有字符变为小写
UPPER(str)	将字符串 str 中所有字符变为大写
LEFT(str,x)	返回字符串 str 最左边的 x 个字符
RIGHT(str,x)	返回字符串 str 最右边的 x 个字符
LPAD(str,n,pad)	用字符串 pad 对 str 最左边进行填充，直到长度为 n 个字符长度
RPAD(str,n,pad)	用字符串 pad 对 str 最右边进行填充，直到长度为 n 个字符长度
LTRIM(str)	去掉字符串 str 左侧的空格
RTRIM(str)	去掉字符串 str 行尾的空格
REPEAT(str,x)	返回 str 重复 x 次的结果
REPLACE(str,a,b)	用字符串 b 替换字符串 str 中所有出现的字符串 a
STRCMP(s1,s2)	比较字符串 s1 和 s2

函　　数	功　　能
TRIM(str)	去掉字符串行尾和行头的空格
SUBSTRING(str,x,y)	返回从字符串 str x 位置起 y 个字符长度的字串

下面通过具体的实例来逐个地研究每个函数的用法，需要注意的是，这里的例子仅仅用于说明各个函数的使用方法，所以函数都是单个出现的，但是在一个具体的应用中通常可能需要综合几个甚至几类函数才能实现相应的应用。

- CONCAT(s1,s2,…,sn)函数：把传入的参数连接成为一个字符串。

下面的例子把"aaa""bbb""ccc"3 个字符串连接成了一个字符串"aaabbbccc"。另外，任何字符串与 NULL 进行连接的结果都将是 NULL。

```
mysql> select concat('aaa', 'bbb', 'ccc') ,concat('aaa',null);
+-----------------------------+--------------------+
| concat('aaa','bbb','ccc')   | concat('aaa',null) |
+-----------------------------+--------------------+
| aaabbbccc                   | NULL               |
+-----------------------------+--------------------+
1 row in set (0.05 sec)
```

- INSERT(str,x,y,instr)函数：将字符串 str 从第 x 位置开始，y 个字符长的子串替换为字符串 instr。

下面的例子把字符串"beijing2008you"中从第 12 个字符开始以后的 3 个字符替换成"me"。

```
mysql> select INSERT('beijing2008you',12,3, 'me') ;
+-------------------------------------+
| INSERT('beijing2008you',12,3, 'me') |
+-------------------------------------+
| beijing2008me                       |
+-------------------------------------+
1 row in set (0.00 sec)
```

- LOWER(str)和 UPPER(str)函数：把字符串转换成小写或大写。

在字符串比较中，通常要将比较的字符串全部转换为大写或者小写，如下例所示：

```
mysql>  select LOWER('BEIJING2008'), UPPER('beijing2008');
+----------------------+----------------------+
| LOWER('BEIJING2008') | UPPER('beijing2008') |
+----------------------+----------------------+
| beijing2008          | BEIJING2008          |
+----------------------+----------------------+
1 row in set (0.00 sec)
```

- LEFT(str,x)和 RIGHT(str,x)函数：分别返回字符串最左边的 x 个字符和最右边的 x 个字符。如果第二个参数是 NULL，那么将不返回任何字符串。

下例中显示了对字符串"beijing2008"应用函数后的结果。

```
mysql> SELECT LEFT('beijing2008',7),LEFT('beijing',null),RIGHT('beijing2008',4);

+-----------------------+----------------------+------------------------+
| LEFT('beijing2008',7) | LEFT('beijing',null) | RIGHT('beijing2008',4) |
+-----------------------+----------------------+------------------------+
| beijing               |                      | 2008                   |
+-----------------------+----------------------+------------------------+
1 row in set (0.00 sec)
```

- LPAD(str,n,pad)和 RPAD(str,n,pad)函数：用字符串 pad 对 str 最左边和最右边进行填充，直到长度为 n 个字符长度。

5.1 字符串函数

下例中显示了对字符串"2008"和"beijing"分别填充后的结果。

```
mysql> select lpad('2008',20,'beijing'),rpad('beijing',20,'2008');
+---------------------------+---------------------------+
| lpad('2008',20,'beijing') | rpad('beijing',20,'2008') |
+---------------------------+---------------------------+
| beijingbeijingbe2008      | beijing2008200820082      |
+---------------------------+---------------------------+
1 row in set (0.00 sec)
```

- LTRIM(str)和 RTRIM(str)函数：去掉字符串 str 左侧和右侧空格。

下例中显示了字符串"beijing"加空格进行过滤后的结果。

```
mysql> select ltrim('  |beijing'),rtrim('beijing |     ');
+---------------------+-------------------------+
| ltrim('  |beijing') | rtrim('beijing|     ')  |
+---------------------+-------------------------+
| |beijing            | beijing|                |
+---------------------+-------------------------+
1 row in set (0.00 sec)
```

- REPEAT(str,x)函数：返回 str 重复 x 次的结果。

下例中对字符串"mysql"重复显示了3次。

```
mysql> select  repeat('mysql ',3);
+--------------------+
| repeat('mysql ',3) |
+--------------------+
| mysql mysql mysql  |
+--------------------+
1 row in set (0.00 sec)
```

- REPLACE(str,a,b)函数：用字符串 b 替换字符串 str 中所有出现的字符串 a。

下例中用字符串"2008"代替了字符串"beijing_2010"中的"_2010"。

```
mysql> select replace('beijing_2010','_2010','2008');
+----------------------------------------+
| replace('beijing_2010','_2010','2008') |
+----------------------------------------+
| beijing2008                            |
+----------------------------------------+
1 row in set (0.00 sec)
```

- STRCMP(s1,s2)函数：比较字符串 s1 和 s2 的 ASCII 码值的大小。

如果 s1 比 s2 小，那么返回-1；如果 s1 与 s2 相等，那么返回 0；如果 s1 比 s2 大，那么返回 1，如下例所示。

```
mysql> select  strcmp('a','b'),strcmp('b','b'),strcmp('c','b');
+-----------------+-----------------+-----------------+
| strcmp('a','b') | strcmp('b','b') | strcmp('c','b') |
+-----------------+-----------------+-----------------+
|              -1 |               0 |               1 |
+-----------------+-----------------+-----------------+
1 row in set (0.00 sec)
```

- TRIM(str)函数：去掉目标字符串的开头和结尾的空格。

下例中对字符串" $ beijing2008 $ "进行了前后空格的过滤。

```
mysql> select trim(' $ beijing2008 $    ');
+------------------------------+
| trim(' $ beijing2008 $    ') |
+------------------------------+
| $ beijing2008 $              |
+------------------------------+
1 row in set (0.00 sec)
```

- SUBSTRING(str,x,y)函数：返回从字符串 str 中的第 x 位置起 y 个字符长度的子串。

此函数经常用来对给定字符串进行字串的提取，如下例所示。

```
mysql> select substring('beijing2008',8,4),substring('beijing2008',1,7);
+------------------------------+------------------------------+
| substring('beijing2008',8,4) | substring('beijing2008',1,7) |
+------------------------------+------------------------------+
| 2008                         | beijing                      |
+------------------------------+------------------------------+
```

5.2 数值函数

MySQL 中另外一类很重要的函数就是数值函数，这些函数能处理很多数值方面的运算。可以想象，如果没有这些函数的支持，用户在编写有关数值运算方面的代码时将会困难重重。举个例子，如果没有 ABS 函数，要取一个数值的绝对值，就需要进行好多次判断才能返回这个值，而数值函数能够大大提高用户的工作效率。表 5-2 中列出了在 MySQL 中会经常使用的数值函数。

表 5-2　　　　　　　　　　　MySQL 中的常用数值函数

函　　数	功　　能
ABS(x)	返回 x 的绝对值
CEIL(x)	返回大于 x 的最小整数值
FLOOR(x)	返回小于 x 的最大整数值
MOD(x,y)	返回 x/y 的模
RAND()	返回 0～1 内的随机值
ROUND(x,y)	返回参数 x 的四舍五入的有 y 位小数的值
TRUNCATE(x,y)	返回数字 x 截断为 y 位小数的结果

下面将结合实例对这些函数进行介绍。

- ABS(x)函数：返回 x 的绝对值。

下例中显示了对正数和负数分别取绝对值之后的结果。

```
mysql> select ABS(-0.8) ,ABS(0.8);
+-----------+----------+
| ABS(-0.8) | ABS(0.8) |
+-----------+----------+
|       0.8 |      0.8 |
+-----------+----------+
1 row in set (0.09 sec)
```

- CEIL(x)函数：返回大于 x 的最小整数。

下例中显示了对 0.8 和-0.8 分别 CEIL 后的结果。

```
mysql> select CEIL(-0.8),CEIL(0.8);
+------------+-----------+
| CEIL(-0.8) | CEIL(0.8) |
+------------+-----------+
|          0 |         1 |
+------------+-----------+
1 row in set (0.03 sec)
```

- FLOOR(x)函数：返回小于 x 的最大整数，和 CEIL 的用法刚好相反。

下例中显示了对 0.8 和-0.8 分别 FLOOR 后的结果。

```
mysql> select FLOOR(-0.8), FLOOR(0.8);
+-------------+------------+
| FLOOR(-0.8) | FLOOR(0.8) |
```

5.2 数值函数

```
+------------+-----------+
|         -1 |         0 |
+------------+-----------+
1 row in set (0.00 sec)
```

- MOD(x,y)函数：返回 x/y 的模。

和 x%y 的结果相同，模数和被模数任何一个为 NULL 结果都为 NULL，如下例所示：

```
mysql> select MOD(15,10),MOD(1,11),MOD(NULL,10);
+------------+-----------+--------------+
| MOD(15,10) | MOD(1,11) | MOD(NULL,10) |
+------------+-----------+--------------+
|          5 |         1 |         NULL |
+------------+-----------+--------------+
1 row in set (0.00 sec)
```

- RAND()函数：返回 0~1 的随机值。

每次执行结果都不一样，如下例所示：

```
mysql> select RAND(),RAND();
+------------------+------------------+
| RAND()           | RAND()           |
+------------------+------------------+
| 0.12090325459922 | 0.83369727882901 |
+------------------+------------------+
1 row in set (0.00 sec)
```

利用此函数可以取任意指定范围内的随机数，比如需要产生 0~100 的任意随机整数，可以进行如下操作：

```
mysql> select ceil(100*rand()),ceil(100*rand());
+------------------+------------------+
| ceil(100*rand()) | ceil(100*rand()) |
+------------------+------------------+
|               91 |               15 |
+------------------+------------------+
1 row in set (0.00 sec)
```

- ROUND(x,y)函数：返回参数 x 的四舍五入的有 y 位小数的值。

如果是整数，将会保留 y 位数量的 0；如果不写 y，则默认 y 为 0，即将 x 四舍五入后取整。适合于将所有数字保留同样小数位的情况，如下例所示：

```
mysql> select ROUND(1.1),ROUND(1.1,2),ROUND(1,2);
+------------+--------------+------------+
| ROUND(1.1) | ROUND(1.1,2) | ROUND(1,2) |
+------------+--------------+------------+
|          1 |         1.10 |       1.00 |
+------------+--------------+------------+
1 row in set (0.00 sec)
```

- TRUNCATE(x,y)函数：返回数字 x 截断为 y 位小数的结果。

注意 TRUNCATE 和 ROUND 的区别在于 TRUNCATE 仅仅是截断，而不进行四舍五入。下例中描述了两者的区别：

```
mysql> select ROUND(1.235,2),TRUNCATE(1.235,2);
+----------------+-------------------+
| ROUND(1.235,2) | TRUNCATE(1.235,2) |
+----------------+-------------------+
|           1.24 |              1.23 |
+----------------+-------------------+
1 row in set (0.00 sec)
```

5.3 日期和时间函数

有时我们可能会遇到这样的需求：当前时间是多少；下个月的今天是星期几；统计截止到当前日期前 3 天的收入总和；等等。这些需求就需要日期和时间函数来实现，表 5-3 列出了 MySQL 中支持的一些常用日期和时间函数。

表 5-3　　　　　　　　　　MySQL 中的常用日期和时间函数

函　　数	功　　能
CURDATE()	返回当前日期
CURTIME()	返回当前时间
NOW()	返回当前的日期和时间
UNIX_TIMESTAMP(date)	返回日期 date 的 UNIX 时间戳
FROM_UNIXTIME	返回 UNIX 时间戳的日期值
WEEK(date)	返回日期 date 为一年中的第几周
YEAR(date)	返回日期 date 的年份
HOUR(time)	返回 time 的小时值
MINUTE(time)	返回 time 的分钟值
MONTHNAME(date)	返回 date 的月份名
DATE_FORMAT(date,fmt)	返回按字符串 fmt 格式化日期 date 值
DATE_ADD(date,INTERVAL expr type)	返回一个日期或时间值加上一个时间间隔的时间值
DATEDIFF(expr,expr2)	返回起始时间 expr 和结束时间 expr2 之间的天数

下面结合一些实例来逐个讲解每个函数的使用方法。

- CURDATE()函数：返回当前日期，只包含年、月、日。

```
mysql> select CURDATE();
+------------+
| CURDATE()  |
+------------+
| 2007-07-11 |
+------------+
1 row in set (0.03 sec)
```

- CURTIME()函数：返回当前时间，只包含时、分、秒。

```
mysql> select CURTIME();
+-----------+
| CURTIME() |
+-----------+
| 14:13:46  |
+-----------+
1 row in set (0.00 sec)
```

- NOW()函数：返回当前的日期和时间，年、月、日、时、分、秒全都包含。

```
mysql> select NOW();
+---------------------+
| NOW()               |
+---------------------+
| 2007-07-11 14:14:06 |
+---------------------+
1 row in set (0.00 sec)
```

- UNIX_TIMESTAMP(date)函数：返回日期 date 的 UNIX 时间戳。

5.3 日期和时间函数

```
mysql> select UNIX_TIMESTAMP(now());
+----------------------+
| UNIX_TIMESTAMP(now()) |
+----------------------+
|           1184134516 |
+----------------------+
1 row in set (0.02 sec)
```

● FROM_UNIXTIME(unixtime)函数：返回 UNIXTIME 时间戳的日期值，和 UNIX_TIMESTAMP(date)互为逆操作。

```
mysql> select FROM_UNIXTIME(1184134516);
+---------------------------+
| FROM_UNIXTIME(1184134516) |
+---------------------------+
| 2007-07-11 14:15:16       |
+---------------------------+
1 row in set (0.00 sec)
```

● WEEK(DATE)和 YEAR(DATE)函数：前者返回所给的日期是一年中的第几周，后者返回所给的日期是哪一年。

```
mysql> select WEEK(now()),YEAR(now());
+-------------+-------------+
| WEEK(now()) | YEAR(now()) |
+-------------+-------------+
|          27 |        2007 |
+-------------+-------------+
1 row in set (0.02 sec)
```

● HOUR(time)和 MINUTE(time)函数：前者返回所给时间的小时，后者返回所给时间的分钟。

```
mysql> select HOUR(CURTIME()),MINUTE(CURTIME());
+-----------------+-------------------+
| HOUR(CURTIME()) | MINUTE(CURTIME()) |
+-----------------+-------------------+
|              14 |                18 |
+-----------------+-------------------+
1 row in set (0.00 sec)
```

● MONTHNAME(date)函数：返回 date 的英文月份名称。

```
mysql> select MONTHNAME(now());
+------------------+
| MONTHNAME(now()) |
+------------------+
| July             |
+------------------+
1 row in set (0.00 sec)
```

● DATE_FORMAT(date,fmt)函数：按字符串 fmt 格式化日期 date 值，此函数能够按指定的格式显示日期，可以用到的格式符如表 5-4 所示。

表 5-4　　　　　　　　　　MySQL 中的日期和时间格式

格 式 符	格 式 说 明
%S 和%s	两位数字形式的秒（00,01,…,59）
%i	两位数字形式的分（00,01,…,59）
%H	两位数字形式的小时，24 小时（00,01,…,23）
%h 和%I	两位数字形式的小时，12 小时（01,02,…,12）
%k	数字形式的小时，24 小时（0,1,…,23）
%l	数字形式的小时，12 小时（1,2,…,12）

续表

格 式 符	格 式 说 明
%T	24 小时的时间形式（hh:mm:ss）
%r	12 小时的时间形式（hh:mm:ssAM 或 hh:mm:ssPM）
%p	AM 或 PM
%W	一周中每一天的名称（Sunday,Monday,…,Saturday）
%a	一周中每一天名称的缩写（Sun,Mon,…,Sat）
%d	两位数字表示月中的天数（00,01,…,31）
%e	数字形式表示月中的天数（1,2,…,31）
%D	英文后缀表示月中的天数（1st,2nd,3rd,…）
%w	以数字形式表示周中的天数（0=Sunday,1=Monday,…,6=Saturday）
%j	以 3 位数字表示年中的天数（001,002,…,366）
%U	周（0,1,52），其中 Sunday 为周中的第一天
%u	周（0,1,52），其中 Monday 为周中的第一天
%M	月名（January,February,…,December）
%b	缩写的月名（January,February,…,December）
%m	两位数字表示的月份（01,02,…,12）
%c	数字表示的月份（1,2,…,12）
%Y	4 位数字表示的年份
%y	两位数字表示的年份
%%	直接值"%"

下面的例子将当前时间显示为"月，日，年"格式：

```
mysql> select DATE_FORMAT(now(),'%M,%D,%Y');
+-------------------------------+
| DATE_FORMAT(now(),'%M,%D,%Y') |
+-------------------------------+
| July,11th,2007                |
+-------------------------------+
1 row in set (0.00 sec)
```

- DATE_ADD(date,INTERVAL expr type)函数：返回与所给日期 date 相差 INTERVAL 时间段的日期。

其中 INTERVAL 是间隔类型关键字，expr 是一个表达式，这个表达式对应后面的类型，type 是间隔类型。MySQL 提供了 13 种间隔类型，如表 5-5 所示。

表 5-5 　　　　　　　　　　MySQL 中的日期间隔类型

表达式类型	描　述	格　式
HOUR	小时	hh
MINUTE	分	mm
SECOND	秒	ss
YEAR	年	YY
MONTH	月	MM
DAY	日	DD
YEAR_MONTH	年和月	YY-MM
DAY_HOUR	日和小时	DD hh

续表

表达式类型	描述	格式
DAY_MINUTE	日和分钟	DD hh:mm
DAY_SECOND	日和秒	DD hh:mm:ss
HOUR_MINUTE	小时和分	hh:mm
HOUR_SECOND	小时和秒	hh:ss
MINUTE_SECOND	分钟和秒	mm:ss

来看一个具体的例子,在这个例子中第一列返回了当前日期时间,第二列返回距离当前日期 31 天后的日期时间,第三列返回距离当前日期一年两个月后的日期时间。

```
mysql> select now() current,date_add(now(),INTERVAL  31 day) after31days,
date_add(now(),INTERVAL  '1_2' year_month) after_oneyear_twomonth;
+---------------------+---------------------+------------------------+
| current             | after31days         | after_oneyear_twomonth |
+---------------------+---------------------+------------------------+
| 2007-09-03 11:30:48 | 2007-10-04 11:30:48 | 2008-11-03 11:30:48    |
+---------------------+---------------------+------------------------+
1 row in set (0.01 sec)
```

同样也可以用负数让它返回之前的某个日期时间,如下面这个例子第一列返回了当前日期时间,第二列返回距离当前日期 31 天前的日期时间,第三列返回距离当前日期一年两个月前的日期时间。

```
mysql> select now() current,date_add(now(),INTERVAL  -31 day) after31days, date_a
dd(now(),INTERVAL  '-1_-2' year_month) after_oneyear_twomonth;
+---------------------+---------------------+------------------------+
| current             | after31days         | after_oneyear_twomonth |
+---------------------+---------------------+------------------------+
| 2007-09-03 11:36:35 | 2007-08-03 11:36:35 | 2006-07-03 11:36:35    |
+---------------------+---------------------+------------------------+
1 row in set (0.00 sec)
```

● DATEDIFF(date1,date2)函数:用来计算两个日期之间相差的天数。

下面的例子计算出当前距离 2008 年 8 月 8 日的奥运会开幕式还有多少天:

```
mysql> select DATEDIFF('2008-08-08',now());
+------------------------------+
| DATEDIFF('2008-08-08',now()) |
+------------------------------+
|                          328 |
+------------------------------+
1 row in set (0.01 sec)
```

5.4 流程函数

流程函数也是常用的一类函数,用户可以使用这类函数在一个 SQL 语句中实现条件选择,这样做能够提高语句的效率。表 5-6 列出了 MySQL 中跟条件选择有关的流程函数。下面将通过具体的实例来讲解每个函数的用法。

表 5-6   MySQL 中的流程函数

函数	功能
IF(value,t f)	如果 value 是真,返回 t;否则返回 f
IFNULL(value1,value2)	如果 value1 不为空,返回 value1,否则返回 value2
CASE WHEN [value1] THEN[result1]...ELSE [default]END	如果 value1 是真,返回 result1,否则返回 default
CASE [expr] WHEN [value1] THEN[result1]... ELSE[default]END	如果 expr 等于 value1,返回 result1,否则返回 default

下面的例子中模拟了对职员薪水进行分类,这里首先创建并初始化一个职员薪水表:

```
mysql> create table salary (userid int,salary decimal(9,2));
Query OK, 0 rows affected (0.06 sec)
```

插入一些测试数据:

```
mysql> insert into salary values(1,1000),(2,2000), (3,3000),(4,4000),(5,5000), (1,null);
Query OK, 6 rows affected (0.00 sec)

mysql> select * from salary;
+--------+---------+
| userid | salary  |
+--------+---------+
| 1      | 1000.00 |
| 2      | 2000.00 |
| 3      | 3000.00 |
| 4      | 4000.00 |
| 5      | 5000.00 |
| 1      | NULL    |
+--------+---------+
6 rows in set (0.00 sec)
```

接下来,通过这个表来介绍各个函数的应用。

- IF(value,t,f)函数:这里认为月薪在 2000 元以上的职员属于高薪,用 "high" 表示;而 2000 元以下的职员属于低薪,用 "low" 表示。

```
mysql> select userid,salary,if(salary>2000, 'high', 'low') as salary_level from salary;
+--------+---------+--------------+
| userid | salary  | salary_level |
+--------+---------+--------------+
|      1 | 1000.00 | low          |
|      2 | 2000.00 | low          |
|      3 | 3000.00 | high         |
|      4 | 4000.00 | high         |
|      5 | 5000.00 | high         |
|      1 |    NULL | low          |
+--------+---------+--------------+
6 rows in set (0.00 sec)
```

- IFNULL(value1,value2)函数:这个函数一般用来替换 NULL 值,我们知道 NULL 值是不能参与数值运算的。下面这个语句就是把 NULL 值用 0 来替换。

```
mysql> select userid,salary,ifnull(salary,0) from salary;
+--------+---------+------------------+
| userid | salary  | ifnull(salary,0) |
+--------+---------+------------------+
|      1 | 1000.00 |          1000.00 |
|      2 | 2000.00 |          2000.00 |
|      3 | 3000.00 |          3000.00 |
|      4 | 4000.00 |          4000.00 |
|      5 | 5000.00 |          5000.00 |
|      1 |    NULL |             0.00 |
+--------+---------+------------------+
6 rows in set (0.03 sec)
```

- CASE [expr] WHEN [value1] THEN[result1]…ELSE[default]END 函数:这是 case 的简单函数用法,case 后面跟列名或者列的表达式,when 后面枚举这个表达式所有可能的值,但不能是值的范围。如果要实现上面例子中高薪低薪的问题,写法如下:

```
mysql> select userid,salary,case salary when 1000 then 'low' when 2000 then 'low' else 'high' end salary_level from salary;
+--------+---------+--------------+
| userid | salary  | salary_level |
+--------+---------+--------------+
|      1 | 1000.00 | low          |
```

```
|    2 | 2000.00 | low    |
|    3 | 3000.00 | high   |
|    4 | 4000.00 | high   |
|    5 | 5000.00 | high   |
|    1 |    NULL | high   |
+--------+---------+-------------+
6 rows in set (0.00 sec)
```

- CASE WHEN [expr] THEN[result1]…ELSE[default]END 函数：这是 case 的搜索函数用法，直接在 when 后面写条件表达式，并且只返回第一个符合条件的值，使用起来更加灵活，上例可以改写如下：

```
mysql> select userid,salary,case when salary<=2000 then 'low' else 'high' end as salary_level
from salary;
+--------+---------+--------------+
| userid | salary  | salary_level |
+--------+---------+--------------+
|    1 | 1000.00 | low          |
|    2 | 2000.00 | low          |
|    3 | 3000.00 | high         |
|    4 | 4000.00 | high         |
|    5 | 5000.00 | high         |
|    1 |    NULL | high         |
+--------+---------+--------------+
6 rows in set (0.00 sec)
```

5.5 JSON 函数

第 3 章已经介绍了 MySQL 5.7.8 之后新引入的 JSON 文档类型，对于 JSON 文档的操作，除了简单读写之外，通常还会有各种各样的查询、修改等需求，为此 MySQL 也提供了很多相应的函数，具体见表 5-7。

表 5-7　　　　　　　　　　　　MySQL 中的 JSON 函数

函数类型	名　称	功　能
创建 JSON	JSON_ARRAY()	创建 JSON 数组
	JSON_OBJECT()	创建 JSON 对象
	JSON_QUOTE()/JSON_UNQUOTE()	加上/去掉 JSON 文档两边的双引号
查询 JSON	JSON_CONTAINS()	查询文档中是否包含指定的元素
	JSON_CONTAINS_PATH()	查询文档中是否包含指定的路径
	JSON_EXTRACT()/->/->>	根据条件提取文档中的数据
	JSON_KEYS()	提取所有 key 的集合
	JSON_SEARCH()	返回所有符合条件的路径集合
修改 JSON	JSON_MERGE() (deprecated 5.7.22)/ JSON_MERGE_PRESERVE	将两个文档合并
	JSON_ARRAY_APPEND()	数组尾部追加元素
	JSON_ARRAY_INSERT()	在数组的指定位置插入元素
	JSON_REMOVE()	删除文档中指定位置的元素
	JSON_REPLACE()	替换文档中指定位置的元素
	JSON_SET()	给文档中指定位置的元素设置新值，如果元素不存在，则进行插入

续表

函数类型	名称	功能
查询 JSON 元数据	JSON_DEPTH()	JSON 文档的深度（元素最大嵌套层数）
	JSON_LENGTH()	JSON 文档的长度（元素个数）
	JSON_TYPE()	JSON 文档类型（数组、对象、标量类型）
	JSON_VALID()	JSON 格式是否合法
其他函数	JSON_PRETTY()	美化 JSON 格式
	JSON_STORAGE_SIZE()	JSON 文档占用的存储空间
	JSON_STORAGE_FREE()	JSON 文档更新操作后剩余的空间，MySQL 8.0 新增
	JSON_TABLE()	将 JSON 文档转换为表格，MySQL 8.0 新增
	JSON_ARRAYAGG()	将聚合后参数中的多个值转换为 JSON 数组
	JSON_OBJECTAGG()	把两个列或者是表达式解释为一个 key 和一个 value，返回一个 JSON 对象

这些函数安装功能可以分为以下几类：

- 创建 JSON 函数；
- 查询 JSON 函数；
- 修改 JSON 函数；
- 查询 JSON 元数据函数；
- 其他函数。

下面将详细介绍这些函数。

5.5.1 创建 JSON 函数

1. JSON_ARRAY([val[, val] ...])

此函数可以返回包含参数中所有值列表的 JSON 数组。

以下示例创建了一个包含数字、字符串、null、布尔、日期类型在内的混合数组，需要注意的是，参数中的 null 和 true/false 大小写不敏感。

```
mysql> SELECT JSON_ARRAY(1, "abc", NULL, TRUE, CURTIME());
+---------------------------------------------+
| JSON_ARRAY(1, "abc", NULL, TRUE, CURTIME()) |
+---------------------------------------------+
| [1, "abc", null, true, "11:30:24.000000"]   |
+---------------------------------------------+
```

2. JSON_OBJECT([key, val[, key, val] ...])

此函数可以返回包含参数中所有键值对的对象列表。参数中的 key 不能为 null，参数个数也不能为奇数，否则报语法错误。

以下示例使用了正确的语法：

```
mysql> SELECT JSON_OBJECT('id', 100, 'name', 'jack');
+----------------------------------------+
| JSON_OBJECT('id', 100, 'name', 'jack') |
+----------------------------------------+
| {"id": 100, "name": "jack"}            |
+----------------------------------------+
```

以下示例则使用了错误的语法：

```
mysql> SELECT JSON_OBJECT('id', 100, 'name');//参数个数为奇数
ERROR 1582 (42000): Incorrect parameter count in the call to native function 'JSON_OBJECT'
mysql> SELECT JSON_OBJECT('id', 100, null,1);//key不能为null
ERROR 3158 (22032): JSON documents may not contain NULL member names.
```

3. JSON_QUOTE(string)

此函数可以将参数中的 JSON 文档转换为双引号引起来的字符串，如果 JSON 文档包含双引号，则转换后的字符串自动加上转义字符"\"，如以下示例：

```
mysql> SELECT JSON_QUOTE('[1,2,3]'), JSON_QUOTE('"null"');
+-----------------------+----------------------+
| JSON_QUOTE('[1,2,3]') | JSON_QUOTE('"null"') |
+-----------------------+----------------------+
| "[1,2,3]"             | "\"null\""           |
+-----------------------+----------------------+
```

如果需要将非 JSON 文档转换为 JSON 文档，或者反过来，可以使用 CONVERT 或者 CAST 函数进行强制转换，这两个函数可以在不同数据类型之间进行强制转换，具体用法请参考官方文档。

5.5.2 查询 JSON 函数

1. JSON_CONTAINS(target, candidate[, path])

此函数可以查询指定的元素（candidate）是否包含在目标 JSON 文档（target）中，包含则返回 1，否则返回 0，path 参数可选。如果有参数为 NULL 或 path 不存在，则返回 NULL。

以下示例分别要查询元素"abc"、1、10 是否包含在 JSON 文档中：

```
mysql> select json_contains('[1, 2, 3, "abc", null]','"abc"') ;
+-------------------------------------------------+
| json_contains('[1, 2, 3, "abc", null]','"abc"') |
+-------------------------------------------------+
|                                               1 |
+-------------------------------------------------+
1 row in set (0.00 sec)

mysql> select json_contains('[1, 2, 3, "abc", null]','1') ;
+---------------------------------------------+
| json_contains('[1, 2, 3, "abc", null]','1') |
+---------------------------------------------+
|                                           1 |
+---------------------------------------------+
1 row in set (0.00 sec)

mysql> select json_contains('[1, 2, 3, "abc", null]','10') ;
+----------------------------------------------+
| json_contains('[1, 2, 3, "abc", null]','10') |
+----------------------------------------------+
|                                            0 |
+----------------------------------------------+
1 row in set (0.00 sec)
```

显然结果符合我们的预期。

元素如果是数组也是可以的：

```
mysql> select json_contains('[1, 2, 3, "abc", null]','[1,3]') ;
+-------------------------------------------------+
| json_contains('[1, 2, 3, "abc", null]','[1,3]') |
+-------------------------------------------------+
|                                               1 |
+-------------------------------------------------+
1 row in set (0.00 sec)
```

path 参数是可选的，可以指定在特定的路径下查询。如果 JSON 文档为对象，则路径格式通常类似于$.a 或者$.a.b 这种格式。$.a 很好理解，表示 key 为 a；$.a.b 通常用在 value 也是对象列表的情况，表示键 a 下层的键 b，比如{"id":{"id1":1,"id2":2}}。如果 JSON 文档为数组，则路径通常写为$[i]这种格式，表示数组中第 i 个元素。

在下例中，要查询 JSON 文档 j 中是否包含 value 为 10 的对象，并指定路径为$.jack（key='jack'），如果包含则返回 1，如果不包含则返回 0。那么 SQL 代码可以这么写：

```
mysql> SET @j = '{"jack": 10, "tom": 20, "lisa": 30}';
Query OK, 0 rows affected (0.00 sec)

mysql> SET @j2='10';
Query OK, 0 rows affected (0.00 sec)

mysql> select json_contains(@j,@j2,'$.jack');
+--------------------------------+
| json_contains(@j,@j2,'$.jack') |
+--------------------------------+
|                              1 |
+--------------------------------+
1 row in set (0.00 sec)
```

返回 1，表示在路径 key='jack'下，存在 value 为 10 的值。将查询路径改为 tom 后，再次查询：

```
mysql> select json_contains(@j,@j2,'$.tom');
+-------------------------------+
| json_contains(@j,@j2,'$.tom') |
+-------------------------------+
|                             0 |
+-------------------------------+
```

此时返回 0，则表示 JSON 文档中不包含{"tom":10}的元素。

2. JSON_CONTAINS_PATH(json_doc, one_or_all, path[, path] ...)

此函数可以查询 JSON 文档中是否存在指定路径，存在则返回 1，否则返回 0。one_or_all 只能取值 one 或 all，one 表示只要有一个存在即可；all 表示所有的都存在才行。如果有参数为 NULL 或 path 不存在，则返回 NULL。

比如，要查询给定的 3 个 path 是否至少一个存在或者必须全部存在，可以分别写 SQL 代码如下：

```
mysql> select json_contains_path('{"k1":"jack","k2":"tom","k3":"lisa"}','one','$.k1','$.k4') one_path ;
+----------+
| one_path |
+----------+
|        1 |
+----------+
1 row in set (0.00 sec)

mysql> select json_contains_path('{"k1":"jack","k2":"tom","k3":"lisa"}','all','$.k1','$.k4') all_path;
+----------+
| all_path |
+----------+
|        0 |
+----------+
1 row in set (0.00 sec)
```

3. JSON_EXTRACT(json_doc, path[, path] ...)

此函数可以从 JSON 文档里抽取数据。如果有参数有 NULL 或 path 不存在，则返回 NULL。如果抽取出多个 path，则返回的数据合并在一个 JSON ARRAY 里。

以下示例从 JSON 文档的第一和第二个元素中提取出对应的 value：

```
mysql> SELECT JSON_EXTRACT('[10, 20, [30, 40]]', '$[0]','$[1]');
+---------------------------------------------------+
| JSON_EXTRACT('[10, 20, [30, 40]]', '$[0]','$[1]') |
+---------------------------------------------------+
| [10, 20]                                          |
+---------------------------------------------------+
1 row in set (0.00 sec)
```

可以看到，返回的两个值以数组的形式进行了合并。如果要取第三个数组值，path 可以写为$2 或者$[2][*]：

```
mysql> SELECT JSON_EXTRACT('[10, 20, [30, 40]]', '$[2]');
+--------------------------------------------+
| JSON_EXTRACT('[10, 20, [30, 40]]', '$[2]') |
+--------------------------------------------+
| [30, 40]                                   |
+--------------------------------------------+
1 row in set (0.00 sec)

mysql> SELECT JSON_EXTRACT('[10, 20, [30, 40]]', '$[2][*]');
+-----------------------------------------------+
| JSON_EXTRACT('[10, 20, [30, 40]]', '$[2][*]') |
+-----------------------------------------------+
| [30, 40]                                      |
+-----------------------------------------------+
1 row in set (0.00 sec)
```

在 MySQL 5.7.9 版本之后，可以用一种更简单的函数 "->" 来替代 JSON_EXTRACT，语法如下：

column->path

注意左边只能是列名，不能是表达式；右边是要匹配的 JSON 路径。上面的例子可以改写为：

```
mysql> insert into t1 values('[10,20,[30,40]]');
Query OK, 1 row affected (0.00 sec)
mysql> select id1,id1->"$[0]",id1->"$[1]" from t1 where id1->"$[0]"=10;
+--------------------+-------------+-------------+
| id1                | id1->"$[0]" | id1->"$[1]" |
+--------------------+-------------+-------------+
| [10, 20, [30, 40]] | 10          | 20          |
+--------------------+-------------+-------------+
1 row in set (0.00 sec)
```

如果 JSON 文档查询的结果是字符串，则显示结果默认会包含双引号，在很多情况下是不需要的，为了解决这个问题，MySQL 提供了另外两个函数 JSON_UNQUOTE 和 "->>"，用法类似于 JSON_EXTRACT 和 "->"，简单举例如下：

```
mysql> select    json_EXTRACT(id1,'$.k1'),json_unquote(id1->'$.k1'), id1->'$.k1',id1->>'$.k1'
from t1 where id1->'$.k1'='jack';
+--------------------------+---------------------------+-------------+--------------+
| json_EXTRACT(id1,'$.k1') | json_unquote(id1->'$.k1') | id1->'$.k1' | id1->>'$.k1' |
+--------------------------+---------------------------+-------------+--------------+
| "jack"                   | jack                      | "jack"      | jack         |
+--------------------------+---------------------------+-------------+--------------+
1 row in set (0.00 sec)
```

即下面 3 种写法效果是一样的：

- ***JSON_UNQUOTE(JSON_EXTRACT(column, path))***
- ***JSON_UNQUOTE(column -> path)***
- ***column->>path***

4. JSON_KEYS(json_doc[, path])

此函数可以获取 JSON 文档在指定路径下的所有键值，返回一个 JSON ARRAY。如果有参数为 NULL 或 path 不存在，则返回 NULL。

参数 path 通常使用在嵌套对象列表中，如下例所示：

```
mysql> select JSON_KEYS('{"a": 1, "b": {"c": 30}}');
+---------------------------------------+
| JSON_KEYS('{"a": 1, "b": {"c": 30}}') |
+---------------------------------------+
| ["a", "b"]                            |
+---------------------------------------+
1 row in set (0.00 sec)

mysql> select JSON_KEYS('{"a": 1, "b": {"c": 30}}','$.b');
+---------------------------------------------+
| JSON_KEYS('{"a": 1, "b": {"c": 30}}','$.b') |
+---------------------------------------------+
| ["c"]                                       |
+---------------------------------------------+
1 row in set (0.00 sec)
```

如果元素中都是数组 ARRAY，则返回为 NULL。

5. JSON_SEARCH(json_doc, one_or_all, search_str[, escape_char[, path] ...])

此函数可以查询包含指定字符串的路径，并作为一个 JSON ARRAY 返回。如果有参数为 NULL 或 path 不存在，则返回 NULL。各参数含义如下。

- one_or_all: one 表示查询到一个即返回；all 表示查询所有。
- search_str: 要查询的字符串，可以用 LIKE 里的'%'或'_'匹配。
- path: 表示在指定 path 下进行查询。

以下示例给出了如何查询 JSON 文档中以字母 t 开头的元素的第一个路径：

```
mysql> select json_search('{"k1":"jack","k2":"tom","k3":"lisa","k4":"tony"}','one','t%');
+----------------------------------------------------------------------------+
| json_search('{"k1":"jack","k2":"tom","k3":"lisa","k4":"tony"}','one','t%') |
+----------------------------------------------------------------------------+
| "$.k2"                                                                     |
+----------------------------------------------------------------------------+
1 row in set (0.00 sec)
```

可以看出，满足条件的第一个元素是"k2":"tom"，path 描述为"$.k2"。

下面把条件"one"改成"all"，再看看结果：

```
mysql> select json_search('{"k1":"jack","k2":"tom","k3":"lisa","k4":"tony"}','all','t%');
+----------------------------------------------------------------------------+
| json_search('{"k1":"jack","k2":"tom","k3":"lisa","k4":"tony"}','all','t%') |
+----------------------------------------------------------------------------+
| ["$.k2", "$.k4"]                                                           |
+----------------------------------------------------------------------------+
1 row in set (0.00 sec)
```

此时，满足条件的所有元素是"k2":"tom"和"k4":"tony"，路径描述为["$.k2", "$.k4"]数组。

如果把 JSON 文档改为数组，则返回路径也将成为数组的描述格式，如下例所示：

```
mysql> select json_search('["tom","lisa","jack",{"name":"tony"}]','all','t%');
+-----------------------------------------------------------------+
| json_search('["tom","lisa","jack",{"name":"tony"}]','all','t%') |
+-----------------------------------------------------------------+
| ["$[0]", "$[3].name"]                                           |
+-----------------------------------------------------------------+
1 row in set (0.00 sec)
```

5.5.3 修改 JSON 的函数

1. JSON_ARRAY_APPEND(json_doc, path, val[, path, val] ...)

此函数可以在指定 path 的 json array 尾部追加 val。如果指定 path 是一个 json object，则将其封装成一个 json array 再追加。如果有参数为 NULL，则返回 NULL。

以下示例在 JSON 文档的不同 path 处分别追加字符 "1"：

```
mysql> SELECT JSON_ARRAY_APPEND('["a", ["b", "c"], "d"]', '$[0]', "1");
+----------------------------------------------------------+
| JSON_ARRAY_APPEND('["a", ["b", "c"], "d"]', '$[0]', "1") |
+----------------------------------------------------------+
| [["a", "1"], ["b", "c"], "d"]                            |
+----------------------------------------------------------+
1 row in set (0.00 sec)

mysql> SELECT JSON_ARRAY_APPEND('["a", ["b", "c"], "d"]', '$[1]', "1");
+----------------------------------------------------------+
| JSON_ARRAY_APPEND('["a", ["b", "c"], "d"]', '$[1]', "1") |
+----------------------------------------------------------+
| ["a", ["b", "c", "1"], "d"]                              |
+----------------------------------------------------------+
1 row in set (0.00 sec)

mysql> SELECT JSON_ARRAY_APPEND('["a", ["b", "c"], "d"]', '$[1][0]', "1");
+-------------------------------------------------------------+
| JSON_ARRAY_APPEND('["a", ["b", "c"], "d"]', '$[1][0]', "1") |
+-------------------------------------------------------------+
| ["a", [["b", "1"], "c"], "d"]                               |
+-------------------------------------------------------------+
1 row in set (0.00 sec)

mysql> SELECT JSON_ARRAY_APPEND('{"a": 1, "b": [2, 3], "c": 4}','$.b', "1");
+---------------------------------------------------------------+
| JSON_ARRAY_APPEND('{"a": 1, "b": [2, 3], "c": 4}','$.b', "1") |
+---------------------------------------------------------------+
| {"a": 1, "b": [2, 3, "1"], "c": 4}                            |
+---------------------------------------------------------------+
1 row in set (0.00 sec)
```

2. JSON_ARRAY_INSERT(json_doc, path, val[, path, val] ...)

此函数可以在 path 指定的 json array 元素插入 val，原位置及以右的元素顺次右移。如果 path 指定的数据非 json array 元素，则略过此 val；如果指定的元素下标超过 json array 的长度，则插入尾部。

将上面例子中的 4 个 SQL 语句改成 JSON_ARRAY_INSERT，看一下结果：

```
mysql> SELECT JSON_ARRAY_INSERT('["a", ["b", "c"], "d"]', '$[0]', "1");
+----------------------------------------------------------+
| JSON_ARRAY_INSERT('["a", ["b", "c"], "d"]', '$[0]', "1") |
+----------------------------------------------------------+
| ["1", "a", ["b", "c"], "d"]                              |
+----------------------------------------------------------+
1 row in set (0.00 sec)

mysql> SELECT JSON_ARRAY_INSERT('["a", ["b", "c"], "d"]', '$[1]', "1");
+----------------------------------------------------------+
| JSON_ARRAY_INSERT('["a", ["b", "c"], "d"]', '$[1]', "1") |
+----------------------------------------------------------+
| ["a", "1", ["b", "c"], "d"]                              |
+----------------------------------------------------------+
1 row in set (0.00 sec)
```

```
mysql> SELECT JSON_ARRAY_INSERT('["a", ["b", "c"], "d"]', '$[1][0]', "1");
+--------------------------------------------------------------+
| JSON_ARRAY_INSERT('["a", ["b", "c"], "d"]', '$[1][0]', "1") |
+--------------------------------------------------------------+
| ["a", ["1", "b", "c"], "d"]                                  |
+--------------------------------------------------------------+
1 row in set (0.00 sec)

mysql> SELECT JSON_ARRAY_INSERT('{"a": 1, "b": [2, 3], "c": 4}','$.b', "1");
ERROR 3165 (42000): A path expression is not a path to a cell in an array.
```

最后一个 SQL 报错，提示路径不对，将 "$.b" 改为 "$[0]" 试一下：

```
mysql> SELECT JSON_ARRAY_INSERT('{"a": 1, "b": [2, 3], "c": 4}','$[0]', "1");
+----------------------------------------------------------------+
| JSON_ARRAY_INSERT('{"a": 1, "b": [2, 3], "c": 4}','$[0]', "1") |
+----------------------------------------------------------------+
| {"a": 1, "b": [2, 3], "c": 4}                                  |
+----------------------------------------------------------------+
1 row in set (0.00 sec)
```

插入路径正确，但字符并没有插入到 JSON 文档中，因为所有元素都是对象，跳过忽略。

3. JSON_REPLACE(json_doc, path, val[, path, val] ...)

此函数可以替换指定路径的数据，如果某个路径不存在，则略过（存在才替换）。如果有参数为 NULL，则返回 NULL。

下例将 JSON 文档中的第一个元素和第二个元素分别替换为 "1" 和 "2"：

```
+-----------------------------------------------------+
| JSON_REPLACE('["a", "b", "c", "d"]', '$[0]', "1",'$[1]',"2") |
+-----------------------------------------------------+
| ["1", "2", "d"]                                     |
+-----------------------------------------------------+
1 row in set (0.00 sec)
```

下例将 JSON 文档中 key 为 a 和 d 的对象的 value 分别替换为 "10" 和 "20"：

```
mysql> SELECT JSON_REPLACE('{"a": 1, "b": [2, 3], "c": 4}','$.a', "10",'$.d',"20");
+----------------------------------------------------------------------+
| JSON_REPLACE('{"a": 1, "b": [2, 3], "c": 4}','$.a', "10",'$.d',"20") |
+----------------------------------------------------------------------+
| {"a": "10", "b": [2, 3], "c": 4}                                     |
+----------------------------------------------------------------------+
1 row in set (0.00 sec)
```

4. JSON_SET(json_doc, path, val[, path, val] ...)

此函数可以设置指定路径的数据（不管是否存在）。如果有参数为 NULL，则返回 NULL。和 JSON_REPLACE 功能有些类似，最主要的区别是指定的路径不存在时，会在文档中自动添加，如下例所示：

```
mysql> SELECT JSON_SET('{"a": 1, "b": [2, 3], "c": 4}','$.a', "10",'$.d',"20");
+------------------------------------------------------------------+
| JSON_SET('{"a": 1, "b": [2, 3], "c": 4}','$.a', "10",'$.d',"20") |
+------------------------------------------------------------------+
| {"a": "10", "b": [2, 3], "c": 4, "d": "20"}                      |
+------------------------------------------------------------------+
1 row in set (0.00 sec)
```

5. JSON_MERGE_PRESERVE (json_doc, json_doc[, json_doc] ...)

此函数可以将多个 JSON 文档进行合并。合并规则如下：
- 如果都是 json array，则结果自动 merge 为一个 json array；
- 如果都是 json object，则结果自动 merge 为一个 json object；

5.5 JSON 函数

- 如果有多种类型，则将非 json array 的元素封装成 json array 再按照规则一进行 merge。

下例中分别将两个数组合并、两个对象合并、数组和对象合并：

```
mysql> SELECT JSON_MERGE_PRESERVE('[1, 2]', '[3,4]');
+----------------------------------------+
| JSON_MERGE_PRESERVE('[1, 2]', '[3,4]') |
+----------------------------------------+
| [1, 2, 3, 4]                           |
+----------------------------------------+
1 row in set (0.00 sec)

mysql> SELECT JSON_MERGE_PRESERVE('{"key1": "tom"}', '{"key2": "lisa"}');
+------------------------------------------------------------+
| JSON_MERGE_PRESERVE('{"key1": "tom"}', '{"key2": "lisa"}') |
+------------------------------------------------------------+
| {"key1": "tom", "key2": "lisa"}                            |
+------------------------------------------------------------+
1 row in set (0.00 sec)

mysql> SELECT JSON_MERGE_PRESERVE('[1,2]','{"key1": "tom"}');
+------------------------------------------------+
| JSON_MERGE_PRESERVE('[1,2]','{"key1": "tom"}') |
+------------------------------------------------+
| [1, 2, {"key1": "tom"}]                        |
+------------------------------------------------+
1 row in set (0.00 sec)
```

6. JSON_REMOVE(json_doc, path[, path] ...)

此函数可以移除指定路径的数据，如果某个路径不存在则略过此路径。如果有参数为 NULL，则返回 NULL。

下例中把 JSON 文档中第二个和第三个元素删除：

```
mysql> SELECT JSON_REMOVE('[1,2,3,4]','$[1]','$[2]');
+----------------------------------------+
| JSON_REMOVE('[1,2,3,4]','$[1]','$[2]') |
+----------------------------------------+
| [1, 3]                                 |
+----------------------------------------+
1 row in set (0.00 sec)
```

结果有些意外，'$[1]','$[2]'分别为 2 和 3，删除后不是应该为[1，4]吗？这里要注意，如果指定了多个 path，则删除操作是串行操作的，即先删除'$[1]'后 JSON 文档变为[1,3,4]，然后在[1,3,4]上删除'$[2]'后变为[1,3]。

5.5.4 查询 JSON 元数据函数

1. JSON_DEPTH(json_doc)

此函数用来获取 JSON 文档的深度。

如果文档是空数组、空对象、null、true/false，则深度为 1；如果非空数组或者非空对象里面包含的都是深度为 1 的对象，则整个文档深度为 2；依次类推，整个文档的深度取决于最大元素的深度。如下例所示：

```
mysql> SELECT JSON_DEPTH('{}'), JSON_DEPTH('[]'), JSON_DEPTH('true');
+------------------+------------------+--------------------+
| JSON_DEPTH('{}') | JSON_DEPTH('[]') | JSON_DEPTH('true') |
+------------------+------------------+--------------------+
|                1 |                1 |                  1 |
+------------------+------------------+--------------------+
1 row in set (0.00 sec)
```

```
mysql> SELECT JSON_DEPTH('[10, 20]'), JSON_DEPTH('[[], {}]');
+------------------------+------------------------+
| JSON_DEPTH('[10, 20]') | JSON_DEPTH('[[], {}]') |
+------------------------+------------------------+
|                      2 |                      2 |
+------------------------+------------------------+
1 row in set (0.00 sec)
mysql> SELECT JSON_DEPTH('[10, {"a": 20}]');
+-------------------------------+
| JSON_DEPTH('[10, {"a": 20}]') |
+-------------------------------+
|                             3 |
+-------------------------------+
1 row in set (0.00 sec)
```

2. JSON_LENGTH(json_doc[, path])

此函数可以获取指定路径下的文档长度。长度的计算规则如下：

- 标量（字符串、数字）的长度为 1；
- json array 的长度为元素的个数；
- json object 的长度为对象的个数；
- 嵌套数组或者嵌套对象不计算长度。

见下例所示：

```
mysql> SELECT JSON_LENGTH('1'),JSON_LENGTH('[1, 2,[3,4]]'),JSON_LENGTH('{"KEY":"TOM"}');
+------------------+-----------------------------+------------------------------+
| JSON_LENGTH('1') | JSON_LENGTH('[1, 2,[3,4]]') | JSON_LENGTH('{"KEY":"TOM"}') |
+------------------+-----------------------------+------------------------------+
|                1 |                           3 |                            1 |
+------------------+-----------------------------+------------------------------+
1 row in set (0.00 sec)
```

3. JSON_TYPE(json_val)

此函数可以获取 JSON 文档的具体类型，可以是数组、对象或者标量类型。

```
mysql> select json_type('[1,3]'),json_type('{"id":"tom"}');
+--------------------+---------------------------+
| json_type('[1,3]') | json_type('{"id":"tom"}') |
+--------------------+---------------------------+
| ARRAY              | OBJECT                    |
+--------------------+---------------------------+
1 row in set (0.00 sec)
mysql> select json_type('1'),json_type('"abc"'),json_type('null'),json_type('true');
+----------------+--------------------+-------------------+-------------------+
| json_type('1') | json_type('"abc"') | json_type('null') | json_type('true') |
+----------------+--------------------+-------------------+-------------------+
| INTEGER        | STRING             | NULL              | BOOLEAN           |
+----------------+--------------------+-------------------+-------------------+
1 row in set (0.00 sec)
```

4. JSON_VALID(val)

此函数判断 val 是否为有效的 JSON 格式，有效为 1，否则为 0。

这个函数在第 3 章做过简单介绍，见下例所示：

```
mysql> select json_valid('abc'),json_valid('"abc"'),json_valid('[1,2]'),json_valid('[1,2');
+-------------------+---------------------+---------------------+--------------------+
| json_valid('abc') | json_valid('"abc"') | json_valid('[1,2]') | json_valid('[1,2') |
+-------------------+---------------------+---------------------+--------------------+
|                 0 |                   1 |                   1 |                  0 |
+-------------------+---------------------+---------------------+--------------------+
1 row in set (0.00 sec)
```

显然，字符串两边不加双引号是无效的 JSON 格式，'[1,2'少了右中括号也是无效的，都返回 0。

5.5.5 JSON 工具函数

1. JSON_PRETTY(json_val)

此函数是在 5.7.22 版本中新增的，用来美化 JSON 的输出格式，使得结果更加易读。对于数组、对象，每一行显示一个元素，多层嵌套的元素会在新行中进行缩进，清楚地显示层次关系，如下例所示：

```
mysql> SELECT JSON_PRETTY('{"a":"10","b":"15","x":{"x1":1,"x2":2,"x3":3}}');
+--------------------------------------------------------------+
| JSON_PRETTY('{"a":"10","b":"15","x":{"x1":1,"x2":2,"x3":3}}') |
+--------------------------------------------------------------+
| {
  "a": "10",
  "b": "15",
  "x": {
    "x1": 1,
    "x2": 2,
    "x3": 3
  }
} |
+--------------------------------------------------------------+
```

2. JSON_STORAGE_SIZE(json_val)/ JSON_STORAGE_FREE(json_val)

JSON_STORAGE_SIZE(json_val)函数可以获取 JSON 文档占用的存储空间（byte），而 JSON_STORAGE_FREE(json_val) 函数可以获取由于 JSON_SET、JSON_REPLACE、JSON_REMOVE 操作导致释放的空间。

其中，JSON_STORAGE_FREE 是 8.0 版本新增的函数，用户可以在 MySQL 8.0 环境下测试以下示例。下面的例子显示了对 JSON 字段 update 操作前和操作后，两个函数的显示结果。

update 前：

```
mysql> SELECT JSON_STORAGE_SIZE(jcol),JSON_STORAGE_FREE(jcol),jcol FROM jtable;
+-------------------------+-------------------------+--------------------------------+
| JSON_STORAGE_SIZE(jcol) | JSON_STORAGE_FREE(jcol) | jcol                           |
+-------------------------+-------------------------+--------------------------------+
|                      40 |                       0 |{"Name": "Homer", "Stupid": "True"}|
+-------------------------+-------------------------+--------------------------------+
1 row in set (0.00 sec)
```

JSON_STORAGE_SIZE 显示了 jcol 列所占用的空间为 40 字节，由于没有字段更新，所以 JSON_STORAGE_FREE 显示为 0。

update 后：

```
mysql> update jtable set jcol=json_set(jcol,'$.Stupid',1);
Query OK, 1 row affected (0.07 sec)
Rows matched: 1  Changed: 1  Warnings: 0

mysql> SELECT JSON_STORAGE_SIZE(jcol),JSON_STORAGE_FREE(jcol),jcol FROM jtable;
+-------------------------+-------------------------+--------------------------------+
| JSON_STORAGE_SIZE(jcol) | JSON_STORAGE_FREE(jcol) | jcol                           |
+-------------------------+-------------------------+--------------------------------+
|                      40 |                       5 |{"Name": "Homer", "Stupid": 1}  |
+-------------------------+-------------------------+--------------------------------+
1 row in set (0.00 sec)
```

从结果上看，update 操作释放了 5 个字节的空间，但 JSON_STORAGE_SIZE(jcol)返回的结果并没有改变，仍然是 40 字节，这是由于 MySQL 规定局部更新（使用 JSON_SET/JSON_REPLACE/JSON_REMOVE 函数进行操作）后的文档存储只能大于等于更新前的 size。如果更新的值大于原值，则 JSON_STORAGE_SIZE 则会大于原文档 size，如下例所示：

```
mysql> update jtable set jcol=json_set(jcol,'$.Stupid',"True123");
Query OK, 1 row affected (0.02 sec)
Rows matched: 1  Changed: 1  Warnings: 0

mysql> SELECT JSON_STORAGE_SIZE(jcol),JSON_STORAGE_FREE(jcol),jcol FROM jtable;
+-------------------------+-------------------------+----------------------------------------+
| JSON_STORAGE_SIZE(jcol) | JSON_STORAGE_FREE(jcol) | jcol                                   |
+-------------------------+-------------------------+----------------------------------------+
|                      43 |                       0 |{"Name": "Homer", "Stupid": "True123"}|
+-------------------------+-------------------------+----------------------------------------+
1 row in set (0.00 sec)
```

由于更新操作没有释放空间，所以 JSON_STORAGE_FREE 返回为 0。但 JSON_STORAGE_SIZE 已经显示为增大后的 size 43。对于非局部更新（即不使用 JSON_SET/JSON_REPLACE/JSON_REMOVE 操作进行更新），上面的函数不满足之前的逻辑，如下例所示：

```
mysql> update jtable set jcol='{"Name": "Homer", "Stupid": "True"}';
Query OK, 1 row affected (0.06 sec)
Rows matched: 1  Changed: 1  Warnings: 0

mysql> SELECT JSON_STORAGE_SIZE(jcol),JSON_STORAGE_FREE(jcol),jcol FROM jtable;
+-------------------------+-------------------------+----------------------------------------+
| JSON_STORAGE_SIZE(jcol) | JSON_STORAGE_FREE(jcol) | jcol                                   |
+-------------------------+-------------------------+----------------------------------------+
|                      40 |                       0 |{"Name": "Homer", "Stupid": "True"}|
+-------------------------+-------------------------+----------------------------------------+
1 row in set (0.00 sec)

mysql> update jtable set jcol='{"Name": "Homer", "Stupid": 1}';
Query OK, 1 row affected (0.01 sec)
Rows matched: 1  Changed: 1  Warnings: 0

mysql> SELECT JSON_STORAGE_SIZE(jcol),JSON_STORAGE_FREE(jcol),jcol FROM jtable;
+-------------------------+-------------------------+----------------------------------------+
| JSON_STORAGE_SIZE(jcol) | JSON_STORAGE_FREE(jcol) | jcol                                   |
+-------------------------+-------------------------+----------------------------------------+
|                      35 |                       0 | {"Name": "Homer", "Stupid": 1}|
+-------------------------+-------------------------+----------------------------------------+
1 row in set (0.00 sec)
```

显然，两个函数的结果和之前的结果都不一样，JSON_STORAGE_SIZE 显示的都是 JSON 文档的实际 size，JSON_STORAGE_FREE 则永远为 0。

3. JSON_TABLE(expr, path COLUMNS (column_list) [AS] alias)

此函数可以将 JSON 文档映射为表格。参数中 expr 可以是表达式或者列；path 是用来过滤的 JSON 路径；COLUMNS 是常量关键字；column_list 是转换后的字段列表。

这个函数是 MySQL 8.0.4 后新增的一个重要的函数，可以将复杂的 JSON 文档转换为表格数据，转换后的表格可以像正常表一样做连接、排序、creat table as select 等各种操作，对 JSON 的数据展示、数据迁移等很多应用领域带来极大的灵活性和便利性。

下面的例子将 JSON 文档中的全部数据转换为表格，并按表格中的 ac 字段进行排序：

```
mysql> SELECT *
    ->        FROM
    ->            JSON_TABLE(
```

```
    ->              '[{"a":"3"},{"a":2},{"b":1},{"a":0},{"a":[1,2]}]',
    ->              "$[*]"
    ->          COLUMNS(
    ->              rowid FOR ORDINALITY,
    ->              ac VARCHAR(100) PATH "$.a" DEFAULT '999' ON ERROR DEFAULT '111' ON EMPTY,
    ->              aj JSON PATH "$.a" DEFAULT '{"x": 333}' ON EMPTY,
    ->              bx INT EXISTS PATH "$.b"
    ->          )
    ->      ) AS tt
    ->      order by ac;
+-------+-----+------------+------+
| rowid | ac  | aj         | bx   |
+-------+-----+------------+------+
|     4 | 0   | 0          |    0 |
|     3 | 111 | {"x": 333} |    1 |
|     2 | 2   | 2          |    0 |
|     1 | 3   | "3"        |    0 |
|     5 | 999 | [1, 2]     |    0 |
+-------+-----+------------+------+
5 rows in set (0.00 sec)
```

对例子中的参数简单介绍一下。

（1）expr，即 JSON 对象数组'[{"a":"3"},{"a":2},{"b":1},{"a":0},{"a":[1,2]}]'。

（2）过滤路径（path），其中"$[*]"表示文档中所有的数据，如果改为"$[0]"，则表示只转换文档中的第一个元素{"a":"3"}。

（3）column list 包含以下 4 个部分的内容。

● rowid FOR ORDINALITY：rowid 是转换后的列名，FOR ORDINALITY 表示按照序列顺序加一，类似于 MySQL 中的自增列。数据类型为 UNSIGNED INT，初始值为 1。

● ac VARCHAR(100) PATH "$.a" DEFAULT '999' ON ERROR DEFAULT '111' ON EMPTY：ac 是转换后的列名；VARCHAR(100)是转换后的列类型；PATH "$.a"说明此字段只记录对象的 key="a"的 value；DEFAULT '999' ON ERROR 说明发生 error，则转换为默认值 999，比如{"a":[1,2]}，value 为 JSON 数组，和 VARCHAR 不匹配，所以此对象转换后为"999"；DEFAULT '111' ON EMPTY 说明对应的 key 不匹配'a'，此对象转换后为"111"，比如{"b":1}。

● aj 和 ac 类似，只是转换后的列类型为 JSON。

● bx INT EXISTS PATH "$.b: bx 是转换后列名，如果存在路径"$.b"，即 key='b'的对象，则转换为 1；否则为 0。

4. JSON_ARRAYAGG(col_or_expr)

此函数可以将聚合后参数中的多个值转换为 JSON 数组。

下面的例子中按照 o_id 聚合后的属性列表转换为一个字符串 JSON 数组：

```
mysql> select * from t;
+------+-----------+-------+
| o_id | attribute | value |
+------+-----------+-------+
|    2 | color     | red   |
|    2 | fabric    | silk  |
|    3 | color     | green |
|    3 | shape     | square|
+------+-----------+-------+
4 rows in set (0.00 sec)

mysql> SELECT o_id, JSON_ARRAYAGG(attribute) AS attributes   FROM t GROUP BY o_id;
+------+---------------------+
| o_id | attributes          |
+------+---------------------+
```

```
|     2 | ["color", "fabric"] |
|     3 | ["color", "shape"]  |
+-------+---------------------+
2 rows in set (0.01 sec)
```

5. JSON_OBJECTAGG(key, value)

此函数可以把两个列或者是表达式解释为一个 key 和一个 value，返回一个 JSON 对象。

还是上例中的数据，这次按照 o_id 聚合后的 attribute/value 作为对象的 key/value 组成一个 JSON 对象文档。

```
mysql> SELECT o_id, JSON_OBJECTAGG(attribute, value) FROM t GROUP BY o_id;
+------+---------------------------------------+
| o_id | JSON_OBJECTAGG(attribute, value)      |
+------+---------------------------------------+
|    2 | {"color": "red", "fabric": "silk"}    |
|    3 | {"color": "green", "shape": "square"} |
+------+---------------------------------------+
2 rows in set (0.00 sec)
```

5.6 窗口函数

日常开发工作中，经常会遇到下面这些需求。

- 去医院看病，怎样知道上次就医距现在的时长？
- 环比如何计算？
- 怎样得到各部门工资排名前 N 名的员工列表？
- 如何查找组内每人工资占总工资的百分比？

如果使用传统的 SQL 来解决这些问题，理论上都是可以的，但逻辑却会相当复杂。这类需求都有一个共同特点，为了得到结果，都需要在某个结果集内做一些特定的函数操作。为了很方便地解决这一类问题，MySQL 8.0 中引入了窗口函数。窗口的概念非常重要，它可以理解为记录集合，窗口函数也就是在满足某种条件的记录集合上执行的特殊函数，对于每条记录都要在此窗口内执行函数。有的函数，随着记录不同，窗口大小都是固定的，这种属于静态窗口；有的函数则相反，不同的记录对应着不同的窗口，这种动态变化的窗口叫滑动窗口。

窗口函数和聚合函数有些类似，两者最大的区别是聚合函数是多行聚合为一行，窗口函数则是多行聚合为相同的行数，每行会多一个聚合后的新列。窗口函数在其他数据库中（比如 Oracle）也称为分析函数，功能也都大体相似。

MySQL 中支持的窗口函数如表 5-8 所示。

表 5-8 MySQL 中的窗口函数

函　　数	功　　能
ROW_NUMBER()	分区中的当前行号
RANK()	当前行在分区中的排名，含序号间隙
DENSE_RANK()	当前行在分区的排名，没有序号间隙
PERCENT_RANK()	百分比等级值
CUME_DIST()	累计分配值
FIRST_VALUE()	窗口中第一行的参数值
LAST_VALUE()	窗口中最后一行的参数值

续表

函　　数	功　　能
LAG()	分区中指定行落后于当前行的参数值
LEAD()	分区中领先当前行的参数值
NTH_VALUE()	从第 N 行窗口框架的参数值
NTILE(N)	分区中当前行的桶号

下面以订单表 order_tab 为例，逐个讲解这些函数的使用。测试表中的数据如下，各字段含义按顺序分别为订单号、用户 id、订单金额、订单创建日期：

```
mysql> select * from order_tab;
+----------+---------+--------+---------------------+
| order_id | user_no | amount | create_date         |
+----------+---------+--------+---------------------+
|        1 | 001     |    100 | 2018-01-01 00:00:00 |
|        2 | 001     |    300 | 2018-01-02 00:00:00 |
|        3 | 001     |    500 | 2018-01-02 00:00:00 |
|        4 | 001     |    800 | 2018-01-03 00:00:00 |
|        5 | 001     |    900 | 2018-01-04 00:00:00 |
|        6 | 002     |    500 | 2018-01-03 00:00:00 |
|        7 | 002     |    600 | 2018-01-04 00:00:00 |
|        8 | 002     |    300 | 2018-01-10 00:00:00 |
|        9 | 002     |    800 | 2018-01-16 00:00:00 |
|       10 | 002     |    800 | 2018-01-22 00:00:00 |
+----------+---------+--------+---------------------+
10 rows in set (0.00 sec)
```

5.6.1　ROW_NUMBER()

如果要查询每个用户最新的一笔订单，我们希望的结果是 order_id 分别为 5 和 10 的记录，此时可以使用 ROW_NUMBER()函数按照用户进行分组并按照订单日期进行由大到小排序，最后查找每组中序号为 1 的记录，SQL 语句如下：

```
mysql> select * from
    -> (
    ->     select row_number()over(partition by user_no order by create_date desc) as row_num,
    ->     order_id,user_no,amount,create_date
    ->     from order_tab
    -> )t where row_num=1;
+---------+----------+---------+--------+---------------------+
| row_num | order_id | user_no | amount | create_date         |
+---------+----------+---------+--------+---------------------+
|       1 |        5 | 001     |    900 | 2018-01-04 00:00:00 |
|       1 |       10 | 002     |    800 | 2018-01-22 00:00:00 |
+---------+----------+---------+--------+---------------------+
2 rows in set (0.00 sec)
```

其中，row_number()后面的 over 是关键字，用来指定函数执行的窗口范围，如果后面的括号中什么都不写，则意味着窗口包含所有行，窗口函数在所有行上进行计算；如果不为空，则支持以下 4 种语法。

- window_name: 给窗口指定一个别名，如果 SQL 中涉及的窗口较多，采用别名则更清晰易读。上面的例子中如果指定一个别名 w，则改写代码如下：

```
select * from
(
    select row_number() over w as row_num,
    order_id,user_no,amount,create_date
    from order_tab
```

```
     WINDOW w AS (partition by user_no order by create_date desc)
    )t where row_num=1;
```

- **partition 子句**：窗口按照哪些字段进行分组，窗口函数在不同的分组上分别执行。上面的例子就按照用户 id 进行分组。在每个用户 id 上，分别执行从 1 开始的顺序编号。
- **order by 子句**：按照哪些字段进行排序，窗口函数将按照排序后的记录顺序进行编号，既可以和 partition 子句配合使用，也可以单独使用。上例中二者同时使用。
- **frame 子句**：frame 是当前分区的一个子集，子句用来定义子集的规则，通常用来作为滑动窗口使用。比如要根据每个订单动态计算包括本订单和按时间顺序前后两个订单的平均订单金额，则可以设置如下 frame 子句来创建滑动窗口：

```
mysql> select * from
    -> (
    ->     select
    ->         order_id,user_no,amount,
    ->         avg(amount) over w as avg_num,
    ->         create_date
    ->     from order_tab
    ->     WINDOW w AS (partition by user_no order by create_date desc ROWS BETWEEN 1 PRECEDING AND 1 FOLLOWING)
    -> )t;
+----------+---------+--------+----------+---------------------+
| order_id | user_no | amount | avg_num  | create_date         |
+----------+---------+--------+----------+---------------------+
|        5 | 001     |    900 | 850.0000 | 2018-01-04 00:00:00 |
|        4 | 001     |    800 | 666.6667 | 2018-01-03 00:00:00 |
|        2 | 001     |    300 | 533.3333 | 2018-01-02 00:00:00 |
|        3 | 001     |    500 | 300.0000 | 2018-01-02 00:00:00 |
|        1 | 001     |    100 | 300.0000 | 2018-01-01 00:00:00 |
|       10 | 002     |    800 | 800.0000 | 2018-01-22 00:00:00 |
|        9 | 002     |    800 | 633.3333 | 2018-01-16 00:00:00 |
|        8 | 002     |    300 | 566.6667 | 2018-01-10 00:00:00 |
|        7 | 002     |    600 | 466.6667 | 2018-01-04 00:00:00 |
|        6 | 002     |    500 | 550.0000 | 2018-01-03 00:00:00 |
+----------+---------+--------+----------+---------------------+
10 rows in set (0.00 sec)
```

从结果可以看出，order_id 为 5 订单属于边界值，没有前一行，因此平均订单金额为（900+800）/2=850；order_id 为 4 的订单前后都有订单，所以平均订单金额为（900+800+300）/3=666.6667，以此类推就可以得到一个基于滑动窗口的动态平均订单值。

对于滑动窗口的范围指定，有如下两种方式。

（1）基于行：通常使用 BETWEEN frame_start AND frame_end 语法来表示行范围，frame_start 和 frame_end 可以支持如下关键字，来确定不同的动态行记录：

```
CURRENT ROW          边界是当前行，一般和其他范围关键字一起使用
UNBOUNDED PRECEDING  边界是分区中的第一行
UNBOUNDED FOLLOWING  边界是分区中的最后一行
expr PRECEDING       边界是当前行减去 expr 的值
expr FOLLOWING       边界是当前行加上 expr 的值
```

比如，下面都是合法的范围：

```
rows BETWEEN 1 PRECEDING AND 1 FOLLOWING              窗口范围是当前行、前一行、后一行一共 3 行记录
rows  UNBOUNDED FOLLOWING                             窗口范围是当前行到分区中的最后一行
rows BETWEEN UNBOUNDED PRECEDING AND UNBOUNDED FOLLOWING  窗口范围是当前分区中所有行，等同于不写
```

（2）基于范围：和基于行类似，但有些范围不是直接可以用行数来表示的，比如希望窗口范围是一周前的订单开始，截止到当前行，则无法使用 rows 来直接表示，此时就可以使用范围来表示窗口：INTERVAL 7 DAY PRECEDING。Linux 中常见的计算最近 1 分钟、5 分钟、15 分钟负载就是一个典型的应用场景。

5.6.2 RANK()/DENSE_RANK()

RANK()和 DENSE_RANK()这两个函数与 row_number()非常类似，只是在出现重复值时处理逻辑有所不同。这里稍微改一下上面的示例，假设需要查询不同用户的订单，按照订单金额进行排序，显示出相应的排名序号，SQL 语句中用 row_number()、rank()、dense_rank()分别显示序号，我们来看一下有什么区别。

```
mysql> select * from
    -> (
    ->   select row_number() over(partition by user_no order by amount desc) as row_num1,
    ->   rank() over(partition by user_no order by amount desc) as row_num2,
    ->   dense_rank() over(partition by user_no order by amount desc) as row_num3,
    ->     order_id,user_no,amount,create_date
    ->     from order_tab
    -> ) t;
+---------+---------+---------+----------+---------+--------+---------------------+
| row_num1| row_num2| row_num3| order_id | user_no | amount | create_date         |
+---------+---------+---------+----------+---------+--------+---------------------+
|       1 |       1 |       1 |        5 |     001 |    900 | 2018-01-04 00:00:00 |
|       2 |       2 |       2 |        4 |     001 |    800 | 2018-01-03 00:00:00 |
|       3 |       3 |       3 |        3 |     001 |    500 | 2018-01-02 00:00:00 |
|       4 |       4 |       4 |        2 |     001 |    300 | 2018-01-02 00:00:00 |
|       5 |       5 |       5 |        1 |     001 |    100 | 2018-01-01 00:00:00 |
|       1 |       1 |       1 |        9 |     002 |    800 | 2018-01-16 00:00:00 |
|       2 |       1 |       1 |       10 |     002 |    800 | 2018-01-22 00:00:00 |
|       3 |       3 |       2 |        7 |     002 |    600 | 2018-01-04 00:00:00 |
|       4 |       4 |       3 |        6 |     002 |    500 | 2018-01-03 00:00:00 |
|       5 |       5 |       4 |        8 |     002 |    300 | 2018-01-10 00:00:00 |
+---------+---------+---------+----------+---------+--------+---------------------+
10 rows in set (0.00 sec)
```

上面的记录中倒数第 3、4、5 行的斜体显示了 3 个函数的区别，row_number()在 amount 都是 800 的两条记录上随机排序，但序号按照 1、2 递增，后面 amount 为 600 的的序号继续递增为 3，中间不会产生序号间隙；rank()/dense_rank()则把 amount 为 800 的两条记录序号都设置为 1，但后续 amount 为 600 的需要则分别设置为 3（rank）和 2（dense_rank）。即 rank()会产生序号相同的记录，同时可能产生序号间隙；而 dense_rank()也会产生序号相同的记录，但不会产生序号间隙。

5.6.3 PERCENT_RANK()/CUME_DIST()

PERCENT_RANK() 和 CUME_DIST() 这两个函数都是计算数据分布的函数，PERCENT_RANK()和之前的 RANK()函数相关，每行按照以下公式进行计算：

```
(rank - 1) / (rows - 1)
```

其中，*rank* 为 RANK()函数产生的序号，*rows* 为当前窗口的记录总行数，上面的例子修改如下：

```
mysql> select * from
    -> (
    ->   select
    -> rank() over w as row_num,
    -> percent_rank() over w as percent,
    ->     order_id,user_no,amount,create_date
    ->     from order_tab
    ->     WINDOW w AS (partition by user_no order by amount desc)
    -> ) t;
+---------+---------+----------+---------+--------+---------------------+
| row_num | percent | order_id | user_no | amount | create_date         |
+---------+---------+----------+---------+--------+---------------------+
```

```
|    1    |    0    |    5    |   001   |   900   | 2018-01-04 00:00:00 |
|    2    |  0.25   |    4    |   001   |   800   | 2018-01-03 00:00:00 |
|    3    |   0.5   |    3    |   001   |   500   | 2018-01-02 00:00:00 |
|    4    |  0.75   |    2    |   001   |   300   | 2018-01-02 00:00:00 |
|    5    |    1    |    1    |   001   |   100   | 2018-01-01 00:00:00 |
|    1    |    0    |    9    |   002   |   800   | 2018-01-16 00:00:00 |
|    1    |    0    |   10    |   002   |   800   | 2018-01-22 00:00:00 |
|    3    |   0.5   |    7    |   002   |   600   | 2018-01-04 00:00:00 |
|    4    |  0.75   |    6    |   002   |   500   | 2018-01-03 00:00:00 |
|    5    |    1    |    8    |   002   |   300   | 2018-01-10 00:00:00 |
+---------+---------+---------+---------+---------+---------------------+
10 rows in set (0.00 sec)
```

从结果中可以看出，percent 列按照公式 (*rank* - 1) / (*rows* - 1) 代入 *rank* 值（row_num 列）和 *rows* 值（user_no 为'001'和'002'的值均为 5）。此函数主要应用在分析领域，日常应用场景较少。

相比 PERCENT_RANK()，CUME_DIST() 函数的应用场景更多，它的作用是分组内小于等于当前 *rank* 值的行数/分组内总行数。上例中，统计大于等于当前订单金额的订单数，占总订单数的比例，SQL 代码如下：

```
mysql> select * from
    -> (
    ->    select
    -> rank() over w as row_num,
    -> cume_dist() over w as cume,
    ->    order_id,user_no,amount,create_date
    ->    from order_tab
    ->    WINDOW w AS (partition by user_no order by amount desc)
    -> ) t;
+---------+------+---------+---------+--------+---------------------+
| row_num | cume | order_id| user_no | amount | create_date         |
+---------+------+---------+---------+--------+---------------------+
|    1    | 0.2  |    5    |   001   |  900   | 2018-01-04 00:00:00 |
|    2    | 0.4  |    4    |   001   |  800   | 2018-01-03 00:00:00 |
|    3    | 0.6  |    3    |   001   |  500   | 2018-01-02 00:00:00 |
|    4    | 0.8  |    2    |   001   |  300   | 2018-01-02 00:00:00 |
|    5    |  1   |    1    |   001   |  100   | 2018-01-01 00:00:00 |
|    1    | 0.4  |    9    |   002   |  800   | 2018-01-16 00:00:00 |
|    1    | 0.4  |   10    |   002   |  800   | 2018-01-22 00:00:00 |
|    3    | 0.6  |    7    |   002   |  600   | 2018-01-04 00:00:00 |
|    4    | 0.8  |    6    |   002   |  500   | 2018-01-03 00:00:00 |
|    5    |  1   |    8    |   002   |  300   | 2018-01-10 00:00:00 |
+---------+------+---------+---------+--------+---------------------+
10 rows in set (0.00 sec)
```

列 cume 显示了预期的结果。

5.6.4 NTILE(N)

NTILE(N) 函数的功能是对一个数据分区中的有序结果集进行划分，将其分为 N 个组，并为每个小组分配一个唯一的组编号。继续上面的例子，对每个用户的订单记录分为 3 组，NTILE(N) 函数记录每组组编号，SQL 代码如下：

```
mysql> select * from
    -> (
    ->    select
    -> ntile(3) over w as nf,
    ->    order_id,user_no,amount,create_date
    ->    from order_tab
    ->    WINDOW w AS (partition by user_no order by amount desc )
    -> ) t;
+------+---------+---------+--------+---------------------+
|  nf  | order_id| user_no | amount | create_date         |
+------+---------+---------+--------+---------------------+
|   1  |    5    |   001   |  900   | 2018-01-04 00:00:00 |
```

```
|   1 |    4 | 001 | 800 | 2018-01-03 00:00:00 |
|   2 |    3 | 001 | 500 | 2018-01-02 00:00:00 |
|   2 |    2 | 001 | 300 | 2018-01-02 00:00:00 |
|   3 |    1 | 001 | 100 | 2018-01-01 00:00:00 |
|   1 |    9 | 002 | 800 | 2018-01-16 00:00:00 |
|   1 |   10 | 002 | 800 | 2018-01-22 00:00:00 |
|   2 |    7 | 002 | 600 | 2018-01-04 00:00:00 |
|   2 |    6 | 002 | 500 | 2018-01-03 00:00:00 |
|   3 |    8 | 002 | 300 | 2018-01-10 00:00:00 |
+-----+------+-----+-----+---------------------+
10 rows in set (0.00 sec)
```

此函数在数据分析中应用较多，比如由于数据量大，需要将数据分配到 N 个并行的进程分别计算，此时就可以用 NTILE(N)对数据进行分组，由于记录数不一定被 N 整除，所以每组记录数不一定完全一致，然后将不同组号的数据再分配。

5.6.5 NTH_VALUE(expr,N)

NTH_VALUE(expr, N)函数可以返回窗口中第 N 个 expr 的值，expr 既可以是表达式，也可以是列名。这个函数不太好理解，来看下面的例子：

```
mysql> select * from
    -> (
    -> select
    -> ntile(3) over w as nf,
    -> nth_value(order_id,3) over w as nth,
    -> order_id,user_no,amount,create_date
    -> from order_tab
    -> WINDOW w AS (partition by user_no order by amount desc )
    -> ) t;
+----+------+----------+---------+--------+---------------------+
| nf | nth  | order_id | user_no | amount | create_date         |
+----+------+----------+---------+--------+---------------------+
|  1 | NULL |        5 | 001     |    900 | 2018-01-04 00:00:00 |
|  1 | NULL |        4 | 001     |    800 | 2018-01-03 00:00:00 |
|  2 |    3 |        3 | 001     |    500 | 2018-01-02 00:00:00 |
|  2 |    3 |        2 | 001     |    300 | 2018-01-02 00:00:00 |
|  3 |    3 |        1 | 001     |    100 | 2018-01-01 00:00:00 |
|  1 | NULL |        9 | 002     |    800 | 2018-01-16 00:00:00 |
|  1 | NULL |       10 | 002     |    800 | 2018-01-22 00:00:00 |
|  2 |    7 |        7 | 002     |    600 | 2018-01-04 00:00:00 |
|  2 |    7 |        6 | 002     |    500 | 2018-01-03 00:00:00 |
|  3 |    7 |        8 | 002     |    300 | 2018-01-10 00:00:00 |
+----+------+----------+---------+--------+---------------------+
10 rows in set (0.00 sec)
```

nth 列返回了分组排序后的窗口中 order_id 的第三个值，'001'用户返回 3，'002'用户返回 7，对于前 N-1 列，本函数返回 NULL。

5.6.6 LAG(expr,N)/LEAD(expr,N)

LAG(expr,N)和 LEAD(expr,N)这两个函数的功能是获取当前数据行按照某种排序规则的上 N 行（LAG）/下 N 行（LEAD）数据的某个字段。比如，每个订单中希望增加一个字段，用来记录本订单距离上一个订单的时间间隔，那么就可以用 LAG 函数来实现，SQL 代码如下：

```
mysql> select order_id,user_no,amount,create_date, last_date,datediff(create_date,last_date) as diff
    -> from
    -> (
    ->   select
```

```
    ->    order_id,user_no,amount,create_date,
    ->    lag(create_date,1) over w as last_date
    ->    from order_tab
    ->    WINDOW w AS (partition by user_no order by create_date )
    -> ) t;
+----------+---------+--------+---------------------+---------------------+------+
| order_id | user_no | amount | create_date         | last_date           | diff |
+----------+---------+--------+---------------------+---------------------+------+
|        1 | 001     |    100 | 2018-01-01 00:00:00 | NULL                | NULL |
|        2 | 001     |    300 | 2018-01-02 00:00:00 | 2018-01-01 00:00:00 |    1 |
|        3 | 001     |    500 | 2018-01-02 00:00:00 | 2018-01-02 00:00:00 |    0 |
|        4 | 001     |    800 | 2018-01-03 00:00:00 | 2018-01-02 00:00:00 |    1 |
|        5 | 001     |    900 | 2018-01-04 00:00:00 | 2018-01-03 00:00:00 |    1 |
|        6 | 002     |    500 | 2018-01-03 00:00:00 | NULL                | NULL |
|        7 | 002     |    600 | 2018-01-04 00:00:00 | 2018-01-03 00:00:00 |    1 |
|        8 | 002     |    300 | 2018-01-10 00:00:00 | 2018-01-04 00:00:00 |    6 |
|        9 | 002     |    800 | 2018-01-16 00:00:00 | 2018-01-10 00:00:00 |    6 |
|       10 | 002     |    800 | 2018-01-22 00:00:00 | 2018-01-16 00:00:00 |    6 |
+----------+---------+--------+---------------------+---------------------+------+
10 rows in set (0.00 sec)
```

内层 SQL 先通过 lag 函数得到上一次订单的日期，外层 SQL 再将本次订单和上次订单日期做差得到时间间隔。

5.6.7　FIRST_VALUE（expr）/LAST_VALUE（expr）

FIRST_VALUE（expr）函数和 LAST_VALUE（expr）函数的功能分别是获得滑动窗口范围内的参数字段中第一个（FIRST_VALUE）和最后一个（LAST_VALUE）的值。下例中，每个用户在每个订单记录中希望看到截止到当前订单为止，按照日期排序最早订单和最晚订单的订单金额，SQL 语句如下：

```
mysql> select *
    -> from
    -> (
    ->    select
    ->    order_id,user_no,amount,create_date,
    ->    first_value(amount) over w as first_amount,
    ->    last_value(amount) over w as last_amount
    ->    from order_tab
    ->    WINDOW w AS (partition by user_no order by create_date )
    -> ) t;
+----------+---------+--------+---------------------+--------------+-------------+
| order_id | user_no | amount | create_date         | first_amount | last_amount |
+----------+---------+--------+---------------------+--------------+-------------+
|        1 | 001     |    100 | 2018-01-01 00:00:00 |          100 |         100 |
|        2 | 001     |    300 | 2018-01-02 00:00:00 |          100 |         500 |
|        3 | 001     |    500 | 2018-01-02 00:00:00 |          100 |         500 |
|        4 | 001     |    800 | 2018-01-03 00:00:00 |          100 |         800 |
|        5 | 001     |    900 | 2018-01-04 00:00:00 |          100 |         900 |
|        6 | 002     |    500 | 2018-01-03 00:00:00 |          500 |         500 |
|        7 | 002     |    600 | 2018-01-04 00:00:00 |          500 |         600 |
|        8 | 002     |    300 | 2018-01-10 00:00:00 |          500 |         300 |
|        9 | 002     |    800 | 2018-01-16 00:00:00 |          500 |         800 |
|       10 | 002     |    800 | 2018-01-22 00:00:00 |          500 |         800 |
+----------+---------+--------+---------------------+--------------+-------------+
10 rows in set (0.00 sec)
```

结果和预期一致，比如 order_id 为 4 的记录，first_amount 和 last_amount 分别记录了用户 '001' 截止到时间 2018-01-03 00:00:00 为止，第一条订单金额 100 和最后一条订单金额 800，注意这里是按时间排序的最早订单和最晚订单，并不是最小金额和最大金额订单。

5.6.8 聚合函数作为窗口函数

除了前面介绍的各类窗口函数外，我们经常使用的各种聚合函数（SUM/AVG/MAX/MIN/COUNT）也可以作为窗口函数来使用。比如要统计每个用户按照订单 id，截止到当前的累计订单金额/平均订单金额/最大订单金额/最小订单金额/订单数是多少，可以用聚合函数作为窗口函数实现如下：

```
mysql> select
    -> order_id,user_no,amount,create_date,
    -> sum(amount) over w as sum1,
    -> avg(amount) over w as avg1,
    -> max(amount) over w as max1,
    -> min(amount) over w as min1,
    -> count(amount) over w as count1
    -> from
    ->  order_tab
    -> WINDOW w AS (partition by user_no order by order_id)
    -> ;
+----------+---------+--------+---------------------+------+----------+------+------+--------+
| order_id | user_no | amount | create_date         | sum1 | avg1     | max1 | min1 | count1 |
+----------+---------+--------+---------------------+------+----------+------+------+--------+
|        1 | 001     |    100 | 2018-01-01 00:00:00 |  100 | 100.0000 |  100 |  100 |      1 |
|        2 | 001     |    300 | 2018-01-02 00:00:00 |  400 | 200.0000 |  300 |  100 |      2 |
|        3 | 001     |    500 | 2018-01-02 00:00:00 |  900 | 300.0000 |  500 |  100 |      3 |
|        4 | 001     |    800 | 2018-01-03 00:00:00 | 1700 | 425.0000 |  800 |  100 |      4 |
|        5 | 001     |    900 | 2018-01-04 00:00:00 | 2600 | 520.0000 |  900 |  100 |      5 |
|        6 | 002     |    500 | 2018-01-03 00:00:00 |  500 | 500.0000 |  500 |  500 |      1 |
|        7 | 002     |    600 | 2018-01-04 00:00:00 | 1100 | 550.0000 |  600 |  500 |      2 |
|        8 | 002     |    300 | 2018-01-10 00:00:00 | 1400 | 466.6667 |  600 |  300 |      3 |
|        9 | 002     |    800 | 2018-01-16 00:00:00 | 2200 | 550.0000 |  800 |  300 |      4 |
|       10 | 002     |    800 | 2018-01-22 00:00:00 | 3000 | 600.0000 |  800 |  300 |      5 |
+----------+---------+--------+---------------------+------+----------+------+------+--------+
10 rows in set (0.07 sec)
```

可以看到 sum1/avg1/max1/min1/count1 的结果完全符合预期。

5.7 其他常用函数

MySQL 提供的函数很丰富，除了前面介绍的字符串函数、数字函数、日期函数、流程函数以外，还有很多其他函数，在此不再一一列举，有兴趣的读者可以参考 MySQL 官方手册。表 5-9 列举了一些其他常用的函数。

表 5-9　　　　　　　　　　　　　　MySQL 中的其他常用函数

函　　数	功　　能
DATABASE()	返回当前数据库名
VERSION()	返回当前数据库版本
USER()	返回当前登录用户名
INET_ATON(IP)	返回 IP 地址的数字表示
INET_NTOA(num)	返回数字代表的 IP 地址
PASSWORD(str)	返回字符串 str 的加密版本
MD5()	返回字符串 str 的 MD5 值

下面结合实例简单介绍一下这些函数的用法。

- DATABASE()函数：返回当前数据库名。

```
mysql> select DATABASE();
+------------+
| DATABASE() |
+------------+
| test       |
+------------+
1 row in set (0.00 sec)
```

- VERSION()函数：返回当前数据库版本。

```
mysql> select VERSION();
+-----------+
| VERSION() |
+-----------+
| 5.0.18-nt |
+-----------+
1 row in set (0.00 sec)
```

- USER()函数：返回当前登录用户名。

```
mysql> select USER();
+----------------+
| USER()         |
+----------------+
| root@localhost |
+----------------+
1 row in set (0.03 sec)
```

- INET_ATON(IP)函数：返回IP地址的网络字节序表示。

```
mysql> select INET_ATON('192.168.1.1');
+--------------------------+
| INET_ATON('192.168.1.1') |
+--------------------------+
|               3232235777 |
+--------------------------+
1 row in set (0.00 sec)
```

- INET_NTOA(num)函数：返回网络字节序代表的IP地址。

```
mysql> select INET_NTOA(3232235777);
+-----------------------+
| INET_NTOA(3232235777) |
+-----------------------+
| 192.168.1.1           |
+-----------------------+
1 row in set (0.00 sec)
```

INET_ATON(IP)和INET_NTOA(num)函数主要的用途是将字符串的IP地址转换为数字表示的网络字节序，这样可以更方便地进行IP或者网段的比较。比如在下面的表t中，想要知道在"192.168.1.3"和"192.168.1.20"之间一共有多少IP地址，可以这么做：

```
mysql> select * from t;
+--------------+
| ip           |
+--------------+
| 192.168.1.1  |
| 192.168.1.3  |
| 192.168.1.6  |
| 192.168.1.10 |
| 192.168.1.20 |
| 192.168.1.30 |
+--------------+
6 rows in set (0.00 sec)
```

按照正常的思维，应该用字符串来进行比较。下面是字符串的比较结果：

```
mysql> select * from t where ip>='192.168.1.3' and ip<='192.168.1.20';
Empty set (0.01 sec)
```

结果没有如我们所愿，竟然是个空集。其原因就在于字符串的比较是一个字符一个字符地比较，当对应字符相同时，就比较下一个，直到遇到能区分出大小的字符才停止比较，后面的字符也将忽略。显然，在此例中，"192.168.1.3"其实比"192.168.1.20"要"大"，因为"3"比"2"大，而不能用我们日常的思维 3<20，所以"ip>='192.168.1.3' and ip<='192.168.1.20'"必然是个空集。

在这里，如果要想实现上面的功能，就可用函数 INET_ATON 来实现，将 IP 地址转换为字节序后再比较，如下所示：

```
mysql> select * from t where inet_aton(ip)>=inet_aton('192.168.1.3') and inet_aton(ip)<= inet_aton('192.168.1.20');
+--------------+
| ip           |
+--------------+
| 192.168.1.3  |
| 192.168.1.6  |
| 192.168.1.10 |
| 192.168.1.20 |
+--------------+
4 rows in set (0.00 sec)
```

结果完全符合我们的要求。

- PASSWORD(str)函数：返回字符串 str 的加密版本，一个 41 位长的字符串。

此函数只用来设置系统用户的密码，但是不能用来对应用的数据加密。如果应用方面有加密的需求，可以使用 MD5 等加密函数来实现。

下例中显示了字符串"123456"的 PASSWORD 加密后的值：

```
mysql> select PASSWORD('123456');
+-------------------------------------------+
| PASSWORD('123456')                        |
+-------------------------------------------+
| *6BB4837EB74329105EE4568DDA7DC67ED2CA2AD9 |
+-------------------------------------------+
1 row in set (0.08 sec)
```

- MD5(str)函数：返回字符串 str 的 MD5 值，常用来对应用中的数据进行加密。

下例中显示了字符串"123456"的 MD5 值：

```
mysql> select MD5('123456');
+----------------------------------+
| MD5('123456')                    |
+----------------------------------+
| e10adc3949ba59abbe56e057f20f883e |
+----------------------------------+
1 row in set (0.06 sec)
```

5.8 小结

本章主要对 MySQL 常用的各类函数通过实例做了介绍。MySQL 有很多内建函数，这些内建函数实现了很多应用需要的功能并且拥有很好的性能，如果用户在工作中需要实现某种功能，最好先查一下 MySQL 官方文档或者帮助，看是否已经有相应的函数实现了用户需要的功能，这样可以大大提高工作效率。由于篇幅所限，本章并没有介绍所有的函数，读者可以进一步查询相关文档。

第二部分
开发篇

第 6 章 表类型（存储引擎）的选择

和大多数数据库不同，MySQL 中有一个存储引擎的概念，针对不同的存储需求可以选择最优的存储引擎。本章将详细介绍存储引擎的概念、分类以及实际应用中的选择原则。

6.1 MySQL 存储引擎概述

插件式存储引擎是 MySQL 数据库最重要的特性之一，用户可以根据应用的需要选择如何存储和索引数据、是否使用事务等。MySQL 默认支持多种存储引擎，以适用于不同领域的数据库应用需要，用户可以通过选择使用不同的存储引擎提高应用的效率，提供灵活的存储，用户甚至可以按照自己的需要定制和使用自己的存储引擎，以实现最大程度的可定制性。

MySQL 5.7 支持的存储引擎包括 InnoDB、MyISAM、MEMORY、CSV、BLACKHOLE、ARCHIVE、MERGE（MRG_MyISAM）、FEDERATED、EXAMPLE、NDB 等，其中 InnoDB 和 NDB 提供事务安全表，其他存储引擎都是非事务安全表。

创建新表时如果不指定存储引擎，那么系统就会使用默认存储引擎，MySQL 5.5 之前的默认存储引擎是 MyISAM，5.5 版本之后改为了 InnoDB。如果要修改默认的存储引擎，可以在参数文件中设置 default_storage_engine。查看当前的默认存储引擎，可以使用以下命令：

```
mysql> show variables like 'default_storage_engine';
+------------------------+--------+
| Variable_name          | Value  |
+------------------------+--------+
| default_storage_engine | InnoDB |
+------------------------+--------+
1 row in set (0.00 sec)
```

可以通过以下方法查询当前数据库支持的存储引擎：

```
mysql> show engines \G
*************************** 1. row ***************************
      Engine: InnoDB
     Support: DEFAULT
     Comment: Supports transactions, row-level locking, and foreign keys
Transactions: YES
          XA: YES
  Savepoints: YES
*************************** 2. row ***************************
      Engine: CSV
     Support: YES
     Comment: CSV storage engine
Transactions: NO
          XA: NO
```

```
      Savepoints: NO
*************************** 3. row ***************************
         Engine: MyISAM
        Support: YES
        Comment: MyISAM storage engine
   Transactions: NO
             XA: NO
     Savepoints: NO
*************************** 4. row ***************************
         Engine: BLACKHOLE
        Support: YES
        Comment: /dev/null storage engine (anything you write to it disappears)
   Transactions: NO
             XA: NO
     Savepoints: NO
*************************** 5. row ***************************
         Engine: PERFORMANCE_SCHEMA
        Support: YES
        Comment: Performance Schema
   Transactions: NO
             XA: NO
     Savepoints: NO
*************************** 6. row ***************************
         Engine: MRG_MYISAM
        Support: YES
        Comment: Collection of identical MyISAM tables
   Transactions: NO
             XA: NO
     Savepoints: NO
*************************** 7. row ***************************
         Engine: ARCHIVE
        Support: YES
        Comment: Archive storage engine
   Transactions: NO
             XA: NO
     Savepoints: NO
*************************** 8. row ***************************
         Engine: MEMORY
        Support: YES
        Comment: Hash based, stored in memory, useful for temporary tables
   Transactions: NO
             XA: NO
     Savepoints: NO
*************************** 9. row ***************************
         Engine: FEDERATED
        Support: NO
        Comment: Federated MySQL storage engine
   Transactions: NULL
             XA: NULL
     Savepoints: NULL
9 rows in set (0.00 sec)
```

通过上面的结果可以查看当前支持哪些存储引擎，其中 Support 不同值的含义分别为：DEFAULT——支持并启用，并且为默认引擎；YES——支持并启用；NO——不支持；DISABLED——支持，但是数据库启动的时候被禁用。

在创建新表的时候，可以通过增加 ENGINE 关键字设置新建表的存储引擎，例如，在下面的例子中，表 ai 的存储引擎是 MyISAM，而 country 表的存储引擎是 InnoDB：

```
CREATE TABLE ai (
  i bigint(20) NOT NULL AUTO_INCREMENT,
  PRIMARY KEY (i)
) ENGINE=MyISAM DEFAULT CHARSET=utf8;

CREATE TABLE country (
  country_id SMALLINT UNSIGNED NOT NULL AUTO_INCREMENT,
  country VARCHAR(50) NOT NULL,
  last_update TIMESTAMP NOT NULL DEFAULT CURRENT_TIMESTAMP ON UPDATE CURRENT_TIMESTAMP,
  PRIMARY KEY  (country_id)
)ENGINE=InnoDB DEFAULT CHARSET=utf8;
```

也可以使用 ALTER TABLE 语句，将一个已经存在的表修改成其他的存储引擎。下面的例子介绍了如何将表 ai 从 MyISAM 存储引擎修改到 InnoDB 存储引擎：

```
mysql> alter table ai engine = innodb;
Query OK, 0 rows affected (0.13 sec)
Records: 0  Duplicates: 0  Warnings: 0

mysql> show create table ai \G
*************************** 1. row ***************************
       Table: ai
Create Table: CREATE TABLE 'ai' (
  'i' bigint(20) NOT NULL AUTO_INCREMENT,
  PRIMARY KEY ('i')
) ENGINE=InnoDB DEFAULT CHARSET=utf8
1 row in set (0.00 sec)
```

这样修改后，表 ai 的存储引擎是 InnoDB，可以使用 InnoDB 存储引擎的相关特性。

注意：修改表的存储引擎需要锁表并复制数据，对于线上环境的表进行这个操作非常危险，除非你非常了解可能造成的影响，否则在线上环境请使用其他方式，例如 percona 的 OSC 工具，后续章节会有详细介绍。

6.2 各种存储引擎的特性

下面重点介绍几种常用的存储引擎，并对比各个存储引擎之间的区别，以帮助读者理解不同存储引擎的使用方式，如表 6-1 所示。

表 6-1 常用存储引擎的对比

特　点	MyISAM	Memory	InnoDB	Archive	NDB
B 树索引	支持	支持	支持	—	—
备份/时间点恢复	支持	支持	支持	支持	支持
支持集群	—	—	—	—	支持
聚簇索引	—	—	支持	—	—
数据压缩	支持	—	支持	支持	—
数据缓存	—	N/A	支持	—	支持
数据加密	支持	支持	支持	支持	支持
支持外键	—	—	支持	—	—
全文索引	支持	—	—	—	—
地理坐标数据类型	支持	—	支持	支持	支持
地理坐标索引	支持	—	支持	—	—
哈希索引	—	支持	—	—	支持
索引缓存	支持	N/A	支持	—	支持
锁粒度	表级	表级	行级	行级	行级
MVCC 多版本控制	—	—	支持	—	—
支持复制	支持	有限支持	支持	支持	支持
存储限制	256TB	RAM	64TB	None	384EB
T 树索引	—	—	—	—	支持
支持事务	—	—	支持	—	—
统计信息	支持	支持	支持	支持	支持

下面将重点介绍常用的 4 种存储引擎：MyISAM、MEMORY、InnoDB 和 Archive。

6.2.1 MyISAM

MyISAM 是 MySQL 5.5 之前版本的默认的存储引擎。MyISAM 既不支持事务，也不支持外键，在 5.5 之前的版本中，MyISAM 在某些场景中相对 InnoDB 的访问速度有明显的优势，对事务完整性没有要求或者以 SELECT、INSERT 为主的应用可以使用这个引擎来创建表。MySQL 5.6 之后，MyISAM 已经越来越少地被使用。

每个 MyISAM 在磁盘上存储成 3 个文件，其文件名都和表名相同，但扩展名分别如下：
- .frm（存储表定义）；
- .MYD（MYData，存储数据）；
- .MYI（MYIndex，存储索引）。

数据文件和索引文件可以放置在不同的目录，平均分布 IO，获得更快的速度。

要指定索引文件和数据文件的路径，需要在创建表的时候通过 DATA DIRECTORY 和 INDEX DIRECTORY 语句指定，也就是说不同 MyISAM 表的索引文件和数据文件可以放置到不同的路径下。文件路径需要是绝对路径，并且具有访问权限。

MyISAM 类型的表可能会损坏，原因可能是多种多样的，损坏后的表可能不能被访问，会提示需要修复或者访问后返回错误的结果。MyISAM 类型的表提供修复的工具，可以用 CHECK TABLE 语句来检查 MyISAM 表的健康，并用 REPAIR TABLE 语句修复一个损坏的 MyISAM 表。表损坏可能导致数据库异常重新启动，需要尽快修复并尽可能地确认损坏的原因。

MyISAM 的表还支持 3 种不同的存储格式，分别如下：
- 静态（固定长度）表；
- 动态表；
- 压缩表。

其中，静态表是默认的存储格式。静态表中的字段都是非变长字段，这样每个记录都是固定长度的，这种存储方式的优点是存储非常迅速，容易缓存，出现故障容易恢复；缺点是占用的空间通常比动态表多。静态表的数据在存储时会按照列的宽度定义补足空格，但是在应用访问的时候并不会得到这些空格，这些空格在返回给应用之前已经去掉。

但是也有些需要特别注意的问题，如果需要保存的内容后面本来就带有空格，那么在返回结果的时候也会被去掉，开发人员在编写程序的时候需要特别注意，因为静态表是默认的存储格式，开发人员可能并没有意识到这一点，从而丢失了尾部的空格。下面的例子演示了插入的记录包含空格时处理的情况：

```
mysql> create table Myisam_char (name char(10)) engine=myisam;
Query OK, 0 rows affected (0.04 sec)

mysql> insert into Myisam_char values('abcde'),('abcde '),(' abcde'),(' abcde ');
Query OK, 4 rows affected (0.00 sec)
Records: 4  Duplicates: 0  Warnings: 0

mysql> select name,length(name) from Myisam_char;
+---------+--------------+
| name    | length(name) |
+---------+--------------+
| abcde   |            5 |
```

```
|   abcde   |      5    |
|   abcde   |      7    |
|   abcde   |      7    |
+-----------+-----------+
4 rows in set (0.01 sec)
```

从上面的例子可以看出，插入记录后面的空格都被去掉了，前面的空格保留了。

动态表中包含变长字段，记录不是固定长度的，这样存储的优点是占用的空间相对较少，但是频繁地更新和删除记录会产生碎片，需要定期执行 OPTIMIZE TABLE 语句或 myisamchk-r 命令来改善性能，并且在出现故障时恢复相对比较困难。

压缩表由 myisampack 工具创建，占用非常小的磁盘空间。因为每个记录是被单独压缩的，所以只有非常小的访问开支。

6.2.2　InnoDB

InnoDB 作为 MySQL 5.5 之后的默认存储引擎，提供了具有提交、回滚和崩溃恢复能力的事务安全保障，同时提供了更小的锁粒度和更强的并发能力，拥有自己独立的缓存和日志，在 MySQL 5.6 和 5.7 版本中性能有较大的提升。

对比 MyISAM 存储引擎，InnoDB 会占用更多的磁盘空间以保留数据和索引。但是在大多数场景下，InnoDB 都会是更好的选择，这也是为何 MySQL 将默认存储引擎改为 InnoDB，并且在最新的 MySQL 8.0 中，将所有系统表也都改为 InnoDB 存储引擎。

下面将重点介绍存储引擎为 InnoDB 的表在使用过程中不同于使用其他存储引擎的表的特点，以及如何更好地使用 InnoDB 存储引擎。

1．自动增长列

InnoDB 表的自动增长列可以手工插入，但是插入的值如果是空，则实际插入的将是自动增长后的值。下面定义新表 autoincre_demo，其中列 i 使用自动增长列，对该表插入记录，然后查看自动增长列的处理情况，可以发现插入空时，实际插入的都将是自动增长后的值：

```
mysql> create table autoincre_demo
    -> (i smallint not null auto_increment,
    -> name varchar(10),primary key(i)
    -> )engine=innodb;
Query OK, 0 rows affected (0.13 sec)

mysql> insert into autoincre_demo values(null,'1'),(2,'2'),(null,'3');
Query OK, 3 rows affected (0.00 sec)
Records: 3  Duplicates: 0  Warnings: 0

mysql> select * from autoincre_demo;
+---+------+
| i | name |
+---+------+
| 1 | 1    |
| 2 | 2    |
| 3 | 3    |
+---+------+
3 rows in set (0.00 sec)
```

可以通过"ALTER TABLE *** AUTO_INCREMENT = n;"语句强制设置自动增长列的初始值，默认从 1 开始，但是在 MySQL 8.0 之前，对于 InnoDB 存储引擎来说，这个值只保留在内存中，如果数据库重新启动，那么这个值就会丢失，数据库会自动将 AUTO_INCREMENT 重置为自增列当前存储的最大值+1，这可能会导致在数据库重启后，自增列记录的值和预期不一致，从而导致数据冲突。以下示例演示了在 MySQL 5.7 中，AUTO_INCREMENT 值在重启前后的表现。

首先，创建测试表 test_auto_incre，id 为自增主键：

```
mysql> show create table test_auto_incre\G
*************************** 1. row ***************************
       Table: test_auto_incre
Create Table: CREATE TABLE `test_auto_incre` (
  `id` int(11) NOT NULL AUTO_INCREMENT,
  `name` varchar(10) COLLATE utf8_unicode_ci DEFAULT NULL,
  PRIMARY KEY (`id`)
) ENGINE=InnoDB  DEFAULT CHARSET=utf8 COLLATE=utf8_unicode_ci
1 row in set (0.00 sec)
```

修改 AUTO_INCREMENT=100：

```
mysql> alter table test_auto_incre auto_increment=100;
Query OK, 0 rows affected (0.04 sec)
Records: 0  Duplicates: 0  Warnings: 0
```

尝试写入一行数据，看一下是否生效：

```
mysql> insert into test_auto_incre values(null,'abc');
Query OK, 1 row affected (0.03 sec)

mysql> select * from test_auto_incre;
+-----+------+
| id  | name |
+-----+------+
| 100 | abc  |
+-----+------+
1 row in set (0.00 sec)
```

id 值为预期的 100，接下来再次修改 AUTO_INCREMENT 的值为 200：

```
mysql> alter table test_auto_incre auto_increment=200;
Query OK, 0 rows affected (0.01 sec)
Records: 0  Duplicates: 0  Warnings: 0

mysql> show create table test_auto_incre\G
*************************** 1. row ***************************
       Table: test_auto_incre
Create Table: CREATE TABLE `test_auto_incre` (
  `id` int(11) NOT NULL AUTO_INCREMENT,
  `name` varchar(10) COLLATE utf8_unicode_ci DEFAULT NULL,
  PRIMARY KEY (`id`)
) ENGINE=InnoDB AUTO_INCREMENT=200 DEFAULT CHARSET=utf8 COLLATE=utf8_unicode_ci
1 row in set (0.00 sec)
```

然后重启 MySQL 实例：

```
mysql> show create table test_auto_incre\G
ERROR 2006 (HY000): MySQL server has gone away
No connection. Trying to reconnect...
Connection id:    3
Current database: employees

*************************** 1. row ***************************
       Table: test_auto_incre
Create Table: CREATE TABLE `test_auto_incre` (
  `id` int(11) NOT NULL AUTO_INCREMENT,
  `name` varchar(10) COLLATE utf8_unicode_ci DEFAULT NULL,
  PRIMARY KEY (`id`)
) ENGINE=InnoDB AUTO_INCREMENT=101 DEFAULT CHARSET=utf8 COLLATE=utf8_unicode_ci
1 row in set (0.21 sec)
```

可以看到，重启之后，AUTO_INCREMENT 的值变成了 101，是当前自增列的最大值+1，而不是重启前设置的 200，这种情况可能导致在某些历史数据归档或者复制环境中发生数据冲突。在 MySQL 8.0 中，这个 BUG 得到了修复，具体实现方式是将自增主键的计数器持久化到 REDO LOG 中，每次计数器发生改变，都会将其写入到 REDO LOG。如果数据库发生重启，InnoDB 会根据 REDO LOG 中的计数器信息来初始化其内存值。

可以使用 LAST_INSERT_ID() 查询当前线程最后插入记录使用的值。如果一次插入了多条记录，那么返回的是第一条记录使用的自动增长值，需要注意的是，如果人为指定自增列的值，那么 LAST_INSERT_ID() 的值不会更新。下面的例子演示了使用 LAST_ INSERT_ID() 的情况：

```
mysql> select LAST_INSERT_ID();
+------------------+
| LAST_INSERT_ID() |
+------------------+
| 1                |
+------------------+
1 row in set (0.00 sec)

mysql> insert into autoincre_demo values(4,'4');
Query OK, 1 row affected (0.04 sec)

mysql> select LAST_INSERT_ID();
+------------------+
| LAST_INSERT_ID() |
+------------------+
| 1                |
+------------------+
1 row in set (0.00 sec)
```

可以看到，手工指定自增列的值为 4，LAST_INSERT_ID() 的值并不会更新。接下来一次插入 3 行，这次自增列的值将自动生成。由于此时自增列的最大值是 4，对于插入的这 3 行，自增列会自动分配 5、6、7 这 3 个值。

```
mysql> insert into autoincre_demo(name) values('5'),('6'),('7');
Query OK, 3 rows affected (0.05 sec)
Records: 3  Duplicates: 0  Warnings: 0

mysql> select LAST_INSERT_ID();
+------------------+
| LAST_INSERT_ID() |
+------------------+
| 5                |
+------------------+
1 row in set (0.00 sec)
```

这时 LAST_INSERT_ID() 的值等于批量插入的第一条记录的值 5，而不是最后插入的 7。

对于 InnoDB 表，自动增长列必须被索引。如果是组合索引，也必须是组合索引的第一列，但是对于 MyISAM 表，自动增长列可以是组合索引的其他列，这样插入记录后，自动增长列是按照组合索引的前面几列进行排序后递增的。例如，创建一个新的 MyISAM 类型的表 autoincre_demo，自动增长列 d1 作为组合索引的第二列，对该表插入一些记录后，可以发现自动增长列是按照组合索引的第一列 d2 进行排序后递增的：

```
mysql> create table autoincre_demo
    -> (d1 smallint not null auto_increment,
    -> d2 smallint not null,
    -> name varchar(10),
    -> index(d2,d1)
    -> )engine=myisam;
Query OK, 0 rows affected (0.03 sec)

mysql> insert into autoincre_demo(d2,name) values(2,'2'),(3,'3'),(4, '4'), (2,'2'), (3,'3'), (4,'4');
Query OK, 6 rows affected (0.00 sec)
Records: 6  Duplicates: 0  Warnings: 0
```

```
mysql> select * from autoincre_demo;
+----+----+------+
| d1 | d2 | name |
+----+----+------+
| 1  | 2  | 2    |
| 1  | 3  | 3    |
| 1  | 4  | 4    |
| 2  | 2  | 2    |
| 2  | 3  | 3    |
| 2  | 4  | 4    |
+----+----+------+
6 rows in set (0.00 sec)
```

2. 外键约束

MySQL 支持外键的常用存储引擎只有 InnoDB，在创建外键的时候，要求父表必须有对应的索引，子表在创建外键的时候也会自动创建对应的索引。

下面是样例数据库中的两个表，country 表是父表，country_id 为主键索引，city 表是子表，country_id 字段为外键，对应于 country 表的主键 country_id。

```
CREATE TABLE country (
  country_id SMALLINT UNSIGNED NOT NULL AUTO_INCREMENT,
  country VARCHAR(50) NOT NULL,
  last_update TIMESTAMP NOT NULL DEFAULT CURRENT_TIMESTAMP ON UPDATE CURRENT_ TIMESTAMP,
  PRIMARY KEY  (country_id)
)ENGINE=InnoDB DEFAULT CHARSET=utf8;

CREATE TABLE city (
  city_id SMALLINT UNSIGNED NOT NULL AUTO_INCREMENT,
  city VARCHAR(50) NOT NULL,
  country_id SMALLINT UNSIGNED NOT NULL,
  last_update TIMESTAMP NOT NULL DEFAULT CURRENT_TIMESTAMP ON UPDATE CURRENT_TIMESTAMP,
  PRIMARY KEY  (city_id),
  KEY idx_fk_country_id (country_id),
  CONSTRAINT fk_city_country FOREIGN KEY (country_id) REFERENCES country (country_id) ON DELETE RESTRICT ON UPDATE CASCADE
)ENGINE=InnoDB DEFAULT CHARSET=utf8;
```

在创建索引时，可以指定在删除、更新父表时，对子表进行的相应操作，包括 RESTRICT、CASCADE、SET NULL 和 NO ACTION。其中 RESTRICT 和 NO ACTION 相同，是指限制在子表有关联记录的情况下父表不能更新；CASCADE 表示父表在更新或者删除时，更新或者删除子表对应记录；SET NULL 则表示父表在更新或者删除的时候，子表的对应字段被 SET NULL。选择后两种方式的时候要谨慎，可能会因为错误的操作导致数据的丢失。

例如，对上面创建的两个表，子表的外键指定是 ON DELETE RESTRICT ON UPDATE CASCADE 方式的，那么在主表删除记录的时候，如果子表有对应记录，则不允许删除；主表在更新记录的时候，如果子表有对应记录，则子表对应更新：

```
mysql> select * from country where country_id = 1;
+------------+-------------+---------------------+
| country_id | country     | last_update         |
+------------+-------------+---------------------+
| 1          | Afghanistan | 2006-02-15 04:44:00 |
+------------+-------------+---------------------+
1 row in set (0.00 sec)

mysql> select * from city where country_id = 1;
+---------+-------+------------+---------------------+
| city_id | city  | country_id | last_update         |
+---------+-------+------------+---------------------+
| 251     | Kabul | 1          | 2006-02-15 04:45:25 |
+---------+-------+------------+---------------------+
1 row in set (0.00 sec)
```

```
mysql> delete from country where country_id=1;
ERROR 1451 (23000): Cannot delete or update a parent row: a foreign key constraint fails
('sakila/city', CONSTRAINT 'fk_city_country' FOREIGN KEY ('country_id') REFERENCES 'country'
('country_id') ON UPDATE CASCADE)

mysql> update country set country_id = 10000 where country_id = 1;
Query OK, 1 row affected (0.04 sec)
Rows matched: 1  Changed: 1  Warnings: 0

mysql> select * from country where country = 'Afghanistan';
+------------+-------------+---------------------+
| country_id | country     | last_update         |
+------------+-------------+---------------------+
| 10000      | Afghanistan | 2007-07-17 09:45:23 |
+------------+-------------+---------------------+
1 row in set (0.00 sec)

mysql> select * from city where city_id = 251;
+---------+-------+------------+---------------------+
| city_id | city  | country_id | last_update         |
+---------+-------+------------+---------------------+
| 251     | Kabul | 10000      | 2006-02-15 04:45:25 |
+---------+-------+------------+---------------------+
1 row in set (0.00 sec)
```

当某个表被其他表创建了外键参照，那么该表的对应索引或者主键禁止被删除。

在导入多个表的数据时，如果需要忽略表之前的导入顺序，可以暂时关闭外键的检查；同样，在执行 LOAD DATA 和 ALTER TABLE 操作的时候，可以通过暂时关闭外键约束来加快处理的速度，关闭的命令是"SET FOREIGN_KEY_CHECKS = 0;"，执行完成之后，通过执行"SET FOREIGN_KEY_CHECKS = 1;"语句改回原状态。

对于 InnoDB 类型的表，外键的信息可以通过使用 show create table 或者 show table status 命令显示。

```
mysql> show table status like 'city' \G
*************************** 1. row ***************************
           Name: city
         Engine: InnoDB
        Version: 10
     Row_format: Compact
           Rows: 427
 Avg_row_length: 115
    Data_length: 49152
Max_data_length: 0
   Index_length: 16384
      Data_free: 0
 Auto_increment: 601
    Create_time: 2007-07-17 09:45:33
    Update_time: NULL
     Check_time: NULL
      Collation: utf8_general_ci
       Checksum: NULL
 Create_options:
        Comment: InnoDB free: 0 kB; ('country_id') REFER 'sakila/country'' ('country_id') ON
UPDATE
1 row in set (0.00 sec)
```

注意：外键需要注意的细节较多，一旦使用不当，可能会带来性能下降或者数据不一致的问题，尤其在 OLTP 类型的应用中，需要慎重使用外键。

3. 主键和索引

不同于其他存储引擎，InnoDB 的数据文件本身就是以聚簇索引的形式保存的，这个聚簇索引也被称为主索引，并且也是 InnoDB 表的主键，InnoDB 表的每行数据都保存在主索引的叶子节点上。因此，所有 InnoDB 表都必须包含主键，如果创建表时候，没有显式指定主键，那么 InnoDB 存储引擎会自动创建一个长度为 6 个字节的 long 类型隐藏字段作为主键。

考虑到聚簇索引的特点和对于查询的优化效果，所有 InnoDB 表都应该显式的指定主键，一般来说，主键应该按照下列原则来选择：

- 满足唯一和非空约束；
- 优先考虑使用最经常被当作查询条件的字段或者自增字段；
- 字段值基本不会被修改；
- 使用尽可能短的字段。

在 InnoDB 表上，除了主键之外的其他索引都叫作辅助索引或者二级索引，二级索引会指向主索引，并通过主索引获取最终数据。因此，主键是否合理的创建，会对所有索引的效率产生影响。

关于索引使用的更多内容，会在第 9 章做更加详细的介绍。

4．存储方式

InnoDB 存储表和索引有以下两种方式。

- 使用共享表空间存储：这种方式创建的表的表结构保存在 .frm 文件中，数据和索引保存在 innodb_data_home_dir 和 innodb_data_file_path 定义的表空间中，可以是多个文件。
- 使用多表空间存储：这种方式创建的表的表结构仍然保存在 .frm 文件中，但是每个表的数据和索引单独保存在 .ibd 中。如果是一个分区表，则每个分区对应单独的 .ibd 文件，文件名是 "表名+分区名"，可以在创建分区的时候指定每个分区的数据文件的位置，以此将表的 IO 均匀分布在多个磁盘上。

使用共享表空间时，随着数据的不断增长，表空间的管理维护会变得越来越困难，所以一般都建议使用多表空间。要使用多表空间的存储方式，需设置参数 innodb_file_per_table，在 MySQL 5.7 中，此参数默认设置为 ON，即新建的表默认都是按照多表空间的方式创建。如果修改此参数为 OFF，则新创建的表都会改为共享表空间存储，但已有的多表空间的表仍然保存原来的访问方式。

一些老版本中的表，很多是共享表空间，可以通过下面的命令改为多表空间：

```
mysql> SET GLOBAL innodb_file_per_table=1;
mysql> ALTER TABLE table_name ENGINE=InnoDB;
```

多表空间的数据文件没有大小限制，既不需要设置初始大小，也不需要设置文件的最大限制、扩展大小等参数。

对于使用多表空间特性的表，可以比较方便地进行单表备份和恢复操作，但是直接复制 .ibd 文件是不行的，因为没有共享表空间的数据字典信息，直接复制的 .ibd 文件和 .frm 文件恢复时是不能被正确识别的，但可以通过以下命令：

```
ALTER TABLE tbl_name DISCARD TABLESPACE;
ALTER TABLE tbl_name IMPORT TABLESPACE;
```

将备份恢复到数据库中。

注意：即便在多表空间的存储方式下，共享表空间仍然是必须的，InnoDB 把内部数据词典和在线重做日志放在这个文件中。

6.2.3 MEMORY

MEMORY 存储引擎使用存在于内存中的内容来创建表。每个 MEMORY 表只实际对应一个磁盘文件，格式是.frm。MEMORY 类型的表访问非常地快，因为它的数据是放在内存中的，并且默认使用 HASH 索引，但是一旦服务关闭，表中的数据就会丢失掉。

下面的例子创建了一个 MEMORY 的表，并从 city 表中获得记录：

```
mysql> CREATE TABLE tab_memory ENGINE=MEMORY
    ->        SELECT city_id,city,country_id
    ->        FROM city GROUP BY city_id;
Query OK, 600 rows affected (0.06 sec)
Records: 600  Duplicates: 0  Warnings: 0

mysql> select count(*) from tab_memory;
+----------+
| count(*) |
+----------+
|      600 |
+----------+
1 row in set (0.00 sec)

mysql> show table status like 'tab_memory' \G
*************************** 1. row ***************************
           Name: tab_memory
         Engine: MEMORY
        Version: 10
     Row_format: Fixed
           Rows: 600
 Avg_row_length: 155
    Data_length: 127040
Max_data_length: 16252835
   Index_length: 0
      Data_free: 0
 Auto_increment: NULL
    Create_time: 2018-05-28 15:10:04
    Update_time: NULL
     Check_time: NULL
      Collation: gbk_chinese_ci
       Checksum: NULL
 Create_options:
        Comment:
1 row in set (0.00 sec)
```

给 MEMORY 表创建索引的时候，可以指定使用 HASH 索引还是 BTREE 索引：

```
mysql> create index mem_hash USING HASH on tab_memory (city_id) ;
Query OK, 600 rows affected (0.04 sec)
Records: 600  Duplicates: 0  Warnings: 0

mysql> SHOW INDEX FROM tab_memory \G
*************************** 1. row ***************************
        Table: tab_memory
   Non_unique: 1
     Key_name: mem_hash
 Seq_in_index: 1
  Column_name: city_id
    Collation: NULL
  Cardinality: 300
     Sub_part: NULL
       Packed: NULL
         Null:
   Index_type: HASH
      Comment:
1 row in set (0.01 sec)

mysql> drop index mem_hash on tab_memory;
```

```
Query OK, 600 rows affected (0.04 sec)
Records: 600  Duplicates: 0  Warnings: 0

mysql> create index mem_hash USING BTREE on tab_memory (city_id) ;
Query OK, 600 rows affected (0.03 sec)
Records: 600  Duplicates: 0  Warnings: 0

mysql> SHOW INDEX FROM tab_memory \G
*************************** 1. row ***************************
        Table: tab_memory
   Non_unique: 1
     Key_name: mem_hash
 Seq_in_index: 1
  Column_name: city_id
    Collation: A
  Cardinality: NULL
     Sub_part: NULL
       Packed: NULL
         Null:
   Index_type: BTREE
      Comment:
1 row in set (0.00 sec)
```

在启动 MySQL 服务的时候使用--init-file 选项，把 INSERT INTO ... SELECT 或 LOAD DATA INFILE 这样的语句放入这个文件中，就可以在服务启动时从持久稳固的数据源装载表。

服务器需要足够的内存来维持所有在同一时间使用的 MEMORY 表，当不再需要 MEMORY 表的内容之时，要释放被 MEMORY 表使用的内存，应该执行 DELETE FROM 或 TRUNCATE TABLE，或者整个删除表（使用 DROP TABLE 操作）。

每个 MEMORY 表中可以放置的数据量的大小，受到 max_heap_table_size 系统变量的约束，这个系统变量的初始值是 16MB，可以根据需要加大。此外，在定义 MEMORY 表的时候，可以通过 MAX_ROWS 子句指定表的最大行数。

MEMORY 类型的存储引擎主要用于那些内容变化不频繁的代码表，或者作为统计操作的中间结果表，便于高效地对中间结果进行分析并得到最终的统计结果。对存储引擎为 MEMORY 的表进行更新操作要谨慎，因为数据并没有实际写入到磁盘中，所以一定要对下次重新启动服务后如何获得这些修改后的数据有所考虑。

6.2.4 MERGE

MERGE 存储引擎也被称为 MRG_MyISAM，是一组 MyISAM 表的组合。这些 MyISAM 表必须结构完全相同，MERGE 表本身并没有数据，对 MERGE 类型的表可以进行查询、更新、删除操作，这些操作实际上是对内部的 MyISAM 表进行的。对于 MERGE 类型表的插入操作，是通过 INSERT_METHOD 子句定义插入的表，可以有 3 个不同的值，使用 FIRST 或 LAST 值使得插入操作被相应地作用在第一个或最后一个表上，不定义这个子句或者定义为 NO，表示不能对这个 MERGE 表执行插入操作。

可以对 MERGE 表进行 DROP 操作，这个操作只是删除 MERGE 的定义，对内部的表没有任何影响。

MERGE 表在磁盘上保留两个文件，文件名以表的名字开始，一个.frm 文件存储表定义，另一个.MRG 文件包含组合表的信息，包括 MERGE 表由哪些表组成、插入新的数据时的依据。可以通过修改.MRG 文件来修改 MERGE 表，但是修改后要通过 FLUSH TABLES 刷新。

下面是一个创建和使用 MERGE 表的示例。

（1）创建 3 个测试表 payment_2006、payment_2007 和 payment_all，其中 payment_all 是前两个表的 MERGE 表：

```
mysql> create table payment_2006(
    -> country_id smallint,
    -> payment_date datetime,
    -> amount DECIMAL(15,2),
    -> KEY idx_fk_country_id (country_id)
    -> )engine=myisam;
Query OK, 0 rows affected (0.03 sec)

mysql> create table payment_2007(
    -> country_id smallint,
    -> payment_date datetime,
    -> amount DECIMAL(15,2),
    -> KEY idx_fk_country_id (country_id)
    -> )engine=myisam;
Query OK, 0 rows affected (0.02 sec)

mysql> CREATE TABLE payment_all(
    -> country_id smallint,
    -> payment_date datetime,
    -> amount DECIMAL(15,2),
    -> INDEX(country_id)
    -> )engine=merge union=(payment_2006,payment_2007) INSERT_METHOD=LAST;
Query OK, 0 rows affected (0.04 sec)
```

（2）分别向 payment_2006 和 payment_2007 表中插入测试数据：

```
mysql> insert into payment_2006 values(1,'2006-05-01',100000),(2, '2006-08-15',150000);
Query OK, 2 rows affected (0.00 sec)
Records: 2  Duplicates: 0  Warnings: 0

mysql> insert into payment_2007 values(1, '2007-02-20',35000),(2, '2007-07-15', 220000);
Query OK, 2 rows affected (0.00 sec)
Records: 2  Duplicates: 0  Warnings: 0
```

（3）分别查看这 3 个表中的记录：

```
mysql> select * from payment_2006;
+------------+---------------------+-----------+
| country_id | payment_date        | amount    |
+------------+---------------------+-----------+
|          1 | 2006-05-01 00:00:00 | 100000.00 |
|          2 | 2006-08-15 00:00:00 | 150000.00 |
+------------+---------------------+-----------+
2 rows in set (0.00 sec)

mysql> select * from payment_2007;
+------------+---------------------+-----------+
| country_id | payment_date        | amount    |
+------------+---------------------+-----------+
|          1 | 2007-02-20 00:00:00 |  35000.00 |
|          2 | 2007-07-15 00:00:00 | 220000.00 |
+------------+---------------------+-----------+
2 rows in set (0.00 sec)

mysql> select * from payment_all;
+------------+---------------------+-----------+
| country_id | payment_date        | amount    |
+------------+---------------------+-----------+
|          1 | 2006-05-01 00:00:00 | 100000.00 |
|          2 | 2006-08-15 00:00:00 | 150000.00 |
|          1 | 2007-02-20 00:00:00 |  35000.00 |
|          2 | 2007-07-15 00:00:00 | 220000.00 |
+------------+---------------------+-----------+
4 rows in set (0.00 sec)
```

可以发现，payment_all 表中的数据是 payment_2006 和 payment_2007 表的记录合并后的

结果集。

下面向 MERGE 表中插入一条记录，由于 MERGE 表的定义是 INSERT_METHOD=LAST，就会向最后一个表中插入记录，所以虽然这里插入的记录是 2006 年的，但仍然会写到 payment_2007 表中。

```
mysql> insert into payment_all values(3, '2006-03-31',112200);
Query OK, 1 row affected (0.00 sec)

mysql> select * from payment_all;
+------------+---------------------+-----------+
| country_id | payment_date        | amount    |
+------------+---------------------+-----------+
| 1          | 2006-05-01 00:00:00 | 100000.00 |
| 2          | 2006-08-15 00:00:00 | 150000.00 |
| 1          | 2007-02-20 00:00:00 | 35000.00  |
| 2          | 2007-07-15 00:00:00 | 220000.00 |
| 3          | 2006-03-31 00:00:00 | 112200.00 |
+------------+---------------------+-----------+
5 rows in set (0.00 sec)

mysql> select * from payment_2007;
+------------+---------------------+-----------+
| country_id | payment_date        | amount    |
+------------+---------------------+-----------+
| 1          | 2007-02-20 00:00:00 | 35000.00  |
| 2          | 2007-07-15 00:00:00 | 220000.00 |
| 3          | 2006-03-31 00:00:00 | 112200.00 |
+------------+---------------------+-----------+
3 rows in set (0.00 sec)
```

这也是 MERGE 表和分区表的区别，MERGE 表并不能智能地将记录写到对应的表中，而分区表是可以的（分区功能在 5.1 版中正式推出，经过多个版本的更新，目前已经比较完善）。通常我们使用 MERGE 表来透明地对多个表进行查询和更新操作，而对这种按照时间记录的操作日志表则可以透明地进行插入操作。

6.2.5 TokuDB

前面介绍的都是 MySQL 自带的存储引擎，除了这些之外，还有一些常见的第三方存储引擎，在某些特定应用中也有广泛使用，比如列式存储引擎 Infobright 以及高写性能、高压缩的 TokuDB 就是其中非常有代表性的两种。本节将简单介绍 TokuDB。

TokuDB 是一个高性能、支持事务处理的存储引擎，在 MySQL 5.6 版本之前，可以安装到 MySQL 和 MariaDB 中，被 Percona 公司收购之后，目前最新版本可以在 Percona Server for MySQL 之中使用。TokuDB 存储引擎具有高扩展性、高压缩率、高效的写入性能，支持大多数在线 DDL 操作。最新版本已经开源，读者可以从 Percona 官方网站中进行下载和安装（*https://www.percona.com/software/mysql-database/percona-server/*）。

针对 TokuDB 存储引擎的主要特性，Tokutek 网站公布了这款优秀存储引擎与经典的 InnoDB 存储引擎的对比结果，如图 6-1 所示。由于本书内容以 MySQL 社区版为主，因此下面的内容是针对 MySQL 5.6 和其对应的 TokuDB 版本。

通过对比，可以看出 TokuDB 主要有以下几项特性：
- 使用 Fractal 树索引保证高效的插入性能；
- 优秀的压缩特性，比 InnoDB 高近 10 倍；
- Hot Schema Changes 特性支持在线创建索引和添加、删除属性列等 DDL 操作；

- 使用 Bulk Loader 达到快速加载大量数据；
- 提供了主从延迟消除技术；
- 支持 ACID 和 MVCC。

TokuDB vs. InnoDB		
	InnoDB	TokuDB®
Index Type	B-tree	Fractal Tree® index [more]
Insertion Rate at Scale	100s / second	10,000s / second [more]
Compression	~2x	5x – 10x typical, up to 25x possible [more]
Hot Indexing	No (hrs+)	Yes (secs to mins) [more]
Hot Column Addition/Deletion/Expansion/Rename	No (hrs+)	Yes (secs to mins) [more]
Fast Loader	No	Yes [more]
Fragmentation Immunity	No	Yes (no dump/reload downtime – no index fragmentation) [more]
Clustering Indexes	Primary Key Only	Multiple [more]
Fast Recovery Time	No	Yes [more]
Eliminates Slave Lag	No	Yes [more]
MariaDB Compatible	Yes	Yes
ACID	Yes	Yes
MVCC	Yes	Yes

图 6-1　TokuDB 与 InnoDB 的比较

关于 TokuDB 最新各特性的具体性能测试数据，读者可以从 Percona 网站获得（*https://www.percona.com/doc/percona-server/LATEST/index.html*）。

通过上面的介绍，可以发现 TokuDB 特别适用以下几种场景：
- 日志数据，因为日志通常插入频繁且存储量大；
- 历史数据，通常不会再有写操作，可以利用 TokuDB 的高压缩特性进行存储；
- 在线 DDL 较频繁的场景，使用 TokuDB 可以大大增加系统的可用性。

6.3　如何选择合适的存储引擎

在选择存储引擎时，应根据应用特点选择合适的存储引擎。对于复杂的应用系统，还可以根据实际情况选择多种存储引擎进行组合。

下面是几种常用的存储引擎的适用环境。

- MyISAM：MySQL 5.5 之前版本默认的存储引擎。如果应用是以读操作和插入操作为主，只有极少的更新和删除操作，并且对事务的完整性没有要求、没有并发写操作，那么选择这个存储引擎是适合的。OLTP 环境一般建议不要再使用 MyISAM。
- InnoDB：MySQL 5.5 之后版本的默认存储引擎，用于事务处理应用程序，支持外键。对于大多数的应用系统，InnoDB 都是合适的选择。如果应用对事务的完整性有比较高的要求，在并发条件下要求数据的一致性。数据操作除了插入和查询以外，还包括很多的更新、删除操作，那么应该优先选择 InnoDB 存储引擎。InnoDB 存储引擎除了有效地降低由于删除和更新导致的锁定，还可以确保事务的完整提交（Commit）和回滚（Rollback）。
- MEMORY：将所有数据保存在 RAM 中，在需要快速定位记录和其他类似数据的环境下，可提供极快的访问速度。MEMORY 的缺陷是对表的大小有限制，太大的表无法缓存在内存中，其次是要确保表的数据可以恢复，数据库异常终止后表中的数据是可以恢复的。

MEMORY 表通常用于更新不太频繁的小表，用以快速得到访问结果。

- MERGE：用于将一系列等同的 MyISAM 表以逻辑方式组合在一起，并作为一个对象引用它们。MERGE 表的优点在于可以突破对单个 MyISAM 表大小的限制，并且通过将不同的表分布在多个磁盘上，可以有效地改善 MERGE 表的访问效率。这对于诸如数据仓储等 VLDB 环境十分适合。

> **注意：** 以上只是我们按照实施经验提出的关于存储引擎选择的一些建议，但是不同应用的特点是千差万别的。选择使用哪种存储引擎才是最佳方案也不是绝对的，这需要根据用户各自的应用进行测试，从而得到最适合自己的结果。

6.4 小结

本章重点介绍了 MySQL 提供的几种主要的存储引擎及其使用特性，以及如何根据应用的需要选择合适的存储引擎。这些存储引擎有各自的优势和适用的场合，正确地选择存储引擎对改善应用的效率可以起到事半功倍的效果。

正确地选择了存储引擎之后，还需要正确选择表中的数据类型。第 7 章将详细介绍如何选择合适的数据类型。

第 7 章 选择合适的数据类型

在使用 MySQL 创建数据表时都会遇到一个问题,即如何为字段选择合适的数据类型。例如,创建一张员工表用来记录员工的信息,这时对员工的各种属性如何来进行定义?也许读者会想,这个问题很简单,每个字段可以使用很多种数据类型来定义,比如 int、float、double、decimal 等。其实正因为可选择的数据类型太多,才需要依据一些原则来"挑选"最适合的数据类型。本章将详细介绍字符、数值、日期数据类型的一些选择原则。

7.1 CHAR 与 VARCHAR

CHAR 和 VARCHAR 类型类似,都用来存储字符串,但它们保存和检索的方式不同。CHAR 属于固定长度的字符类型,而 VARCHAR 属于可变长度的字符类型。

表 7-1 显示了将各种字符串值保存到 CHAR(4)和 VARCHAR(4)列后的结果,说明了 CHAR 和 VARCHAR 之间的差别。

表 7-1   CHAR 和 VARCHAR 的对比

值	CHAR(4)	存储需求	VARCHAR(4)	存储需求
''	' '	4 个字节	''	1 个字节
'ab'	'ab '	4 个字节	'ab'	3 个字节
'abcd'	'abcd'	4 个字节	'abcd'	5 个字节
'abcdefgh'	'abcd'	4 个字节	'abcd'	5 个字节

注意表 7-1 中最后一行的值只适用 MySQL 运行在非"严格模式"时,如果 MySQL 运行在严格模式,超过列长度的值将不会保存,并且会出现错误提示。关于"严格模式",将在第 13 章中详细介绍。VARCHAR(4)列显示的存储需求比实际字符长度多 1 是因为 VARCHAR 类型要用一到两个字节来记录字节长度,如果数据位占用字节数小于 255 时,用一个字节记录;大于 255 时,用两个字节记录。

从 CHAR(4)和 VARCHAR(4)列检索的值并不总是相同,因为检索时从 CHAR 列删除了尾部的空格。下面通过一个例子来说明该差别:

```
mysql> CREATE TABLE vc (v VARCHAR(4), c CHAR(4));
Query OK, 0 rows affected (0.02 sec)

mysql> INSERT INTO vc VALUES ('ab  ', 'ab  ');
```

```
Query OK, 1 row affected (0.00 sec)

mysql> SELECT CONCAT(v, '+'), CONCAT(c, '+') FROM vc;
+----------------+----------------+
| CONCAT(v, '+') | CONCAT(c, '+') |
+----------------+----------------+
| ab +           | ab+            |
+----------------+----------------+
1 row in set (0.00 sec)
```

由于 CHAR 是固定长度的，所以它的处理速度比 VARCHAR 快，但是其缺点是浪费存储空间，程序需要对行尾空格进行处理，所以对于那些长度变化不大并且对查询速度有较高要求的数据可以考虑使用 CHAR 类型来存储。

在使用 VARCHAR 类型的时候，不能因为 VARCHAR 类型长度可变就都为 VARCHAR 定义一个很大的长度，仍然需要按需定义长度，定义一个远超实际需求长度的 VARCHAR 字段可能影响应用程序的效率，并且有更大的概率触发 MySQL 在 VARCHAR 上存在的一些 BUG。

在 MySQL 中，不同的存储引擎对 CHAR 和 VARCHAR 的使用原则有所不同，这里简单概括如下。

- MyISAM 存储引擎：建议使用固定长度的数据列代替可变长度的数据列。
- MEMORY 存储引擎：目前都使用固定长度的数据行存储，因此无论使用 CHAR 或 VARCHAR 列都没有关系，两者都是作为 CHAR 类型处理。
- InnoDB 存储引擎：建议使用 VARCHAR 类型。对于 InnoDB 数据表，内部的行存储格式没有区分固定长度和可变长度列（所有数据行都使用指向数据列值的头指针），因此在本质上，使用固定长度的 CHAR 列不一定比使用可变长度 VARCHAR 列性能要好。因而，主要的性能因素是数据行使用的存储总量。由于 CHAR 平均占用的空间多于 VARCHAR，因此使用 VARCHAR 来最小化需要处理的数据行的存储总量和磁盘 I/O 是比较好的。

7.2 TEXT 与 BLOB

一般在保存少量字符串的时候，我们会选择 CHAR 或者 VARCHAR；而在保存较大文本时，通常会选择使用 TEXT 或者 BLOB。二者之间的主要差别是 BLOB 能用来保存二进制数据，比如照片；而 TEXT 只能保存字符数据，比如一篇文章或者日记。TEXT 和 BLOB 中又分别包括 TINYTEXT、TEXT、MEDIUMTEXT、LONGTEXT 和 TINYBLOB、BLOB、MEDIUMBLOB、LONGBLOB 等不同的类型，它们之间的主要区别是存储文本长度不同和存储字节不同，用户应该根据实际情况选择能够满足需求的最小存储类型。本节主要对 BLOB 和 TEXT 存在的一些常见问题进行介绍。

（1）BLOB 和 TEXT 值会引起一些性能问题，特别是在执行大量的删除操作时。

删除操作会在数据表中留下很大的"空洞"，以后填入这些"空洞"的记录在插入的性能上会有影响。为了提高性能，建议定期使用 OPTIMIZE TABLE 功能对这类表进行碎片整理，避免因为"空洞"导致性能问题。

下面的例子描述了 OPTIMIZE TABLE 的碎片整理功能。首先创建测试表 t，字段 id 和 context 的类型分别为 varchar(100)和 text：

```
mysql> create table t (id varchar(100),context  text);
Query OK, 0 rows affected (0.01 sec)
```

然后往 t 中插入大量记录，这里使用 repeat 函数插入大量字符串：

```
mysql> insert into t values(1,repeat('haha',100));
Query OK, 1 row affected (0.00 sec)

mysql> insert into t values(2,repeat('haha',100));
Query OK, 1 row affected (0.00 sec)

mysql> insert into t values(3,repeat('haha',100));
Query OK, 1 row affected (0.00 sec)

mysql> insert into t select * from t;
…
mysql> insert into t select * from t;
Query OK, 196608 rows affected (4.86 sec)
Records: 196608  Duplicates: 0  Warnings: 0

mysql> exit
Bye
```

退出到操作系统下，查看表 t 的物理文件大小：

```
[root@hz_10_120_240_251 employees]$ du -sh t.*
12K     t.frm
155M    t.ibd
```

这里数据文件显示为 155MB。从表 t 中删除 id 为 "1" 的数据，这些数据占总数据量的 1/3：

```
mysql> delete from t where id=1;
Query OK, 131072 rows affected (4.33 sec)
```

再次退出到操作系统下，查看表 t 的物理文件大小：

```
[root@hz_10_120_240_251 employees]$ du -sh t.*
12K     t.frm
155M    t.ibd
```

可以发现，表 t 的数据文件仍然为 155MB，并没有因为数据删除而减少。接下来对表进行 OPTIMIZE（优化）操作：

```
mysql> OPTIMIZE TABLE   t\G
*************************** 1. row ***************************
   Table: employees.t
      Op: optimize
Msg_type: note
Msg_text: Table does not support optimize, doing recreate + analyze instead
*************************** 2. row ***************************
   Table: employees.t
      Op: optimize
Msg_type: status
Msg_text: OK
2 rows in set (2.09 sec)
```

再次查看表 t 的物理文件大小：

```
[root@hz_10_120_240_251 employees]$ du -sh t.*
12K     t.frm
104M    t.ibd
```

可以发现，表的数据文件大大缩小，"空洞" 空间已经被回收。另外，注意到，对于 InnoDB 引擎，OPTIMIZE 语句被自动转换为 recreate+analyze 语句。

（2）可以使用合成的（Synthetic）索引来提高大文本字段（BLOB 或 TEXT）的查询性能。

简单来说，合成索引就是根据大文本字段的内容建立一个散列值，并把这个值存储在单独的数据列中，接下来就可以通过检索散列值找到数据行了。但是，要注意这种技术只能用于精确匹配的查询（散列值对于类似 "<" 或 ">=" 等范围搜索操作符是没有用处的）。

可以使用 MD5() 函数生成散列值，也可以使用 SHA1() 或 CRC32()，或者使用自己的应用

7.2 TEXT 与 BLOB

程序逻辑来计算散列值。请记住数值型散列值可以高效率地存储。

同样，如果散列算法生成的字符串带有尾部空格，就不要把它们存储在 CHAR 字段中，而是使用 VARCHAR 字段，因为 CHAR 字段会受到尾部空格去除的影响。合成的散列索引对于那些 BLOB 或 TEXT 数据列特别有用。用散列标识符值查找的速度比搜索 BLOB 列本身的速度快很多。

下面通过实例介绍一下合成索引的使用方法。首先创建测试表 t，字段 id、context、hash_value 字段类型分别为 varchar(100)、blob、varchar(40)，并且在 hash_value 列上创建索引：

```
mysql> create table t (id varchar(100),context  blob,hash_value varchar(40));
Query OK, 0 rows affected (0.03 sec)
mysql> create index idx_hash_value on t(hash_value);
Query OK, 0 rows affected (0.04 sec)
Records: 0  Duplicates: 0  Warnings: 0
```

然后往 t 中插入测试数据，其中 hash_value 用来存放 context 列的 MD5 散列值：

```
mysql> insert into t values(1,repeat('beijing',2),md5(context));
Query OK, 1 row affected (0.00 sec)

mysql> insert into t values(2,repeat('beijing',2),md5(context));
Query OK, 1 row affected (0.00 sec)

mysql> insert into t values(3,repeat('beijing 2008',2),md5(context));
Query OK, 1 row affected (0.00 sec)
mysql> select * from t;
+------+-------------------------+----------------------------------+
| id   | context                 | hash_value                       |
+------+-------------------------+----------------------------------+
| 1    | beijingbeijing          | 09746eef633dbbccb7997dfd795cff17 |
| 2    | beijingbeijing          | 09746eef633dbbccb7997dfd795cff17 |
| 3    | beijing 2008beijing 2008| 1c0ddb82cca9ed63e1cacbddd3f74082 |
+------+-------------------------+----------------------------------+
3 rows in set (0.00 sec)
```

如果要查询 context 值为 "beijing 2008beijing 2008" 的记录，则可以通过相应的散列值来查询：

```
mysql> select * from t where hash_value=md5(repeat('beijing 2008',2));
+------+-------------------------+----------------------------------+
| id   | context                 | hash_value                       |
+------+-------------------------+----------------------------------+
| 3    | beijing 2008beijing 2008| 1c0ddb82cca9ed63e1cacbddd3f74082 |
+------+-------------------------+----------------------------------+
1 row in set (0.00 sec)
```

上面的例子展示了合成索引的用法，由于这种技术只能用于精确匹配，在一定程度上减少了 I/O，从而提高了查询效率。如果需要对 BLOB 或者 CLOB 字段进行模糊查询，MySQL 提供了前缀索引，也就是只为字段的前 n 列创建索引，举例如下：

```
mysql> create index idx_blob on t(context(100));
Query OK, 0 rows affected (0.04 sec)
Records: 0  Duplicates: 0  Warnings: 0

mysql> desc select * from t where context like 'beijing%' \G
*************************** 1. row ***************************
           id : 1
  select_type : SIMPLE
        table : t
   partitions : NULL
         type : ALL
possible_keys : idx_blob
          key : NULL
      key_len : NULL
          ref : NULL
```

```
        rows: 3
    filtered: 100.00
       Extra: Using where
1 row in set, 1 warning (0.00 sec)
```

可以发现，对 context 前 100 个字符进行模糊查询，就可能用到前缀索引。注意，由于这个表只有 3 行数据，虽然索引可用，但优化器最终并没有选择使用索引，关于索引是否被使用的更多内容，将会在后面的章节中介绍。另外，这里的查询条件中，"%" 不能放在最前面，否则索引将不会被使用。

（3）在不必要的时候避免检索大型的 BLOB 或 TEXT 值。

例如，SELECT * 查询就不是很好的想法，除非能够确定作为约束条件的 WHERE 子句只会找到所需要的数据行。否则，很可能毫无目的地在网络上传输大量的值。这也是 BLOB 或 TEXT 标识符信息存储在合成的索引列中对用户有所帮助的例子。用户可以搜索索引列，决定需要哪些数据行，然后从符合条件的数据行中检索 BLOB 或 TEXT 值。

（4）把 BLOB 或 TEXT 列分离到单独的表中。

在某些环境中，确实有使用 BLOB 或 TEXT 类型的需求，这时建议将 BLOB 或 TEXT 类型的字段分离到单独的表中存储。这会减少主表中的碎片，显著减小主表的数据量从而获得性能优势，主数据表在运行 SELECT * 查询的时候也不会再需要通过网络传输大量的 BLOB 或 TEXT 值。

例如，在 user_info 表中需要一个 BLOB 字段来保存用户的身份证图像信息，更好的做法是新建一个 user_id_pic 的表，包含 user_id 和 id_pic 两列，大多数的查询需求通过访问 user_info 表就可以完成，只有需要访问身份证图像时候才关联 user_id_pic 表进行查询。

注意：尽可能在 OLTP 环境避免使用 BLOB 或者 TEXT 类型，优先使用 VARCHAR。VARCHAR 类型最长可以支持 65533 字节的长度，已经可以满足绝大多数的需求。

7.3 浮点数与定点数

浮点数一般用于表示含有小数部分的数值。当一个字段被定义为浮点类型后，如果插入数据的精度超过该列定义的实际精度，则插入值会被四舍五入到实际定义的精度值，然后插入，四舍五入的过程不会报错。在 MySQL 中，float 和 double（或 real）用来表示浮点数。

定点数不同于浮点数，定点数实际上是以字符串形式存放的，所以定点数可以更精确地保存数据。如果实际插入的数值精度大于实际定义的精度，则 MySQL 会进行警告（默认的 SQLMode 下），但是数据按照实际精度四舍五入后插入；如果 SQLMode 是在 TRADITIONAL（传统模式）下，则系统会直接报错，导致数据无法插入。在 MySQL 中，decimal（或 numeric）用来表示定点数。

在简单了解了浮点数和定点数的区别之后，来看一个例子，回顾一下前面讲到的浮点数精确性问题。

```
mysql> create table t (f float( 8,1));
Query OK, 0 rows affected (0.03 sec)

mysql> desc t;
+-------+----------+------+-----+---------+-------+
| Field | Type     | Null | Key | Default | Extra |
+-------+----------+------+-----+---------+-------+
| f     | float(8,1) | YES |     | NULL    |       |
```

7.3 浮点数与定点数

```
+-------+------------+------+-----+---------+-------+
1 row in set (0.00 sec)

mysql> insert into t values (1.23456);
Query OK, 1 row affected (0.00 sec)

mysql> select * from t;
+------+
| f    |
+------+
| 1.2  |
+------+
1 row in set (0.00 sec)

mysql> insert into t values (1.25456);
Query OK, 1 row affected (0.00 sec)

mysql> select * from t;
+------+
| f    |
+------+
| 1.2  |
| 1.3  |
+------+
2 rows in set (0.00 sec)
```

从上面的例子中,可以发现对于第一次插入值 1.23456 到 float(8,1)时,该值被截断,并保存为 1.2,而第二次插入值 1.25456 到 float(8,1)时,该值进行了四舍五入然后被截断,并保存为 1.3,所以在选择浮点型数据保存小数时,要注意四舍五入的问题,并尽量保留足够的小数位,避免存储的数据不准确。

为了能够让读者了解浮点数与定点数的区别,再来看一个例子:

```
mysql> CREATE TABLE test (c1 float(10,2),c2 decimal(10,2));
Query OK, 0 rows affected (0.29 sec)
mysql> insert into test values(131072.32,131072.32);
Query OK, 1 row affected (0.07 sec)
mysql> select * from test;
+-----------+-----------+
| c1        | c2        |
+-----------+-----------+
| 131072.31 | 131072.32 |
+-----------+-----------+
1 row in set (0.00 sec)
```

从上面的例子中可以看到,c1 列的值由 131072.32 变成了 131072.31,这是上面的数值在使用单精度浮点数表示时,产生了误差。这是浮点数特有的问题。因此在精度要求比较高的应用中(比如货币)要使用定点数而不是浮点数来保存数据。

另外,浮点数的比较也是一个普遍存在的问题,下面的程序片断中对两个浮点数做减法运算:

```
public class Test {
            public static void main(String[] args) throws Exception {
                System.out.print("7.22-7.0=" + (7.22f-7.0f));
            }
        }
```

对上面 Java 程序的输出结果可能会想当然地认为是 0.22,但是,实际结果却是 7.22-7.0=0.21999979,因此,在编程中应尽量避免浮点数的比较。如果非要使用浮点数的比较,则最好使用范围比较,而不要使用"=="比较。

下面使用定点数来实现上面的例子:

```
import java.math.BigDecimal;
/*
```

```
 * 提供精确的减法运算。
 * @param v1
 * @param v2
 */
public class Test {
    public static void main(String[] args) throws Exception {
        System.out.print("7.22-7.0=" + subtract(7.22,7.0));
    }
    public static double subtract(double v1, double v2)      {
        BigDecimal b1 = new BigDecimal(Double.toString(v1));
        BigDecimal b2 = new BigDecimal(Double.toString(v2));
        return b1.subtract(b2).doubleValue();
    }
}
```

上面的实例使用 Java 的 BigDecimal 类实现了定点数的精确计算，所以 7.22 减 7.0 的结果和预想的相同，为 7.22-7.0=0.22。

注意：在今后关于浮点数和定点数的应用中，用户要考虑到以下几个原则：
- 浮点数存在误差问题；
- 对货币等对精度敏感的数据，应该用定点数表示或存储；
- 在编程中，如果用到浮点数，要特别注意误差问题，并尽量避免做浮点数比较；
- 要注意浮点数中一些特殊值的处理。

7.4 日期类型选择

MySQL 提供的常用日期类型有 DATE、TIME、DATETIME 和 TIMESTAMP，它们之间的区别在第 3 章中已经进行过详细论述，这里就不再赘述。下面主要总结一下选择日期类型的原则。

- 根据实际需要选择能够满足应用的最小存储的日期类型。如果应用只需要记录"年份"，那么用 1 个字节来存储的 YEAR 类型完全可以满足，而不需要用 4 个字节来存储的 DATE 类型。这样不仅仅能节约存储，更能够提高表的操作效率。
- 如果要记录年月日时分秒，并且记录的年份比较久远，那么最好使用 DATETIME，而不要使用 TIMESTAMP。因为 TIMESTAMP 表示的日期范围比 DATETIME 要短得多。
- 如果记录的日期需要让不同时区的用户使用，那么最好使用 TIMESTAMP，因为日期类型中只有它能够和实际时区相对应。

7.5 小结

本章主要介绍了常见数据类型的选择原则，简单归纳如下。
- 对于字符类型，要根据实际需求来选择类型和长度。
- 对精度要求较高的应用中，建议使用定点数来存储数值，以保证结果的准确性。
- 尽可能减少使用 TEXT 和 BLOB 字段，对含有这两种字段的表，如果经常做删除和修改记录的操作，要定时执行 OPTIMIZE TABLE 对表进行碎片整理。
- 日期类型要根据实际需要选择能够满足应用的最小存储的日期类型。

第 8 章 字符集

从本质上来说，计算机只能识别二进制代码，因此，不论是计算机程序还是其处理的数据，最终都必须转换成二进制码，计算机才能认识。为了使计算机不仅能做科学计算，也能处理文字信息，人们想出了给每个文字符号编码，以便于计算机识别处理的办法，这就是计算机字符集的由来。本章将详细介绍字符集的发展历程和 MySQL 中字符集的使用。

8.1 字符集概述

简单地说，字符集就是一套文字符号及其编码、比较规则的集合。20 世纪 60 年代初期，美国标准化组织 ANSI 发布了第一个计算机字符集——ASCII（American Standard Code for Information Interchange），后来进一步变成了国际标准 ISO-646。这个字符集采用 7 位编码，定义了包括大小写英文字母、阿拉伯数字和标点符号，以及 33 个控制符号等。虽然现在看来，这个美式的字符集很简单，包括的符号也很少，但直到今天它依然是计算机世界里奠基性的标准，其后制定的各种字符集基本都兼容 ASCII 字符集。

自 ASCII 之后，为了处理不同的文字，各大计算机公司、各国政府、标准化组织等先后发明了几百种字符集，如人们熟悉的 ISO-8859 系列、GB 2312-80、GBK、BIG 5 等。这些五花八门的字符集，从收录的字符到编码规则各不相同，给计算机软件开发和移植带来了很大困难。一个软件要在使用不同文字的国家或地区发布，必须进行本地化开发！基于这个原因，统一字符编码，成了 20 世纪 80 年代计算机业的迫切需要和普遍共识。

8.2 Unicode 简述

为了统一字符编码，国际标准化组织（International Organization for Standardization，ISO）的一些成员国于 1984 年发起制定新的国际字符集标准，以容纳全世界各种语言文字和符号。这个标准最后叫做 "Universal Multiple-Octet Coded Character Set"，简称 UCS，标准编号则定为 ISO-10646。ISO-10646 标准采用 4 字节（32bit）编码，因此简称 UCS-4。

具体编码规则是：将代码空间划分为组（group）、面（plane）、行（row）和格（ceil）；第一个字节代表组（group），第二个字节代表面（plane），第三个字节代表行（row），第四个字节代表格（ceil），并规定字符编码的第 32 位必须为 0，且每个面（plane）的最后两个码位

FFFEh 和 FFFFh 保留不用；因此，ISO-10646 共有 128 个群组（0～0x7F），每个群组有 256 个字面（00～0xFF），每个字面有 256 行（00～0xFF），每行包括 256 格（0～0xFF），共有 256×128=32 768 个字面，每个字面有 256×256-2=65 534 个码位，合计 65 534×32 768= 2 147 418 112 个码位。

ISO-10646 发布以后，遭到了部分美国计算机公司的反对。1988 年 Xerox 公司提议制定新的以 16 位编码的统一字符集 Unicode，并联合 Apple、IBM、DEC、Sun、Microsoft、Novell 等公司成立 Unicode 协会（The Unicode Consortium），并成立 Unicode 技术委员会（Unicode Technical Committee），专门负责 Unicode 文字的搜集、整理和编码，并于 1991 年推出了 Unicode 1.0。

都是为了解决字符编码统一问题，ISO 和 Unicode 协会却推出了两个不同的编码标准，这显然是不利的。后来，大家都认识到了这一点，经过双方谈判，1991 年 10 月达成协议，ISO 将 Unicode 编码并入 ISO-10646 的 0 组 0 字面，称之为基本多语言文字面（Basic Multi-lingual Plane，BMP），共有 65 534 个码位，并根据不同用途分为若干区域。

除 BMP 外的 32 767 个字面又分为辅助字面（Supplementary Plane）和专用字面（Private Use Plane）两部分，辅助字面用以收录 ISO-10646 后续搜集的各国文字，专用字面供使用者自定义收录 ISO-10646 未收录的文字符号。

其实，大部分用户使用 BMP 字面就足够了，早期的 ISO-10646-1 标准也只要求实现 BMP 字面，这样只需要 2 字节来编码就足够了，Unicode 也正是这么做的，这叫做 ISO-10646 编码的基本面形式，简称为 UCS-2 编码。UCS-2 编码转换成 UCS-4 编码也很容易，只要在前面加两个取值为 0 的字节即可。

ISO-10646 的编码空间足以容纳人类从古至今使用过的所有文字和符号，但其实许多文字符号都已经很少使用了，超过 99%的在用文字符号都编入了 BMP，因此，绝大部分情况下，Unicode 的双字节编码方式都能满足需求，而这种双字节编码方式比起 ISO-10646 的 4 字节原始编码来说，在节省内存和处理时间上都具有优势，这也是 Unicode 编码方式更流行的原因。

但如果万一要使用 ISO-10646 BMP 字面以外的文字怎么办呢？Unicode 提出了名为 UTF-16 或代理法（Surrogates）的解决方案。UTF 是 UCS/Unicode Transformation Format 的缩写。UTF-16 的解决办法是：对 BMP 字面的编码保持 2 字节不变，对其他字面的文字按一定规则将其 32 位编码转换为两个 16 位的 Unicode 编码，其两个字节的取值范围分别限定为 0xD800～0xDBFF 和 0xDC00～0xDFFF，因此，UTF-16 共有（4×256）×（4×256）= 1 048 576 个码位。

虽然 UTF-16 解决了 ISO-10646 除 BMP 外第 1～15 字面的编码问题，但当时的计算机和网络世界还是 ASCII 的天下，只能处理单字节数据流，UTF-16 离开 Unicode 环境后，在传输和处理中都存在问题。

于是 Unicode 又提出了名为 UTF-8 的解决方案，UTF-8 按一定规则将一个 ISO-10646 或 Unicode 字元码转换成 1～4 个字节的编码，其中将 ASCII 码（0～0x7F）转换成单字节编码，也就是严格兼容 ASCII 字符集；UTF-8 的 2 字节编码，用于转换 ISO-10646 标准 0x0080～ 0x07FF 的 UCS-4 原始码；UTF-8 的 3 字节编码，用于转换 ISO-10646 标准 0x0800～0xFFFF 的 UCS-4 原始码；UTF-8 的 4 字节编码，用于转换 ISO-10646 标准 0x00010000～0001FFFF 的 UCS-4 原始码。

上述各种编码方式，看起来有点让人迷惑。其实，ISO-10646 只是给每一个文字符号分配了一个 4 字节无符号整数编号（UCS-4），并未规定在计算机中如何去表示这个无符号整数编号。UTF-16 和 UTF-8 就是其两种变通表示方式。

ISO-10646 与 Unicode 统一以后，两个组织虽然都继续发布各自的标准，但两者之间是一致的。由于 Unicode 最早投入应用，其编码方式更加普及，因此，许多人都知道 Unicode，但对 ISO-10646 却了解不多。但由于两者是一致的，因此，区分 ISO-10646 和 Unicode 的意义也就不大了。现在，人们说 Unicode 和 ISO-10646，一般指的是同一个东西，只是 Unicode 更直接、更普及罢了。两者不同版本的对应关系如下：

- Unicode 2.0 等同于 ISO/IEC 10646-1:1993；
- Unicode 3.0 等同于 ISO/IEC 10646-1:2000；
- Unicode 4.0 等同于 ISO/IEC 10646:2003。

最后要说的是，UTF-16 和 UTF-32 因字节序的不同，又有了 UTF-16BE（Big Endian）、UTF-16LE（Little Endian）和 UTF-32BE（Big Endian）、UTF-32LE（Little Endian）等，在此不做进一步介绍。

8.3 汉字及一些常见字符集

在计算机发展的不同阶段，我国也参照当时的国际标准和实际需要，制定了一些汉字字符集编码标准，主要内容如下。

- GB 2312-80：全称《信息交换用汉字编码字符集 基本集》，于 1980 年发布。根据 ISO/IEC 2022 提供的字符编码扩充规范，形成双字节编码的字符集，收录了 6 763 个常用汉字和 682 个非汉字图形符号。
- GB 13000：全称《信息技术 通用多八位编码字符集（UCS）第一部分：体系结构与基本多文种平面》，于 1993 年发布。根据 ISO/IEC 10646-1:1993，在 CJK（中、日、韩简称）统一汉字区和 CJK 统一汉字扩充区 A，除收录 GB 2312-80 外，还收录了第 1、3、5、7 辅助集的全部汉字，共 27 484 个，以及一些偏旁部首等。但 GB 13000 推出后，几乎没有得到业界的支持，也就成了一个形式上的标准。
- GBK：全称《汉字内码扩展规范》1.0 版，发布于 1995 年。GBK 在 GB 2312 内码系统的基础上进行了扩充，收录了 GB 13000.1-1993 的全部 20 902 个 CJK 统一汉字，包括 GB 2312 全部的 6 763 个汉字。此外，它增补编码了 52 个汉字，13 个汉字结构符（在 ISO/IEC 10646.1: 2000 中称为表意文字描述符）和一些常用部首与汉字部件。在 GBK 的内码系统中，GB 2312 汉字所在码位保持不变，这样，保证了 GBK 对 GB 2312 的完全兼容。同时，GBK 内码与 GB 13000.1 代码一一对应，为 GBK 向 GB 13000.1 的转换提供了解决办法。有意思的是，GBK 并不是一个强制性的国家标准，只是一个行业指导规范，并没有强制力，但由于得到了 Microsoft Windows 95 的支持而大为流行。
- GB 18030：全称《信息技术信息交换用汉字编码字符集、基本集的扩充》，发布于 2000 年。根据 ISO/IEC 10646-1:2000，收录了 ISO/IEC 10646.1: 2000 全部 27484 个 CJK 统一汉字，13 个表意文字描述符、部分汉字部首和部件、欧元符号等。GB 18030 采用 2 字节或 4 字节编码，其 2 字节编码部分与 GBK 保持一致，因此，GB 18030 是 GBK 的超集，也完全与 GB 13000

向上兼容，制定 GB 18030 也是为了解决 GBK 强制力不够的问题。

以上简要介绍了几种汉字字符集，下面将一些常用字符集的特点归纳如表 8-1 所示。

表 8-1　　　　　　　　　　　　　　常用字符集比较

字　符　集	是否定长	编码方式	其他说明
ACSII	是	单字节 7 位编码	最早的奠基性字符集
ISO-8859-1/latin1	是	单字节 8 位编码	西欧字符集，经常被一些程序员用来转码
GB 2312-80	是	双字节编码	早期标准，不推荐使用
GBK	是	双字节编码	虽然不是国标，但支持的系统不少
GB 18030	否	2 字节或 4 字节编码	开始有一些支持，但数据库支持的还少见
UTF-32	是	4 字节编码	UCS-4 原始编码，目前很少采用
UCS-2	是	2 字节编码	Windows 2000 内部用 UCS-2
UTF-16	否	2 字节或 4 字节编码	Java 和 Windows XP/NT 等内部使用 UTF-16
UTF-8	否	1～4 字节编码	互联网和 UNIX/Linux 广泛支持的 Unicode 字符集；MySQLServer 也使用 UTF-8。包含两个子集 utf8 和 utf8mb4。

8.4　怎样选择合适的字符集

对数据库来说，字符集更加重要，因为数据库存储的数据大部分都是各种文字，字符集对数据库的存储、处理性能，以及日后系统的移植、推广都会有影响。

MySQL 5.7 目前支持几十种字符集，包括 UCS-2、UTF-16、UTF-16LE、UTF-32、UTF-8 等 Unicode 字符集。面对众多的字符集，我们该如何选择呢？

虽然没有一定之规，但在选择数据库字符集时，可以根据应用的需求，结合上面介绍的一些字符集的特点来权衡，主要考虑以下几方面的因素。

（1）满足应用支持语言的需求，如果应用要处理各种各样的文字，或者将发布到使用不同语言的国家或地区，就应该选择 Unicode 字符集。对 MySQL 来说，最常用的字符集就是 UTF-8。更严谨的说法是字符的编码规则是 UTF-8，其中 utf8mb3 和 utf8mb4 是这种编码规则下最常用的两种字符集，后者是前者的超集。我们常说的 utf8 其实是 utf8mb3 的别名，其中的 3 表明这种字符集由 1～3 个字节组成。顾名思义，utf8mb4 表明每个字符由 1～4 个字节组成，如果需要支持 emoji 表情，通常需要选择 utf8mb4。随着互联网和手机应用的飞速发展，即使英文应用，也越来越多的需要 utf8mb4 的字符集来支持。在最新的 MySQL 8.0 中，默认字符集已经由 latin1 变为 utf8mb4。

（2）如果应用中涉及已有数据的导入，就要充分考虑数据库字符集对已有数据的兼容性。假如已有数据是 GBK 文字，如果选择 GB 2312-80 为数据库字符集，就很可能出现某些文字无法正确导入的问题。

（3）如果数据库只需要支持一般中文，数据量很大，性能要求也很高，那就应该选择双字节定长编码的中文字符集，比如 GBK。因为，相对于 UTF-8 而言，GBK 比较"小"，每个汉字只占 2 个字节，而 UTF-8（包括 utf8 和 utf8mb4）汉字编码需要 3 个字节，这样可以减少磁盘 I/O、数据库 Cache 以及网络传输的时间，从而提高性能。相反，如果应用主要处理英文字符，仅有少量汉字数据，那么选择 UTF-8 更好，因为 GBK、UCS-2、UTF-16 的西文字符编码都是 2 个字节，会造成很多不必要的开销。

(4)如果数据库需要做大量的字符运算,如比较、排序等,那么选择定长字符集可能更好,因为定长字符集的处理速度要比变长字符集的处理速度快。

(5)如果所有客户端程序都支持相同的字符集,则应该优先选择该字符集作为数据库字符集。这样可以避免因字符集转换带来的性能开销和数据损失。

8.5 MySQL 支持的字符集简介

MySQL 服务器可以支持多种字符集,在同一台服务器、同一个数据库甚至同一个表的不同字段都可以指定使用不同的字符集,相比 Oracle 等其他数据库管理系统,在同一个数据库只能使用相同的字符集,MySQL 明显存在更大的灵活性。

查看所有可用的字符集的命令是 show character set:

```
mysql> show character set;
+----------+-----------------------+---------------------+--------+
| Charset  | Description           | Default collation   | Maxlen |
+----------+-----------------------+---------------------+--------+
| big5     | Big5 Traditional Chinese | big5_chinese_ci  |      2 |
| dec8     | DEC West European     | dec8_swedish_ci     |      1 |
| cp850    | DOS West European     | cp850_general_ci    |      1 |
| hp8      | HP West European      | hp8_english_ci      |      1 |
......
```

或者查看 information_schema.character_set,可以显示所有的字符集和该字符集默认的排序规则。

```
mysql> desc information_schema.character_sets;
+----------------------+-------------+------+-----+---------+-------+
| Field                | Type        | Null | Key | Default | Extra |
+----------------------+-------------+------+-----+---------+-------+
| CHARACTER_SET_NAME   | varchar(32) | NO   |     |         |       |
| DEFAULT_COLLATE_NAME | varchar(32) | NO   |     |         |       |
| DESCRIPTION          | varchar(60) | NO   |     |         |       |
| MAXLEN               | bigint(3)   | NO   |     | 0       |       |
+----------------------+-------------+------+-----+---------+-------+
4 rows in set (0.00 sec)
```

MySQL 的字符集包括字符集(CHARACTER)和排序规则(COLLATION)两个概念。其中字符集用来定义 MySQL 存储字符串的方式,排序规则用来定义比较字符串的方式。字符集和排序规则是一对多的关系,MySQL 支持 30 多种字符集的 70 多种排序规则。

每个字符集至少对应一个排序规则。可以用 "SHOW COLLATION LIKE '***';" 命令或者通过系统表 information_schema.COLLATIONS 来查看相关字符集的排序规则。

```
mysql> SHOW COLLATION LIKE 'utf8%';
+--------------------------+---------+----+---------+----------+---------+
| Collation                | Charset | Id | Default | Compiled | Sortlen |
+--------------------------+---------+----+---------+----------+---------+
| utf8_general_ci          | utf8    | 33 | Yes     | Yes      |       1 |
| utf8_bin                 | utf8    | 83 |         | Yes      |       1 |
......
```

排序规则命名约定:它们以其相关的字符集名开始,通常包括一个语言名,并且以_ci(大小写不敏感)、_cs(大小写敏感)或_bin(二元,即比较是基于字符编码的值而与 language 无关)结束。

例如,上面例子中 utf8 的排序规则,其中 utf8_general_ci 是默认的排序规则,对大小写不敏感;而 utf8_bin 按照编码的值进行比较,对大小写敏感。

下面的这个例子中,如果指定"A"和"a"按照 utf8_general_ci 排序规则进行比较,则认为两个字符是相同的,如果按照 utf8_bin 排序规则进行比较,则认为两个字符是不同的。我们事先需要确认应用的需求,是需要按照什么样的排序方式,是否需要区分大小写,以确定排序规则的选择。

```
mysql> select case when 'A' COLLATE utf8_general_ci = 'a' collate utf8_general_ci then 1 else 0 end as CaseInsensitive;
+-----------------+
| CaseInsensitive |
+-----------------+
|               1 |
+-----------------+
1 row in set (0.00 sec)

mysql> select case when 'A' COLLATE utf8_bin = 'a' collate utf8_bin then 1 else 0 end as CaseInsensitive;
+-----------------+
| CaseInsensitive |
+-----------------+
|               0 |
+-----------------+
1 row in set (0.00 sec)
```

8.6 MySQL 字符集的设置

MySQL 的字符集和排序规则有 4 个级别的默认设置:服务器级、数据库级、表级和字段级。它们分别在不同的地方设置,作用也不相同。

8.6.1 服务器字符集和排序规则

服务器字符集和排序规则,可以在 MySQL 服务启动的时候确定。

- 可以在 my.cnf 中设置:

```
[mysqld]
character-set-server=utf8
```

- 或者在启动选项中指定:

```
mysqld --character-set-server=utf8
```

- 或者在编译时指定:

```
shell> cmake . -DDEFAULT_CHARSET=utf8
```

如果没有特别的指定服务器字符集,那么在 MySQL 5.7 中默认使用 latin1 作为服务器字符集。上面 3 种设置的方式都只指定了字符集,没有指定排序规则,这样意味着使用该字符集默认的排序规则。如果要使用该字符集的非默认排序规则,则需要在指定字符集的同时指定排序规则。

注意:在最新的 MySQL 8.0 中,默认字符集已经变为 utf8mb4。

可以用"show variables like 'character_set_server';"命令查询当前服务器的字符集和排序规则。

```
mysql> show variables like 'character_set_server';
+----------------------+-------+
| Variable_name        | Value |
+----------------------+-------+
| character_set_server | utf8  |
```

```
+--------------------+-------+
1 row in set (0.00 sec)

mysql> show variables like 'collation_server';
+------------------+-----------------+
| Variable_name    | Value           |
+------------------+-----------------+
| collation_server | utf8_unicode_ci |
+------------------+-----------------+
1 row in set (0.00 sec)
```

8.6.2 数据库字符集和排序规则

数据库的字符集和排序规则既可以在创建数据库的时候指定，也可以在创建完数据库后通过"alter database"命令进行修改。需要注意的是，如果数据库里已经存在数据，因为修改字符集并不能将已有的数据按照新的字符集进行存放，所以不能通过修改数据库的字符集直接修改数据的内容。8.7 节会通过一个具体的例子介绍字符集的修改方法。

设置数据库字符集的规则如下：
- 如果指定了字符集和排序规则，则使用指定的字符集和排序规则；
- 如果指定了字符集没有指定排序规则，则使用指定字符集的默认排序规则；
- 如果指定了排序规则但未指定字符集，则字符集使用与该排序规则关联的字符集；
- 如果没有指定字符集和排序规则，则使用服务器字符集和排序规则作为数据库的字符集和排序规则。

推荐在创建数据库时明确指定字符集和排序规则，避免受到默认值的影响。要显示当前数据库的字符集和排序规则，可以使用"show variables like ' character_set_ database '"和"show variables like ' collation_database '"命令查看：

```
mysql> show variables like 'character_set_database';
+------------------------+-------+
| Variable_name          | Value |
+------------------------+-------+
| character_set_database | utf8  |
+------------------------+-------+
1 row in set (0.00 sec)

mysql> show variables like 'collation_database';
+--------------------+-----------------+
| Variable_name      | Value           |
+--------------------+-----------------+
| collation_database | utf8_unicode_ci |
+--------------------+-----------------+
1 row in set (0.00 sec)
```

8.6.3 表字符集和排序规则

表的字符集和排序规则在创建表的时候指定，可以通过 alter table 命令进行修改，同样，如果表中已有记录，修改字符集对原有的记录并没有影响，不会按照新的字符集进行存放。表的字段仍然使用原来的字符集。

设置表的字符集的规则和上面基本类似：
- 如果指定了字符集和排序规则，使用指定的字符集和排序规则；
- 如果指定了字符集没有指定排序规则，使用指定字符集的默认排序规则；

- 如果指定了排序规则但未指定字符集，则字符集使用与该排序规则关联的字符集；
- 如果没有指定字符集和排序规则，使用数据库字符集和排序规则作为表的字符集和排序规则。

推荐在创建表的时候明确指定字符集和排序规则，以避免受到默认值的影响。要显示表的字符集和排序规则，可以使用 show create table 命令查看：

```
mysql>  show create table person\G
*************************** 1. row ***************************
       Table: person
Create Table: CREATE TABLE `person` (
  `id` smallint(5) unsigned NOT NULL AUTO_INCREMENT,
  `name` char(60) COLLATE utf8_unicode_ci NOT NULL,
  PRIMARY KEY (`id`)
) ENGINE=InnoDB DEFAULT CHARSET=utf8 COLLATE=utf8_unicode_ci
1 row in set (0.00 sec)
```

8.6.4　列字符集和排序规则

MySQL 可以定义列级别的字符集和排序规则，主要是针对相同的表不同字段需要使用不同的字符集的情况，应该说一般遇到这种情况的概率比较小，这只是 MySQL 提供给我们一个灵活设置的手段。

列字符集和排序规则的定义可以在创建表时指定，或者在修改表时调整。如果在创建表的时候没有特别指定字符集和排序规则，则默认使用表的字符集和排序规则。

8.6.5　连接字符集和排序规则

上面 4 种设置方式，确定的是数据保存的字符集和排序规则，对于实际的应用访问来说，还存在客户端和服务器之间交互的字符集和排序规则的设置。

对于客户端和服务器的交互操作，MySQL 提供了 3 个不同的参数：character_set_client、character_set_connection 和 character_set_results，分别代表客户端、连接和返回结果的字符集。通常情况下，这 3 个字符集应该是相同的，才可以确保用户写入的数据可以正确地读出，特别是对于中文字符，不同的写入字符集和返回结果字符集将导致写入的记录不能正确读出。

通常情况下，不会单个地设置这 3 个参数，可以通过以下命令：

```
SET NAMES ***;
```

来设置连接的字符集和排序规则，这个命令可以同时修改这 3 个参数的值。使用这个方法修改连接的字符集和排序规则，需要应用每次连接数据库后都执行这个命令。

另一个更简便的办法，是在 my.cnf 中设置以下语句：

```
[mysql]
default-character-set=utf8
```

这样服务器启动后，所有连接默认就是使用 utf8 字符集进行连接的，而不需要在程序中再执行 set names 命令。另外，字符串常量的字符集也是由 character_set_connection 参数来指定的。

可以通过 "[_charset_name]'string' [COLLATE collation_name]" 命令强制字符串的字符集和排序规则。例如：

```
select _utf8 '字符集';
select _latin1 '字符集';
```

通常情况下，基本不需要用户强制指定字符串字符集。

8.7 字符集的修改步骤

如果在应用开始阶段没有正确地设置字符集，在运行一段时间以后才发现存在不能满足要求需要调整，又不想丢弃这段时间的数据，那么就需要进行字符集的修改。字符集的修改不能直接通过"alter database character set ***"或者"alter table tablename character set ***"命令进行，这两个命令都没有更新已有记录的字符集，而只是对新创建的表或者记录生效。已有记录的字符集调整，需要先将数据导出，经过适当的调整重新导入后才可完成。

以下模拟的是将 latin1 字符集的数据库修改成 utf8 字符集的数据库的过程。

（1）导出表结构。

```
mysqldump -uroot -p --default-character-set=utf8 -d databasename> createtab.sql
```

其中--default-character-set=utf8 表示设置以什么字符集连接；-d 表示只导出表结构，不导出数据。

（2）手工修改 createtab.sql 中表结构定义中的字符集为新的字符集。

（3）确保记录不再更新，导出所有记录。

```
mysqldump -uroot -p --quick --no-create-info --extended-insert --default-character-set= latin1 databasename> data.sql
```

- --quick: 该选项用于转储大的表。它强制 mysqldump 从服务器一次一行地检索表中的行而不是检索所有行，并在输出前将它缓存到内存中。
- --extended-insert: 使用包括几个 VALUES 列表的多行 INSERT 语法。这样使转储文件更小，重载文件时可以加速插入。
- --no-create-info: 不导出每个转储表的 CREATE TABLE 语句。
- --default-character-set=latin1: 按照原有的字符集导出所有数据，这样导出的文件中，所有中文都是可见的，不会保存成乱码。

（4）打开 data.sql，将 SET NAMES latin1 修改成 SET NAMES utf8。

（5）使用新的字符集创建新的数据库。

```
create database databasename default charset utf8;
```

（6）创建表，执行 createtab.sql。

```
mysql -uroot -p databasename < createtab.sql
```

（7）导入数据，执行 data.sql。

```
mysql -uroot -p databasename < data.sql
```

注意：选择目标字符集的时候，要注意最好是源字符集的超集，或者确定比源字符集的字库更大，否则如果目标字符集的字库小于源字符集的字库，那么目标字符集中不支持的字导入后会变成乱码，丢失一部分数据。例如，gbk 字符集的字库大于 gb2312 字符集，那么 gbk 字符集的数据，如果导入 gb2312 数据库中，就会丢失 gb2312 中不支持的那部分汉字的数据。

8.8 小结

本章主要介绍了 MySQL 中字符集和排序规则的概念、设置方法，以及推荐读者使用的字符集。最后，举例介绍了字符集修改的步骤和修改过程中遇到的问题，希望会对读者有所帮助。

第 9 章 索引的设计和使用

索引是数据库中用来提高性能的常用工具。本章主要介绍了 MySQL 5.7 支持的索引类型,并简单介绍了索引的设计原则。在后面的优化篇中,将会对索引做更多的介绍。

9.1 索引概述

所有 MySQL 列类型都可以被索引,对相关列使用索引是提高 SELECT 操作性能的最佳途径。根据存储引擎可以定义每个表的最大索引数和最大索引长度,每种存储引擎(如 MyISAM、InnoDB、BDB、MEMORY 等)对每个表至少支持 16 个索引,总索引长度至少为 256 字节。大多数存储引擎有更高的限制。

MyISAM 和 InnoDB 存储引擎的表默认创建的都是 BTREE 索引。除了直接在单列或者多列上直接创建索引外,MySQL 5.7 之后可以通过虚拟列索引来实现函数索引的功能,同时 MySQL 也支持前缀索引,即对索引字段的前 N 个字符创建索引。前缀索引的长度跟存储引擎相关,对于 MyISAM 存储引擎的表,索引的前缀长度可以达到 1 000 字节长,而对于 InnoDB 存储引擎的表,索引的前缀长度最长是 3 072 字节。请注意前缀的限制应以字节为单位进行测量,而 CREATE TABLE 语句中的前缀长度解释为字符数。在为使用多字节字符集的列指定前缀长度时一定要加以考虑。

MySQL 中还支持全文本(FULLTEXT)索引,该索引可以用于全文搜索。在 MySQL 5.6 之后,InnoDB 和 MyISAM 存储引擎都可以支持 FULLTEXT 索引,但只限于 CHAR、VARCHAR 和 TEXT 列。索引总是对整个列进行的,不支持局部(前缀)索引。

MySQL 也可以为空间列类型创建索引,MySQL 5.7 之前只有 MyISAM 存储引擎支持空间类型索引,且索引的字段必须是非空的。MySQL 5.7 中,InnoDB 存储引擎也开始支持空间类型索引,索引以 R-Trees 的数据结构保存。

默认情况下,MEMORY 存储引擎使用 HASH 索引,但也支持 BTREE 索引。

索引在创建表的时候可以同时创建,也可以随时增加新的索引。创建新索引的语法如下:

```
CREATE [UNIQUE|FULLTEXT|SPATIAL] INDEX index_name
  [USING index_type]
  ON tbl_name (index_col_name,...)
 [index_option]
[algorithm_option | lock_option]…

index_col_name:
  col_name [(length)] [ASC | DESC]
```

也可以使用 ALTER TABLE 的语法来增加索引，语法与 CREATE INDEX 类似，可以查询帮助获得详细的语法，这里不再复述。

例如，要为 city 表创建 10 字节的前缀索引，代码如下：

```
mysql> create index cityname on city (city(10));
Query OK, 600 rows affected (0.26 sec)
Records: 600  Duplicates: 0  Warnings: 0
```

如果以 city 为条件进行查询，可以发现索引 cityname 被使用：

```
mysql> explain select * from city where city = 'Fuzhou' \G
*************************** 1. row ***************************
           id: 1
  select_type: SIMPLE
        table: city
         type: ref
possible_keys: cityname
          key: cityname
      key_len: 32
          ref: const
         rows: 1
        Extra: Using where
1 row in set (0.00 sec)
```

索引的删除语法如下：

```
DROP INDEX index_name ON tbl_name
```

例如，想要删除 city 表上的索引 cityname，可以操作如下：

```
mysql> drop index cityname on city;
Query OK, 600 rows affected (0.23 sec)
Records: 600  Duplicates: 0  Warnings: 0
```

9.2 设计索引的原则

索引的设计可以遵循一些已有的原则，创建索引的时候请尽量考虑符合这些原则，便于提升索引的使用效率，更高效地使用索引。

- 要在条件列上创建索引，而不是查询列。换句话说，最适合索引的列是出现在 WHERE 子句中的列，或连接子句中指定的列，而不是出现在 SELECT 关键字后的选择列表中的列。

- 尽量使用高选择度索引。考虑某列中值的分布。索引的列的基数越大，索引的效果越好。例如，存放出生日期的列具有不同值，很容易区分各行。而用来记录性别的列，只含有"M"和"F"，则对此列进行索引没有多大用处，因为不管搜索哪个值，都会得出大约一半的行。

- 使用短索引。如果对字符串列进行索引，应该指定一个前缀长度，只要有可能就应该这样做。例如，有一个 CHAR(200) 列，如果在前 10 个或 20 个字符内，多数值是唯一的，那么就不要对整个列进行索引。对前 10 个或 20 个字符进行索引能够节省大量索引空间，也可能会使查询更快。较小的索引涉及的磁盘 IO 较少，较短的值比较起来更快。更为重要的是，对于较短的键值，索引高速缓存中的块能容纳更多的键值，因此，MySQL 也可以在内存中容纳更多的值。这样就增加了找到行而不用读取索引中较多块的可能性。

- 利用最左前缀。在创建一个 n 列的索引时，实际相当于创建了 MySQL 可利用的 n 个索引。多列索引可起几个索引的作用，因为可利用索引中最左边的列集来匹配行。这样的列集称为最左前缀。例如以 a、b、c 的顺序在 3 列上创建一个组合索引之后，利用 a = ?或者 a=?

and b=?或者 a=? and b=? and c=?这 3 种条件的查询，都可以使用这个索引。通过这种方式，可以有效的降低索引的数量，提高索引的使用效率。

- 对于 InnoDB 存储引擎的表，尽量手工指定主键。记录默认会按照一定的顺序保存，如果有明确定义的主键，则按照主键顺序保存。如果没有主键，但是有唯一索引，那么就是按照唯一索引的顺序保存。如果既没有主键又没有唯一索引，那么表中会自动生成一个内部列，按照这个列的顺序保存。按照主键或者内部列进行的访问是最快的，所以 InnoDB 表尽量自己指定主键。当表中同时有几个列都是唯一的，都可以作为主键的时候，要选择最常作为访问条件的列作为主键，提高查询的效率。另外，还需要注意，InnoDB 表的普通索引都会保存主键的键值，所以主键要尽可能选择较短的数据类型，有效地减少索引的磁盘占用，提高索引的缓存效果。

9.3 索引设计的误区

设计索引时，有一些常见的误区，总结如下。

- 不是所有的表都需要创建索引。通常来说，常见的代码表、配置表等数据量很小的表，除了主键外，再创建索引没有太大的意义，索引扫描和全表扫描相比，并不会带来性能的大幅提升。而大表的查询、更新、删除操作则要尽可能通过索引。对于大表来说，任何全表扫描对于系统来说都会是非常大的冲击，因此每个操作都尽可能通过索引进行。这类表要经常统计操作频率较高的 SQL，然后对这些 SQL 进行分析，提取最常用的一些选择性高的列来创建索引。

- 不要过度索引。不要以为索引"越多越好"，什么东西都用索引是错误的。每个额外的索引都要占用额外的磁盘空间，并降低写操作的性能。在修改表的内容时，索引必须进行更新，有时可能需要重构，因此，索引越多，所花的时间越长。如果有一个索引很少利用或从不使用，那么会不必要地减缓表的修改速度。此外，MySQL 在生成一个执行计划时，要考虑各个索引，这也要花费时间。创建多余的索引给查询优化带来了更多的工作。索引太多，也可能会使 MySQL 选择不到所要使用的最好索引。因此，只保持所需的索引有利于查询优化。

- 谨慎创建低选择度索引。对于选择性低并且数据分布均衡的列，因为过滤的结果集大，创建索引的效果通常不好；但如果列的选择性低且数据分布不均衡，比如男女比例为 99%:1%，那么此时创建索引对于查询条件为'女'的过滤结果集就比较小，索引的效率较高，此时创建索引就比较合适。在 MySQL 8.0 之后也可以使用直方图取得类似的优化效果。

- 慎重使用唯一索引。使用唯一索引，尽管在特定查询的速度上有稍许提升，同时能够满足业务上对于唯一性约束的需求，但是在 MySQL 5.6 之后，InnoDB 引擎新增了 Change Buffer 特性来大幅提升写入性能，而除主键外的唯一索引会导致 Change Buffer 无法被使用，对写入性能影响较大。

9.4 索引设计的一般步骤

通过上面的介绍，当对一个大表做索引设计时，一般可以采用下面的步骤。

（1）整理表上的所有 SQL，重点包括 select、update、delete 操作的 where 条件所用到的列的组合、关联查询的关联条件等。

（2）整理所有查询 SQL 的预期执行频率。

（3）整理所有涉及的列的选择度，列的不同值相比总非空行数的比例越大，选择度越好，比如全部都是唯一值的主键列选择度最高。当然，上面所提到的查询频率、选择度，都是估算的值，能够在设计索引时作为参考即可。

（4）遵照之前提到的设计原则，给表选择合适的主键，没有特别合适的列时，建议使用自增列作为主键。

（5）优先给那些执行频率最高的 SQL 创建索引，执行频率很高的 SQL，使用到的索引的效率对整体性能影响也会比较大，选择其中选择度最高的列来创建索引，如果选择度都不够好，那么应该考虑是否可以使用其他选择度更好的条件，或者选择创建联合索引。

（6）按执行频率排序，依次检查是否需要为每个 SQL 创建索引，可以复用之前已经创建的索引的 SQL，无须再重复创建索引，除非 SQL 执行频率很高，新创建的索引，对选择度提升也很大。

（7）索引合并，利用复合索引来降低索引的总数，充分利用最左前缀的原则，让索引可以被尽可能多地复用，同时在保证复用率的情况下，把选择度更高的列放到索引的更左侧。

（8）上线之后，通过慢查询分析、执行计划分析、索引使用统计，来确定索引的实际使用情况，并根据情况做出调整。

9.5　BTREE 索引与 HASH 索引

MEMORY 存储引擎的表可以选择使用 BTREE 索引或者 HASH 索引，两种不同类型的索引各有其不同的适用范围。HASH 索引有一些重要的特征在使用时需特别注意，如下所示。

- 只用于使用=或<=>操作符的等式比较。
- 优化器不能使用 HASH 索引来加速 ORDER BY 操作。
- MySQL 不能确定在两个值之间大约有多少行。如果将一个 MyISAM 表改为 HASH 索引的 MEMORY 表，会影响一些查询的执行效率。
- 只能使用整个关键字来搜索一行。

而对于 BTREE 索引，当使用>、<、>=、<=、BETWEEN、!=或者<>，或者 LIKE 'pattern'（其中'pattern'不以通配符开始）操作符时，都可以使用相关列上的索引。下列范围查询适用于 BTREE 索引和 HASH 索引：

```
SELECT * FROM t1 WHERE key_col = 1 OR key_col IN (15,18,20);
```

下列范围查询只适用于 BTREE 索引：

```
SELECT * FROM t1 WHERE key_col > 1 AND key_col < 10;
SELECT * FROM t1 WHERE key_col LIKE 'ab%' OR key_col BETWEEN 'lisa' AND 'simon';
```

例如，创建一个和 city 表完全相同的 MEMORY 存储引擎的表 city_memory：

```
mysql> CREATE TABLE city_memory (
    ->   city_id SMALLINT UNSIGNED NOT NULL AUTO_INCREMENT,
    ->   city VARCHAR(50) NOT NULL,
    ->   country_id SMALLINT UNSIGNED NOT NULL,
    ->   last_update TIMESTAMP NOT NULL DEFAULT CURRENT_TIMESTAMP ON UPDATE CURRENT_ TIMESTAMP,
    ->   PRIMARY KEY  (city_id),
    ->   KEY idx_fk_country_id (country_id)
    -> )ENGINE=Memory DEFAULT CHARSET=utf8;
Query OK, 0 rows affected (0.03 sec)

mysql> insert into city_memory select * from city;
Query OK, 600 rows affected (0.00 sec)
Records: 600  Duplicates: 0  Warnings: 0
```

当对索引字段进行范围查询的时候，只有 BTREE 索引可以通过索引访问：

```
mysql> explain SELECT * FROM city WHERE country_id > 1 and country_id < 10 \G
*************************** 1. row ***************************
           id: 1
  select_type: SIMPLE
        table: city
         type: range
possible_keys: idx_fk_country_id
          key: idx_fk_country_id
      key_len: 2
          ref: NULL
         rows: 24
        Extra: Using where
1 row in set (0.00 sec)
```

而 HASH 索引实际上是全表扫描的:

```
mysql> explain SELECT * FROM city_memory WHERE country_id > 1 and country_id < 10 \G
*************************** 1. row ***************************
           id: 1
  select_type: SIMPLE
        table: city_memory
         type: ALL
possible_keys: idx_fk_country_id
          key: NULL
      key_len: NULL
          ref: NULL
         rows: 600
        Extra: Using where
1 row in set (0.00 sec)
```

了解了 BTREE 索引和 HASH 索引不同后,当使用 MEMORY 表时,如果是默认创建的 HASH 索引,就要注意 SQL 语句的编写,确保可以使用上索引;如果一定要使用范围查询,那么在创建索引时,就应该选择创建成 BTREE 索引。

9.6 索引在 MySQL 8.0 中的改进

索引的正确使用,对于 MySQL 的性能优化,起着非常关键的作用。在 MySQL 8.0 中,索引也引入了不少新的特性。下面介绍几个比较重点的改进。

9.6.1 不可见索引

在 MySQL 8.0 中,增加了对于不可见索引(invisible index)的支持,这也是一个从 Oracle 数据库借鉴而来的新特性。所谓不可见,指的是对于查询优化器不可见,SQL 在执行时自然也就不会选择,但在查看表结构时候索引仍然能看到,也可以通过 information_schema.statistics 或者 show index 来查看索引是否可见的状态。

索引默认是可见的,可以通过在创建索引时指定 invisible 关键字来创建不可见索引:

```
CREATE TABLE t1 (
  i INT,
  j INT,
  k INT,
  INDEX i_idx (i) INVISIBLE
) ENGINE = InnoDB;
```

也可以通过命令来单独添加不可见索引:

```
CREATE INDEX j_idx ON t1 (j) INVISIBLE;
ALTER TABLE t1 ADD INDEX k_idx (k) INVISIBLE;
```

可以通过 alter table 命令来修改索引是否可见：

```
ALTER TABLE t1 ALTER INDEX i_idx INVISIBLE;
ALTER TABLE t1 ALTER INDEX i_idx VISIBLE;
```

为什么数据库中要设计这么一种消耗资源，却又不能够对 SQL 起到任何优化左右的索引呢？实际上，引入不可见索引的目的，主要是为了减小对于表上的索引进行调整时的潜在风险。

随着表的数据量增大，达到了几百 GB，几 TB 甚至更大的时候，如果此时对表上的索引进行调整，往往面临着很大的风险。例如，当删除一个认为不再需要的索引时，一旦系统中还存在个别使用这个索引的 SQL，那么这些 SQL 的执行计划有可能会变成对这个大表的全表扫描，这会对数据库服务器造成巨大冲击，很有可能直接导致服务不可用。而由于表的数据量大，重建索引需要的时间和消耗的系统资源也会很大，很难马上通过重建索引解决问题。

有了不可见索引，当需要删除一个表上的冗余索引时，可以先将索引设置为不可见，而不是直接删除，一旦发现没有这个索引之后，对系统性能产生了负面影响，可以很方便地恢复这个索引，而不再需要重建索引。

同样，当增加一个索引之后，如果发现对系统带来了负面影响，可以首先将索引设置为不可见，待系统负载恢复正常后，再做索引的删除，避免了系统压力大的时候雪上加霜。

9.6.2 倒序索引

在 MySQL 8.0 中，正式增加了对于倒序索引（descending index）的支持，在之前的版本中，虽然在创建索引的时候可以指定 desc 关键字，但是实际上 MySQL 仍然会保存为正序索引。

```
mysql 5.7> CREATE TABLE t1 (a INT, b INT, INDEX a_desc_b_asc (a DESC, b ASC));
Query OK, 0 rows affected (0.47 sec)

mysql 5.7> SHOW CREATE TABLE t1\G
*************************** 1. row ***************************
       Table: t1
Create Table: CREATE TABLE `t1` (
  `a` int(11) DEFAULT NULL,
  `b` int(11) DEFAULT NULL,
  KEY `a_desc_b_asc` (`a`,`b`) <-- Order is not preserved
) ENGINE=InnoDB DEFAULT CHARSET=latin1
1 row in set (0.00 sec)
```

在 MySQL 8.0 中，倒序索引能够被正确创建：

```
mysql 8.0> CREATE TABLE t1 (a INT, b INT, INDEX a_desc_b_asc (a DESC, b ASC));
Query OK, 0 rows affected (0.47 sec)

mysql 8.0> show create table t1;
+-------+-----------------------------------------------------------------+
| Table | Create Table |
+-------+-----------------------------------------------------------------+
| t1    | CREATE TABLE `t1` (
 `a` int(11) DEFAULT NULL,
 `b` int(11) DEFAULT NULL,
 KEY `a_desc_b_asc` (`a` DESC,`b`)
) ENGINE=InnoDB DEFAULT CHARSET=latin1 |
+-------+-----------------------------------------------------------------+
1 row in set (0.00 sec)
```

倒序索引在某些情况下，可以起到更好的作用。但是相比于 Oracle 倒序索引对于查询的优化效果，MySQL 倒序索引起到的作用还是比较弱的，有待未来的版本继续加强。但是需要注意的是，由于倒序索引的引入，MySQL 8.0 取消了对于 group by 操作的隐式排序和显式排序，如果业务中有依赖于此特性的，在升级数据库版本的时候要谨慎。

9.7 小结

索引用于快速找出在某个列中有某个特定值的行。如果不使用索引，MySQL 必须从第 1 条记录开始然后读完整个表直到找出相关的行。表越大，花费的时间越多。如果表中查询的列有一个索引，MySQL 能快速到达一个位置去搜寻数据文件的中间，没有必要看所有数据。如果一个表有 1 000 行，这比顺序读取至少快 100 倍。注意如果需要访问大部分行，顺序读取要快得多，因为此时应避免磁盘搜索。

大多数 MySQL 索引（如 PRIMARY KEY、UNIQUE、INDEX 和 FULLTEXT 等）在 BTREE 中存储。只是空间列类型的索引使用 RTREE，并且 MEMORY 表还支持 HASH 索引。

本节简单地介绍了在设计索引时需要注意的一些常见问题，至于数据库何时会使用索引，何时不会使用索引，可参见优化篇的相关章节，这里不再赘述。

第 10 章 开发常用数据库对象

10.1 视图

MySQL 从 5.0.1 版本开始提供视图功能,本节将对 MySQL 中的视图进行介绍。

10.1.1 什么是视图

视图(View)是一种虚拟存在的表,对于使用视图的用户来说基本上是透明的。视图并不在数据库中实际存在,行和列数据来自定义视图的查询中使用的表,并且是在使用视图时动态生成的。

视图相对于普通的表的优势主要包括以下几项。

- 简单:使用视图的用户完全不需要关心后面对应的表的结构、关联条件和筛选条件,对用户来说已经是过滤好的复合条件的结果集。
- 安全:使用视图的用户只能访问他们被允许查询的结果集,对表的权限管理并不能限制到某个行某个列,但是通过视图就可以简单地实现。
- 数据独立:一旦视图的结构确定了,可以屏蔽表结构变化对用户的影响,源表增加列对视图没有影响;源表修改列名,则可以通过修改视图来解决,不会造成对访问者的影响。

10.1.2 视图操作

视图的操作包括创建或者修改视图、删除视图,以及查看视图定义。

10.1.3 创建或者修改视图

创建视图需要有 CREATE VIEW 的权限,并且对于查询涉及的列有 SELECT 权限。如果使用 CREATE OR REPLACE 或者 ALTER 修改视图,那么还需要该视图的 DROP 权限。

创建视图的语法如下:

```
CREATE [OR REPLACE] [ALGORITHM = {UNDEFINED | MERGE | TEMPTABLE}]
    VIEW view_name [(column_list)]
    AS select_statement
    [WITH [CASCADED | LOCAL] CHECK OPTION]
```

修改视图的语法如下：

```
ALTER [ALGORITHM = {UNDEFINED | MERGE | TEMPTABLE}]
    VIEW view_name [(column_list)]
    AS select_statement
    [WITH [CASCADED | LOCAL] CHECK OPTION]
```

例如，要创建视图 staff_list_view，可以使用以下命令：

```
mysql> CREATE OR REPLACE VIEW staff_list_view AS
    -> SELECT s.staff_id,s.first_name,s.last_name,a.address
    -> FROM staff AS s,address AS a
    -> where s.address_id = a.address_id ;
Query OK, 0 rows affected (0.00 sec)
```

MySQL 视图的定义有一些限制，例如，在 5.7.7 版本之前，FROM 关键字后面不能包含子查询，这和其他数据库是不同的，如果视图是从其他数据库迁移过来的，那么可能需要因此做一些改动。

视图的可更新性和视图中查询的定义有关，以下类型的视图是不可更新的。

- 包含以下关键字的 SQL 语句：聚合函数（SUM、MIN、MAX、COUNT 等）、DISTINCT、GROUP BY、HAVING、UNION 或者 UNION ALL。
- 常量视图。
- SELECT 中包含子查询。
- JOIN。
- FROM 一个不能更新的视图。
- WHERE 字句的子查询引用了 FROM 字句中的表。

例如，以下的视图都是不可更新的：

```
--包含聚合函数
mysql> create or replace view payment_sum as
    -> select staff_id,sum(amount) from payment group by staff_id;
Query OK, 0 rows affected (0.00 sec)

--常量视图
mysql> create or replace view pi as select 3.1415926 as pi;
Query OK, 0 rows affected (0.00 sec)

--select 中包含子查询
mysql> create view city_view as
    -> select (select city from city where city_id = 1) ;
Query OK, 0 rows affected (0.00 sec)
```

WITH [CASCADED | LOCAL] CHECK OPTION 决定了是否允许更新数据使记录不再满足视图的条件。这个选项与 Oracle 数据库中的选项是类似的，其中：

- LOCAL 只要满足本视图的条件就可以更新；
- CASCADED 则必须满足所有针对该视图的所有视图的条件才可以更新。

如果没有明确是 LOCAL 还是 CASCADED，则默认是 CASCADED。

例如，对 payment 表创建两层视图，并进行更新操作：

```
mysql> create or replace view payment_view as
    -> select payment_id,amount from payment
    -> where amount < 10 WITH CHECK OPTION;
Query OK, 0 rows affected (0.00 sec)

mysql>
mysql> create or replace view payment_view1 as
    -> select payment_id,amount from payment_view
    -> where amount > 5 WITH LOCAL CHECK OPTION;
```

```
Query OK, 0 rows affected (0.00 sec)

mysql>
mysql> create or replace view payment_view2 as
    -> select payment_id,amount from payment_view
    -> where amount > 5 WITH CASCADED CHECK OPTION;
Query OK, 0 rows affected (0.00 sec)

mysql> select * from payment_view1 limit 1;
+------------+--------+
| payment_id | amount |
+------------+--------+
| 3          | 5.99   |
+------------+--------+
1 row in set (0.00 sec)

mysql> update payment_view1 set amount=10
    -> where payment_id = 3;
Query OK, 1 row affected (0.03 sec)
Rows matched: 1  Changed: 1  Warnings: 0

mysql> update payment_view2 set amount=10
    -> where payment_id = 3;
ERROR 1369 (HY000): CHECK OPTION failed 'sakila.payment_view2'
```

从测试结果可以看出，payment_view1 是 WITH LOCAL CHECK OPTION 的，所以只要满足本视图的条件就可以更新，但是 payment_view2 是 WITH CASCADED CHECK OPTION 的，必须满足针对该视图的所有视图才可以更新，因为更新后记录不再满足 payment_view 的条件，所以更新操作提示错误退出。

10.1.4 删除视图

用户可以一次删除一个或者多个视图，前提是必须有该视图的 DROP 权限。

DROP VIEW [IF EXISTS] view_name [, view_name] ...[RESTRICT | CASCADE]

例如，删除 staff_list 视图：

```
mysql> drop view staff_list;
Query OK, 0 rows affected (0.00 sec)
```

10.1.5 查看视图

从 MySQL 5.1 版本开始，使用 SHOW TABLES 命令的时候不仅显示表的名字，同时也会显示视图的名字，而不存在单独显示视图的 SHOW VIEWS 命令。

```
mysql> use sakila
Database changed
mysql> show tables;
+----------------------------+
| Tables_in_sakila           |
+----------------------------+
......
| staff                      |
| staff_list                 |
| store                      |
+----------------------------+
26 rows in set (0.00 sec)
```

同样，在使用 SHOW TABLE STATUS 命令的时候，不但可以显示表的信息，同时也可以显示视图的信息。所以，可以通过下面的命令显示视图的信息：

SHOW TABLE STATUS [FROM db_name] [LIKE 'pattern']

下面演示的是查看 staff_list 视图信息的操作：

```
mysql> show table status like 'staff_list' \G
*************************** 1. row ***************************
           Name : staff_list
         Engine : NULL
        Version : NULL
     Row_format : NULL
           Rows : NULL
 Avg_row_length : NULL
    Data_length : NULL
Max_data_length : NULL
   Index_length : NULL
      Data_free : NULL
 Auto_increment : NULL
    Create_time : NULL
    Update_time : NULL
     Check_time : NULL
      Collation : NULL
       Checksum : NULL
 Create_options : NULL
        Comment : VIEW
1 row in set (0.01 sec)
```

如果需要查询某个视图的定义，可以使用 SHOW CREATE VIEW 命令进行查看：

```
mysql> show create view staff_list \G
*************************** 1. row ***************************
       View: staff_list
Create View: CREATE ALGORITHM=UNDEFINED DEFINER='root'@'localhost' SQL SECURITY DEFINER VIEW
'staff_list' AS select 's'.'staff_id' AS 'ID',concat ('s'.'first_ name',_utf8 ',','s'.'last_name')
AS 'name','a'.'address' AS 'address','a'.'postal_ code' AS 'zip code','a'.'phone' AS 'phone',
'city'.'city' AS 'city','country'. 'country' AS 'country','s''.'store_id' AS 'SID' from ((('staff'
's' join 'address' 'a' on(('s'.'address_id' = 'a'.'address_id'))) join 'city' on(('a'.'city_id'
= 'city'.'city_id'))) join 'country' on(('city'.'country_id' = 'country'.'country_id')))
CHARACTER_SET_CLIENT: utf8
COLLATION_CONNECTION: utf8_general_ci
1 row in set (0.00 sec)
```

最后，通过查看系统表 information_schema.views 也可以查看视图的相关信息：

```
mysql> select * from views where table_name = 'staff_list' \G
*************************** 1. row ***************************
   TABLE_CATALOG: NULL
    TABLE_SCHEMA: sakila
      TABLE_NAME: staff_list
 VIEW_DEFINITION: select 's'.'staff_id' AS 'ID',concat('s'.'first_name',_ utf8 ',','s'.'last_name')
AS 'name','a'.'address' AS 'address','a'.'postal_code' AS 'zip code','a'.'phone' AS 'phone',
'sakila'.'city'.'city' AS 'city','sakila'. 'country'. 'country' AS 'country','s'.'store_id' AS '
SID' from ((('sakila'.'staff' 's' join 'sakila'.'address' 'a' on(('s'.'address_id' = 'a'.'address_
id'))) join 'sakila'.'city' on(('a'.'city_id' = 'sakila'.'city'.'city_id'))) join 'sakila'.'country'
on(('sakila'. 'city'.'country_id' = 'sakila'.'country'.'country_id')))
    CHECK_OPTION: NONE
    IS_UPDATABLE: YES
         DEFINER: root@localhost
   SECURITY_TYPE: DEFINER
CHARACTER_SET_CLIENT: utf8
COLLATION_CONNECTION: utf8_general_ci
1 row in set (0.00 sec)
```

10.2 存储过程和函数

MySQL 从 5.0 版本开始支持存储过程和函数。

10.2.1 什么是存储过程和函数

存储过程和函数是事先经过编译并存储在数据库中的一段 SQL 语句的集合，调用存储过程和函数可以简化应用开发人员的很多工作，减少数据在数据库和应用服务器之间的传输，对于提高数据处理的效率是有好处的。

存储过程和函数的区别在于函数必须有返回值，而存储过程没有，存储过程的参数可以使用 IN、OUT、INOUT 类型，而函数的参数只能是 IN 类型的。如果有函数从其他类型的数据库迁移到 MySQL，那么就可能因此需要将函数改造成存储过程。

10.2.2 存储过程和函数的相关操作

在对存储过程或函数进行操作时，需要首先确认用户是否具有相应的权限。例如，创建存储过程或者函数需要 CREATE ROUTINE 权限，修改或者删除存储过程或者函数需要 ALTER ROUTINE 权限，执行存储过程或者函数需要 EXECUTE 权限。

10.2.3 创建、修改存储过程或者函数

创建、修改存储过程或者函数的语法如下：

```
CREATE
    [DEFINER = { user | CURRENT_USER }]
    PROCEDURE sp_name ([proc_parameter[,...]])
    [characteristic ...] routine_body

CREATE
    [DEFINER = { user | CURRENT_USER }]
    FUNCTION sp_name ([func_parameter[,...]])
    RETURNS type
    [characteristic ...] routine_body

proc_parameter:
    [ IN | OUT | INOUT ] param_name type

func_parameter:
    param_name type

type:
    Any valid MySQL data type

characteristic:
    COMMENT 'string'
    | LANGUAGE SQL
    | [NOT] DETERMINISTIC
    | { CONTAINS SQL | NO SQL | READS SQL DATA | MODIFIES SQL DATA }
    | SQL SECURITY { DEFINER | INVOKER }

routine_body:
    Valid SQL routine statement
```

调用过程的语法如下：

```
CALL sp_name([parameter[,...]])
```

MySQL 的存储过程和函数中允许包含 DDL 语句，也允许在存储过程中执行提交（Commit，即确认之前的修改）或者回滚（Rollback，即放弃之前的修改），但是存储过程和函数中不允许执行 LOAD DATA INFILE 语句。此外，存储过程和函数中可以调用其他的过程

或者函数。

下面创建了一个新的过程 film_in_stock：

```
mysql> DELIMITER $$
mysql>
mysql> CREATE PROCEDURE film_in_stock(IN p_film_id INT, IN p_store_id INT, OUT p_film_count INT)
    -> READS SQL DATA
    -> BEGIN
    ->     SELECT inventory_id
    ->     FROM inventory
    ->     WHERE film_id = p_film_id
    ->     AND store_id = p_store_id
    ->     AND inventory_in_stock(inventory_id);
    ->
    ->     SELECT FOUND_ROWS() INTO p_film_count;
    -> END $$
Query OK, 0 rows affected (0.00 sec)

mysql>
mysql> DELIMITER ;
```

上面是在使用的样例数据库中创建的一个过程，该过程用来检查 film_id 和 store_id 对应的 inventory 是否满足要求，并且返回满足要求的 inventory_id 以及满足要求的记录数。

通常我们在执行创建过程和函数之前，都会通过"DELIMITER $$"命令将语句的结束符从";"修改成其他符号，这里使用的是"$$"，这样在过程和函数中的";"就不会被 MySQL 解释成语句的结束而提示错误。存储过程或者函数创建完毕，通过"DELIMITER ;"命令再将结束符改回成";"。

可以看到在这个过程中调用了函数 inventory_in_stock()，并且这个过程有两个输入参数和一个输出参数。下面可以通过调用这个过程来看看返回的结果。

如果需要检查 film_id=2 store_id=2 对应的 inventory 的情况，则首先手工执行过程中的 SQL 语句，以查看执行的效果：

```
mysql> SELECT inventory_id
    -> FROM inventory
    -> WHERE film_id = 2
    -> AND store_id = 2
    -> AND inventory_in_stock(inventory_id);
+--------------+
| inventory_id |
+--------------+
| 10           |
| 11           |
+--------------+
2 rows in set (0.00 sec)
```

满足条件的记录应该是两条，inventory_id 分别是 10 和 11。如果将这个查询封装在存储过程中调用，那么调用过程的执行情况如下：

```
mysql> CALL film_in_stock(2,2,@a);
+--------------+
| inventory_id |
+--------------+
| 10           |
| 11           |
+--------------+
2 rows in set (0.00 sec)

Query OK, 0 rows affected (0.00 sec)

mysql> select @a;
+------+
| @a   |
```

```
+------+
| 2    |
+------+
1 row in set (0.00 sec)
```

可以看到调用存储过程与直接执行 SQL 的效果是相同的，但是存储过程的好处在于处理逻辑都封装在数据库端，调用者不需要了解中间的处理逻辑。一旦处理逻辑发生变化，只需要修改存储过程即可，而对调用者的程序完全没有影响。

另外，和视图的创建语法稍有不同，存储过程和函数的 CREATE 语法不支持使用 CREATE OR REPLACE 对存储过程和函数进行修改，如果需要对已有的存储过程或者函数进行修改，需要执行 ALTER 语法。

下面对 characteristic 特征值的部分进行简单的说明。

- LANGUAGE SQL：说明下面过程的 BODY 是使用 SQL 编写，这条是系统默认的，为今后 MySQL 会支持的除 SQL 外的其他语言支持的存储过程而准备。

- [NOT] DETERMINISTIC：DETERMINISTIC 确定的，即每次输入一样输出也一样的程序，NOT DETERMINISTIC 非确定的，默认是非确定的。当前，这个特征值还没有被优化程序使用。

- { CONTAINS SQL | NO SQL | READS SQL DATA | MODIFIES SQL DATA }：这些特征值提供子程序使用数据的内在信息，这些特征值目前只是提供给服务器，并没有根据这些特征值来约束过程实际使用数据的情况。CONTAINS SQL 表示子程序不包含读或写数据的语句。NO SQL 表示子程序不包含 SQL 语句。READS SQL DATA 表示子程序包含读数据的语句，但不包含写数据的语句。MODIFIES SQL DATA 表示子程序包含写数据的语句。如果这些特征没有明确给定，默认使用的值是 CONTAINS SQL。

- SQL SECURITY { DEFINER | INVOKER }：可以用来指定子程序该用创建子程序者的许可来执行，还是使用调用者的许可来执行，默认值是 DEFINER。

- COMMENT 'string'：存储过程或者函数的注释信息。

下面的例子对比了 SQL SECURITY 特征值的不同，使用 root 用户创建了两个相似的存储过程，分别指定使用创建者的权限执行和调用者的权限执行，然后使用一个普通用户调用这两个存储过程，对比执行的效果。

首先用 root 用户创建以下两个存储过程 film_in_stock_definer 和 film_in_stock_invoker：

```
mysql> DELIMITER $$
mysql>
mysql> CREATE PROCEDURE film_in_stock_definer(IN p_film_id INT, IN p_store_id INT, OUT p_film
_count INT)
    -> SQL SECURITY DEFINER
    -> BEGIN
    ->     SELECT inventory_id
    ->     FROM inventory
    ->     WHERE film_id = p_film_id
    ->     AND store_id = p_store_id
    ->     AND inventory_in_stock(inventory_id);
    ->
    ->     SELECT FOUND_ROWS() INTO p_film_count;
    -> END $$
Query OK, 0 rows affected (0.00 sec)

mysql>
mysql> CREATE PROCEDURE film_in_stock_invoker(IN p_film_id INT, IN p_store_id INT, OUT p_film
_count INT)
    -> SQL SECURITY INVOKER
```

```
    -> BEGIN
    ->     SELECT inventory_id
    ->     FROM inventory
    ->     WHERE film_id = p_film_id
    ->     AND store_id = p_store_id
    ->     AND inventory_in_stock(inventory_id);
    ->
    ->     SELECT FOUND_ROWS() INTO p_film_count;
    -> END $$
Query OK, 0 rows affected (0.00 sec)

mysql>
mysql> DELIMITER ;
```

给普通用户 lisa 赋予可以执行存储过程的权限，但是不能查询 inventory 表：

```
mysql> GRANT EXECUTE ON sakila.* TO 'lisa'@'localhost';
Query OK, 0 rows affected (0.00 sec)
```

使用 lisa 登录后，直接查询 inventory 表会提示查询被拒绝：

```
mysql> select count(*) from inventory;
ERROR 1142 (42000): SELECT command denied to user 'lisa'@'localhost' for table 'inventory'
```

lisa 用户分别调用 film_in_stock_definer 和 film_in_stock_invoker：

```
mysql> CALL film_in_stock_definer(2,2,@a);
+--------------+
| inventory_id |
+--------------+
| 10           |
| 11           |
+--------------+
2 rows in set (0.03 sec)

Query OK, 0 rows affected (0.03 sec)

mysql> CALL film_in_stock_invoker(2,2,@a);
ERROR 1142 (42000): SELECT command denied to user 'lisa'@'localhost' for table 'inventory'
```

从上面的例子可以看出，film_in_stock_definer 是以创建者的权限执行的，因为是 root 用户创建的，所以可以访问 inventory 表的内容，film_in_stock_invoker 是以调用者的权限执行的，lisa 用户没有访问 inventory 表的权限，所以会提示权限不足。

10.2.4 删除存储过程或者函数

一次只能删除一个存储过程或者函数，删除存储过程或者函数需要有该过程或者函数的 ALTER ROUTINE 权限，具体语法如下：

```
DROP {PROCEDURE | FUNCTION} [IF EXISTS] sp_name
```

例如，使用 DROP 语法删除 film_in_stock 过程：

```
mysql> DROP PROCEDURE film_in_stock;
Query OK, 0 rows affected (0.00 sec)
```

10.2.5 查看存储过程或者函数

存储过程或者函数被创建后，用户可能需要查看存储过程、函数的状态、定义等信息，便于了解存储过程或者函数的基本情况。下面介绍如何查看存储过程或函数相关信息。

1. 查看存储过程或者函数的状态

```
SHOW {PROCEDURE | FUNCTION} STATUS [LIKE 'pattern']
```

下面演示的是查看过程 film_in_stock 的信息:

```
mysql> show procedure status like 'film_in_stock'\G
*************************** 1. row ***************************
                  Db: sakila
                Name: film_in_stock
                Type: PROCEDURE
             Definer: root@localhost
            Modified: 2007-07-06 09:29:00
             Created: 2007-07-06 09:29:00
       Security_type: DEFINER
             Comment:
1 row in set (0.00 sec)
```

2. 查看存储过程或者函数的定义

SHOW CREATE {PROCEDURE | FUNCTION} sp_name

下面演示的是查看过程 film_in_stock 的定义:

```
mysql> show create procedure film_in_stock \G
*************************** 1. row ***************************
           Procedure: film_in_stock
            sql_mode:
    Create Procedure: CREATE DEFINER='root'@'localhost' PROCEDURE 'film_in_stock'(IN p_film_id INT,
IN p_store_id INT, OUT p_film_count INT)
        READS SQL DATA
BEGIN
    SELECT inventory_id
    FROM inventory
    WHERE film_id = p_film_id
    AND store_id = p_store_id
    AND inventory_in_stock(inventory_id);
    SELECT FOUND_ROWS() INTO p_film_count;
END
1 row in set (0.00 sec)
```

3. 通过查看 information_schema.Routines 了解存储过程和函数的信息

除了以上两种方法,我们还可以查看系统表来了解存储过程和函数的相关信息,通过查看 information_schema.Routines 就可以获得存储过程和函数的名称、类型、语法、创建人等信息。

例如,通过查看 information_schema.Routines 得到过程 film_in_stock 的定义:

```
mysql> select *  from routines where ROUTINE_NAME = 'film_in_stock' \G
*************************** 1. row ***************************
       SPECIFIC_NAME: film_in_stock
     ROUTINE_CATALOG: NULL
      ROUTINE_SCHEMA: sakila
        ROUTINE_NAME: film_in_stock
        ROUTINE_TYPE: PROCEDURE
      DTD_IDENTIFIER: NULL
        ROUTINE_BODY: SQL
  ROUTINE_DEFINITION: BEGIN
    SELECT inventory_id
    FROM inventory
    WHERE film_id = p_film_id
    AND store_id = p_store_id
    AND inventory_in_stock(inventory_id);
    SELECT FOUND_ROWS() INTO p_film_count;
END
       EXTERNAL_NAME : NULL
   EXTERNAL_LANGUAGE : NULL
      PARAMETER_STYLE : SQL
     IS_DETERMINISTIC : NO
     SQL_DATA_ACCESS : READS SQL DATA
            SQL_PATH : NULL
       SECURITY_TYPE : DEFINER
```

```
            CREATED: 2007-07-06 09:29:00
       LAST_ALTERED: 2007-07-06 09:29:00
           SQL_MODE:
    ROUTINE_COMMENT:
            DEFINER: root@localhost
1 row in set (0.00 sec)
```

10.2.6 变量的使用

存储过程和函数中可以使用变量，而且在 MySQL 5.1 版本中，变量是不区分大小写的。

1. 变量的定义

通过 DECLARE 可以定义一个局部变量，该变量的作用范围只能在 BEGIN…END 块中，可以用在嵌套的块中。变量的定义必须写在复合语句的开头，并且在任何其他语句的前面。可以一次声明多个相同类型的变量。如果需要，可以使用 DEFAULT 赋默认值。

定义一个变量的语法如下：

```
DECLARE var_name[,...] type [DEFAULT value]
```

例如，定义一个 DATE 类型的变量，名称是 last_month_start：

```
DECLARE last_month_start DATE;
```

2. 变量的赋值

变量可以直接赋值，或者通过查询赋值。直接赋值使用 SET，可以赋常量或者赋表达式，具体语法如下：

```
SET var_name = expr [, var_name = expr] ...
```

给刚才定义的变量 last_month_start 赋值，具体语法如下：

```
SET last_month_start = DATE_SUB(CURRENT_DATE(), INTERVAL 1 MONTH);
```

也可以通过查询将结果赋给变量，这要求查询返回的结果必须只有一行，具体语法如下：

```
SELECT col_name[,...] INTO var_name[,...] table_expr
```

通过查询将结果赋值给变量 v_payments：

```
CREATE FUNCTION get_customer_balance(p_customer_id INT,
p_effective_date DATETIME)
RETURNS DECIMAL(5,2)
DETERMINISTIC
READS SQL DATA
BEGIN
    …
    DECLARE v_payments DECIMAL(5,2); #SUM OF PAYMENTS MADE PREVIOUSLY
    …
    SELECT IFNULL(SUM(payment.amount),0) INTO v_payments
    FROM payment
    WHERE payment.payment_date <= p_effective_date
    AND payment.customer_id = p_customer_id;
    …
    RETURN v_rentfees + v_overfees - v_payments;
END $$
```

10.2.7 定义条件和处理

条件的定义和处理可以用来定义在处理过程中遇到问题时将如何进行相应的处理。

1. 条件的定义

```
DECLARE condition_name CONDITION FOR condition_value

condition_value:
        mysql_error_code
    | SQLSTATE [VALUE] sqlstate_value
```

2. 条件的处理

```
DECLARE handler_type HANDLER
 FOR condition_value[,...]
 statement

handler_type:
        CONTINUE
    | EXIT
    | UNDO

condition_value:
    mysql_error_code
    | SQLSTATE [VALUE] sqlstate_value
    | condition_name
    | SQLWARNING
    | NOT FOUND
    | SQLEXCEPTION
```

下面将通过两个例子来说明：在向 actor 表中插入记录时，如果没有进行条件的处理，那么在主键冲突的时候会抛出异常并退出；如果对条件进行了处理，那么就不会再抛出异常。

（1）当没有进行条件处理时，执行结果如下：

```
mysql> select max(actor_id) from actor;
+---------------+
| max(actor_id) |
+---------------+
| 200           |
+---------------+
1 row in set (0.00 sec)

mysql> delimiter $$
mysql>
mysql> CREATE PROCEDURE actor_insert ()
    -> BEGIN
    ->   SET @x = 1;
    ->   INSERT INTO actor(actor_id,first_name,last_name) VALUES (201,'Test','201');
    ->   SET @x = 2;
    ->   INSERT INTO actor(actor_id,first_name,last_name) VALUES (1,'Test','1');
    ->   SET @x = 3;
    -> END;
    -> $$
Query OK, 0 rows affected (0.00 sec)

mysql> delimiter ;
mysql> call actor_insert();
ERROR 1062 (23000): Duplicate entry '1' for key 'PRIMARY'
mysql> select @x;
+------+
| @x   |
+------+
| 2    |
+------+
1 row in set (0.00 sec)
```

从上面的例子可以看出，执行到插入 actor_id=1 的记录时，会主键冲突并退出，没有执行到下面其他的语句。

（2）当对主键冲突的异常进行处理时，执行结果如下：

```
mysql> delimiter $$
mysql>
mysql> CREATE PROCEDURE actor_insert ()
    -> BEGIN
    ->   DECLARE CONTINUE HANDLER FOR SQLSTATE '23000' SET @x2 = 1;
    ->   SET @x = 1;
    ->   INSERT INTO actor(actor_id,first_name,last_name) VALUES (201,'Test','201');
    ->   SET @x = 2;
    ->   INSERT INTO actor(actor_id,first_name,last_name) VALUES (1,'Test','1');
    ->   SET @x = 3;
    -> END;
    -> $$
Query OK, 0 rows affected (0.00 sec)

mysql> delimiter ;
mysql> call actor_insert();
Query OK, 0 rows affected (0.06 sec)

mysql> select @x,@x2;
+------+------+
| @x   | @x2  |
+------+------+
| 3    | 1    |
+------+------+
1 row in set (0.00 sec)
```

调用条件处理的过程，再遇到主键冲突的错误时，会按照定义的处理方式进行处理，由于例子中定义的是 CONTINUE，所以会继续执行下面的语句。

handler_type 现在还只支持 CONTINUE 和 EXIT 两种，CONTINUE 表示继续执行下面的语句，EXIT 则表示执行终止，UNDO 现在还不支持。

condition_value 的值既可以是通过 DECLARE 定义的 condition_name，也可以是 SQLSTATE 的值或者 mysql-error-code 的值或者 SQLWARNING、NOT FOUND、SQLEXCEPTION，这 3 个值是 3 种定义好的错误类别，分别代表不同的含义。

● SQLWARNING 是对所有以 01 开头的 SQLSTATE 代码的速记。
● NOT FOUND 是对所有以 02 开头的 SQLSTATE 代码的速记。
● SQLEXCEPTION 是对所有没有被 SQLWARNING 或 NOT FOUND 捕获的 SQLSTATE 代码的速记。

因此，上面的例子还可以写成以下几种方式：

```
--捕获mysql-error-code:
DECLARE CONTINUE HANDLER FOR 1062 SET @x2 = 1;
--事先定义condition_name:
DECLARE DuplicateKey CONDITION FOR SQLSTATE '23000';
DECLARE CONTINUE HANDLER FOR DuplicateKey SET @x2 = 1;
--捕获SQLEXCEPTION
DECLARE CONTINUE HANDLER FOR SQLEXCEPTION SET @x2 = 1;
```

10.2.8 光标的使用

在存储过程和函数中，可以使用光标对结果集进行循环的处理。光标的使用包括光标的声明、OPEN、FETCH 和 CLOSE，其语法分别如下。

● 声明光标：

```
DECLARE cursor_name CURSOR FOR select_statement
```

● OPEN 光标：

```
OPEN cursor_name
```

- FETCH 光标：

`FETCH [[NEXT] FROM] cursor_name INTO var_name [, var_name] ...`

- CLOSE 光标：

`CLOSE cursor_name`

以下例子是一个简单的使用光标的过程，对 payment 表按照行进行循环的处理，按照 staff_id 值的不同累加 amount 的值，判断循环结束的条件是捕获 NOT FOUND 的条件，当 FETCH 光标找不到下一条记录的时候，就会关闭光标然后退出过程。

```
mysql> delimiter $$
mysql>
mysql> CREATE PROCEDURE payment_stat ()
    -> BEGIN
    ->   DECLARE i_staff_id int;
    ->   DECLARE d_amount decimal(5,2);
    ->   DECLARE cur_payment cursor for select staff_id,amount from payment;
    ->   DECLARE EXIT HANDLER FOR NOT FOUND CLOSE cur_payment;
    ->
    ->   set @x1 = 0;
    ->   set @x2 = 0;
    ->
    ->   OPEN cur_payment;
    ->
    ->   REPEAT
    ->     FETCH cur_payment INTO i_staff_id, d_amount;
    ->         if i_staff_id = 2 then
    ->             set @x1 = @x1 + d_amount;
    ->         else
    ->             set @x2 = @x2 + d_amount;
    ->         end if;
    ->   UNTIL 0 END REPEAT;
    ->
    ->   CLOSE cur_payment;
    ->
    -> END;
    -> $$
Query OK, 0 rows affected (0.00 sec)

mysql> delimiter ;
mysql>
mysql> call payment_stat();
Query OK, 0 rows affected (0.11 sec)

mysql> select @x1,@x2;
+----------+----------+
| @x1      | @x2      |
+----------+----------+
| 33927.04 | 33489.47 |
+----------+----------+
1 row in set (0.00 sec)
```

注意：变量、条件、处理程序、光标都是通过 DECLARE 定义的，它们之间是有先后顺序要求的。变量和条件必须在最前面声明，然后才能是光标的声明，最后才可以是处理程序的声明。

10.2.9 流程控制

下面将逐一进行说明如何使用 IF、CASE、LOOP、LEAVE、ITERATE、REPEAT 及 WHILE 语句来控制流程。

1. IF 语句

IF 实现条件判断，满足不同的条件执行不同的语句列表，具体语法如下：

```
IF search_condition THEN statement_list
    [ELSEIF search_condition THEN statement_list] ...
    [ELSE statement_list]
END IF
```

10.2.8 节中使用光标的例子中已经涉及了 IF 语句的使用，这里就不再举例说明了。

2. CASE 语句

CASE 实现比 IF 更复杂一些的条件构造，具体语法如下：

```
CASE case_value
    WHEN when_value THEN statement_list
    [WHEN when_value THEN statement_list] ...
    [ELSE statement_list]
END CASE
```

或者：

```
CASE
    WHEN search_condition THEN statement_list
    [WHEN search_condition THEN statement_list] ...
    [ELSE statement_list]
END CASE
```

在上文光标的使用例子中，IF 语句也可以使用 CASE 语句来完成：

```
case
  when i_staff_id = 2 then
    set @x1 = @x1 + d_amount;
  else
    set @x2 = @x2 + d_amount;
end case;
```

或者：

```
case i_staff_id
  when 2 then
    set @x1 = @x1 + d_amount;
  else
    set @x2 = @x2 + d_amount;
end case;
```

3. LOOP 语句

LOOP 实现简单的循环，退出循环的条件需要使用其他的语句定义，通常可以使用 LEAVE 语句实现，具体语法如下：

```
[begin_label:] LOOP
    statement_list
END LOOP [end_label]
```

如果不在 statement_list 中增加退出循环的语句，那么 LOOP 语句可以用来实现简单的死循环。

4. LEAVE 语句

LEAVE 语句用来从标注的流程构造中退出，通常和 BEGIN ... END 或者循环一起使用。

下面是一个使用 LOOP 和 LEAVE 的简单例子，循环 100 次向 actor 表中插入记录，当插入 100 条记录后，退出循环：

```
mysql> CREATE PROCEDURE actor_insert ()
    -> BEGIN
    ->   set @x = 0;
    ->   ins: LOOP
    ->     set @x = @x + 1;
    ->     IF @x = 100 then
    ->       leave ins;
    ->     END IF;
```

```
        ->         INSERT INTO actor(first_name,last_name) VALUES ('Test','201');
        ->     END LOOP ins;
        -> END;
        -> $$
Query OK, 0 rows affected (0.00 sec)

mysql> call actor_insert();
Query OK, 0 rows affected (0.01 sec)

mysql> select count(*) from actor where first_name='Test';
+----------+
| count(*) |
+----------+
|      100 |
+----------+
1 row in set (0.00 sec)
```

5. ITERATE 语句

ITERATE 语句必须用在循环中，作用是跳过当前循环的剩下的语句，直接进入下一轮循环。

下面的例子使用了 ITERATE 语句，当@x 变量是偶数的时候，不再执行循环中剩下的语句，而直接进行下一轮循环：

```
mysql> CREATE PROCEDURE actor_insert ()
    -> BEGIN
    ->   set @x = 0;
    ->   ins: LOOP
    ->     set @x = @x + 1;
    ->     IF @x = 10 then
    ->     leave ins;
    ->     ELSEIF mod(@x,2) = 0 then
    ->     ITERATE ins;
    ->     END IF;
    ->     INSERT INTO actor(actor_id,first_name,last_name) VALUES (@x+200, 'Test',@x);
    ->   END LOOP ins;
    -> END;
    -> $$
Query OK, 0 rows affected (0.00 sec)

mysql> call actor_insert();
Query OK, 0 rows affected (0.00 sec)

mysql> select actor_id,first_name,last_name from actor where first_name='Test';
+----------+------------+-----------+
| actor_id | first_name | last_name |
+----------+------------+-----------+
|      201 | Test       | 1         |
|      203 | Test       | 3         |
|      205 | Test       | 5         |
|      207 | Test       | 7         |
|      209 | Test       | 9         |
+----------+------------+-----------+
5 rows in set (0.00 sec)
```

6. REPEAT 语句

有条件的循环控制语句，当满足条件的时候退出循环，具体语法如下：

```
[begin_label:] REPEAT
        statement_list
UNTIL search_condition
END REPEAT [end_label]
```

10.2.8 节中的示例使用了 REPEAT 语句来实现光标的循环获得，下面节选的代码就是其中使用 REPEAT 语句的部分，详细的执行过程可以参照 10.2.8 节，这里不再赘述。

```
    -> REPEAT
    ->   FETCH cur_payment INTO i_staff_id, d_amount;
    ->     if i_staff_id = 2 then
    ->       set @x1 = @x1 + d_amount;
    ->     else
    ->       set @x2 = @x2 + d_amount;
    ->     end if;
    -> UNTIL 0 END REPEAT;
```

7. WHILE 语句

WHILE 语句实现的也是有条件的循环控制语句，即当满足条件时执行循环的内容，具体语法如下：

```
[begin_label:] WHILE search_condition DO
        statement_list
END WHILE [end_label]
```

WHILE 循环和 REPEAT 循环的区别在于：WHILE 是满足条件才执行循环，REPEAT 是满足条件退出循环；WHILE 在首次循环执行之前就判断条件，所以循环最少执行 0 次，而 REPEAT 是在首次执行循环之后才判断条件，所以循环最少执行 1 次。

以下例子用来对比 REPEAT 和 WHILE 语句的功能：

```
mysql> delimiter $$
mysql> CREATE PROCEDURE loop_demo ()
    -> BEGIN
    ->     set @x = 1 , @x1 = 1;
    ->     REPEAT
    ->        set @x = @x + 1;
    ->     until @x > 0 end repeat;
    ->
    ->     while @x1 < 0 do
    ->        set @x1 = @x1 + 1;
    ->     end while;
    -> END;
    -> $$
Query OK, 0 rows affected (0.00 sec)

mysql> delimiter ;
mysql> call loop_demo();
Query OK, 0 rows affected (0.00 sec)

mysql> select @x,@x1;
+------+------+
| @x   | @x1  |
+------+------+
| 2    | 1    |
+------+------+
1 row in set (0.00 sec)
```

从判断的条件上看，初始值都是满足退出循环的条件的，但是 REPEAT 循环仍然执行了一次以后才退出循环，而 WHILE 循环则一次都没有执行。

10.2.10 事件调度器

事件调度器是 MySQL 5.1 后新增的功能，可以将数据库按自定义的时间周期触发某种操作，可以理解为时间触发器，类似 Linux 系统下的任务调度器 crontab。

下面是一个最简单的事件调度器：

```
CREATE EVENT myevent
    ON SCHEDULE AT CURRENT_TIMESTAMP + INTERVAL 1 HOUR
```

```
    DO
        UPDATE myschema.mytable SET mycol = mycol + 1;
```

其中：
- 事件名称在 create event 关键字后指定；
- 通过 ON SCHEDULE 子句指定事件在何时执行及执行频次；
- 通过 DO 子句指定要执行的具体操作或事件。

上述创建的调度事件首先创建了 myevent 调度事件，然后执行更新操作，起始执行时间为调度器创建时间，后续在起始时间的基础上每隔 1 小时触发一次。

下面通过一个完整的实例来熟悉事件调度器的使用。

（1）创建测试表 test：

```
mysql> create table test(id1 varchar(10),create_time datetime);
Query OK, 0 rows affected (0.19 sec)
```

（2）创建事件调度器 test_event_1，每隔 5s 向 test 表插入一条记录：

```
mysql> CREATE EVENT test_event_1
    -> ON SCHEDULE
    -> EVERY 5 SECOND
    -> DO
    -> INSERT INTO test.test(id1,create_time)
    -> VALUES ('test',now());
Query OK, 0 rows affected (0.05 sec)
```

（3）查看调度器状态：

```
      mysql> show events \G;
*************************** 1. row ***************************
              Db: test
            Name: test_event_1
         Definer: root@localhost
       Time zone: SYSTEM
            Type: RECURRING
      Execute at: NULL
  Interval value: 5
  Interval field: SECOND
          Starts: 2013-07-26 14:02:02
            Ends: NULL
          Status: ENABLED
      Originator: 8306
character_set_client: latin1
collation_connection: latin1_swedish_ci
  Database Collation: gbk_chinese_ci
1 row in set (0.00 sec)
```

（4）隔几秒后，查看 test 表，发现并没有数据插入：

```
mysql> select * from test;
Empty set (0.00 sec)
```

（5）查看事件调度器状态，发现默认是关闭的：

```
mysql> show variables like '%scheduler%';
+-----------------+-------+
| Variable_name   | Value |
+-----------------+-------+
| event_scheduler | OFF   |
+-----------------+-------+
1 row in set (0.01 sec)
```

（6）通过下面的命令打开调度器，同时 show processslist 发现新产生一个后台进程：

```
mysql> SET GLOBAL event_scheduler = 1;
Query OK, 0 rows affected (0.00 sec)

mysql> show variables like '%scheduler%';
```

```
+----------------+-------+
| Variable name  | Value |
+----------------+-------+
| event_scheduler | ON   |
+----------------+-------+
1 row in set (0.01 sec)

mysql> show processlist \G;
…（前面省略）
*************************** 4. row ***************************
     Id: 464905
   User: event_scheduler
   Host: localhost
     db: NULL
Command: Daemon
   Time: 1
  State: Waiting for next activation
   Info: NULL
Rows_sent: 0
Rows_examined: 0
Rows_read: 1
4 rows in set (0.00 sec)
```

（7）隔几秒后，再次查看 test 表，发现已经有了一些数据，且日期间隔都为 5s：

```
mysql> select * from test;
+------+---------------------+
| id1  | create_time         |
+------+---------------------+
| test | 2013-07-29 05:28:48 |
| test | 2013-07-29 05:28:53 |
| test | 2013-07-29 05:28:58 |
| test | 2013-07-29 05:29:03 |
```

（8）为了防止表变得很大，创建一个新的调度器，每隔 1min 清空一次 test 表：

```
CREATE EVENT trunc_test
ON SCHEDULE every 1 MINUTE
DO TRUNCATE TABLE test;
```

隔一段时间后，可以发现，test 表中数据会定期清空，这类触发器非常适合去定期清空临时表或者日志表。

（9）如果事件调度器不再使用，可以禁用（disable）或者删除（drop）掉：

```
--禁用 event
mysql> alter event test_event_1 disable;
Query OK, 0 rows affected (0.00 sec)
--删除 event
mysql> drop event test_event_1;
Query OK, 0 rows affected (0.05 sec)
```

对于事件调度器，还有很多选项，比如指定事件开始时间和结束时间，或者指定某个时间执行一次而不是循环执行，详细信息可以参考事件调度器的相关帮助，这里不再详述。

最后，总结一下事件调度器的优势、适用场景及使用中的注意事项，如表 10-1 所示。

表 10-1　　　　　　　　事件调度器的优势、适用场景及注意事项

事件调度器	说　　明
优势	MySQL 事件调度器部署在数据库内部由 DBA 或专人统一维护和管理，避免将一些数据库相关的定时任务部署在操作系统层，减少操作系统管理员产生误操作的风险，对后续的管理和维护也非常有益。例如，后续进行数据库迁移时无须再迁移操作系统层的定时任务，数据库迁移本身已经包含了调度事件的迁移
适用场景	事件调度器适用于定期收集统计信息、定期清理历史数据、定期数据库检查（例如，自动监控和恢复 Slave 失败进程）

事件调度器	说　明
注意事项	在繁忙且要求性能的数据库服务器上要慎重部署和启用调度器 过于复杂的处理更适合用程序实现 开启和关闭事件调度器需要具有超级用户权限

10.3　触发器

MySQL 从 5.0.2 版本开始支持触发器的功能。触发器是与表有关的数据库对象，在满足定义条件时触发，并执行触发器中定义的语句集合。触发器的这种特性可以协助应用在数据库端确保数据的完整性。本章将详细介绍 MySQL 中触发器的使用方法。

10.3.1　创建触发器

创建触发器的语法如下：

```
CREATE TRIGGER trigger_name trigger_time trigger_event
    ON tbl_name FOR EACH ROW [trigger_order] trigger_body
```

注意：触发器只能创建在永久表（Permanent Table）上，不能对临时表（Temporary Table）创建触发器。

其中 trigger_time 是触发器的触发时间，可以是 BEFORE 或者 AFTER，BEFORE 的含义指在检查约束前触发，而 AFTER 是在检查约束后触发。

而 trigger_event 就是触发器的触发事件，可以是 INSERT、UPDATE 或者 DELETE。

使用别名 OLD 和 NEW 来引用触发器中发生变化的记录内容，这与其他的数据库是相似的。现在触发器还只支持行级触发，不支持语句级触发。

在样例数据库中，为 film 表创建了 AFTER INSERT 的触发器，具体如下：

```
DELIMITER $$
CREATE TRIGGER ins_film
AFTER INSERT ON film FOR EACH ROW BEGIN
    INSERT INTO film_text (film_id, title, description)
        VALUES (new.film_id, new.title, new.description);
END;
$$
delimiter ;
```

插入 film 表记录的时候，会向 film_text 表中也插入相应的记录。

```
mysql> INSERT INTO film VALUES
    -> (1001,'ACADEMY DINOSAUR',
    -> 'A Epic Drama of a Feminist And a Mad Scientist who must Battle a Teacher in The Canadian Rockies',
    -> 2006,1,NULL,6,'0.99',86,'20.99','PG','Deleted Scenes,Behind the Scenes', '2006-02-15 05:03:42');
Query OK, 1 row affected (0.05 sec)

mysql> select * from film_text where film_id=1001 \G
*************************** 1. row ***************************
    film_id: 1001
      title: ACADEMY DINOSAUR
description: A Epic Drama of a Feminist And a Mad Scientist who must Battle a Teacher in The Canadian Rockies
1 row in set (0.00 sec)
```

对于 INSERT INTO...ON DUPLICATE KEY UPDATE...语句来说，触发触发器的顺序可

能会造成疑惑。下面对 film 表分别创建 BEFORE INSERT、AFTER INSERT、BEFORE UPDATE、AFTER UPDATE 触发器，然后插入记录，观察触发器的触发情况：

```
--创建 BEFORE INSERT、AFTER INSERT、BEFORE UPDATE、AFTER UPDATE 触发器
mysql> create table tri_demo(id int AUTO_INCREMENT,note varchar(20),PRIMARY KEY (id));
Query OK, 0 rows affected (0.03 sec)

mysql> CREATE TRIGGER ins_film_bef
    -> BEFORE INSERT ON film FOR EACH ROW BEGIN
    ->     INSERT INTO tri_demo (note) VALUES ('before insert');
    -> END;
    -> $$
Query OK, 0 rows affected (0.00 sec)

mysql> CREATE TRIGGER ins_film_aft
    -> AFTER INSERT ON film FOR EACH ROW BEGIN
    ->     INSERT INTO tri_demo (note) VALUES ('after insert');
    -> END;
    -> $$
Query OK, 0 rows affected (0.00 sec)

mysql> CREATE TRIGGER upd_film_bef
    -> BEFORE update ON film FOR EACH ROW BEGIN
    ->     INSERT INTO tri_demo (note) VALUES ('before update');
    -> END;
    -> $$
Query OK, 0 rows affected (0.00 sec)

mysql> CREATE TRIGGER upd_film_aft
    -> AFTER update ON film FOR EACH ROW BEGIN
    ->     INSERT INTO tri_demo (note) VALUES ('after update');
    -> END;
    -> $$
Query OK, 0 rows affected (0.00 sec)

--插入记录已经存在的情况
mysql> select film_id,title from film where film_id = 1001;
+---------+------------------+
| film_id | title            |
+---------+------------------+
| 1001    | ACADEMY DINOSAUR |
+---------+------------------+
1 row in set (0.00 sec)

mysql> INSERT INTO film VALUES
    -> (1001,'Only test',
    -> 'Only test',2006,1,NULL,6,'0.99',86,'20.99','PG',
    -> 'Deleted Scenes,Behind the Scenes','2006-02-15 05:03:42')
    -> ON DUPLICATE KEY
    -> UPDATE title='update record';
Query OK, 2 rows affected (0.05 sec)

mysql> select * from tri_demo;
+----+---------------+
| id | note          |
+----+---------------+
| 1  | before insert |
| 2  | before update |
| 3  | after update  |
+----+---------------+
3 rows in set (0.00 sec)

--插入新记录的情况
mysql> delete from tri_demo;
Query OK, 3 rows affected (0.00 sec)

mysql> select film_id,title from film where film_id = 1002;
Empty set (0.00 sec)
```

```
mysql> INSERT INTO film VALUES
    -> (1002,'Only test',
    -> 'Only test',2006,1,NULL,6,'0.99',86,'20.99','PG',
    -> 'Deleted Scenes,Behind the Scenes','2006-02-15 05:03:42')
    -> ON DUPLICATE KEY
    -> UPDATE title='update record';
Query OK, 1 row affected (0.05 sec)

mysql>
mysql> select * from tri_demo;
+----+---------------+
| id | note          |
+----+---------------+
| 4  | before insert |
| 5  | after insert  |
+----+---------------+
2 rows in set (0.00 sec)
```

从上面的例子可以知道，对于有重复记录、需要进行 UPDATE 操作的 INSERT，触发器触发的顺序是 BEFORE INSERT、BEFORE UPDATE、AFTER UPDATE；对于没有重复记录的 INSERT，就是简单地执行 INSERT 操作，触发器触发的顺序是 BEFORE INSERT、AFTER INSERT。对于那些实际执行 UPDATE 操作的记录，仍然会执行 BEFORE INSERT 触发器的内容，在设计触发器的时候一定要考虑这种情况，避免错误地触发触发器。

10.3.2 删除触发器

一次可以删除一个触发程序，如果没有指定 schema_name，默认为当前数据库，具体语法如下：

```
DROP TRIGGER [schema_name.]trigger_name
```

例如，要删除 film 表上的触发器 ins_film，可以使用以下命令：

```
mysql> drop trigger ins_film;
Query OK, 0 rows affected (0.00 sec)
```

10.3.3 查看触发器

可以通过执行 SHOW TRIGGERS 命令查看触发器的状态、语法等信息，但是因为不能查询指定的触发器，所以每次都返回所有的触发器的信息，使用起来不是很方便，具体语法如下：

```
mysql> show triggers \G
*************************** 1. row ***************************
             Trigger: customer_create_date
               Event: INSERT
               Table: customer
           Statement: SET NEW.create_date = NOW()
              Timing: BEFORE
             Created: NULL
            sql_mode: STRICT_TRANS_TABLES,STRICT_ALL_TABLES,NO_ZERO_IN_DATE,NO_ZERO_DATE,ERROR_FOR_DIVISION_BY_ZERO,TRADITIONAL,NO_AUTO_CREATE_USER
             Definer: root@localhost
*************************** 2. row ***************************
……
```

另一个查看方式是查询系统表的 information_schema.triggers 表，这个方式可以查询指定触发器的指定信息，操作起来明显很方便：

```
mysql> desc triggers;
+---------------------------+--------------+------+-----+---------+-------+
| Field                     | Type         | Null | Key | Default | Extra |
+---------------------------+--------------+------+-----+---------+-------+
| TRIGGER_CATALOG           | varchar(512) | YES  |     |         |       |
| TRIGGER_SCHEMA            | varchar(64)  | NO   |     |         |       |
| TRIGGER_NAME              | varchar(64)  | NO   |     |         |       |
| EVENT_MANIPULATION        | varchar(6)   | NO   |     |         |       |
| EVENT_OBJECT_CATALOG      | varchar(512) | YES  |     |         |       |
| EVENT_OBJECT_SCHEMA       | varchar(64)  | NO   |     |         |       |
| EVENT_OBJECT_TABLE        | varchar(64)  | NO   |     |         |       |
| ACTION_ORDER              | bigint(4)    | NO   |     | 0       |       |
| ACTION_CONDITION          | longtext     | YES  |     |         |       |
| ACTION_STATEMENT          | longtext     | NO   |     |         |       |
| ACTION_ORIENTATION        | varchar(9)   | NO   |     |         |       |
| ACTION_TIMING             | varchar(6)   | NO   |     |         |       |
| ACTION_REFERENCE_OLD_TABLE| varchar(64)  | YES  |     |         |       |
| ACTION_REFERENCE_NEW_TABLE| varchar(64)  | YES  |     |         |       |
| ACTION_REFERENCE_OLD_ROW  | varchar(3)   | NO   |     |         |       |
| ACTION_REFERENCE_NEW_ROW  | varchar(3)   | NO   |     |         |       |
| CREATED                   | datetime     | YES  |     |         |       |
| SQL_MODE                  | longtext     | NO   |     |         |       |
| DEFINER                   | longtext     | NO   |     |         |       |
+---------------------------+--------------+------+-----+---------+-------+
19 rows in set (0.00 sec)

mysql> select * from triggers where trigger_name = 'ins_film_bef' \G
*************************** 1. row ***************************
           TRIGGER_CATALOG: NULL
            TRIGGER_SCHEMA: sakila
              TRIGGER_NAME: ins_film_bef
        EVENT_MANIPULATION: INSERT
      EVENT_OBJECT_CATALOG: NULL
       EVENT_OBJECT_SCHEMA: sakila
        EVENT_OBJECT_TABLE: film
              ACTION_ORDER: 0
          ACTION_CONDITION: NULL
          ACTION_STATEMENT: BEGIN
    INSERT INTO tri_demo (note) VALUES ('before insert');
END
        ACTION_ORIENTATION: ROW
             ACTION_TIMING: BEFORE
ACTION_REFERENCE_OLD_TABLE: NULL
ACTION_REFERENCE_NEW_TABLE: NULL
  ACTION_REFERENCE_OLD_ROW: OLD
  ACTION_REFERENCE_NEW_ROW: NEW
                   CREATED: NULL
                  SQL_MODE:
                   DEFINER: root@localhost
1 row in set (0.01 sec)
```

10.3.4 触发器的使用

触发器执行的语句有以下两个限制。

● 触发程序既不能调用将数据返回客户端的存储程序，也不能使用采用 CALL 语句的动态 SQL 语句，但是允许存储程序通过参数将数据返回触发程序。也就是存储过程或者函数通过 OUT 或者 INOUT 类型的参数将数据返回触发器是可以的，但是不能调用直接返回数据的过程。

● 不能在触发器中使用以显式或隐式方式开始或结束事务的语句，如 START TRANSACTION、COMMIT 或 ROLLBACK。

MySQL 的触发器是按照 BEFORE 触发器、行操作、AFTER 触发器的顺序执行的，其中任何一步操作发生错误都不会继续执行剩下的操作。如果是对事务表进行的操作，那么会整个作为一个事务被回滚（Rollback），但是如果是对非事务表进行的操作，那么已经更新的记录将无法回滚，这也是设计触发器的时候需要注意的问题。

10.4 小结

本章主要介绍了 MySQL 提供的视图、存储过程、函数、触发器的创建、维护等相关语法，也介绍了它们分别适用的场合，但是由于篇幅问题，本章并没有对这部分内容进行深入，读者如果有兴趣，可以查询在线的 MySQL 文档获得帮助。

关于触发器这部分，需要特别注意的是触发器是行触发的，每次增加、修改或者删除记录都会触发进行处理，编写过于复杂的触发器或者增加过多的触发器对记录的插入、更新、删除操作肯定会有比较严重的影响，因此在设计数据库的时候要有所考虑，不要将应用的处理逻辑过多地依赖于触发器来处理。

第 11 章 事务控制和锁定语句

MySQL 支持对 MyISAM 和 MEMORY 存储引擎的表进行表级锁定，对 BDB 存储引擎的表进行页级锁定，对 InnoDB 存储引擎的表进行行级锁定。默认情况下，表锁和行锁都是自动获得的，不需要额外的命令。但是在有的情况下，用户需要明确地进行锁表或者进行事务的控制，以便确保整个事务的完整性，这样就需要使用事务控制和锁定语句来完成。

有关锁机制、不同存储引擎对锁的处理、死锁等内容，将会在后面的优化篇中进行更详细的介绍，有兴趣的读者可以参见相关的章节。

11.1 LOCK TABLES 和 UNLOCK TABLES

LOCK TABLES 可以锁定用于当前线程的表。如果表被其他线程锁定，则当前线程会等待，直到可以获取所有锁定为止。

UNLOCK TABLES 可以释放当前线程获得的任何锁定。当前线程执行另一个 LOCK TABLES 时，或当与服务器的连接被关闭时，所有由当前线程锁定的表被隐含地解锁，具体语法如下：

```
LOCK TABLES
    tbl_name [AS alias] {READ [LOCAL] | [LOW_PRIORITY] WRITE}
    [, tbl_name [AS alias] {READ [LOCAL] | [LOW_PRIORITY] WRITE}] ...
UNLOCK TABLES
```

表 11-1 给出了一个获得表锁和释放表锁的简单例子，其中 film_text 表获得 read 锁的情况，其他 session 更新该表记录会等待锁，film_text 表释放锁以后，其他 session 可以进行更新操作。其中 session1 和 session2 表示两个同时打开的 session，表格中的每一行表示同一时刻两个 session 的运行状况，后面的例子也都是同样格式，这里不再赘述。

表 11-1　　　　　　　　　　　一个获得表锁和释放表锁的简单例子

session_1	session_2
获得表 film_text 的 READ 锁定： mysql> lock table film_text read; Query OK, 0 rows affected (0.00 sec)	

session_1	session_2
当前 session 可以查询该表记录： mysql> select film_id,title from film_text where film_id = 1001; +---------+-----------------+ | film_id | title | +---------+-----------------+ | 1001 | ACADEMY DINOSAUR | +---------+-----------------+ 1 row in set (0.00 sec)	其他 session 也可以查询该表的记录： mysql> select film_id,title from film_text where film_id = 1001; +---------+-----------------+ | film_id | title | +---------+-----------------+ | 1001 | ACADEMY DINOSAUR | +---------+-----------------+ 1 row in set (0.00 sec)
	其他 session 更新锁定表会等待获得锁： mysql> update film_text set title = 'Test' where film_id = 1001; 等待
释放锁： mysql> unlock tables; Query OK, 0 rows affected (0.00 sec)	等待
	Session 获得锁，更新操作完成： mysql> update film_text set title = 'Test' where film_id = 1001; Query OK, 1 row affected (1 min 0.71 sec) Rows matched: 1 Changed: 1 Warnings: 0

有关表锁的使用，可以参见 16.2 节以获得更详细的信息。

注意：LOCK TABLES/UNLOCK TABLES 有时也写为 LOCK TABLE/UNLOCK TABLE，两种写法含义一致。

11.2 事务控制

MySQL 通过 SET AUTOCOMMIT、START TRANSACTION、COMMIT 和 ROLLBACK 等语句支持本地事务，具体语法如下：

```
START TRANSACTION | BEGIN [WORK]
COMMIT [WORK] [AND [NO] CHAIN] [[NO] RELEASE]
ROLLBACK [WORK] [AND [NO] CHAIN] [[NO] RELEASE]
SET AUTOCOMMIT = {0 | 1}
```

默认情况下，MySQL 是自动提交（Autocommit）的。如果需要通过明确的 Commit 和 Rollback 来提交和回滚事务，那么就需要通过明确的事务控制命令来开始事务，这是和 Oracle 的事务管理明显不同的地方。如果应用是从 Oracle 数据库迁移到 MySQL 数据库，则需要确保应用中是否对事务进行了明确的管理。

- START TRANSACTION 或 BEGIN 语句可以开始一项新的事务。
- COMMIT 和 ROLLBACK 用来提交或者回滚事务。
- CHAIN 和 RELEASE 子句分别用来定义在事务提交或者回滚之后的操作，CHAIN 会立即启动一个新事务，并且和刚才的事务具有相同的隔离级别，RELEASE 则会断开和客户端的连接。
- SET AUTOCOMMIT 可以修改当前连接的提交方式，如果设置了 SET AUTOCOMMIT=0，则设置之后的所有事务都需要通过明确的命令进行提交或者回滚。

如果只是对某些语句需要进行事务控制，则使用 START TRANSACTION 语句开始一个事务比较方便，这样，事务结束之后可以自动回到自动提交的方式。如果希望所有的事务都不是自动提交的，那么通过修改 AUTOCOMMIT 来控制事务比较方便，这样不用在每个事务开始的时候再执行 START TRANSACTION 语句。

表 11-2 中的例子演示了使用 START TRANSACTION 开始的事务在提交后自动回到自动提交的方式；如果在提交时使用 COMMIT AND CHAIN，那么会在提交后立即开始一个新的事务。

表 11-2　　START TRANSACTION 和 COMMIT AND CHAIN 的使用例子

session_1	session_2
从表 actor 中查询 actor_id=201 的记录，结果为空： mysql> select * from actor where actor_id = 201; Empty set (0.00 sec)	从表 actor 中查询 actor_id=201 的记录，结果为空： mysql> select * from actor where actor_id = 201; Empty set (0.00 sec)
用 start transaction 命令启动一个事务，往表 actor 中插入一条记录，没有 commit： mysql> start transaction; Query OK, 0 rows affected (0.00 sec) mysql> insert into actor (actor_id,first_name,last_name) values(201,'Lisa','Tom'); Query OK, 1 row affected (0.00 sec)	
	查询表 actor，结果仍然为空： mysql> select * from actor where actor_id= 201; Empty set (0.00 sec)
执行提交： mysql> commit; Query OK, 0 rows affected (0.04 sec)	
	再次查询表 actor，可以查询到结果： mysql> select actor_id,last_name from actor where actor_id in (201,202); +----------+-----------+ | actor_id | last_name | +----------+-----------+ | 201 | Tom | +----------+-----------+ 1 row in set (0.00 sec)
这个事务是按照自动提交执行的： mysql> insert into actor (actor_id,first_name,last_name) values(202,'Lisa','Lan'); Query OK, 1 row affected (0.04 sec)	
	可以从 actor 表中查询到 session1 刚刚插入的数据： mysql> select actor_id,last_name from actor where actor_id in (201,202); +----------+-----------+ | actor_id | last_name | +----------+-----------+ | 201 | Tom | | 202 | Lan | +----------+-----------+ 2 rows in set (0.00 sec)

11.2 事务控制

续表

session_1	session_2

重新用 start transaction 启动一个事务：
mysql> start transaction;
Query OK, 0 rows affected (0.00 sec)
往表 actor 中插入一条记录：
mysql> insert into actor (actor_id,first_name,last_name) values(203,'Lisa','TT');
Query OK, 1 row affected (0.00 sec)
用 commit and chain 命令提交：
mysql> commit and chain;
Query OK, 0 rows affected (0.03 sec)
此时自动开始一个新的事务：
mysql> insert into actor (actor_id,first_name, last_name) values(204,'Lisa','Mou');
Query OK, 1 row affected (0.00 sec)	

	session1 刚插入的记录无法看到： mysql> select actor_id,last_name from actor where first_name = 'Lisa'; +----------+-----------+ | actor_id | last_name | +----------+-----------+ | 178 | MONROE T | | 201 | Tom | | 202 | Lan | | 203 | TT | +----------+-----------+ 4 rows in set (0.00 sec)
用 commit 命令提交： mysql> commit; Query OK, 0 rows affected (0.06 sec)	
	session1 插入的新记录可以看到： mysql> select actor_id,last_name from actor where first_name= 'Lisa'; +----------+-----------+ | actor_id | last_name | +----------+-----------+ | 178 | MONROE T | | 201 | Tom | | 202 | Lan | | 203 | TT | | 204 | Mou | +----------+-----------+ 5 rows in set (0.00 sec)

如果在锁表期间，用 start transaction 命令开始一个新事务，则会造成一个隐含的 UNLOCK TABLES 被执行，如表 11-3 所示。

表 11-3　　　　　start transaction 导致的 UNLOCK TABLES

session_1	session_2
从表 actor 中查询 actor_id=201 的记录，结果为空： mysql> select * from actor where actor_id = 201; Empty set (0.00 sec)	从表 actor 中查询 actor_id=201 的记录，结果为空： mysql> select * from actor where actor_id = 201; Empty set (0.00 sec)
对表 actor 加写锁： mysql> lock table actor write; Query OK, 0 rows affected (0.00 sec)	

续表

session_1	session_2
	对表 actor 的读操作被阻塞： mysql> select actor_id,last_name from actor where actor_id = 201; 等待
插入一条记录： mysql> insert into actor (actor_id,first_name, last_name) values(201,'Lisa','Tom'); Query OK, 1 row affected (0.04 sec)	等待
回滚刚才的记录： mysql> rollback; Query OK, 0 rows affected (0.00 sec)	等待
用 start transaction 命令重新开始一个事务： mysql> start transaction; Query OK, 0 rows affected (0.00 sec)	等待
	session1 开始一个事务时，表锁被释放，可以查询： mysql> select actor_id,last_name from actor where actor_id = 201; +----------+-----------+ | actor_id | last_name | +----------+-----------+ | 201 | Tom | +----------+-----------+ 1 row in set (17.78 sec) 对 lock 方式加的表锁，不能通过 rollback 进行回滚

因此，在同一个事务中，最好使用相同存储引擎的表，否则 ROLLBACK 时需要对非事务类型的表进行特别的处理，因为 COMMIT、ROLLBACK 只能对事务类型的表进行提交和回滚。

通常情况下，只对提交的事务记录到二进制的日志中，但是如果一个事务中包含非事务类型的表，那么回滚操作也会被记录到二进制日志中，以确保非事务类型表的更新可以被复制到从数据库（Slave）中。

和 Oracle 的事务管理相同，所有的 DDL 语句是不能回滚的，并且部分的 DDL 语句会造成隐式的提交。

在事务中可以通过定义 SAVEPOINT，指定回滚事务的一个部分，但是不能指定提交事务的一个部分。对于复杂的应用，可以定义多个不同的 SAVEPOINT，满足不同的条件时，回滚不同的 SAVEPOINT。需要注意的是，如果定义了相同名字的 SAVEPOINT，则后面定义的 SAVEPOINT 会覆盖之前的定义。对于不再需要使用的 SAVEPOINT，可以通过 RELEASE SAVEPOINT 命令删除 SAVEPOINT，删除后的 SAVEPOINT 不能再执行 ROLLBACK TO SAVEPOINT 命令。

表 11-4 中的例子就是模拟回滚事务的一个部分，通过定义 SAVEPOINT 来指定需要回滚的事务的位置。

表 11-4　　　　　　　　　　　　　模拟回滚事务

session_1	session_2
从表 actor 中查询 first_name='Simon'的记录，结果为空： mysql> select * from actor where first_name = 'Simon'; Empty set (0.00 sec)	从表 actor 中查询 first_name='Simon'的记录，结果为空： mysql> select * from actor where first_name = 'Simon'; Empty set (0.00 sec)

11.2 事务控制

续表

session_1	session_2
启动一个事务，往表 actor 中插入一条记录： mysql> start transaction; Query OK, 0 rows affected (0.02 sec) mysql> insert into actor (actor_id,first_name, last_name) values(301,'Simon','Tom'); Query OK, 1 row affected (0.00 sec)	
可以查询到刚插入的记录： mysql> select actor_id,last_name from actor where first_name = 'Simon'; +----------+-----------+ | actor_id | last_name | +----------+-----------+ | 301 | Tom | +----------+-----------+ 1 row in set (0.00 sec)	无法从 actor 表中查到 session1 刚插入的记录： mysql> select * from actor where first_name = 'Simon'; Empty set (0.00 sec)
定义 savepoint，名称为 test： mysql> savepoint test; Query OK, 0 rows affected (0.00 sec) 继续插入一条记录： mysql> insert into actor (actor_id,first_name,last_name) values(302,'Simon','Cof'); Query OK, 1 row affected (0.00 sec)	
可以查询到两条记录： mysql> select actor_id,last_name from actor where first_name = 'Simon'; +----------+-----------+ | actor_id | last_name | +----------+-----------+ | 301 | Tom | | 302 | Cof | +----------+-----------+ 2 rows in set (0.00 sec)	仍然无法查询到结果： mysql> select * from actor where first_name = 'Simon'; Empty set (0.00 sec)
回滚到刚才定义的 savepoint： mysql> rollback to savepoint test; Query OK, 0 rows affected (0.00 sec)	
只能从表 actor 中查询到第一条记录，因为第二条已经被回滚： mysql> select actor_id,last_name from actor where first_name = 'Simon'; +----------+-----------+ | actor_id | last_name | +----------+-----------+ | 301 | Tom | +----------+-----------+ 1 row in set (0.00 sec)	仍然无法查询到结果： mysql> select * from actor where first_name = 'Simon'; Empty set (0.00 sec)
用 commit 命令提交： mysql> commit; Query OK, 0 rows affected (0.05 sec)	
只能从 actor 表中查询到第一条记录： mysql> select actor_id,last_name from actor where first_name = 'Simon'; +----------+-----------+ | actor_id | last_name | +----------+-----------+ | 301 | Tom | +----------+-----------+ 1 row in set (0.00 sec)	只能从 actor 表中查询到 session1 插入的第一条记录： mysql> select actor_id,last_name from actor where first_name = 'Simon'; +----------+-----------+ | actor_id | last_name | +----------+-----------+ | 301 | Tom | +----------+-----------+ 1 row in set (0.00 sec)

11.3 分布式事务的使用

MySQL 从 5.0.3 版本起开始支持分布式事务，当前分布式事务只支持 InnoDB 存储引擎。一个分布式事务会涉及多个行动，这些行动本身是事务性的。所有行动都必须一起成功完成，或者一起被回滚。

11.3.1 分布式事务的原理

在 MySQL 中，使用分布式事务的应用程序涉及一个或多个资源管理器和一个事务管理器。

● 资源管理器（RM）用于提供通向事务资源的途径。数据库服务器是一种资源管理器。该管理器必须可以提交或回滚由 RM 管理的事务。例如，多台 MySQL 数据库作为多台资源管理器或者几台 MySQL 服务器和几台 Oracle 服务器作为资源管理器。

● 事务管理器（TM）用于协调作为一个分布式事务一部分的事务。TM 与管理每个事务的 RMs 进行通信。在一个分布式事务中，各个单个事务均是分布式事务的"分支事务"。分布式事务和各分支通过一种命名方法进行标识。

MySQL 执行 XA MySQL 时，MySQL 服务器相当于一个用于管理分布式事务中的 XA 事务的资源管理器。与 MySQL 服务器连接的客户端相当于事务管理器。

要执行一个分布式事务，必须知道这个分布式事务涉及哪些资源管理器，并且把每个资源管理器的事务执行到事务可以被提交或回滚时。根据每个资源管理器报告的有关执行情况的内容，这些分支事务必须作为一个原子性操作全部提交或回滚。要管理一个分布式事务，必须要考虑任何组件或连接网络可能会出现故障。

用于执行分布式事务的过程使用两阶段提交，发生时间在由分布式事务的各个分支需要进行的行动已经被执行之后。

● 在第一阶段中，所有的分支被预备好。即它们被 TM 告知要准备提交。通常，这意味着用于管理分支的每个 RM 会记录对于被稳定保存的分支的行动。分支指示是否它们可以这么做。这些结果被用于第二阶段。

● 在第二阶段中，TM 告知 RMs 是否要提交或回滚。如果在预备分支时，所有的分支指示它们将能够提交，则所有的分支被告知要提交。如果在预备时，有任何分支指示它将不能提交，则所有分支被告知回滚。

在有些情况下，一个分布式事务可能会使用一阶段提交。例如，当一个事务管理器发现，一个分布式事务只由一个事务资源组成（即单一分支），则该资源可以被告知同时进行预备和提交。

11.3.2 分布式事务的语法

分布式事务（XA 事务）的 SQL 语法如下：

```
XA {START|BEGIN} xid [JOIN|RESUME]
```

XA START xid 用于启动一个带给定 xid 值的 XA 事务。每个 XA 事务必须有一个唯一的 xid 值，因此该值当前不能被其他的 XA 事务使用。xid 是一个 XA 事务标识符，用来唯一标识一个分布式事务。xid 值由客户端提供，或由 MySQL 服务器生成。xid 值包含 1~3 个部分：

```
xid: gtrid [, bqual [, formatID ]]
```

- gtrid 是一个分布式事务标识符，相同的分布式事务应该使用相同的 gtrid，这样可以明确知道 XA 事务属于哪个分布式事务。
- bqual 是一个分支限定符，默认值是空串。对于一个分布式事务中的每个分支事务，bqual 值必须是唯一的。
- formatID 是一个数字，用于标识由 gtrid 和 bqual 值使用的格式，默认值是 1。

下面其他 XA 语法中用到的 xid 值都必须和 START 操作使用的 xid 值相同，也就是表示对这个启动的 XA 事务进行操作。

```
XA END xid [SUSPEND [FOR MIGRATE]]
XA PREPARE xid
```

使事务进入 PREPARE 状态，也就是两阶段提交的第一个提交阶段。

```
XA COMMIT xid [ONE PHASE]
XA ROLLBACK xid
```

这两个命令用来提交或者回滚具体的分支事务。也就是两阶段提交的第二个提交阶段：分支事务被实际提交或者回滚。

```
XA RECOVER
```

XA RECOVER 返回当前数据库中处于 PREPARE 状态的分支事务的详细信息。

分布式的关键在于如何确保分布式事务的完整性，以及在某个分支出现问题时的故障解决。XA 的相关命令就是提供给应用如何在多个独立的数据库之间进行分布式事务的管理，包括启动一个分支事务、使事务进入准备阶段以及事务的实际提交回滚操作等。表 11-5 中的例子演示了一个简单的分布式事务的执行，事务的内容是在 DB1 中插入一条记录，同时在 DB2 中更新一条记录，两个操作作为同一事务提交或者回滚。

表 11-5　　　　　　　　　　　　　　分布式事务例子

session_1 in DB1	session_2 in DB2
在数据库 DB1 中启动一个分布式事务的一个分支事务，xid 的 gtrid 为 "test"，bqual 为 "db1"： mysql> xa start 'test','db1'; Query OK, 0 rows affected (0.00 sec) 分支事务 1 在表 actor 中插入一条记录： mysql> insert into actor (actor_id,first_name,last_name) values(301,'Simon','Tom'); Query OK, 1 row affected (0.00 sec) 对分支事务 1 进行第一阶段提交，进入 prepare 状态： mysql> xa end 'test','db1'; Query OK, 0 rows affected (0.00 sec) mysql> xa prepare 'test','db1'; Query OK, 0 rows affected (0.02 sec)	在数据库 DB2 中启动分布式事务 "test" 的另外一个分支事务，xid 的 gtrid 为 "test"，bqual 为 "db2"： mysql> xa start 'test','db2'; Query OK, 0 rows affected (0.00 sec) 分支事务 2 在表 film_actor 中更新了 23 条记录： mysql> update film_actor set last_update=now() where actor_id = 178; Query OK, 23 rows affected (0.04 sec) Rows matched: 23　Changed: 23　Warnings: 0 对分支事务 2 进行第一阶段提交，进入 prepare 状态： mysql> xa end 'test','db2'; Query OK, 0 rows affected (0.00 sec) mysql> xa prepare 'test','db2'; Query OK, 0 rows affected (0.02 sec)
用 xa recover 命令查看当前分支事务状态： mysql> xa recover \G *************************** 1. row *************************** formatID: 1 gtrid_length: 4 bqual_length: 3 data: testdb1 1 row in set (0.00 sec)	用 xa recover 命令查看当前分支事务状态： mysql> xa recover \G *************************** 1. row *************************** formatID: 1 gtrid_length: 4 bqual_length: 3 data: testdb2 1 row in set (0.00 sec)

session_1 in DB1	session_2 in DB2
两个事务都进入准备提交阶段,如果之前遇到任何错误,都应该回滚所有的分支,以确保分布式事务的正确	
提交分支事务1: mysql> xa commit 'test','db1'; Query OK, 0 rows affected (0.03 sec) 两个事务都到达准备提交阶段后,一旦开始进行提交操作,就需要确保全部的分支都提交成功	提交分支事务2: mysql> xa commit 'test','db2'; Query OK, 0 rows affected (0.03 sec)

11.3.3 存在的问题

虽然 MySQL 支持分布式事务,但是仍然存在一些问题。

在 MySQL 5.5 之前的版本,如果分支事务在达到 prepare 状态时,数据库异常重新启动,服务器重新启动以后,可以选择对分支事务进行提交或者回滚操作,但是即使选择提交事务,该事务也不会被写入 BINLOG。这就存在一定的隐患,可能导致使用 BINLOG 恢复时丢失部分数据。如果存在复制的从库,则有可能导致主从数据库的数据不一致。以下演示了这个过程。

(1)从表 actor 中查询 first_name = 'Simon'的记录,显示有一条。

```
mysql> select actor_id,last_name from actor where first_name = 'Simon';
+----------+-----------+
| actor_id | last_name |
+----------+-----------+
| 301      | Tom       |
+----------+-----------+
1 row in set (0.00 sec)
```

(2)启动分布式事务"test",删除刚才查询的记录。

```
mysql> xa start 'test';
Query OK, 0 rows affected (0.00 sec)

mysql> delete from actor where actor_id = 301;
Query OK, 1 row affected (0.00 sec)

mysql> select actor_id,last_name from actor where first_name = 'Simon';
Empty set (0.00 sec)
```

(3)完成第一阶段提交,进入 prepare 状态。

```
mysql> xa end 'test';
Query OK, 0 rows affected (0.00 sec)

mysql>  xa prepare 'test';
Query OK, 0 rows affected (0.03 sec)
```

(4)此时,数据库异常终止,查询出错。

```
mysql> select actor_id,last_name from actor where first_name = 'Simon';
ERROR 2006 (HY000): MySQL server has gone away
No connection. Trying to reconnect...
ERROR 2002 (HY000): Can't connect to local MySQL server through socket '/mnt/db/mysqld.sock' (2)
ERROR:
Can't connect to the server
```

(5)启动数据库后,分支事务依然存在。

```
mysql> xa recover \G
*************************** 1. row ***************************
    formatID: 1
gtrid_length: 4
bqual_length: 0
```

11.3 分布式事务的使用

```
        data: test
1 row in set (0.00 sec)
```

（6）表中记录并没有被删除。

```
mysql> select actor_id,last_name from actor where first_name = 'Simon';
+----------+-----------+
| actor_id | last_name |
+----------+-----------+
| 301      | Tom       |
+----------+-----------+
1 row in set (0.00 sec)
```

（7）可以在 MySQL 的数据库日志中看到分布式事务的处理情况，数据库启动的时候发现有一个 prepare 状态的事务，提示需要进行处理：

```
InnoDB: Transaction 0 117471044 was in the XA prepared state.
InnoDB: 1 transaction(s) which must be rolled back or cleaned up
InnoDB: in total 0 row operations to undo
InnoDB: Trx id counter is 0 117471488
070710 16:55:41  InnoDB: Started; log sequence number 29 2758352865
070710 16:55:41  InnoDB: Starting recovery for XA transactions...
070710 16:55:41  InnoDB: Transaction 0 117471044 in prepared state after recovery
070710 16:55:41  InnoDB: Transaction contains changes to 1 rows
070710 16:55:41  InnoDB: 1 transactions in prepared state after recovery
070710 16:55:41  [Note] Found 1 prepared transaction(s) in InnoDB
070710 16:55:41  [Warning] Found 1 prepared XA transactions
```

可以进行提交或者回滚。

```
mysql>  xa commit 'test';
Query OK, 0 rows affected (0.02 sec)

mysql> select actor_id,last_name from actor where first_name = 'Simon';
Empty set (0.00 sec)
```

提交后，使用 mysqlbinlog 查看 BINLOG，可以确认最后提交的这个分支事务并没有记录到 BINLOG 中，因为复制和灾难恢复都是依赖于 BINLOG 的，所以 BINLOG 的缺失会导致复制环境的不同步，以及使用 BINLOG 恢复丢失部分数据。由于这个 BUG 的存在，在 MySQL 5.7 之前，对于数据库实例死机，官方的建议是选择回滚 prepare 的事务。

此外，如果分支事务的客户端连接异常中止，例如执行 prepare 之后退出连接，那么数据库会自动回滚未完成的分支事务，但是这种做法实际上仍然存在问题，因为如果此时分支事务已经执行到 prepare 状态，那么这个分布式事务的其他分支可能已经成功提交，如果这个分支回滚，可能导致分布式事务的不完整，丢失部分分支事务的内容，如表 11-6 所示。

表 11-6　　　　　　　　客户端连接中止导致分布式事务失败例子

session_1	session_2
从表 actor 中查询 first_name='Simon'的记录，结果为空： mysql> select * from actor where first_name = 'Simon'; Empty set (0.00 sec)	从表 actor 中查询 first_name='Simon'的记录，结果为空： mysql> select * from actor where first_name = 'Simon'; Empty set (0.00 sec)
启动分布式事务 test： mysql> xa start 'test'; Query OK, 0 rows affected (0.00 sec) 往 actor 表中插入一条记录： mysql> insert into actor (actor_id,first_name,last_name) values(301,'Simon','Tom'); Query OK, 1 row affected (0.00 sec) 事务结束： mysql> xa end 'test'; Query OK, 0 rows affected (0.00 sec)	

session_1	session_2
查询刚插入的记录，可以显示结果： mysql> select actor_id,last_name from actor where first_name = 'Simon'; +----------+-----------+ | actor_id | last_name | +----------+-----------+ | 301 | Tom | +----------+-----------+ 1 row in set (0.00 sec)	查询刚插入的记录，显示结果为空： mysql> select * from actor where first_name = 'Simon'; Empty set (0.00 sec)
完成第一阶段提交，进入 prepare 状态： mysql> xa prepare 'test'; Query OK, 0 rows affected (0.02 sec)	
	查询分布式事务 "test" 的状态： mysql> xa recover \G ****************** 1. row ****************** 　　formatID: 1 gtrid_length: 4 bqual_length: 3 　　　　data: test 1 row in set (0.00 sec)
session_1 异常中止	
session_1 被回滚	session1 异常中止后，分布式事务被回滚，session2 中无法查询到 session1 插入的记录，如果此时 session2 存在分支事务并且被成功提交，则会导致分布式事务的不完整 mysql> select * from actor where first_name = 'Simon'; Empty set (0.00 sec)

而上面也已经提到，当发现部分分支已经提交成功，需要使用备份和 BINLOG 来恢复数据的时候，那些在 prepare 状态的分支事务因为并没有记录到 BINLOG，所以不能通过 BINLOG 进行恢复，在数据库恢复后，将丢失这部分的数据。

在 MySQL 5.7 中，已经解决了 XA 事务的严格持久化问题，在 session 断开和实例崩溃的情况下，事务都不会自动回滚，同时在 XA PREPARE 时，之前的事务信息就会被写入 BINLOG 并同步到备库。最终再由用户决定将悬挂事务回滚或者是提交。下面测试一下 XA 事务在 MySQL 5.7 中的改进。

首先，开启一个 XA 事务：

```
mysql> xa start 'test';
Query OK, 0 rows affected (0.00 sec)

mysql> insert into actor (actor_id,first_name,last_name) values(301,'Simon','Tom');
Query OK, 1 row affected (0.00 sec)

mysql> xa end 'test';
Query OK, 0 rows affected (0.00 sec)
```

在 MySQL 5.7 中，XA 事务在结束之后，提交之前，不允许进行查询：

```
mysql> select * from actor;
ERROR 1399 (XAE07): XAER_RMFAIL: The command cannot be executed when global transaction is in the  IDLE state
mysql> xa prepare 'test';
Query OK, 0 rows affected (0.00 sec)

mysql> select * from actor;
ERROR 1399 (XAE07): XAER_RMFAIL: The command cannot be executed when global transaction is in the  PREPARED state
```

此时查看 BINLOG，可以看到执行 XA PREPARE 后，BINLOG 已经有相应的记录：

```
Query event : XA START X'74657374',X'',1
Table_map event
Write_rows event
Query event: XA END X'74657374',X'',1
XA_prepare event:  XA PREPARE X'74657374',X'',1
```

断开后重新连接 MySQL，可以看到，事务没有被自动回滚，可以手动进行回滚或提交：

```
mysql> exit
Bye
[root ~]# mysql
Welcome to the MySQL monitor.  Commands end with ; or \g.
Your MySQL connection id is 17
Server version: 5.7.22-log MySQL Community Server (GPL)

Copyright (c) 2000, 2018, Oracle and/or its affiliates. All rights reserved.

Oracle is a registered trademark of Oracle Corporation and/or its
affiliates. Other names may be trademarks of their respective
owners.

Type 'help;' or '\h' for help. Type '\c' to clear the current input statement.

mysql> xa recover;
+----------+--------------+--------------+------+
| formatID | gtrid_length | bqual_length | data |
+----------+--------------+--------------+------+
|        1 |            4 |            0 | test |
+----------+--------------+--------------+------+
1 row in set (0.00 sec)
mysql>
mysql> mysql> xa commit 'test';
Query OK, 0 rows affected (0.00 sec)
```

总之，MySQL 的分布式事务还存在一些问题，在数据库或者应用异常的情况下，可能会导致分布式事务的不完整或者需要人工介入处理。如果需要使用分布式事务，建议尽量采用 MySQL 5.7 或者更新的版本。

11.4　小结

事务控制和锁定是 MySQL 的重要特点之一。本章介绍了 MySQL 提供的事务控制和锁定语法，并对分布式事务进行了简单的介绍。MySQL 中锁的管理涉及的内容很广泛，在后面的优化篇中我们将会对锁机制、死锁和应用中需要注意的其他问题进行更深入的讨论。

第 12 章 SQL 中的安全问题

在日常开发过程中，程序员一般只关心 SQL 是否能实现预期的功能，而对于 SQL 的安全问题一般都不太重视。实际上，如果 SQL 语句写作不当，将会给应用系统造成很大的安全隐患，其中最重要的隐患就是 SQL 注入。本章以 MySQL 为例，将会对 SQL 注入以及相应的防范措施进行详细的介绍。

12.1 SQL 注入简介

结构化查询语言（SQL）是一种用来和数据库交互的文本语言。SQL 注入（SQL Injection）就是利用某些数据库的外部接口将用户数据插入到实际的数据库操作语言（SQL）当中，从而达到入侵数据库乃至操作系统的目的。它的产生主要是由于程序对用户输入的数据没有进行严格的过滤，导致非法数据库查询语句的执行。

SQL 注入攻击具有很大的危害，攻击者可以利用它读取、修改或者删除数据库内的数据，获取数据库中的用户名和密码等敏感信息，甚至可以获得数据库管理员的权限，而且，SQL 注入也很难防范。网站管理员无法通过安装系统补丁或者进行简单的安全配置进行自我保护，一般的防火墙也无法拦截 SQL 注入攻击。

下面的用户登录验证程序就是 SQL 注入的一个例子（以 PHP 程序举例）。

（1）创建用户表 user：

```
CREATE TABLE user (
userid int(11) NOT NULL auto_increment,
username varchar(20) NOT NULL default '',
password varchar(20) NOT NULL default '',
PRIMARY KEY (userid)
) TYPE=MyISAM AUTO_INCREMENT=3 ;
```

（2）给用户表 user 添加一条用户记录：

```
INSERT INTO 'user' VALUES (1, 'angel', 'mypass');
```

（3）验证用户 root 登录 localhost 服务器：

```
<?php
    $servername = "localhost";
    $dbusername = "root";
    $dbpassword = "";
    $dbname = "injection";
    mysql_connect($servername,$dbusername,$dbpassword) or die ("数据库连接失败");
    $sql = "SELECT * FROM user WHERE username='$username' AND password= '$password'";
    $result = mysql_db_query($dbname, $sql);
```

```
$userinfo = mysql_fetch_array($result);
if (empty($userinfo))
{
echo "登录失败";
} else {
echo "登录成功";
}
echo "<p>SQL Query:$sql<p>";
?>
```

(4) 然后提交如下 URL：

```
http://127.0.0.1/injection/user.php?username=angel' or '1=1
```

结果发现，这个 URL 可以成功登录系统，但是很显然这并不是我们预期的结果。同样也可以利用 SQL 的注释语句实现 SQL 注入，如下面的例子：

```
http://127.0.0.1/injection/user.php?username=angel'/*
http://127.0.0.1/injection/user.php?username=angel'#
```

因为在 SQL 语句中，"/*" 或者 "#" 都可以将后面的语句注释掉。这样上述语句就可以通过这两个注释符中任意一个将后面的语句给注释掉了，结果导致只根据用户名而没有密码的 URL 都成功进行了登录。利用 "or" 和注释符的不同之处在于，前者是利用逻辑运算，而后者则是根据 MySQL 的特性，这个比逻辑运算简单多了。虽然这两种情况实现的原理不同，但是达到了同样的 SQL 注入效果，都是我们应该关注的。

12.2 应用开发中可以采取的应对措施

对于上面提到的 SQL 注入隐患，后果可想而知是很严重的，轻则获得数据信息，重则可以将数据进行非法更改。那么对这种情况有没有防范措施呢？答案是肯定的。本节将介绍一些常用的防范方法。

12.2.1 PrepareStatement+Bind-Variable

MySQL 服务器端并不存在共享池的概念，所以在 MySQL 上使用绑定变量（Bind Variable）最大的好处主要是为了避免 SQL 注入，增加安全性。下面以 Java 语言为例，同样是根据 username 来访问 user 表：

```
…
Class.forName("com.mysql.jdbc.Driver").newInstance();
String connectionUrl = "jdbc:mysql://localhost:3331/test";
String connectionUser = "test_user";
String connectionPassword = "test_passwd";
conn = DriverManager.getConnection(connectionUrl, connectionUser, connectionPassword);
String sqlStmt = " select * from user where username = ? and password = ?";
System.out.println("Source SQL Statement:" + sqlStmt);
prepStmt = conn.prepareStatement(sqlStmt);
System.out.println("Before Bind Value:" + prepStmt.toString());
prepStmt.setString(1, "angel' or 1=1'");
prepStmt.setString(2, "test");
System.out.println("After Bind Value:" + prepStmt.toString());
rs = prepStmt.executeQuery();
while (rs.next()) {
String ename = rs.getString("username");
String job = rs.getString("password");
System.out.println("username: " + username + ", password: " + password);
}
…
```

输出日志如下：

```
Source SQL Statement: select * from user where username = ? and password = ?
Before Bind Value:com.mysql.jdbc.JDBC4PreparedStatement@6910fe28:     select * from user where username = ** NOT SPECIFIED ** and password = ** NOT SPECIFIED **
After Bind Value:com.mysql.jdbc.JDBC4PreparedStatement@6910fe28:     select * from user where username = 'angel\' or 1=1\'' and password = 'test'
```

可以注意到，虽然传入的变量中带了"angel' or 1=1"的条件，企图蒙混过关，但是由于使用了绑定变量（Java 驱动中采用 PreparedStatement 语句来实现），输入的参数中的单引号被正常转义，导致后续的"or 1=1"作为 username 条件的内容出现，而不会被作为 SQL 的一个单独条件被解析，避免了 SQL 注入的风险。

同样的，在使用绑定变量的情况下，企图通过注释"/*"或"#"让后续条件失效也是会失败的：

```
prepStmt.setString(1, "angel '/*");
After Bind Value:com.mysql.jdbc.JDBC4PreparedStatement@6910fe28:     select * from user where username = 'angel \'/*' and password = 'test'

prepStmt.setString(1, "angel '#");
After Bind Value:com.mysql.jdbc.JDBC4PreparedStatement@5a9e29fb:     select * from user where username = 'angel \'#' and password = 'test'
```

需要注意，PreparedStatement 语句是由 JDBC 驱动来支持的，在使用 PreparedStatement 语句的时候，仅仅做了简单的替换和转义，并不是 MySQL 提供了 PreparedStatement 的特性。

对 Java、JSP 开发的应用，可以使用 PrepareStatement+Bind-variable 来防止 SQL 注入，另外从 PHP 5 开始，也在扩展的 MySQLI 中支持 PrepareStatement，所以在使用这类语言作数据库开发时，强烈建议使用 PrepareStatement+Bind-variable 来实现。下面是 PHP 的例子：

```
...
$stmt = $dbh->prepare("SELECT * FROM users WHERE USERNAME = ? AND PASSWORD = ?");
$stmt->execute(array($username, $password));
...
```

12.2.2 使用应用程序提供的转换函数

很多应用程序接口都提供了对特殊字符进行转换的函数。恰当地使用这些函数，可以防止应用程序用户输入使应用程序生成不期望的语句。

- MySQL C API：使用 mysql_real_escape_string() API 调用。
- MySQL++：使用 escape 和 quote 修饰符。
- PHP：使用 mysql_real_escape_string()函数（适用于 PHP 4.3.0 版本）。从 PHP 5 开始，可以使用扩展的 MySQLI，这是对 MySQL 新特性的一个扩展支持，其中的一个优点就是支持 PrepareStatement。
- Perl DBI：使用 placeholders 或者 quote()方法。
- Ruby DBI：使用 placeholders 或者 quote()方法。

12.2.3 自己定义函数进行校验

如果现有的转换函数仍然不能满足要求，则需要自己编写函数进行输入校验。输入验证是一个很复杂的问题。输入验证的途径可以分为以下几种：

- 整理数据使之变得有效；
- 拒绝已知的非法输入；

- 只接受已知的合法输入。

因此，如果想要获得最好的安全状态，目前最好的解决办法就是，对用户提交或者可能改变的数据进行简单分类，分别应用正则表达式来对用户提供的输入数据进行严格的检测和验证。

下面采用正则表达式的方法提供一个验证函数，以供读者参考。

已知非法符号有：" ' " " ; " " = " " (" ") " " /* " " */ " " % " " + " " " " > " " < " " -- " " [" 和 "] "。

其实只需要过滤非法的符号组合就可以阻止已知形式的攻击，并且如果发现更新的攻击符号组合，也可以将这些符号组合增添进来，继续防范新的攻击。特别是空格符号和与其产生相同作用的分隔关键字的符号，例如 "/**/"，如果能成功过滤这种符号，那么有很多注入攻击将不能发生，并且同时也要过滤它们的十六进制表示 "%XX"。

由此，可以构造如下正则表达式：

```
(|\'|(\%27)|\;|(\%3b)|\=|(\%3d)|\(|(\%28)|\)|(\%29)|(\/*)|(\%2f%2a)|(\*/)|(\%2a%2f)|\+|(\%2b)
|\<|(\%3c)|\>|(\%3e)|\(--)|\[|(\%5b|\]|\%5d)
```

根据上述的正则表达式，可以提供一个函数（以 PHP 举例），可以防范大多数的 SQL 注入，具体函数如下：

```
function SafeRequest ($ParaName, $ParaType)
{
    /* ---传入参数--- */
    /* ParaName：参数名称-字符型 */
    /* ParaType：参数类型-数字型（1 表示参数是数字或字符，0 表示参数为其他）*/

    if ($ParaType == 1)
    {
        $re = "/[^\w+$]/";
    }
    else
    {
        $re = "/(|\' |(\%27)|\;|(\%3b)|\=|(\%3d)|\(|(\%28)|\)|(\%29)|(\/*) |(\%2f%2a)|(\ */)|(\%2a%2f)|\+|(\%2b)|\<|(\%3c)|\>|(\%3e)|\(--))\[|\%5b|\]|\%5d)/";
    }

    if (preg_match($re, $ParaName) > 0)
    {
        echo("参数不符合要求，请重新输入!");
        return 0;
    }
    else
    {
        return 1;
    }
}
```

12.3 小结

本章主要从 SQL 注入的角度讨论了 SQL 的安全问题，阐述了 SQL 注入的原理以及防范措施，最后通过一个 PHP 函数例子给出了类似问题解决方法的参考。

本章的内容不仅仅适用于 MySQL 数据库，一些原理以及解决方案同样适用于其他数据库系统，因为 SQL 注入问题是一个数据库应用普遍存在的安全问题。

第 13 章　SQL Mode 及相关问题

与其他数据库不同，MySQL 可以运行在不同的 SQL Mode（SQL 模式）下。SQL Mode 定义了 MySQL 应支持的 SQL 语法、数据校验等，这样可以更容易地在不同的环境中使用 MySQL。本章将详细介绍常用的 SQL Mode 及其在实际中的应用。

13.1　MySQL SQL Mode 简介

在 MySQL 中，SQL Mode 常用来解决下面几类问题。
- 通过设置 SQL Mode，可以完成不同严格程度的数据校验，有效地保障数据准确性。
- 通过设置 SQL Mode 为 ANSI 模式，来保证大多数 SQL 符合标准的 SQL 语法，这样应用在不同数据库之间进行迁移时，则不需要对业务 SQL 进行较大的修改。
- 在不同数据库之间进行数据迁移之前，通过设置 SQL Mode 可以使 MySQL 上的数据更方便地迁移到目标数据库中。

在 MySQL 5.7 中，SQL Mode 有了较大的变化，查询默认的 SQL Mode（sql_mode 参数）为 ONLY_FULL_GROUP_BY、STRICT_TRANS_TABLES、NO_ZERO_IN_DATE、NO_ZERO_DATE、ERROR_FOR_DIVISION_BY_ZERO、NO_AUTO_CREATE_USER 和 NO_ENGINE_SUBSTITUTION（不同的小版本可能略有区别）。这些 SQL MODE 的含义如表 13-1 所示。

表 13-1　　　　　　　　　　MySQL 5.7 中默认的 SQL Mode

sql_mode 值	描　　述
ONLY_FULL_GROUP_BY	在 group by 子句中没有出现的列，出现在 select 列表、having 条件、order by 条件中时会被拒绝
STRICT_TRANS_TABLES	非法日期，超过字段长度的值插入时，直接报错，拒绝执行
NO_ZERO_IN_DATE	日期中针对月份和日期部分，如果为 0，比如'2018-00-00'，有不同的执行逻辑 ● disable：可以正常插入，实际插入值还是'2018-00-00'没有警告 ● enable：可以插入，有警告，实际插入值变为'0000-00-00'；如果 mode 中包含 STRICT_TRANS_TABLES，则日期被拒绝写入，但可以通过加 ignore 关键字写入'0000-00-00'
NO_ZERO_DATE	针对日期'0000-00-00'，执行逻辑如下 ● disable：可以正常插入，没有警告 ● enable：可以插入，有警告，实际插入值变为'0000-00-00'；如果 mode 中包含 STRICT_TRANS_TABLES，则日期被拒绝写入，但可以通过加 ignore 关键字写入'0000-00-00'，有警告

13.1 MySQL SQL Mode 简介

续表

sql_mode 值	描 述
ERROR_FOR_DIVISION_BY_ZERO	除数为 0（包括 MOD(N,0)），执行逻辑如下 ● disable：插入 NULL，没有警告 ● enable：插入 NULL，有警告；如果 mode 中包含 STRICT_TRANS_TABLES，则数据被拒绝写入，但可以通过加 ignore 关键字写入 NULL，有警告
NO_AUTO_CREATE_USER	防止使用不带密码子句的 grant 语句来创建一个用户
NO_ENGINE_SUBSTITUTION	执行 create table 或者 alter table 语句时，如果指定了不支持（包括 disable 或未编译）的存储引擎，是否自动替换为默认的存储引擎 ● disabe：create table 会自动替换后执行，alter table 不会执行，两个命令都有警告 ● enable：两个命令直接报错

相比之前的版本，MySQL 5.7.5 之后的版本最大的区别是在 SQL Mode 的默认设置中，增加了严格的事物表模式（STRICT_TRANS_TABLES），在这种模式下不允许插入字段类型不一致的值，不允许插入超过字段长度的值，这在绝大多数场景下都更加合理。如果不设置 STRICT_TRANS_TABLES，那么上述操作会被允许，只是在插入后，MySQL 会返回一个 warning，从而导致表中写入异常数据。

NO_ZERO_IN_DATE、NO_ZERO_DATE、ERROR_FOR_DIVISION_BY_ZERO 这几种 SQL Mode 很少单独使用，通常和 STRICT_TRANS_TABLES 一起来用，官网宣称这几种 SQL Mode 将来可能会合并。下面来看个具体的例子。

（1）查看默认 SQL Mode 的命令如下：

```
mysql> select @@sql_mode;
+-------------------------------------------------------------+
| @@sql_mode                                                  |
+-------------------------------------------------------------+
| ONLY_FULL_GROUP_BY,STRICT_TRANS_TABLES,NO_ZERO_IN_DATE……    |
+-------------------------------------------------------------+
1 row in set (0.00 sec)
```

（2）查看测试表 t 的表结构的命令如下：

```
mysql> desc t_sql_mode_strict;
+-------+-------------+------+-----+---------+-------+
| Field | Type        | Null | Key | Default | Extra |
+-------+-------------+------+-----+---------+-------+
| id    | int(11)     | YES  |     | NULL    |       |
| name  | varchar(10) | YES  |     | NULL    |       |
+-------+-------------+------+-----+---------+-------+
2 rows in set (0.00 sec)
```

（3）首先取消 SQL Mode 的严格模式：

```
mysql> set session sql_mode='';
Query OK, 0 rows affected, 1 warning (0.00 sec)
```

（4）在表中插入一条记录，其中 name 故意超出了实际的定义值 varchar(10)：

```
mysql> insert into t_sql_mode_strict values(1,'beijing@126.com');
Query OK, 1 row affected, 1 warning (0.02 sec)
```

（5）可以发现，记录可以插入，但是显示了一个 warning，查看 warning 内容：

```
mysql> show warnings;
+---------+------+------------------------------------------------+
| Level   | Code | Message                                        |
+---------+------+------------------------------------------------+
| Warning | 1265 | Data truncated for column 'name' at row 1      |
+---------+------+------------------------------------------------+
1 row in set (0.00 sec)
```

（6）warning 提示对插入的 name 值进行了截断，从表 t_sql_mode_strict 中查看实际插入值：

```
mysql> select * from t_sql_mode_strict;
+------+------------+
| id   | name       |
+------+------------+
|    1 | beijing@12 |
+------+------------+
1 row in set (0.00 sec)
```

果然，记录虽然插入进去，但是只截取了前 10 位字符。

（7）接下来设置 SQL Mode 为 STRICT_TRANS_TABLES（严格的事物表模式）：

```
mysql> set session sql_mode='STRICT_TRANS_TABLES';
Query OK, 0 rows affected (0.01 sec)

mysql> select @@sql_mode;
+---------------------+
| @@sql_mode          |
+---------------------+
| STRICT_TRANS_TABLES |
+---------------------+
1 row in set (0.01 sec)
```

（8）再次尝试插入上面的测试记录：

```
mysql> insert into t_sql_mode_strict values(1,'beijing@126.com');
ERROR 1406 (22001): Data too long for column 'name' at row 1
```

结果发现，这次记录没有插入成功，给出了一个 ERROR，而不仅仅是 warning。

上面的例子中，给出了 sql_mode 的一种修改方法，即 SET [SESSION|GLOBAL] sql_mode='modes'，其中 SESSION 选项表示只在本次连接中生效；而 GLOBAL 选项表示在本次连接中并不生效，而对于新的连接则生效，这种方法在 MySQL 4.1 开始有效。另外，也可以通过使用 "--sql-mode="modes"" 选项，在 MySQL 启动时设置 sql_mode。

13.2　SQL Mode 的常见功能

下面介绍一下 SQL Mode 的常见功能。

（1）校验日期数据合法性，这是 SQL Mode 的一项常见功能。

在下面的例子中，观察一下非法日期 "2007-04-31"（因为 4 月没有 31 日）在不同 SQL Mode 下能否正确插入。

```
mysql> set session sql_mode='ANSI';
Query OK, 0 rows affected (0.00 sec)

mysql> create table t_sql_mode_ansi (d datetime);
Query OK, 0 rows affected (0.03 sec)

mysql> insert into t_sql_mode_ansi values('2007-04-31');
Query OK, 1 row affected, 1 warning (0.00 sec)
mysql> select * from t;
+---------------------+
| d                   |
+---------------------+
| 0000-00-00 00:00:00 |
+---------------------+
1 row in set (0.00 sec)

mysql> set session sql_mode='TRADITIONAL';
Query OK, 0 rows affected (0.00 sec)
```

```
mysql> insert into t_sql_mode_ansi values('2007-04-31');
ERROR 1292 (22007): Incorrect datetime value: '2007-04-31' for column 'd' at row 1
```

很显然，在 ANSI 模式下，非法日期可以插入，但是插入值却变为"0000-00-00 00:00:00"，并且系统给出了 warning；而在 TRADITIONAL 模式下，会直接提示日期非法，拒绝插入。

（2）启用 NO_BACKSLASH_ESCAPES 模式，使反斜线"\"成为普通字符。在导入数据时，如果数据中含有反斜线字符，那么启用 NO_BACKSLASH_ESCAPES 模式保证数据的正确性，是一个不错的选择。

以下示例说明了启用 NO_BACKSLASH_ESCAPES 模式前后对反斜线"\"插入的变化。

```
mysql> set sql_mode='ansi';
Query OK, 0 rows affected (0.00 sec)

mysql> select @@sql_mode;
+--------------------------------------------------------------------------+
| @@sql_mode                                                               |
+--------------------------------------------------------------------------+
| REAL_AS_FLOAT,PIPES_AS_CONCAT,ANSI_QUOTES,IGNORE_SPACE,ONLY_FULL_GROUP_BY,ANSI |
+--------------------------------------------------------------------------+
1 row in set (0.00 sec)

mysql> create table t_sql_mode_bs (context varchar(20));
Query OK, 0 rows affected (0.04 sec)

mysql> insert into t_sql_mode_bs  values('\beijing');
Query OK, 1 row affected (0.00 sec)

mysql> select * from t_sql_mode_bs;
+---------+
| context |
+---------+
|eijing   |
+---------+
1 row in set (0.00 sec)

mysql> insert into t_sql_mode_bs  values('\\beijing');
Query OK, 1 row affected (0.00 sec)

mysql> select * from t_sql_mode_bs;
+----------+
| context  |
+----------+
|eijing    |
| \beijing |
+----------+
2 rows in set (0.00 sec)

mysql> set sql_mode = 'REAL_AS_FLOAT,PIPES_AS_CONCAT,ANSI_QUOTES,IGNORE_SPACE,ONLY_FULL_GROUP_BY,ANSI,NO_BACKSLASH_ESCAPES';
Query OK, 0 rows affected (0.00 sec)

mysql> mysql> select @@sql_mode;
+------------------------------------------------------------------------------------------+
| @@sql_mode                                                                               |
+------------------------------------------------------------------------------------------+
| REAL_AS_FLOAT,PIPES_AS_CONCAT,ANSI_QUOTES,IGNORE_SPACE,ONLY_FULL_GROUP_BY,ANSI,NO_BACKSLASH_ESCAPES |
+------------------------------------------------------------------------------------------+
1 row in set (0.00 sec)

mysql> insert into t_sql_mode_bs  values('\\beijing');
Query OK, 1 row affected (0.00 sec)

mysql> select * from t_sql_mode_bs;
+----------+
| context  |
```

```
+----------+
|eijing    |
| \beijing |
| \\beijing|
+----------+
3 rows in set (0.00 sec)
```

通过上面的示例可以看到，当在 ANSI 模式中增加了 NO_BACKSLASH_ESCAPES 模式后，反斜线变为了普通字符。如果导入的数据存在反斜线，可以设置此模式，保证导入数据的正确性。

（3）启用 PIPES_AS_CONCAT 模式。将"||"视为字符串连接操作符，在 Oracle 等数据库中，"||"被视为字符串的连接操作符，所以，在其他数据库中含有"||"操作符的 SQL 在 MySQL 中将无法执行。为了解决这个问题，MySQL 提供了 PIPES_AS_CONCAT 模式。

下面通过示例来介绍一下 PIPES_AS_CONCAT 模式的作用。

```
mysql> set sql_mode='ansi';
Query OK, 0 rows affected (0.00 sec)

mysql> select @@sql_mode;
+-------------------------------------------------------------------------------+
| @@sql_mode                                                                    |
+-------------------------------------------------------------------------------+
| REAL_AS_FLOAT,PIPES_AS_CONCAT,ANSI_QUOTES,IGNORE_SPACE,ONLY_FULL_GROUP_BY,ANSI |
+-------------------------------------------------------------------------------+
1 row in set (0.00 sec)

mysql> select 'beijing'||'2018'  ;
+------------------+
| 'beijing'||'2018'|
+------------------+
| beijing2018      |
+------------------+
1 row in set (0.01 sec)
```

通过上面的示例可以看到，ANSI 模式中包含了 PIPES_AS_CONCAT 模式，所以默认情况下，MySQL 版本支持将"||"视为字符串连接操作符。

需要注意的是，在分区表和主从复制环境中，要谨慎修改 SQL Mode。前者可能导致数据的写入逻辑发生变化，新的逻辑可能导致同一条数据在不同的 SQL Mode 下写入不同的分区；如果主从服务器的 SQL Mode 不同，后者会导致复制的数据在主从服务器上写入逻辑不同。这两种情况都可能导致数据的混乱。

13.3 常用的 SQL Mode

熟悉并了解经常使用的 SQL Mode 会帮助用户更好地使用它。表 13-2 总结了常用的 SQL Mode 值及其说明。

表 13-2　　　　　　　　　　MySQL 中的 SQL Mode

sql_mode 值	描　　述
ANSI	等同于 REAL_AS_FLOAT、PIPES_AS_CONCAT、ANSI_QUOTES、IGNORE_SPACE、ONLY_FULL_GROUP_BY 和 ANSI 组合模式，这种模式遇到异常时倾向于警告而不是立刻返回错误
STRICT_TRANS_TABLES	STRICT_TRANS_TABLES 适用于事务表和非事务表，它是严格模式，不允许非法日期，也不允许超过字段长度的值插入字段中，对于插入不正确的值给出错误而不是警告。MySQL 5.7 版本后添加到默认的 SQL Mode 中

13.3 常用的 SQL Mode

续表

sql_mode 值	描 述
TRADITIONAL	TRADITIONAL 模式等同于 STRICT_TRANS_TABLES、STRICT_ALL_TABLES、NO_ZERO_IN_DATE、NO_ZERO_DATE、ERROR_FOR_DIVISION_BY_ZERO、NO_AUTO_CREATE_USER 和 NO_ENGINE_SUBSTITUTION 的组合模式，所以它也是严格模式，对于插入不正确的值会给出错误而不是警告。可以应用在事务表和非事务表，用在事务表时，只要出现错误就会立即回滚

可以发现，表格中第一列 SQL Mode 的值大都是一些原子模式的组合，类似于角色和权限的关系。这样当实际应用时，只需要设置一个模式组合，就可以设置很多的原子模式，大大简化了用户的工作。

其中 TRADITIONAL 和 MySQL 5.7 的默认模式很相似，都属于严格模式。两者的主要区别在于 TRADITIONAL 包含了原子模式 STRICT_ALL_TABLES。STRICT_ALL_TABLES 和 STRICT_TRANS_TABLES 非常类似，对于事物表（比如 InnoDB）的写入规则完全一致，但对于非事务表有细微的差别，如下例所示：

（1）创建非事物表 t1_myisam，存储引擎设置为 MyISAM。

```
mysql> create table t1_myisam (id int,name varchar(4)) engine=myisam;
Query OK, 0 rows affected (0.05 sec)
```

（2）设置 SQL Mode 为 STRICT_TRANS_TABLES。

```
mysql> set sql_mode='STRICT_TRANS_TABLES';
Query OK, 0 rows affected, 1 warning (0.00 sec)
```

（3）分别写入下面的数据，发现第一条正常，第二条有警告，第三条被拒绝。

```
mysql> insert into t1_myisam values(1,'z1');
Query OK, 1 row affected (0.01 sec)

mysql> insert into t1_myisam values(10,'z1'),(20,'z1'),(30,'z111111111111');
Query OK, 3 rows affected, 1 warning (0.00 sec)
Records: 3  Duplicates: 0  Warnings: 1

mysql> insert into t1_myisam values(10,'z10000000000'),(20,'z1'),(30,'z111111111111');
ERROR 1406 (22001): Data too long for column 'name' at row 1
```

（4）查询表发现全部写入，但 name 为 "z111111111111" 的被截断为 "z111"。

```
mysql> select * from t1_myisam;
+------+------+
| id   | name |
+------+------+
|    1 | z1   |
|   10 | z1   |
|   20 | z1   |
|   30 | z111 |
+------+------+
4 rows in set (0.00 sec)
```

（5）更改 SQL Mode 为 STRICT_ALL_TABLES。

```
mysql> set sql_mode='STRICT_ALL_TABLES';
Query OK, 0 rows affected, 1 warning (0.00 sec)
```

（6）写入刚才同样的数据，发现第一条正常，第二条和第三条被拒绝。

```
mysql> insert into t1_myisam values(1,'z1');
Query OK, 1 row affected (0.00 sec)

mysql> insert into t1_myisam values(10,'z1'),(20,'z1'),(30,'z111111111111');
ERROR 1406 (22001): Data too long for column 'name' at row 3
```

```
mysql> insert into t1_myisam values(10,'z10000000000'),(20,'z1'),(30,'z1111111111111');
ERROR 1406 (22001): Data too long for column 'name' at row 1
```

从上例可以看出，对于非事物表，如果一次写入多行记录，在 STRICT_TRANS_TABLES 模式下，只要多行中的第一条写入成功，那么后面的记录即使违反了严格模式的约束，也会自动转换为最接近的数据写入成功；而在 STRICT_ALL_TABLES 模式下则相反，只要多行记录中的任意一行违反严格模式的约束，本次的所有记录都会写入失败。显而易见，两者各有利弊，前者可能导致数据的异常，后者可能导致事物的不一致。避免这种问题的最好办法是让数据逐条写入，读者在开发中一定要注意。

13.4 SQL Mode 在迁移中如何使用

如果 MySQL 与其他异构数据库之间有数据迁移的需求，那么 MySQL 中提供的数据库组合模式就会对数据迁移过程有所帮助。

从表 13-3 中可以看出，MySQL 提供了很多数据库的组合模式名称，例如 "ORACLE" "DB2" 等。这些模式组合是由很多小的 sql_mode 组合而成，在异构数据库之间迁移数据时可以尝试使用这些模式来导出适合于目标数据库格式的数据，这样就使得导出数据更容易导入目标数据库。

表 13-3　　　　　　　　　　MySQL 中的常用数据库 Mode

组合后的模式名称	组合模式中的各个 sql_mode
DB2	PIPES_AS_CONCAT、ANSI_QUOTES、IGNORE_SPACE、NO_KEY_OPTIONS、NO_TABLE_OPTIONS、NO_FIELD_OPTIONS
MAXDB	PIPES_AS_CONCAT、ANSI_QUOTES、IGNORE_SPACE、NO_KEY_OPTIONS、NO_TABLE_OPTIONS、NO_FIELD_OPTIONS、NO_AUTO_CREATE_USER
MSSQL	PIPES_AS_CONCAT、ANSI_QUOTES、IGNORE_SPACE、NO_KEY_OPTIONS、NO_TABLE_OPTIONS、NO_FIELD_OPTIONS
ORACLE	PIPES_AS_CONCAT、ANSI_QUOTES、IGNORE_SPACE、NO_KEY_OPTIONS、NO_FIELD_OPTIONS、NO_AUTO_CREATE_USER
POSTGRESQL	PIPES_AS_CONCAT、ANSI_QUOTES、IGNORE_SPACE、NO_KEY_OPTIONS、NO_TABLE_OPTIONS、NO_FIELD_OPTIONS

在数据迁移过程中，可以设置 SQL Mode 为 NO_TABLE_OPTIONS 模式，这样将去掉 show create table 中的 "engine" 关键字，获得通用的建表脚本。

测试示例如下：

```
mysql> show create table emp \G;
*************************** 1. row ***************************
       Table: emp
Create Table: CREATE TABLE 'emp' (
  'ename' varchar(20) DEFAULT NULL,
  'hiredate' date DEFAULT NULL,
  'sal' decimal(10,2) DEFAULT NULL,
  'deptno' int(2) DEFAULT NULL
) ENGINE=InnoDB DEFAULT CHARSET=utf8
1 row in set (0.00 sec)

mysql> set session sql_mode='NO_TABLE_OPTIONS';
Query OK, 0 rows affected (0.00 sec)

mysql>  show create table emp \G;
```

```
*************************** 1. row ***************************
       Table: emp
Create Table: CREATE TABLE 'emp' (
  'ename' varchar(20) DEFAULT NULL,
  'hiredate' date DEFAULT NULL,
  'sal' decimal(10,2) DEFAULT NULL,
  'deptno' int(2) DEFAULT NULL
)
1 row in set (0.00 sec)
```

13.5 小结

本章介绍了 SQL Mode 的含义以及实际用途，重点讨论了以下内容。

- SQL Mode 的"严格模式"为 MySQL 提供了很好的数据校验功能，保证了数据的准确性，TRADITIONAL 和 STRICT_TRANS_TABLES 是常用的两种严格模式，要注意两者的区别。
- SQL Mode 的多种模式可以灵活组合，组合后的模式可以更好地满足应用程序的需求。尤其在数据迁移中，SQL Mode 的使用更为重要。

第 14 章 MySQL 分区

MySQL 从 5.1 版本起开始支持分区的功能。分区是指根据一定的规则，数据库把一个表分解成多个更小的、更容易管理的部分。就访问数据库的应用而言，逻辑上只有一个表或一个索引，但是实际上这个表可能由数十个物理分区对象组成，每个分区都是一个独立的对象，可以独自处理，可以作为表的一部分进行处理。分区对应用来说是完全透明的，不影响应用的业务逻辑。

14.1 分区概述

分区有利于管理非常大的表，它采用了"分而治之"的逻辑。分区引入了分区键（Partition Key）的概念，分区键用于根据某个区间值（或者范围值）、特定值列表或者 HASH 函数值执行数据的聚集，让数据根据规则分布在不同的分区中，让一个大对象变成一些小对象。

MySQL 分区的优点主要包括以下 4 个方面。

- 和单个磁盘或者文件系统分区相比，可以存储更多数据。
- 优化查询。在 Where 子句中包含分区条件时，可以只扫描必要的一个或多个分区来提高查询效率；同时在涉及 SUM()和 COUNT()这类聚合函数的查询时，可以容易地在每个分区上并行处理，最终只需要汇总所有分区得到的结果。在 MySQL 5.7 中，还可以通过类似 SELECT * FROM t PARTITION (p0,p1)这样的方式来显式查询指定分区的数据。
- 对于已经过期或者不需要保存的数据，可以通过删除与这些数据有关的分区来快速删除数据。
- 跨多个磁盘来分散数据查询，以获得更大的查询吞吐量。

在 MySQL 5.7 中，通过二进制安装会默认包含分区支持，如果通过源码编译安装，那么需要在编译时指定参数 DWITH_PARTITION_STORAGE_ENGINE。

可以通过 SHOW PLUGINS 命令或者查询 PLUGINS 字典表来确定当前的 MySQL 是否支持分区，例如：

```
mysql> select * from information_schema.plugins where PLUGIN_NAME='partition'\G
*************************** 1. row ***************************
           PLUGIN_NAME: partition
        PLUGIN_VERSION: 1.0
         PLUGIN_STATUS: ACTIVE
           PLUGIN_TYPE: STORAGE ENGINE
   PLUGIN_TYPE_VERSION: 50722.0
        PLUGIN_LIBRARY: NULL
```

```
        PLUGIN_LIBRARY_VERSION: NULL
                 PLUGIN_AUTHOR: Mikael Ronstrom, MySQL AB
            PLUGIN_DESCRIPTION: Partition Storage Engine Helper
                PLUGIN_LICENSE: GPL
                   LOAD_OPTION: ON
1 row in set (0.00 sec)
```

通过查到的 partition 插件信息，如果 PLUGIN_STATUS 列的状态是 ACTIVE，说明 MySQL 已经开启了分区功能。

MySQL 支持大部分存储引擎（比如 MyISAM、InnoDB、Memory 等）来创建分区表；但不支持 MERGE、CSV 和 FEDERATED 这 3 类存储引擎。在 MySQL 5.7 版本中，同一个分区表的所有分区必须使用同一个存储引擎，且分区数量不能超过 8192（NDB 存储引擎除外），即同一个表上，不能对一个分区使用 MyISAM 引擎，对另一个分区使用 InnoDB。但是，可以在同一个 MySQL 服务器中，甚至同一个数据库中，对于不同的分区表使用不同的存储引擎。

和非分区表设置存储引擎一样，分区表设置存储引擎，只能用[STORAGE]ENGINE 子句。[STORAGE]ENGINE 子句必须列在 CREATE TABLE 语句中的其他任何分区选项之前。例如，下面的例子创建了一个使用 InnoDB 引擎并有 6 个 HASH 分区的表：

```
mysql> CREATE TABLE emp (empid INT, salary DECIMAL(7,2), birth_date DATE)
    ->        ENGINE=INNODB
    ->        PARTITION BY HASH( MONTH(birth_date) )
    ->        PARTITIONS 6;
Query OK, 0 rows affected (0.11 sec)
```

> 注意：MySQL 的分区适用于一个表的所有数据和索引，不能只对表数据分区而不对索引分区；反过来也是一样的，不能只对索引分区而不对表分区，同时也不能只对表的一部分数据进行分区。MySQL 的分区表上创建的索引一定是本地 LOCAL 索引。

14.2 分区类型

本节主要讨论在 MySQL 5.7 中可用的分区类型，主要有以下 6 种。

- RANGE 分区：基于一个给定连续区间范围，把数据分配到不同的分区。
- LIST 分区：类似 RANGE 分区，区别在 LIST 分区是基于枚举出的值列表分区，RANGE 是基于给定的连续区间范围分区。
- COLUMNS 分区：类似于 RANGE 和 LIST，区别在于分区键既可以是多列，又可以是非整数。
- HASH 分区：基于给定的分区个数，把数据取模分配到不同的分区。
- KEY 分区：类似于 HASH 分区，但使用 MySQL 提供的哈希函数。
- 子分区：也叫作复合分区或者组合分区，即在主分区下再做一层分区，将数据再次分割。

在 MySQL 5.7 中，RANGE 分区、LIST 分区、HASH 分区都要求分区键必须是 INT 类型，或者通过表达式返回 INT 类型，但 KEY 和 COLUMNS 分区例外，可以使用其他类型的列（BLOB 或 TEXT 列类型除外）作为分区键。如果希望在 RANGE 和 LIST 类型的分区中使用非 INT 列作为分区键，可以选择 COLUMNS 分区。

无论是哪种 MySQL 分区类型，要么分区表上没有主键/唯一键，要么分区表的主键/唯一键都必须包含分区键，也就是说不能使用主键/唯一键字段之外的其他字段分区，例如 emp 表的主键为 id 字段，在尝试通过 store_id 字段分区的时候，MySQL 会提示返回失败：

```
mysql> CREATE TABLE emp (
    ->         id INT NOT NULL,
    ->         ename VARCHAR(30),
    ->         hired DATE NOT NULL DEFAULT '1970-01-01',
    ->         separated DATE NOT NULL DEFAULT '9999-12-31',
    ->         job VARCHAR(30) NOT NULL,
    ->         store_id INT NOT NULL,
    ->         PRIMARY KEY (id)
    ->     )
    ->     PARTITION BY RANGE (store_id) (
    ->         PARTITION p0 VALUES LESS THAN (10),
    ->         PARTITION p1 VALUES LESS THAN (20),
    ->         PARTITION p2 VALUES LESS THAN (30)
    ->     );
ERROR 1503 (HY000): A PRIMARY KEY must include all columns in the table's partitioning function
```

去掉主键约束后，创建表就会成功：

```
mysql> CREATE TABLE emp (
    ->         id INT NOT NULL,
    ->         ename VARCHAR(30),
    ->         hired DATE NOT NULL DEFAULT '1970-01-01',
    ->         separated DATE NOT NULL DEFAULT '9999-12-31',
    ->         job VARCHAR(30) NOT NULL,
    ->         store_id INT NOT NULL
    ->     )
    ->     PARTITION BY RANGE (store_id) (
    ->         PARTITION p0 VALUES LESS THAN (10),
    ->         PARTITION p1 VALUES LESS THAN (20),
    ->         PARTITION p2 VALUES LESS THAN (30)
    ->     );
Query OK, 0 rows affected (0.05 sec)
```

分区的名字基本上遵循 MySQL 标识符的原则。说到命名，顺便介绍一下 MySQL 命名中的大小写敏感：在 MySQL 中，对于库名和表名是否大小写敏感，可以通过下面的方法来查看：

```
mysql> show variables like '%lower_case%';
+------------------------+-------+
| Variable_name          | Value |
+------------------------+-------+
| lower_case_file_system | OFF   |
| lower_case_table_names | 1     |
+------------------------+-------+
2 rows in set (0.00 sec)
```

这里 lower_case_file_system 是一个不可以修改的变量，代表了操作系统是否大小写敏感，OFF 代表敏感，ON 代表不敏感。lower_case_table_names 代表数据库的库名和表名是否大小写敏感，1 代表不敏感，0 代表敏感。如果操作系统大小写敏感（例如 Linux），那么库名和表名大小写是否敏感就由 lower_case_table_names 参数来决定，如果操作系统大小写不敏感（例如 Windows），那么 lower_case_table_names 应该设置为 1。否则可能会遇到数据库实例挂起、崩溃、启动报错等问题。

但是需要注意的是，列名、别名、分区名这些是不区分大小写的。例如，无论 lower_case_table_names 的值如何设置，下面的 CREATE TABLE 语句都将会产生错误：

```
mysql> CREATE TABLE t2 (val INT)
    ->     PARTITION BY LIST(val)(
    ->         PARTITION mypart VALUES IN (1,3,5),
    ->         PARTITION MyPart VALUES IN (2,4,6)
    ->     );
ERROR 1517 (HY000): Duplicate partition name mypart
```

这是因为 MySQL 认为分区名字 mypart 和 MyPart 没有区别。

14.2.1 RANGE 分区

按照 RANGE 分区的表是利用取值范围将数据分成分区，区间要连续并且不能互相重叠，使用 VALUES LESS THAN 操作符进行分区定义。

例如，雇员表 emp 中按商店 ID store_id 进行 RANGE 分区：

```
mysql> CREATE TABLE emp (
    ->     id INT NOT NULL,
    ->     ename VARCHAR(30),
    ->     hired DATE NOT NULL DEFAULT '1970-01-01',
    ->     separated DATE NOT NULL DEFAULT '9999-12-31',
    ->     job VARCHAR(30) NOT NULL,
    ->     store_id INT NOT NULL
    -> )
    -> PARTITION BY RANGE (store_id) (
    ->     PARTITION p0 VALUES LESS THAN (10),
    ->     PARTITION p1 VALUES LESS THAN (20),
    ->     PARTITION p2 VALUES LESS THAN (30)
    -> );
Query OK, 0 rows affected (0.05 sec)
```

按照这种分区方案，在商店 1~9 工作的雇员相对应的所有行被保存在分区 P0 中，商店 10~19 的雇员保存在 P1 中，依次类推。注意，每个分区都是按顺序进行定义的，从最低到最高。这是 PARTITION BY RANGE 语法的要求；类似 Java 或者 C 中的 "switch case" 语句。

这时，如果增加了商店 ID 大于等于 30 的行，会出现错误，因为没有规则包含商店 ID 大于等于 30 的行，服务器不知道应该把记录保存在哪里。

```
mysql> insert into emp(id, ename, hired, job, store_id) values ('7934', 'MILLER', '1982-01-23', 'CLERK', 50);
ERROR 1526 (HY000): Table has no partition for value 50
```

可以在设置分区的时候使用 VALUES LESS THAN MAXVALUE 子句，该子句提供给所有大于明确指定的最高值的值，MAXVALUE 表示最大的可能的整数值。例如，增加 P3 分区存储所有商店 ID 大于等于 30 的行之后再执行插入语句就没有问题了：

```
mysql> alter table emp add partition (partition p3 VALUES LESS THAN MAXVALUE);
Query OK, 0 rows affected (0.21 sec)
Records: 0  Duplicates: 0  Warnings: 0

mysql> insert into emp(id, ename, hired, job, store_id) values ('7934', 'MILLER', '1982-01-23', 'CLERK', 50);
Query OK, 1 row affected (0.04 sec)
```

MySQL 支持在 VALUES LESS THAN 子句中使用表达式，比如，以日期作为 RANGE 分区的分区列：

```
mysql> CREATE TABLE emp_date (
    ->     id INT NOT NULL,
    ->     ename VARCHAR(30),
    ->     hired DATE NOT NULL DEFAULT '1970-01-01',
    ->     separated DATE NOT NULL DEFAULT '9999-12-31',
    ->     job VARCHAR(30) NOT NULL,
    ->     store_id INT NOT NULL
    -> )
    -> PARTITION BY RANGE (YEAR(separated)) (
    ->     PARTITION p0 VALUES LESS THAN (1995),
    ->     PARTITION p1 VALUES LESS THAN (2000),
    ->     PARTITION p2 VALUES LESS THAN (2005)
    -> );
Query OK, 0 rows affected (0.08 sec)
```

注意：在 RANGE 分区中，分区键如果是 NULL 值会被当作一个最小值来处理，在 14.2.7 节中有详细的说明。

MySQL 从 5.5 版本开始，改进了 RANGE 分区功能，提供了 RANGE COLUMNS 分区支持非整数分区，例如：

```
mysql> CREATE TABLE emp_date(
    ->     id INT NOT NULL,
    ->     ename VARCHAR(30),
    ->     hired DATE NOT NULL DEFAULT '1970-01-01',
    ->     separated DATE NOT NULL DEFAULT '9999-12-31',
    ->     job VARCHAR(30) NOT NULL,
    ->     store_id INT NOT NULL
    -> )
    -> PARTITION BY RANGE COLUMNS (separated) (
    ->     PARTITION p0 VALUES LESS THAN ('1996-01-01'),
    ->     PARTITION p1 VALUES LESS THAN ('2001-01-01'),
    ->     PARTITION p2 VALUES LESS THAN ('2006-01-01')
    -> );
Query OK, 0 rows affected (0.07 sec)
```

RANGE 分区功能特别适用于以下两种情况。

● 当需要删除过期的数据时，只需要简单的 ALTER TABLE emp DROP PARTITION p0 来删除 p0 分区中的数据。对于具有上百万条记录的表来说，删除分区要比运行一个 DELETE 语句有效得多。

● 经常运行包含分区键的查询，MySQL 可以很快地确定只有某一个或者某些分区需要扫描，因为其他分区不可能包含有符合该 WHERE 子句的任何记录。例如，检索商店 ID 大于等于 25 的记录数，MySQL 只需要扫描 p2 分区即可。

```
mysql> explain partitions select count(1) from emp where store_id >= 25\G
*************************** 1. row ***************************
           id: 1
  select_type: SIMPLE
        table: emp
   partitions: p2
         type: ALL
possible_keys: NULL
          key: NULL
      key_len: NULL
          ref: NULL
         rows: 5
        Extra: Using where
1 row in set (0.00 sec)
```

14.2.2 LIST 分区

LIST 分区是建立离散的值列表告诉数据库特定的值属于哪个分区，LIST 分区在很多方面类似于 RANGE 分区，区别在于 LIST 分区是从属于一个枚举列表的值的集合，RANGE 分区是从属于一个连续区间值的集合。

LIST 分区通过使用 PARTITION BY LIST(expr)子句来实现，expr 是某列值或一个基于某列值返回一个整数值的表达式，然后通过 VALUES IN(value_list)的方式来定义分区，其中 value_list 是一个逗号分隔的整数列表。与 RANGE 分区不同，LIST 分区不必声明任何特定的顺序，例如：

```
mysql> CREATE TABLE expenses (
    ->     expense_date DATE NOT NULL,
    ->     category INT,
```

14.2 分区类型

```
    ->     amount DECIMAL (10,3)
    -> )PARTITION BY LIST(category) (
    ->     PARTITION p0 VALUES IN (3, 5),
    ->     PARTITION p1 VALUES IN (1, 10),
    ->     PARTITION p2 VALUES IN (4, 9),
    ->     PARTITION p3 VALUES IN (2),
    ->     PARTITION p4 VALUES IN (6)
    -> );
Query OK, 0 rows affected (0.09 sec)
```

如果试图插入的列值（或者分区表达式的返回值）不包含分区值列表中时，那么 INSERT 操作会失败并报错。要重点注意的是，LIST 分区不存在类似 VALUES LESS THAN MAXVALUE 这样包含其他值在内的定义方式。将要匹配的任何值都必须在值列表中找得到。

如果要使用非整数分区，则可以创建 LIST COLUMNS 分区：

```
mysql> CREATE TABLE expenses (
    ->     expense_date DATE NOT NULL,
    ->     category VARCHAR(30),
    ->     amount DECIMAL (10,3)
    -> )PARTITION BY LIST COLUMNS (category) (
    ->     PARTITION p0 VALUES IN ( 'lodging', 'food'),
    ->     PARTITION p1 VALUES IN ( 'flights', 'ground transportation'),
    ->     PARTITION p2 VALUES IN ( 'leisure', 'customer entertainment'),
    ->     PARTITION p3 VALUES IN ( 'communications'),
    ->     PARTITION p4 VALUES IN ( 'fees')
    -> );
Query OK, 0 rows affected (0.07 sec)
```

14.2.3 COLUMNS 分区

COLUMNS 分区是在 MySQL 5.5 引入的分区类型。在 MySQL 5.5 版本之前，RANGE 分区和 LIST 分区只支持整数分区，从而需要额外的函数计算来得到整数或者通过额外的转换表来转换为整数再分区。COLUMNS 分区解决了这一问题。

COLUMNS 分区可以细分为 RANGE COLUMNS 分区和 LIST COLUMNS 分区，两者都支持整数、日期时间、字符串三大数据类型。

- 整数类型：tinyint、smallint、mediumint、int 和 bigint；其他数值类型都不支持，例如不支持 Decimal 和 Float。
- 日期时间类型：date 和 datetime。
- 字符类型：char、varchar、binary 和 varbinary；不支持 text 和 blob 类型作为分区键。

注意：在 MySQL 5.7 中，COLUMNS 分区仅支持一个或者多个字段名作为分区键，不支持表达式作为分区键，区别于 RANGE 分区和 LIST 分区。

对比 RANGE 分区和 LIST 分区，COLUMNS 分区的亮点除了支持数据类型增加之外，还支持多列分区。例如，创建了一个根据字段 a、b 组合的 RANGE COLUMNS 分区：

```
mysql> CREATE TABLE rc3 (
    ->     a INT,
    ->     b INT
    -> )
    -> PARTITION BY RANGE COLUMNS(a, b) (
    ->     PARTITION p01 VALUES LESS THAN (0,10),
    ->     PARTITION p02 VALUES LESS THAN (10,10),
    ->     PARTITION p03 VALUES LESS THAN (10,20),
    ->     PARTITION p04 VALUES LESS THAN (10,35),
    ->     PARTITION p05 VALUES LESS THAN (10,MAXVALUE),
    ->     PARTITION p06 VALUES LESS THAN (MAXVALUE, MAXVALUE)
```

```
    -> );
Query OK, 0 rows affected (0.07 sec)
```

需要注意的是，RANGE COLUMNS 分区键的比较是基于元组的比较，也就是基于字段组的比较，这和之前 RANGE 分区键的比较有些差异，我们写入几条测试数据并观察测试数据的分区分布情况来看一看。

- 写入 a=1, b=10 的记录，从 information_schema.partitions 表发现数据实际上写入了 p02 分区，也就是说元组(1, 10) < (10, 10)，如图 14-1 所示。

```
mysql> insert into rc3 (a, b) values (1, 10);
Query OK, 1 row affected (0.00 sec)

mysql> select (1,10) < (10,10) from dual;
+-------------------+
| (1,10) < (10,10)  |
+-------------------+
|                 1 |
+-------------------+
1 row in set (0.01 sec)

mysql> SELECT
    ->   partition_name part,
    ->   partition_expression expr,
    ->   partition_description descr,
    ->   table_rows
    -> FROM
    ->   INFORMATION_SCHEMA.partitions
    -> WHERE
    ->   TABLE_SCHEMA = schema()
    ->   AND TABLE_NAME='rc3';
+------+----------+-------------------+------------+
| part | expr     | descr             | table_rows |
+------+----------+-------------------+------------+
| p01  | `a`,`b`  | 0,10              |          0 |
| p02  | `a`,`b`  | 10,10             |          1 |
| p03  | `a`,`b`  | 10,20             |          0 |
| p04  | `a`,`b`  | 10,35             |          0 |
| p05  | `a`,`b`  | 10,MAXVALUE       |          0 |
| p06  | `a`,`b`  | MAXVALUE,MAXVALUE |          0 |
+------+----------+-------------------+------------+
```

- 写入 a=10, b=9 的记录，从 information_schema.partitions 表能够发现数据实际上写入了 p02 分区，也就是说元组(10, 9) < (10, 10)，元组(10, 9)的大小判断不是简单地通过元组的首字段进行的，如图 14-2 所示。

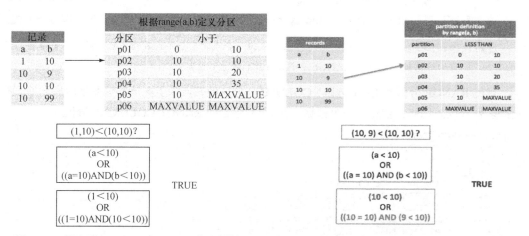

图 14-1　写入 RANGE COLUMNS 分区测试一　　图 14-2　写入 RANGE COLUMNS 分区测试二

14.2 分区类型

```
mysql> insert into rc3(a, b) values(10, 9);
Query OK, 1 row affected (0.00 sec)

mysql> select (10,9) < (10,10) from dual;
+-----------------+
| (10,9) < (10,10) |
+-----------------+
|               1 |
+-----------------+
1 row in set (0.00 sec)

mysql> SELECT
    ->    partition_name part,
    ->    partition_expression expr,
    ->    partition_description descr,
    ->    table_rows
    -> FROM
    ->    INFORMATION_SCHEMA.partitions
    -> WHERE
    ->    TABLE_SCHEMA = schema()
    ->    AND TABLE_NAME='rc3';
+------+---------+-------------------+------------+
| part | expr    | descr             | table_rows |
+------+---------+-------------------+------------+
| p01  | `a`,`b` | 0,10              |          0 |
| p02  | `a`,`b` | 10,10             |          2 |
| p03  | `a`,`b` | 10,20             |          0 |
| p04  | `a`,`b` | 10,35             |          0 |
| p05  | `a`,`b` | 10,MAXVALUE       |          0 |
| p06  | `a`,`b` | MAXVALUE,MAXVALUE |          0 |
+------+---------+-------------------+------------+
6 rows in set (0.01 sec)
```

- 写入 a=10, b=10 的记录，从 information_schema.partitions 表能够发现数据实际上写入了 p03 分区，也就是说元组(10, 10) <= (10, 10) <(10, 20)，如图 14-3 所示。

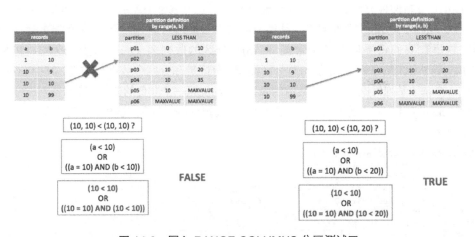

图 14-3 写入 RANGE COLUMNS 分区测试三

```
mysql> insert into rc3(a, b) values(10,10);
Query OK, 1 row affected (0.00 sec)

mysql> select (10,10) < (10,10) from dual;
+-----------------+
| (10,10) < (10,10) |
+-----------------+
|                0 |
+-----------------+
1 row in set (0.00 sec)
```

```
mysql> SELECT
    ->     partition_name part,
    ->     partition_expression expr,
    ->     partition_description descr,
    ->     table_rows
    -> FROM
    ->     INFORMATION_SCHEMA.partitions
    -> WHERE
    ->     TABLE_SCHEMA = schema()
    ->     AND TABLE_NAME='rc3';
+------+---------+-------------------+------------+
| part | expr    | descr             | table_rows |
+------+---------+-------------------+------------+
| p01  | `a`,`b` | 0,10              |          0 |
| p02  | `a`,`b` | 10,10             |          2 |
| p03  | `a`,`b` | 10,20             |          1 |
| p04  | `a`,`b` | 10,35             |          0 |
| p05  | `a`,`b` | 10,MAXVALUE       |          0 |
| p06  | `a`,`b` | MAXVALUE,MAXVALUE |          0 |
+------+---------+-------------------+------------+
6 rows in set (0.00 sec)
```

其实，RANGE COLUMNS 分区键的比较（元组的比较）其实就是多列排序，先根据 a 字段排序再根据 b 字段排序，根据排序结果来分区存放数据。和 RANGE 单字段分区排序的规则实际上是一致的。

14.2.4 HASH 分区

HASH 分区主要用来分散热点读，确保数据在预先确定个数的分区中尽可能平均分布。对一个表执行 HASH 分区时，MySQL 会对分区键应用一个散列函数，以此确定数据应当放在 N 个分区中的哪个分区中。

MySQL 支持两种 HASH 分区，即常规 HASH 分区和线性 HASH 分区（LINEAR HASH 分区）。常规 HASH 使用的是取模算法，线性 HASH 分区使用的是一个线性的 2 的幂的运算法则。

在这里，我们将要创建一个常规 HASH 分区的散列表 emp，使用 PARTITION BY HASH(*expr*) PARTITIONS num 子句对分区类型、分区键和分区个数进行定义，其中 *expr* 是某列值或一个基于某列值返回一个整数值的表达式；num 是一个非负的整数，表示分割成分区的数量，默认 num 为 1。下面的 SQL 语句创建了一个基于 store_id 列 HASH 分区的表，表被分成了 4 个分区。

```
mysql> CREATE TABLE emp (
    ->     id INT NOT NULL,
    ->     ename VARCHAR(30),
    ->     hired DATE NOT NULL DEFAULT '1970-01-01',
    ->     separated DATE NOT NULL DEFAULT '9999-12-31',
    ->     job VARCHAR(30) NOT NULL,
    ->     store_id INT NOT NULL
    -> )
    -> PARTITION BY HASH (store_id) PARTITIONS 4;
Query OK, 0 rows affected (0.05 sec)
```

其实对于一个表达式"expr"，我们是可以计算出它会被保存在哪个分区中。假设将要保存记录的分区编号为 N，那么"N = MOD(expr, num)"。例如，emp 表有 4 个分区，插入一个 store_id 列值为 234 的记录到 emp 表中：

```
mysql> insert into emp values (1, 'Tom', '2010-10-10', '9999-12-31', 'Clerk', 234);
Query OK, 0 rows affected (0.05 sec)
```

234 取模运算如下：

```
MOD(234,4) = 2
```

也就是 store_id = 234 这条记录将会被保存到第二个分区，图 14-4 显示了 MySQL 会检查 store_id 中的值、计算散列、确定给定行会出现在哪个分区。

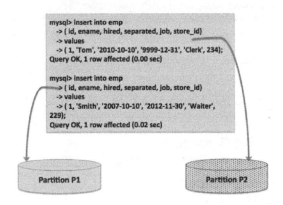

图 14-4　MySQL 写入 Hash 分区

通过执行计划可以确定 store_id = 234 这条记录存储在第二个分区内：

```
sql> explain partitions select * from emp where store_id = 234\G
*************************** 1. row ***************************
           id: 1
  select_type: SIMPLE
        table: emp
   partitions: p2
         type: ALL
possible_keys: NULL
          key: NULL
      key_len: NULL
          ref: NULL
         rows: 2
        Extra: Using where
1 row in set (0.00 sec)
```

表达式"expr"可以是 MySQL 中有效的任何函数或者其他表达式，只要它们返回一个既非常数也非随机数的整数。每当插入/更新/删除一行数据时，这个表达式都需要计算一次，这意味着非常复杂的表达式可能会引起性能问题，MySQL 也不推荐使用涉及多列的哈希表达式。

常规 HASH 分区方式看上去挺不错的，通过取模的方式来使数据尽可能平均分布在每个分区中，让每个分区管理的数据都减少了，提高了查询的效率；可是当我们需要增加分区或者合并分区的时候，问题就出现了。假设原来是 5 个常规 HASH 分区，现在需要新增一个常规 HASH 分区，原来的取模算法是 MOD(expr,5)，根据余数 0～4 分布在 5 个分区中，现在新增一个分区后，取模算法变成 MOD(expr,6)，根据余数 0～5 分区在 6 个分区中，原来 5 个分区中的数据大部分都需要通过重新计算重新分区。常规 HASH 在分区管理上带来的代价太大了，不适合需要灵活变动分区的需求。为了降低分区管理上的代价，MySQL 提供了线性 HASH 分区，分区函数是一个线性的 2 的幂的运算法则。

线性 HASH 分区和常规 HASH 分区在语法上的唯一区别是在"PARTITION BY"子句中添加"LINEAR"关键字，例如：

```
mysql> CREATE TABLE emp (
    ->           id INT NOT NULL,
```

```
    ->          ename VARCHAR(30),
    ->          hired DATE NOT NULL DEFAULT '1970-01-01',
    ->          separated DATE NOT NULL DEFAULT '9999-12-31',
    ->          job VARCHAR(30) NOT NULL,
    ->          store_id INT NOT NULL
    ->      )
    ->      PARTITION BY LINEAR HASH (store_id) PARTITIONS 4;
Query OK, 0 rows affected (0.02 sec)
```

同样的，使用线性 HASH 时，指定记录保存在哪个分区是可以计算出来的，假设将要保存记录的分区编号为 N，num 是一个非负的整数，表示分割成分区的数量，那么 N 可以通过以下算法得到。

（1）首先，找到下一个大于等于 num 的 2 的幂，这个值设为 V，V 可以通过下面的公式得到：

$V = Power\ (2, Ceiling\ (Log\ (2, num)))$

例如，刚才创建的 emp 表预先设定了 4 个分区，也就是 num = 4。

$V = Power\ (2, Ceiling\ (Log\ (2, num)))$

　 $= Power\ (2, Ceiling\ (Log\ (2, 4)))$

　 $= Power\ (2, Ceiling\ (2))$

　 $= Power\ (2, 2)$

　 $= 4$

（2）其次，设置 $N = F(column\_list)\ \&\ (V - 1)$。

例如，我们刚才计算出 $V = 4$，现在计算 store_id = 234 对应的 N 值。

$N = F\ (column\_list)\ \&\ (V - 1)$

　 $= 234\ \&\ (4 - 1)$

　 $= 2$

（3）当 $N >=$ num 时，设置 $V = Ceiling(V/2)$，设置 $N = N\ \&\ (V - 1)$。

对于 store_id = 234 这条记录，由于 $N = 2 < 4$，所以直接就能够判断这条记录会被存储在第二个分区中。

有意思的是，当线性 HASH 的分区个数是 2 的 N 次幂时，线性 HASH 的分区结果和常规 HASH 的分区结果是一致的。

注意：由于负数取模较复杂，仅以非负整数 A 举例，模数 num 为 2 幂次方，那么数值 A 对 num 取模能够转换为位与运算：MOD (A, num) = A & (num - 1)。

假设分区个数 num 为 2 的幂次方，数值 A（A 为非负整数）的所在分区为 N(A)。

- 常规 HASH 分区时，保存数值 A 所在分区 N(A) = MOD(A, num) = A & (num - 1)。
- 线性 HASH 分区时，找到大于等于 num 的幂 V = Power (2, Ceiling (Log (2, num))) = num（num 本身就是 2 的幂次方）；其次，计算 N = A & (num - 1)，考虑到 A 为非负整数，N = A & (num - 1) = MOD(A, num)，也就是 N 为数值 A 对分区个数 num 取模的结果，容易判定 N < num，最终数值 A 所在的分区 N(A) = N = A & (num -1)。这与常规 HASH 分区计算得到结果一致。

线性 HASH 分区的优点是，在分区维护（包含增加、删除、合并、拆分分区）时，MySQL 能够处理得更加迅速；缺点是，对比常规 HASH 分区（取模）的时候，线性 HASH 各个分区之间数据的分布不太均衡。

14.2.5 KEY 分区

按照 KEY 进行分区非常类似于按照 HASH 进行分区，只不过 HASH 分区允许使用用户自定义的表达式，而 KEY 分区则不允许，需要使用 MySQL 服务器提供的 HASH 函数；同时 HASH 分区只支持整数分区，而 KEY 分区支持使用除 BLOB 和 Text 外其他类型的列作为分区键。

我们同样可以用 PARTITION BY KEY(expr)子句来创建一个 KEY 分区表，expr 是零个或者多个字段名名的列表。下面语句创建了一个基于 job 字段进行 KEY 分区的表，表被分成了 4 个分区：

```
mysql> CREATE TABLE emp (
    ->         id INT NOT NULL,
    ->         ename VARCHAR(30),
    ->         hired DATE NOT NULL DEFAULT '1970-01-01',
    ->         separated DATE NOT NULL DEFAULT '9999-12-31',
    ->         job VARCHAR(30) NOT NULL,
    ->         store_id INT NOT NULL
    -> )
    -> PARTITION BY KEY (job) PARTITIONS 4;
Query OK, 0 rows affected (0.04 sec)
```

与 HASH 分区不同，创建 Key 分区表的时候，可以不指定分区键，默认会首先选择使用主键作为分区键，例如：

```
mysql> CREATE TABLE emp (
    ->         id INT NOT NULL,
    ->         ename VARCHAR(30),
    ->         hired DATE NOT NULL DEFAULT '1970-01-01',
    ->         separated DATE NOT NULL DEFAULT '9999-12-31',
    ->         job VARCHAR(30) NOT NULL,
    ->         store_id INT NOT NULL,
    ->         PRIMARY KEY (id)
    -> )
    -> PARTITION BY KEY ( ) PARTITIONS 4;
Query OK, 0 rows affected (0.02 sec)
```

在没有主键的情况，会选择非空唯一键作为分区键：

```
mysql> drop table emp;
Query OK, 0 rows affected (0.02 sec)

mysql> CREATE TABLE emp (
    ->         id INT NOT NULL,
    ->         ename VARCHAR(30),
    ->         hired DATE NOT NULL DEFAULT '1970-01-01',
    ->         separated DATE NOT NULL DEFAULT '9999-12-31',
    ->         job VARCHAR(30) NOT NULL,
    ->         store_id INT NOT NULL,
    ->         UNIQUE KEY (id)
    -> )
    -> PARTITION BY KEY ( ) PARTITIONS 4;
Query OK, 0 rows affected (0.01 sec)
```

注意：作为分区键的唯一键必须是非空的，如果不是非空的，依然会报错。

```
mysql> CREATE TABLE emp (
    ->         id INT,
    ->         ename VARCHAR(30),
    ->         hired DATE NOT NULL DEFAULT '1970-01-01',
    ->         separated DATE NOT NULL DEFAULT '9999-12-31',
    ->         job VARCHAR(30) NOT NULL,
    ->         store_id INT NOT NULL,
    ->         UNIQUE KEY (id,ename)
    -> )
```

```
    -> PARTITION BY KEY ( ) PARTITIONS 4;
ERROR 1488 (HY000): Field in list of fields for partition function not found in table
```

在没有主键,也没有唯一键的情况下,就不能不指定分区键了:

```
mysql> drop table emp;
Query OK, 0 rows affected (0.01 sec)

mysql> CREATE TABLE emp (
    ->         id INT NOT NULL,
    ->         ename VARCHAR(30),
    ->         hired DATE NOT NULL DEFAULT '1970-01-01',
    ->         separated DATE NOT NULL DEFAULT '9999-12-31',
    ->         job VARCHAR(30) NOT NULL,
    ->         store_id INT NOT NULL
    ->     )
    ->     PARTITION BY KEY ( ) PARTITIONS 4;
ERROR 1488 (HY000): Field in list of fields for partition function not found in table
```

> **注意**:在按照 KEY 分区的分区表上,不能够执行 "ALTER TABLE DROP PRIMARY KEY;" 语句来删除主键,MySQL 会返回错误 "Field in list of fields for partition function not found in table"。

和 HASH 分区类似,在 KEY 分区中使用关键字 LINEAR 具有同样的作用,也就是 LINEAR KEY 分区时,分区的编号是通过 2 的幂算法得到的,而不是通过取模得到的。

KEY 分区和 HASH 分区类似,在处理大量数据记录时,能够有效地分散热点。

14.2.6 子分区

子分区(Subpartitioning)是分区表中对每个分区的再次分割,又被称为复合分区(Composite Partitioning)。MySQL 5.7 支持对已经通过 RANGE 或者 LIST 分区了的表再进行子分区。子分区既可以使用 HASH 分区,也可以使用 KEY 分区。例如:

```
mysql> CREATE TABLE ts (id INT, purchased DATE)
    ->     PARTITION BY RANGE(YEAR(purchased))
    ->     SUBPARTITION BY HASH(TO_DAYS(purchased))
    ->     SUBPARTITIONS 2
    ->     (
    ->         PARTITION p0 VALUES LESS THAN (1990),
    ->         PARTITION p1 VALUES LESS THAN (2000),
    ->         PARTITION p2 VALUES LESS THAN MAXVALUE
    ->     );
Query OK, 0 rows affected (0.11 sec)
```

表 ts 有 3 个 RANGE 分区,这 3 个分区中的每个分区(p0、p1、p2)又被进一步分成两个子分区,实际上,整个表被分成了 3×2 = 6 个分区,由于 PARTITION BY RANGE 子句的作用,第一和第二个分区只保存 purchased 列中值小于 1 990 的记录。

复合分区适用于保存非常大量的数据记录。在实际使用中,要注意以下几点。

● 每个分区必须具有相同数量的子分区。

● 如果要显式指定子分区,则每个分区都要显式指定,比如下面的语句中,p1 没有显式指定子分区,执行会失败:

```
CREATE TABLE ts (id INT, purchased DATE)
    PARTITION BY RANGE( YEAR(purchased) )
    SUBPARTITION BY HASH( TO_DAYS(purchased) ) (
        PARTITION p0 VALUES LESS THAN (1990) (
            SUBPARTITION s0,
            SUBPARTITION s1
        ),
```

```
        PARTITION p1 VALUES LESS THAN (2000),
        PARTITION p2 VALUES LESS THAN MAXVALUE (
            SUBPARTITION s2,
            SUBPARTITION s3
        )
    );
```

- 子分区的名称在整个表中是唯一的。下面的语句执行会报分区名重复错误：

```
CREATE TABLE ts (id INT, purchased DATE)
    PARTITION BY RANGE( YEAR(purchased) )
    SUBPARTITION BY HASH( TO_DAYS(purchased) ) (
        PARTITION p0 VALUES LESS THAN (1990) (
            SUBPARTITION s0,
            SUBPARTITION s1
        ),
        PARTITION p1 VALUES LESS THAN (2000) (
            SUBPARTITION s0,
            SUBPARTITION s1
        ),
        PARTITION p2 VALUES LESS THAN MAXVALUE (
            SUBPARTITION s0,
            SUBPARTITION s1
        )
    );
ERROR 1517 (HY000): Duplicate partition name s0
```

将 p1、p2 中的子分区名改一下即执行正常：

```
mysql> CREATE TABLE ts (id INT, purchased DATE)
    ->     PARTITION BY RANGE( YEAR(purchased) )
    ->     SUBPARTITION BY HASH( TO_DAYS(purchased) ) (
    ->         PARTITION p0 VALUES LESS THAN (1990) (
    ->             SUBPARTITION s0,
    ->             SUBPARTITION s1
    ->         ),
    ->         PARTITION p1 VALUES LESS THAN (2000) (
    ->             SUBPARTITION s2,
    ->             SUBPARTITION s3
    ->         ),
    ->         PARTITION p2 VALUES LESS THAN MAXVALUE (
    ->             SUBPARTITION s4,
    ->             SUBPARTITION s5
    ->         )
    ->     );
Query OK, 0 rows affected (0.04 sec)
```

14.2.7 MySQL 分区处理 NULL 值的方式

MySQL 不禁止在分区键值上使用 NULL，分区键可能是一个字段或者一个用户定义的表达式。一般情况下，MySQL 的分区把 NULL 当作零值，或者一个最小值进行处理。

注意：RANGE 分区中，NULL 值会被当作最小值来处理；LIST 分区中，NULL 值必须出现在枚举列表中，否则不被接受；HASH/KEY 分区中，NULL 值会被当作零值来处理。

例如，创建 tb_range 表，按照 id 进行 RANGE 分区，在 RANGE 分区中写入 NULL 值：

```
mysql>CREATE TABLE tb_range (
    -> id INT,
    -> name VARCHAR(5)
    -> )
    -> PARTITION BY RANGE(id) (
    -> PARTITION p0 VALUES LESS THAN (-6),
    -> PARTITION p1 VALUES LESS THAN (0),
    -> PARTITION p2 VALUES LESS THAN (1),
    -> PARTITION p3 VALUES LESS THAN MAXVALUE
    -> );
```

```
Query OK, 0 rows affected (0.06 sec)

mysql>insert into tb_range values (null, 'NULL');
Query OK, 1 row affected (0.00 sec)
```

查询 INFORMATION_SCHEMA.PARTITIONS 表确认写入的 NULL 值被当作最小值处理，NULL 值被分配在分区 p0 内：

```
mysql>SELECT
    ->    partition_name part,
    ->    partition_expression expr,
    ->    partition_description descr,
    ->    table_rows
    -> FROM
    ->    INFORMATION_SCHEMA.partitions
    -> WHERE
    ->    TABLE_SCHEMA = schema()
    ->    AND TABLE_NAME='tb_range';
+------+------+----------+------------+
| part | expr | descr    | table_rows |
+------+------+----------+------------+
| p0   | id   | -6       |          1 |
| p1   | id   | 0        |          0 |
| p2   | id   | 1        |          0 |
| p3   | id   | MAXVALUE |          0 |
+------+------+----------+------------+
4 rows in set (0.00 sec)
```

例如，在 LIST 分区中写入 NULL 值，分区定义不包含 NULL 值的时候，会返回一个错误 "ERROR 1526 (HY000): Table has no partition for value NULL"。

```
mysql>CREATE TABLE tb_list (
    -> id INT,
    -> name VARCHAR(5)
    -> )
    -> PARTITION BY LIST(id) (
    -> PARTITION p1 VALUES IN (0),
    -> PARTITION p2 VALUES IN (1)
    -> );
Query OK, 0 rows affected (0.01 sec)

mysql>insert into tb_list values (null, 'NULL');
ERROR 1526 (HY000): Table has no partition for value NULL
```

在 LIST 分区中增加 NULL 的定义之后，就能够成功写入 NULL 值：

```
mysql>CREATE TABLE tb_list (
    -> id INT,
    -> name VARCHAR(5)
    -> )
    -> PARTITION BY LIST(id) (
    -> PARTITION p1 VALUES IN (0, NULL),
    -> PARTITION p2 VALUES IN (1)
    -> );
Query OK, 0 rows affected (0.01 sec)

root@localhost:test 16:43>insert into tb_list values (NULL, 'NULL');
Query OK, 1 row affected (0.00 sec)

root@localhost:test 16:43>SELECT
    ->    partition_name part,
    ->    partition_expression expr,
    ->    partition_description descr,
    ->    table_rows
    -> FROM
    ->    INFORMATION_SCHEMA.partitions
    -> WHERE
    ->    TABLE_SCHEMA = schema()
    ->    AND TABLE_NAME='tb_list';
+------+------+--------+------------+
```

```
| part | expr | descr  | table_rows |
+------+------+--------+------------+
| p1   | id   | NULL,0 |     1      |
| p2   | id   | 1      |     0      |
+------+------+--------+------------+
2 rows in set (0.00 sec)
```

例如，创建 tb_hash 表，按照 id 列 HASH 分区，在 HASH 分区中写入 NULL 值：

```
mysql>CREATE TABLE tb_hash (
    ->     id INT,
    ->     name VARCHAR(5)
    -> )
    -> PARTITION BY HASH(id)
    -> PARTITIONS 2;
Query OK, 0 rows affected (0.04 sec)

mysql>insert into tb_hash values (null, 'NULL');
Query OK, 1 row affected (0.00 sec)

mysql>SELECT
    ->     partition_name part,
    ->     partition_expression expr,
    ->     partition_description descr,
    ->     table_rows
    -> FROM
    ->     INFORMATION_SCHEMA.partitions
    -> WHERE
    ->     TABLE_SCHEMA = schema()
    ->     AND TABLE_NAME='tb_hash';
+------+------+--------+------------+
| part | expr | descr  | table_rows |
+------+------+--------+------------+
| p0   | id   | NULL   |     1      |
| p1   | id   | NULL   |     0      |
+------+------+--------+------------+
2 rows in set (0.00 sec)
```

由于针对不同的分区类型，NULL 值时而被当作零值处理，时而被当作最小值处理，为了避免在处理 NULL 值时出现误判，更推荐通过设置字段非空和默认值来绕开 MySQL 默认对 NULL 值的处理。

14.3 分区管理

MySQL 提供了添加、删除、重定义、合并、拆分、交换分区的命令，这些操作都可以通过 ALTER TABLE 命令来进行实现。

14.3.1 RANGE 与 LIST 分区管理

在添加、删除、重新定义分区的处理上，RANGE 分区和 LIST 分区非常相似，所以合并一起来说。

从一个 RANGE 或者 LIST 分区的表中删除一个分区，可以使用 ALTER TABLE DROP PARTITION 语句来实现，以之前创建的按照表达式 YEAR(seperated)的值进行 RANGE 分区的 emp_date 表为例，执行如下语句创建 emp_date 表：

```
mysql> CREATE TABLE emp_date (
    ->     id INT NOT NULL,
    ->     ename VARCHAR(30),
    ->     hired DATE NOT NULL DEFAULT '1970-01-01',
    ->     separated DATE NOT NULL DEFAULT '9999-12-31',
    ->     job VARCHAR(30) NOT NULL,
```

```
    ->     store_id INT NOT NULL
    -> )
    -> PARTITION BY RANGE (YEAR(separated)) (
    ->     PARTITION p0 VALUES LESS THAN (1995),
    ->     PARTITION p1 VALUES LESS THAN (2000),
    ->     PARTITION p2 VALUES LESS THAN (2005),
    ->     PARTITION p3 VALUES LESS THAN (2015)
    -> );
Query OK, 0 rows affected (0.08 sec)
```

写入测试数据：

```
mysql>insert into emp_date(id, ename, hired, separated, job, store_id) values     -> (7499, 'ALLEN', '1981-02-20', '2003-08-03', 'SALESMAN', 30 ),
    -> (7521, 'WARD',  '1981-02-22', '1993-09-01', 'SALESMAN', 30),
    -> (7566, 'JONES', '1981-04-02', '2000-08-01', 'MANAGER', 20),
    -> (7654, 'MARTIN','1981-09-28', '2012-12-31', 'SALESMAN', 30),
    -> (7698, 'BLAKE', '1981-05-01', '1998-09-08', 'MANAGER', 30),
    -> (7782, 'CLARK', '1981-06-09', '2007-08-01', 'MANAGER', 10),
    -> (7788, 'SCOTT', '1987-04-19', '2012-05-01', 'ANALYST', 20),
    -> (7839, 'KING',  '1981-11-17', '2011-03-09', 'PRESIDENT', 10),
    -> (7844, 'TURNER','1981-09-08', '2010-12-31', 'SALESMAN', 30),
    -> (7876, 'ADAMS', '1987-05-23', '2000-01-01', 'CLERK', 20),
    -> (7900, 'JAMES', '1981-12-03', '2004-09-02', 'CLERK', 30),
    -> (7902, 'FORD',  '1981',       '2010-12-31', 'ANALYST', 20),
    -> (7934, 'MILLER','1982-01-23', '2011-12-31', 'CLERK', 10);
Query OK, 13 rows affected (0.01 sec)
Records: 13  Duplicates: 0  Warnings: 0
```

通过下面的查询语句查看哪些记录在分区 p2 中（LESS THAN 2005）：

```
mysql>select * from emp_date where separated between '2000-01-01' and '2004-12-31';
+------+--------+------------+------------+----------+----------+
| id   | ename  | hired      | separated  | job      | store_id |
+------+--------+------------+------------+----------+----------+
| 7499 | ALLEN  | 1981-02-20 | 2003-08-03 | SALESMAN |       30 |
| 7566 | JONES  | 1981-04-02 | 2000-08-01 | MANAGER  |       20 |
| 7876 | ADAMS  | 1987-05-23 | 2000-01-01 | CLERK    |       20 |
| 7900 | JAMES  | 1981-12-03 | 2004-09-02 | CLERK    |       30 |
+------+--------+------------+------------+----------+----------+
4 rows in set (0.00 sec)
```

执行下面的语句删除 p2 分区：

```
mysql>alter table emp_date drop partition p2;
Query OK, 0 rows affected (0.01 sec)
Records: 0  Duplicates: 0  Warnings: 0
```

注意：删除分区的命令执行之后，并不显示实际从表中删除的行数，并不是真的没有记录被删除。

从表结构定义上，可以观察到 p2 分区确实被删除了：

```
mysql>show create table emp_date\G
*************************** 1. row ***************************
       Table: emp_date
Create Table: CREATE TABLE `emp_date` (
  `id` int(11) NOT NULL,
  `ename` varchar(30) DEFAULT NULL,
  `hired` date NOT NULL DEFAULT '1970-01-01',
  `separated` date NOT NULL DEFAULT '9999-12-31',
  `job` varchar(30) NOT NULL,
  `store_id` int(11) NOT NULL
) ENGINE=InnoDB DEFAULT CHARSET=latin1
/*!50100 PARTITION BY RANGE (YEAR(separated))
(PARTITION p0 VALUES LESS THAN (1995) ENGINE = InnoDB,
 PARTITION p1 VALUES LESS THAN (2000) ENGINE = InnoDB,
 PARTITION p3 VALUES LESS THAN (2015) ENGINE = InnoDB) */
1 row in set (0.00 sec)
```

删除了 p2 分区，那么也同时删除了该分区中所有数据，重新执行前面的查询来确认一下：

14.3 分区管理

```
mysql>select * from emp_date where separated between '2000-01-01' and '2004-12-31';
Empty set (0.00 sec)
```

同时检查 emp_date 表的记录数，确认受影响的仅是原来在 p2 分区中的记录：

```
mysql>select count(1) from emp_date;
+----------+
| count(11) |
+----------+
|        9 |
+----------+
1 row in set (0.00 sec)
```

再次写入 separated 日期在 '2000-01-01'和'2004-12-31'之间的新记录到 emp_date 表时，这些行会被保存在 p3 分区中，确认一下，首先检查 emp_date 表的记录分布，p3 分区中仅有 7 条记录：

```
mysql>SELECT
    ->    partition_name part,
    ->    partition_expression expr,
    ->    partition_description descr,
    ->    table_rows
    -> FROM
    ->    INFORMATION_SCHEMA.partitions
    -> WHERE
    ->    TABLE_SCHEMA = schema()
    ->    AND TABLE_NAME='emp_date';
+------+-----------------+-------+------------+
| part | expr            | descr | table_rows |
+------+-----------------+-------+------------+
| p0   | YEAR(separated) | 1995  |          0 |
| p1   | YEAR(separated) | 2000  |          0 |
| p3   | YEAR(separated) | 2015  |          7 |
+------+-----------------+-------+------------+
3 rows in set (0.00 sec)
```

在 emp_date 表中写入一条 separated 日期在'2000-01-01'和'2004-12-31'之间的新记录：

```
mysql>insert into emp_date(id, ename, hired, separated, job, store_id) values
    -> (7566, 'JONES', '1981-04-02', '2000-08-01', 'MANAGER', 20);
Query OK, 1 row affected (0.00 sec)
```

再次检查 emp_date 表的记录分布，发现 p3 分区的记录数增加为 8 条，确认新写入的记录写入 p3 分区：

```
mysql>SELECT
    ->    partition_name part,
    ->    partition_expression expr,
    ->    partition_description descr,
    ->    table_rows
    -> FROM
    ->    INFORMATION_SCHEMA.partitions
    -> WHERE
    ->    TABLE_SCHEMA = schema()
    ->    AND TABLE_NAME='emp_date';
+------+-----------------+-------+------------+
| part | expr            | descr | table_rows |
+------+-----------------+-------+------------+
| p0   | YEAR(separated) | 1995  |          0 |
| p1   | YEAR(separated) | 2000  |          0 |
| p3   | YEAR(separated) | 2015  |          8 |
+------+-----------------+-------+------------+
3 rows in set (0.00 sec)
```

删除 LIST 分区和删除 RANGE 分区使用的语句完全相同，只不过删除 LIST 分区之后，由于在 LIST 分区的定义中不再包含已经被删除了的分区的值列表，所以后续无法写入包含有已经删除了的分区的值列表的数据。

为一个 RANGE 或者 LIST 分区的表增加一个分区,可以使用 ALTER TABLE ADD PARTITION 语句来实现。对于 RANGE 分区来说,只能通过 ADD PARTITION 方式添加新的分区到分区列表的最大一端,例如,给 emp_date 表增加 p4 分区,存放 separated 日期在 '2015-01-01'和'2029-12-31'之间的记录:

```
mysql>alter table emp_date add partition (partition p4 values less than (2030));
Query OK, 0 rows affected (0.01 sec)
Records: 0  Duplicates: 0  Warnings: 0

mysql>show create table emp_date\G
*************************** 1. row ***************************
       Table: emp_date
Create Table: CREATE TABLE `emp_date` (
  `id` int(11) NOT NULL,
  `ename` varchar(30) DEFAULT NULL,
  `hired` date NOT NULL DEFAULT '1970-01-01',
  `separated` date NOT NULL DEFAULT '9999-12-31',
  `job` varchar(30) NOT NULL,
  `store_id` int(11) NOT NULL
) ENGINE=InnoDB DEFAULT CHARSET=latin1
/*!50100 PARTITION BY RANGE (YEAR(separated))
(PARTITION p0 VALUES LESS THAN (1995) ENGINE = InnoDB,
 PARTITION p1 VALUES LESS THAN (2000) ENGINE = InnoDB,
 PARTITION p3 VALUES LESS THAN (2015) ENGINE = InnoDB,
 PARTITION p4 VALUES LESS THAN (2030) ENGINE = InnoDB) */
1 row in set (0.01 sec)
```

只能从 RANGE 分区列表最大端增加分区,否则会出现如下错误:

```
mysql>alter table emp_date add partition (partition p5 values less than (2025));
ERROR 1493 (HY000): VALUES LESS THAN value must be strictly increasing for each partition
```

给 LIST 分区增加新分区的方式也类似,以之前创建的 expenses 表为例,当前的 expenses 表结构如下:

```
mysql>CREATE TABLE expenses (
    ->     expense_date DATE NOT NULL,
    ->     category INT,
    ->     amount DECIMAL (10,3)
    ->  )PARTITION BY LIST(category) (
    ->     PARTITION p0 VALUES IN (3, 5),
    ->     PARTITION p1 VALUES IN (1, 10),
    ->     PARTITION p2 VALUES IN (4, 9),
    ->     PARTITION p3 VALUES IN (2),
    ->     PARTITION p4 VALUES IN (6)
    -> );
Query OK, 0 rows affected (0.04 sec)
```

为 expenses 表新增 p5 分区存储 category 分类为 7 和 8 的记录:

```
mysql>alter table expenses add partition (partition p5 values in (7,8));
Query OK, 0 rows affected (0.01 sec)
Records: 0  Duplicates: 0  Warnings: 0
```

增加 LIST 分区时,不能添加一个包含现有分区值列表中的任意值的分区,也就是说对一个固定的分区键值,必须指定并且只能指定一个唯一的分区,否则会出现错误:

```
mysql>alter table expenses add partition (partition p6 values in (6,11));
ERROR 1495 (HY000): Multiple definition of same constant in list partitioning
```

MySQL 也提供了在不丢失数据的情况下,通过重新定义分区的语句 ALTER TABLE REORGANIZE PARTITION INTO 重定义分区。仍以 RANGE 分区的 emp_date 表为例,当前 emp_date 的表结构如下:

```
mysql>show create table emp_date\G
*************************** 1. row ***************************
```

```
        Table: emp_date
Create Table: CREATE TABLE `emp_date` (
  `id` int(11) NOT NULL,
  `ename` varchar(30) DEFAULT NULL,
  `hired` date NOT NULL DEFAULT '1970-01-01',
  `separated` date NOT NULL DEFAULT '9999-12-31',
  `job` varchar(30) NOT NULL,
  `store_id` int(11) NOT NULL
) ENGINE=InnoDB DEFAULT CHARSET=latin1
/*!50100 PARTITION BY RANGE (YEAR(separated))
(PARTITION p0 VALUES LESS THAN (1995) ENGINE = InnoDB,
 PARTITION p1 VALUES LESS THAN (2000) ENGINE = InnoDB,
 PARTITION p3 VALUES LESS THAN (2015) ENGINE = InnoDB,
 PARTITION p4 VALUES LESS THAN (2030) ENGINE = InnoDB) */
1 row in set (0.00 sec)
```

计划拆分 p3 分区（2000～2015）为两个分区 p2（2000～2005）和 p3（2005～2015）：

```
mysql>alter table emp_date reorganize partition p3 into (
    ->      partition p2 values less than (2005),
    ->      partition p3 values less than (2015)
    -> );
Query OK, 8 rows affected (0.03 sec)
Records: 8  Duplicates: 0  Warnings: 0
```

确认拆分之后 emp_date 表的表结构：

```
mysql>show create table emp_date\G
*************************** 1. row ***************************
       Table: emp_date
Create Table: CREATE TABLE `emp_date` (
  `id` int(11) NOT NULL,
  `ename` varchar(30) DEFAULT NULL,
  `hired` date NOT NULL DEFAULT '1970-01-01',
  `separated` date NOT NULL DEFAULT '9999-12-31',
  `job` varchar(30) NOT NULL,
  `store_id` int(11) NOT NULL
) ENGINE=InnoDB DEFAULT CHARSET=latin1
/*!50100 PARTITION BY RANGE (YEAR(separated))
(PARTITION p0 VALUES LESS THAN (1995) ENGINE = InnoDB,
 PARTITION p1 VALUES LESS THAN (2000) ENGINE = InnoDB,
 PARTITION p2 VALUES LESS THAN (2005) ENGINE = InnoDB,
 PARTITION p3 VALUES LESS THAN (2015) ENGINE = InnoDB,
 PARTITION p4 VALUES LESS THAN (2030) ENGINE = InnoDB) */
1 row in set (0.00 sec)
```

重新定义分区可以用来拆分一个 RANGE 分区为多个 RANGE 分区，也可以用来合并多个相邻 RANGE 分区为一个 RANGE 分区或者多个 RANGE 分区：

```
mysql>alter table emp_date reorganize partition p1,p2,p3 into (
    ->      partition p1 values less than (2015)
    -> );
Query OK, 9 rows affected (0.05 sec)
Records: 9  Duplicates: 0  Warnings: 0

mysql>show create table emp_date\G
*************************** 1. row ***************************
       Table: emp_date
Create Table: CREATE TABLE `emp_date` (
  `id` int(11) NOT NULL,
  `ename` varchar(30) DEFAULT NULL,
  `hired` date NOT NULL DEFAULT '1970-01-01',
  `separated` date NOT NULL DEFAULT '9999-12-31',
  `job` varchar(30) NOT NULL,
  `store_id` int(11) NOT NULL
) ENGINE=InnoDB DEFAULT CHARSET=latin1
/*!50100 PARTITION BY RANGE (YEAR(separated))
(PARTITION p0 VALUES LESS THAN (1995) ENGINE = InnoDB,
 PARTITION p1 VALUES LESS THAN (2015) ENGINE = InnoDB,
```

```
  PARTITION p4 VALUES LESS THAN (2030) ENGINE = InnoDB) */
1 row in set (0.00 sec)
```

注意：重新定义 RANGE 分区时，只能够重新定义相邻的分区，不能跳过某个 RANGE 分区进行重新定义，同时重新定义的分区区间必须和原分区区间覆盖相同的区间；也不能使用重新定义分区来改变表分区的类型，例如，不能把 RANGE 分区变为 HASH 分区，也不能把 HASH 分区变成 RANGE 分区。

同样的，对 LIST 分区，也可以使用 ALTER TABLE REORGANIZE PARTITION INTO 语句重定义分区，例如，当前 expenses 表的分区如下：

```
mysql>show create table expenses\G
*************************** 1. row ***************************
       Table: expenses
Create Table: CREATE TABLE `expenses` (
  `expense_date` date NOT NULL,
  `category` int(11) DEFAULT NULL,
  `amount` decimal(10,3) DEFAULT NULL
) ENGINE=InnoDB DEFAULT CHARSET=latin1
/*!50100 PARTITION BY LIST (category)
(PARTITION p0 VALUES IN (3,5) ENGINE = InnoDB,
 PARTITION p1 VALUES IN (1,10) ENGINE = InnoDB,
 PARTITION p2 VALUES IN (4,9) ENGINE = InnoDB,
 PARTITION p3 VALUES IN (2) ENGINE = InnoDB,
 PARTITION p4 VALUES IN (6) ENGINE = InnoDB,
 PARTITION p5 VALUES IN (7,8) ENGINE = InnoDB) */
1 row in set (0.00 sec)
```

现在需要调整 p4 分区，使得 p4 分区包含值为 6 和 11 的记录，即 p4 分区的定义为 PARTITION p4 VALUES IN (6,11)，之前单纯通过 ADD PARTITION 的方式是不可以的：

```
mysql>alter table expenses add partition (partition p6 values in (6,11));
ERROR 1495 (HY000): Multiple definition of same constant in list partitioning
```

可以变通地通过增加分区和重定义分区来实现。首先，先增加不重复值列表的 p6 分区，包含值 11：

```
mysql>alter table expenses add partition (partition p6 values in (11));
Query OK, 0 rows affected (0.02 sec)
Records: 0  Duplicates: 0  Warnings: 0

mysql>show create table expenses\G;
*************************** 1. row ***************************
       Table: expenses
Create Table: CREATE TABLE `expenses` (
  `expense_date` date NOT NULL,
  `category` int(11) DEFAULT NULL,
  `amount` decimal(10,3) DEFAULT NULL
) ENGINE=InnoDB DEFAULT CHARSET=latin1
/*!50100 PARTITION BY LIST (category)
(PARTITION p0 VALUES IN (3,5) ENGINE = InnoDB,
 PARTITION p1 VALUES IN (1,10) ENGINE = InnoDB,
 PARTITION p2 VALUES IN (4,9) ENGINE = InnoDB,
 PARTITION p3 VALUES IN (2) ENGINE = InnoDB,
 PARTITION p4 VALUES IN (6) ENGINE = InnoDB,
 PARTITION p5 VALUES IN (7,8) ENGINE = InnoDB,
 PARTITION p6 VALUES IN (11) ENGINE = InnoDB) */
1 row in set (0.00 sec)
```

之后，通过 REORGANIZE PARTITION 方式重定义 p4、p5、p6 这 3 个分区，合并 p4 和 p6 两个分区为新的 p4 分区，包含值 6 和 11：

```
myql>alter table expenses reorganize partition p4,p5,p6 into (
    ->     partition p4 values in (6,11),
    ->     partition p5 values in (7,8)
```

```
    -> );
Query OK, 0 rows affected (0.05 sec)
Records: 0  Duplicates: 0  Warnings: 0

mysql>show create table expenses\G
*************************** 1. row ***************************
       Table: expenses
Create Table: CREATE TABLE `expenses` (
  `expense_date` date NOT NULL,
  `category` int(11) DEFAULT NULL,
  `amount` decimal(10,3) DEFAULT NULL
) ENGINE=InnoDB DEFAULT CHARSET=latin1
/*!50100 PARTITION BY LIST (category)
(PARTITION p0 VALUES IN (3,5) ENGINE = InnoDB,
 PARTITION p1 VALUES IN (1,10) ENGINE = InnoDB,
 PARTITION p2 VALUES IN (4,9) ENGINE = InnoDB,
 PARTITION p3 VALUES IN (2) ENGINE = InnoDB,
 PARTITION p4 VALUES IN (6,11) ENGINE = InnoDB,
 PARTITION p5 VALUES IN (7,8) ENGINE = InnoDB) */
1 row in set (0.00 sec)
```

通过重定义分区之后，p4 分区的值包含了 6 和 11。类似的重定义 RANGE 分区，重新定义 LIST 分区时，只能够重新定义相邻的分区，不能跳过 LIST 分区进行重新定义，否则提示以下信息：

```
mysql>alter table expenses reorganize partition p4,p6 into (partition p4 values in (6,11));
ERROR 1519 (HY000): When reorganizing a set of partitions they must be in consecutive order
```

注意：类似重新定义 RANGE 分区，重新定义 LIST 分区时，只能够重新定义相邻的分区，不能跳过 LIST 分区进行重新定义，同时重新定义的分区区间必须和原分区区间覆盖相同的区间；也不能使用重新定义分区来改变表分区的类型，例如，不能把 LIST 分区变为 RANGE 分区，也不能把 RANGE 分区变成 LIST 分区。

14.3.2 HASH 与 KEY 分区管理

在改变分区设置方面，HASH 分区和 KEY 分区的表非常类似，所以这两种分区的管理合并在一起讨论。

不能以 RANGE 或者 LIST 分区表中删除分区的相同方式，来从 HASH 或者 KEY 分区的表中删除分区，而可以通过 ALTER TABLE COALESCE PARTITION 语句来合并 HASH 分区或者 KEY 分区。例如，emp 表按照 store_id 分成了 4 个分区：

```
mysql> CREATE TABLE emp (
    ->     id INT NOT NULL,
    ->     ename VARCHAR(30),
    ->     hired DATE NOT NULL DEFAULT '1970-01-01',
    ->     separated DATE NOT NULL DEFAULT '9999-12-31',
    ->     job VARCHAR(30) NOT NULL,
    ->     store_id INT NOT NULL
    -> )
    -> PARTITION BY HASH (store_id) PARTITIONS 4;
Query OK, 0 rows affected (0.05 sec)
```

要减少 HASH 分区的数量，从 4 个分区变为 2 个分区，可以执行下面的 ALTER TABLE 命令：

```
mysql>alter table emp coalesce  partition 2;
Query OK, 0 rows affected (0.03 sec)
Records: 0  Duplicates: 0  Warnings: 0

mysql>show create table emp\G
*************************** 1. row ***************************
       Table: emp
Create Table: CREATE TABLE `emp` (
```

```
  `id` int(11) NOT NULL,
  `ename` varchar(30) DEFAULT NULL,
  `hired` date NOT NULL DEFAULT '1970-01-01',
  `separated` date NOT NULL DEFAULT '9999-12-31',
  `job` varchar(30) NOT NULL,
  `store_id` int(11) NOT NULL
) ENGINE=InnoDB DEFAULT CHARSET=latin1
/*!50100 PARTITION BY HASH (store_id)
PARTITIONS 2 */
1 row in set (0.00 sec)
```

COALESCE 不能用来增加分区的数量,否则会出现以下错误:

```
mysql>alter table emp coalesce  partition 8;
ERROR 1508 (HY000): Cannot remove all partitions, use DROP TABLE instead
```

要增加分区,可以通过 ALTER TABLE ADD PARTITION 语句来实现,例如,当前 emp 表有两个 HASH 分区,现在增加 8 个分区,最终 emp 表一共有 10 个 HASH 分区:

```
mysql>alter table emp add partition partitions 8;
Query OK, 0 rows affected (0.05 sec)
Records: 0  Duplicates: 0  Warnings: 0

root@localhost:test 22:34>show create table emp\G
*************************** 1. row ***************************
       Table: emp
Create Table: CREATE TABLE `emp` (
  `id` int(11) NOT NULL,
  `ename` varchar(30) DEFAULT NULL,
  `hired` date NOT NULL DEFAULT '1970-01-01',
  `separated` date NOT NULL DEFAULT '9999-12-31',
  `job` varchar(30) NOT NULL,
  `store_id` int(11) NOT NULL
) ENGINE=InnoDB DEFAULT CHARSET=latin1
/*!50100 PARTITION BY HASH (store_id)
PARTITIONS 10 */
1 row in set (0.01 sec)
```

> 注意:通过 ALTER TABLE ADD PARTITION PARTITIONS n 语句新增 HASH 分区或者 KEY 分区时,其实是对原表新增 n 个分区,而不是增加到 n 个分区。

14.3.3 交换分区

MySQL 5.6 增加了交换分区的功能,使用 ALTER TABLE pt EXCHANGE PARTITION p WITH TABLE nt 命令,可以实现将分区表 pt 中的一个分区或者子分区 p 中的数据和普通表 nt 中的数据进行交换。交换分区需要满足下列条件。

- 表 nt 不能是分区表,由于交换分区不能通过分区对分区的方式进行,如果有这种需求,可以用一个普通表作为中间表,通过交换两次分区来实现。
- 表 nt 不能是临时表。
- 表 pt 和 nt 的结构,除了分区之外,应该完全一致,包括索引的名称和索引列都要一致。
- 表 nt 上不能有外键,也不能有其他表的外键依赖 nt。
- nt 表的所有数据,应该都在分区 p 定义的范围内,在 MySQL 5.7 中,如果确定表中数据都在界限内,可以通过增加 WITHOUT VALIDATION 来跳过这个逐行验证的过程。

下面对交换分区做一下测试。

(1)首先创建一个分区表,并插入几行测试数据:

```
mysql> CREATE TABLE e (
    ->     id INT NOT NULL,
    ->     fname VARCHAR(30),
```

14.3 分区管理

```
    ->     lname VARCHAR(30)
    -> )
    ->     PARTITION BY RANGE (id) (
    ->         PARTITION p0 VALUES LESS THAN (50),
    ->         PARTITION p1 VALUES LESS THAN (100),
    ->         PARTITION p2 VALUES LESS THAN (150),
    ->         PARTITION p3 VALUES LESS THAN (MAXVALUE)
    -> );
Query OK, 0 rows affected (0.03 sec)

mysql> INSERT INTO e VALUES
    ->     (1669, "Jim", "Smith"),
    ->     (337, "Mary", "Jones"),
    ->     (16, "Frank", "White"),
    ->     (2005, "Linda", "Black");
Query OK, 4 rows affected (0.00 sec)
Records: 4  Duplicates: 0  Warnings: 0
```

（2）接下来创建一个普通表，表结构和上面的分区表一致：

```
mysql> CREATE TABLE e2 LIKE e;
Query OK, 0 rows affected (0.01 sec)
```

（3）通过 REMOVE PARTITIONING 子句将分区表改为非分区表：

```
mysql> ALTER TABLE e2 REMOVE PARTITIONING;
Query OK, 0 rows affected (0.02 sec)
Records: 0  Duplicates: 0  Warnings: 0
```

（4）查看分区表 e 中的数据分布：

```
mysql> SELECT PARTITION_NAME, TABLE_ROWS FROM INFORMATION_SCHEMA.PARTITIONS
    -> WHERE TABLE_NAME = 'e';
+----------------+------------+
| PARTITION_NAME | TABLE_ROWS |
+----------------+------------+
| p0             |          1 |
| p1             |          0 |
| p2             |          0 |
| p3             |          3 |
+----------------+------------+
4 rows in set (0.00 sec)
```

（5）执行交换分区命令，并观察交换之后的数据分布：

```
mysql> ALTER TABLE e EXCHANGE PARTITION p0 WITH TABLE e2;
Query OK, 0 rows affected (0.00 sec)

mysql> SELECT PARTITION_NAME, TABLE_ROWS FROM INFORMATION_SCHEMA.PARTITIONS
    -> WHERE TABLE_NAME = 'e';
+----------------+------------+
| PARTITION_NAME | TABLE_ROWS |
+----------------+------------+
| p0             |          0 |
| p1             |          0 |
| p2             |          0 |
| p3             |          3 |
+----------------+------------+
4 rows in set (0.00 sec)

mysql>
mysql> select * from e2;
+----+-------+-------+
| id | fname | lname |
+----+-------+-------+
| 16 | Frank | White |
+----+-------+-------+
1 row in set (0.00 sec)
```

可以看到，数据已经成功地从分区表 e 的分区 p0 交换到了表 e2 中。

（6）在 e2 的 name 字段上创建索引，此时 e 和 e2 定义不一致，交换命令报错：

```
mysql> create index idx_e2 on e2(fname);
Query OK, 0 rows affected (0.00 sec)
Records: 0  Duplicates: 0  Warnings: 0

mysql>  ALTER TABLE e EXCHANGE PARTITION p0 WITH TABLE e2;
ERROR 1736 (HY000): Tables have different definitions
```

此时 e 上 name 列创建索引，但索引名和 e2 的不同，依然报错：

```
mysql> create index idx_e on e(fname);
Query OK, 0 rows affected (0.02 sec)
Records: 0  Duplicates: 0  Warnings: 0

mysql>  ALTER TABLE e EXCHANGE PARTITION p0 WITH TABLE e2;
ERROR 1736 (HY000): Tables have different definitions
```

将索引名称改为和 e2 一致，再次执行交换命令成功：

```
mysql> alter table e rename index idx_e to idx_e2;
Query OK, 0 rows affected (0.00 sec)
Records: 0  Duplicates: 0  Warnings: 0

mysql>  ALTER TABLE e EXCHANGE PARTITION p0 WITH TABLE e2;
Query OK, 0 rows affected (0.15 sec)
```

使用交换分区，可以方便地完成对包含大量数据的分区、子分区的备份，迁移等工作。但要特别注意以下几点，交换前尽量提前最好备份，避免交换后可能带来的数据问题：

- 交换分区不会触发任何被交换的表或分区上的触发器；
- 表中自增列的值会被重置；
- 交换分区的命令中，IGNORE 关键字不会产生影响。

14.4 小结

本章重点介绍了 MySQL 的集中主要的分区类型、适用场景以及常规的管理维护命令，分区通过"分而治之"的方法管理数据库表，提高了数据处理的并行度，从而能够提升性能。

第三部分
优化篇

第 15 章　SQL 优化

在应用的开发过程中，由于初期数据量小，开发人员写 SQL 语句时更重视功能上的实现，但是当应用系统正式上线后，随着生产数据量的急剧增长，很多 SQL 语句开始逐渐显露出性能问题，对生产的影响也越来越大，此时这些有问题的 SQL 语句就成为整个系统性能的瓶颈，因此我们必须要对它们进行优化。本章将详细介绍在 MySQL 中优化 SQL 语句的方法。

15.1　优化 SQL 语句的一般步骤

当面对一个有 SQL 性能问题的数据库时，我们应该从何处入手来进行系统的分析，使得能够尽快定位问题 SQL 并尽快解决问题。本节将向读者介绍这个过程。

本章大部分涉及的案例表位于 MySQL 的案例库 Sakila 上。Sakila 是一个 MySQL 官方提供的模拟电影出租厅信息管理系统的数据库，可在 MySQL 官网页面上搜索并下载。

压缩包包括 3 个文件：sakila-schema.sql、sakila-data.sql 和 sakila.mwb，分别是 Sakila 库的表结构创建、数据灌入、Sakila 的 MySQL Workbench 数据模型（可以在 MySQL 工作台打开查看数据库模型）。

15.1.1　通过 show status 命令了解各种 SQL 的执行频率

MySQL 客户端连接成功后，通过 show [session|global]status 命令可以提供服务器状态信息，也可以在操作系统上使用 mysqladmin extended-status 命令获得这些消息。show [session|global] status 可以根据需要加上参数 "session" 或者 "global" 来显示 session 级（当前连接）的统计结果和 global 级（自数据库上次启动至今）的统计结果。如果不写，默认使用的参数是 "session"。

下面的命令显示了当前 session 中部分统计参数的值：

```
mysql> show status like 'Com_%';
+-------------------------------+-------+
| Variable_name                 | Value |
+-------------------------------+-------+
| Com_admin_commands            | 0     |
| Com_assign_to_keycache        | 0     |
| Com_alter_db                  | 0     |
| Com_alter_db_upgrade          | 0     |
| Com_alter_event               | 0     |
```

```
| Com_alter_function    | 0 |
| Com_alter_instance    | 0 |
| Com_alter_procedure   | 0 |
| Com_alter_server      | 0 |
| Com_alter_table       | 0 |
| Com_alter_tablespace  | 0 |
...
```

Com_xxx 表示每个 xxx 语句执行的次数，我们通常比较关心的是以下几个统计参数。

- Com_select：执行 SELECT 操作的次数，一次查询只累加 1。
- Com_insert：执行 INSERT 操作的次数，对于批量插入的 INSERT 操作，只累加一次。
- Com_update：执行 UPDATE 操作的次数。
- Com_delete：执行 DELETE 操作的次数。

上面这些参数对于所有存储引擎的表操作都会进行累计。下面这几个参数只是针对 InnoDB 存储引擎的，累加的算法也略有不同。

- Innodb_rows_read：SELECT 查询返回的行数。
- Innodb_rows_inserted：执行 INSERT 操作插入的行数。
- Innodb_rows_updated：执行 UPDATE 操作更新的行数。
- Innodb_rows_deleted：执行 DELETE 操作删除的行数。

通过以上几个参数，可以很容易地了解当前数据库的应用是以插入更新为主还是以查询操作为主，以及各种类型的 SQL 大致的执行比例是多少。对于更新操作的计数，是对执行次数的计数，不论提交还是回滚都会进行累加。

对于事务型的应用，通过 Com_commit 和 Com_rollback 可以了解事务提交和回滚的情况，对于回滚操作非常频繁的数据库，可能意味着应用编写存在问题。

此外，以下几个参数便于用户了解数据库的基本情况。

- Connections：试图连接 MySQL 服务器的次数。
- Uptime：服务器工作时间。
- Slow_queries：慢查询的次数。

15.1.2 定位执行效率较低的 SQL 语句

可以通过以下两种方式定位执行效率较低的 SQL 语句。

- 通过慢查询日志定位那些执行效率较低的 SQL 语句，将 slow-query-log 参数设置为 1 之后，MySQL 会将所有执行时间超过 long_query_time 参数所设定阈值的 SQL，写入 slow_query_log_file 参数所指定的文件中。

- 慢查询日志在查询结束以后才记录，所以在应用反映执行效率出现问题的时候查询慢查询日志并不能定位问题，可以使用 show processlist 命令查看当前 MySQL 在进行的线程，包括线程的状态、是否锁表等，可以实时地查看 SQL 的执行情况，同时对一些锁表操作进行优化。

15.1.3 通过 EXPLAIN 分析低效 SQL 的执行计划

通过以上步骤查询到效率低的 SQL 语句后，可以通过 EXPLAIN 或者 DESC 命令获取 MySQL 如何执行 SELECT 语句的信息，包括在 SELECT 语句执行过程中表如何连接和连接的顺序，比如想统计某个 email 为租赁电影拷贝所支付的总金额，需要关联客户表 customer 和付款表 payment，并且对付款金额 amount 字段做求和（sum）操作，相应 SQL 的执行计划如下：

```
mysql> explain select sum(amount) from customer a, payment b where 1=1 and a.customer_id = b.customer_id and email = 'JANE.BENNETT@sakilacustomer.org'\G
*************************** 1. row ***************************
           id: 1
  select_type: SIMPLE
        table: a
         type: ALL
possible_keys: PRIMARY
          key: NULL
      key_len: NULL
          ref: NULL
         rows: 583
        Extra: Using where
*************************** 2. row ***************************
           id: 1
  select_type: SIMPLE
        table: b
         type: ref
possible_keys: idx_fk_customer_id
          key: idx_fk_customer_id
      key_len: 2
          ref: sakila.a.customer_id
         rows: 12
        Extra:
2 rows in set (0.00 sec)
```

对每个列简单地进行一下说明。

- select_type: 表示 SELECT 的类型，常见的取值有 SIMPLE（简单表，即不使用表连接或者子查询）、PRIMARY（主查询，即外层的查询）、UNION（UNION 中的第二个或者后面的查询语句）、SUBQUERY（子查询中的第一个 SELECT）等。
- table: 输出结果集的表。
- type: 表示 MySQL 在表中找到所需行的方式，或者叫访问类型，常见的类型如图 15-1 所示。

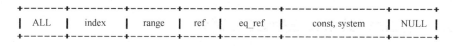

图 15-1　常见访问类型

从左至右，性能由最差到最好。

（1）type=ALL，全表扫描，MySQL 遍历全表来找到匹配的行：

```
mysql> explain select * from film where rating >9\G
*************************** 1. row ***************************
           id: 1
  select_type: SIMPLE
        table: film
         type: ALL
possible_keys: NULL
          key: NULL
      key_len: NULL
          ref: NULL
         rows: 916
        Extra: Using where
1 row in set (0.00 sec)
```

（2）type=index，索引全扫描，MySQL 遍历整个索引来查询匹配的行：

```
mysql> explain select title from film\G
*************************** 1. row ***************************
           id: 1
  select_type: SIMPLE
        table: film
         type: index
possible_keys: NULL
          key: idx_title
      key_len: 767
          ref: NULL
         rows: 916
        Extra: Using index
1 row in set (0.00 sec)
```

（3）type=range，索引范围扫描，常见于<、<=、>、>=、between 等操作符：

```
mysql> explain select * from payment where customer_id >= 300 and customer_id <= 350\G
*************************** 1. row ***************************
           id: 1
  select_type: SIMPLE
        table: payment
         type: range
possible_keys: idx_fk_customer_id
          key: idx_fk_customer_id
      key_len: 2
          ref: NULL
         rows: 1349
        Extra: Using where
1 row in set (0.00 sec)
```

（4）type=ref，使用非唯一索引扫描或唯一索引的前缀扫描，返回匹配某个单独值的记录行，例如：

```
mysql> explain select * from payment where customer_id =350\G
*************************** 1. row ***************************
           id: 1
  select_type : SIMPLE
         table: payment
          type: ref
possible_keys: idx_fk_customer_id
          key: idx_fk_customer_id
      key_len : 2
          ref: const
         rows: 23
        Extra:
1 row in set (0.00 sec)
```

索引 idx_fk_customer_id 是非唯一索引，查询条件为等值查询条件 customer_id=35，所以扫描索引的类型为 ref。ref 还经常出现在 join 操作中：

```
mysql> explain select b.*, a.* from payment a, customer b where a.customer_id = b.customer_id\G
*************************** 1. row ***************************
           id: 1
  select_type: SIMPLE
        table: b
         type: ALL
possible_keys: PRIMARY
          key: NULL
      key_len: NULL
          ref: NULL
         rows: 505
        Extra:
*************************** 2. row ***************************
           id: 1
  select_type: SIMPLE
        table: a
         type: ref
possible_keys: idx_fk_customer_id
          key: idx_fk_customer_id
      key_len: 2
```

```
         ref: sakila.b.customer_id
        rows: 12
       Extra:
2 rows in set (0.00 sec)
```

（5）type=eq_ref，类似 ref，区别就在使用的索引是唯一索引，对于每个索引键值，表中只有一条记录匹配；简单来说，就是多表连接中使用 primary key 或者 unique index 作为关联条件。

```
mysql> explain select * from film a, film_text b where a.film_id = b.film_id\G
*************************** 1. row ***************************
           id: 1
  select_type: SIMPLE
        table: a
         type: ALL
possible_keys: PRIMARY
          key: NULL
      key_len: NULL
          ref: NULL
         rows: 916
        Extra:
*************************** 2. row ***************************
           id: 1
  select_type: SIMPLE
        table: b
         type: eq_ref
possible_keys: PRIMARY
          key: PRIMARY
      key_len: 2
          ref: sakila.a.film_id
         rows: 1
        Extra: Using where
2 rows in set (0.00 sec)
```

（6）type=const/system，单表中最多有一个匹配行，查询起来非常迅速，所以这个匹配行中的其他列的值可以被优化器在当前查询中当作常量来处理，例如，根据主键 primary key 或者唯一索引 unique index 进行的查询。

构造一个查询：

```
mysql> alter table customer drop index idx_email;
Query OK, 0 rows affected (0.08 sec)
Records: 0  Duplicates: 0  Warnings: 0

mysql> alter table customer add unique index uk_email (email);
Query OK, 0 rows affected (0.15 sec)
Records: 0  Duplicates: 0  Warnings: 0

mysql> explain select * from (select * from customer where email = 'AARON.SELBY@sakilacustomer.org')a\G
*************************** 1. row ***************************
           id: 1
  select_type: PRIMARY
        table: <derived2>
         type: system
possible_keys: NULL
          key: NULL
      key_len: NULL
          ref: NULL
         rows: 1
        Extra:
*************************** 2. row ***************************
           id: 2
  select_type: DERIVED
        table: customer
         type: const
possible_keys: uk_email
          key: uk_email
      key_len: 153
```

15.1 优化 SQL 语句的一般步骤

```
            ref:
           rows: 1
          Extra:
2 rows in set (0.00 sec)
```

通过唯一索引 uk_email 访问的时候，类型 type 为 const；而从我们构造的仅有一条记录的 a 表中检索时，类型 type 就为 system。

（7）type=NULL，MySQL 不用访问表或者索引，直接就能够得到结果，例如：

```
mysql> explain select 1 from dual where 1\G
*************************** 1. row ***************************
           id: 1
  select_type: SIMPLE
        table: NULL
         type: NULL
possible_keys: NULL
          key: NULL
      key_len: NULL
          ref: NULL
         rows: NULL
        Extra: No tables used
1 row in set (0.00 sec)
```

类型 type 还有其他值，如 ref_or_null（与 ref 类似，区别在于条件中包含对 NULL 的查询）、index_merge（索引合并优化）、unique_subquery（in 的后面是一个查询主键字段的子查询）、index_subquery（与 unique_subquery 类似，区别在于 in 的后面是查询非唯一索引字段的子查询）等。

- possible_keys: 表示查询时可能使用的索引。
- key: 表示实际使用的索引。
- key_len: 使用到索引字段的长度。
- rows: 扫描行的数量。
- Extra: 执行情况的说明和描述，包含不适合在其他列中显示但是对执行计划非常重要的额外信息。

MySQL 从 5.1 版本开始支持分区功能，同时 explain 命令也增加了对分区的支持。可以通过 explain partitions 命令查看 SQL 所访问的分区。例如，创建一个 Hash 分区的 customer_part 表，根据分区键查询的时候，能够看到 explain partitions 的输出结果中有一列 partitions，其中显示了 SQL 所需要访问的分区名字 p2：

```
mysql> create table customer_part (
    ->  `customer_id` smallint(5) unsigned NOT NULL AUTO_INCREMENT,
…
    ->   PRIMARY KEY (`customer_id`)
    -> )partition by hash(customer_id) partitions 8;
Query OK, 0 rows affected (0.56 sec)

mysql> insert into customer_part select * from customer;
Query OK, 599 rows affected (0.08 sec)
Records: 599  Duplicates: 0  Warnings: 0

mysql> explain partitions select * from customer_part where customer_id = 130\G
*************************** 1. row ***************************
           id: 1
  select_type: SIMPLE
        table: customer_part
   partitions: p2
         type: const
possible_keys: PRIMARY
          key: PRIMARY
      key_len: 2
```

```
        ref: const
       rows: 1
      Extra:
1 row in set (0.00 sec)
```

有时，仅仅通过 explain 分析执行计划并不能很快地定位 SQL 的问题，那么这时我们还可以选择 profile 联合分析。

15.1.4　通过 show profile 分析 SQL

MySQL 从 5.0.37 版本开始增加了对 show profiles 和 show profile 语句的支持。通过 have_profiling 参数，能够看到当前 MySQL 是否支持 profile：

```
mysql> select @@have_profiling;
+------------------+
| @@have_profiling |
+------------------+
| YES              |
+------------------+
1 row in set (0.00 sec)
```

默认 profiling 是关闭的，可以通过 set 语句在 Session 级别开启 profiling：

```
mysql> select @@profiling;
+-------------+
| @@profiling |
+-------------+
|           0 |
+-------------+
1 row in set (0.02 sec)

mysql> set profiling=1;
Query OK, 0 rows affected (0.00 sec)
```

通过 profile，用户能够更清楚地了解 SQL 执行的过程。例如，我们知道 MyISAM 表有表元数据的缓存（例如行数，即 COUNT(*)值），那么对一个 MyISAM 表的 COUNT(*)是不需要消耗太多资源的，而对于 InnoDB 来说，就没有这种元数据缓存，COUNT(*)执行得较慢。下面来做个实验验证一下。

首先，在一个 innodb 引擎的付款表 payment 上，执行一个 COUNT(*)查询：

```
mysql> select count(*) from payment;
+----------+
| count(*) |
+----------+
|    16049 |
+----------+
1 row in set (0.01 sec)
```

执行完毕后，通过 show profiles 语句，看到当前 SQL 的 Query ID 为 4：

```
mysql> show profiles;
+----------+------------+-------------------------------+
| Query_ID | Duration   | Query                         |
+----------+------------+-------------------------------+
|        1 | 0.00019300 | SELECT DATABASE()             |
|        2 | 0.00049000 | show databases                |
|        3 | 0.00281600 | show tables                   |
|        4 | 0.00774175 | select count(*) from payment  |
+----------+------------+-------------------------------+
4 rows in set (0.00 sec)
```

通过 show profile for query 语句能够看到执行过程中线程的每个状态和消耗的时间：

```
mysql> show profile for query 4;
```

15.1 优化 SQL 语句的一般步骤

```
| Status                              | Duration |
+-------------------------------------+----------+
| starting                            | 0.000026 |
| Waiting for query cache lock        | 0.000006 |
| checking query cache for query      | 0.000057 |
| checking permissions                | 0.000011 |
| Opening tables                      | 0.000300 |
| System lock                         | 0.000016 |
| Waiting for query cache lock        | 0.000024 |
| init                                | 0.000018 |
| optimizing                          | 0.000009 |
| statistics                          | 0.000016 |
| preparing                           | 0.000014 |
| executing                           | 0.000009 |
| Sending data                        | 0.007143 |
| end                                 | 0.000011 |
| query end                           | 0.000012 |
| closing tables                      | 0.000015 |
| freeing items                       | 0.000012 |
| Waiting for query cache lock        | 0.000004 |
| freeing items                       | 0.000020 |
| Waiting for query cache lock        | 0.000004 |
| freeing items                       | 0.000004 |
| storing result in query cache       | 0.000006 |
| logging slow query                  | 0.000004 |
| cleaning up                         | 0.000005 |
+-------------------------------------+----------+
24 rows in set (0.00 sec)
```

注意：Sending data 状态表示 MySQL 线程开始访问数据行并把结果返回给客户端，而不仅仅是返回结果给客户端。由于在 Sending data 状态下，MySQL 线程往往需要做大量的磁盘读取操作，所以经常是整个查询中耗时最长的状态。

通过仔细检查 show profile for query 的输出，能够发现在执行 COUNT(*) 的过程中，时间主要消耗在 Sending data 这个状态上。为了更清晰地看到排序结果，可以查询 INFORMATION_SCHEMA.PROFILING 表并按照时间做个 DESC 排序：

```
mysql> SET @query_id := 4;
Query OK, 0 rows affected (0.00 sec)

mysql> SELECT STATE, SUM(DURATION) AS Total_R,
    ->        ROUND(
    ->            100 * SUM(DURATION) /
    ->             (SELECT SUM(DURATION)
    ->              FROM INFORMATION_SCHEMA.PROFILING
    ->              WHERE QUERY_ID = @query_id
    ->            ), 2) AS Pct_R,
    ->        COUNT(*) AS Calls,
    ->        SUM(DURATION) / COUNT(*) AS "R/Call"
    ->    FROM INFORMATION_SCHEMA.PROFILING
    ->    WHERE QUERY_ID = @query_id
    ->    GROUP BY STATE
    ->    ORDER BY Total_R DESC;
+----------------------+----------+-------+-------+----------------+
| STATE                | Total_R  | Pct_R | Calls | R/Call         |
+----------------------+----------+-------+-------+----------------+
| Sending data         | 0.007143 | 92.22 |     1 | 0.0071430000   |
| Opening tables       | 0.000300 |  3.87 |     1 | 0.0003000000   |
...
| logging slow query   | 0.000004 |  0.05 |     1 | 0.0000040000   |
+----------------------+----------+-------+-------+----------------+
19 rows in set (0.04 sec)
```

在获取到最消耗时间的线程状态后，MySQL 支持进一步选择 all、cpu、block io、context switch、page faults 等明细类型来查看 MySQL 在使用什么资源上耗费了过高的时间，例如，选择查看 CPU 的耗费时间：

```
mysql> show profile cpu for query 4;
+----------------------+----------+----------+------------+
| Status               | Duration | CPU_user | CPU_system |
+----------------------+----------+----------+------------+
| starting             | 0.000036 | 0.000000 | 0.000000   |
...
| executing            | 0.000009 | 0.000000 | 0.000000   |
| Sending data         | 0.007143 | 0.006999 | 0.000000   |
| end                  | 0.000011 | 0.000000 | 0.000000   |
...
| logging slow query   | 0.000002 | 0.000000 | 0.000000   |
| cleaning up          | 0.000003 | 0.000000 | 0.000000   |
+----------------------+----------+----------+------------+
24 rows in set (0.00 sec)
```

能够发现 Sending data 状态下,时间主要消耗在 CPU 上了。

对比 MyISAM 表的 COUNT(*)操作,也创建一个同样表结构的 MyISAM 表,数据量也完全一致:

```
mysql> create table payment_myisam like payment;
Query OK, 0 rows affected (0.11 sec)

mysql> alter table payment_myisam engine=myisam;
Query OK, 0 rows affected (0.24 sec)
Records: 0  Duplicates: 0  Warnings: 0

mysql> insert into payment_myisam select * from payment;
Query OK, 16049 rows affected (0.37 sec)
Records: 16049  Duplicates: 0  Warnings: 0
```

同样执行 COUNT(*)操作,检查 profile:

```
mysql> select count(*) from payment_myisam;
+----------+
| count(*) |
+----------+
|    16049 |
+----------+
1 row in set (0.00 sec)
...
mysql> show profiles;
...
mysql> show profile for query 10;
+----------------------+----------+
| Status               | Duration |
+----------------------+----------+
| starting             | 0.000029 |
...
| executing            | 0.000015 |
| end                  | 0.000007 |
| query end            | 0.000009 |
...
| cleaning up          | 0.000006 |
+----------------------+----------+
21 rows in set (0.00 sec)
```

从 profile 的结果能够看出,InnoDB 引擎的表在 COUNT(*)时经历了 Sending data 状态,存在访问数据的过程,而 MyISAM 引擎的表在 executing 之后直接就结束查询,完全不需要访问数据。

读者如果对 MySQL 源码感兴趣,还可以通过 show profile source for query 查看 SQL 解析执行过程中每个步骤对应的源码的文件、函数名以及具体的源文件行数:

```
mysql> show profile source for query 4\G
...
*************************** 4. row ***************************
    Status: checking permissions
```

```
            Duration: 0.000015
     Source_function: check_access
         Source_file: sql_parse.cc
         Source_line: 4627
...
```

show profile 能够在做 SQL 优化时帮助我们了解时间都耗费到哪里去了。而 MySQL 5.6 之后则通过 trace 文件进一步向我们展示了优化器是如何选择执行计划的。

注意：在 MySQL 5.7 中，profile 已经不建议使用，而使用 performance schema 中的一系列性能视图来替代，详细内容请参考第 20 章。

15.1.5 通过 trace 分析优化器如何选择执行计划

MySQL 从 5.6 版本开始提供了对 SQL 的跟踪 trace，通过 trace 文件能够进一步了解为什么优化器选择 A 执行计划而不选择 B 执行计划，帮助我们更好地理解优化器的行为。

使用方式：首先打开 trace，设置格式为 JSON，设置 trace 最大能够使用的内存大小，避免解析过程中因为默认内存过小而不能够完整显示。

```
mysql> SET OPTIMIZER_TRACE="enabled=on",END_MARKERS_IN_JSON=on;
Query OK, 0 rows affected (0.03 sec)

mysql> SET OPTIMIZER_TRACE_MAX_MEM_SIZE=1000000;
Query OK, 0 rows affected (0.00 sec)
```

接下来执行想做 trace 的 SQL 语句，例如，想了解租赁表 rental 中库存编号 inventory_id 为 4466 的电影拷贝在出租日期 rental_date 为 2005-05-25 4:00:00~5:00:00 出租的记录：

```
mysql> select rental_id from rental where 1=1 and rental_date >= '2005-05-25 04:00:00' and
rental_date <= '2005-05-25 05:00:00' and inventory_id=4466;
+-----------+
| rental_id |
+-----------+
|        39 |
+-----------+
1 row in set (0.00 sec)
```

然后，检查 INFORMATION_SCHEMA.OPTIMIZER_TRACE 就可以知道 MySQL 是如何执行 SQL 语句的：

```
mysql> SELECT * FROM INFORMATION_SCHEMA.OPTIMIZER_TRACE\G
```

最后会输出一个格式如下的跟踪文件（截取部分内容）：

```
*************************** 1. row ***************************
                            QUERY: select rental_id from rental where 1=1 and rental_date >=
'2005-05-25 04:00:00' and rental_date <= '2005-05-25 05:00:00' and inventory_id=4466
                            TRACE: {
    "steps": [
      {
        "join_preparation": {
          "select#": 1,
          "steps": [
            {
              "expanded_query": "/* select#1 */ select `rental`.`rental_id` AS `rental_id`
from `rental` where ((1 = 1) and (`rental`.`rental_date` >= '2005-05-25 04:00:00') and (`rental
`.`rental_date` <= '2005-05-25 05:00:00') and (`rental`.`inventory_id` = 4466))"
            }
          ] /* steps */
        } /* join_preparation */
      },
      ……
   MISSING_BYTES_BEYOND_MAX_MEM_SIZE: 0
```

```
        INSUFFICIENT_PRIVILEGES: 0
1 row in set (0.00 sec)
```

文件里面记录了很多信息,包括访问表的路径、行数、成本等,来帮助读者对执行计划的选择过程进行分析,后面会有一些使用的例子。

15.1.6 确定问题并采取相应的优化措施

经过以上步骤,基本就可以确认问题出现的原因。此时用户可以根据情况采取相应的措施,进行优化以提高执行的效率。

在 15.1.3 节的例子中,已经可以确认是由于对客户表 customer 的全表扫描导致效率不理想,那么对客户表 customer 的 email 字段创建索引,具体如下:

```
mysql> create index idx_email on customer(email);
Query OK, 0 rows affected (0.37 sec)
Records: 0  Duplicates: 0  Warnings: 0
```

创建索引后,再看一下这条语句的执行计划,具体如下:

```
mysql> explain select sum(amount) from customer a, payment b where 1=1 and a.customer_id =
b.customer_id and email = 'JANE.BENNETT@sakilacustomer.org'\G
*************************** 1. row ***************************
           id: 1
  select_type: SIMPLE
        table: a
         type: ref
possible_keys: PRIMARY,idx_email
          key: idx_email
      key_len: 153
          ref: const
         rows: 1
        Extra: Using where; Using index
*************************** 2. row ***************************
           id: 1
  select_type: SIMPLE
        table: b
         type: ref
possible_keys: idx_fk_customer_id
          key: idx_fk_customer_id
      key_len: 2
          ref: sakila.a.customer_id
         rows: 14
        Extra:
2 rows in set (0.00 sec)
```

可以发现,建立索引后对客户表 customer 需要扫描的行数明显减少(从 583 行减少到 1 行),可见索引的使用可以大大提高数据库的访问速度,尤其在表很庞大的时候这种优势更为明显。

15.2 索引问题

索引是数据库优化中最常用也是最重要的手段之一,通过索引通常可以帮助用户解决大多数的 SQL 性能问题。本节将详细讨论 MySQL 中索引的分类、存储和使用方法。

15.2.1 索引的存储分类

索引是在 MySQL 的存储引擎层中实现的,而不是在服务器层实现的。所以每种存储引

15.2 索引问题

擎的索引都不一定完全相同,也不是所有的存储引擎都支持所有的索引类型。MySQL 目前提供了以下 4 种索引。

- **B-Tree** 索引:最常见的索引类型,大部分引擎都支持 B 树索引。
- **HASH** 索引:只有 Memory/NDB 引擎支持,使用场景简单。
- **R-Tree** 索引(空间索引):空间索引是 MyISAM 的一个特殊索引类型,主要用于地理空间数据类型,通常使用较少,不做特别介绍。
- **Full-text**(全文索引):全文索引也是 MyISAM 的一个特殊索引类型,主要用于全文索引,InnoDB 从 MySQL 5.6 版本开始提供对全文索引的支持。

MySQL 目前版本(8.0.11)还不支持函数索引(本书截稿前官方发布的最新版本 8.0.13 已经支持),但可以通过两种方式实现函数索引的功能。

(1)前缀索引,即对列的前面某一部分进行索引。例如标题 title 字段,可以只取 title 的前 10 个字符进行索引,这个特性可以大大缩小索引文件的大小,但前缀索引也有缺点,在排序 Order By 和分组 Group By 操作的时候无法使用。用户在设计表结构的时候也可以对文本列根据此特性进行灵活设计。

下面是创建前缀索引的一个例子:

```
mysql> create index idx_title on film(title(10));
Query OK, 0 rows affected (0.06 sec)
Records: 0  Duplicates: 0  Warnings: 0
```

(2)虚拟列索引。在 Oracle 等大多商业数据库中,早已支持函数索引,但 MySQL 一直没有实现这个功能。在 MySQL 5.7 之后,可以通过创建虚拟列索引的方式来实现函数索引的功能,如下例所示:

在表 salaries 中执行如下 SQL:

```
Select * from salaries where round(salary/1000)<10
```

表的定义如下:

```
mysql> desc salaries;
+-----------+---------+------+-----+---------+-------+
| Field     | Type    | Null | Key | Default | Extra |
+-----------+---------+------+-----+---------+-------+
| emp_no    | int(11) | NO   | PRI | NULL    |       |
| salary    | int(11) | NO   |     | NULL    |       |
| from_date | date    | NO   | PRI | NULL    |       |
| to_date   | date    | NO   |     | NULL    |       |
+-----------+---------+------+-----+---------+-------+
4 rows in set (0.00 sec)
```

此时,直接在 salary 上创建索引并不会被这个 SQL 所使用。这里创建一个虚拟列 salary_by_1k:

```
mysql> alter table salaries add column salary_by_1k int generated always as (round(salary/1000));
Query OK, 0 rows affected (0.06 sec)
Records: 0  Duplicates: 0  Warnings: 0
```

然后在这个虚拟列上创建索引:

```
mysql> alter table salaries add key idx_salary_by_1k(salary_by_1k);
Query OK, 0 rows affected (11.18 sec)
Records: 0  Duplicates: 0  Warnings: 0
```

此时观察执行计划,显示新创建的虚拟列索引已经被使用:

```
mysql> desc  Select count(1) from salaries where round(salary/1000)<10 \G;
*************************** 1. row ***************************
           id: 1
```

```
        select_type: SIMPLE
              table: salaries
         partitions: NULL
               type: range
      possible_keys: idx_salary_by_1k
                key: idx_salary_by_1k
            key_len: 5
                ref: NULL
               rows: 1
           filtered: 100.00
              Extra: Using where
1 row in set, 1 warning (0.00 sec)
```

实际执行后，SQL 的执行时间从 8s 降到了 0.2s，性能大幅提升。

表 15-1 主要对比了 MyISAM、InnoDB、Memory 这 3 个常用引擎支持的索引类型。

表 15-1　MyISAM、InnoDB、Memory 这 3 个常用引擎支持的索引类型比较

索引	MyISAM 引擎	InnoDB 引擎	Memory 引擎
B-Tree 索引	支持	支持	支持
HASH 索引	不支持	不支持	支持
R-Tree 索引	支持	不支持	不支持
Full-text 索引	支持	不支持	不支持

比较常用到的索引就是 B-Tree 索引和 Hash 索引。Hash 索引相对简单，只有 Memory/NDB 引擎支持完全的 Hash 索引，InnoDB 存储引擎在 MySQL 5.7 中支持自适应的 Hash 索引。所谓自适应，就是 MySQL 根据数据的访问频率和模式为某些热点页自动创建 Hash 索引，索引由 buffer pool 中的 B-tree 来自动生成，效率很高，这个特性由参数 innodb_adaptive_hash_index 来控制，默认是打开的。

Hash 索引适用于 Key-Value 查询，通过 Hash 索引要比通过 B-Tree 索引查询更迅速；Hash 索引不适用范围查询，例如<、>、<=、>=这类操作。如果使用 Memory/NDB 引擎并且 where 条件中不使用"="进行索引列，那么不会用到索引。Memory/Heap 引擎只有在"="和"<=>"的条件下才会使用索引。

B-Tree 索引比较复杂，下面将详细分析 MySQL 是如何利用 B-Tree 索引的。

15.2.2　MySQL 如何使用索引

B-Tree 索引是最常见的索引，构造类似二叉树，能根据键值提供一行或者一个行集的快速访问，通常只需要很少的读操作就可以找到正确的行。不过，需要注意 B-Tree 索引中的 B 不代表二叉树（binary），而是代表平衡树（balanced）。B-Tree 索引并不是一棵二叉树。

图 15-2 展示了经典的 B-Tree 的结构：根节点 root 下面有多个分支（Branch 节点），Branch 节点下面就是明细的叶子（Leaf 节点）。

可以利用 B-Tree 索引进行全关键字、关键字范围和关键字前缀查询，以下例子如果没有特别说明都能够在 MySQL 5.7 上执行通过。

为了避免混淆，重命名租赁表 rental 上的索引 rental_date 为 idx_rental_date：

```
mysql> alter table rental rename index rental_date to idx_rental_date;
Query OK, 0 rows affected (0.25 sec)
Records: 0  Duplicates: 0  Warnings: 0
```

15.2 索引问题

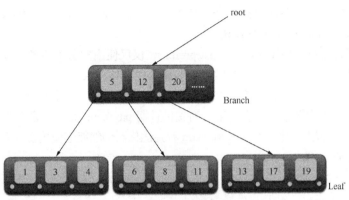

图 15-2 B-Tree 索引的结构

1. MySQL 中能够使用索引的典型场景

（1）匹配全值（Match the full value），对索引中所有列都指定具体值，即是对索引中的所有列都有等值匹配的条件。例如，租赁表 rental 中通过指定出租日期 rental_date + 库存编号 inventory_id + 客户编号 customer_id 的组合条件进行查询，从执行计划的 key 和 extra 两字段的值看到优化器选择了复合索引 idx_rental_date：

```
mysql> explain select * from rental where rental_date='2005-05-25 17:22:10' and inventory_id
=373 and customer_id=343\G
*************************** 1. row ***************************
           id: 1
  select_type: SIMPLE
        table: rental
   partitions: NULL
         type: const
possible_keys: idx_rental_date,idx_fk_inventory_id,idx_fk_customer_id
          key: idx_rental_date
      key_len: 10
          ref: const,const,const
         rows: 1
     filtered: 100.00
        Extra: NULL
1 row in set, 1 warning (0.08 sec)
```

explain 输出结果中字段 type 的值为 const，表示是常量；字段 key 的值为 idx_rental_date，表示优化器选择索引 idx_rental_date 进行扫描。

（2）匹配值的范围查询（Match a range of values），对索引的值能够进行范围查找。例如，检索租赁表 rental 中客户编号 customer_id 在指定范围内的记录：

```
mysql> explain select * from rental where customer_id >= 373 and customer_id < 400\G
*************************** 1. row ***************************
           id: 1
  select_type: SIMPLE
        table: rental
   partitions: NULL
         type: range
possible_keys: idx_fk_customer_id
          key: idx_fk_customer_id
      key_len: 2
          ref: NULL
         rows: 718
     filtered: 100.00
```

```
    Extra: Using index condition
1 row in set (0.00 sec)
```

类型 type 为 range 说明优化器选择范围查询，索引 key 为 idx_fk_customer_id 说明优化器选择索引 idx_fk_customer_id 来加速访问。

（3）匹配最左前缀（Match a leftmost prefix），仅仅使用索引中的最左边列进行查找，比如在 col1 + col2 + col3 字段上的联合索引能够被包含 col1、(col1 + col2)、(col1 + col2 + col3) 的等值查询利用到，可是不能够被 col2、(col2 + col3) 的等值查询利用到；以支付表 payment 为例，如果查询条件中仅包含索引的第一列支付日期 payment_date 和索引的第三列更新时间 last_update 的时候，从执行计划的 key 和 extra 看到优化器仍然能够使用复合索引 idx_payment_date 进行条件过滤：

```
mysql> alter table payment add index idx_payment_date (payment_date, amount, last_update);
Query OK, 0 rows affected (0.21 sec)
Records: 0  Duplicates: 0  Warnings: 0

mysql> explain select * from payment where payment_date = '2006-02-14 15:16:03' and last_update ='2006-02-15 22:12:32'\G
*************************** 1. row ***************************
           id: 1
  select_type: SIMPLE
        table: payment
   partitions: NULL
         type: ref
possible_keys: idx_payment_date
          key: idx_payment_date
      key_len: 5
          ref: const
         rows: 182
     filtered: 10.00
        Extra: Using index condition
1 row in set (0.00 sec)
```

但是，如果仅仅选择复合索引 idx_payment_date 的第二列支付金额 amount 和第三列更新时间 last_update 进行查询时，那么执行计划显示并不会利用到索引 idx_payment_date：

```
mysql> explain select * from payment where amount = 3.98 and last_update='2006-02-15 22:12:32'\G
*************************** 1. row ***************************
           id: 1
  select_type: SIMPLE
        table: payment
   partitions: NULL
         type: ALL
possible_keys: NULL
          key: NULL
      key_len: NULL
          ref: NULL
         rows: 16086
     filtered: 1.00
        Extra: Using where
1 row in set (0.01 sec)
```

最左匹配原则可以算是 MySQL 中 B-Tree 索引使用的首要原则。

（4）仅仅对索引进行查询（Index only query），当查询的列都在索引的字段中时，查询的效率更高；对比上一个例子使用 Select *，本次选择查询的字段都包含在索引 idx_payment_date 中时，能够看到查询计划有了一点变动：

```
mysql> explain select last_update from payment where payment_date = '2006-02-14 15:16:03' and amount = 3.98\G
*************************** 1. row ***************************
           id: 1
  select_type: SIMPLE
        table: payment
```

```
        partitions: NULL
              type: ref
     possible_keys: idx_payment_date
               key: idx_payment_date
           key_len: 8
               ref: const,const
              rows: 8
          filtered: 100.00
             Extra: Using index
1 row in set (0.00 sec)
```

Extra 部分变成了 Using index,也就意味着,现在直接访问索引就足够获取到所需要的数据,不需要通过索引回表,Using index 也就是平常说的覆盖索引扫描。只访问必须访问的数据,在一般情况下,减少不必要的数据访问能够提升效率。

(5)匹配列前缀(match a column prefix),仅仅使用索引中的第一列,并且只包含索引第一列的开头一部分进行查找。例如,现在需要查询出标题 title 是以 AFRICAN 开头的电影信息,从执行计划能够清楚看到,idx_title_desc_part 索引被利用上了:

```
mysql> create index idx_title_desc_part on film_text (title(10), description(20));
Query OK, 1000 rows affected (0.08 sec)
Records: 1000  Duplicates: 0  Warnings: 0

mysql> explain select title from film_text where title like 'AFRICAN%'\G
*************************** 1. row ***************************
                id: 1
       select_type: SIMPLE
             table: film_text
        partitions: NULL
              type: range
     possible_keys: idx_title_desc_part,idx_title_description
               key: idx_title_desc_part
           key_len: 32
               ref: NULL
              rows: 1
          filtered: 100.00
             Extra: Using where
1 row in set (0.02 sec)
```

Extra 值为 Using where 表示优化器需要通过索引回表查询数据。

(6)能够实现索引匹配部分精确而其他部分进行范围匹配(match one part exactly and match a range on another part)。例如,需要查询出租日期 rental_date 为指定日期且客户编号 customer_id 为指定范围的库存:

```
mysql> explain select inventory_id from rental where rental_date='2006-02-14 15:16:03' and
customer_id >= 300 and customer_id <= 400\G
*************************** 1. row ***************************
                id: 1
       select_type: SIMPLE
             table: rental
        partitions: NULL
              type: ref
     possible_keys: idx_rental_date,idx_fk_customer_id
               key: idx_rental_date
           key_len: 5
               ref: const
              rows: 182
          filtered: 16.85
             Extra: Using where; Using index
1 row in set (0.00 sec)
```

类型 type 为 ref 说明优化器选择等值非唯一方式进行查询,索引 key 为 idx_rental_date 说明优化器选择索引 idx_rental_date 帮助加速查询,同时由于只查询索引字段 inventory_id 的值,所以在 Extra 部分能看到 Using index,表示查询使用了覆盖索引扫描。

（7）如果列名是索引，那么使用 column_name is null 就会使用索引（区别于 Oracle）。例如，查询支付表 payment 的租赁编号 rental_id 字段为空的记录就用到了索引：

```
mysql> explain select * from payment where rental_id is null\G
*************************** 1. row ***************************
           id: 1
  select_type: SIMPLE
        table: payment
   partitions: NULL
         type: ref
possible_keys: fk_payment_rental
          key: fk_payment_rental
      key_len: 5
          ref: const
         rows: 5
     filtered: 100.00
        Extra: Using index condition
1 row in set (0.00 sec)
```

（8）MySQL 5.6 引入了 Index Condition Pushdown（ICP）的特性，进一步优化了查询。Pushdown 表示操作下放，某些情况下的条件过滤操作下放到存储引擎。

例如，查询租赁表 rental 中租赁时间 rental_date 在某一指定时间点且客户编号 customer_id 在指定范围内的数据，MySQL 5.5/5.1 的执行计划显示：优化器首先使用复合索引 idx_rental_date 的首字段 rental_date 过滤出符合条件 rental_date='2006-02-14 15:16:03' 的记录（执行计划中 key 字段值显示为 idx_rental_date），然后根据复合索引 idx_rental_date 回表获取记录后，最终根据条件 customer_id >= 300 and customer_id <= 400 来过滤出最后的查询结果（执行计划中 Extra 字段值显示为 Using where）。

```
mysql> select version();
+------------+
| version()  |
+------------+
| 5.5.32-log |
+------------+
1 row in set (0.00 sec)

mysql> explain select * from rental where rental_date='2006-02-14 15:16:03' and customer_id >= 300 and customer_id <= 400\G
*************************** 1. row ***************************
           id: 1
  select_type: SIMPLE
        table: rental
         type: ref
possible_keys: idx_fk_customer_id,idx_rental_date
          key: idx_rental_date
      key_len: 8
          ref: const
         rows: 182
        Extra: Using where
1 row in set (0.00 sec)
```

复合索引检索如图 15-3 所示。

在 MySQL 5.7 上做同样的案例，能够发现 Explain 执行计划的 Extra 部分从 Using where 变成了 Using index condition 的提示：

```
mysql> select version();
+------------+
| version()  |
+------------+
| 5.7.22-log |
+------------+
1 row in set (0.00 sec)
```

15.2 索引问题

```
mysql> explain select * from rental where rental_date='2006-02-14 15:16:03' and customer_id
>= 300 and customer_id <= 400\G
*************************** 1. row ***************************
           id: 1
  select_type: SIMPLE
        table: rental
   partitions: NULL
         type: ref
possible_keys: idx_rental_date,idx_fk_customer_id
          key: idx_rental_date
      key_len: 5
          ref: const
         rows: 182
     filtered: 16.85
        Extra: Using index condition
1 row in set (0.00 sec)
```

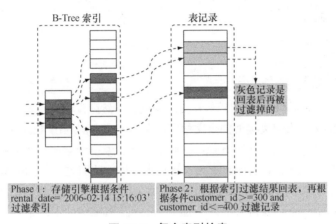

图 15-3　复合索引检索

Using index condition 就表示 MySQL 使用了 ICP 来进一步优化查询，在检索的时候，把条件 customer_id 的过滤操作下推到存储引擎层来完成，这样能够降低不必要的 IO 访问，如图 15-4 所示。

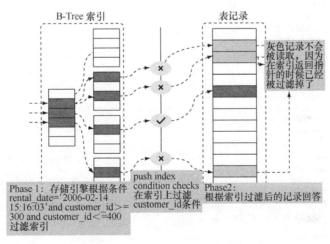

图 15-4　Index Condition Pushdown

2. 存在索引但不能使用索引的典型场景

有些时候虽然有索引,但是并不被优化器选择使用。下面列举几个不能使用索引的常见场景。

(1) 以%开头的 LIKE 查询不能够利用 B-Tree 索引,执行计划中 key 的值为 NULL 表示没有使用索引:

```
mysql> explain select * from actor where last_name like '%NI%'\G
*************************** 1. row ***************************
           id: 1
  select_type: SIMPLE
        table: actor
   partitions: NULL
         type: ALL
possible_keys: NULL
          key: NULL
      key_len: NULL
          ref: NULL
         rows: 200
     filtered: 11.11
        Extra: Using where
1 row in set (0.00 sec)
```

因为 B-Tree 索引的结构,所以以%开头的查询很自然就没法利用索引了,一般都推荐使用全文索引(Fulltext)来解决类似的全文检索问题。或者考虑利用 InnoDB 的表都是聚簇表的特点,采取一种轻量级别的解决方式:一般情况下,索引都会比表小,扫描索引要比扫描表更快(某些特殊情况下,索引比表更大,不在本例讨论范围内),而 InnoDB 表上二级索引 idx_last_name 实际上存储字段 last_name 还有主键 actor_id,那么理想的访问方式应该是首先扫描二级索引 idx_last_name 获得满足条件 last_name like '%NI%'的主键 actor_id 列表,之后根据主键回表去检索记录,这样访问避开了全表扫描演员表 actor 产生的大量 IO 请求。验证一下:

```
mysql> explain select * from (select actor_id from actor where last_name like '%NI%')a, actor b where a.actor_id = b.actor_id\G
*************************** 1. row ***************************
           id: 1
  select_type: SIMPLE
        table: actor
   partitions: NULL
         type: index
possible_keys: PRIMARY
          key: idx_actor_last_name
      key_len: 137
          ref: NULL
         rows: 200
     filtered: 11.11
        Extra: Using where; Using index
*************************** 2. row ***************************
           id: 1
  select_type: SIMPLE
        table: b
   partitions: NULL
         type: eq_ref
possible_keys: PRIMARY
          key: PRIMARY
      key_len: 2
          ref: sakila.actor.actor_id
         rows: 1
     filtered: 100.00
        Extra: NULL
2 rows in set (0.00 sec)
```

从执行计划中能够看到,内层查询的 Using index 代表索引覆盖扫描,之后通过主键 join

操作去演员表 actor 中获取最终查询结果，理论上是能够比直接全表扫描更快一些。

（2）数据类型出现隐式转换的时候也不会使用索引，特别是当列类型是字符串，那么一定记得在 where 条件中把字符常量值用引号引起来，否则即便这个列上有索引，MySQL 也不会用到，因为 MySQL 默认把输入的常量值进行转换以后才进行检索。例如，演员表 actor 中的姓氏字段 last_name 是字符型的，但是 SQL 语句中的条件值 1 是一个数值型值，因此即便存在索引 idx_last_name，MySQL 也不能正确地用上索引，而是继续进行全表扫描：

```
mysql> explain select * from actor where last_name = 1\G
*************************** 1. row ***************************
           id: 1
  select_type: SIMPLE
        table: actor
   partitions: NULL
         type: ALL
possible_keys: idx_actor_last_name
          key: NULL
      key_len: NULL
          ref: NULL
         rows: 200
     filtered: 10.00
        Extra: Using where
1 row in set (0.00 sec)
```

加上引号之后，再次检查执行计划，就发现使用上索引了：

```
mysql> explain select * from actor where last_name = '1'\G
*************************** 1. row ***************************
           id: 1
  select_type: SIMPLE
        table: actor
   partitions: NULL
         type: ref
possible_keys: idx_actor_last_name
          key: idx_actor_last_name
      key_len: 137
          ref: const
         rows: 1
     filtered: 100.00
        Extra: NULL
1 row in set (0.00 sec)
```

（3）复合索引的情况下，假如查询条件不包含索引列最左边部分，即不满足最左原则 Leftmost，是不会使用复合索引的：

```
mysql> explain select * from payment where amount = 3.98 and last_update='2006-02-15 22:12:32'\G
*************************** 1. row ***************************
           id: 1
  select_type: SIMPLE
        table: payment
   partitions: NULL
         type: ALL
possible_keys: NULL
          key: NULL
      key_len: NULL
          ref: NULL
         rows: 16086
     filtered: 1.00
        Extra: Using where
1 row in set (0.01 sec)
```

（4）如果 MySQL 估计使用索引比全表扫描更慢，则不使用索引。例如，查询以"S"开头的电影标题，需要返回的记录比例较大，MySQL 就预估索引扫描还不如全表扫描更快：

```
mysql> update film_text set title = concat('S', title);
Query OK, 1000 rows affected (0.19 sec)
Rows matched: 1000  Changed: 1000  Warnings: 0
```

```
mysql> explain select * from film_text where title like 'S%'\G
*************************** 1. row ***************************
           id: 1
  select_type: SIMPLE
        table: film_text
   partitions: NULL
         type: ALL
possible_keys: idx_title_desc_part,idx_title_description
          key: NULL
      key_len: NULL
          ref: NULL
         rows: 1000
     filtered: 100.00
        Extra: Using where
1 row in set (0.00 sec)
```

在 MySQL 5.7 版本中，能够通过 Trace 清晰地看到优化器选择的过程，全表扫描 table scan 需要访问的记录 rows 为 1000，代价 cost 计算为 233.53：

```
"table_scan": {
    "rows": 1000,
    "cost": 233.53
} /* table_scan */,
```

而对应 idx_title_desc_part 索引过滤条件时，优化器预估需要返回 998 条记录，访问代价 cost 为 1198.6，远高于全表扫描的代价，索引优化器倾向于选择全表扫描：

```
{
    "index": "idx_title_desc_part",
    "ranges": [
        "S\u0000\u0000\u0000\u0000\u0000\u0000\u0000\u0000\u0000 <= title <= S?????????"
    ] /* ranges */,
    "index_dives_for_eq_ranges": true,
    "rowid_ordered": false,
    "using_mrr": false,
    "index_only": false,
    "rows": 998,
    "cost": 1198.6,
    "chosen": false,
    "cause": "cost"
}
```

更换查询的值为一个选择率更高的值，就能发现优化器更倾向于选择索引扫描：

```
mysql> explain select * from film_text where title like 'SW%'\G
*************************** 1. row ***************************
           id: 1
  select_type: SIMPLE
        table: film_text
         type: range
possible_keys: idx_title_desc_part,idx_title_description
          key: idx_title_desc_part
      key_len: 32
          ref: NULL
         rows: 66
     filtered: 100.00
        Extra: Using where
1 row in set (0.00 sec)
```

同样通过 trace 能够看到，title like 'SW%' 优化器预估需要返回 66 条记录，代价 cost 为 80.21，远小于全表扫描的代价，所以优化器倾向于选择索引扫描：

```
{
    "index": "idx_title_desc_part",
    "ranges": [
        "SW\u0000\u0000\u0000\u0000\u0000\u0000\u0000\u0000 <= title <= SW????????"
    ] /* ranges */,
    "index_dives_for_eq_ranges": true,
```

```
        "rowid_ordered": false,
        "using_mrr": false,
        "index_only": false,
        "rows": 66,
        "cost": 80.21,
        "chosen": true
}
```

也就是在查询时，筛选性越高越容易使用到索引，筛选性越低越不容易使用索引。

（5）用 or 分割开的条件，如果 or 前的条件中的列有索引，而后面的列中没有索引，那么涉及的索引都不会被用到，例如：

```
mysql> explain select * from payment where customer_id = 203 or amount = 3.96\G
*************************** 1. row ***************************
           id: 1
  select_type: SIMPLE
        table: payment
         type: ALL
possible_keys: idx_fk_customer_id
          key: NULL
      key_len: NULL
          ref: NULL
         rows: 16451
     filtered: 10.15
        Extra: Using where
1 row in set (0.00 sec)
```

因为 or 后面的条件列中没有索引，那么后面的查询肯定要走全表扫描，在存在全表扫描的情况下，就没有必要多一次索引扫描增加 I/O 访问，一次全表扫描过滤条件就足够了。

15.2.3　查看索引使用情况

如果索引正在工作，Handler_read_key 的值将很高，这个值代表了一个行被索引值读的次数，很低的值表明增加索引得到的性能改善不高，因为索引并不经常使用。

Handler_read_rnd_next 的值高则意味着查询运行低效，并且应该建立索引补救。这个值的含义是在数据文件中读下一行的请求数。如果正进行大量的表扫描，Handler_read_rnd_next 的值较高，则通常说明表索引不正确或写入的查询没有利用索引，具体如下：

```
mysql> show status like 'Handler_read%';
+-----------------------+-------+
| Variable_name         | Value |
+-----------------------+-------+
| Handler_read_first    | 0     |
| Handler_read_key      | 5     |
| Handler_read_next     | 0     |
| Handler_read_prev     | 0     |
| Handler_read_rnd      | 0     |
| Handler_read_rnd_next | 2055  |
+-----------------------+-------+
6 rows in set (0.00 sec)
```

从上面的例子中可以看出，目前使用的 MySQL 数据库的索引情况并不理想。

15.3　两个简单实用的优化方法

对于大多数开发人员来说，可能只希望掌握一些简单实用的优化方法；对于更多、更复杂的优化，更倾向于交给专业 DBA 来做。本节将向读者介绍两个简单适用的优化方法。

15.3.1 定期分析表和检查表

分析表的语法如下：

```
ANALYZE [LOCAL | NO_WRITE_TO_BINLOG] TABLE tbl_name [, tbl_name] ...
```

本语句用于分析和存储表的关键字分布，分析的结果将可以使得系统得到准确的统计信息，使得 SQL 能够生成正确的执行计划。如果用户感觉实际执行计划并不是预期的执行计划，执行一次分析表可能会解决问题。在分析期间，使用一个读取锁定对表进行锁定。这对于 MyISAM、BDB 和 InnoDB 表有作用。对于 MyISAM 表，本语句与使用 myisamchk -a 相当，下例中对表 sales 做了表分析：

```
mysql> analyze table payment;
+----------------+---------+----------+----------+
| Table          | Op      | Msg_type | Msg_text |
+----------------+---------+----------+----------+
| sakila.payment | analyze | status   | OK       |
+----------------+---------+----------+----------+
1 row in set (0.00 sec)
```

检查表的语法如下：

```
CHECK TABLE tbl_name [, tbl_name] ... [option] ... option = {QUICK | FAST | MEDIUM | EXTENDED | CHANGED}
```

检查表的作用是检查一个或多个表是否有错误。CHECK TABLE 对 MyISAM 和 InnoDB 表有作用。对于 MyISAM 表，关键字统计数据被更新，例如：

```
mysql> check table payment_myisam;
+-----------------------+-------+----------+----------+
| Table                 | Op    | Msg_type | Msg_text |
+-----------------------+-------+----------+----------+
| sakila.payment_myisam | check | status   | OK       |
+-----------------------+-------+----------+----------+
1 row in set (0.03 sec)
```

CHECK TABLE 也可以检查视图是否有错误，比如在视图定义中被引用的表已不存在，下面给出一个示例。

（1）首先创建一个视图：

```
mysql> create view v_payment_myisam as select * from payment_myisam;
Query OK, 0 rows affected (0.05 sec)
```

（2）然后 CHECK 一下该视图，发现没有问题：

```
mysql> check table v_payment_myisam;
+-------------------------+-------+----------+----------+
| Table                   | Op    | Msg_type | Msg_text |
+-------------------------+-------+----------+----------+
| sakila.v_payment_myisam | check | status   | OK       |
+-------------------------+-------+----------+----------+
1 row in set (0.05 sec)
```

（3）现在删除掉视图依赖的表：

```
mysql> drop table payment_myisam;
Query OK, 0 rows affected (0.00 sec)
```

（4）再来 CHECK 一下刚才的视图，发现报错了，并且提示出错的原因是 Table sakila.payment_myisam 不存在了：

```
mysql> check table v_payment_myisam\G
*************************** 1. row ***************************
    Table: sakila.v_payment_myisam
```

```
        Op: check
  Msg_type: Error
  Msg_text: Table 'sakila.payment_myisam' doesn't exist
*************************** 2. row ***************************
     Table: sakila.v_payment_myisam
        Op: check
  Msg_type: Error
  Msg_text: View 'sakila.v_payment_myisam' references invalid table(s) or column(s) or function
(s) or definer/invoker of view lack rights to use them
*************************** 3. row ***************************
     Table: sakila.v_payment_myisam
        Op: check
  Msg_type: error
  Msg_text: Corrupt
3 rows in set (0.00 sec)
```

15.3.2 定期优化表

优化表的语法如下：

```
OPTIMIZE [LOCAL | NO_WRITE_TO_BINLOG] TABLE tbl_name [, tbl_name] ...
```

如果已经删除了表的一大部分，或者如果已经对含有可变长度行的表（含有 VARCHAR、BLOB 或 TEXT 列的表）进行了很多更改，则应使用 OPTIMIZE TABLE 命令来进行表优化。这个命令可以将表中的空间碎片进行合并，并且可以消除由于删除或者更新造成的空间浪费，但 OPTIMIZE TABLE 命令只对 MyISAM、BDB 和 InnoDB 表起作用。命令执行期间 MyISAM 表会全程锁表，而 InnoDB 表则会将优化命令转换为重建表和分析表两个操作，加锁时间也仅仅在整个工作的 prepare 和 commit 期间做短暂的加锁工作，对于表的读写几乎没有影响。

以下例子显示了优化表 payment_myisam 的过程：

```
mysql> optimize table payment_myisam;
+-----------------------+----------+----------+----------+
| Table                 | Op       | Msg_type | Msg_text |
+-----------------------+----------+----------+----------+
| sakila.payment_myisam | optimize | status   | OK       |
+-----------------------+----------+----------+----------+
1 row in set (0.02 sec)
```

对于 InnoDB 引擎的表来说，通过设置 innodb_file_per_table 参数，设置 InnoDB 为独立表空间模式，这样每个数据库的每个表都会生成一个独立的 ibd 文件，用于存储表的数据和索引，这样可以一定程度上减轻 InnoDB 表的空间回收问题。另外，在删除大量数据后，InnoDB 表可以通过 alter table 但是不修改引擎的方式来回收不用的空间：

```
mysql> alter table payment engine=innodb;
Query OK, 16049 rows affected (2.56 sec)
Records: 16049  Duplicates: 0  Warnings: 0
```

注意：ANALYZE、CHECK、OPTIMIZE、ALTER TABLE 执行期间将对表进行锁定，因此一定注意要在数据库不繁忙的时候执行相关的操作。

15.4 常用 SQL 的优化

前面已经介绍了 MySQL 中是如何通过索引来优化查询的。在日常开发中，除了使用查询外，我们还会使用一些其他的常用 SQL，比如 INSERT、GROUP BY 等。对于这些 SQL 语句，哪又该如何进行优化呢？本节将针对这些 SQL 语句介绍一些优化的方法。

15.4.1 大批量插入数据

当用 load 命令导入数据的时候，适当的设置可以提高导入的速度。

对于 MyISAM 存储引擎的表，可以通过以下方式快速地导入大量的数据。

```
ALTER TABLE tbl_name DISABLE KEYS;
loading the data
ALTER TABLE tbl_name ENABLE KEYS;
```

DISABLE KEYS 和 ENABLE KEYS 用来打开或者关闭 MyISAM 表非唯一索引的更新。在导入大量的数据到一个非空的 MyISAM 表时，通过设置这两个命令，可以提高导入的效率。对于导入大量数据到一个空的 MyISAM 表，默认就是先导入数据然后才创建索引的，所以不用进行设置。

以下示例用 LOAD 语句导入数据耗时 115.12s：

```
mysql> load data infile '/home/mysql/film_test.txt' into table film_test2;
Query OK, 529056 rows affected (1 min 55.12 sec)
Records: 529056  Deleted: 0  Skipped: 0  Warnings: 0
```

而用 alter table tbl_name disable keys 方式总耗时 6.34 + 12.25 = 18.59s，提高了 6 倍多。

```
mysql> alter table film_test2 disable keys;
Query OK, 0 rows affected (0.00 sec)

mysql> load data infile '/home/mysql/film_test.txt' into table film_test2;
Query OK, 529056 rows affected (6.34 sec)
Records: 529056  Deleted: 0  Skipped: 0  Warnings: 0

mysql> alter table film_test2 enable keys;
Query OK, 0 rows affected (12.25 sec)
```

上面是对 MyISAM 表进行数据导入时的优化措施，对于 InnoDB 类型的表，这种方式并不能提高导入数据的效率，可以有以下几种方式提高 InnoDB 表的导入效率。

（1）因为 InnoDB 类型的表是按照主键的顺序保存的，所以将导入的数据按照主键的顺序排列，可以有效地提高导入数据的效率。

例如，下面的文本 film_test3.txt 是按表 film_test4 的主键存储的，那么导入时共耗时 27.92s。

```
mysql> load data infile '/home/mysql/film_test3.txt' into table film_test4;
Query OK, 1587168 rows affected (22.92 sec)
Records: 1587168  Deleted: 0  Skipped: 0  Warnings: 0
```

而下面的 film_test4.txt 是没有任何顺序的文本，那么导入时共耗时 31.16s。

```
mysql> load data infile '/home/mysql/film_test4.txt' into table film_test4;
Query OK, 1587168 rows affected (31.16 sec)
Records: 1587168  Deleted: 0  Skipped: 0  Warnings: 0
```

从上面的例子可以看出，当被导入的文件按表主键顺序存储时比不按主键顺序存储时快 1.12 倍。

（2）在导入数据前执行 SET UNIQUE_CHECKS=0，关闭唯一性校验；在导入结束后执行 SET UNIQUE_CHECKS=1，恢复唯一性校验，可以提高导入的效率。

例如，当 UNIQUE_CHECKS=1 时：

```
mysql> load data infile '/home/mysql/film_test3.txt' into table film_test4;
Query OK, 1587168 rows affected (22.92 sec)
Records: 1587168  Deleted: 0  Skipped: 0  Warnings: 0
```

当 SET UNIQUE_CHECKS=0 时：

```
mysql> load data infile '/home/mysql/film_test3.txt' into table film_test4;
Query OK, 1587168 rows affected (19.92 sec)
Records: 1587168  Deleted: 0  Skipped: 0  Warnings: 0
```

可见 UNIQUE_CHECKS=0 时比 SET UNIQUE_CHECKS=1 时要快一些。

（3）如果应用使用自动提交的方式，建议在导入前执行 SET AUTOCOMMIT=0，关闭自动提交，导入结束后再执行 SET AUTOCOMMIT=1，打开自动提交，也可以提高导入的效率。

例如，当 AUTOCOMMIT=1 时：

```
mysql> load data infile '/home/mysql/film_test3.txt' into table film_test4;
Query OK, 1587168 rows affected (22.92 sec)
Records: 1587168  Deleted: 0  Skipped: 0  Warnings: 0
```

当 AUTOCOMMIT=0 时：

```
mysql> load data infile '/home/mysql/film_test3.txt' into table film_test4;
Query OK, 1587168 rows affected (20.87 sec)
Records: 1587168  Deleted: 0  Skipped: 0  Warnings: 0
```

对比一下可以知道，当 AUTOCOMMIT=0 时比 AUTOCOMMIT=1 时导入数据要快一些。

15.4.2 优化 INSERT 语句

当进行数据 INSERT 的时候，可以考虑采用以下几种优化方式。

● 如果同时从同一客户插入很多行，应尽量使用多个值表的 INSERT 语句，这种方式将大大缩减客户端与数据库之间的连接、关闭等消耗，使得效率比分开执行的单个 INSERT 语句快（在大部分情况下，使用多个值表的 INSERT 语句能比单个 INSERT 语句快上好几倍）。下面是一次插入多值的一个例子：

```
insert into test values(1,2),(1,3),(1,4)…
```

● 如果从不同客户插入很多行，可以通过使用 INSERT DELAYED 语句得到更高的速度。DELAYED 的含义是让 INSERT 语句马上执行，其实数据都被放在内存的队列中，并没有真正写入磁盘，这比每条语句分别插入要快得多；LOW_PRIORITY 刚好相反，在所有其他用户对表的读写完成后才进行插入。

● 将索引文件和数据文件分在不同的磁盘上存放（利用建表中的选项）。

● 如果进行批量插入，可以通过增加 bulk_insert_buffer_size 变量值的方法来提高速度，但是，这只能对 MyISAM 表使用。

● 当从一个文本文件装载一个表时，使用 LOAD DATA INFILE。这通常比使用很多 INSERT 语句快 20 倍。

15.4.3 优化 ORDER BY 语句

优化 ORDER BY 语句之前，首先来了解一下 MySQL 中的排序方式。先看 customer 表上的索引情况：

```
mysql> show index from customer\G
*************************** 1. row ***************************
        Table: customer
   Non_unique: 0
     Key_name: PRIMARY
 Seq_in_index: 1
  Column_name: customer_id
    Collation: A
```

```
        Cardinality: 541
          Sub_part: NULL
           Packed: NULL
             Null:
        Index_type: BTREE
          Comment:
*************************** 2. row ***************************
            Table: customer
       Non_unique: 1
         Key_name: idx_fk_store_id
     Seq_in_index: 1
      Column_name: store_id
        Collation: A
      Cardinality: 3
         Sub_part: NULL
           Packed: NULL
             Null:
        Index_type: BTREE
          Comment:
*************************** 3. row ***************************
            Table: customer
       Non_unique: 1
         Key_name: idx_fk_address_id
     Seq_in_index: 1
      Column_name: address_id
        Collation: A
      Cardinality: 541
         Sub_part: NULL
           Packed: NULL
             Null:
        Index_type: BTREE
          Comment:
*************************** 4. row ***************************
            Table: customer
       Non_unique: 1
         Key_name: idx_last_name
     Seq_in_index: 1
      Column_name: last_name
        Collation: A
      Cardinality: 541
         Sub_part: NULL
           Packed: NULL
             Null:
        Index_type: BTREE
          Comment:
4 rows in set (1.05 sec)
```

1. MySQL 中有两种排序方式

第一种通过有序索引顺序扫描直接返回有序数据，这种方式在使用 explain 分析查询的时候显示为 Using Index，不需要额外的排序，操作效率较高，例如：

```
mysql> explain select customer_id from customer order by store_id\G
*************************** 1. row ***************************
           id: 1
  select_type: SIMPLE
        table: customer
         type: index
possible_keys: NULL
          key: idx_fk_store_id
      key_len: 1
          ref: NULL
         rows: 541
        Extra: Using index
1 row in set (0.41 sec)
```

第二种是通过对返回数据进行排序，也就是通常说的 Filesort 排序，所有不是通过索引直接返回排序结果的排序都叫 Filesort 排序。Filesort 并不代表通过磁盘文件进行排序，而只是说明进行了一个排序操作,至于排序操作是否使用了磁盘文件或临时表等,则取决于 MySQL 服务器对排序参数的设置和需要排序数据的大小。例如,按照商店 store_id 排序返回所有客户记录时,出现了对全表扫描的结果的排序:

```
mysql> explain select * from customer order by store_id\G
*************************** 1. row ***************************
           id: 1
  select_type: SIMPLE
        table: customer
         type: ALL
possible_keys: NULL
          key: NULL
      key_len: NULL
          ref: NULL
         rows: 541
        Extra: Using filesort
1 row in set (0.68 sec)
```

又如,只需要获取商店 store_id 和顾客 email 信息时,对表 customer 的扫描就被对覆盖索引 idx_storeid_email 扫描替代,此时虽然只访问了索引就足够,但是在索引 idx_storeid_email 上发生了一次排序操作,所以执行计划中仍然有 Using Filesort。

```
mysql> alter table customer add index idx_storeid_email (store_id, email);
Query OK, 599 rows affected (1.17 sec)
Records: 599  Duplicates: 0  Warnings: 0

mysql> explain select store_id, email, customer_id from customer  order by email\G
*************************** 1. row ***************************
           id: 1
  select_type: SIMPLE
        table: customer
         type: index
possible_keys: NULL
          key: idx_storeid_email
      key_len: 154
          ref: NULL
         rows: 590
        Extra: Using index; Using filesort
1 row in set (0.10 sec)
```

Filesort 是通过相应的排序算法,将取得的数据在 sort_buffer_size 系统变量设置的内存排序区中进行排序,如果内存装载不下,它就会将磁盘上的数据进行分块,再对各个数据块进行排序,然后将各个块合并成有序的结果集。sort_buffer_size 设置的排序区是每个线程独占的,所以同一个时刻,MySQL 中存在多个 sort buffer 排序区。

了解了 MySQL 排序的方式,优化目标就清晰了:尽量减少额外的排序,通过索引直接返回有序数据。WHERE 条件和 ORDER BY 使用相同的索引,并且 ORDER BY 的顺序和索引顺序相同,并且 ORDER BY 的字段都是升序或者都是降序,否则肯定需要额外的排序操作,这样就会出现 Filesort。

例如,查询商店编号 store_id 为 1,按照 email 逆序排序的记录主键 customer_id 时,优化器使用扫描索引 idx_storeid_email 直接返回排序完毕的记录:

```
mysql> explain select store_id, email, customer_id from customer where store_id = 1 order by email desc \G
*************************** 1. row ***************************
           id: 1
  select_type: SIMPLE
        table: customer
         type: ref
possible_keys: idx_fk_store_id,idx_storeid_email
          key: idx_storeid_email
      key_len: 1
```

```
            ref: const
           rows: 295
          Extra: Using where; Using index
1 row in set (1.93 sec)
```

而查询商店编号 store_id 大于等于 1 小于等于 3，按照 email 排序的记录主键 customer_id 的时候，由于优化器评估使用索引 idx_storeid_email 进行范围扫描代价 cost 最低，所以最终是对索引扫描的结果，进行了额外的按照 email 逆序排序操作：

```
mysql> explain select store_id, email, customer_id from customer where store_id >= 1 and store_
id <= 3 order by email desc \G
*************************** 1. row ***************************
           id: 1
  select_type: SIMPLE
        table: customer
         type: range
possible_keys: idx_fk_store_id,idx_storeid_email
          key: idx_storeid_email
      key_len: 1
          ref: NULL
         rows: 295
        Extra: Using where; Using index; Using filesort
1 row in set (1.53 sec)
```

总结，下列 SQL 可以使用索引：

```
SELECT * FROM tabname ORDER BY key_part1,key_part2,... ;
SELECT * FROM tabname WHERE key_part1=1 ORDER BY key_part1 DESC, key_part2 DESC;
SELECT * FROM tabname ORDER BY key_part1 DESC, key_part2 DESC;
```

但是在以下几种情况下则不使用索引：

```
SELECT * FROM tabname ORDER BY key_part1 DESC, key_part2 ASC;
--order by 的字段混合 ASC 和 DESC
SELECT * FROM tabname WHERE key2=constant ORDER BY key1;
--用于查询行的关键字与 ORDER BY 中所使用的不相同
SELECT * FROM tabname ORDER BY key1, key2;
--对不同的关键字使用 ORDER BY；
```

2．Filesort 的优化

通过创建合适的索引能够减少 Filesort 出现，但是在某些情况下，条件限制不能让 Filesort 消失，那就需要想办法加快 Filesort 的操作。对于 Filesort，MySQL 有两种排序算法。

- 两次扫描算法（Two Passes）：首先根据条件取出排序字段和行指针信息，之后在排序区 sort buffer 中排序。如果排序区 sort buffer 不够，则在临时表 Temporary Table 中存储排序结果。完成排序后根据行指针回表读取记录。该算法是 MySQL 4.1 之前采用的算法，需要两次访问数据，第一次获取排序字段和行指针信息，第二次根据行指针获取记录，尤其是第二次读取操作可能导致大量随机 I/O 操作；优点是排序的时候内存开销较少。

- 一次扫描算法（Single Pass）：一次性取出满足条件的行的所有字段，然后在排序区 sort buffer 中排序后直接输出结果集。排序的时候内存开销比较大，但是排序效率比两次扫描算法要高。

MySQL 通过比较系统变量 max_length_for_sort_data 的大小和 Query 语句取出的字段总大小来判断使用哪种排序算法。如果 max_length_for_sort_data 更大，那么使用第二种优化之后的算法；否则使用第一种算法。

适当加大系统变量 max_length_for_sort_data 的值，能够让 MySQL 选择更优化的 Filesort 排序算法。当然，假如 max_length_for_sort_data 设置过大，会造成 CPU 利用率过低和磁盘 I/O 过高，CPU 和 I/O 利用平衡就足够了。

15.4 常用 SQL 的优化

适当加大 sort_buffer_size 排序区,尽量让排序在内存中完成,而不是通过创建临时表放在文件中进行;当然也不能无限制加大 sort_buffer_size 排序区,因为 sort_buffer_size 参数是每个线程独占的,设置过大,会导致服务器 SWAP 严重,要考虑数据库活动连接数和服务器内存的大小来适当设置排序区。

尽量只使用必要的字段,SELECT 具体的字段名称,而不是 SELECT * 选择所有字段,这样可以减少排序区的使用,提高 SQL 性能。

15.4.4 优化 GROUP BY 语句

默认情况下,MySQL 对所有 GROUP BY col1,col2,…的字段进行排序。这与在查询中指定 ORDER BY col1,col2,…类似。因此,如果显式包括一个包含相同列的 ORDER BY 子句,则对 MySQL 的实际执行性能没有什么影响。

如果查询包括 GROUP BY 但用户想要避免排序结果的消耗,则可以指定 ORDER BY NULL 禁止排序,如下面的例子:

```
mysql> explain select payment_date, sum(amount) from payment group by payment_date\G
*************************** 1. row ***************************
           id: 1
  select_type: SIMPLE
        table: payment
         type: ALL
possible_keys: NULL
          key: NULL
      key_len: NULL
          ref: NULL
         rows: 16310
        Extra: Using temporary; Using filesort
1 row in set (0.01 sec)

mysql> explain select payment_date, sum(amount) from payment group by payment_date order by null\G
*************************** 1. row ***************************
           id: 1
  select_type: SIMPLE
        table: payment
         type: ALL
possible_keys: NULL
          key: NULL
      key_len: NULL
          ref: NULL
         rows: 16310
        Extra: Using temporary
1 row in set (0.06 sec)
```

从上面的例子可以看出,第一个 SQL 语句需要进行"Filesort",而第二个 SQL 由于 ORDER BY NULL 不需要进行"Filesort",而上文提过 Filesort 往往非常耗费时间。

15.4.5 优化 JOIN 操作

MySQL 对于多表 JOIN 在目前只支持一种算法——Nested-Loop Join(NLJ)。NLJ 的原理非常简单,就是内外两层循环,对于外循环中的每条记录,都要在内循环中做一次检索,如下面的伪代码所示:

```
for each row in t1 matching range {
    for each row in t2 matching reference key {
        if row satisfies join conditions, send to client
```

```
    }
}
```

其中 t1 和 t2 表进行 join，t1 通过范围扫描取每条记录作为外循环，t2 通过关联字段在表中做扫描，满足条件则返回客户端；不断重复这个过程直到外循环结束。外循环的表通常也称为驱动表。

从这个流程来看，NLJ 的性能高低主要取决于两方面：一是外循环的结果集大小，二是内循环扫描数据的效率。常见的优化方案是在驱动表上加上尽可能的 where 条件并创建合适索引，使得外循环的结果集更小，读取效率更高；内循环为了提高扫描效率，通常需要在关联字段上加索引。

通过上面的优化，在大多数情况下，NLJ 的性能是可以满足需求的，尤其是关联字段在内循环是主键或者唯一索引时效率尤其高。但有两种情况，NLJ 的性能会有比较明显地下降。

- 外循环结果集大，导致访问内循环表的 io 次数非常多。
- 内循环的关联字段并不是唯一索引，而是普通的辅助索引。如果访问的数据列不在辅助索引上，此时通常需要再次回表，通过辅助索引的主键找到聚集索引的实际数据，而回表会导致大量的随机 io 产生，导致性能下降明显。

为了优化这两个问题，MySQL 先后推出了 NLJ 的变种 BNL（Block Nested-Loop Join）和 BKA（Batched Key Access）。

1. BNL

BNL 在 MySQL 较早版本就引入，算法伪代码如下：

```
for each row in t1 matching range {
  for each row in t2 matching reference key {
    store used columns from t1, t2 in join buffer
    if buffer is full {
      for each row in t3 {
        for each t1, t2 combination in join buffer {
          if row satisfies join conditions, send to client
        }
      }
      empty join buffer
    }
  }
}

if buffer is not empty {
  for each row in t3 {
    for each t1, t2 combination in join buffer {
      if row satisfies join conditions, send to client
    }
  }
}
```

通过缓存外层循环读的行，来降低内层表的读取次数。例如，10 行数据读入到 buffer 中，然后 buffer 被传递到内层循环，内层表读出的每一行都要跟这个缓存的 10 行依次做对比，这样就降低了内层表数据的读取次数。

在 MySQL 5.7 中，BNL 优化器特性默认是打开的，以下示例将 customer 和 payment 表进行 join，关联字段上均无索引：

```
mysql> show variables like 'optimizer_switch' \G;
*************************** 1. row ***************************
Variable_name: optimizer_switch
        Value: index_merge=on,index_merge_union=on,index_merge_sort_union=on,index_merge_
```

```
intersection=on,engine_condition_pushdown=on,index_condition_pushdown=on,mrr=on,mrr_cost_based=
on,block_nested_loop=on,batched_key_access=off,materialization=on,semijoin=on,loosescan=on,
firstmatch=on,duplicateweedout=on,subquery_materialization_cost_based=on,use_index_extensions=
on,condition_fanout_filter=on,derived_merge=on
1 row in set (0.00 sec)

mysql> desc  select count(1) from customer a,payment b where a.create_date=b.payment_date;
+----+-------------+-------+------------+------+---------------+------+---------+------+-------+----------+-------------------------------------------------+
| id | select_type | table | partitions | type | possible_keys | key  | key_len | ref  | rows  | filtered | Extra                                           |
+----+-------------+-------+------------+------+---------------+------+---------+------+-------+----------+-------------------------------------------------+
|  1 | SIMPLE      | a     | NULL       | ALL  | NULL          | NULL | NULL    | NULL |   599 |   100.00 | NULL                                            |
|  1 | SIMPLE      | b     | NULL       | ALL  | NULL          | NULL | NULL    | NULL | 16086 |    10.00 | Using where; Using join buffer (Block Nested Loop) |
+----+-------------+-------+------------+------+---------------+------+---------+------+-------+----------+-------------------------------------------------+
2 rows in set, 1 warning (0.01 sec)
```

如执行计划的 Extra 部分显示,连接使用了 BNL。实际执行上面的 SQL,完成时间为 1.54s:

```
mysql>  select count(1) from customer a,payment b where a.create_date=b.payment_date;
+----------+
| count(1) |
+----------+
|        0 |
+----------+
1 row in set (1.54 sec)
```

关闭 BNL 特性,再次观察一下:

```
mysql>  set optimizer_switch='block_nested_loop=off';
Query OK, 0 rows affected (0.01 sec)

mysql> desc  select count(1) from customer a,payment b where a.create_date=b.payment_date;
+----+-------------+-------+------------+------+---------------+------+---------+------+-------+----------+-------------+
| id | select_type | table | partitions | type | possible_keys | key  | key_len | ref  | rows  | filtered | Extra       |
+----+-------------+-------+------------+------+---------------+------+---------+------+-------+----------+-------------+
|  1 | SIMPLE      | a     | NULL       | ALL  | NULL          | NULL | NULL    | NULL |   599 |   100.00 | NULL        |
|  1 | SIMPLE      | b     | NULL       | ALL  | NULL          | NULL | NULL    | NULL | 16086 |    10.00 | Using where |
+----+-------------+-------+------------+------+---------------+------+---------+------+-------+----------+-------------+
2 rows in set, 1 warning (0.00 sec)
```

执行计划中的 BNL 部分消失,再次执行 SQL:

```
mysql>  select count(1) from customer a,payment b where a.create_date=b.payment_date;
+----------+
| count(1) |
+----------+
|        0 |
+----------+
1 row in set (4.25 sec)
```

完成时间为 4.25s,比使用 BNL 特性慢了近两倍。

BNL 性能虽然大幅提高,但使用条件较为苛刻,只有当 join 类型是 all/index/range 时才可以,也就是内表不使用索引或者索引效率很低时才不得已使用。buffer 的大小由参数 join_buffer_size 进行设置,buffer 中保存参与连接的所有列信息,join 完成后 buffer 释放。对于使用到 BNL 特性且性能较差的 SQL,建议在 session 级别将 join_buffer_size 临时增大来提高性能。

2. MRR & BKA

从上面的介绍知道，BNL 的使用场景较为苛刻，最重要的条件是内表关联字段没有索引或者索引效率很低，此时使用 BNL 可以较明显的降低访问内表的次数，同时降低回表的 IO 次数，以此来达到优化的目的。但在大多数情况下，表的 join 操作通常是通过效率较高的索引来做 ref 或者 eq_ref 方式进行连接，这种情况下，BNL 是无法使用的。为了优化这种更常见的 join，MySQL 引入了 MRR 和 BKA。

MRR（Multi Range Read）是 MySQL 5.6 引入的特性。MRR 优化的目的就是为了减少磁盘的随机访问，InnoDB 由于聚集索引的特性，如果查询使用辅助索引，并且用到表中非索引列，那么需要回表读取数据做后续处理，过于随机的回表会伴随着大量的随机 IO。而 MRR 的优化并不是每次通过辅助索引读取到数据就回表，而是通过范围扫描将数据存入 read_rnd_buffer_size，然后对其按照 Primary Key（RowID）排序，最后使用排序好的数据进行顺序回表，因为 InnoDB 中叶子节点数据是按照 Primary Key（RowID）进行排列的，这样就转换随机 IO 为顺序 IO 了，对于瓶颈为 IO 的 SQL 查询语句将带来极大的性能提升。

MRR 特性在单表和多表 join 查询中都可以使用。其中，单表通常通过范围查询（range access）；多表 join 方式如果是 ref/eq_ref，则先通过 BKA 算法（后面介绍）批量提取 key 到 join buffer，然后将 buffer 中的 key 作为参数传入 MRR 的调用接口，MRR 高效读取需要的数据返回，过程如图 15-5 所示。

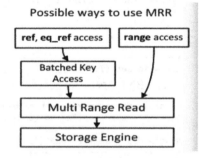

图 15-5　MRR 的使用方式

如果要打开 MRR 特性，则需要设置以下两个优化器参数：

```
set optimizer_switch='mrr=on,mrr_cost_based=off';
```

其中 mrr 参数控制 MRR 特性是否打开，默认为 on；mrr_cost_based 控制是否根据优化器的计算成本来决定使用 MRR 特性，默认是 on；如果希望尽可能使用 MRR，可以将此参数设置为 off。

要查看是否使用了 MRR 特性，需要观察在执行计划的 Extra 部分是否存在 "Using MRR" 字符串，下例 SQL 就使用了 MRR 特性：

```
mysql> desc select * from  payment where customer_id between 1 and 200 \G;
*************************** 1. row ***************************
           id: 1
  select_type: SIMPLE
        table: payment
   partitions: NULL
         type: range
possible_keys: idx_fk_customer_id
          key: idx_fk_customer_id
      key_len: 2
          ref: NULL
         rows: 5444
     filtered: 100.00
        Extra: Using index condition; Using MRR
1 row in set, 1 warning (0.00 sec)
```

BKA（Batched Key Access）是 MySQL 5.6 引入的新算法，结合 MRR 特性进行高效 JOIN

操作，算法具体工作步骤如下。

- 将外循环表中相关的列放入 Join Buffer 中。
- 批量的将 Key（索引键值）发送到 MRR 接口。
- MRR 通过收到的 Key，根据其对应的 Primary Key(RowID)进行排序，然后再根据排序后的 Primary Key(RowID)顺序的读取聚集索引，得到需要的列数据。
- 返回结果集给客户端。

MySQL 5.7 以后，BKA 默认是打开的，由优化器中的参数 batched_key_access 来控制。如果要使用 BKA，则先需打开 MRR 特性，通常一起设置如下参数：

```
mysql> SET optimizer_switch='mrr=on,mrr_cost_based=off,batched_key_access=on';
         Query OK, 0 rows affected (0.00 sec)
```

判断是否使用了 BKA 算法，需要查看执行计划中 extra 部分是否含有"Using join buffer (Batched Key Access)"字符串，如下例中倒数第 2 行所示：

```
mysql> create index em_diredate on employess(hire_date); desc    select count(*) from employees a,salaries b where a.hire_date=b.to_date and b.salary>5000 and a.gender=1 \G;
*************************** 1. row ***************************
           id: 1
  select_type: SIMPLE
        table: b
   partitions: NULL
         type: ALL
possible_keys: test_salaries
          key: NULL
      key_len: NULL
          ref: NULL
         rows: 2790144
     filtered: 33.33
        Extra: Using where
*************************** 2. row ***************************
           id: 1
  select_type: SIMPLE
        table: a
   partitions: NULL
         type: ref
possible_keys: emp_hiredate
          key: emp_hiredate
      key_len: 3
          ref: employees.b.to_date
         rows: 53
     filtered: 50.00
        Extra: Using where; Using join buffer (Batched Key Access)
2 rows in set, 1 warning (0.00 sec)
```

通过 BKA 来做 JOIN，很多情况下可以提高连接的效率，但对 JOIN 也有一定的条件限制，一个条件是连接的列要求是唯一索引或者普通索引，但不能是主键；另一个是要有对非主键列的查询操作，否则优化器就可以通过覆盖索引等方式直接得到需要的数据，不需要回表，也就不需要用到 MRR 接口。

15.4.6 优化嵌套查询

MySQL 4.1 开始支持 SQL 的子查询。这个技术可以使用 SELECT 语句来创建一个单列的查询结果，然后把这个结果作为过滤条件用在另一个查询中。使用子查询可以一次性地完成很多逻辑上需要多个步骤才能完成的 SQL 操作，同时也可以避免事务或者表锁死，并且写起来也很容易。但是，有些情况下，子查询可以被更有效率的连接（JOIN）替代。

在下面的例子中，要从客户表 customer 中找到不在支付表 payment 中的所有客户信息：

```
mysql> explain select * from customer where customer_id not in (select customer_id from payment )\G
*************************** 1. row ***************************
           id: 1
  select_type: PRIMARY
        table: customer
   partitions: NULL
         type: ALL
possible_keys: NULL
          key: NULL
      key_len: NULL
          ref: NULL
         rows: 599
     filtered: 100.00
        Extra: Using where
*************************** 2. row ***************************
           id: 2
  select_type: DEPENDENT SUBQUERY
        table: payment
   partitions: NULL
         type: index_subquery
possible_keys: idx_fk_customer_id
          key: idx_fk_customer_id
      key_len: 2
          ref: func
         rows: 26
     filtered: 100.00
        Extra: Using index
2 rows in set (0.00 sec)
```

如果使用连接（JOIN）来完成这个查询工作，速度将会快很多。尤其是当 payment 表中对 customer_id 建有索引，性能将会更好，具体查询如下：

```
mysql> explain select * from customer a left join payment b on a.customer_id = b.customer_id where b.customer_id is null\G
*************************** 1. row ***************************
           id: 1
  select_type: SIMPLE
        table: a
   partitions: NULL
         type: ALL
possible_keys: NULL
          key: NULL
      key_len: NULL
          ref: NULL
         rows: 599
     filtered: 100.00
        Extra: NULL
*************************** 2. row ***************************
           id: 1
  select_type: SIMPLE
        table: b
   partitions: NULL
         type: ref
possible_keys: idx_fk_customer_id
          key: idx_fk_customer_id
      key_len: 2
          ref: sakila.a.customer_id
         rows: 26
     filtered: 100.00
        Extra: Using where; Not exists
2 rows in set (0.00 sec)
```

从执行计划中可以看出查询关联的类型从 index_subquery 调整为了 ref，在 MySQL 5.5 以下版本（包括 MySQL 5.5），子查询的效率还是不如关联查询（JOIN）。

连接（JOIN）之所以更有效率一些，是因为 MySQL 不需要在内存中创建临时表来完成这个逻辑上需要两个步骤的查询工作。

15.4.7 MySQL 如何优化 OR 条件

对于含有 OR 的查询子句，如果要利用索引，则 OR 之间的每个条件列都必须用到索引；如果没有索引，则应该考虑增加索引。

例如，首先使用 show index 命令查看表 sales2 的索引，可知它有 3 个索引，在 id 和 year 两个字段上分别有 1 个独立的索引，在 company_id 和 year 字段上有 1 个复合索引。

```
mysql> show index from sales2\G;
*************************** 1. row ***************************
        Table: sales2
   Non_unique: 1
     Key_name: ind_sales2_id
 Seq_in_index: 1
  Column_name: id
    Collation: A
  Cardinality: 1000
     Sub_part: NULL
       Packed: NULL
         Null: YES
   Index_type: BTREE
      Comment:
*************************** 2. row ***************************
        Table: sales2
   Non_unique: 1
     Key_name: ind_sales2_year
 Seq_in_index: 1
  Column_name: year
    Collation: A
  Cardinality: 250
     Sub_part: NULL
       Packed: NULL
         Null: YES
   Index_type: BTREE
      Comment:
*************************** 3. row ***************************
        Table: sales2
   Non_unique: 1
     Key_name: ind_sales2_companyid_moneys
 Seq_in_index: 1
  Column_name: company_id
    Collation: A
  Cardinality: 1000
     Sub_part: NULL
       Packed: NULL
         Null: YES
   Index_type: BTREE
      Comment:
*************************** 4. row ***************************
        Table: sales2
   Non_unique: 1
     Key_name: ind_sales2_companyid_moneys
 Seq_in_index: 2
  Column_name: year
    Collation: A
  Cardinality: 1000
     Sub_part: NULL
       Packed: NULL
         Null: YES
   Index_type: BTREE
      Comment:
4 rows in set (0.00 sec)
```

然后在两个独立索引上面做 OR 操作，具体如下：

```
mysql> explain select * from sales2 where id = 2 or year = 1998\G;
*************************** 1. row ***************************
           id: 1
  select_type: SIMPLE
        table: sales2
         type: index_merge
possible_keys: ind_sales2_id,ind_sales2_year
          key: ind_sales2_id,ind_sales2_year
      key_len: 5,2
          ref: NULL
         rows: 2
        Extra: Using union(ind_sales2_id,ind_sales2_year); Using where
1 row in set (0.00 sec)
```

可以发现查询正确地用到了索引，并且从执行计划的描述中，发现 MySQL 在处理含有 OR 字句的查询时，实际是对 OR 的各个字段分别查询后的结果进行了 UNION 操作。

但是当在建有复合索引的列 company_id 和 moneys 上面做 OR 操作时，却不能用到索引，具体结果如下：

```
mysql> explain select * from sales2 where company_id = 3 or moneys = 100\G;
*************************** 1. row ***************************
           id: 1
  select_type: SIMPLE
        table: sales2
         type: ALL
possible_keys: ind_sales2_companyid_moneys
          key: NULL
      key_len: NULL
          ref: NULL
         rows: 1000
        Extra: Using where
1 row in set (0.00 sec)
)
```

15.4.8 优化分页查询

一般分页查询时，通过创建覆盖索引能够比较好地提高性能。一个常见又非常头痛的分页场景是："limit 1000,20"，此时 MySQL 排序出前 1020 条记录后仅仅需要返回第 1001 到 1020 条记录，前 1000 条记录都会被抛弃，查询和排序的代价非常高。

1. 第一种优化思路

在索引上完成排序分页的操作，最后根据主键关联回原表查询所需要的其他列内容。例如，对电影表 film 根据标题 title 排序后取某一页数据，直接查询的时候，能够从 explain 的输出结果中看到优化器实际上做了全表扫描，处理效率不高：

```
mysql> explain select film_id, description from film order by title limit 50,5\G
*************************** 1. row ***************************
           id: 1
  select_type: SIMPLE
        table: film
         type: ALL
possible_keys: NULL
          key: NULL
      key_len: NULL
          ref: NULL
         rows: 919
        Extra: Using filesort
1 row in set (0.00 sec)
```

而按照索引分页后回表方式改写 SQL 后，从 explain 的输出结果中已经看不到全表扫描了：

```
mysql> explain select a.film_id, a.description from film a inner join (select film_id from
film order by title limit 50,5)b on a.film_id = b.film_id\G
*************************** 1. row ***************************
           id: 1
  select_type: PRIMARY
        table: <derived2>
         type: ALL
possible_keys: NULL
          key: NULL
      key_len: NULL
          ref: NULL
         rows: 5
        Extra:
*************************** 2. row ***************************
           id: 1
  select_type: PRIMARY
        table: a
         type: eq_ref
possible_keys: PRIMARY
          key: PRIMARY
      key_len: 2
          ref: b.film_id
         rows: 1
        Extra:
*************************** 3. row ***************************
           id: 2
  select_type: DERIVED
        table: film
         type: index
possible_keys: NULL
          key: idx_title
      key_len: 767
          ref: NULL
         rows: 55
        Extra: Using index
3 rows din set (0.00 sec)
```

这种方式让 MySQL 扫描尽可能少的页面来提高分页效率。

2. 第二种优化思路

把 LIMIIT 查询转换成某个位置的查询，例如，假设每页 10 条记录，查询支付表 payment 中按照租赁编号 rental_id 逆序排序的第 42 页记录，能够看到执行计划走了全表扫描：

```
mysql> explain select * from payment order by rental_id desc limit 410,10\G
*************************** 1. row ***************************
           id: 1
  select_type: SIMPLE
        table: payment
         type: ALL
possible_keys: NULL
          key: NULL
      key_len: NULL
          ref: NULL
         rows: 16451
        Extra: Using filesort
1 row in set (0.00 sec)
```

和开发人员协商一下，翻页的过程中通过增加一个参数 last_page_record，用来记录上一页最后一行的租赁编号 rental_id，例如第 41 页最后一行的租赁编号 rental_id=15640：

```
mysql> select payment_id, rental_id from payment order by rental_id desc limit 400,10;
+------------+-----------+
| payment_id | rental_id |
+------------+-----------+
|       1669 |     15649 |
|       2193 |     15648 |
|       6785 |     15647 |
```

```
|      3088 |       15646 |
|      5831 |       15645 |
|      1201 |       15644 |
|      8105 |       15643 |
|      4369 |       15642 |
|      6499 |       15641 |
|      7095 |       15640 |
+-----------+-------------+
10 rows in set (0.00 sec)
```

那么在翻页到第 42 页时,可以根据第 41 页最后一条记录向后追溯,相应的 SQL 可以改写为:

```
mysql> explain select * from payment where rental_id < 15640 order by rental_id desc limit 10\G
*************************** 1. row ***************************
           id: 1
  select_type: SIMPLE
        table: payment
         type: range
possible_keys: fk_payment_rental
          key: fk_payment_rental
      key_len: 5
          ref: NULL
         rows: 8225
        Extra: Using where
1 row in set (0.00 sec)
```

注意:这样把 LIMIT m,n 转换成 LIMIT n 的查询,只适合在排序字段不会出现重复值的特定环境,能够减轻分页翻页的压力;如果排序字段出现大量重复值,而仍进行这种优化,那么分页结果可能会丢失部分记录,不适用这种方式进行优化。

15.4.9 使用 SQL 提示

SQL 提示(SQL HINT)是优化数据库的一个重要手段,简单来说就是在 SQL 语句中加入一些人为的提示来达到优化操作的目的。下面是一个使用 SQL 提示的示例:

```
SELECT SQL_BUFFER_RESULTS * FROM...
```

这个语句将强制 MySQL 生成一个临时结果集。只要临时结果集生成后,所有表上的锁定均被释放。这能在遇到表锁定问题时或要花很长时间将结果传给客户端时有所帮助,因为可以尽快释放锁资源。

下面是一些在 MySQL 中常用的 SQL 提示。

1. USE INDEX

在查询语句中表名的后面,添加 USE INDEX 来提供希望 MySQL 去参考的索引列表,就可以让 MySQL 不再考虑其他可用的索引。

```
mysql> explain select count(*) from rental use index (idx_rental_date)\G
*************************** 1. row ***************************
           id: 1
  select_type: SIMPLE
        table: rental
         type: index
possible_keys: NULL
          key: idx_rental_date
      key_len: 13
          ref: NULL
         rows: 16291
        Extra: Using index
1 row in set (0.00 sec)
```

2. IGNORE INDEX

如果用户只是单纯地想让 MySQL 忽略一个或者多个索引,则可以使用 IGNORE INDEX 作为 HINT。同样是上面的例子,这次来看一下查询过程忽略索引 ind_sales2_id 的情况:

```
mysql> explain select count(*) from rental ignore index (idx_rental_date)\G
*************************** 1. row ***************************
           id: 1
  select_type: SIMPLE
        table: rental
         type: index
possible_keys: NULL
          key: idx_fk_staff_id
      key_len: 1
          ref: NULL
         rows: 16291
        Extra: Using index
1 row in set (0.00 sec)
```

从执行计划可以看出,系统忽略了指定的索引,使用索引 idx_fk_staff_id。

3. FORCE INDEX

为强制 MySQL 使用一个特定的索引,可在查询中使用 FORCE INDEX 作为 HINT。例如,当不强制使用索引的时候,因为大部分库存 inventory_id 的值都是大于 1 的,因此 MySQL 会默认进行全表扫描,而不使用索引,如下所示:

```
mysql> explain select * from rental  where inventory_id > 1\G
*************************** 1. row ***************************
           id: 1
  select_type: SIMPLE
        table: rental
         type: ALL
possible_keys: idx_fk_inventory_id
          key: NULL
      key_len: NULL
          ref: NULL
         rows: 16291
        Extra: Using where
1 row in set (0.00 sec)
```

尝试使用 use index 的 hint 看看:

```
mysql> explain select * from rental use index (idx_fk_inventory_id) where inventory_id > 1\G
*************************** 1. row ***************************
           id: 1
  select_type: SIMPLE
        table: rental
         type: ALL
possible_keys: idx_fk_inventory_id
          key: NULL
      key_len: NULL
          ref: NULL
         rows: 16291
        Extra: Using where
1 row in set (0.00 sec)
```

发现仍然不行,MySQL 还是选择走全表扫描。但是,当使用 FORCE INDEX 进行提示时,即便使用索引的效率不是最高,MySQL 还是选择使用了索引,这是 MySQL 留给用户的一个自行选择执行计划的权力。加入 FORCE INDEX 提示后再次执行以上 SQL 语句:

```
mysql> explain select * from rental force index (idx_fk_inventory_id) where inventory_id > 1\G
*************************** 1. row ***************************
          id : 1
 select_type : SIMPLE
```

```
            table : rental
             type : range
    possible_keys : idx_fk_inventory_id
              key : idx_fk_inventory_id
          key_len : 3
              ref : NULL
             rows : 8145
            Extra : Using where
1 row in set (0.01 sec)
```

果然,执行计划中使用了 FORCE INDEX 后的索引。

15.5 直方图

直方图是 MySQL 8.0 引入的新功能。利用直方图,用户可以对一张表的一列做数据分布的统计,特别是针对没有索引的字段。这可以帮助优化器找到更优的执行计划。统计直方图的主要使用场景是用来计算字段选择性,即过滤效率。

15.5.1 什么是直方图

在数据库中,查询优化器负责将 SQL 转换成最有效的执行计划。但有时,由于一些字段的数据分布不均衡,导致优化器针对某些值不会选择最优的执行计划,从而使得执行效率降低。为了能做出更准确的选择,优化器需要了解条件列中具体的数据分布情况,而直方图的引入就是为了统计这些信息。

直方图的主要操作命令有以下两个。

- 生成直方图:

```
ANALYZE TABLE tbl_name UPDATE HISTOGRAM ON col_name [, col_name] WITH N BUCKETS;
```

- 删除直方图:

```
ANALYZE TABLE tbl_name DROP HISTOGRAM ON col_name [, col_name];
```

其中,BUCKETS 表示生成桶的个数,桶用来存放列中不同值的分布情况,默认值为 100,最大到 1024。

举一个简单的例子,员工表 emp1 的性别字段 gender 数据分布如下:

```
mysql> select gender,count(1) from emp1 group by gender;
+--------+----------+
| gender | count(1) |
+--------+----------+
| M      |   299025 |
| F      |      999 |
+--------+----------+
```

如果没有直方图,查询 gender 为 "M" 或者 "F",执行计划如下:

```
mysql> desc select count(1) from emp1 where gender='F' \G;
*************************** 1. row ***************************
           id: 1
  select_type: SIMPLE
        table: emp1
   partitions: NULL
         type: ALL
possible_keys: NULL
          key: NULL
      key_len: NULL
          ref: NULL
         rows: 299556
```

```
        filtered: 50.00
           Extra: Using where
1 row in set, 1 warning (0.00 sec)
```

可以发现，执行计划中的 filtered 值都是 50%，即优化器不知道数据实际分布情况，只是按照值的个数进行平均分配。如果在 gender 上创建了直方图，则执行计划会按照实际的数据分布进行过滤，如下所示：

```
mysql> analyze table emp1 update histogram on gender  \G;
*************************** 1. row ***************************
   Table: employees.emp1
      Op: histogram
Msg_type: status
Msg_text: Histogram statistics created for column 'gender'.
1 row in set (0.32 sec)

mysql> desc select count(1) from emp1 where gender='F' \G;
*************************** 1. row ***************************
           id: 1
  select_type: SIMPLE
        table: emp1
   partitions: NULL
         type: ALL
possible_keys: NULL
          key: NULL
      key_len: NULL
          ref: NULL
         rows: 299556
     filtered: 0.33
        Extra: Using where
1 row in set, 1 warning (0.00 sec)
```

上面的 analyze 命令在 emp1 的 gender 字段上创建了直方图，后面的执行计划显示 filtered 属性已经从 50% 改为 0.33%，符合实际分布情况。这个信息使得优化器对 gender 字段查询的代价计算更为精确，从而在某些 gender 查询相关的 SQL 中生成更为高效的执行计划。

15.5.2 直方图的分类

MySQL 目前支持两种直方图类型：等宽直方图（singleton）和等高直方图（equi-height）。它们的共同点是，都将数据分到了一系列的 buckets 中；区别在于如果列中不同值的个数小于等于 bucket 数，则为等宽直方图；反之则为等高直方图。MySQL 会自动将数据划到不同的 bucket 中，也会自动决定创建哪种类型的直方图。

直方图的统计信息存放在 information_schema 库的 column_statistics 视图中，以 json 格式保存。上例中 gender 列的直方图信息内容如下：

```
mysql> select * from information_schema.column_statistics \G;
*************************** 1. row ***************************
SCHEMA_NAME: employees
 TABLE_NAME: emp1
COLUMN_NAME: gender
  HISTOGRAM: {"buckets": [[1, 0.9966702663786897], [2, 1.0]], "data-type": "enum", "null-values"
: 0.0, "collation-id": 192, "last-updated": "2018-10-22 02:58:20.035927", "sampling-rate": 1.0,
"histogram-type": "singleton", "number-of-buckets-specified": 100}
1 row in set (0.00 sec)
```

其中，HISTOGRAM 列记录了 gender 字段的直方图信息，内容包括如下几项：

- buckets: [[1, 0.9966702663786897], [2, 1.0]] 表示 gender 上创建了两个 bucket，bucket 为 1 的值包含了 99.67% 的数据；bucket 为 2 的值包含了 1−99.67%=0.33% 的数据，即 2 后面的 1.0 是个累积的数据分布。由于 enum 保存的并不是字面值，这里的 1 和 2 是 enum 类型的

"M"和"F"实际保存的值。
- data-type:"enum"表示 gender 字段是枚举类型。
- null-values:0.0 表示 gender 列中没有空值。
- collation-id:192 表示排序方式,对应 information_schema 下 collations 中的 id 字段。
- last-updated:表示直方图的最后更新日期。直方图只能手工 update,而不会随着数据的更新而更新;当数据量发生大的变更时,要重新手工生成新的直方图。
- sampling-rate:1.0,表示数据的采样率。对于数据量巨大的表,MySQL 出于性能考虑,不会全部扫描,只会采样部分数据来生成直方图。采样大小由参数 histogram_generation_max_mem_size 进行控制,这个值是控制最大多少内存能允许被使用。
- histogram-type:生成的直方图类型,singleton 表示等宽直方图,equi-height 表示等高直方图。
- number-of-buckets-specified:表示指定了多少个 bucket。对于等宽直方图来说,如果创建直方图时指定的 buckets 的个数大于列中唯一值的个数,则实际只需要创建唯一值的个数的 buckets 就可以了。本例中的值为 100,而实际创建的 bucket 是两个。

对于等高直方图,bucket 的显示内容有所不同,由于列中不同值的个数大于 bucket 的数量,因此每个 bucket 上对应的不是一个值,而是一个具有上下限的列值范围。下例是一个具有 8 个 buckets 的等高直方图:

```
{"buckets": [["1952-02-01", "1953-09-17", 0.12507666053382396, 595], ["1953-09-18", "1955-04
-30", 0.24993667173292802, 590], ["1955-05-01", "1956-12-16", 0.375, 596], ["1956-12-17", "1958-
08-01", 0.5001199904007679, 593], ["1958-08-02", "1960-03-12", 0.625059995200384, 589], ["1960-
03-13", "1961-10-27", 0.749953337066368, 594], ["1961-10-28", "1963-06-16", 0.87506666133376,
597], ["1963-06-17", "1965-02-01", 1.0, 596]], "data-type": "date", "null-values": 0.0, "collation-
id": 8, "last-updated": "2018-10-22 14:36:53.303565", "sampling-rate": 1.0, "histogram-type":
"equi-height", "number-of-buckets-specified": 8}
```

可以看出,每个 bucket 由 4 个值组成,第一、二个值表示列值上下限范围,第三个值表示列值在此范围内的记录数占总记录的百分比,第四个值表示 bucket 内列上唯一值的估算数量,通常也称为 Cardinality。通常来说,对于等高直方图,bucket 的个数越多,统计的分布越精确,但很多情况下,单纯提高 bucket 数量对数据分布的准确性不会有明显提高,建议初始时先设置较小的值,如果达不到效果再逐渐增加,重新设置 bucket 的命令和创建命令一样。

15.5.3 直方图实例应用

上面简单介绍了直方图的概念和分类,本节给出一个示例来验证一下直方图在 SQL 优化中的作用。

两个表 emp1 和 titles1 进行 join,SQL 如下:

```
select count(1) from  emp1 a,titles1 b  where a.emp_no=b.emp_no and a.gender='M' ;
```

两表的定义和数据量分别如下:

```
mysql> desc emp1
    -> ;
+------------+--------+------+-----+---------+-------+
| Field      | Type   | Null | Key | Default | Extra |
+------------+--------+------+-----+---------+-------+
| emp_no     | int(11)| NO   | PRI | NULL    |       |
| birth_date | date   | NO   |     | NULL    |       |
```

15.5 直方图

```
| first_name | varchar(14)   | NO  |     | NULL    |       |
| last_name  | varchar(16)   | NO  |     | NULL    |       |
| gender     | enum('M','F') | NO  |     | NULL    |       |
| hire_date  | date          | NO  |     | NULL    |       |
| gender1    | varchar(10)   | YES |     | NULL    |       |
+------------+---------------+-----+-----+---------+-------+
7 rows in set (0.01 sec)

mysql> select count(1) from emp1;
+----------+
| count(1) |
+----------+
|   300024 |
+----------+

mysql> desc titles1;
+-----------+-------------+------+-----+---------+-------+
| Field     | Type        | Null | Key | Default | Extra |
+-----------+-------------+------+-----+---------+-------+
| emp_no    | int(11)     | NO   | PRI | NULL    |       |
| title     | varchar(50) | NO   | PRI | NULL    |       |
| from_date | date        | NO   | PRI | NULL    |       |
| to_date   | date        | YES  |     | NULL    |       |
+-----------+-------------+------+-----+---------+-------+
4 rows in set (0.00 sec)

mysql> select count(1) from titles1;
+----------+
| count(1) |
+----------+
|   262408 |
+----------+
```

gender 字段的数据分布不均衡，如下所示：

```
mysql> select gender,count(1) from emp1 group by gender;
+--------+----------+
| gender | count(1) |
+--------+----------+
| F      |      999 |
| M      |   299025 |
+--------+----------+
2 rows in set (0.41 sec)
```

gender 字段创建直方图之前，执行计划如下：

```
mysql> desc  select count(1) from  emp1 a,titles1 b  where a.emp_no=b.emp_no and a.gender='M' \G;
*************************** 1. row ***************************
           id: 1
  select_type: SIMPLE
        table: a
   partitions: NULL
         type: ALL
possible_keys: PRIMARY
          key: NULL
      key_len: NULL
          ref: NULL
         rows: 299550
     filtered: 50.00
        Extra: Using where
*************************** 2. row ***************************
           id: 1
  select_type: SIMPLE
        table: b
   partitions: NULL
         type: ref
possible_keys: PRIMARY
          key: PRIMARY
      key_len: 4
          ref: employees.a.emp_no
```

```
            rows: 1
        filtered: 100.00
           Extra: Using index
2 rows in set, 1 warning (0.01 sec)
```

SQL 实际执行时间为 1.54s：

```
mysql> select count(1) from  emp1 a,titles1 b  where a.emp_no=b.emp_no and a.gender='M' ;
+----------+
| count(1) |
+----------+
|   260939|
+----------+
1 row in set (1.42 sec)
```

现在在 gender 字段上创建直方图：

```
mysql> analyze table emp1 update histogram on gender;
+--------------+----------+----------+------------------------------------------------+
|Table         |Op        |Msg_type  | Msg_text                                       |
+--------------+----------+----------+------------------------------------------------+
|employees.emp1| histogram| status   |Histogram statistics created for column 'gender'.|
+--------------+----------+----------+------------------------------------------------+
1 row in set (0.34 sec)
```

再次观察执行计划，连接顺序发生了变化，titles1 成为新的驱动表：

```
mysql> desc select count(1) from  emp1 a,titles1 b  where a.emp_no=b.emp_no and a.gender='M'
\G;
*************************** 1. row ***************************
           id: 1
  select_type: SIMPLE
        table: b
   partitions: NULL
         type: index
possible_keys: PRIMARY
          key: PRIMARY
      key_len: 159
          ref: NULL
         rows: 261872
     filtered: 100.00
        Extra: Using index
*************************** 2. row ***************************
           id: 1
  select_type: SIMPLE
        table: a
   partitions: NULL
         type: eq_ref
possible_keys: PRIMARY
          key: PRIMARY
      key_len: 4
          ref: employees.b.emp_no
         rows: 1
     filtered: 99.67
        Extra: Using where
2 rows in set, 1 warning (0.00 sec)
```

执行时间也大大缩短，仅仅用了 0.46s，降了近 3 倍：

```
mysql> select count(1) from  emp1 a,titles1 b  where a.emp_no=b.emp_no and a.gender='M';
+----------+
| count(1) |
+----------+
|   260939|
+----------+
1 row in set (0.17 sec)
```

此时，利用前面介绍的 trace 来分析一下执行计划的差异，如表 15-2 所示。

15.5 直方图

表 15-2 直方图使用前后 trace 对比

	直方图创建前	直方图创建后
执行计划 1，选择驱动表为 titles1，cost 二者都为 118 027	`"rest_of_plan": [` `{` `"plan_prefix": [` `"`titles1``b`"` `] /* plan_prefix */,` `"table": "`emp1``a`",` `"filtering_effect": [` `] /* filtering_effect */,` `...` `"final_filtering_effect": 0.5,` `...` `"rows_for_plan": 261872,` `"cost_for_plan": 118027,` `"chosen": true` `}` `] /* rest_of_plan */`	`"rest_of_plan": [` `{` `"plan_prefix": [` `"`titles1``b`"` `] /* plan_prefix */,` `"table": "`emp1``a`",` `"filtering_effect": [` `{` `"condition":` `"(`a`.`gender` = 'M')",` `"histogram_selectivity": 0.9967` `}` `] /* filtering_effect */,` `"final_filtering_effect": 0.9967,` `...` `"rows_for_plan": 261872,` `"cost_for_plan": 118027,` `"chosen": true` `}` `] /* rest_of_plan */`
执行计划 2，选择驱动表为 emp1，cost 分别为 90 311 和 150 018	`"rest_of_plan": [` `{` `"plan_prefix": [` `"`emp1``a`"` `] /* plan_prefix */,` `"table": "`titles1``b`",` `"best_access_path": {` `"considered_access_paths": [` `{` `"access_type": "ref",` `"index": "PRIMARY",` `"rows": 1.4682,` `"cost": 60108,` `"chosen": true` `}` `...` `"rows_for_plan": 219897,` `"cost_for_plan": 90311,` `"chosen": true` `}` `]` `/* rest_of_plan */`	`"rest_of_plan": [` `{` `"plan_prefix": [` `"`emp1``a`"` `] /* plan_prefix */,` `"table": "`titles1``b`",` `"best_access_path": {` `"considered_access_paths": [` `{` `"access_type": "ref",` `"index": "PRIMARY",` `"rows": 1.4682,` `"cost": 119815,` `"chosen": true` `},` `...` `"rows_for_plan": 438329,` `"cost_for_plan": 150018,` `"pruned_by_cost": true` `}` `]` `/* rest_of_plan */`
选择结果	由于 90311<118027，选择执行计划 2，即连接顺序为 emp1、titles1	由于 150018>118027，选择执行计划 1，即连接顺序为 titles1、emp1

可以看出，由于直方图生成前后，过滤因子发生了变化，导致优化器在选择执行计划时的 cost 随着发生了改变，从而选择了更优的执行计划，达到了优化目的。

15.5.4 直方图小结

直方图的引入给 SQL 优化提供了一个新的思路，但并不是所有大表的字段都需要创建直方图。通常在一些唯一值较少、数据分布不均衡、查询较为频繁、没有创建索引的字段上考虑创建直方图。虽然在创建索引有时也可以达到优化效果，但由于这类字段索引使用率低、索引维护成本高，因此通常不会 在这些字段上单独创建索引。而直方图只需要创建一次，对数据的变更不需要实时进行维护，代价较小，更适合于此类条件的查询。

15.6 使用查询重写

使用过 Oracle 数据库的同学可能知道，如果某个 SQL 的执行计划出了问题，可以使用 sql_profile，在不修改 SQL 本身的情况下，为 SQL 绑定更好的执行计划，这样做的好处是可以不依赖于代码的调整，第一时间解决因为执行计划选择错误而带来的问题。

而 MySQL 一直以来，都缺少能够便捷地干预执行计划的方式，经常需要通过修改 SQL 的写法或者调整索引的方式来达到改变执行计划的目的，而修改 SQL 写法依赖于应用程序的修改和发布，不一定能够及时作出调整；通过修改索引的方式，代价往往也比较大，而且无法保证修改索引后执行计划就能够符合预期，这是 DBA 在面对 MySQL 中的 SQL 优化时经常会遇到的一个问题。

在 MySQL 5.7 中，提供了 Query Rewrite Plugin，可以通过规则匹配的方式，将符合条件的 SQL 进行重写，从而达到调整执行计划或者其他的目标。下面来测试一下这个插件的安装和使用方法。

在$mysqlhome/share 目录下执行安装脚本，创建 query_rewrite 数据库和 rewrite_rules 规则表：

```
mysql -uroot -p < install_rewriter.sql
mysql> show databases like 'query%';
+------------------+
| Database (query%) |
+------------------+
| query_rewrite    |
+------------------+
1 row in set (0.00 sec)

mysql> use query_rewrite
Database changed

mysql> show tables;
+-------------------------+
| Tables_in_query_rewrite |
+-------------------------+
| rewrite_rules           |
+-------------------------+
1 row in set (0.00 sec)

mysql> SHOW GLOBAL VARIABLES LIKE 'rewriter_enabled';
+------------------+-------+
| Variable_name    | Value |
+------------------+-------+
| rewriter_enabled | ON    |
+------------------+-------+
1 row in set (0.01 sec)
```

15.6 使用查询重写

通过上面的查询,可以看到 SQL 重写插件已经打开。需要注意的是,SQL 重写插件安装之后,即使关闭插件,仍然会有一定的额外开销,考虑到 SQL 重写插件可以动态安装和打开,因此,如果不是确定要使用这一插件,没有必要提前安装。

下面通过一个示例来演示一下查询重写是如何发挥作用的。

(1)增加匹配规则,将"SELECT ?"全部重写为"SELECT ? + 1",其中"?"是通配符,用来匹配数据值:

```
mysql> INSERT INTO query_rewrite.rewrite_rules (pattern, replacement)
    -> VALUES('SELECT ?', 'SELECT ? + 1');
Query OK, 1 row affected (0.00 sec)
```

(2)刷新使规则生效:

```
mysql> CALL query_rewrite.flush_rewrite_rules();
Query OK, 0 rows affected, 1 warning (0.01 sec)
```

(3)看一下重新的效果:

```
mysql> select 1;
+-------+
| 1 + 1 |
+-------+
|     2 |
+-------+
1 row in set, 1 warning (0.00 sec)

mysql> show warnings;
+-------+------+----------------------------------------------------------------+
| Level | Code | Message                                                        |
+-------+------+----------------------------------------------------------------+
| Note  | 1105 | Query 'select 1' rewritten to 'SELECT 1 + 1' by a query rewrite plugin |
+-------+------+----------------------------------------------------------------+
1 row in set (0.00 sec)
```

可以看到 select 1 被重写为 select1+1,同时 MySQL 也返回一个 warning,提示原 SQL 被重写了:

```
mysql> show warnings;
+-------+------+----------------------------------------------------------------+
| Level | Code | Message                                                        |
+-------+------+----------------------------------------------------------------+
| Note  | 1105 | Query 'select 1' rewritten to 'SELECT 1 + 1' by a query rewrite plugin |
+-------+------+----------------------------------------------------------------+
1 row in set (0.00 sec)
```

需要注意的是,改写对函数无效:

```
mysql> select PI();
+----------+
| PI()     |
+----------+
| 3.141593 |
+----------+
1 row in set (0.00 sec)
```

这里 PI()就没有没改写为 PI()+1。

接下来,再给出一个示例,来说明如何通过查询重写来优化执行计划:

```
mysql> create table `tab_test_rewrite` (
    -> `order_id` varchar(20) not null,
    -> `user_id` varchar(40) not null,
    -> `status` smallint(5) not null,
    -> primary key (`order_id`),
    -> key `idx_user_id` (`user_id`),
    -> key `idx_status` (`status`));
Query OK, 0 rows affected (0.01 sec)
```

假设这是一个订单表,其中保存了订单 id、用户 id 和订单状态,导入数据后,总共有 100 个用户,每个用户有 1 000 个订单,并且这些订单的状态都是正常,用 status=1 来表示,最后再加入 1 条状态为 0 的数据,表示异常。

```
mysql> DELIMITER //
mysql>
mysql> CREATE PROCEDURE `p_test1` ()
    -> BEGIN
    -> declare v_user int;
    -> declare v_order int;
    -> set @v_user=1;
    -> while @v_user<101
    -> do
    -> set @v_order=1;
    ->    while @v_order<1001
    ->    do
    ->      insert into tab_test_rewrite(order_id,user_id,status)
    ->      values (concat('order_',@v_user,'_',@v_order),concat('user_',@v_user),1);
    ->      set @v_order=@v_order+1;
    ->    END while ;
    ->    set @v_user=@v_user+1;
    -> END while ;
    ->    insert into tab_test_rewrite(order_id,user_id,status) values ('order_1_1001','user_1',0);
    ->
    -> END //
Query OK, 0 rows affected (0.00 sec)

mysql>
mysql> DELIMITER ;
mysql> call p_test1();
Query OK, 1 row affected (1 min 30.17 sec)
```

现在要查询某个用户的状态为 1 的数据,首先来看一下默认的执行计划:

```
mysql> explain select * from tab_test_rewrite where user_id = 'user_1' and status = 1\G
*************************** 1. row ***************************
           id: 1
  select_type: SIMPLE
        table: tab_test_rewrite
   partitions: NULL
         type: index_merge
possible_keys: idx_user_id,idx_status
          key: idx_user_id,idx_status
      key_len: 122,2
          ref: NULL
         rows: 500
     filtered: 100.00
        Extra: Using intersect(idx_user_id,idx_status); Using where; Using index
1 row in set, 1 warning (0.00 sec)
```

可以看到,默认的执行计划是扫描了 idx_user_id 和 idx_status 两个索引,并做了 index_merge,而实际上,绝大多数记录的 status 都等于 1,所以 idx_status 对于状态是 1 的数据的选择度很差,一般没有必要去扫描这个索引,再来看一下如果使用 force index 之后的执行计划:

```
mysql> explain select * from tab_test_rewrite force index (idx_user_id) where user_id = 'user_1' and status = 1\G
*************************** 1. row ***************************
           id: 1
  select_type: SIMPLE
        table: tab_test_rewrite
   partitions: NULL
         type: ref
possible_keys: idx_user_id
          key: idx_user_id
      key_len: 122
          ref: const
         rows: 1001
     filtered: 100.00
```

```
        Extra: Using where
1 row in set, 1 warning (0.00 sec)
```

这个执行计划更加符合我们的预期,为了优化这个查询,下面对这个查询进行改写:

```
mysql> INSERT INTO query_rewrite.rewrite_rules
    -> (pattern_database, pattern, replacement)
    -> VALUES('employees',
    -> 'select * from tab_test_rewrite where user_id = ? and status = 1',
    -> 'select * from tab_test_rewrite force index (idx_user_id) where user_id = ? and status = 1');
Query OK, 1 row affected (0.00 sec)

mysql> CALL query_rewrite.flush_rewrite_rules();
Query OK, 0 rows affected (0.01 sec)
```

重新执行以下查询:

```
mysql> select * from tab_test_rewrite where user_id = 'user_1' and status = 1;
……

mysql> show warnings \G;
*************************** 1. row ***************************
  Level: Note
   Code: 1105
Message: Query 'select * from tab_test_rewrite where user_id = 'user_1' and status = 1' rewritten
to 'select * from tab_test_rewrite force index (idx_user_id) where user_id = 'user_1' and status
= 1' by a query rewrite plugin
1 row in set (0.00 sec)
```

可以看到,查询已经正确被改写。

当然,MySQL 对于查询改写的处理相比 Oracle 来说仍然比较简单,而这种通过字符串匹配和替换的方式来做的改写,也存在一些问题,例如,可能造成一些查询被错误的修改。因此在使用这一特性时,需要做好充分的测试。

15.7 常用 SQL 技巧

15.7.1 正则表达式的使用

正则表达式(regular expression)是指一个用来描述或者匹配一系列符合某个句法规则的字符串的单个字符串。在很多文本编辑器或其他工具里,正则表达式通常被用来检索或替换那些符合某个模式的文本内容。许多程序设计语言都支持利用正则表达式进行字符串操作。例如,在 Perl 中就内建了一个功能强大的正则表达式引擎。正则表达式这个概念最初是由 UNIX 中的工具软件(例如 SED 和 GREP)普及开的,通常缩写成"RegEx"或者"RegExp"。

MySQL 利用 RegExp 命令提供给用户扩展的正则表达式功能,RegExp 实现的功能类似 UNIX 上 GREP 和 SED 的功能,并且 RegExp 在进行模式匹配时是区分大小写的。熟悉并掌握 RegExp 的功能可以使模式匹配工作事半功倍。

MySQL 5.7 中可以使用的模式序列如表 15-3 所示。

表 15-3   正则表达式中的模式

序 列	序 列 说 明
^	在字符串的开始处进行匹配
$	在字符串的末尾处进行匹配
.	匹配任意单个字符,包括换行符
[...]	匹配出括号内的任意字符

序　列	序 列 说 明
[^...]	匹配不出括号内的任意字符
a*	匹配零个或多个 a（包括空串）
a+	匹配 1 个或多个 a（不包括空串）
a?	匹配 1 个或零个 a
a1|a2	匹配 a1 或 a2
a(m)	匹配 m 个 a
a(m,)	匹配 m 个或更多个 a
a(m,n)	匹配 m 到 n 个 a
a(,n)	匹配 0 到 n 个 a
(...)	将模式元素组成单一元素

下面举一些例子来介绍常用正则表达式的使用方法。

● "^" 在字符串的开始处进行匹配，返回结果为 1 表示匹配，返回结果为 0 表示不匹配。下例中尝试匹配字符串 "abcdefg" 是否以字符 "a" 开始。

```
mysql> select 'abcdefg' REGEXP '^a';
+-----------------------+
| 'abcdefg' REGEXP '^a' |
+-----------------------+
|                     1 |
+-----------------------+
1 row in set (0.39 sec)
```

● "$" 在字符串的末尾处进行匹配。下例中尝试匹配字符串 "abcdefg" 是否以字符 "g" 结束。

```
mysql> select 'abcdefg' REGEXP 'g$';
+-----------------------+
| 'abcdefg' REGEXP 'g$' |
+-----------------------+
|                     1 |
+-----------------------+
1 row in set (0.01 sec)
```

● "." 匹配任意单个字符，包括换行符。下例中字符串 "abcdefg" 尝试匹配单字符 "h" 和 "f"。

```
mysql> select 'abcdefg' REGEXP '.h', 'abcdefg' REGEXP '.f';
+-----------------------+-----------------------+
| 'abcdefg' REGEXP '.h' | 'abcdefg' REGEXP '.f' |
+-----------------------+-----------------------+
|                     0 |                     1 |
+-----------------------+-----------------------+
1 row in set (0.00 sec)
```

● "[...]" 匹配出括号内的任意字符。下例中字符串 "abcdefg" 尝试匹配 "fhk" 中的任意一个字符，如果有一个字符能匹配上，则返回 1。

```
mysql> select 'abcdefg' REGEXP "[fhk]";
+--------------------------+
| 'abcdefg' REGEXP "[fhk]" |
+--------------------------+
|                        1 |
+--------------------------+
1 row in set (0.01 sec)
```

● "[^...]" 匹配不出括号内的任意字符，和 "[...]" 刚好相反。下例中字符串 "efg" 和 "X" 中如果有任何一个字符匹配不上 "[XYZ]" 中的任意一个字符，则返回 0；如果全部都能匹配上，则返回 1。

```
mysql> select 'efg' REGEXP "[^XYZ]",'X' REGEXP "[^XYZ]";
+-----------------------+---------------------+
| 'efg' REGEXP "[^XYZ]" | 'X' REGEXP "[^XYZ]" |
+-----------------------+---------------------+
|                     1 |                   0 |
+-----------------------+---------------------+
1 row in set (0.00 sec)
```

上文介绍了正则表达式的常见使用方法。但是，在实际工作中，正则表达式到底会在什么地方用到呢？下面给出一个实例，使用正则表达式查询出使用 163.com 邮箱的用户和邮箱。

（1）创建测试数据：

```
mysql> insert into customer (store_id, first_name, last_name, address_id, email) values
(1, '188mail', 'beijing', 605, 'beijing@188.com');
Query OK, 1 row affected, 1 warning (0.06 sec)

mysql> insert into customer (store_id, first_name, last_name, address_id, email) values
(1, '126mail', 'beijing', 605, 'beijing@126.com');
Query OK, 1 row affected, 1 warning (0.05 sec)

mysql> insert into customer (store_id, first_name, last_name, address_id, email) values
(1, '163mail', 'beijing', 605, 'beijing@163.com');
Query OK, 1 row affected, 1 warning (0.03 sec)
```

（2）使用正则表达式"$"和"[...]"进行匹配：

```
mysql> select first_name, email from customer where email regexp "@163[,.]com$";
+------------+-----------------+
| first_name | email           |
+------------+-----------------+
| 163mail    | beijing@163.com |
+------------+-----------------+
1 row in set (0.00 sec)
```

从以上可以看出，如果不使用正则表达式而使用普通的 LIKE 语句，则 WHERE 条件需要写成如下格式：

```
email like "%@163.com" or  email like "%@163,com"
```

显然，采用正则表达式可以使得代码更加简单易读。

15.7.2 巧用 RAND()提取随机行

大多数数据库都会提供产生随机数的包或者函数，通过这些包或者函数可以产生用户需要的随机数，也可以用来从数据表中抽取随机产生的记录，这对一些抽样分析统计是非常有用的。例如 ORACLE 中用 DBMS_RANDOM 包产生随机数，而在 MySQL 中，产生随机数的方法是 RAND()函数。可以利用这个函数与 ORDER BY 子句一起完成随机抽取某些行的功能。它的原理其实就是 ORDER BY RAND()能够把数据随机排序。

例如，可按照随机顺序检索数据行：

```
mysql> select * from category order by rand();
+-------------+----------+---------------------+
| category_id | name     | last_update         |
+-------------+----------+---------------------+
|          12 | Music    | 2006-02-15 04:46:27 |
|           9 | Foreign  | 2006-02-15 04:46:27 |
|          13 | New      | 2006-02-15 04:46:27 |
|           7 | Drama    | 2006-02-15 04:46:27 |
|           1 | Action   | 2006-02-15 04:46:27 |
|           3 | Children | 2006-02-15 04:46:27 |
...
```

这样，如果想随机抽取一部分样本的时候，把数据随机排序后再抽取前 n 条记录就可以了，比如：

```
mysql> select * from category order by rand() limit 5;
+-------------+-----------+---------------------+
| category_id | name      | last_update         |
+-------------+-----------+---------------------+
|           4 | Classics  | 2006-02-15 04:46:27 |
|           9 | Foreign   | 2006-02-15 04:46:27 |
|           2 | Animation | 2006-02-15 04:46:27 |
|          16 | Travel    | 2006-02-15 04:46:27 |
|           7 | Drama     | 2006-02-15 04:46:27 |
+-------------+-----------+---------------------+
5 rows in set (0.00 sec)
```

上面的例子从类别表 category 中随机抽取了 5 个样本，随机抽取样本对总体的统计具有十分重要的意义，因此这个函数非常有用。

15.7.3 利用 GROUP BY 的 WITH ROLLUP 子句

在 SQL 语句中，使用 GROUP BY 的 WITH ROLLUP 字句可以检索出更多的分组聚合信息，它不仅仅能像一般的 GROUP BY 语句那样检索出各组的聚合信息，还能检索出本组类的整体聚合信息，具体如下例所示。

在支付表 payment 中，按照支付时间 payment_date 的年月、经手员工编号 staff_id 列分组对支付金额 amount 列进行聚合计算如下：

```
mysql> select date_format(payment_date, '%Y-%m'), staff_id, sum(amount) from payment group by date_format(payment_date, '%Y-%m'), staff_id;
+------------------------------------+----------+-------------+
| date_format(payment_date, '%Y-%m') | staff_id | sum(amount) |
+------------------------------------+----------+-------------+
| 2005-05                            |        1 |     2621.83 |
| 2005-05                            |        2 |     2202.60 |
| 2005-06                            |        1 |     4776.36 |
| 2005-06                            |        2 |     4855.52 |
| 2005-07                            |        1 |    14003.54 |
| 2005-07                            |        2 |    14370.35 |
| 2005-08                            |        1 |    11853.65 |
| 2005-08                            |        2 |    12218.48 |
| 2006-02                            |        1 |      234.09 |
| 2006-02                            |        2 |      280.09 |
+------------------------------------+----------+-------------+
10 rows in set (0.06 sec)

mysql> select date_format(payment_date, '%Y-%m'), IFNULL(staff_id,''), sum(amount) from payment group by date_format(payment_date, '%Y-%m'), staff_id with rollup;
+------------------------------------+---------------------+-------------+
| date_format(payment_date, '%Y-%m') | IFNULL(staff_id,'') | sum(amount) |
+------------------------------------+---------------------+-------------+
| 2005-05                            | 1                   |     2621.83 |
| 2005-05                            | 2                   |     2202.60 |
| 2005-05                            |                     |     4824.43 |
| 2005-06                            | 1                   |     4776.36 |
| 2005-06                            | 2                   |     4855.52 |
| 2005-06                            |                     |     9631.88 |
| 2005-07                            | 1                   |    14003.54 |
| 2005-07                            | 2                   |    14370.35 |
| 2005-07                            |                     |    28373.89 |
| 2005-08                            | 1                   |    11853.65 |
| 2005-08                            | 2                   |    12218.48 |
| 2005-08                            |                     |    24072.13 |
| 2006-02                            | 1                   |      234.09 |
| 2006-02                            | 2                   |      280.09 |
| 2006-02                            |                     |      514.18 |
| NULL                               |                     |    67416.51 |
+------------------------------------+---------------------+-------------+
16 rows in set (0.05 sec)
```

从上面的例子中可以看到，第二个 SQL 语句的结果比第一个 SQL 语句的结果多出了很多行，而这些行反映出了更多的信息。例如，第二个 SQL 语句的结果的前两行表示 2005-05 月份各个员工（1、2）的经手的支付金额，而第三行表示 2005-05 月份总支付金额为 4 824.43，这个信息在第一个 SQL 语句中是不能反映出来的，第 16 行表示总支付金额为 67 416.51，这个信息在第一个 SQL 语句中是没有的。

其实 WITH ROLLUP 反映的是一种 OLAP 思想，也就是说这一个 GROUP BY 语句执行完成后可以满足用户想要得到的任何一个分组以及分组组合的聚合信息值。

注意：当使用 ROLLUP 时，不能同时使用 ORDER BY 子句进行结果排序。换言之，ROLLUP 和 ORDER BY 是互相排斥的。此外，LIMIT 用在 ROLLUP 后面。

15.7.4 用 BIT GROUP FUNCTIONS 做统计

本节主要介绍如何共同使用 GROUP BY 语句和 BIT_AND、BIT_OR 函数完成统计工作。这两个函数的一般用途就是做数值之间的逻辑位运算，但是，当把它们与 GROUP BY 子句联合使用的时候就可以做一些其他的任务。

假设现在有这样一个任务：一个超市需要记录每个用户每次来超市都购买了哪些商品。为了将问题简单化，假设该超市只有面包、牛奶、饼干、啤酒 4 种商品。那么通常该怎么做呢？一般先建立一个购物单表，里面记录购物发生的时间、顾客信息等；然后再建立一个购物单明细表，里面记录该顾客所购买的商品。这样设计表结构的优点是顾客所购买的商品的详细信息可以记录下来，比如数量、单价等，但是如果目前的这个任务只需要知道用户购买商品的种类和每次购物总价等信息，那么这种数据库结构的设计就显得太复杂了。一般还可能会想到用一个表实现这个功能，并且用一个字段以字符串的形式记录顾客所购买的所有商品的商品号，这也是一种方法，但是如果顾客一次购买商品比较多，需要很大的存储空间，而且将来做各种统计的时候也会捉襟见肘。

下面给出一种新的解决办法，类似于上面讲到的第二种方案，仍然用一个字段表示顾客购买商品的信息，但是这个字段是数值型的而不是字符型的，该字段存储一个十进制数字，当它转换成二进制的时候，那么每一位代表一种商品，而且如果所在位是 "1" 那么表示顾客购买了该种商品，"0" 表示没有购买该种商品。比如数值的第一位代表面包（规定从右向左开始计算）、第二位代表牛奶、第三位代表饼干、第 4 位代表啤酒，这样如果一个用户购物单的商品列的数值为 5，那么二进制表示为 0101，这样从右向左第一位和第三位是 1，那么就可以知道这个用户购买了面包和饼干，而如果这个客户有多个这样的购物单（在数据库中就是有多条记录），把这些购物单按用户分组做 BIT_OR() 操作就可以知道这个用户都购买过什么商品。

下面举例说明一下这个操作，首先初始化一组数据：

```
mysql> create table order_rab (id int,customer_id int,kind int);
Query OK, 0 rows affected (0.05 sec)
mysql> insert into order_rab values (1,1,5),(2,1,4);
Query OK, 2 rows affected (0.00 sec)
mysql> insert into order_rab values (3,2,3),(4,2,4);
Query OK, 2 rows affected (0.00 sec)
mysql> select * from order_rab;
+------+-------------+------+
| id   | customer_id | kind |
+------+-------------+------+
|  1   |  1          |  5   |
|  2   |  1          |  4   |
```

```
|   3 |    2 |    3 |
|   4 |    2 |    4 |
+-----+------+------+
4 rows in set (0.00 sec)
```

其中 customerid 是顾客编号，kind 是所购买的商品，初始化了两个顾客 1 和 2 的数据，他们每人购物两次，前者购买的商品数值是 5 和 4，转化为二进制分别为 0101、0100，表示这个顾客第一次购买了牛奶和啤酒，第二次购买了牛奶；后者购买的商品数值是 3 和 4，转化为二进制分别为 0011、0100，表示这个顾客第一次购买了饼干和啤酒，第二次购买了牛奶。

下面用 BIT_OR() 函数与 GROUP BY 子句联合起来，统计一下这两个顾客在这个超市一共都购买过什么商品，如下例：

```
mysql> select customer_id,bit_or(kind) from order_rab group by customer_id;
+-------------+--------------+
| customer_id | bit_or(kind) |
+-------------+--------------+
|           1 |            5 |
|           2 |            7 |
+-------------+--------------+
2 rows in set (0.00 sec)
```

可以看到顾客 1 的 BIT_OR() 结果是 5，即 0101，表示这个顾客在本超市购买过牛奶和啤酒；顾客 2 的 BIT_OR() 结果是 7，即 0111，表示这个顾客在本超市购买过牛奶、饼干、啤酒。

下面解释一下数据库在处理这个逻辑时的计算过程，以第一个顾客举例，BIT_OR(kind) 就相当于把 kind 的各个值做了一个"或"操作，最终结果是十进制的 5。逻辑计算公式如下：

```
#       ..0101
#       ..0100
#  OR   ..0000
#      ---------
#       ..0101
```

同理，可以用 BIT_AND() 统计每个顾客每次来本超市都会购买的商品，具体如下：

```
mysql> select customer_id,bit_and(kind) from order_rab group by customer_id;
+-------------+---------------+
| customer_id | bit_and(kind) |
+-------------+---------------+
|           1 |             4 |
|           2 |             0 |
+-------------+---------------+
2 rows in set (0.01 sec)
```

顾客 1 的 BIT_AND() 结果是 4，即 0100，表示顾客 1 每次来本超市都会购买牛奶；顾客 2 的 BIT_AND() 结果是 0，即 0000，表示顾客 2 不是每次来本超市都会购买的商品。

数据库在处理 BIT_AND() 的时候就是把 kind 的各个值做了一个"与"操作，拿顾客 1 举例说明一下，逻辑计算公式如下：

```
#        ..0101
#        ..0100
#  AND   ..1111
#       ---------
#        ..0100
```

从上面的例子可以看出，这种数据库结构设计的好处就是能用很简洁的数据表示很丰富的信息，这种方法能够大大地节省存储空间，而且能够提高部分统计计算的速度。不过需要注意的是，这种设计其实损失了顾客购买商品的详细信息，比如购买商品的数量、当时单价、是否有折扣、是否有促销等，因此还要根据应用的实际情况有选择地考虑数据库的结构设计。

15.7.5 数据库名、表名大小写问题

在 MySQL 中，数据库对应操作系统下的数据目录。数据库中的每个表至少对应数据库目录中的一个文件（也可能是多个，这取决于存储引擎）。因此，所使用操作系统的大小写敏感性决定了数据库名和表名的大小写敏感性。在大多数 UNIX 环境中，由于操作系统对大小写的敏感性导致了数据库名和表名对大小写敏感性，而在 Windows 中，由于操作系统本身对大小写不敏感，因此在 Windows 下的 MySQL 数据库名和表名对大小写也不敏感。

列、索引、存储子程序和触发器名在任何平台上对大小写不敏感。默认情况下，表别名在 UNIX 中对大小写敏感，但在 Windows 或 macOS X 中对大小写不敏感。下面的查询在 UNIX 中会报错，因为它同时引用了别名 a 和 A：

```
mysql> select id from order_rab a where A.id = 1;
ERROR 1054 (42S22): Unknown column 'A.id' in 'where clause'
```

然而，该查询在 Windows 中是可以的。要想避免出现差别，最好采用一致的转换，例如，总是用小写创建并引用数据库名和表名。

在 MySQL 中，如何在硬盘上保存、使用表名和数据库名是由 lower_case_table_names 系统变量决定的，用户可以在启动 MySQL 服务时设置这个系统变量。lower_case_table_names 可以采用如表 15-4 所示的任一值。

表 15-4 lower_case_table_names 的取值范围

值	含 义
0	使用 CREATE TABLE 或 CREATE DATABASE 语句指定的大写和小写在硬盘上保存表名和数据库名。名称对大小写不敏感。在 UNIX 系统中的默认设置就是这个值
1	表名在硬盘上以小写保存，名称对大小写敏感。MySQL 将所有表名转换为小写以便存储和查找。该值为 Windows 和 macOS X 系统中的默认值
2	表名和数据库名在硬盘上使用 CREATE TABLE 或 CREATE DATABASE 语句指定的大小写进行保存，但 MySQL 将它们转换为小写以便查找。此值只在对大小写不敏感的文件系统上适用

如果只在一个平台上使用 MySQL，通常不需要更改 lower_case_table_names 变量。然而，如果用户想要在对大小写敏感性不同的文件系统的平台之间转移表，就会遇到困难。例如，在 UNIX 中，my_tables 和 MY_tables 是两个不同的表，但在 Windows 中，这两个表名相同。

在 UNIX 中使用 lower_case_table_names=0，而在 Windows 中使用 lower_case_table_names=2，这样可以保留数据库名和表名的大小写。不利之处是必须确保在 Windows 中的所有 SQL 语句总是正确地使用大小写来引用数据库名和表名，如果 SQL 语句中没有正确引用数据库名和表名的大小写，那么虽然在 Windows 中能正确执行，但是如果将查询转移到 UNIX 中，大小写不正确，将会导致查询失败。

> **注意**：在 UNIX 中将 lower_case_table_names 设置为 1 并且重启 mysqld 之前，必须先将旧的数据库名和表名转换为小写。尽管在某些平台中数据库名和表名对大小写不敏感，但是最好养成在同一查询中使用相同的大小写来引用给定的数据库名或表名的习惯。

15.7.6 使用外键需要注意的问题

在 MySQL 中，InnoDB 存储引擎支持对外部关键字约束条件的检查。而对于其他类型存储引擎的表，当使用 REFERENCES tbl_name(col_name) 子句定义列时可以使用外部关键字，但是该子

句没有实际的效果，只作为备忘录或注释来提醒用户目前正定义的列指向另一个表中的一个列。

例如，下面的 myisam 表外键就没有起作用：

```
mysql> create table users(id int,name varchar(10),primary key(id)) engine=myisam;
Query OK, 0 rows affected (0.03 sec)

mysql> create table books(id int,bookname varchar(10),userid int ,primary key(id),constraint
fk_userid_id foreign key(userid) references users(id)) engine=myisam;
Query OK, 0 rows affected (0.03 sec)

mysql> insert into books values(1,'book1',1);
Query OK, 1 row affected (0.00 sec)
```

如果用 InnoDB 存储引擎建表的话，外键就会起作用，具体如下：

```
mysql> create table users2(id int,name varchar(10),primary key(id)) engine=innodb;
Query OK, 0 rows affected (0.14 sec)

mysql> create table books2(id int,bookname varchar(10),userid int ,primary key(id),constraint
fk_userid_id foreign key(userid) references users2(id)) engine=innodb;
Query OK, 0 rows affected (0.18 sec)

mysql> insert into books2 values(1,'book1',1);
ERROR 1452 (23000): Cannot add or update a child row: a foreign key constraint fails ('sakila/
books2', CONSTRAINT 'fk_userid_id' FOREIGN KEY ('userid') REFERENCES 'users2' ('id'))
```

而且，用 show create table 命令查看建表语句的时候，发现 MyISAM 存储引擎并不显示外键的语句，而 InnoDB 存储引擎就会显示外键语句，具体如下：

```
mysql> show create table books\G;
*************************** 1. row ***************************
       Table: books
Create Table: CREATE TABLE 'books' (
  'id' int(11) NOT NULL DEFAULT '0',
  'bookname' varchar(10) DEFAULT NULL,
  'userid' int(11) DEFAULT NULL,
  PRIMARY KEY ('id'),
  KEY 'fk_userid_id' ('userid')
) ENGINE=MyISAM DEFAULT CHARSET=gbk
1 row in set (0.00 sec)

mysql> show create table books2\G;
*************************** 1. row ***************************
       Table: books2
Create Table: CREATE TABLE 'books2' (
  'id' int(11) NOT NULL DEFAULT '0',
  'bookname' varchar(10) DEFAULT NULL,
  'userid' int(11) DEFAULT NULL,
  PRIMARY KEY ('id'),
  KEY 'fk_userid_id' ('userid'),
  CONSTRAINT 'fk_userid_id' FOREIGN KEY ('userid') REFERENCES 'users2' ('id')
) ENGINE=InnoDB DEFAULT CHARSET=gbk
1 row in set (0.00 sec)
```

15.8 小结

SQL 优化问题是数据库性能优化最基础也是最重要的一个问题，实践表明很多数据库性能问题都是由不合适的 SQL 语句造成。本章通过实例描述了 SQL 优化的一般过程，从定位一个有性能问题的 SQL 语句到分析产生性能问题的原因，最后到采取什么措施优化 SQL 语句的性能。另外，针对特定的 SQL（比如排序、join 等）和 MySQL 8.0 引入的一些新功能（比如直方图）也做了一些介绍，希望能帮助读者拓宽优化的思路。

第 16 章 锁问题

锁是计算机协调多个进程或线程并发访问某一资源的机制。在数据库中，除传统的计算资源（如 CPU、RAM、I/O 等）的争用以外，数据也是一种供许多用户共享的资源。如何保证数据并发访问的一致性、有效性是所有数据库必须解决的一个问题，锁冲突也是影响数据库并发访问性能的一个重要因素。从这个角度来说，锁对数据库而言显得尤其重要，也更加复杂。本章我们着重讨论 MySQL 锁机制的特点、常见的锁问题，以及解决 MySQL 锁问题的一些方法或建议。

16.1 MySQL 锁概述

相对其他数据库而言，MySQL 的锁机制比较简单，其最显著的特点是不同的存储引擎支持不同的锁机制。比如，MyISAM 和 MEMORY 存储引擎采用的是表级锁（table-level locking）；BDB 存储引擎采用的是页面锁（page-level locking），但也支持表级锁；InnoDB 存储引擎既支持行级锁（row-level locking），也支持表级锁，但默认情况下是采用行级锁。

MySQL 这 3 种锁的特性可大致归纳如下。

- 表级锁：开销小，加锁快；不会出现死锁；锁定粒度大，发生锁冲突的概率最高,并发度最低。
- 行级锁：开销大，加锁慢；会出现死锁；锁定粒度最小，发生锁冲突的概率最低,并发度也最高。
- 页面锁：开销和加锁时间界于表锁和行锁之间；会出现死锁；锁定粒度界于表锁和行锁之间，并发度一般。

从上述特点可见，很难笼统地说哪种锁更好，只能就具体应用的特点来说哪种锁更合适！仅从锁的角度来说：表级锁更适合于以查询为主，只有少量按索引条件更新数据的应用，如 Web 应用；而行级锁则更适合于有大量按索引条件并发更新少量不同数据，同时又有并发查询的应用，如一些在线事务处理（OLTP）系统。这一点在本书的"开发篇"介绍表类型的选择时，也曾提到过。下面重点介绍 MySQL 表锁和 InnoDB 行锁的问题。由于 BDB 已经被 InnoDB 取代，即将成为历史，这里就不进一步讨论了。

16.2 MyISAM 表锁

MyISAM 存储引擎只支持表锁，这也是 MySQL 开始几个版本中唯一支持的锁类型。随着应用对事务完整性和并发性要求的不断提高，MySQL 才开始开发基于事务的存储引擎，后来慢慢出现了支持页锁的 BDB 存储引擎和支持行锁的 InnoDB 存储引擎。但是，MyISAM 的表锁依然是使用最为广泛的锁类型。本节将详细介绍 MyISAM 表锁的使用。

16.2.1 查询表级锁争用情况

可以通过检查 table_locks_waited 和 table_locks_immediate 状态变量来分析系统上的表锁定争用情况：

```
mysql> show status like 'table%';
+-----------------------+-------+
| Variable_name         | Value |
+-----------------------+-------+
| Table_locks_immediate | 2979  |
| Table_locks_waited    | 0     |
+-----------------------+-------+
2 rows in set (0.00 sec)
```

如果 Table_locks_waited 的值比较高，则说明存在着较严重的表级锁争用情况。

16.2.2 MySQL 表级锁的锁模式

MySQL 的表级锁有两种模式：表共享读锁（Table Read Lock）和表独占写锁（Table Write Lock）。锁模式的兼容性如表 16-1 所示。

表 16-1　　　　　　　　　　MySQL 中的表锁兼容性

是否兼容　　请求锁模式 当前锁模式	None	读锁	写锁
读锁	是	是	否
写锁	是	否	否

可见，对 MyISAM 表的读操作，不会阻塞其他用户对同一表的读请求，但会阻塞对同一表的写请求；对 MyISAM 表的写操作，则会阻塞其他用户对同一表的读和写操作；MyISAM 表的读操作与写操作之间，以及写操作之间是串行的！根据如表 16-2 所示的示例可以知道，当一个线程获得对一个表的写锁后，只有持有锁的线程可以对表进行更新操作。其他线程的读、写操作都会等待，直到锁被释放为止。

表 16-2　　　　　　　　MyISAM 存储引擎的写阻塞读示例

session_1	session_2
获得表 film_text 的 WRITE 锁定： mysql> lock table film_text write; Query OK, 0 rows affected (0.00 sec)	

16.2　MyISAM 表锁

session_1	session_2

当前 session 对锁定表的查询、更新、插入操作都可以执行：
mysql> select film_id,title from film_text where film_id = 1001;
+---------+---------------+
| film_id | title |
+---------+---------------+
| 1001 | Update Test |
+---------+---------------+
1 row in set (0.00 sec)
mysql> insert into film_text (film_id,title) values(1003, 'Test');
Query OK, 1 row affected (0.00 sec)
mysql> update film_text set title = 'Test' where film_id = 1001;
Query OK, 1 row affected (0.00 sec)
Rows matched: 1 Changed: 1 Warnings: 0
	其他 session 对锁定表的查询被阻塞，需要等待锁被释放：
	mysql> select film_id,title from film_text where film_id = 1001;
	等待

释放锁： mysql> unlock tables; Query OK, 0 rows affected (0.00 sec)	等待
	Session2 获得锁，查询返回： mysql> select film_id,title from film_text where film_id = 1001; +---------+---------+ | film_id | title | +---------+---------+ | 1001 | Test | +---------+---------+ 1 row in set (57.59 sec)

16.2.3　如何加表锁

　　MyISAM 在执行查询语句（SELECT）前，会自动给涉及的所有表加读锁；在执行更新操作（UPDATE、DELETE、INSERT 等）前，会自动给涉及的表加写锁，这个过程并不需要用户干预，因此，用户一般不需要直接用 LOCK TABLE 命令给 MyISAM 表显式加锁。在本书的示例中，显式加锁基本上都是为了方便说明问题，并非必须如此。

　　给 MyISAM 表显式加锁，一般是为了在一定程度模拟事务操作，实现对某一时间点多个表的一致性读取。例如，有一个订单表 orders，其中记录有各订单的总金额 total，同时还有一个订单明细表 order_detail，其中记录有各订单每一产品的金额小计 subtotal，假设我们需要检查这两个表的金额合计是否相符，可能就需要执行如下两条 SQL 语句：

```
Select sum(total) from orders;
Select sum(subtotal) from order_detail;
```

　　这时，如果不先给两个表加锁，就可能产生错误的结果，因为第一条语句执行过程中，order_detail 表可能已经发生了改变。因此，正确的方法应该是：

```
Lock tables orders read local, order_detail read local;
Select sum(total) from orders;
Select sum(subtotal) from order_detail;
Unlock tables;
```

　　要特别说明以下两点内容。

●　上面的示例在 LOCK TABLES 时加了 "local" 选项，其作用就是在满足 MyISAM 表并发插入条件的情况下，允许其他用户在表尾并发插入记录，有关 MyISAM 表的并发插入问题，在后面的章节中还会进一步介绍。

- 在用 LOCK TABLES 给表显式加表锁时，必须同时取得所有涉及表的锁，并且 MySQL 不支持锁升级。也就是说，在执行 LOCK TABLES 后，只能访问显式加锁的这些表，不能访问未加锁的表；同时，如果加的是读锁，那么只能执行查询操作，而不能执行更新操作。在自动加锁的情况下也是如此，MyISAM 总是一次获得 SQL 语句所需要的全部锁。这也正是 MyISAM 表不会出现死锁（Deadlock Free）的原因。

在如表 16-3 所示的示例中，一个 session 使用 LOCK TABLE 命令给表 film_text 加了读锁，这个 session 可以查询锁定表中的记录，但更新或访问其他表都会提示错误；同时，另外一个 session 可以查询表中的记录，但更新就会出现锁等待。

表 16-3　　　　　　　　　　MyISAM 存储引擎的读阻塞写示例

session_1	session_2
获得表 film_text 的 READ 锁定： mysql> lock table film_text read; Query OK, 0 rows affected (0.00 sec)	
当前 session 可以查询该表记录： mysql> select film_id,title from film_text where film_id = 1001; +---------+---------------+ | film_id | title | +---------+---------------+ | 1001 |ACADEMY DINOSAUR| +---------+---------------+ 1 row in set (0.00 sec)	其他 session 也可以查询该表的记录： mysql> select film_id,title from film_text where film_id = 1001; +---------+---------------+ | film_id | title | +---------+---------------+ | 1001 |ACADEMY DINOSAUR| +---------+---------------+ 1 row in set (0.00 sec)
当前 session 不能查询没有锁定的表： mysql> select film_id,title from film where film_id= 1001; ERROR 1100 (HY000): Table 'film' was not locked with LOCK TABLES	其他 session 可以查询或者更新未锁定的表： mysql> select film_id,title from film where film_id = 1001; +---------+---------------+ | film_id | title | +---------+---------------+ | 1001 | update record | +---------+---------------+ 1 row in set (0.00 sec) mysql> update film set title = 'Test' where film_id = 1001; Query OK, 1 row affected (0.04 sec) Rows matched: 1　Changed: 1　Warnings: 0
当前 session 中插入或者更新锁定的表都会提示错误： mysql> insert into film_text (film_id,title) values (1002,'Test'); ERROR 1099 (HY000): Table 'film_text' was locked with a READ lock and can't be updated mysql>update film_text set title = 'Test' where film_id= 1001; ERROR 1099 (HY000): Table 'film_text' was locked with a READ lock and can't be updated	其他 session 更新锁定表会等待获得锁： mysql>update film_text set title ='Test'where film_id=1001; 等待
释放锁： mysql> unlock tables; Query OK, 0 rows affected (0.00 sec)	等待
	session 获得锁，更新操作完成： mysql>update film_text set title ='Test' where film_id=1001; Query OK, 1 row affected (1 min 0.71 sec) Rows matched: 1　Changed: 1　Warnings: 0

当使用 LOCK TABLES 时，不仅需要一次锁定用到的所有表，而且，同一个表在 SQL 语句中出现多少次，就要通过与 SQL 语句中相同的别名锁定多少次，否则也会出错！举例说明如下。

（1）对 actor 表获得读锁：

```
mysql> lock table actor read;
Query OK, 0 rows affected (0.00 sec)
```

（2）但是通过别名访问会提示错误：

```
mysql> select a.first_name,a.last_name,b.first_name,b.last_name from actor a,actor b where a.first_name = b.first_name and a.first_name = 'Lisa' and a.last_name = 'Tom' and a.last_name <> b.last_name;
ERROR 1100 (HY000): Table 'a' was not locked with LOCK TABLES
```

（3）需要对别名分别锁定：

```
mysql> lock table actor as a read,actor as b read;
Query OK, 0 rows affected (0.00 sec)
```

（4）按照别名的查询可以正确执行：

```
mysql> select a.first_name,a.last_name,b.first_name,b.last_name from actor a,actor b where a.first_name = b.first_name and a.first_name = 'Lisa' and a.last_name = 'Tom' and a.last_name <> b.last_name;
+------------+-----------+------------+-----------+
| first_name | last_name | first_name | last_name |
+------------+-----------+------------+-----------+
| Lisa       | Tom       | LISA       | MONROE    |
+------------+-----------+------------+-----------+
1 row in set (0.00 sec)
```

16.2.4 并发插入（Concurrent Inserts）

上文提到过 MyISAM 表的读和写是串行的，但这是就总体而言的。在一定条件下，MyISAM 表也支持查询和插入操作的并发进行。

MyISAM 存储引擎有一个系统变量 concurrent_insert，专门用以控制其并发插入的行为，其值分别可以为 0、1 或 2。

- 当 concurrent_insert 设置为 0 时，不允许并发插入。
- 当 concurrent_insert 设置为 1 时，如果 MyISAM 表中没有空洞（表的中间没有被删除的行），MyISAM 允许在一个进程读表的同时，另一个进程从表尾插入记录。这也是 MySQL 的默认设置。
- 当 concurrent_insert 设置为 2 时，无论 MyISAM 表中有没有空洞，都允许在表尾并发插入记录。

在如表 16-4 所示的示例中，session_1 获得了一个表的 READ LOCAL 锁，该线程可以对表进行查询操作，但不能对表进行更新操作；其他的线程（session_2），虽然不能对表进行删除和更新操作，但却可以对该表进行并发插入操作，这里假设该表中间不存在空洞。

表 16-4   MyISAM 存储引擎的读写（INSERT）并发示例

session_1	session_2
获得表 film_text 的 READ LOCAL 锁定： mysql> lock table film_text read local; Query OK, 0 rows affected (0.00 sec)	

续表

session_1	session_2
当前 session 不能对锁定表进行更新或者插入操作： mysql> insert into film_text (film_id,title) values (1002,'Test'); ERROR 1099 (HY000): Table 'film_text' was locked with a READ lock and can't be updated mysql> update film_text set title = 'Test' where film_id = 1001; ERROR 1099 (HY000): Table 'film_text' was locked with a READ lock and can't be updated	其他 session 可以进行插入操作，但是更新会等待： mysql> insert into film_text (film_id,title) values (1002,'Test'); Query OK, 1 row affected (0.00 sec) mysql> update film_text set title = 'Update Test' where film_id = 1001; 等待
当前 session 不能访问其他 session 插入的记录： mysql> select film_id,title from film_text where film_id = 1002; Empty set (0.00 sec)	
释放锁： mysql> unlock tables; Query OK, 0 rows affected (0.00 sec)	等待
当前 session 解锁后可以获得其他 session 插入的记录： mysql> select film_id,title from film_text where film_id = 1002; +---------+-------+ | film_id | title | +---------+-------+ | 1002 | Test | +---------+-------+ 1 row in set (0.00 sec)	session2 获得锁，更新操作完成： mysql> update film_text set title = 'Update Test' where film_id = 1001; Query OK, 1 row affected (1 min 17.75 sec) Rows matched: 1 Changed: 1 Warnings: 0

可以利用 MyISAM 存储引擎的并发插入特性来解决应用中对同一表查询和插入的锁争用。例如，将 concurrent_insert 系统变量设为 2，总是允许并发插入；同时，通过定期在系统空闲时段执行 OPTIMIZE TABLE 语句来整理空间碎片，收回因删除记录而产生的中间空洞。有关 OPTIMIZE TABLE 语句的详细介绍，可以参见 15.3 节的内容。

16.2.5 MyISAM 的锁调度

前面讲过，MyISAM 存储引擎的读锁和写锁是互斥的，读写操作是串行的。那么，一个进程请求某个 MyISAM 表的读锁，同时另一个进程也请求同一表的写锁，MySQL 如何处理呢？答案是写进程先获得锁。不仅如此，即使读请求先到锁等待队列，写请求后到，写锁也会插到读锁请求之前！这是因为 MySQL 认为写请求一般比读请求要重要。这也正是 MyISAM 表不太适合于有大量更新操作和查询操作应用的原因，因为大量的更新操作会造成查询操作很难获得读锁，从而可能永远阻塞。这种情况有时可能会变得非常糟糕！幸好我们可以通过一些设置来调节 MyISAM 的调度行为。

- 通过指定启动参数 low-priority-updates，使 MyISAM 引擎默认给予读请求以优先的权利。
- 通过执行命令 SET LOW_PRIORITY_UPDATES=1，使该连接发出的更新请求优先级降低。
- 通过指定 INSERT、UPDATE、DELETE 语句的 LOW_PRIORITY 属性，降低该语句的优先级。

虽然上面 3 种方法都是要么更新优先，要么查询优先的方法，但还是可以用其来解决查询相对重要的应用（如用户登录系统）中读锁等待严重的问题。

另外，MySQL 也提供了一种折中的办法来调节读写冲突，即给系统参数 max_write_lock_count 设置一个合适的值，当一个表的写锁达到这个值后，MySQL 就暂时将写请求的优先级降低，给读进程一些获得锁的机会。

上面已经讨论了写优先调度机制带来的问题和解决办法。这里还要强调一点：一些需要长时间运行的查询操作，也会使写进程"饿死"！因此，应用中应尽量避免出现长时间运行的查询操作，不要总想用一条 SELECT 语句来解决问题，因为这种看似巧妙的 SQL 语句，往往比较复杂，执行时间较长，在可能的情况下可以通过使用中间表等措施对 SQL 语句做一定的"分解"，使每一步查询都能在较短时间完成，从而减少锁冲突。如果复杂查询不可避免，应尽量安排在数据库空闲时段执行，比如一些定期统计可以安排在夜间执行。

16.3 InnoDB 锁问题

InnoDB 与 MyISAM 的最大不同有两点：一是支持事务（TRANSACTION）；二是采用了行级锁。行级锁与表级锁本来就有许多不同之处，另外，事务的引入也带来了一些新问题。下面我们先介绍一点背景知识，然后详细讨论 InnoDB 的锁问题。

16.3.1 背景知识

1．事务（Transaction）及其 ACID 属性

事务是由一组 SQL 语句组成的逻辑处理单元，具有以下 4 个属性，通常简称为事务的 ACID 属性。

- 原子性（Atomicity）：事务是一个原子操作单元，其对数据的修改，要么全都执行，要么全都不执行。
- 一致性（Consistent）：在事务开始和完成时，数据都必须保持一致状态。这意味着所有相关的数据规则都必须应用于事务的修改，以保持数据的完整性；事务结束时，所有的内部数据结构（如 B 树索引或双向链表）也都必须是正确的。
- 隔离性（Isolation）：数据库系统提供一定的隔离机制，保证事务在不受外部并发操作影响的"独立"环境执行。这意味着事务处理过程中的中间状态对外部是不可见的，反之亦然。
- 持久性（Durable）：事务完成之后，它对于数据的修改是永久性的，即使出现系统故障也能够保持。

银行转账就是事务的一个典型示例。

2．并发事务处理带来的问题

相对于串行处理来说，并发事务处理能大大增加数据库资源的利用率，提高数据库系统的事务吞吐量，从而可以支持更多的用户。但并发事务处理也会带来一些问题，主要包括以下几种情况。

- 更新丢失（Lost Update）：当两个或多个事务选择同一行，然后基于最初选定的值更新该行时，由于每个事务都不知道其他事务的存在，就会发生丢失更新问题——最后的更新覆盖了由其他事务所做的更新。例如，两个编辑人员制作了同一文档的电子副本。每个编辑人员独立地更改其副本，然后保存更改后的副本，这样就覆盖了原始文档。最后保存其更改副本的编辑人员覆盖另一个编辑人员所做的更改。如果在一个编辑人员完成并提交事务之前，另一个编辑人员不能访问同一文件，则可避免此问题。
- 脏读（Dirty Read）：一个事务正在对一条记录做修改，在这个事务完成并提交前，这条记录的数据就处于不一致状态；这时，另一个事务也来读取同一条记录，如果不加控制，第二个事务读取了这些"脏"数据，并据此做进一步的处理，就会产生未提交的数据依赖关系。这种现象被形象地叫做"脏读"。

- 不可重复读（Non-Repeatable Read）：一个事务在读取某些数据后的某个时间，再次读取以前读过的数据，却发现其读出的数据已经发生了改变或某些记录已经被删除了！这种现象就叫做"不可重复读"。

- 幻读（Phantom Read）：一个事务按相同的查询条件重新读取以前检索过的数据，却发现其他事务插入了满足其查询条件的新数据，这种现象就称为"幻读"。

3．事务隔离级别

在上面讲到的并发事务处理带来的问题中，"更新丢失"通常是应该完全避免的。但防止更新丢失，并不能单靠数据库事务控制器来解决，需要应用程序对要更新的数据加必要的锁来解决，因此，防止更新丢失应该是应用的责任。

"脏读"、"不可重复读"和"幻读"，其实都是数据库读一致性问题，必须由数据库提供一定的事务隔离机制来解决。数据库实现事务隔离的方式，基本上可分为以下两种。

- 一种是在读取数据前，对其加锁，阻止其他事务对数据进行修改。

- 另一种是不用加任何锁，通过一定机制生成一个数据请求时间点的一致性数据快照（Snapshot），并用这个快照来提供一定级别（语句级或事务级）的一致性读取。从用户的角度来看，好像是数据库可以提供同一数据的多个版本，因此，这种技术叫做数据多版本并发控制（MultiVersion Concurrency Control，MVCC 或 MCC），也经常称为多版本数据库。

数据库的事务隔离越严格，并发副作用越小，但付出的代价也就越大，因为事务隔离实质上就是使事务在一定程度上"串行化"进行，这显然与"并发"是矛盾的。同时，不同的应用对读一致性和事务隔离程度的要求也是不同的，比如许多应用对"不可重复读"和"幻读"并不敏感，可能更关心数据并发访问的能力。

为了解决"隔离"与"并发"的矛盾，ISO/ANSI SQL92 定义了 4 个事务隔离级别，每个级别的隔离程度不同，允许出现的副作用也不同，应用可以根据自己的业务逻辑要求，通过选择不同的隔离级别来平衡"隔离"与"并发"的矛盾。表 16-5 很好地概括了这 4 个隔离级别的特性。

表 16-5 4 种隔离级别比较

隔离级别	读数据一致性及允许的并发副作用	读数据一致性	脏读	不可重复读	幻读
未提交读（Read uncommitted）		最低级别，只能保证不读取物理上损坏的数据	是	是	是
已提交读（Read committed）		语句级	否	是	是
可重复读（Repeatable read）		事务级	否	否	是
可序列化（Serializable）		最高级别，事务级	否	否	否

最后要说明的是：各具体数据库并不一定完全实现了上述 4 个隔离级别，例如，Oracle 只提供 Read committed 和 Serializable 两个标准隔离级别，另外还提供自己定义的 Read only 隔离级别；SQL Server 除支持上述 ISO/ANSI SQL92 定义的 4 个隔离级别外，还支持一个叫做"快照"的隔离级别，但严格来说，它是一个用 MVCC 实现的 Serializable 隔离级别。MySQL

支持全部 4 个隔离级别,但在具体实现时,有一些 MySQL 自有的特点(区别于其他数据库的实现),比如在一些隔离级别下是采用 MVCC 一致性读,但某些情况下又不是,在后面的章节中将会对这些内容做进一步介绍。

16.3.2 获取 InnoDB 行锁争用情况

可以通过检查 InnoDB_row_lock 状态变量来分析系统上的行锁的争夺情况:

```
mysql> show status like 'innodb_row_lock%';
+-------------------------------+-------+
| Variable_name                 | Value |
+-------------------------------+-------+
| InnoDB_row_lock_current_waits | 0     |
| InnoDB_row_lock_time          | 0     |
| InnoDB_row_lock_time_avg      | 0     |
| InnoDB_row_lock_time_max      | 0     |
| InnoDB_row_lock_waits         | 0     |
+-------------------------------+-------+
5 rows in set (0.01 sec)
```

如果发现锁争用比较严重,如 InnoDB_row_lock_waits 和 InnoDB_row_lock_time_avg 的值比较高,可以通过设置 InnoDB Monitors 来进一步观察发生锁冲突的表、数据行等,并分析锁争用的原因。

通过设置下面两个参数可以来设置监视器,MySQL 定期将包含锁冲突信息的日志写入 error log:

```
SET GLOBAL innodb_status_output=ON;
SET GLOBAL innodb_status_output_locks=ON;
```

也可以用以下语句来查看最新的状态信息(粗体部分显示为锁冲突信息):

```
mysql> Show engine innodb status\G;
| InnoDB |    |
=====================================
2018-11-05 17:55:31 0x7f3b909a7700 INNODB MONITOR OUTPUT
=====================================
Per second averages calculated from the last 20 seconds
-----------------
LATEST FOREIGN KEY ERROR
------------------------
2018-11-02 15:59:59 0x7f3b90966700 Transaction:
TRANSACTION 2285382, ACTIVE 0 sec inserting
mysql tables in use 1, locked 1
2 lock struct(s), heap size 1136, 0 row lock(s)
MySQL thread id 10, OS thread handle 139893805573888, query id 108 localhost root update
insert into titles values(111111111,1,now(),now())
Foreign key constraint fails for table `employees`.`titles`:

   CONSTRAINT `titles_ibfk_1` FOREIGN KEY (`emp_no`) REFERENCES `employees` (`emp_no`) ON DELETE CASCADE
 ……
Trying to add in child table, in index PRIMARY tuple:
MySQL thread id 8124, OS thread handle 139893341038336, query id 2451665 localhost root
TABLE LOCK table `sakila`.`customer` trx id 3811967 lock mode IX
RECORD LOCKS space id 72 page no 7 n bits 256 index PRIMARY of table `sakila`.`customer` trx id 3811967 lock_m
ode X locks rec but not gap
Record lock, heap no 2 PHYSICAL RECORD: n_fields 11; compact format; info bits 0
 ……
```

可以在 MySQL 命令行中通过发出下列语句来停止监视器:

```
SET GLOBAL innodb_status_output=OFF;

SET GLOBAL innodb_status_output_locks=OFF;
```

设置监视器后,在 SHOW ENGINE INNODB STATUS 的显示内容中,会有当前锁等待的详细信息,包括表名、锁类型、锁定记录的情况等,以便进一步分析和确定问题。

打开监视器以后,默认情况下每 15s 会向日志中记录监控的内容,如果长时间打开会导致 error log 文件变得非常巨大,所以用户在确认问题原因之后,要记得关闭监视器。用户也可以在启动选项中加入--innodb-status-file 选项使得监控信息写入指定的 innodb_status.*pid* 文件,以便和 error 日志进行隔离。

16.3.3 InnoDB 的行锁模式及加锁方法

InnoDB 实现了以下两种类型的行锁。

- 共享锁(S):允许一个事务读一行,阻止其他事务获得相同数据集的排他锁。
- 排他锁(X):允许获得排他锁的事务更新数据,阻止其他事务取得相同数据集的共享读锁和排他写锁。

另外,为了允许行锁和表锁共存,实现多粒度锁机制,InnoDB 还有两种内部使用的意向锁(Intention Lock),这两种意向锁都是表锁。

- 意向共享锁(IS):事务打算给数据行加行共享锁,事务在给一个数据行加共享锁前必须先取得该表的 IS 锁。
- 意向排他锁(IX):事务打算给数据行加行排他锁,事务在给一个数据行加排他锁前必须先取得该表的 IX 锁。

上述锁模式的兼容情况具体如表 16-6 所示。

表 16-6   InnoDB 行锁模式兼容性列表

是否兼容 \ 当前锁模式 \ 请求锁模式	X	IX	S	IS
X	冲突	冲突	冲突	冲突
IX	冲突	兼容	冲突	兼容
S	冲突	冲突	兼容	兼容
IS	冲突	兼容	兼容	兼容

如果一个事务请求的锁模式与当前的锁兼容,InnoDB 就将请求的锁授予该事务;反之,如果两者不兼容,该事务就要等待锁释放。

意向锁是 InnoDB 自动加的,不需用户干预。对于 UPDATE、DELETE 和 INSERT 语句,InnoDB 会自动给涉及数据集加排他锁(X);对于普通 SELECT 语句,InnoDB 不会加任何锁;事务可以通过以下语句显式给记录集加共享锁或排他锁。

(1)在 MYSQL 5.7 中。

- 共享锁(S):SELECT * FROM table_name WHERE ... LOCK IN SHARE MODE。
- 排他锁(X):SELECT * FROM table_name WHERE ... FOR UPDATE。

(2)在 MySQL 8.0 中。

- 共享锁(S):SELECT * FROM table_name WHERE ... FOR SHARE
- 排他锁(X):SELECT * FROM table_name WHERE ... FOR UPDATE [NOWAIT|SKIP

16.3 InnoDB 锁问题

LOCKED]

在 MySQL 5.7 及之前的版本，用 SELECT ... IN SHARE MODE 获得共享锁，MySQL 8.0 后改为了 SELECT ... FOR SHARE 语句，为了向下兼容，IN SHARE MODE 命令仍然可以在 MySQL 8.0 中使用。共享锁主要用在需要数据依存关系时来确认某行记录是否存在，并确保没有人对这个记录进行 UPDATE 或者 DELETE 操作。但是如果当前事务也需要对该记录进行更新操作，则很有可能造成死锁，对于锁定行记录后需要进行更新操作的应用，应该使用 SELECT... FOR UPDATE 方式获得排他锁。

在 MySQL 5.7 中，SELECT... FOR UPDATE 语句用来显式加排他锁。如果遇到锁等待，那么 session 默认会等待 50s，这在高并发的应用系统中，一旦出现对于热点行的争用，将会造成连接数的快速增加，甚至超过最大连接数。为了解决并发问题，在 MySQL 8.0 中，FOR UPDATE 后增加了两个选项：SKIP LOCKED 和 NOWAIT，使用这两个选项在一定程度上会避免这种情况。其中，NOWAIT 选项发现有锁等待后会立即返回错误，不用等到锁超时后报错；某些特殊场景下，也可以使用 SKIP LOCKED 来跳过被锁定的行，直接更新其他行，但这样要注意是否会造成更新结果不符合预期。

在如表 16-7 所示的示例中，使用了 SELECT ... IN SHARE MODE 加锁后再更新记录，看看会出现什么情况，其中 actor 表的 actor_id 字段为主键。

表 16-7　　　　　　　　　　InnoDB 存储引擎的共享锁示例

session_1	session_2
mysql> set autocommit = 0; Query OK, 0 rows affected (0.00 sec)	mysql> set autocommit = 0; Query OK, 0 rows affected (0.00 sec)
mysql> select actor_id,first_name,last_name from actor where actor_id = 178; +----------+------------+-----------+ | actor_id | first_name | last_name | +----------+------------+-----------+ | 178 | LISA | MONROE | +----------+------------+-----------+ 1 row in set (0.00 sec)	mysql> select actor_id,first_name,last_name from actor where actor_id = 178; +----------+------------+-----------+ | actor_id | first_name | last_name | +----------+------------+-----------+ | 178 | LISA | MONROE | +----------+------------+-----------+ 1 row in set (0.00 sec)
当前 session 对 actor_id=178 的记录加 share mode 的共享锁： mysql> select actor_id,first_name,last_name from actor where actor_id = 178 lock in share mode; +----------+------------+-----------+ | actor_id | first_name | last_name | +----------+------------+-----------+ | 178 | LISA | MONROE | +----------+------------+-----------+ 1 row in set (0.01 sec)	
	其他 session 仍然可以查询记录，并也可以对该记录加 share mode 的共享锁： mysql> select actor_id,first_name,last_name from actor where actor_id = 178 lock in share mode; +----------+------------+-----------+ | actor_id | first_name | last_name | +----------+------------+-----------+ | 178 | LISA | MONROE | +----------+------------+-----------+ 1 row in set (0.01 sec)

session_1	session_2
当前 session 对锁定的记录进行更新操作，等待锁： mysql> update actor set last_name = 'MONROE T' where actor_id = 178; 等待	
	其他 session 也对该记录进行更新操作，则会导致死锁退出： mysql> update actor set last_name = 'MONROE T' where actor_id = 178; ERROR 1213 (40001): Deadlock found when trying to get lock; try restarting transaction
获得锁后，可以成功更新： mysql> update actor set last_name = 'MONROE T' where actor_id = 178; Query OK, 1 row affected (17.67 sec) Rows matched: 1　Changed: 1　Warnings: 0	

当使用 SELECT...FOR UPDATE 加锁后再更新记录，出现如表 16-8 所示的情况。

表 16-8　　　　　　　　　　InnoDB 存储引擎的排他锁示例

session_1	session_2
mysql> set autocommit = 0; Query OK, 0 rows affected (0.00 sec) mysql> select actor_id,first_name,last_name from actor where actor_id = 178; +----------+------------+-----------+ | actor_id | first_name | last_name | +----------+------------+-----------+ | 178 | LISA | MONROE | +----------+------------+-----------+ 1 row in set (0.00 sec)	mysql> set autocommit = 0; Query OK, 0 rows affected (0.00 sec) mysql> select actor_id,first_name,last_name from actor where actor_id = 178; +----------+------------+-----------+ | actor_id | first_name | last_name | +----------+------------+-----------+ | 178 | LISA | MONROE | +----------+------------+-----------+ 1 row in set (0.00 sec)
当前 session 对 actor_id=178 的记录加 for update 的排他锁： mysql> select actor_id,first_name,last_name from actor where actor_id = 178 for update; +----------+------------+-----------+ | actor_id | first_name | last_name | +----------+------------+-----------+ | 178 | LISA | MONROE | +----------+------------+-----------+ 1 row in set (0.00 sec)	
	其他 session 可以查询该记录，但是不能对该记录加共享锁，会等待获得锁： mysql> select actor_id,first_name,last_name from actor where actor_id = 178; +----------+------------+-----------+ | actor_id | first_name | last_name | +----------+------------+-----------+ | 178 | LISA | MONROE | +----------+------------+-----------+ 1 row in set (0.00 sec) mysql> select actor_id,first_name,last_name from actor where actor_id = 178 for update; 等待

session_1	session_2
当前 session 可以对锁定的记录进行更新操作，更新后释放锁： mysql> update actor set last_name = 'MONROE T' where actor_id = 178; Query OK, 1 row affected (0.00 sec) Rows matched: 1　Changed: 1　Warnings: 0 mysql> commit; Query OK, 0 rows affected (0.01 sec)	
	其他 session 获得锁，得到其他 session 提交的记录： mysql> select actor_id,first_name,last_name from actor where actor_id = 178 for update; +----------+------------+----------+ | actor_id | first_name | last_name | +----------+------------+----------+ | 178　　| LISA　　| MONROE T | +----------+------------+----------+ 1 row in set (9.59 sec)

16.3.4　InnoDB 行锁实现方式

InnoDB 行锁是通过给索引上的索引项加锁来实现的，如果没有索引，InnoDB 将通过隐藏的聚簇索引来对记录加锁。InnoDB 行锁分为 3 种情形。

● Record lock：对索引项加锁。

● Gap lock：对索引项之间的"间隙"、第一条记录前的"间隙"或最后一条记录后的"间隙"加锁。

● Next-key lock：前两种的组合，对记录及其前面的间隙加锁。

InnoDB 这种行锁实现特点意味着：如果不通过索引条件检索数据，那么 InnoDB 将对表中的所有记录加锁，实际效果跟表锁一样！

在实际应用中，要特别注意 InnoDB 行锁的这一特性，否则可能导致大量的锁冲突，从而影响并发性能。下面通过一些实际示例来加以说明。

（1）在不通过索引条件查询时，InnoDB 会锁定表中的所有记录。

在表 16-9 中的示例中，tab_no_index 表开始没有索引：

```
mysql> create table tab_no_index(id int,name varchar(10)) engine=innodb;
Query OK, 0 rows affected (0.15 sec)

mysql> insert into tab_no_index values(1,'1'),(2,'2'),(3,'3'),(4,'4');
Query OK, 4 rows affected (0.00 sec)
Records: 4  Duplicates: 0  Warnings: 0
```

表 16-9　　InnoDB 存储引擎的表在不使用索引时对全部记录加锁的示例

session_1	session_2
mysql> set autocommit=0; Query OK, 0 rows affected (0.00 sec) mysql> select * from tab_no_index where id = 1 ; +------+------+ | id　| name | +------+------+ | 1　| 1　| +------+------+ 1 row in set (0.00 sec)	mysql> set autocommit=0; Query OK, 0 rows affected (0.00 sec) mysql> select * from tab_no_index where id = 2 ; +------+------+ | id　| name | +------+------+ | 2　| 2　| +------+------+ 1 row in set (0.00 sec)

session_1	session_2
mysql> select * from tab_no_index where id = 1 for update; +------+------+ | id | name | +------+------+ | 1 | 1 | +------+------+ 1 row in set (0.00 sec)	
	mysql> select * from tab_no_index where id = 2 for update; 等待

在表 16-9 中，看起来 session_1 只给一行加了排他锁，但 session_2 在请求其他行的排他锁时，却出现了锁等待！原因是在没有索引的情况下，InnoDB 会对所有记录都加锁。当给其增加一个索引后，InnoDB 就只锁定了符合条件的行，如表 16-10 所示。

表 16-10　　　　InnoDB 存储引擎的表在使用索引时使用行锁的示例

session_1	session_2
mysql> set autocommit=0; Query OK, 0 rows affected (0.00 sec) mysql> select * from tab_with_index where id = 1 ; +------+------+ | id | name | +------+------+ | 1 | 1 | +------+------+ 1 row in set (0.00 sec)	mysql> set autocommit=0; Query OK, 0 rows affected (0.00 sec) mysql> select * from tab_with_index where id = 2 ; +------+------+ | id | name | +------+------+ | 2 | 2 | +------+------+ 1 row in set (0.00 sec)
mysql> select * from tab_with_index where id = 1 for update; +------+------+ | id | name | +------+------+ | 1 | 1 | +------+------+ 1 row in set (0.00 sec)	
	mysql> select * from tab_with_index where id = 2 for update; +------+------+ | id | name | +------+------+ | 2 | 2 | +------+------+ 1 row in set (0.00 sec)

创建 tab_with_index 表，id 字段有普通索引：

```
mysql> create table tab_with_index(id int,name varchar(10)) engine=innodb;
Query OK, 0 rows affected (0.15 sec)
mysql> insert into tab_with_index values(1,'1'),(2,'2'),(3,'3'),(4,'4');
Query OK, 4 rows affected (0.00 sec)
Records: 4  Duplicates: 0  Warnings: 0
mysql> alter table tab_with_index add index id(id);
Query OK, 4 rows affected (0.24 sec)
Records: 4  Duplicates: 0  Warnings: 0
```

（2）由于 MySQL 的行锁是针对索引加的锁，不是针对记录加的锁，所以虽然是访问不同行的记录,但是如果是使用相同的索引键,是会出现锁冲突的。应用设计的时候要注意这一点。

在如表 16-11 所示的示例中,表 tab_with_index 的 id 字段有索引,name 字段没有索引:

```
mysql> insert into tab_with_index  values(1,'4');
Query OK, 1 row affected (0.00 sec)

mysql> select * from tab_with_index where id = 1;
+------+------+
| id   | name |
+------+------+
| 1    | 1    |
| 1    | 4    |
+------+------+
2 rows in set (0.00 sec)
```

表 16-11　InnoDB 存储引擎使用相同索引键的阻塞示例

session_1	session_2
mysql> set autocommit=0; Query OK, 0 rows affected (0.00 sec)	mysql> set autocommit=0; Query OK, 0 rows affected (0.00 sec)
mysql> select * from tab_with_index where id = 1 and name = '1' for update; +------+------+ | id | name | +------+------+ | 1 | 1 | +------+------+ 1 row in set (0.00 sec)	
	虽然 session_2 访问的是和 session_1 不同的记录,但是因为使用了相同的索引,所以需要等待锁: mysql> select * from tab_with_index where id = 1 and name = '4' for update; 等待

(3)当表有多个索引的时候,不同的事务可以使用不同的索引锁定不同的行,不论是使用主键索引、唯一索引或普通索引,InnoDB 都会使用行锁来对数据加锁。

在如表 16-12 所示的示例中,表 tab_with_index 的 id 字段有主键索引,name 字段有普通索引:

```
mysql> alter table tab_with_index add index name(name);
Query OK, 5 rows affected (0.23 sec)
Records: 5  Duplicates: 0  Warnings: 0
```

表 16-12　InnoDB 存储引擎的表使用不同索引的阻塞示例

session_1	session_2
mysql> set autocommit=0; Query OK, 0 rows affected (0.00 sec)	mysql> set autocommit=0; Query OK, 0 rows affected (0.00 sec)
mysql> select * from tab_with_index where id = 1 for update; +------+------+ | id | name | +------+------+ | 1 | 1 | | 1 | 4 | +------+------+ 2 rows in set (0.00 sec)	

session_1	session_2
	session_2 使用 name 的索引访问记录，因为记录没有被加锁，所以可以获得锁： `mysql> select * from tab_with_index where name = '2' for update;` `+------+------+` `| id	name |` `+------+------+` `| 2	2 |` `+------+------+` `1 row in set (0.00 sec)`
	由于访问的记录已经被 session_1 锁定，所以等待获得锁： `mysql> select * from tab_with_index where name = '4' for update;`

（4）即便在条件中使用了索引字段，但是否使用索引来检索数据是由 MySQL 通过判断不同执行计划的代价来决定的，如果 MySQL 认为全表扫描效率更高，比如对一些很小的表，它就不会使用索引，这种情况下 InnoDB 也会对所有记录加锁。因此，在分析锁冲突时，别忘了检查 SQL 的执行计划，以确认是否真正使用了索引。关于 MySQL 在什么情况下不使用索引的详细讨论，可以参考 15.2 节。

在下面的示例中，检索值的数据类型与索引字段不同，虽然 MySQL 能够进行数据类型转换，却不会使用索引，从而导致 InnoDB 对所有记录加锁。通过用 explain 检查两条 SQL 的执行计划，读者可以清楚地看到这一点。

在示例中，tab_with_index 表的 name 字段有索引，但是 name 字段是 varchar 类型的，如果 where 条件中 name 字段不是 varchar 类型，则会对 name 进行隐式的数据类型转换，从而导致索引无法使用，走了全表扫描。

```
mysql> alter table tab_with_index add index name(name);
Query OK, 4 rows affected (8.06 sec)
Records: 4  Duplicates: 0  Warnings: 0

mysql> explain select * from tab_with_index where name = 1 \G
*************************** 1. row ***************************
           id: 1
  select_type: SIMPLE
        table: tab_with_index
         type: ALL
possible_keys: name
          key: NULL
      key_len: NULL
          ref: NULL
         rows: 4
        Extra: Using where
1 row in set (0.00 sec)
mysql> explain select * from tab_with_index where name = '1' \G
*************************** 1. row ***************************
           id: 1
  select_type: SIMPLE
        table: tab_with_index
         type: ref
possible_keys: name
          key: name
      key_len: 23
          ref: const
         rows: 1
        Extra: Using where
1 row in set (0.00 sec)
```

16.3.5 Next-Key 锁

当我们使用范围条件而不是相等条件检索数据，并请求共享或排他锁时，InnoDB 会给符合条件的已有数据记录的索引项加锁；对于键值在条件范围内但并不存在的记录，叫做间隙（GAP），InnoDB 也会对这个"间隙"加锁，这种锁机制就是所谓的 **Next-Key** 锁。

举例来说，假如 emp 表中只有 101 条记录，其 empid 的值分别是 1、2、…、100、101，SQL 语句如下：

```
Select * from emp where empid > 100 for update;
```

这是一个范围条件的检索，InnoDB 不仅会对符合条件的 empid 值为 101 的记录加锁，也会对 empid 大于 101（这些记录并不存在）的"间隙"加锁。

InnoDB 使用 Next-Key 锁的目的，一方面是为了防止幻读，以满足相关隔离级别的要求，对于上面的示例，要是不使用间隙锁，如果其他事务插入了 empid 大于 100 的任何记录，那么本事务如果再次执行上述语句，就会发生幻读；另一方面，是为了满足其恢复和复制的需要。有关其恢复和复制对锁机制的影响，以及不同隔离级别下 InnoDB 使用 Next-Key 锁的情况，在后续的章节中会做进一步介绍。

很显然，在使用范围条件检索并锁定记录时，InnoDB 这种加锁机制会阻塞符合条件范围内键值的并发插入，这往往会造成严重的锁等待。因此，在实际应用开发中，尤其是并发插入比较多的应用，要尽量优化业务逻辑，尽量使用相等条件来访问更新数据，避免使用范围条件。

还要特别说明的是，InnoDB 除了通过范围条件加锁时使用 Next-Key 锁外，如果使用相等条件请求给一个不存在的记录加锁，InnoDB 也会使用 Next-Key 锁！

在如表 16-13 所示的示例中，假如 emp 表中只有 101 条记录，其 empid 的值分别是 1、2、…、100、101。

表 16-13   InnoDB 存储引擎的 Next-Key 锁阻塞示例

session_1	session_2
mysql> select @@tx_isolation; +-----------------------------+ | @@tx_isolation | +-----------------------------+ | REPEATABLE-READ | +-----------------------------+ 1 row in set (0.00 sec) mysql> set autocommit = 0; Query OK, 0 rows affected (0.00 sec)	mysql> select @@tx_isolation; +-----------------------------+ | @@tx_isolation | +-----------------------------+ | REPEATABLE-READ | +-----------------------------+ 1 row in set (0.00 sec) mysql> set autocommit = 0; Query OK, 0 rows affected (0.00 sec)
当前 session 对不存在的记录加 for update 的锁： mysql> select * from emp where empid = 102 for update; Empty set (0.00 sec)	
	这时，如果其他 session 插入 empid 为 201 的记录（注意：这条记录并不存在），也会出现锁等待： mysql>insert into emp(empid,...) values(201,...); 阻塞等待
Session_1 执行 rollback： mysql> rollback; Query OK, 0 rows affected (13.04 sec)	

session_1	session_2
	由于其他 session_1 回退后释放了 Next-Key 锁，当前 session 可以获得锁并成功插入记录： mysql>insert into emp(empid,...) values(201,...); Query OK, 1 row affected (13.35 sec)

16.3.6　恢复和复制的需要，对 InnoDB 锁机制的影响

MySQL 通过 BINLOG 记录执行成功的 INSERT、UPDATE、DELETE 等更新数据的 SQL 语句，并由此实现 MySQL 数据库的恢复和主从复制（参见本书的"管理维护篇"）。MySQL 5.7 支持 3 种日志格式，即基于语句的日志格式 SBL、基于行的日志格式 RBL 和混合格式。它还支持 4 种复制模式。

- 基于 SQL 语句的复制 SBR：这也是 MySQL 最早支持的复制模式。
- 基于行数据的复制 RBR：这是 MySQL 5.1 以后开始支持的复制模式，主要优点是支持对非安全 SQL 的复制。
- 混合复制模式：对安全的 SQL 语句采用基于 SQL 语句的复制模式，对于非安全的 SQL 语句采用居于行的复制模式。
- 使用全局事务 ID（GTIDs）的复制：主要是解决主从自动同步一致问题。

对基于语句日志格式（SBL）的恢复和复制而言，由于 MySQL 的 BINLOG 是按照事务提交的先后顺序记录的，因此要正确恢复或复制数据，就必须满足：在一个事务未提交前，其他并发事务不能插入满足其锁定条件的任何记录，也就是不允许出现幻读。这已经超过了 ISO/ANSI SQL92 "可重复读"隔离级别的要求，实际上是要求事务要串行化。这也是许多情况下，InnoDB 要用到 Next-Key 锁的原因，比如在用范围条件更新记录时，无论在 Read Committed 或是 Repeatable Read 隔离级别下，InnoDB 都要使用 Next-Key 锁，但这并不是隔离级别要求的，有关 InnoDB 在不同隔离级别下加锁的差异在 16.3.7 节还会介绍。

对于 "insert into target_tab select * from source_tab where ..." 和 "create table new_tab ...select ... From source_tab where ...（CTAS）"这种 SQL 语句，用户并没有对 source_tab 做任何更新操作，但 MySQL 对这种 SQL 语句做了特别处理。先来看表 16-14 中的示例。

表 16-14　　　　　　　　CTAS 操作给原表加锁的示例

session_1	session_2
mysql> set autocommit = 0; Query OK, 0 rows affected (0.00 sec) mysql> select * from target_tab; Empty set (0.00 sec) mysql> select * from source_tab where name = '1'; +----+-------+----+ | d1 | name | d2 | +----+-------+----+ | 4 | 1 | 1 | | 5 | 1 | 1 | | 6 | 1 | 1 | | 7 | 1 | 1 | | 8 | 1 | 1 | +----+-------+----+ 5 rows in set (0.00 sec)	mysql> set autocommit = 0; Query OK, 0 rows affected (0.00 sec) mysql> select * from target_tab; Empty set (0.00 sec) mysql> select * from source_tab where name = '1'; +----+-------+----+ | d1 | name | d2 | +----+-------+----+ | 4 | 1 | 1 | | 5 | 1 | 1 | | 6 | 1 | 1 | | 7 | 1 | 1 | | 8 | 1 | 1 | +----+-------+----+ 5 rows in set (0.00 sec)

16.3 InnoDB 锁问题

续表

session_1	session_2
mysql> insert into target_tab select d1,name from source_tab where name = '1'; Query OK, 5 rows affected (0.00 sec) Records: 5 Duplicates: 0 Warnings: 0	
	mysql> update source_tab set name = '1' where name = '8'; 等待
commit;	
	返回结果 commit;

在上面的示例中,只是简单地读 source_tab 表的数据,相当于执行一个普通的 SELECT 语句,用一致性读就可以了。Oracle 正是这么做的,它通过 MVCC 技术实现的多版本数据来实现一致性读,不需要给 source_tab 加任何锁。我们知道 InnoDB 也实现了多版本数据,对普通的 SELECT 一致性读,也不需要加任何锁;但这里 InnoDB 却给 source_tab 加了共享锁,并没有使用多版本数据一致性读技术!

MySQL 为什么要这么做呢?其原因还是为了保证恢复和复制的正确性。因为在不加锁的情况下,如果在上述语句执行过程中,其他事务对 source_tab 做了更新操作,就可能导致数据恢复的结果错误。为了演示这一点,我们再重复一下前面的示例,不同的是在 session_1 执行事务前,先将系统变量 innodb_locks_unsafe_for_binlog 的值设置为 "on"(其默认值为 off),具体结果如表 16-15 所示。

表 16-15　　　　CTAS 操作不给原表加锁带来的安全问题示例

session_1	session_2
mysql> set autocommit = 0; Query OK, 0 rows affected (0.00 sec) mysql>set innodb_locks_unsafe_for_binlog='on' Query OK, 0 rows affected (0.00 sec) mysql> select * from target_tab; Empty set (0.00 sec) mysql> select * from source_tab where name = '1'; +------+--------+--------+ | d1 | name | d2 | +------+--------+--------+ | 4 | 1 | 1 | | 5 | 1 | 1 | | 6 | 1 | 1 | | 7 | 1 | 1 | | 8 | 1 | 1 | +------+--------+--------+ 5 rows in set (0.00 sec) mysql> insert into target_tab select d1,name from source_tab where name = '1'; Query OK, 5 rows affected (0.00 sec) Records: 5 Duplicates: 0 Warnings: 0	mysql> set autocommit = 0; Query OK, 0 rows affected (0.00 sec) mysql> select * from target_tab; Empty set (0.00 sec) mysql> select * from source_tab where name = '1'; +------+--------+------+ | d1 | name | d2 | +------+--------+------+ | 4 | 1 | 1 | | 5 | 1 | 1 | | 6 | 1 | 1 | | 7 | 1 | 1 | | 8 | 1 | 1 | +------+--------+------+ 5 rows in set (0.00 sec)

session_1	session_2
	session_1 未提交，可以对 session_1 的 select 的记录进行更新操作： mysql> update source_tab set name = '8' where name = '1'; Query OK, 5 rows affected (0.00 sec) Rows matched: 5　Changed: 5　Warnings: 0 mysql> select * from source_tab where name = '8'; +----+------+----+ | d1 | name | d2 | +----+------+----+ | 4 | 8 | 1 | | 5 | 8 | 1 | | 6 | 8 | 1 | | 7 | 8 | 1 | | 8 | 8 | 1 | +----+------+----+ 5 rows in set (0.00 sec)
	更新操作先提交 mysql> commit; Query OK, 0 rows affected (0.05 sec)
插入操作后提交： mysql> commit; Query OK, 0 rows affected (0.07 sec)	
此时查看数据，target_tab 中可以插入 source_tab 更新前的结果，这符合应用逻辑： mysql> select * from source_tab where name = '8'; +----+------+----+ | d1 | name | d2 | +----+------+----+ | 4 | 8 | 1 | | 5 | 8 | 1 | | 6 | 8 | 1 | | 7 | 8 | 1 | | 8 | 8 | 1 | +----+------+----+ 5 rows in set (0.00 sec) mysql> select * from target_tab; +------+------+ | id | name | +------+------+ | 4 | 1.00 | | 5 | 1.00 | | 6 | 1.00 | | 7 | 1.00 | | 8 | 1.00 | +------+------+ 5 rows in set (0.00 sec)	mysql> select * from tt1 where name = '1'; Empty set (0.00 sec) mysql> select * from source_tab where name = '8'; +----+------+----+ | d1 | name | d2 | +----+------+----+ | 4 | 8 | 1 | | 5 | 8 | 1 | | 6 | 8 | 1 | | 7 | 8 | 1 | | 8 | 8 | 1 | +----+------+----+ 5 rows in set (0.00 sec) mysql> select * from target_tab; +------+------+ | id | name | +------+------+ | 4 | 1.00 | | 5 | 1.00 | | 6 | 1.00 | | 7 | 1.00 | | 8 | 1.00 | +------+------+ 5 rows in set (0.00 sec)

从上可见，设置系统变量 innodb_locks_unsafe_for_binlog 的值为"on"后，InnoDB 不再对 source_tab 加锁，结果也符合应用逻辑，但是如果分析 BINLOG 的内容：

```
...
SET TIMESTAMP=1169175130;
BEGIN;
# at 274
```

```
    #070119 10:51:57 server id 1   end_log_pos 105    Query    thread_id=1    exec_time=0
error_code=0
    SET TIMESTAMP=1169175117;
    update source_tab set name = '8' where name = '1';
    # at 379
    #070119 10:52:10 server id 1   end_log_pos 406    xid = 5
    COMMIT;
    # at 406
    #070119 10:52:14 server id 1   end_log_pos 474    Query    thread_id=2    exec_time=0
error_code=0
    SET TIMESTAMP=1169175134;
    BEGIN;
    # at 474
    #070119 10:51:29 server id 1   end_log_pos 119    Query    thread_id=2    exec_time=0
error_code=0
    SET TIMESTAMP=1169175089;
    insert into target_tab select d1,name from source_tab where name = '1';
    # at 593
    #070119 10:52:14 server id 1   end_log_pos 620    xid = 7
    COMMIT;
    ...
```

可以发现，在 BINLOG 中，更新操作的位置在 INSERT...SELECT 之前，如果使用这个 BINLOG 进行数据库恢复，恢复的结果与实际的应用逻辑不符；如果进行复制，就会导致主从数据库不一致！

通过上面的示例，读者就不难理解为什么 MySQL 在处理 "Insert into target_tab select * from source_tab where ..." 和 "create table new_tab ...select ... From source_tab where ..." 时要给 source_tab 加锁，而不是使用对并发影响最小的多版本数据来实现一致性读。还要特别说明的是，如果上述语句的 SELECT 是范围条件，InnoDB 还会给源表加 Next-Key 锁。

因此，INSERT...SELECT...和 CREATE TABLE...SELECT...语句，可能会阻止对源表的并发更新。如果查询比较复杂，会造成严重的性能问题，读者在应用中应尽量避免使用。实际上，MySQL 将这种 SQL 称为不确定（non-deterministic）的 SQL，属于"Unsafe SQL"，不推荐使用。

如果应用中一定要用这种 SQL 来实现业务逻辑，又不希望对源表的并发更新产生影响，可以采取以下 3 种措施。

- 采取上面示例中的做法，将 innodb_locks_unsafe_for_binlog 的值设置为 "on"，强制 MySQL 使用多版本数据一致性读。但付出的代价是可能无法用 BINLOG 正确地恢复或复制数据，因此，不推荐使用这种方式。
- 通过使用 "select * from source_tab ... Into outfile" 和 "load data infile ..." 语句组合来间接实现，采用这种方式 MySQL 不会给 source_tab 加锁。
- 使用基于行的 BINLOG 格式和基于行数据的复制。

16.3.7　InnoDB 在不同隔离级别下的一致性读及锁的差异

前面讲过，锁和多版本数据是 InnoDB 实现一致性读和 ISO/ANSI SQL92 隔离级别的手段，因此，在不同的隔离级别下，InnoDB 处理 SQL 时采用的一致性读策略和需要的锁是不同的。同时，数据恢复和复制机制的特点，也对一些 SQL 的一致性读策略和锁策略有很大影响。将这些特性归纳成如表 16-16 所示的内容，以便读者查阅。

表 16-16　InnoDB 存储引擎中不同 SQL 在不同隔离级别下锁的比较

一致性读和锁 \ 隔离级别 \ SQL	条件	Read Uncommited	Read Commited	Repeatable Read	Serializable
select	相等	none locks	consistent read/none lock	consistent read/none lock	share locks
	范围	none locks	consistent read/none lock	consistent read/none lock	share next-key
update	相等	exclusive locks	exclusive locks	exclusive locks	exclusive locks
	范围	exclusive next-key	exclusive next-key	exclusive next-key	exclusive next-key
Insert	N/A	exclusive locks	exclusive locks	exclusive locks	exclusive locks
replace	无键冲突	exclusive locks	exclusive locks	exclusive locks	exclusive locks
	键冲突	exclusive next-key	exclusive next-key	exclusive next-key	exclusive next-key
delete	相等	exclusive locks	exclusive locks	exclusive locks	exclusive locks
	范围	exclusive next-key	exclusive next-key	exclusive next-key	exclusive next-key
Select ... from ... Lock in share mode	相等	share locks	share locks	share locks	share locks
	范围	share locks	share locks	share next-key	share next-key
Select * from ... For update	相等	exclusive locks	exclusive locks	exclusive locks	exclusive locks
	范围	exclusive locks	share locks	exclusive next-key	exclusive next-key
Insert into ... Select ...（指源表锁）	innodb_locks_unsafe_for_binlog=off	share next-key	share next-key	share next-key	share next-key
	innodb_locks_unsafe_for_binlog=on	none locks	consistent read/none lock	consistent read/none lock	share next-key
create table ... Select ...（指源表锁）	innodb_locks_unsafe_for_binlog=off	share next-key	share next-key	share next-key	share next-key
	innodb_locks_unsafe_for_binlog=on	none locks	consistent read/none lock	consistent read/none lock	share next-key

从表 16-16 中可以看出，对于许多 SQL，隔离级别越高，InnoDB 给记录集加的锁就越严格（尤其是使用范围条件的时候），产生锁冲突的可能性也就越高，从而对并发性事务处理性能的影响也就越大。因此，在应用中，应该尽量使用较低的隔离级别，以减少锁争用的机率。实际上，通过优化事务逻辑，大部分应用使用 Read Committed 隔离级别就足够了。对于一些确实需要更高隔离级别的事务，可以通过在程序中执行 SET SESSION TRANSACTION ISOLATION LEVEL REPEATABLE READ 或 SET SESSION TRANSACTION ISOLATION LEVEL SERIALIZABLE 动态改变隔离级别的方式来满足需求。

16.3.8　什么时候使用表锁

对于 InnoDB 表，在绝大部分情况下都应该使用行级锁，因为事务和行锁往往是我们选择 InnoDB 表的理由。但在个别特殊事务中，也可以考虑使用表级锁。

- 第一种情况是：事务需要更新大部分或全部数据，表又比较大，如果使用默认的行锁，不仅这个事务执行效率低，而且可能造成其他事务长时间锁等待和锁冲突，这种情况下可以考虑使用表锁来提高该事务的执行速度。
- 第二种情况是：事务涉及多个表，比较复杂，很可能引起死锁，造成大量事务回滚。这

16.3 InnoDB 锁问题

种情况也可以考虑一次性锁定事务涉及的表,从而避免死锁,减少数据库因事务回滚带来的开销。

当然,应用中这两种事务不能太多,否则,就应该考虑使用 MyISAM 表了。在 InnoDB 下,使用表锁要注意以下两点。

(1)使用 LOCK TABLES 虽然可以给 InnoDB 加表级锁,但必须说明的是,表锁不是由 InnoDB 存储引擎层管理的,而是由其上一层—MySQL Server 负责的,仅当 autocommit=0、innodb_table_locks=1(默认设置)时,InnoDB 层才能知道 MySQL 加的表锁,MySQL Server 也才能感知 InnoDB 加的行锁,这种情况下,InnoDB 才能自动识别涉及表级锁的死锁;否则,InnoDB 将无法自动检测并处理这种死锁。有关死锁,16.3.9 节还会继续讨论。

(2)在用 LOCK TABLES 对 InnoDB 表加锁时要注意,要将 AUTOCOMMIT 设为 0,否则 MySQL 不会给表加锁;事务结束前,不要用 UNLOCK TABLES 释放表锁,因为 UNLOCK TABLES 会隐含地提交事务;COMMIT 或 ROLLBACK 并不能释放用 LOCK TABLES 加的表级锁,必须用 UNLOCK TABLES 释放表锁。正确的方式如下所示。

例如,如果需要写表 t1 并从表 t 读,可以按如下做:

```
SET AUTOCOMMIT=0;
LOCK TABLES t1 WRITE, t2 READ, ...;
[do something with tables t1 and t2 here];
COMMIT;
UNLOCK TABLES;
```

16.3.9 关于死锁

上文讲过,MyISAM 表锁是 deadlock free 的,这是因为 MyISAM 总是一次获得所需的全部锁,要么全部满足,要么等待,因此不会出现死锁。但在 InnoDB 中,除单个 SQL 组成的事务外,锁是逐步获得的,这就决定了在 InnoDB 中发生死锁是可能的。表 16-17 就是一个发生死锁的示例。

表 16-17　　　　　　　　　InnoDB 存储引擎中的死锁示例

session_1	session_2
mysql> set autocommit = 0; Query OK, 0 rows affected (0.00 sec)	mysql> set autocommit = 0; Query OK, 0 rows affected (0.00 sec)
mysql> select * from table_1 where where id=1 for update; …… 做一些其他处理	mysql> select * from table_2 where id=1 for update; …… 做一些其他处理
select * from table_2 where id =1 for update; 因 session_2 已取得排他锁,等待	
	mysql> select * from table_1 where where id=1 for update; 死锁

在上面的示例中,两个事务都需要获得对方持有的排他锁才能继续完成事务,这种循环锁等待就是典型的死锁。

发生死锁后,InnoDB 一般都能自动检测到,并使一个事务释放锁并回退,另一个事务获得锁,继续完成事务。但在涉及外部锁或涉及表锁的情况下,InnoDB 并不能完全自动检测到死锁,这需要通过设置锁等待超时参数 innodb_lock_wait_timeout 来解决。需要说明的是,这个参数并不是只用来解决死锁问题,在并发访问比较高的情况下,如果大量事务因无法立即获得所需的锁而挂起,会占用大量计算机资源,造成严重性能问题,甚至拖垮数据库。通过设置合适的锁等待超时阈值，可以避免这种情况发生。

通常来说，死锁都是应用设计的问题，通过调整业务流程、数据库对象设计、事务大小，以及访问数据库的 SQL 语句，绝大部分死锁都可以避免。下面通过几个实例来介绍几种避免死锁的常用方法。

（1）在应用中，如果不同的程序会并发存取多个表，应尽量约定以相同的顺序来访问表，这样可以大大降低产生死锁的机会。在表 16-18 所示的示例中，由于两个 session 访问两个表的顺序不同，发生死锁的机会就非常高！但如果以相同的顺序来访问，死锁就可以避免。

表 16-18　　　　InnoDB 存储引擎中表顺序造成的死锁示例

session_1	session_2
mysql> set autocommit=0; Query OK, 0 rows affected (0.00 sec)	mysql> set autocommit=0; Query OK, 0 rows affected (0.00 sec)
mysql> select first_name,last_name from actor where actor_id = 1 for update; +------------------+-----------+ | first_name | last_name | +------------------+-----------+ | PENELOPE | GUINESS | +------------------+-----------+ 1 row in set (0.00 sec)	
	mysql> insert into country (country_id,country) values(110,'Test'); Query OK, 1 row affected (0.00 sec)
mysql>　insert into country (country_id,country) values(110,'Test'); 等待	
	mysql> select first_name,last_name from actor where actor_id = 1 for update; +------------------+-----------+ | first_name | last_name | +------------------+-----------+ | PENELOPE | GUINESS | +------------------+-----------+ 1 row in set (0.00 sec)
mysql>　insert into country (country_id,country) values (110,'Test'); ERROR 1213 (40001): Deadlock found when trying to get lock; try restarting transaction	

（2）在程序以批量方式处理数据的时候，如果事先对数据排序，保证每个线程按固定的顺序来处理记录，也可以大大降低出现死锁的可能。表 16-19 是一个数据操作顺序不一致而造成死锁的示例。

表 16-19　　　　InnoDB 存储引擎中表数据操作顺序不一致造成的死锁示例

session_1	session_2
mysql> set autocommit=0; Query OK, 0 rows affected (0.00 sec)	mysql> set autocommit=0; Query OK, 0 rows affected (0.00 sec)
mysql> select first_name,last_name from actor where actor_id = 1 for update; +------------------+-----------+ | first_name | last_name | +------------------+-----------+ | PENELOPE | GUINESS | +------------------+-----------+ 1 row in set (0.00 sec)	

16.3 InnoDB 锁问题

续表

session_1	session_2
	mysql> select first_name,last_name from actor where actor_id = 3 for update; +------------+-----------+ | first_name | last_name | +------------+-----------+ | ED | CHASE | +------------+-----------+ 1 row in set (0.00 sec)
mysql> select first_name,last_name from actor where actor_id = 3 for update; 等待	
	mysql> select first_name,last_name from actor where actor_id = 1 for update; ERROR 1213 (40001): Deadlock found when trying to get lock; try restarting transaction
mysql> select first_name,last_name from actor where actor_id = 3 for update; +------------+-----------+ | first_name | last_name | +------------+-----------+ | ED | CHASE | +------------+-----------+ 1 row in set (4.71 sec)	

（3）在事务中，如果要更新记录，应该直接申请足够级别的锁，即排他锁，而不应先申请共享锁，更新时再申请排他锁，因为当用户申请排他锁时，其他事务可能又已经获得了相同记录的共享锁，从而造成锁冲突，甚至死锁。具体演示可参见 16.3.3 节中的示例。

（4）前面讲过，在 REPEATABLE-READ 隔离级别下，如果两个线程同时对相同条件记录用 SELECT...FOR UPDATE 加排他锁，在没有符合该条件记录情况下，两个线程都会加锁成功。程序发现记录尚不存在，就试图插入一条新记录，如果两个线程都这么做，就会出现死锁。这种情况下，将隔离级别改成 READ COMMITTED，就可避免问题，如表 16-20 所示。

表 16-20　InnoDB 存储引擎中隔离级别引起的死锁示例 1

session_1	session_2
mysql> select @@tx_isolation; +-----------------+ | @@tx_isolation | +-----------------+ | REPEATABLE-READ | +-----------------+ 1 row in set (0.00 sec) mysql> set autocommit = 0; Query OK, 0 rows affected (0.00 sec)	mysql> select @@tx_isolation; +-----------------+ | @@tx_isolation | +-----------------+ | REPEATABLE-READ | +-----------------+ 1 row in set (0.00 sec) mysql> set autocommit = 0; Query OK, 0 rows affected (0.00 sec)
当前 session 对不存在的记录加 for update 的锁： mysql> select actor_id,first_name,last_name from actor where actor_id = 201 for update; Empty set (0.00 sec)	

续表

session_1	session_2
	其他 session 也可以对不存在的记录加 for update 的锁： mysql> select actor_id,first_name,last_name from actor where actor_id = 201 for update; Empty set (0.00 sec)
因为其他 session 也对该记录加了锁，所以当前的插入会等待： mysql> insert into actor (actor_id , first_name , last_name) values(201,'Lisa','Tom'); 等待	
	因为其他 session 已经对记录进行了更新，这时候再插入记录就会提示死锁并退出： mysql> insert into actor (actor_id, first_name , last_name) values(201,'Lisa','Tom'); ERROR 1213 (40001): Deadlock found when trying to get lock; try restarting transaction
由于其他 session 已经退出，当前 session 可以获得锁并成功插入记录： mysql> insert into actor (actor_id , first_name , last_name) values(201,'Lisa','Tom'); Query OK, 1 row affected (13.35 sec)	

（5）当隔离级别为 READ COMMITTED 时，如果两个线程都先执行 SELECT...FOR UPDATE，判断是否存在符合条件的记录，如果没有，就插入记录。此时，只有一个线程能插入成功，另一个线程会出现锁等待，当第一个线程提交后，第二个线程会因主键重出错，但虽然这个线程出错了，却会获得一个排他锁！这时如果有第三个线程又来申请排他锁，也会出现死锁。

对于这种情况，可以直接做插入操作，然后再捕获主键重异常，或者在遇到主键重错误时，总是执行 ROLLBACK 释放获得的排他锁，如表 16-21 所示。

表 16-21　　InnoDB 存储引擎中隔离级别引起的死锁示例 2

session_1	session_2	session_3
mysql> select @@tx_isolation; +----------------------------+ | @@tx_isolation | +----------------------------+ | READ-COMMITTED | +----------------------------+ 1 row in set (0.00 sec) mysql> set autocommit=0; Query OK, 0 rows affected (0.01 sec)	mysql> select @@tx_isolation; +----------------------------+ | @@tx_isolation | +----------------------------+ | READ-COMMITTED | +----------------------------+ 1 row in set (0.00 sec) mysql> set autocommit=0; Query OK, 0 rows affected (0.01 sec)	mysql> select @@tx_isolation; +----------------------------+ | @@tx_isolation | +----------------------------+ | READ-COMMITTED | +----------------------------+ 1 row in set (0.00 sec) mysql> set autocommit=0; Query OK, 0 rows affected (0.01 sec)
session_1 获得 for update 的共享锁： mysql> select actor_id, first_name, last_name from actor where actor_id = 201 for update; Empty set (0.00 sec)	由于记录不存在，session_2 也可以获得 for update 的共享锁： mysql> select actor_id, first_name, last_name from actor where actor_id = 201 for update; Empty set (0.00 sec)	

16.3 InnoDB 锁问题

续表

session_1	session_2	session_3
session_1 可以成功插入记录： mysql> insert into actor (actor_id,first_name, last_name) values(201,'Lisa','Tom'); Query OK, 1 row affected (0.00 sec)		
	session_2 插入申请等待获得锁： mysql> insert into actor (actor_id, first_name,last_name) values (201,'Lisa','Tom'); 等待	
session_1 成功提交： mysql> commit; Query OK, 0 rows affected (0.04 sec)		
	session_2 获得锁，发现插入记录主键重，这个时候抛出了异常，但是并没有释放共享锁： mysql> insert into actor (actor_id, irst_name,last_name) values (201,'Lisa','Tom'); ERROR 1062 (23000): Duplicate entry '201' for key 'PRIMARY'	
		session_3 申请获得共享锁，因为 session_2 已经锁定该记录，所以 session_3 需要等待： mysql> select actor_id, first_name, last_name from actor where actor_id= 201 for update; 等待
	这个时候，如果 session_2 直接对记录进行更新操作，则会抛出死锁的异常： mysql> update actor set last_name='Lan' where actor_id = 201; ERROR 1213 (40001): Deadlock found when trying to get lock; try restarting transaction	
		session_2 释放锁后，session_3 获得锁： mysql> select first_name, last_name from actor where actor_id = 201 for update; +------------+-----------+ | first_name | last_name | +------------+-----------+ | Lisa | Tom | +------------+-----------+ 1 row in set (31.12 sec)

尽管通过上面介绍的设计和 SQL 优化等措施，可以大大减少死锁，但死锁很难完全避免。因此，在程序设计中总是捕获并处理死锁异常是一个很好的编程习惯。

如果出现死锁，可以用 SHOW ENGINE INNODB STATUS 命令来确定最后一个死锁产生的原因。返回结果中包括死锁相关事务的详细信息，如引发死锁的 SQL 语句，事务已经获得的锁，正在等待什么锁，以及被回滚的事务等。据此可以分析死锁产生的原因和改进措施。

下面是一段 SHOW INNODB STATUS 输出的示例：

```
mysql> show engine innodb status \G
…
------------------------
LATEST DETECTED DEADLOCK
------------------------
070710 14:05:16
*** (1) TRANSACTION:
TRANSACTION 0 117470078, ACTIVE 117 sec, process no 1468, OS thread id 1197328736 inserting
mysql tables in use 1, locked 1
LOCK WAIT 5 lock struct(s), heap size 1216
MySQL thread id 7521657, query id 673468054 localhost root update
insert into country (country_id,country) values(110,'Test')
…

*** (2) TRANSACTION:
TRANSACTION 0 117470079, ACTIVE 39 sec, process no 1468, OS thread id 1164048736 starting index
read, thread declared inside InnoDB 500
mysql tables in use 1, locked 1
4 lock struct(s), heap size 1216, undo log entries 1
MySQL thread id 7521664, query id 673468058 localhost root statistics
select first_name,last_name from actor where actor_id = 1 for update
*** (2) HOLDS THE LOCK(S):
…

*** (2) WAITING FOR THIS LOCK TO BE GRANTED:
…

*** WE ROLL BACK TRANSACTION (1)
…
```

16.4 小结

本章重点介绍了 MySQL 中 MyISAM 表级锁和 InnoDB 行级锁的实现特点，并讨论了两种存储引擎经常遇到的锁问题和解决办法。

对于 MyISAM 的表锁，主要讨论了以下几点。

（1）共享读锁（S）之间是兼容的，但共享读锁（S）与排他写锁（X）之间，以及排他写锁（X）之间是互斥的，也就是说读和写是串行的。

（2）在一定条件下，MyISAM 允许查询和插入并发执行，可以利用这一点来解决应用中对同一表查询和插入的锁争用问题。

（3）MyISAM 默认的锁调度机制是写优先，这并不一定适合所有应用，用户可以通过设置 LOW_PRIORITY_UPDATES 参数，或在 INSERT、UPDATE、DELETE 语句中指定 LOW_PRIORITY 选项来调节读写锁的争用。

（4）由于表锁的锁定粒度大，读写之间又是串行的，因此，如果更新操作较多，MyISAM 表可能会出现严重的锁等待，可以考虑采用 InnoDB 表来减少锁冲突。

对于 InnoDB 表，本章主要讨论了以下几个内容。

- InnoDB 的行锁是基于索引实现的，如果不通过索引访问数据，InnoDB 会对所有数据加锁。
- 介绍了 InnoDB Next-Key 锁机制，以及 InnoDB 使用 Next-Key 锁的原因。
- 在不同的隔离级别下，InnoDB 的锁机制和一致性读策略不同。
- MySQL 的恢复和复制对 InnoDB 锁机制和一致性读策略也有较大影响。

16.4 小结

- 锁冲突甚至死锁很难完全避免。

在了解 InnoDB 锁特性后，用户可以通过设计和 SQL 调整等措施减少锁冲突和死锁，包括以下几项：

- 尽量使用较低的隔离级别；
- 精心设计索引，并尽量使用索引访问数据，使加锁更精确，从而减少锁冲突的机会；
- 选择合理的事务大小，小事务发生锁冲突的概率也更小；
- 给记录集显式加锁时，最好一次性请求足够级别的锁。比如要修改数据，最好直接申请排他锁，而不是先申请共享锁，修改时再请求排他锁，这样容易产生死锁；
- 不同的程序访问一组表时，应尽量约定以相同的顺序访问各表，对一个表而言，尽可能以固定的顺序存取表中的行。这样可以大大减少死锁的机会；
- 尽量用相等的条件访问数据，这样可以避免 Next-Key 锁对并发插入的影响；
- 不要申请超过实际需要的锁级别；除非必须，查询时不要显示加锁；
- 对于一些特定的事务，可以使用表锁来提高处理速度或减少发生死锁的概率。

第 17 章 优化 MySQL Server

MySQL 是一个可高度定制化的数据库系统，提供了很多可配置的参数。除非是做简单测试，我们一般都需要根据应用特性和硬件情况对 MySQL 做配置优化。本章在简单说明 MySQL 体系结构和内部机制的基础上，着重介绍 MySQL Server 的性能优化和调整。

17.1 MySQL 体系结构概览

MySQL 实例由一组后台线程、一些内存块和若干服务线程组成，如图 17-1 所示。

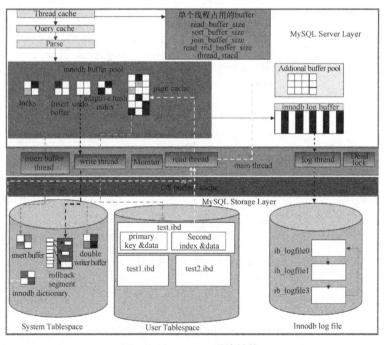

图 17-1　MySQL 系统结构

在默认情况下，MySQL 有 7 组后台线程，分别是 1 个主线程、4 组 IO 线程、1 个锁线程和 1 个错误监控线程。MySQL 5.5 和 MySQL 5.6 之后又分别新增了一个 purge 线程和一个 page cleaner 线程。这些后台线程的主要功能如下。

17.1 MySQL 体系结构概览

- master thread：主要负责将脏缓存页刷新到数据文件，执行 purge 操作，触发检查点，合并插入缓冲区等。
- insert buffer thread：主要负责插入缓冲区的合并操作。
- read thread：负责数据库读取操作，可配置多个读线程。
- write thread：负责数据库写操作，可配置多个写线程。
- log thread：用于将重做日志刷新到 logfile 中。
- purge thread：MySQL 5.5 之后用单独的 purge thread 执行 purge 操作。
- page cleaner thread：MySQL 5.6 之后，用来执行 buffer pool 中脏页的 flush 操作。
- lock thread：负责锁控制和死锁检测等。
- 错误监控线程：主要负责错误监控和错误处理。

我们可以通过 show engine innodb status 命令来查看这些线程的状态：

```
show engine innodb status\G;
=====================================
2018-10-10 16:22:39 0x7f6f18355700 INNODB MONITOR OUTPUT
=====================================
Per second averages calculated from the last 3 seconds
-----------------
BACKGROUND THREAD
-----------------
srv_master_thread loops: 3424 srv_active, 0 srv_shutdown, 6297767 srv_idle
srv_master_thread log flush and writes: 5955867
----------
SEMAPHORES
----------
OS WAIT ARRAY INFO: reservation count 806944
OS WAIT ARRAY INFO: signal count 844536
RW-shared spins 0, rounds 338076, OS waits 14683
RW-excl spins 0, rounds 1522728, OS waits 7608
RW-sx spins 26111, rounds 672324, OS waits 4511
Spin rounds per wait: 338076.00 RW-shared, 1522728.00 RW-excl, 25.75 RW-sx
------------------------
LATEST FOREIGN KEY ERROR
------------------------
2018-09-29 16:56:21 0x7f6f20135700 Error in foreign key constraint of table sakila/#sql-6efd_2d3:
foreign key fk_payment_customer(customer_id) references customer(id):
Cannot resolve column name close to:
)
------------------------
LATEST DETECTED DEADLOCK
------------------------
2018-10-09 11:39:36 0x7f6f20072700
*** (1) TRANSACTION:
TRANSACTION 2110061, ACTIVE 0 sec starting index read
......
*** WE ROLL BACK TRANSACTION (1)
------------
TRANSACTIONS
------------
Trx id counter 2255786
Purge done for trx's n:o < 2255762 undo n:o < 0 state: running but idle
History list length 75
LIST OF TRANSACTIONS FOR EACH SESSION:
---TRANSACTION 421596340903536, not started
0 lock struct(s), heap size 1136, 0 row lock(s)
---TRANSACTION 421596340901712, not started
0 lock struct(s), heap size 1136, 0 row lock(s)
---TRANSACTION 421596340902624, not started
0 lock struct(s), heap size 1136, 0 row lock(s)
--------
FILE I/O
--------
```

```
I/O thread 0 state: waiting for completed aio requests (insert buffer thread)
I/O thread 1 state: waiting for completed aio requests (log thread)
I/O thread 2 state: waiting for completed aio requests (read thread)
……
I/O thread 9 state: waiting for completed aio requests (read thread)
I/O thread 10 state: waiting for completed aio requests (write thread)
……
I/O thread 17 state: waiting for completed aio requests (write thread)
Pending normal aio reads: [0, 0, 0, 0, 0, 0, 0, 0] , aio writes: [0, 0, 0, 0, 0, 0, 0, 0] ,
 ibuf aio reads:, log i/o's:, sync i/o's:
Pending flushes (fsync) log: 0; buffer pool: 0
25172 OS file reads, 734032 OS file writes, 522428 OS fsyncs
0.00 reads/s, 0 avg bytes/read, 0.00 writes/s, 0.00 fsyncs/s
…
----------------------------
END OF INNODB MONITOR OUTPUT
============================
```

17.2 MySQL 内存管理及优化

内存是影响数据库性能的重要资源，也是 MySQL 性能优化的一个重要方面。本节主要介绍 MyISAM 和 InnoDB 的内存管理及优化。

17.2.1 内存优化原则

在调整 MySQL 内存分配时，要注意以下 3 点。

● 将尽量多的内存分配给 MySQL 做缓存，但要给操作系统和其他程序的运行预留足够的内存，否则如果产生 SWAP 页交换，将严重影响系统性能。

● MyISAM 的数据文件读取依赖于操作系统自身的 IO 缓存，因此，如果有 MyISAM 表，就要预留更多的内存给操作系统做 IO 缓存。

● 排序区、连接区等缓存是分配给每个数据库会话（session）专用的，其默认值的设置要根据最大连接数合理分配，如果设置太大，不但浪费内存资源，而且在并发连接较高时会导致物理内存耗尽。

17.2.2 MyISAM 内存优化

MyISAM 存储引擎使用 key buffer 缓存索引块，以加速 MyISAM 索引的读写速度。对于 MyISAM 表的数据块，MySQL 没有特别的缓存机制，完全依赖于操作系统的 IO 缓存。

1. key_buffer_size 设置

key_buffer_size 决定 MyISAM 索引块缓存区的大小，它直接影响 MyISAM 表的存取效率。可以在 MySQL 的参数文件中设置 key_buffer_size 的值，对于一般 MyISAM 数据库，建议至少将 1/4 可用内存分配给 key_buffer_size：

```
…
Key_buffer_size = 4G
…
```

可以通过检查 key_read_requests、key_reads、key_write_requests 和 key_writes 等 MySQL 状态变量来评估索引缓存的效率。一般来说，索引块物理读比率 key_reads / key_read_requests 应小于 0.01。索引块写比率 key_writes / key_write_requests 也应尽可能小，但这与应用特点有

关，对于更新和删除操作特别多的应用，key_writes / key_write_requests 可能会接近 1，而对于每次更新很多行记录的应用，key_writes / key_write_requests 就会比较小。

除通过索引块的物理读写比率衡量 key buffer 的效率外，我们也可以通过评估 key buffer 的使用率来判断索引缓存设置是否合理。key buffer 使用率计算公式如下：

$$1 - ((key\_blocks\_unused \times key\_cache\_block\_size) / key\_buffer\_size)$$

一般来说，使用率在 80% 左右比较合适，大于 80%，可能因索引缓存不足而导致性能下降；小于 80%，会导致内存浪费。

2. 使用多个索引缓存

MySQL 通过各 session 共享的 key buffer 提高了 MyISAM 索引存取的性能，但它并不能消除 session 间对 key buffer 的竞争。比如，一个 session 如果对某个很大的索引进行扫描，就可能将其他的索引数据块挤出索引缓存区，而这些索引块可能是其他 session 要用的热数据。为减少 session 间对 key buffer 的竞争，MySQL 从 5.1 版本开始引入了多索引缓存的机制，从而可以将不同表的索引缓存到不同的 key buffer 中。

可以通过下述命令创建新的 key buffer：

```
mysql> set global hot_cache.key_buffer_size=128*1024;
Query OK, 0 rows affected (0.01 sec)
```

其中，hot_cache 是新建索引缓存的名称，global 关键字表示新建的缓存对每一个新的连接都有效。

创建的索引缓存可以使用以下命令删除：

```
mysql> set global hot_cache.key_buffer_size=0;
Query OK, 0 rows affected (0.00 sec)
```

但我们不能删除默认的 key_buffer：

```
mysql> show variables like 'key_buffer_size';
+-----------------+---------+
| Variable_name   | Value   |
+-----------------+---------+
| key_buffer_size | 8388600 |
+-----------------+---------+
1 row in set (0.00 sec)
mysql> set global key_buffer_size=0;
Query OK, 0 rows affected, 1 warning (0.00 sec)

mysql> show warnings;
+---------+------+------------------------------+
| Level   | Code | Message                      |
+---------+------+------------------------------+
| Warning | 1438 | Cannot drop default keycache |
+---------+------+------------------------------+
1 row in set (0.01 sec)
```

在默认情况下，MySQL 将使用默认 key buffer 缓存 MyISAM 表的索引，我们可以用 cache index 命令指定表的索引缓存：

```
mysql> cache index sales,sales2 in hot_cache;
+---------------+--------------------+----------+----------+
| Table         | Op                 | Msg_type | Msg_text |
+---------------+--------------------+----------+----------+
| sakila.sales  | assign_to_keycache | status   | OK       |
| sakila.sales2 | assign_to_keycache | status   | OK       |
+---------------+--------------------+----------+----------+
2 rows in set (0.04 sec)
```

除上述通过命令动态创建并分配辅助索引缓存外，更常见的做法是通过配置文件在 MySQL 启动时自动创建并加载索引缓存：

```
key_buffer_size = 4G
hot_cache.key_buffer_size = 2G
cold_cache.key_buffer_size = 1G
init_file=/path/to/data-directory/mysqld_init.sql
```

在 mysqld_init.sql 中，可以通过 cache index 命令分配索引缓存，并用 load index into cache 命令来进行索引预加载：

```
cache index sales in hot_cache;
cache index sales2 in cold_cache;
load index into cache sales, sales2;
```

3．调整"中点插入策略"

在默认情况下，MySQL 使用简单的 LRU（Least Recently Used）策略来选择要淘汰的索引数据块。但这种算法不是很精细，在某些情况下会导致真正的热块被淘汰。

如果出现这种情况，除了使用上面介绍的多个索引缓存机制外，还可以利用中点插入策略（Midpoint Insertion Strategy）来优化索引块淘汰算法。所谓"中点插入策略"，是对简单 LRU 淘汰算法的改进，它将 LRU 链分成两部分：hot 子表和 warm 子表，当一个索引块读入内存时，先被放到 LRU 链表的"中点"，即 warm 子表的尾部，当达到一定的命中次数后，该索引块会被晋升到 hot 子表的尾部；此后，该数据块在 hot 子表流转，如果其到达 hot 子表的头部并超过一定时间，它将由 hot 子表的头部降级到 warm 子表的头部；当需要淘汰索引块时，缓存管理程序会选择优先淘汰 warm 表头部的内存块。不难理解，这种算法能够避免偶尔被访问的索引块将访问频繁的热块淘汰。

可以通过调节 key_cache_division_limit 来控制多大比例的缓存用做 warm 子表，key_cache_division_limit 的默认值是 100，意思是全部缓存块都放在 warm 子表，其实也就是不启用"中点插入策略"。如果希望将大致 30%的缓存用来 cache 最热的索引块，可以对 key_cache_division_limit 做如下设置：

```
Set global key_cache_division_limit = 70
Set global hot_cache.key_cache_division_limit = 70;
```

除调节 warm 子表的比例外，还可以通过 key_cache_age_threshold 控制数据块由 hot 子表向 warm 子表降级的时间，值越小，数据块将越快被降级。对于有 N 个块的索引缓存来说，如果一个在 hot 子表头部的索引块，在最后 *N×key_cache_age_threshold / 100* 次缓存命准内未被访问过，就会被降级到 warm 子表。

4．调整 read_buffer_size 和 read_rnd_buffer_size

如果需要经常顺序扫描 MyISAM 表，可以通过增大 read_buffer_size 的值来改善性能。但需要注意的是，read_buffer_size 是每个 session 独占的，如果默认值太大，就会造成内存浪费，甚至导致物理内存耗尽。

对于需要做排序的 MyISAM 表查询，如带有 order by 子句的 SQL，适当增大 read_rnd_buffer_size 的值，也可以改善此类 SQL 的性能。但同样要注意的是，read_rnd_buffer_size 也是按 session 分配的，默认值不能设置得太大。

17.2.3 InnoDB 内存优化

InnoDB 的缓存机制与 MyISAM 不尽相同，本节重点介绍其缓存内部机制和优化方法。

1. InnoDB 缓存机制

InnoDB 用一块内存区做 IO 缓存池，该缓存池不仅用来缓存 InnoDB 的索引块，而且也用来缓存 InnoDB 的数据块，这一点与 MyISAM 不同。

在内部，InnoDB 缓存池逻辑上由 free list、flush list 和 LRU list 组成。顾名思义，free list 是空闲缓存块列表，flush list 是需要刷新到磁盘的缓存块列表，而 LRU list 是 InnoDB 正在使用的缓存块，它是 InnoDB buffer pool 的核心。

InnoDB 使用的 LRU 算法与 MyISAM 的 "中点插入策略" LRU 算法很类似，大致原理是：将 LRU list 分为 young sublist 和 old sublist，数据从磁盘读入时，会将该缓存块插入到 LRU list 的 "中点"，即 old sublist 的头部；经过一定时间的访问（由 innodb_old_blocks_time 系统参数决定），该数据块将会由 old sublist 转移到 young sublist 的头部，也就是整个 LRU list 的头部；随着时间的推移，young sublist 和 old sublist 中较少被访问的缓存块将从各自链表的头部逐渐向尾部移动；需要淘汰数据块时，优先从链表尾部淘汰。这种设计同样是为了防止偶尔被访问的索引块将访问频繁的热块淘汰。

脏页的刷新存在于 FLUSH list 和 LRU list 这两个链表，LRU 上也存在可以刷新的脏页，这里是直接可以刷新的，默认 BP（INNODB_BUFFER_POOL）中不存在可用的数据页的时候，会扫描 LRU list 尾部的 innodb_lru_scan_depth 个数据页（默认是 1024 个数据页），进行相关刷新操作。从 LRU list 淘汰的数据页会立刻放入到 free list 中去。

可以通过调整 InnoDB buffer pool 的大小、改变 young sublist 和 old sublist 的分配比例、控制脏缓存的刷新活动、使用多个 InnoDB 缓存池等方法来优化 InnoDB 的性能。

2. innodb_buffer_pool_size 的设置

innodb_buffer_pool_size 决定 InnoDB 存储引擎表数据和索引数据的最大缓存区大小。和 MyISAM 存储引擎不同，Innod buffer pool 同时为数据块和索引块提供数据缓存，这与 Oracle 的缓存机制很相似。在保证操作系统及其他程序有足够内存可用的情况下，innodb_buffer_pool_size 的值越大，缓存命中率越高，访问 InnoDB 表需要的磁盘 I/O 就越少，性能也就越高。在一个专用的数据库服务器上，可以将 80% 的物理内存分配给 InnoDB buffer pool，但一定要注意避免设置过大而导致页交换。

通过以下命令查看 buffer pool 的使用情况：

```
mysqladmin --socket=/tmp/mysql_3307.sock ext|grep -i innodb_buffer_pool
| Innodb_buffer_pool_dump_status         | Dumping of buffer pool not started           |
| Innodb_buffer_pool_load_status         | Buffer pool(s) load completed at 180729 17:44:29 |
| Innodb_buffer_pool_resize_status       |                                              |
| Innodb_buffer_pool_pages_data          | 119623                                       |
| Innodb_buffer_pool_bytes_data          | 1959903232                                   |
| Innodb_buffer_pool_pages_dirty         | 0                                            |
| Innodb_buffer_pool_bytes_dirty         | 0                                            |
| Innodb_buffer_pool_pages_flushed       | 205650                                       |
| Innodb_buffer_pool_pages_free          | 206703                                       |
| Innodb_buffer_pool_pages_misc          | 1334                                         |
| Innodb_buffer_pool_pages_total         | 327660                                       |
| Innodb_buffer_pool_read_ahead_rnd      | 0                                            |
```

```
| Innodb_buffer_pool_read_ahead         | 548       |
| Innodb_buffer_pool_read_ahead_evicted | 0         |
| Innodb_buffer_pool_read_requests      | 627668161 |
| Innodb_buffer_pool_reads              | 17113     |
| Innodb_buffer_pool_wait_free          | 0         |
| Innodb_buffer_pool_write_requests     | 78539424  |
```

可用以下公式计算 InnoDB 缓存池的命中率：

（*1 − innodb_buffer_pool_reads/innodb_buffer_pool_read_request*）×*100*

如果命中率太低，则应考虑扩充内存、增加 innodb_buffer_pool_size 的值。

在 MySQL 8.0 中，增加了一个自动分配内存的参数——innodb_dedicated_server，如果一台服务器上面只运行了一个 MySQL 实例，那么将这个参数设置为 1 之后，innodb_buffer_pool_size、innodb_log_file_size 和 innodb_flush_method 这 3 个参数会根据物理内存的大小自动分配，具体规则如下。

- 物理内存小于 1GB：

innodb_buffer_pool_size=128MB

innodb_log_file_size=48MB

innodb_flush_method=O_DIRECT_NO_FSYNC（如果系统允许）

- 物理内存为 1GB～4GB：

innodb_buffer_pool_size=物理内存*0.5

innodb_log_file_size=128MB

innodb_flush_method=O_DIRECT_NO_FSYNC（如果系统允许）

- 物理内存大于 4GB：

innodb_buffer_pool_size=物理内存*0.75

innodb_log_file_size=1024MB

innodb_flush_method=O_DIRECT_NO_FSYNC（如果系统允许）

其实，即使使用的是 MySQL 5.6 或 MySQL 5.7，也可以参考上面的规则，来分配实例的这几个内存参数。

3．调整 old sublist 大小

在 LRU list 中，old sublist 的比例由系统参数 innodb_old_blocks_pct 决定，其取值范围是 5～95，默认值是 37。通过以下命令可以查看其当前设置：

```
mysql> show global variables like '%innodb_old_blocks_pct%';
+-----------------------+-------+
| Variable_name         | Value |
+-----------------------+-------+
| innodb_old_blocks_pct | 75    |
+-----------------------+-------+
```

可以根据 InnoDB Monitor 的输出信息来调整 innodb_old_blocks_pct 的值。例如，在没有较大表扫描或索引扫描的情况下，如果 young/s 的值很低，可能就需要适当增大 innodb_old_blocks_pct 的值或减小 innodb_old_blocks_time 的值。

4．调整 innodb_old_blocks_time 的设置

innodb_old_blocks_time 参数决定了缓存数据块由 old sublist 转移到 young sublist 的快慢，当一个缓存数据块被插入到 midpoint（old sublist）后，至少要在 old sublist 停留超过

innodb_old_blocks_time（ms）后，才有可能被转移到 new sublist。例如，将 innodb_old_blocks_time 设置为 1000（即 1s），当出现 table scan 时，InnoDB 先将数据块载入到 midpoint（old sublist）上，程序读取数据块，因为这时，数据块在 old sublist 中的停留时间还不到 innodb_old_blocks_time，所以不会被转移到有 young sublist 中，这样就避免了表扫描污染 buffer pool 的情况。

可以根据 InnoDB Monitor 的输出信息来调整 innodb_old_blocks_time 的值。在进行表扫描时，如果 non-youngs/s 很低，young/s 很高，就应考虑将 innodb_old_blocks_time 适当调大，以防止表扫描将真正的热数据淘汰。更酷的是，这个值可以动态设置，如果要进行大的表扫描操作，可以很方便地临时做调整。

5．调整缓存池数量，减少内部对缓存池数据结构的争用

MySQL 内部不同线程对 InnoDB 缓存池的访问在某些阶段是互斥的，这种内部竞争也会产生性能问题，尤其在高并发和 buffer pool 较大的情况下。为解决这个问题，InnoDB 的缓存系统引入了 innodb_buffer_pool_instances 配置参数，默认为 8 个，对于较大的缓存池，适当增大此参数的值，可以降低并发导致的内部缓存访问冲突，改善性能。InnoDB 缓存系统会将参数 innodb_buffer_pool_size 指定大小的缓存平分为 innodb_buffer_pool_instances 个 buffer pool。

6．控制 innodb buffer 刷新，延长数据缓存时间，减缓磁盘 I/O

在 InnoDB 找不到干净的可用缓存页或检查点被触发等情况下，InnoDB 的后台线程就会开始把"脏的缓存页"回写到磁盘文件中，这个过程叫缓存刷新。

我们通常都希望 buffer pool 中的数据在缓存中停留的时间尽可能长，以备重用，从而减少磁盘 IO 的次数。同时，磁盘 IO 较慢，是数据库系统最主要的性能瓶颈，我们往往也希望通过延迟缓存刷新来减轻 IO 系统的压力。InnoDB buffer pool 的刷新快慢主要取决于两个参数。

- 一个是 innodb_max_dirty_pages_pct，它控制缓存池中脏页的最大比例，默认值是 50%，如果脏页的数量达到或超过该值，InnoDB 的后台线程将开始缓存刷新。
- 另一个是 innodb_io_capacity，它代表磁盘系统的 IO 能力，其值在一定程度上代表磁盘每秒可完成 I/O 的次数。innodb_io_capacity 的默认值是 200，对于转速较低的磁盘，如 7200RPM 的磁盘，可将 innodb_io_capacity 的值降低到 100，而对于固态硬盘和由多个磁盘组成的盘阵，innodb_io_capacity 的值可以适当增大，尤其是对于固态硬盘来说，建议设置为 2000 或者更高。

innodb_io_capacity 决定一批刷新脏页的数量，当缓存池脏页的比例达到 innodb_max_dirty_pages_pct 时，InnoDB 大约将 innodb_io_capacity 个已改变的缓存页刷新到磁盘；当脏页比例小于 innodb_max_dirty_pages_pct 时，如果 innodb_adaptive_flushing 的设置为 true，InnoDB 将根据函数 buf_flush_get_desired_flush_rate 返回的重做日志产生速度来确定要刷新的脏页数。在合并插入缓存时，InnoDB 每次合并的页数是 0.05×innodb_io_capacity。

可以根据一些 InnoDB Monitor 的值来调整 innodb_max_dirty_pages_pct 和 innodb_io_capacity。例如，若 innodb_buffer_pool_wait_free 的值增长较快，则说明 InnoDB 经常在等待空闲缓存页，如果无法增大缓存池，那么应将 innodb_max_dirty_pages_pct 的值调小，或将 innodb_io_capacity 的值提高，以加快脏页的刷新。

7．InnoDB doublewrite

在进行脏页刷新时，InnoDB 采用了双写（doublewrite）策略，这么做的原因是：MySQL

的数据页大小（一般是 16KB）与操作系统的 IO 数据页大小（一般是 4KB）不一致，无法保证 InnoDB 缓存页被完整、一致地刷新到磁盘，而 InnoDB 的 redo 日志只记录了数据页改变的部分，并未记录数据页的完整前像，当发生部分写或断裂写时（比如将缓存页的第一个 4KB 写入磁盘后，服务器突然断电），就会出现页面无法恢复的问题。为解决这个问题，InnoDB 引入了 doublewrite 技术。

InnoDB doublewrite 机制的实现原理是：用系统表空间中的一块连续磁盘空间（100 个连续数据页，大小为 2MB）作为 doublewrite buffer，当进行脏页刷新时，首先将脏页的副本写到系统表空间的 doublewrite buffer 中，然后调用 fsync() 刷新操作系统 IO 缓存，确保副本被真正写入磁盘，最后 InnoDB 后台 IO 线程将脏页刷新到磁盘数据文件。其原理示意图如图 17-2 所示。

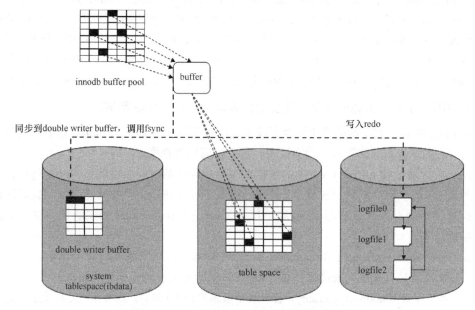

图 17-2　InnoDB doublewrite 原理示意图

在做恢复时，如果发现不一致的页，InnoDB 会用系统表空间 doublewrite buffer 区的相应副本来恢复数据页。

默认情况下，InnoDB doublewrite 是开启的，可以用以下命令查看：

```
mysql> show global variables like '%doublewrite%';
+--------------------+-------+
| Variable_name      | Value |
+--------------------+-------+
| innodb_doublewrite | ON    |
+--------------------+-------+
1 row in set (0.01 sec)
```

由于同步到 doublewrite buffer 是对连续磁盘空间的顺序写，因此开启双写对性能的影响并不太大。对于要求超高性能，又能容忍极端情况下少量数据丢失的应用，可以通过在配置文件中增加 innodb_doublewrite=0 参数设置来关闭 doublewrite，以尽量满足性能方面的要求。

17.2.4 调整用户服务线程排序缓存区

如果通过 show global status 看到 sort_merge_passes 的值很大,可以考虑通过调整参数 sort_buffer_size 的值来增大排序缓存区,以改善带有 order by 子句或 group 子句 SQL 的性能。

对于无法通过索引进行连接操作的查询,可以尝试通过增大 join_buffer_size 的值来改善性能。

不过需要注意的是,sort buffer 和 join buffer 都是面向客户服务线程分配的,如果设置过大可能造成内存浪费,甚至导致内存交换。尤其是 join buffer,如果是多表关联的复杂查询,还可能会分配多个 join buffer,因此,最好的策略是设置较小的全局 join_buffer_size,而对需要做复杂连接操作的 session 单独设置较大的 join_buffer_size。

17.3 InnoDB log 机制及优化

支持事务的数据库系统都需要有一套机制来保证事务更新的一致性和持久性,InnoDB 与 Oracle 等支持事务的关系数据库一样,也采用 redo log 机制来保证事务更新的一致性和持久性。本节将介绍 InnoDB 重做日志的内部机制,并讨论相关的性能优化问题。

17.3.1 InnoDB 重做日志

redo log 是 InnoDB 保证事务 ACID 属性的重要机制,其工作原理图如图 17-3 所示。

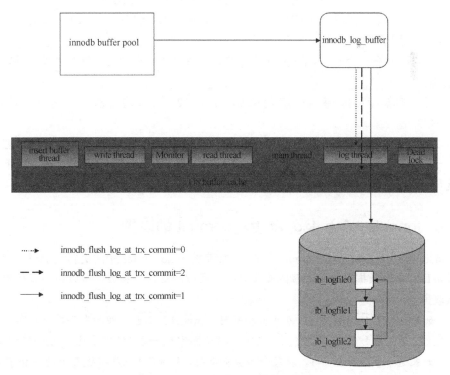

图 17-3 redo 日志回写磁盘示意图

当更新数据时，InnoDB 内部的操作流程大致如下。

（1）将数据读入 InnoDB buffer pool，并对相关记录加独占锁。

（2）将 UNDO 信息写入 undo 表空间的回滚段中。

（3）更改缓存页中的数据，并将更新记录写入 redo buffer 中。

（4）提交时，根据 innodb_flush_log_at_trx_commit 的设置，用不同的方式将 redo buffer 中的更新记录刷新到 InnoDB redo log file 中，然后释放独占锁。

（5）最后，后台 IO 线程根据需要择机将缓存中更新过的数据刷新到磁盘文件中。

可以通过 show engine innodb status 命令查看当前日志的写入情况，如图 17-4 所示。

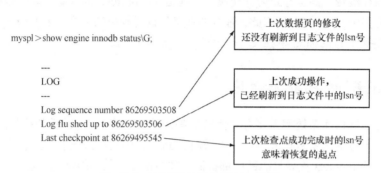

图 17-4　查看当前日志的写入情况

其中，LSN（Log Sequence Number）称为日志序列号，它实际上对应日志文件的偏移量，其生成公式是：

$$新的 LSN = 旧的 LSN + 写入的日志大小$$

例如，日志文件大小为 600MB，目前的 LSN 是 1GB，现在要将 512 字节的更新记录写入 redo log，则实际写入过程如下。

- 求出偏移量：由于 LSN 数值远大于日志文件大小，因此通过取余方式，得到偏移量为 400MB。
- 写入日志：找到偏移 400MB 的位置，写入 512 字节日志内容，下一个事务的 LSN 就是 1073742336。

由以上介绍可知，除 InnoDB buffer pool 外，InnoDB log buffer 的大小、redo 日志文件的大小以及 innodb_flush_log_at_trx_commit 参数的设置等，都会影响 InnoDB 的性能。下面我们介绍这几个参数的优化调整。

17.3.2　innodb_flush_log_at_trx_commit 的设置

innodb_flush_log_at_trx_commit 参数可以控制将 redo buffer 中的更新记录写入到日志文件以及将日志文件数据刷新到磁盘的操作时机。通过调整这个参数，可以在性能和数据安全之间做取舍。

- 如果这个参数设置为 0，在事务提交时，InnoDB 不会立即触发将缓存日志写到磁盘文件的操作，而是每秒触发一次缓存日志回写磁盘操作，并调用操作系统 fsync 刷新 IO 缓存。
- 如果这个参数设置为 1，在每个事务提交时，InnoDB 立即将缓存中的 redo 日志回写到日志文件，并调用操作系统 fsync 刷新 IO 缓存。

- 如果这个参数设置为 2，在每个事务提交时，InnoDB 立即将缓存中的 redo 日志回写到日志文件，但并不马上调用 fsync 来刷新 IO 缓存，而是每秒只做一次磁盘 IO 缓存刷新操作。

innodb_flush_log_at_trx_commit 参数的默认值是 1，即每个事务提交时都会从 log buffer 写更新记录到日志文件，而且会实际刷新磁盘缓存，显然，这完全能满足事务的持久化要求，是最安全的，但这样会有较大的性能损失。

在某些情况下，我们需要尽量提高性能，并且可以容忍在数据库崩溃时丢失小部分数据，那么通过将参数 innodb_flush_log_at_trx_commit 设置成 0 或 2 都能明显减少日志同步 IO，加快事务提交，从而改善性能。

将此参数设置成 0，如果数据库崩溃，最后 1 秒钟的事务重做日志可能会由于未及写入磁盘文件而丢失，这种方式是效率最高的，但也是最不安全的。

将此参数设置成 2，如果数据库崩溃，由于已执行重做日志写入磁盘操作，只是没有做磁盘 IO 刷新操作，因此，只要不发生操作系统崩溃，数据就不会丢失，这种方式是对性能和数据安全的折中，其性能和数据安全性介于其他两种方式之间。

17.3.3 设置 innodb_log_file_size，控制检查点

当一个日志文件写满后，InnoDB 会自动切换到另一个日志文件，但切换时会触发数据库检查点（checkpoint），这将导致 InnoDB 缓存脏页的小批量刷新，会明显降低 InnoDB 的性能。

那是不是将 innodb_log_file_size 设得越大越好呢？理论上是，但如果日志文件设置得太大，恢复时将需要更长时间，同时也不便于管理。一般来说，平均每半小时写满 1 个日志文件比较合适。我们可以通过下面的方法来计算 InnoDB 每小时产生的日志量并估算合适的 innodb_log_file_size 值。

首先，计算 InnoDB 每分钟产生的日志量：

```
mysql> pager grep -i "Log sequence number"
PAGER set to 'grep -i "Log sequence number"'
mysql> SHOW ENGINE INNODB STATUS\G SELECT SLEEP(60); SHOW ENGINE INNODB STATUS\G
Log sequence number 90176272406
1 row in set (0.00 sec)

1 row in set (59.99 sec)

Log sequence number 90196407469
1 row in set (0.00 sec)

mysql> nopager
PAGER set to stdout
mysql> select ROUND((90196407469 - 90176272406)/ 1024 / 1024) as MB;
+------+
| MB   |
+------+
|   19 |
+------+
1 row in set (0.00 sec)
```

这一步也可以通过查询 INFORMATION_SCHEMA.GLOBAL_STATUS 表来计算：

```
SELECT @a1 := variable_value AS a1
FROM information_schema.global_status
WHERE variable_name = 'innodb_os_log_written'
UNION ALL
SELECT Sleep(60)
UNION ALL
SELECT @a2 := variable_value AS a2
```

```
FROM information_schema.global_status
WHERE variable_name = 'innodb_os_log_written';
+--------------+
| a1           |
+--------------+
| 90176272406  |
| 0            |
| 90196407469  |
+--------------+
1 row in set (0.01 sec)

SELECT ROUND((@a2-@a1) / 1024 / 1024 / @@innodb_log_files_in_group) as MB;
+------+
| MB   |
+------+
| 19   |
+------+
1 row in set (0.01 sec)
```

通过上述操作得到 InnoDB 每分钟产生的日志量是 19MB。然后，计算每半小时的日志量：

半小时日志量 = 30 × 19MB = 570MB

这样，就可以得出 innodb_log_file_size 的大小至少应该是 512MB。

17.3.4 调整 innodb_log_buffer_size

innodb_log_buffer_size 决定 InnoDB 重做日志缓存池的大小，默认值是 16MB。对于可能产生大量更新记录的大事务，增加 innodb_log_buffer_size 的大小，可以避免 InnoDB 在事务提交前就执行不必要的日志写入磁盘操作。因此，对于会在一个事务中更新、插入或删除大量记录的应用，我们可以通过增大 innodb_log_buffer_size 来减少日志写磁盘操作，从而提高事务处理的性能。

17.4 调整 MySQL 并发相关的参数

从实现上来说，MySQL Server 是多线程结构，包括后台线程和客户服务线程。多线程可以有效利用服务器资源，提高数据库的并发性能。在 MySQL 中，控制并发连接和线程的主要参数包括 max_connections、back_log、thread_cache_size 以及 table_open_cache 等。

17.4.1 调整 max_connections，提高并发连接

参数 max_connections 控制允许连接到 MySQL 数据库的最大数量，默认值是 151。如果状态变量 connection_errors_max_connections 不为零，并且一直在增长，就说明不断有连接请求因数据库连接数已达到最大允许的值而失败，应考虑增大 max_connections 的值。

MySQL 最大可支持的数据库连接取决于很多因素，包括给定操作系统平台线程库的质量、内存大小、每个连接的负荷以及期望的响应时间等。在 Linux 平台下，MySQL 支持 5000～10000 个连接不是难事，如果内存足够、不考虑响应时间，甚至能达到上万个连接。而在 Windows 平台下，受其所用线程库的影响，最大连接数有以下限制：

(open tables × 2 + open connections) < 2048

每一个 session 操作 MySQL 数据库表都需要占用文件描述符，数据库连接本身也要占用文件描述符，因此，在增大 max_connections 时，也要注意评估 open_files-limit 的设置是否够用。

17.4.2 调整 back_log

back_log 参数控制 MySQL 监听 TCP 端口时设置的积压请求栈大小，5.6.6 版本以前的默认值是 50，5.6.6 版本以后的默认值是 50+（*max_connections* / 5），但最大不能超过 900。

如果需要数据库在较短时间内处理大量连接请求，可以考虑适当增大 back_log 的值。

17.4.3 调整 table_open_cache

每一个 SQL 执行线程至少都要打开 1 个表缓存，参数 table_open_cache 控制所有 SQL 执行线程可打开表缓存的数量。这个参数的值应根据最大连接数 max_connections 以及每个连接执行关联查询中所涉及表的最大个数（用 N 表示）来设定：

$$max\_connections \times N$$

在未执行 flush tables 命令的情况下，如果 MySQL 状态变量 opened_tables 的值较大，就说明 table_open_cache 设置得太小，应适当增大。增大 table_open_cache 的值，会增加 MySQL 对文件描述符的使用量，因此，也要注意评估 open-files-limit 的设置是否够用。

17.4.4 调整 thread_cache_size

为加快连接数据库的速度，MySQL 会缓存一定数量的客户服务线程以备重用，通过参数 thread_cache_size 可控制 MySQL 缓存客户服务线程的数量。

可以通过计算线程 cache 的失效率 threads_created/connections 来衡量 thread_cache_size 的设置是否合适。该值越接近 1，说明线程 cache 命中率越低，应考虑适当增加 thread_cache_size 的值。

17.4.5 innodb_lock_wait_timeout 的设置

参数 innodb_lock_wait_timeout 可以控制 InnoDB 事务等待行锁的时间，默认值是 50ms，可以根据需要动态设置。对于需要快速反馈的交互式 OLTP 应用，可以将行锁等待超时时间调小，以避免事务长时间挂起；对于后台运行的批处理操作，可以将行锁等待超时时间调大，以避免发生大的回滚操作。

17.5 持久化全局变量

前面几节介绍了通过设置 MySQL 参数来优化数据库实例的方法，在 MySQL 8.0 之前，一般都是通过 set global 命令来设置 MySQL 参数，一旦数据库实例重启，那么实例会从配置文件中重新加载旧的参数配置，之前做过的修改就会失效。因此，在使用 set global 命令修改了实例参数后，还要记得去修改配置文件中的配置。操作需要两步完成，因此容易产生遗漏，从而带来潜在的问题。

在 MySQL 8.0 中，提供了全新的 set persist/set persist only 命令，可以对数据库参数做持久化的修改，通过 set persist 方式做过的修改，数据库实例在重启后，就能够正确加载修改后的参数。其中，set persist 命令会同时修改当前环境并持久化参数修改，set persist only 命令则不修改当前环境，仅仅对修改做持久化，当下次实例启动时，参数修改才会生效。set persist 命令的实现原理如下：

当使用 set persist 命令来修改 MySQL 参数时，会在 datadir 中创建一个名为 mysqld-auto.cnf 的文件，并在文件中以 json 的格式，保存被修改的参数的值。

数据库实例启动时，在加载了默认的 my.cnf 配置文件之后，会再读取 mysqld-auto.cnf 文件中的配置。如果发现文件中有修改的参数值，那么会使用后读到的 mysqld-auto.cnf 文件中的参数值来启动数据库。

下面来测试一下这个过程，将参数 max_connections 从 10050 持久化修改为 11000：

```
root@localhost:mysql_3488.sock   [(none)]>show variables like 'max_connections';
+-----------------+-------+
| Variable_name   | Value |
+-----------------+-------+
| max_connections | 10050 |
+-----------------+-------+
1 row in set (0.00 sec)

root@localhost:mysql_3488.sock   [(none)]>set persist max_connections = 11000;
Query OK, 0 rows affected (0.09 sec)
```

此时，发现 datadir 下生成了新文件 mysqld-auto.cnf：

```
[mysql3488@hz_10_120_240_251 mysqlhome]$ ll /data2/mysql3488/data/mysqld-auto.cnf
-rw-r----- 1 mysql3488 mysql3488 173 Sep 23 11:15 /data2/mysql3488/data/mysqld-auto.cnf
```

文件内容如下：

```
[mysql3488@hz_10_120_240_251 mysqlhome]$ more /data2/mysql3488/data/mysqld-auto.cnf
{ "Version" : 1 , "mysql_server" : { "max_connections" : { "Value" : "11000" , "Metadata" :
{ "Timestamp" : 1537672549953953 , "User" : "root" , "H
ost" : "localhost" } } } }
```

可以使用 reset persist 命令来清除 mysqld-auto.cnf 文件中的所有配置，也可以通过 reset persist + 参数名的方式，来清除某个制定的配置。下面命令完全清除了刚才的配置：

```
root@localhost:mysql_3488.sock   [(none)]>reset persist max_connections;
Query OK, 0 rows affected (0.07 sec)

root@localhost:mysql_3488.sock   [(none)]>exit
Bye
[mysql3488@hz_10_120_240_251 mysqlhome]$ more /data2/mysql3488/data/mysqld-auto.cnf
{ "Version" : 1 , "mysql_server" : { } }
```

通过增加持久化参数修改命令，MySQL 8.0 在一定程度上降低了 DBA 的维护复杂度，减少了因为忘记修改配置文件从而引发问题的可能，但是对比 Oracle 使用 pfile+spfile 的方式来实现参数持久化的方案，MySQL 8.0 版本的参数持久化方式还存在一些不足，比如不能很方便的合并参数，需要查看两个文件，才能确定参数的实际值。希望未来 MySQL 可以继续改进，提供更加方便的参数管理方案。

17.6　使用资源组

在 MySQL 8.0 中，新增了资源组（Resource Group）特性。通过资源组，可以限制某个任务或者查询对于系统资源的使用。目前 MySQL 8.0 中，能够调整的是对 CPU 资源的使用限制和线程的优先级。

设置资源组可以使数据库更加有针对性地应对一些业务场景，例如数据库在执行一些批量任务的时候，如果希望能够控制对数据库产生的压力，从而减小对其他更重要服务的影响。我们就可以选择使用资源组，将批量任务能够使用的 CPU 减少，线程优先级降低。

17.6 使用资源组

在 CPU 层面，MySQL 控制的粒度为 vCPU，也就是逻辑 CPU 的个数，逻辑 CPU 包括了超线程技术虚拟出来的 CPU，可以通过命令 more /proc/cpuinfo | grep processor | wc –l 来查看服务器总共的 vCPU 数量。

在线程优先级层面，默认的线程优先级都是 0，数字越小，线程优先级越高。对于系统资源组，线程优先级可以直接设置在-20～0 之间；对于用户资源组，线程优先级可以设置在 0 到 19 之间。

可以通过 INFORMATION_SCHEMA.RESOURCE_GROUPS 来查看资源组的设置。

```
root@localhost> select * from INFORMATION_SCHEMA.RESOURCE_GROUPS;
+---------------------+---------------------+-----------------------+---------+-----------------+
|RESOURCE_GROUP_NAME  |RESOURCE_GROUP_TYPE  |RESOURCE_GROUP_ENABLED |VCPU_IDS |THREAD_PRIORITY  |
+---------------------+---------------------+-----------------------+---------+-----------------+
| USR_default         | USER                |                    1  | 0-7     |               0 |
| SYS_default         | SYSTEM              |                    1  | 0-7     |               0 |
+---------------------+---------------------+-----------------------+---------+-----------------+
2 rows in set (0.38 sec)
```

可以看到，MySQL 默认创建了两个资源组，都可以使用全部的 vCPU，线程优先级都是 0，这两个资源组的属性是不可修改的。新创建的线程默认都会被分配到这两个资源组中的一个。

接下来，创建一个用于执行批量任务的资源组：

```
root@localhost>CREATE RESOURCE GROUP Batch
    -> TYPE = USER
    -> VCPU = 2-3
    -> THREAD_PRIORITY = 10;
Query OK, 0 rows affected, 1 warning (0.13 sec)

root@localhost>show warnings;
+---------+------+----------------------------------------------------------------------+
| Level   | Code | Message                                                              |
+---------+------+----------------------------------------------------------------------+
| Warning | 3659 | Attribute thread_priority is ignored (using default value).          |
+---------+------+----------------------------------------------------------------------+
1 row in set (0.00 sec)

root@localhost>select * from INFORMATION_SCHEMA.RESOURCE_GROUPS;
+---------------------+---------------------+-----------------------+---------+-----------------+
|RESOURCE_GROUP_NAME  |RESOURCE_GROUP_TYPE  |RESOURCE_GROUP_ENABLED |VCPU_IDS |THREAD_PRIORITY  |
+---------------------+---------------------+-----------------------+---------+-----------------+
| USR_default         | USER                |                    1  | 0-7     |               0 |
| SYS_default         | SYSTEM              |                    1  | 0-7     |               0 |
| Batch               | USER                |                    1  | 2-3     |               0 |
+---------------------+---------------------+-----------------------+---------+-----------------+
3 rows in set (0.00 sec)
```

可以看到，Batch 资源组创建成功，可以使用 vCPU2 和 3，但是对于资源组的线程优先级设置并没有成功，显示依然为 0。对于 Linux 系统，需要首先用 root 用户或者 sudo 权限执行下面的操作并重启数据库实例：

```
[root@hz_10_120_240_251 mysqlhome]# pwd
/home/mysql3488/mysqlhome
[root@hz_10_120_240_251 mysqlhome]# setcap cap_sys_nice+ep ./bin/mysqld
[root@hz_10_120_240_251 mysqlhome]# getcap ./bin/mysqld
./bin/mysqld = cap_sys_nice+ep
```

重启实例后，我们重新设置一下资源组的线程优先级：

```
root@localhost>alter resource group Batch thread_priority = 19;
Query OK, 0 rows affected (0.08 sec)
```

这次设置成功：

```
root@localhost>select * from INFORMATION_SCHEMA.RESOURCE_GROUPS;
+---------------------+---------------------+-----------------------+---------+-----------------+
```

```
|RESOURCE_GROUP_NAME|RESOURCE_GROUP_TYPE|RESOURCE_GROUP_ENABLED|VCPU_IDS|THREAD_PRIORITY|
+-------------------+-------------------+----------------------+--------+---------------+
| USR_default       | USER              |                    1 | 0-7    |             0 |
| SYS_default       | SYSTEM            |                    1 | 0-7    |             0 |
| Batch             | USER              |                    1 | 2-3    |            19 |
+-------------------+-------------------+----------------------+--------+---------------+
3 rows in set (0.00 sec)
```

对于正常执行的线程，可以使用以下几种方法来设置资源组。

（1）第一种方法是通过线程 id 来设置：

首先使用 show porcesslist 或者 information_schema.processlist 获得 session 的 id，再通过 performance_schema.threads 获得 session 对应的线程 id。

例如，获得的线程 id 为 62，那么接下来执行。

```
root@localhost>set resource group Batch for 62;
Query OK, 0 rows affected (0.00 sec)
```

（2）第二种方法是在任务开始之前，设置 session 的资源组。

在 session 中直接执行：

```
root@localhost>set resource group Batch;
Query OK, 0 rows affected (0.00 sec)
```

这样，当前 session 的所有操作都会被放到 Batch 资源组中。

（3）第三种方法是通过注释的方式，将某个具体的 SQL，放到 Batch 资源组中进行：

```
INSERT /*+ RESOURCE_GROUP(Batch) */ INTO t2 VALUES(2);
```

17.7 小结

本章在简要说明 MySQL 系统架构、缓存体系和 redo log 机制的基础上，介绍了影响 MySQL 数据库性能的一些重要配置参数，讨论了缓存、InnoDB 日志和并发相关的性能优化问题。最后还介绍了 MySQL 5.7 和 8.0 中新增的一些实例优化方法。

第 18 章 磁盘 I/O 问题

作为应用系统的持久化层，不管数据库采取了什么样的 Cache 机制，但数据库最终总是要将数据储存到可以长久保存的 I/O 设备——磁盘上，但磁盘的存取速度显然要比 CPU、RAM 的速度慢很多，因此，对于比较大的数据库，磁盘 I/O 一般总会成为数据库的一个性能瓶颈！

实际上，前面提到的 SQL 优化、数据库对象优化、数据库参数优化以及应用程序优化等，大部分都是想通过减少或延缓磁盘读写来减轻磁盘 I/O 的压力及其对性能的影响。解决磁盘 I/O 问题，减少或延缓磁盘操作肯定是一个重要方面，但磁盘 I/O 是不可避免的，因此，增强磁盘 I/O 本身的性能和吞吐量也是一个重要方面。本章将从磁盘类型、磁盘阵列、符号链接、裸设备等更底层的方面来介绍提高磁盘 I/O 能力的一些技术和方法。

18.1 使用固态硬盘

固态硬盘（Solid State Disk，SSD）是近年来提升数据库 I/O 性能最常见、性价比最高的手段之一，如果存在 I/O 瓶颈的数据库仍然使用的是 SAS 甚至 SATA 硬盘，那么应该优先考虑将硬盘升级为 SSD 固态硬盘。

传统机械硬盘的原理是依靠磁头来读写高速旋转的盘片，一般来说，性能受到以下 3 个方面的影响。

- 盘片转速：相对来说，转速越高的磁盘读写速度也越快，但是受到机械性能和发热的限制，机械硬盘的转速为 4 200~15 000RPM。
- 存储密度：同样面积大小的盘片，如果容量更大，那么存储密度也更大，同样转速下，磁头每秒钟能够读写的数据量也更大。
- 接口类型：服务器上使用的机械硬盘常用的接口一般有 SATA2、SATA3、SAS 等类型，带宽从 3Gbit/s 到 12Gbit/s 不等，硬盘的性能自然会受到接口速度的限制。

对于机械硬盘来说，性能上最大的瓶颈来自于随机 I/O。发生随机 I/O 时，硬盘需要不断地通过转动盘片加移动磁头这种机械操作，来定位到需要读写的数据块，效率比较低。

固态硬盘由主控芯片加闪存芯片组成，不存在机械操作过程，对比机械硬盘，固态硬盘的优点主要有以下 3 点。

- 读写速度快：普通 SATA 接口的 SSD，连续 I/O 能力一般是机械硬盘的数倍，随机 I/O

能力更是可以达到机械硬盘的几十倍以上，采用了 PCI-E 接口的 SSD，I/O 能力还会更高。

- 可靠性高：早期的 SSD 闪存颗粒的擦写次数比较低，因此当时 SSD 的寿命也比较短，但是现在企业级 SSD 的擦写次数已经比较高，寿命对于大多数应用场景来说已经不成问题，反而由于不存在易损坏的机械结构，相对来说更加稳定可靠。
- 功耗低、体积小、重量轻、无噪音。

在使用固态硬盘的过程中，也有一些需要注意的方面。

- 合理规划成本：虽然现在 SSD 的价格已经比较低，但是相比 SAS 硬盘仍然贵不少，因此，对于某些对 I/O 资源需求不大，数据量又非常大的场景，例如日志类型的数据，也需要认真评估是否需要使用 SSD。
- 选择适合的 SSD 型号：不同型号的 SSD，由于采用了不同的芯片和接口类型，相互之间 I/O 能力差距很大，尤其是写入的 IOPS，差距更加明显。在实际使用 SSD 的时候，需要根据业务的类型，来选择适合当前业务的 SSD 型号。
- 合理预留空间：目前企业级 SSD 的寿命（一般通过 PE 来计算）已经比较大，但是如果是针对写密集型应用，可以按 SSD 的 PE 计算一下预期的寿命，如果希望加大 SSD 的寿命，在 SSD 型号不变的情况下，可以通过配置预留空间来解决。

18.2 使用磁盘阵列

RAID 是 Redundant Array of Independent Disks 的缩写，翻译成中文就是"独立磁盘冗余阵列"，通常就叫做磁盘阵列。RAID 就是按照一定策略将数据分布到若干物理磁盘上，这样不仅增强了数据存储的可靠性，而且可以提高数据读写的整体性能，因为通过分布实现了数据的"并行"读写。

RAID 最早是用来取代大型计算机上高档存储设备的，相对于那些高档存储设备而言，RAID 的价格很便宜，这也是曾经其名称中带有"廉价"（Inexpensive）一词的原因。但其实很长时间，对于 PC 而言，其价格远远谈不上"廉价"！近几年，随着存储技术的发展，RAID 开始变得真正的物美价廉了。

18.2.1 常见 RAID 级别及其特性

根据数据分布和冗余方式，RAID 分为许多级别，不同存储厂商提供的 RAID 卡或设备支持的 RAID 级别也不尽相同，这里只介绍最常见也是最基本的几种，如表 18-1 所示，其他 RAID 级别基本上都是在这几种基础上的改进。

表 18-1　　　　　　　　　　常见的 RAID 级别的比较

RAID 级别	特　　性	优　　点	缺　　点
RAID 0	也称之为条带化（Stripe），按一定的条带大小（Chunk Size）将数据依次分布到各个磁盘，没有数据冗余	数据并发读写速度快，无额外磁盘空间开销，投资省	数据无冗余保护，可靠性差
RAID 1	也称之为磁盘镜像（Mirror），两个磁盘一组，所有数据都同时写入两个磁盘，但读时从任一磁盘读都可以	数据有完全冗余保护，只要不出现两块镜像磁盘同时损坏，不会影响使用；可以提高并发读性能	在容量一定的情况下，需要 2 倍的磁盘，投资比较大

续表

RAID 级别	特　性	优　点	缺　点
RAID 10	是 RAID 1 和 RAID 0 的结合，也称之为 RAID 1+0。先对磁盘做镜像，再条带化，使其兼具 RAID 1 的可靠性和 RAID 0 的优良并发读写性能	可靠性高，并发读写性能优良	在容量一定的情况下，需要 2 倍的磁盘，投资比较大
RAID 4	像 RAID 0 一样对磁盘组条带化，不同的是：需要额外增加一个磁盘，用来写各 Stripe 的校验纠错数据	RAID 中的一个磁盘损坏，其数据可以通过校验纠错数据计算出来，具有一定容错保护能力；读数据速度快	每个 Stripe 上数据的修改都要写校验纠错块，写性能受影响；所有纠错数据都在同一磁盘上，风险大，也会形成一个性能瓶颈；在出现坏盘时，读性能会下降
RAID 5	是对 RAID 4 的改进：将每一个条带（Stripe）的校验纠错数据块也分别写到各个磁盘，而不是写到一个特定的磁盘	基本同 RAID 4，只是其写性能和数据保护能力要更强一点	写性能不及 RAID 0、RAID 1 和 RAID 10，容错能力也不及 RAID 1；在出现坏盘时，读性能会下降
RAID6	在 RAID 5 基础上，为了进一步加强数据保护而设计的一种 RAID 方式，实际上是一种扩展 RAID 5 等级。与 RAID 5 的不同之处于除了每个硬盘上都有同级数据 XOR 校验区外，还有一个针对每个数据块的 XOR 校验区	每个数据块有了两个校验保护屏障（一个分层校验，一个是总体校验），因此 RAID 6 的数据冗余性能更好，每组 RAID 允许两块磁盘同时出现故障	写性能相比 RAID5 更差；需要消耗两块硬盘的容量作为数据校验盘

18.2.2 如何选择 RAID 级别

了解各种 RAID 级别的特性后，就可以根据数据读写的特点、可靠性要求以及投资预算等来选择合适的 RAID 级别，比如：

- 数据读写都很频繁，可靠性要求也很高，最好选择 RAID 10；
- 数据读很频繁，写相对较少，对可靠性有一定要求，可以选择 RAID 5 或者 RAID 6；
- 数据读写都很频繁，但可以接受全部数据丢失，可以选择 RAID 0。

18.3 虚拟文件卷或软 RAID

最初，RAID 都是由硬件实现的，要使用 RAID，至少需要有一个 RAID 卡。但现在，一些操作系统中提供的软件包，也模拟实现了一些 RAID 的特性。虽然性能上不如硬 RAID，但相比单个磁盘，性能和可靠性都有所改善。比如，Linux 下的逻辑卷（Logical Volume）系统 lvm2，支持条带化（Stripe）；Linux 下的 MD（Multiple Device）驱动，支持 RAID 0、RAID 1、RAID 4、RAID 5、RAID 6 等。在不具备硬件条件的情况下，可以考虑使用上述虚拟文件卷或软 RAID 技术，具体配置方法可参见 Linux 帮助文档。

18.4 使用 Symbolic Links 分布 I/O

MySQL 的数据库名和表名是与文件系统的目录名和文件名对应的，在默认情况下，创建的数据库和表都存放在参数 datadir 定义的目录下。这样如果不使用 RAID 或逻辑卷，所有的表都存放在一个磁盘设备上，无法发挥多磁盘并行读写的优势！在这种情况下，就可以利用操作系统的符号连接（Symbolic Links）将不同的数据库、表或索引指向不同的物理磁盘，从而达到分布磁盘 I/O 的目的。

（1）将一个数据库指向其他物理磁盘。

其方法是先在目标磁盘上创建目录，然后再创建从 MySQL 数据目录（/path/to/datadir）到目标目录的符号连接：

```
shell> mkdir /otherdisk/databases/test
shell> ln -s /otherdisk/databases/test /path/to/datadir
```

（2）将表的数据文件或索引文件指向其他物理磁盘。

- 对于新建的表，可以通过在 CREATE TABLE 语句中增加 DATA DIRECTORY 选项来完成，例如：

```
Create table test(id int primary key,
Name varchar(20))
ENGINE = innodb
DATA DIRECTORY = '/data1/data';
```

- 对于已有的表，可以先将其数据文件（.ibd）转移到目标磁盘，然后再建立符号连接即可。需要说明的是，表定义文件（.frm）必须位于 MySQL 数据文件目录下，不能用符号连接。

（3）在 Windows 下使用符号连接。

以上介绍的是 Linux/UNIX 下符号连接的使用方法，在 Windows 下，是通过在 MySQL 数据文件目录下创建包含目标路径并以".sym"结尾的文本文件来实现的。例如，假设 MySQL 的数据文件目录是 C:\mysql\data，要把数据库 foo 存放到 D:\data\foo，可以执行以下步骤。

- 创建目录 D:\data\foo。
- 创建文件 C:\mysql\data\foo.sym，在其中输入 D:\data\foo。

这样在数据库 foo 创建的表都会存储到 D:\data\foo 目录下。

注意：使用 Symbolic Links 存在一定的安全风险，如果不使用 Symbolic Links，应通过启动参数 skip-symbolic-links 禁用这一功能。

18.5 禁止操作系统更新文件的 atime 属性

atime 是 Linux/UNIX 系统下的一个文件属性，每当读取文件时，操作系统都会将读操作发生的时间回写到磁盘上。对于读写频繁的数据库文件来说，记录文件的访问时间一般没有任何用处，还会增加磁盘系统的负担，影响 I/O 的性能！因此，可以通过设置文件系统的 mount 属性，阻止操作系统写 atime 信息，以减轻磁盘 I/O 的负担。在 Linux 下的具体做法如下。

（1）修改文件系统配置文件/etc/fstab，指定 noatime 选项：

```
LABEL=/home /home ext3 noatime 1 2
```

（2）重新 mount 文件系统：

```
#mount -oremount /home
```

完成上述操作，以后读/home 下文件就不会再写磁盘了。

18.6 调整 I/O 调度算法

目前来说，传统的磁盘仍然是主流的存储设备，从传统的硬盘上读取数据分为以下 3 个步骤。

（1）将磁头移动到磁盘表面的正确位置，花费的时间称为寻道时间。

（2）等待磁盘旋转，需要的数据会移动到磁头下面，花费的时间取决于磁盘的转速，转速越高的磁盘需要的时间越短。

（3）磁盘继续旋转，直到所有需要的数据都经过磁头。

磁盘在做这样动作时的快慢可以归结为两个因素：访问时间（步骤1和步骤2）和传输速度，这两个因素也叫作延迟和吞吐量。

传统的磁盘结构如图18-1所示。

I/O 请求处理的快慢有很大程度上取决于磁盘的寻道时间。为了减少寻道时间，操作系统不对每次 I/O 请求都直接寻道处理，而是将 I/O 请求放入队列，对请求进行合并和排序，来减少磁盘寻道操作次数。

I/O 请求合并是指将两个或者多个 I/O 请求合并成一个新请求，例如，当新来的请求和当前请求队列中的某个请求需要访问的是相

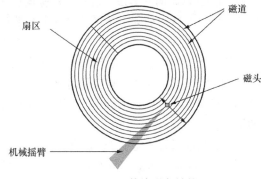

图 18-1 传统磁盘结构

同或者相邻扇区时，那么就可以把两个请求合并为对同一个或者多个相邻扇区的请求，这样只需要一次寻道就足够。通过合并，多个 I/O 请求被压缩为一次 I/O，最后只需要一次寻址就可以完成多次寻址的效果。

对于相邻扇区的访问通过合并处理，对于非相邻扇区的访问则通过排序处理。机械臂的转动是朝着扇区增长方向的，如果把 I/O 请求按照扇区增长排序，一次旋转就可以访问更多的扇区，能够缩短所有请求的实际寻道时间。

为此 Linux 实现了 4 种 I/O 调度算法，分别是 NOOP 算法（No Operation）、最后期限算法（Deadline）、预期算法（Anticipatory）和完全公平队列算法（CFQ）。用户既可以在内核引导时指定一种 I/O 调度算法，也可以在系统运行时动态修改 I/O 调度算法。从内核 2.5 开始，默认的 I/O 调度算法是 Deadline，之后默认 I/O 调度算法为 Anticipatory，直到内核 2.6.17 为止，从内核 2.6.18 开始，CFQ 成为默认的 I/O 调度算法。4 种算法详细介绍如下：

● NOOP 算法（No Operation）不对 I/O 请求排序，除了合并请求也不会进行其他任何优化，用最简单的先进先出（FIFO）队列顺序提交 I/O 请求。NOOP 算法面向的主要是随机访问设备，例如 SSD 等。NOOP 算法更适合随机访问设备的原因主要是，随机访问设备不存在传统机械磁盘的机械臂移动（也就是磁头移动）造成的寻道时间，那么就没有必要做多余的事情。

● 最后期限算法（Deadline）除了维护一个拥有合并和排序功能的请求队列之外，还额外维护了两个队列，分别是读请求队列和写请求队列，它们都是带有超时的 FIFO 队列。当新来一个 I/O 请求时，会被同时插入普通队列和读/写队列，然后 I/O 调度器正常处理普通队列中的请求。当调度器发现读/写请求队列中的请求超时的时候，会优先处理这些请求，保证尽可能不产生饥饿请求。Deadline 在全局吞吐量和延迟方面做了权衡，牺牲一定的全局吞吐量来避免出现饥饿请求的可能。当系统存在大量顺序请求的时候，Deadline 可能导致请求无法被很好地排序，引发频繁寻道。

● 预期算法（Anticipatory）是基于预测的 I/O 算法，它和 Deadline 很类似，也维护了 3 个请求队列。两者的区别在于，Anticipatory 处理完一个 I/O 请求之后并不会直接返回处理下一个请求，而是等待片刻（默认 6ms），等待期间如果有新来的相邻扇区的请求，会直接处理新来的请求，当等待时间结束后，调度才返回处理下一个队列请求。Anticipatory 适合写入较

多的环境，例如文件服务器等，不适合 MySQL 等随机读取较多的数据库环境。

● 完全公平队列（Complete Fair Queuing，CFQ）把 I/O 请求按照进程分别放入进程对应的队列中。CFQ 的公平是针对进程而言的，每一个提交 I/O 请求的进程都会有自己的 I/O 队列，CFQ 以时间片算法为前提，轮转调动队列，默认从当前队列中取出 4 个请求处理，然后处理下一个队列的 4 个请求，确保每个进程享有的 I/O 资源是均衡的。

从上面的算法中可以看到，在不同的场景下选择不同的 I/O 调度器是十分必要的。在完全随机的访问环境下，CFQ 和 Deadline 性能差异很小，但是在有大的连续 I/O 出现的情况下，CFQ 可能会造成小 I/O 的响应延时增加，所以建议 MySQL 数据库环境设置为 Deadline 算法，这样更稳定。对于 SSD 等设备，采用 NOOP 或者 Deadline 通常也可以获取比默认调度器更好的性能。

查看当前系统支持的 I/O 调度算法：

```
shell> dmesg | grep -i scheduler
```

查看当前设备（/dev/sda）使用的 I/O 调度算法：

```
shell> more /sys/block/sda/queue/scheduler
noop anticipatory deadline [cfq]
```

修改当前块设备（/dev/sda）使用的 I/O 调度算法，修改 I/O 调度算法后直接生效：

```
shell> echo "deadline" > /sys/block/sda/queue/scheduler
```

永久地修改 I/O 调度算法，可以通过修改内核引导参数，增加 elevator=调度程序名：

```
shell> vi /boot/grub/menu.lst
```

更改内容：

```
kernel /boot/vmlinuz-2.6.18-308.el5 ro root=LABEL=/ elevator=deadline
```

18.7 RAID 卡电池充放电问题

RAID 卡的写缓存设置对于系统的 I/O 性能影响极大，而影响写缓存的最重要因素是 RAID 卡电池的充放电问题。本节将详细介绍 RAID 卡充放电的原理和可能导致的各种问题。

18.7.1 什么是 RAID 卡电池充放电

RAID 卡都有写缓存（Battery Backed Write Cache），写缓存对 I/O 性能的提升非常明显，为了避免掉电丢失写缓存中的数据，所以 RAID 卡都有电池（Battery Backup Unit，BBU）来提供掉电后将写缓存中的数据写入磁盘。

为了记录 RAID 卡电池的放电曲线，便于 RAID 卡控制器了解电池的状态，同时为了延长电池的使用寿命，默认会定期启动自动校准模式（Auto Learn Mode），在 Learn Cycle 期间，RAID 卡控制器不会启用 BBU 直到完成校准。通俗地说，RAID 卡电池会定期充放电，定期充放电的操作叫作电池 Relearn 或者电池校准。

查看 RAID 卡 BBU 的状态：

```
shell> MegaCli64 -AdpBbuCmd -GetBbuStatus -aALL
BBU status for Adapter: 0
BatteryType: BBU
```

```
Voltage: 3945 mV
Current: 0 mA
Temperature: 47 C
Battery State: Optimal
BBU Firmware Status:

  Charging Status                          : None
  Voltage                                  : OK
  Temperature                              : OK
  Learn Cycle Requested                    : No
  Learn Cycle Active                       : No
  Learn Cycle Status                       : OK
  Learn Cycle Timeout                      : No
  I2c Errors Detected                      : No
  Battery Pack Missing                     : No
  Battery Replacement required             : No
  Remaining Capacity Low                   : No
  Periodic Learn Required                  : No
  Transparent Learn                        : No
  No space to cache offload                : No
  Pack is about to fail & should be replaced : No
  Cache Offload premium feature required   : No
  Module microcode update required         : No

BBU GasGauge Status: 0x0228
Relative State of Charge: 100 %
Charger Status: Complete
Remaining Capacity: 442 mAh
Full Charge Capacity: 446 mAh
isSOHGood: Yes

Exit Code: 0x00
```

其中，粗体显示的几个主要属性含义如下。

- Charging Status：None、Charging、Discharging 分别代表 BBU 处于不充放电状态、充电状态和放电状态。
- Learn Cycle Requested：Yes 代表当前有 Learn Cycle 请求，正在处于校准中。
- Learn Cycle Active：Yes 代表处于 Learn Cycle 校准阶段，控制器开始校准。
- Battery Replacement Required：Yes 代表电池需要更换。
- Remaining Capacity Low：Yes 代表电池容量过低，需要更换电池。

电池校准一般会经历 3 个阶段：首先 RAID 卡控制器会将 BBU 充满到最大程度，然后开始放电，放电完毕后重新将 BBU 充满到最大程度，一次 BBU 校准完成。整个过程一般为 3 个小时或者更长，期间 RAID 卡会自动禁用 WriteBack 策略，以保证数据完整性，而系统 I/O 性能会出现较大的波动。

默认 DELL 服务器 90 天执行一次校准，而 IBM 服务器是 30 天。DELL 和 IBM 都不推荐关闭 BBU 电池的 Auto Learn 模式，不做校准的 RAID 卡电池寿命会从正常的两年降低到正常寿命的 1/3，也就是 8 个月。

18.7.2 RAID 卡缓存策略

可以通过 MegaCli64 -LDInfo -Lall -aALL 命令来查看当前 RAID 卡设置的缓存策略。

```
shell> MegaCli64 -LDInfo -Lall -aALL
…
Adapter 0 -- Virtual Drive Information:
Virtual Drive: 0 (Target Id: 0)
Name                :
RAID Level          : Primary-1, Secondary-0, RAID Level Qualifier-0
Size                : 1.089 TB
```

```
Mirror Data          : 1.089 TB
State                : Optimal
Strip Size           : 64 KB
Number Of Drives per span:2
Span Depth           : 2
Default Cache Policy : WriteBack, ReadAdaptive, Direct, No Write Cache if Bad BBU
Current Cache Policy : WriteBack, ReadAdaptive, Direct, No Write Cache if Bad BBU
Default Access Policy: Read/Write
Current Access Policy: Read/Write
Disk Cache Policy    : Disk's Default
Encryption Type      : None
…
```

- Default Cache Policy：默认的缓存策略。
- Current Cache Policy：当前生效的缓存策略。

下面对缓存策略进行详细的说明。

1. 缓存策略第 1 段

写缓存策略，包括 WriteBack 和 WriteThrough。

- WriteBack：进行写入操作的时候，将数据写入 RAID 卡缓存后直接返回，RAID 卡控制器将在系统负载低或者 RAID 缓存满的情况下把数据写入磁盘，减少了磁盘操作的频次，大大提升了 RAID 卡写入性能，在大多数情况下能够有效降低系统 I/O 负载。写入 RAID 卡缓存的数据的可靠性由 RAID 卡的 BBU（Battery Backup Unit）保证。
- WriteThrough：进行写入操作的时候，不使用 RAID 卡缓存，数据直接写入磁盘才返回。也就是 RAID 卡写缓存被穿透，每次写入都直接写入磁盘。大多数情况下，WriteThrough 的策略设置会造成系统 I/O 负载上升。和 WriteBack 策略相比，WriteThrough 策略则不需要 BBU 电池来保证数据的完整性，但会造成传统机械硬盘写性能的大幅下降，但是对于 SSD 来说，使用 WriteThrough 反而是更好的选择。

2. 缓存策略第 2 段

是否开启预读，包括 ReadAheadNone、ReadAhead 和 ReadAdaptive。

- ReadAheadNone：不开启预读。
- ReadAhead：开启预读，在读操作的时候，预先把后面顺序的数据加载入缓存，在顺序读取的时候，能提供性能，但是在随机读的时候，开启预读做了不必要的操作，会降低随机读的性能。
- ReadAdaptive：自适应预读，在缓存和 I/O 空闲的时候，选择顺序预读，需要消耗一些计算能力，是默认的策略。

3. 缓存策略第 3 段

读操作是否缓存到 RAID 卡缓存中，包括 Direct 和 Cached。

- Direct：读操作不缓存到 RAID 卡缓存中。
- Cached：读操作缓存到 RAID 卡缓存中。

4. 缓存策略第 4 段

如果 BBU 出问题，是否启用 Write Cache，包括 Write Cache OK if Bad BBU 和 No Write Cache if Bad BBU。

- No Write Cache if Bad BBU：如果 BBU 出问题，则不再使用 Write Cache，从 WriteBack

策略自动切换到 WriteThrough 模式。这是默认配置，确保在没有 BBU 电池支持的情况，直接写入磁盘而不是 RAID 卡缓存，以确保数据安全。

- Write Cache OK if Bad BBU：如果 BBU 出问题，仍然启用 Write Cache。通常不推荐这么配置，因为如果 BBU 出问题，将无法保证意外断电后数据能够完整写回磁盘。除非有 UPS 后备电源或者其他类似方案做电源方面额外的保证。

RAID 卡缓存策略可以通过 MegaCli64 -LDSetProp 命令进行修改。常用的策略主要有下面 4 种：WriteBack、WriteThrough、Write Cache OK if Bad BBU 和 No Write Cache if Bad BBU。

```
shell> MegaCli64 -LDSetProp -WB -Lall -aALL
shell> MegaCli64 -LDSetProp -WT -Lall -aALL
shell> MegaCli64 -LDSetProp -CachedBadBBU -Lall -aALL
shell> MegaCli64 -LDSetProp -NoCachedBadBBU -Lall -aALL
```

在 RAID 卡电池校准期间（RAID 卡电池电量在降低到特定阈值之后）或者电池故障期间，默认的 RAID 卡写缓存策略会自动发生变动，从 WriteBack 变为 WriteThrough，造成系统写入性能下降，此时如果正好是业务高峰时间，会引起系统负载大幅度上升、响应时间变长。

此时，可以通过临时修改 RAID 卡缓存策略为 Write Cache OK if Bad BBU 来解决，修改后立即生效，无须重启系统等额外的配置：

```
shell> MegaCli64  -LDSetProp CachedBadBBU -Lall -aALL

Set Write Cache OK if bad BBU on Adapter 0, VD 0 (target id: 0) success

Exit Code: 0x00
# MegaCli64 -LDInfo -Lall -aALL
…
Default Cache Policy: WriteBack, ReadAdaptive, Direct, Write Cache OK if Bad BBU
Current Cache Policy: WriteBack, ReadAdaptive, Direct, Write Cache OK if Bad BBU
…
Current Access Policy: Read/Write
```

注意，临时修改 RAID 卡缓存策略度过业务高峰期之后，仍然应该及时恢复缓存策略为 No Write Cache if Bad BBU，避免断电可能导致的数据损失：

```
shell> MegaCli64  -LDSetProp NoCachedBadBBU -Lall -aALL

Set No Write Cache if bad BBU on Adapter 0, VD 0 (target id: 0) success

Exit Code: 0x00
shell> MegaCli64 -LDInfo -Lall -aALL
…
Default Cache Policy: WriteBack, ReadAdaptive, Direct, No Write Cache if Bad BBU
Current Cache Policy: WriteBack, ReadAdaptive, Direct, No Write Cache if Bad BBU
…
```

对于 RAID 卡缓存策略的自动变动带来的 I/O 性能波动，一般解决方案有两种。

18.7.3 如何应对 RAID 卡电池充放电带来的 I/O 性能波动

1. 解决方案 1

根据 RAID 卡电池（BBU 电池）下次充放电的时间，定期在业务量较低的时候，提前进行充放电，避免在业务高峰时发生 RAID 卡写入策略从 Write Back 到 Write Through 的更改。在手工触发电池充放电之后，下一次充放电的时间会往后顺延。例如，DELL 服务器电池 Relearn 周期一般为 90 天，在手工触发 DELL 服务器 BBU 电池校准之后，下一次电池 Relearn 的时间会往后顺延

90 天；而 IBM 服务器电池 Relearn 的周期一般为 30 天，手工触发 IBM 服务器 BBU 电池校准之后，下一次电池 Relearn 的时间就会往后顺延 30 天。具体时间周期可以在系统中查看。

可以从 BBU 电池的日志中获取到下次电池 Relearn 时间：

```
shell> MegaCli64 -fwtermlog -dsply -a0 -nolog
...
11/02/12 21:18:33: Next Learn will start on 01 31 2013
11/02/12 21:18:33:      *** BATTERY FEATURE PROPERTIES ***
11/02/12 21:18:33:      _____
11/02/12 21:18:33:      Auto Learn Period       : 90  days
11/02/12 21:18:33:      Next Learn Time         : 412982313
11/02/12 21:18:33:      Delayed Learn Interval  : 0  hours from scheduled time
11/02/12 21:18:33:      Next Learn scheduled on : 01 31 2013  21:18:33
11/02/12 21:18:33:      _____
...
```

或者通过命令直接得到下一次电池 Relearn 的时间：

```
shell> MegaCli -AdpBbuCmd -GetBbuProperties -aall

BBU Properties for Adapter: 0

  Auto Learn Period: 90 Days
  Next Learn time: Thu Jan 31 21:18:33 2013
  Learn Delay Interval:0 Hours
  Auto-Learn Mode: Transparent

Exit Code: 0x00
```

手工触发电池 Relearn（电池校准）的操作：

```
shell> MegaCli64 -AdpBbuCmd -BbuLearn -aALL
```

2. 解决方案 2

设置 Forced WriteBack 写策略，也就是说即使在电池电量低于警戒值甚至电池放电完毕的情况下，强制使用 WriteBack 写缓存策略，避免写入性能波动，此时一定要有 UPS 之类的后备电源，否则当电池放电完毕时服务器恰好断电，就会导致写入 RAID 卡缓存中的数据丢失。

18.8 NUMA 架构优化

从系统架构来看，目前的商用服务器大体可以分为 3 类：对称多处理器结构（Symmetric Multi-Processor，SMP）、非一致存储访问结构（Non-Uniform Memory Access，NUMA）和海量并行处理结构（Massive Parallel Processing，MPP）。一般服务器是 SMP 或者 NUMA 架构的较多。

SMP 架构是指在一台计算机上汇集了一组处理器（多 CPU），各个 CPU 之间共享内存子系统和总线结构，如图 18-2 所示。SMP 架构同时使用多个 CPU，系统将任务队列对称地分布于多个 CPU 上，从而极大地提高了整个系统的数据处理能力。所有的 CPU 都可以平等地访问内存、I/O 和外设。架构中多个 CPU 没有区别，共享相同的物理内存，每个 CPU 访问内存中的任何地址所需时间是相同的，因此 SMP 也被称为一致存储器访问（Uniform Memory Access，UMA）结构。对 SMP 服务器进行扩展的方式有增加内存、使用更快的 CPU、增加 CPU、扩充 I/O，以及增加更多的磁盘等。

SMP 服务器的主要特征是共享，系统中所有的资源（CPU、内存、I/O 等）都是共享的。正是由于共享，导致了 SMP 服务器的扩展能力非常有限。对 SMP 服务器来说，扩展最受限制的是内存，由于每个 CPU 都必须通过相同的总线访问相同的内存资源，如果两个 CPU 同

18.8 NUMA 架构优化

时请求访问一个内存资源（例如同一段内存地址），由硬件、软件的锁机制来解决资源争用的问题。所以随着 CPU 数量的增加，CPU 之间内存访问冲突加剧，最终造成 CPU 资源的浪费，使得 CPU 性能的有效性大幅度降低。

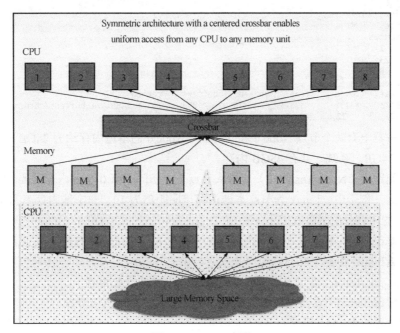

图 18-2　SMP 架构

由于 SMP 架构在扩展能力上的限制，人们开始探究如何进行有效地扩展以构建大型系统的技术，这时 NUMA 架构出现了。NUMA 把一台计算机分成多个节点（Node），每个节点内部拥有多个 CPU，节点内部使用共有的内存控制器，节点之间是通过互联模块进行连接和信息交互，如图 18-3 所示。因此节点的所有内存对于本节点所有的 CPU 都是等同的，而对于其他节点中的所有 CPU 都是不同的。因此每个 CPU 可以访问整个系统内存，但是访问本地节点的内存速度最快（不需要经过互联模块），访问非本地节点的内存的速度较慢（需要经过互联模块），即 CPU 访问内存的速度与节点的距离有关，距离称为 Node Distance。

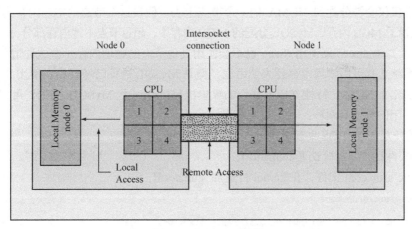

图 18-3　NUMA 架构

当前 NUMA 的节点情况如下所示：

```
shell> numactl --hardware
available: 2 nodes (0-1)
node 0 size: 24194 MB
node 0 free: 117 MB
node 1 size: 24240 MB
node 1 free: 17 MB
node distances:
node   0   1
  0:  10  21
  1:  21  10
shell> free -m
           total    used     free    shared   buffers   cached
Mem:       48273    48138    135     0        343       14891-/+ buffers/cache:   32903   15369
Swap:      2047     22       2025
```

当前服务器上有两个节点 Node 0 和 Node 1，Node 0 的本地内存约为 24GB，Node 1 的本地内存约为 24GB。系统一共有 48GB 内存。

节点之间距离（Node Distance）是指从节点 1 上访问节点 0 上的内存需要付出的代价的一种表现形式。例子中，Linux 为节点本地内存声明距离为 10，非本地内存声明距离为 21。

NUMA 的内存分配策略有以下 4 种。

- 缺省 default：总是在本地节点上分配（分配在当前进程运行的节点上）。
- 绑定 bind：强制分配到指定节点上。
- 交叉 interleave：在所有节点或者指定节点上交叉分配内存。
- 优先 preferred：在指定节点上分配，失败则在其他节点上分配。

显示当前系统 NUMA 策略：

```
shell> numactl --show
policy: default
preferred node: current
physcpubind: 0 1 2 3 4 5 6 7 8 9 10 11 12 13 14 15
cpubind: 0 1
nodebind: 0 1
membind: 0 1
```

因为 NUMA 默认的内存分配策略是优先在进程所在 CPU 的本地内存中分配，会导致 CPU 节点之间内存分配不均衡，当某个 CPU 节点内存不足时，会导致 Swap 产生，而不是从远程节点分配内存，这就是 Swap Insanity 现象。

MySQL 是单进程多线程架构的数据库，当 NUMA 采用默认内存分配策略时，MySQL 进程会被并且仅仅会被分配到 NUMA 的一个节点上去。假设这个节点的本地内存为 8GB，而 MySQL 配置了 14GB 内存，MySQL 分配的 14GB 内存中，超过节点本地内存部分（14GB-8GB = 6GB 内存）。Linux 宁愿使用 Swap 也不会使用其他节点的物理内存。在这种情况下，能观察到虽然系统总共可用物理内存还未使用完，但是 MySQL 进程已经开始在使用 Swap 了。

MySQL 对 NUMA 特性支持不好，如果单机只运行一个 MySQL 实例，可以选择关闭 NUMA。关闭的方式有两种。

- 硬件层，在 BIOS 中设置关闭；
- OS 内核，启动时设置 numa=off。

修改/etc/grub.conf 文件，在 kernel 行追加 numa=off：

```
shell> vi /etc/grub.conf
...
title Red Hat Enterprise Linux (2.6.32-279.el6.x86_64)
    root (hd0,0)
```

```
kernel /boot/vmlinuz-2.6.32-279.el6.x86_64 ro root=UUID=7971e5ab-ee55-4848-80f6-33
f811702a51 rd_NO_LUKS rd_NO_LVM LANG=en_US.UTF-8 rd_NO_MD SYSFONT=latarcyrheb-sun16 crashke
rnel=auto KEYBOARDTYPE=pc KEYTABLE=us rd_NO_DM rhgb quiet numa=off
```

保存后重启服务器,再次检查 NUMA 只剩下一个节点就成功了:

```
shell> numactl --hardware
available: 1 nodes (0)
node 0 cpus: 0 1 2 3 4 5 6 7 8 9 10 11 12 13 14 15 16 17 18 19 20 21 22 23
node 0 size: 65490 MB
node 0 free: 60968 MB
node distances:
node   0
  0:  10
```

或者通过 numactl 命令将 NUMA 内存分配策略修改为 interleave。

修改 mysqld_safe 启动脚本,添加 "cmd="/usr/bin/numactl --interleave all $cmd"" 一行即可,启动 MySQL 时指定内存分配策略为 interleave:

```
shell> vi $MYSQL_HOME/bin/mysqld_safe
...
cmd="/usr/bin/numactl --interleave all $cmd"
```

```
for i in "$ledir/$MYSQLD" "$defaults" "--basedir=$MY_BASEDIR_VERSION" \
  "--datadir=$DATADIR" "--plugin-dir=$plugin_dir" "$USER_OPTION"
do
  cmd="$cmd `shell_quote_string "$i"`"
done
```

保存后重启 MySQL 进程即可。

如果单机运行多个 MySQL 实例,可以将 MySQL 绑定到不同的 CPU 节点上,同时配置合适的 MySQL 内存参数,并且采用绑定的内存分配策略,强制在本节点分配内存。

NUMA 技术可以很好地解决 SMP 架构的扩展问题,但是 NUMA 一样存在缺陷,由于访问远程内存的延时远远超过访问本地内存,因此随着 CPU 数量增加,系统性能并不能够线性增加。

MPP 解决了 NUMA 架构增加 CPU 并不能线性提升性能的问题,MPP 由多个 SMP 服务器通过一定的节点互联网络进行连接,每个节点只访问自己的本地资源(内存、存储等),不访问其他节点的资源,是一种 Share Nothing 的架构,因而理论上可以无限扩展。

在 MPP 架构中,每个节点的 CPU 不能访问其他节点的内存,此时节点之间的信息交互是通过节点互联网络来实现的,这个过程称为 Data Redistribution。但是 MPP 服务器需要复杂的机制来调度和平衡各个节点的负载和并行处理。目前一些基于 MPP 技术的服务器会通过例如数据库软件等系统级别的软件来屏蔽底层的复杂性。例如,Teradata 就是基于 MPP 技术的一个关系数据库软件。

18.9 小结

本章站在操作系统的角度介绍了如何对 MySQL 数据库进行优化,主要讨论了 I/O 的优化问题、文件系统分布的优化问题等。在大多数的数据库系统中,磁盘 I/O 都会成为影响系统性能的瓶颈。希望读者通过本章能够掌握一些减少磁盘 I/O 以提高系统性能的方法。

第 19 章 应用优化

前面章节介绍了很多数据库的优化措施。但是在实际生产环境中，由于数据库服务器本身的性能局限，数据库的前期设计和应用访问设计就显得非常重要。好的设计在节点瓶颈到来时，可以通过简单的扩容或者配置变更来缩短服务不可用时间，甚至可以达到无缝扩容。数据库领域有一句流传很广的名言"数据库的性能是设计出来的，而非调优调出来的"，可见设计的重要性。本章将介绍一些常用的数据库设计和应用优化方法。

19.1 优化数据表的设计

在数据库设计过程中，用户可能会经常遇到这种问题：是否应该把所有表都按照第三范式来设计？表里面的字段到底该设置为多大长度合适？这些问题虽然很小，但是如果设计不当，则会给将来的应用带来很多的性能问题。本节将介绍 MySQL 中一些数据表的优化方法，其中一些方法不仅仅适用于 MySQL，也适用于其他类型的数据库管理系统。

19.1.1 优化表的数据类型

表需要使用何种数据类型是需要根据应用来判断的。虽然应用设计的时候需要考虑字段的长度留有一定的冗余，但是不推荐让很多字段都留有大量的冗余，这样既浪费磁盘存储空间，同时在应用程序操作时也浪费物理内存。

在 MySQL 中，可以使用函数 PROCEDURE ANALYSE() 对当前应用的表进行分析，该函数可以对数据表中列的数据类型提出优化建议，用户可以根据应用的实际情况酌情考虑是否实施优化。

以下是函数 PROCEDURE ANALYSE() 的使用方法：

```
SELECT * FROM tbl_name PROCEDURE ANALYSE();
SELECT * FROM tbl_name PROCEDURE ANALYSE(16,256);
```

输出的每一列信息都会对数据表中的列的数据类型提出优化建议。以上第二个语句告诉 PROCEDURE ANALYSE() 不要为那些包含的值多于 16 个或者 256 个字节的 ENUM 类型提出建议。如果没有这样的限制，输出信息可能很长；ENUM 定义通常很难阅读。

根据 PROCEDURE ANALYSE() 函数的输出信息，用户可能会发现，一些表中的字段可以修改为效率更高的数据类型。如果决定改变某个字段的类型，则需要使用 ALTER TABLE 语句。

19.1 优化数据表的设计

下面分析一下表 duck_cust 的数据类型是否需要优化。

（1）首先创建测试表 duck_cust，duck_cust 表中记录了客户的一些基本信息：

```sql
drop table duck_cust;
CREATE TABLE duck_cust(
cust_num MEDIUMINT  AUTO_INCREMENT, --客户编号
cust_title TINYINT, --客户标题号
cust_last CHAR(20) NOT NULL,  --客户姓氏
cust_first CHAR(15) NOT NULL, --客户名
cust_suffix ENUM('Jr.', 'II', 'III','IV', 'V', 'M.D.','PhD'), --附加码
cust_add1 CHAR(30) NOT NULL, --客户地址
cust_add2 CHAR(10), --客户地址
cust_city CHAR(18) NOT NULL, --客户所在城市
cust_state CHAR(2) NOT NULL, --客户所在州
cust_zip1 CHAR(5)NOT NULL, --客户邮编
cust_zip2 CHAR(4), --客户邮编
cust_duckname CHAR(25) NOT NULL, --客户名称
cust_duckbday DATE, --客户生日
PRIMARY KEY (cust_num)
) ENGINE=InnoDB;
```

（2）然后生成一些测试数据：

```sql
INSERT INTO duck_cust VALUES(NULL, 1, 'Irishlord', 'Red', 'III', '1022 N.E. Sea of Rye', 'A207', 'Seacouver', 'WA', '98601', '3464', 'Netrek Rules', '1967:10:21');
INSERT INTO duck_cust VALUES(NULL, 4, 'Thegreat', 'Vicki', NULL, '2004 Singleton Dr.', 0, 'Freedom', 'KS', '67209', '4321', 'Frida Kahlo de Tomayo', '1948:03:21');
INSERT INTO duck_cust VALUES(NULL, 9, 'Montgomery', 'Chantel', NULL, '1567 Terra Cotta Way', 0, 'Chicago', 'IL', '89129', '4444', 'Bianca', '1971:07:29');
INSERT INTO duck_cust VALUES(NULL, 7, 'Robert', 'David', NULL, '20113 Open Road Highway', '#6', 'Blacktop', 'AZ', '00606', '1952', 'Harley', '1949:08:00');
INSERT INTO duck_cust VALUES(NULL, 5, 'Kazui', 'Wonko', 'PhD', '42 Cube Farm Lane', 'Gatehouse', 'Vlimpt', 'CA', '45362', 0, 'Fitzwhistle', '1961:12:04');
INSERT INTO duck_cust VALUES(NULL, 6, 'Gashlycrumb', 'Karen', NULL, '3113 Picket Fence Lane', 0, 'Fedora', 'VT', '41927', '5698', 'Tess D''urberville', '1948:08:19');
```

这时，查看一下表结构：

```
desc duck_cust;
+---------------+-----------------------------------------+------+-----+---------+----------------+
| Field         | Type                                    | Null | Key | Default | Extra          |
+---------------+-----------------------------------------+------+-----+---------+----------------+
| cust_num      | mediumint(9)                            | NO   | PRI | NULL    | auto_increment |
| cust_title    | tinyint(4)                              | YES  |     | NULL    |                |
| cust_last     | char(20)                                | NO   |     | NULL    |                |
| cust_first    | char(15)                                | NO   |     | NULL    |                |
| cust_suffix   | enum('Jr.','II','III','IV','V','M.D.','PhD')| YES |     | NULL    |                |
| cust_add1     | char(30)                                | NO   |     | NULL    |                |
| cust_add2     | char(10)                                | YES  |     | NULL    |                |
| cust_city     | char(18)                                | NO   |     | NULL    |                |
| cust_state    | char(2)                                 | NO   |     | NULL    |                |
| cust_zip1     | char(5)                                 | NO   |     | NULL    |                |
| cust_zip2     | char(4)                                 | YES  |     | NULL    |                |
| cust_duckname | char(25)                                | NO   |     | NULL    |                |
| cust_duckbday | date                                    | YES  |     | NULL    |                |
+---------------+-----------------------------------------+------+-----+---------+----------------+
```

（3）使用 PROCEDURE ANALYSE() 函数确定要优化的列：

```
    mysql> SELECT * FROM duck_cust PROCEDURE ANALYSE()\G;
*************************** 1. row ***************************
            Field_name: sakila.duck_cust.cust_num
             Min_value: 1
             Max_value: 6
            Min_length: 1
            Max_length: 1
      Empties_or_zeros : 0
                 Nulls : 0
```

```
      Avg_value_or_avg_length: 3.5000
                         Std: 1.7078
           Optimal_fieldtype: ENUM('1','2','3','4','5','6') NOT NULL
*************************** 2. row ***************************
...
```

从结果中可以看到 test.duck_cust.cust_num 列的 Min_length、Max_length、Avg_value_or_avg_length，根据这些统计值，可以对列进行优化，例如，插入的数据最大长度和最小长度都是 1，所以，可以优化字段 cust_num 为 mediumint(2)；同时，上面的结果也给出了优化建议"Optimal_fieldtype: ENUM('1','2','3','4','5','6') NOT NULL"。

看到这个建议读者可能会觉得很奇怪，怎么给出了枚举类型，而不是我们预期的整型？因为这时分析的测试表记录数太少，使得 cust_name 的唯一值太少，因此函数觉得用枚举类型会更合理。如果是对一个大表进行分析，提出的建议会更准确。

根据给出的统计信息和优化建议，可以使用如下语句进行字段类型的更改：

```
mysql> alter table  duck_cust  modify cust_num mediumint(2);
Query OK, 6 rows affected (0.03 sec)
Records: 6  Duplicates: 0  Warnings: 0
```

19.1.2 通过拆分提高表的访问效率

这里所说的"拆分"，是指对数据表进行拆分。如果针对 InnoDB 类型的表进行，那么有两种拆分方法。

（1）第一种方法是垂直拆分，即把主键和一些列放到一个表，然后把主键和另外的列放到另一个表中。

如果一个表中某些列常用，而另一些列不常用，则可以采用垂直拆分。另外，垂直拆分可以使得数据行变小，一个数据页就能存放更多的数据，在查询时就会减少 I/O 次数。其缺点是需要管理冗余列，查询所有数据需要联合（JOIN）操作。

（2）第二种方法是水平拆分，即根据一列或多列数据的值把数据行放到多个独立的表或者分区中。

水平拆分通常在以下几种情况下使用。

● 表很大，分割后可以降低在查询时需要读的数据和索引的页数，同时也降低了索引的层数，提高查询速度。

● 表中的数据本来就有独立性，例如，表中分别记录各个地区的数据或不同时期的数据，特别是有些数据常用，而另外一些数据不常用。

● 需要把数据存放到多个介质上。

例如，移动电话的账单表就可以分成多个表或者多个分区。分表的话，可以把最近 3 个月的账单数据存在一个表中，3 个月前的历史账单存放在另外一个表中，超过 1 年的历史账单可以存储到单独的存储介质上。如果采用分区的话，可以按月分区，并且在查询的时候都带上时间条件，这样大多数查询都只会用到表中一个分区的数据。

用分表的方式做水平拆分会给应用增加复杂度，例如拆表通常在查询时需要多个表名，查询所有数据需要 UNION 操作。在许多数据库应用中，这种复杂性会超过它带来的优点，因为只要索引关键字不大，则在索引用于查询时，表中增加 2～3 倍数据量，查询时也就增加读一个索引层的磁盘次数，所以水平拆分要考虑数据量的增长速度，根据实际情况决定是否需要对表进行水平拆分。而分区相对来说带来的不便会小很多,所以一般来说,如果数据量不是特别大,可以优先采用分区的方式来做水平拆分。

19.1.3 逆规范化

数据库设计时要满足规范化这个道理大家都非常清楚,但是否数据的规范化程度越高越好呢?这是由实际需求来决定的。因为规范化越高,那么产生的关系就越多,关系过多的直接结果就是导致表之间的连接操作越频繁,而表之间的连接操作是性能较低的操作,直接影响到查询的速度,所以,对于查询较多的应用,就需要根据实际情况运用逆规范化对数据进行设计,通过逆规范化来提高查询的性能。

例如,移动电话的用户每月都会查询自己的账单。账单信息一般包含用户的名字和本月消费总金额,设想一下,如果用户的姓名和属性信息存放在一个表中,假设表名为 A,而用户的编号和他对应的账单信息存放在另外一张 B 表中,那么,用户每次查询自己的月账单时,数据库查询时都要进行表连接,因为账单表 B 中并不包含用户的名字,所以必须通过关联 A 表取过来,如果在数据库设计时考虑到这一点,就可以在 B 表增加一个冗余字段存放用户的名字,这样在查询账单时就不用再做表关联,可以使查询有更好的性能。

反规范的好处是降低连接操作的需求、降低外码和索引的数目,还可能减少表的数目,相应带来的问题是可能出现数据的完整性问题。加快查询速度,但会降低修改速度。因此,决定做反规范时,一定要权衡利弊,仔细分析应用的数据存取需求和实际的性能特点,好的索引和其他方法经常能够解决性能问题,而不必采用反规范这种方法。

在进行反规范操作之前,要充分考虑数据的存取需求、常用表的大小、一些特殊的计算(例如合计)、数据的物理存储位置等。常用的反规范技术有增加冗余列、增加派生列、重新组表和分割表。

- 增加冗余列:指在多个表中具有相同的列,常用来在查询时避免连接操作。
- 增加派生列:指增加的列来自其他表中的数据,由其他表中的数据经过计算生成。增加派生列的作用是在查询时减少连接操作,避免使用集函数。
- 重新组表:指如果许多用户需要查看两个表连接出来的结果数据,则把这两个表重新组成一个表来减少连接而提高性能。
- 分割表:可以参见 19.1.2 节的内容。

另外,逆规范技术需要维护数据的完整性。无论使用何种反规范技术,都需要一定的管理来维护数据的完整性。常用的方法是批处理维护、应用逻辑和触发器。

- 批处理维护是指对复制列或派生列的修改积累一定的时间后,运行一批处理作业或存储过程对复制或派生列进行修改,这只能在对实时性要求不高的情况下使用。
- 数据的完整性也可由应用逻辑来实现,这就要求必须在同一事务中对所有涉及的表进行增、删、改操作。在应用逻辑中控制,可以根据对数据完整性的实际需求来灵活控制,但是对于应用开发来说也会增加一定的复杂度,需要从系统整体需求综合考虑。
- 另一种方式就是使用触发器,对数据的任何修改立即触发对复制列或派生列的相应修改。触发器是实时的,而且相应的处理逻辑只在一个地方出现,易于维护。这也是解决这类问题的一种思路。

19.2 数据库应用优化

19.2.1 使用连接池

对于访问数据库来说，建立连接的代价比较昂贵，因此，我们有必要建立"连接池"以提高访问的性能。从名字上理解，"连接池"是一个存放"连接"的"池子"，再具体一些，我们可以把连接当作对象或者设备，统一放在一个"池子"里面，以前需要直接访问数据库的地方，现在都改为从这个"池子"里面获取连接来使用。因为"池子"中的连接都已经预先创建好，可以直接分配给应用使用，因此大大减少了创建新连接所耗费的资源。连接返回后，本次访问将连接交还给"连接池"，以供新的访问使用。

19.2.2 减少对 MySQL 的访问

在实际应用中，我们的硬件资源通常是有限的、无法扩充的。这种情况下，应用有什么措施能减少对数据库的访问呢？本节将向大家介绍一些简单的方法。

1. 避免对同一数据做重复检索

应用中需要厘清对数据库的访问逻辑。能够一次连接就能够提取出所有结果的，就不用两次连接，这样可以大大减少对数据库无谓的重复访问。

例如，在某应用中需要检索某人的年龄和性别，那么就可以执行以下查询：

```
Select old,gender from users where userid = 231;
```

之后又需要这个人的家庭住址，可以执行：

```
Select address from users where userid = 231;
```

这样，就需要向数据库提交两次请求，数据库就要做两次查询操作，其实完全可以用一句 SQL 语句得到想要的结果，然后把得到的结果放到变量中已备后用，比如：

```
Select old,gender,address from users where userid = 231;
```

不管读者是否相信，由于上面的原因导致的性能问题，在很多应用系统中都存在，因此在厘清应用逻辑并向数据库提交请求前进行深思熟虑是很有必要的。

2. 增加 CACHE 层

在应用中，我们可以在应用端加 CACHE 层来达到减轻数据库负担的目的。CACHE 层有很多种，也有很多种实现的方式，只要能达到降低数据库的负担又能满足应用就可以，这就需要根据应用的实际情况进行特殊处理。

比如，可以把部分数据从数据库中抽取出来放到应用端以文本方式存储，然后如果有查询需求，可以直接从这个"CACHE"中检索。由于这里的数据量小，所以能够达到很高的查询效率，而且也减轻了数据库的负担。当然这种方案还涉及很多其他问题，比如如果有数据更新怎么办、多长时间刷新一次"CACHE"等，都需要根据具体应用环境进行相应的处理。

再比如用户可以在应用端建立一个二级数据库，把访问频度非常大的数据放到二级库上，然后设定一个机制与主数据库进行同步，这样用户的主要操作都在二级数据库上进行，大大地降低了主数据库的压力，各种 NoSQL 数据库，例如 Redis 等，都可以很方便地充当这种二级数据库的角色。

各种"CACHE"层的实现方式不同,这里只是抛砖引玉,给读者一个解决问题的思路,不可千篇一律地照搬照抄。

19.2.3 负载均衡

负载均衡(Load Balance)是实际应用中使用非常普遍的一种优化方法,它的机制就是利用某种均衡算法,将固定的负载量分布到不同的服务器上,以此来减轻单台服务器的负载,达到优化的目的。负载均衡可以用在系统中的各个层面中,从前台的 Web 服务器到中间层的应用服务器,最后到数据层的数据库服务器,都可以使用。本节主要介绍 MySQL 数据库端的一些负载均衡方法。

1. 利用 MySQL 复制分流查询操作

利用 MySQL 的主从复制(具体介绍见第 30 章)可以有效地分流更新操作和查询操作。具体的实现是一个主服务器承担更新操作,而多台从服务器承担查询操作,主从之间通过复制实现数据的同步。多台从服务器一方面用来确保可用性,另一方面可以创建不同的索引,以满足不同查询的需要。

对于主从之间不需要复制全部表的情况,可以通过在主服务器上搭建一个虚拟的从服务器,将需要复制到从服务器的表设置成 BLACKHOLE 引擎,然后定义 replicate-do-table 参数只复制这些表,这样就过滤出需要复制的 BINLOG,减少了传输 BINLOG 的带宽。因为搭建的虚拟从服务器只起到过滤 BINLOG 的作用,并没有实际记录任何数据,所以对主数据库服务器的性能影响也非常有限。

通过复制来分流查询是减少主数据库负载的一个常用方法,但是这种办法也存在一些问题,最主要的问题是当主数据库上更新频繁或者网络出现问题的时候,主从之间的数据可能存在比较大的延迟更新,从而造成查询结果和主数据库上有所差异。因此应用在设计的时候需要有所考虑。

2. 采用分布式数据库架构

分布式的数据库架构适合大数据量、负载高的情况,它具有良好的扩展性和高可用性。通过在多台服务器之间分布数据,可以实现在多台服务器之间的负载平均,提高了访问的执行效率。具体实现的时候,可以使用 MySQL 的 CLUSTER 或者第三方中间件来实现读写分离、数据分片、全局事务等功能。第 31 章和第 32 章将详细介绍常用的高可用架构和中间件产品。

19.3 小结

本章介绍了在数据库设计和应用层面的优化。数据表的设计是一个数据库设计的基础,一旦数据库设计完毕并投入使用,将来再进行修改就比较麻烦,因此,在数据库设计时一定要尽可能地考虑周到。

数据库使用方式的优化也是数据库优化的重要组成部分,数据库本身的优化有一定的局限性,到了一定程度就很难再有大的提升,但是如果从使用方式的角度进行考虑,那么还有很多可以优化的内容。我们从连接池、减少数据库访问、负载均衡以及 CACHE 层等方面进行了讨论。

应用层优化的方面还有很多与应用本身密切相关,这些内容需要读者在实际应用环境中不断总结、不断积累经验才能达到更好的优化效果。

第 20 章 PS/SYS 数据库

MySQL Performance Schema（PS）和 SYS Schema 是 MySQL 官方提供的，可以用来监控性能和诊断故障。

PS 库主要用于收集数据库运行时的性能数据，通过充分利用 PS 库表的数据，让 DBA 更了解数据库的运行状态，也有助于排查定位问题。PS 数据库最早出现在 MySQL 5.5 中，默认关闭信息收集，需要手动开启。PS 库自出现以来，在每一个 MySQL 版本都有增强。在 MySQL 5.6 中，默认打开 PS 库信息收集。而 MySQL 5.7 的 PS 库中添加了更多的监控项，如内存监控、复制信息、metadata lock 监控等，统计信息更加丰富，功能跟 Oracle 的 awr 也越来越像。到了 MySQL 8.0 版本，PS 库表又添加了索引，极大地加快表的查询速度。

MySQL 从 5.7.7 版本开始提供 SYS 库，DBA 和开发同学可以使用 SYS 库进行数据库的性能调优和问题诊断。SYS 库除了 sys_config 表，其他都是视图，视图的数据来源于 PS 库和 IS（information_schema）库，也就是说，SYS 库本身既不采集数据也不存储数据（sys_config 表除外）。引入 SYS 库是为了以更直观、更易懂的方式展示 PS 库和 IS 库的信息。

20.1 Performance Schema 库

本节主要介绍 Performance Schema（PS）库的使用和表用途。

20.1.1 如何开启 PS 库

PS 库在 MySQL 5.6 之后的版本是默认开启的。如果需要手工开启或者关闭 PS 库，可以在 my.cnf 文件修改相应配置，例如添加以下参数，开启 PS 库：

```
[mysqld]
performance_schema=ON
```

在 MySQL 实例中确认 PS 库是否已开启：

```
mysql>SHOW VARIABLES LIKE 'performance_schema';
+--------------------+-------+
| Variable_name      | Value |
+--------------------+-------+
| performance_schema | ON    |
+--------------------+-------+
1 row in set (0.00 sec)
```

从查询结果上看，performance_schema 参数的值为 ON，表明当前实例已经开启 PS 库。

PS 库的开启或者关闭，需要修改配置文件后重启实例生效。如果想在线修改 performance_schema 参数的值，则会报错：

```
mysql>set global performance_schema=off;
ERROR 1238 (HY000): Variable 'performance_schema' is a read only variable
```

20.1.2　PS 库的表

PS 库的表使用 PERFORMANCE_SCHEMA 存储引擎，可以使用 show create table tablename 命令查看表的结构，例如查看 accounts 表结构：

```
mysql>show create table accounts\G
*************************** 1. row ***************************
       Table: accounts
Create Table: CREATE TABLE `accounts` (
  `USER` char(32) CHARACTER SET utf8 COLLATE utf8_bin DEFAULT NULL,
  `HOST` char(60) CHARACTER SET utf8 COLLATE utf8_bin DEFAULT NULL,
  `CURRENT_CONNECTIONS` bigint(20) NOT NULL,
  `TOTAL_CONNECTIONS` bigint(20) NOT NULL
) ENGINE=PERFORMANCE_SCHEMA DEFAULT CHARSET=utf8
1 row in set (0.00 sec)
```

在 MySQL 5.7 中，PS 库一共有 87 个表。PS 库的 setup 表保存了监控的配置数据：

```
mysql>SELECT TABLE_NAME FROM INFORMATION_SCHEMA.TABLES WHERE TABLE_SCHEMA = 'performance_schema' AND TABLE_NAME LIKE 'setup%';
+-------------------+
| TABLE_NAME        |
+-------------------+
| setup_actors      |
| setup_consumers   |
| setup_instruments |
| setup_objects     |
| setup_timers      |
+-------------------+
5 rows in set (0.00 sec)
```

setup_actors 表用于配置新的线程的监控状态，默认监控所有用户：

```
mysql>select * from setup_actors;
+------+------+------+---------+---------+
| HOST | USER | ROLE | ENABLED | HISTORY |
+------+------+------+---------+---------+
| %    | %    | %    | YES     | YES     |
+------+------+------+---------+---------+
1 row in set (0.00 sec)
```

setup_consumers 表存放所有的事件消息的消费者类型以及事件的开启状态：

```
mysql>select * from setup_consumers;
+----------------------------------+---------+
| NAME                             | ENABLED |
+----------------------------------+---------+
| events_stages_current            | NO      |
| events_stages_history            | NO      |
| events_stages_history_long       | NO      |
| events_statements_current        | YES     |
| events_statements_history        | YES     |
| events_statements_history_long   | NO      |
| events_transactions_current      | NO      |
| events_transactions_history      | NO      |
| events_transactions_history_long | NO      |
| events_waits_current             | NO      |
| events_waits_history             | NO      |
| events_waits_history_long        | NO      |
| global_instrumentation           | YES     |
| thread_instrumentation           | YES     |
| statements_digest                | YES     |
```

```
+----------------------------------+--------+
15 rows in set (0.00 sec)
```

setup_instruments 表存放 6 种类型（idle、wait、stage、statement、transaction 和 memory）相关的 instrument 对象以及对象的开启状态：

```
mysql>select * from setup_instruments;
+-----------------------------------------------+---------+-------+
| NAME                                          | ENABLED | TIMED |
+-----------------------------------------------+---------+-------+
| wait/synch/mutex/sql/TC_LOG_MMAP::LOCK_tc     | NO      | NO    |
| wait/synch/mutex/sql/LOCK_des_key_file        | NO      | NO    |
……
| memory/sql/servers_cache                      | NO      | NO    |
| memory/sql/udf_mem                            | NO      | NO    |
| memory/sql/Relay_log_info::mts_coor           | NO      | NO    |
| wait/lock/metadata/sql/mdl                    | YES     | YES   |
+-----------------------------------------------+---------+-------+
1030 rows in set (0.01 sec)
```

setup_objects 存放对象和所属 db 的监控列表，setup_timers 存放的是 instrument 对象所使用的 timer 类型。

对于上面的配置信息，除非有特殊的需求，一般来说使用默认项即可。

PS 库按照各个维度对数据库进行性能监控，包括事件、文件使用、内存使用、复制相关、会话统计、socket 使用、表相关和锁等维度，其中事件包括阶段事件、语句事件、事务事件和等待事件，如表 20-1 所示。

表 20-1　　　　　PS 库用来对数据库进行性能监控的统计维度

统计维度	关　键　字	主　要　功　能
阶段事件	%stages%	记录线程监视的阶段事件的状态
语句事件	%statements%	记录线程监视的语句事件的状态
事务事件	%transactions%	记录线程监视的事务事件的状态
等待事件	%waits%	记录线程监视的等待事件的状态
文件使用	%file%	汇总有关 I/O 操作的信息
内存使用	%memory%	检测内存使用和聚合内存使用统计
复制统计	%replication%	记录多源复制和 MGR 的信息
会话统计	%session%	记录会话的属性和连接的状态参数
socket 使用	%socket%	记录活跃会话对象实例
表使用	%table%	按照表和索引聚合每个表的 I/O 操作
锁统计	%lock%	记录持有读写锁的记录和汇总表的锁等待信息

使用下面的 SQL 结合表 20-1 中的关键字可以找到对应统计维度的表，例如查看阶段事件相关的表：

```
mysql> select table_name from information_schema.tables
    where table_schema = 'performance_schema' and table_name like '%stages%';
+----------------------------------------------------+
| TABLE_NAME                                         |
+----------------------------------------------------+
| events_stages_current                              |
| events_stages_history                              |
| events_stages_history_long                         |
| events_stages_summary_by_account_by_event_name     |
| events_stages_summary_by_host_by_event_name        |
| events_stages_summary_by_thread_by_event_name      |
| events_stages_summary_by_user_by_event_name        |
| events_stages_summary_global_by_event_name         |
```

```
8 rows in set (0.00 sec)
```

查询结果包含 3 种不同维度的数据。

- 当前事件表：表名含有"current"，当前表包含每个线程的最新事件。
- 历史事件表：表名含有"history"，历史表与当前表结构相同，但包含更多行。例如，events_stages_history 表包含每个线程最近的 10 个事件。events_stages_history_long 包含最近 10 000 个事件。
- 汇总表：表名含有"summary"，汇总表包含通过事件聚合的信息，包括已经从历史记录表中丢弃的事件。

20.2 SYS 库

本节主要介绍 SYS 库的对象和用途。

20.2.1 SYS 库的对象

SYS 库中的对象按照名称划分，一种是以字母开头，另一种是以 x$开头。比如 host_summary 和 x$host_summary，两者的底层数据是一样的，区别是 host_summary 对某些字段进行了格式化处理，方便阅读；x$host_summary 直接展示了原始采集的数据，方便对数据进行二次加工。

SYS 库的对象主要基于 PS 库，所以 SYS 库的监控维度跟 PS 库类似，主要包括主机相关、innodb 相关、io 相关、内存相关、连接数和会话、db 相关、SQL 语句相关和等待事件相关，如表 20-2 所示。

表 20-2　　　　　　　　　　　SYS 库的统计维度

统计维度	关 键 字	主 要 功 能
主机	host%和 x$host%	按照主机维度汇总全局信息、文件 IO 信息和语句信息
innodb 相关	innodb%和 x$innodb%	统计 innodb 缓冲池信息和锁等待信息
io 相关	io%和 x$io%	按照线程、文件和等待事件维护统计 IO 操作信息
内存	memory%和 x$memory%	按照主机、线程、用户和全局维度统计内存信息
连接和会话	%processlist%和%session%	记录连接和会话信息
db	schema%和 x$schema%	从 db 角度统计索引、锁和表的信息
SQL 语句	statement%和 x$statement%	从表的维护汇总排序、全表扫描等信息
等待事件	wait%和 x$wait%	按照主机、用户、用户和全局维度统计内存信息

同上类似，使用下面的 SQL 结合表 20-2 中的关键字可以找到对应统计维度的对象，例如主机相关的表：

```
mysql> select table_name from information_schema.tables where table_schema = 'sys' and (table_name like 'host%' or table_name like 'x$host%');
+----------------------------------+
| table_name                       |
+----------------------------------+
| host_summary                     |
| host_summary_by_file_io          |
| host_summary_by_file_io_type     |
| host_summary_by_stages           |
```

```
| host_summary_by_statement_latency   |
| host_summary_by_statement_type      |
| x$host_summary                      |
| x$host_summary_by_file_io           |
| x$host_summary_by_file_io_type      |
| x$host_summary_by_stages            |
| x$host_summary_by_statement_latency |
| x$host_summary_by_statement_type    |
+-------------------------------------+
12 rows in set (0.01 sec)
```

查询结果中如果包含"summary"字样，都属于汇总表，即通过特定事件来聚合信息。

20.2.2 SYS 对象的实际应用

SYS 的对象基于 PS 库和 IS 库进行了整合，并格式化输出，所以直接查询 SYS 库将会更方便，结果也更直观。在实际工作中，通过查询 SYS 库的对象获取信息，可以对连接会话、表的使用、内存等进行优化。本节主要介绍 SYS 库的实际应用。

1. 主机相关

以连接 mysql 实例的 host 为第一维度，分析文件的 io 等待时间、文件 io 的等待事件、SQL 语句的查询时间等信息。例如 host_summary：

```
mysql>select * from host_summary\G
*************************** 1. row ***************************
                  host: 192.168.1.62
            statements: 29760
     statement_latency: 8.76 w
 statement_avg_latency: 2.97 m
           table_scans: 5424
              file_ios: 139382
        file_io_latency: 12.22 s
    current_connections: 1
      total_connections: 83
           unique_users: 2
         current_memory: 0 bytes
 total_memory_allocated: 0 bytes
```

该类对象比较重要的字段如下。

- host：连接到 MySQL 实例的主机 IP。
- statements：主机 IP 执行总的 SQL 语句数量。
- statement_latency：主机 IP 执行 SQL 语句消耗的总时间。
- file_io_latency：主机 IP 发生文件等待事件消耗的总时间。
- current_connections：主机 IP 当前总的连接数。
- current_memory：实例为主机 IP 当前分配的内存。

通过上面的信息，我们了解到主机 IP 为 192.168.1.62 的连接，执行了 29 760 条 SQL 语句，执行这些语句一共消耗了 8.76 w（w 为 week 的简称），文件等待事件消耗总时间是 12.22s，当前连接数只有一个，消耗了 0 byte 的内存。如果发现上面某个维度的信息有异常，可以再进一步进行分析。

2. innodb 相关

该类对象分为两部分，其中 innodb_buffer_stats_by_%用于分析哪些对象占用缓冲池的内存比较多；另一类 innodb_lock_waits 用于行锁等待分析，得到阻塞和被阻塞的会话信息。

20.2 SYS 库

下面从表的维度，分析缓冲池使用情况。

```
mysql>select * from innodb_buffer_stats_by_table  order by pages desc limit 1\G
*************************** 1. row ***************************
object_schema: employees
  object_name: salaries
    allocated: 143.61 MiB
         data: 132.74 MiB
        pages: 9191
 pages_hashed: 75
    pages_old: 5403
  rows_cached: 2841496
```

通过上面的结果，可以看到 employees.salaries 表占用了缓冲池 143MB 的内存，这样就找到了占用缓冲池内存最多的对象。

通过下面的例子，看看 innodb_lock_waits 怎么帮助用户分析行锁等待。

第一个会话，开始事务，根据主键更新 employees.departments 表的一行记录。

```
mysql>use employees
Database changed
mysql>start transaction;
Query OK, 0 rows affected (0.01 sec)

mysql>update departments set dept_name='Repair Service' where dept_no='d009';
Query OK, 1 row affected (0.14 sec)
Rows matched: 1  Changed: 1  Warnings: 0
```

第一个会话的事务未提交，接着开启第二个事务，更新同一行记录。为了方便观察，设置事务锁超时时间为 1 000s。

```
mysql>set session innodb_lock_wait_timeout=1000;
Query OK, 0 rows affected (0.00 sec)

mysql>start transaction;
Query OK, 0 rows affected (0.00 sec)

mysql>update departments set dept_name='Customer Service' where dept_no='d009';
```

第二个会话被第一个会话阻塞，查询 innodb_lock_waits 得到详细的会话信息：

```
mysql>select * from innodb_lock_waits\G
*************************** 1. row ***************************
              wait_started: 2018-09-25 12:27:19
              wait_age_secs: 19
               locked_table: `employees`.`departments`
               locked_index: PRIMARY
                locked_type: RECORD
             waiting_trx_id: 571091
        waiting_trx_started: 2018-09-25 12:27:19
            waiting_trx_age: 00:00:19
……
               waiting_pid: 691
             waiting_query: update departments set dept_na ...  Service' where dept_no='d009'
……
            blocking_trx_id: 571090
              blocking_pid: 690
……
      sql_kill_blocking_query: KILL QUERY 690
 sql_kill_blocking_connection: KILL 690
1 row in set, 3 warnings (0.01 sec)
```

结果显示，第一个会话（blocking）阻塞第二个会话（waiting）。第一个会话的 ID（blocking_pid）是 690，第二个会话的 ID（waiting_pid）是 691。提交第一个事务，或者执行命令 KILL QUERY 690 把第一个事务会话杀掉，第二个会话的事务就可以正常执行了。

3. 连接数和会话相关

对象 sys.processlist 和 sys.session 的信息比 information_schema.processlist 全面，其中 sys.session 去除了后台线程的信息，只保留了用户会话的信息。

```
mysql>select * from session\G
*************************** 1. row ***************************
                thd_id: 727
               conn_id: 693
                  user: root@localhost
                    db: employees
               command: Sleep
                 state: NULL
                  time: 2
     current_statement: NULL
     statement_latency: NULL
              progress: NULL
          lock_latency: 160.00 us
         rows_examined: 2834339
             rows_sent: 1
         rows_affected: 0
            tmp_tables: 0
       tmp_disk_tables: 0
              full_scan: NO
        last_statement: select count(0) from salaries
last_statement_latency: 991.52 ms
        current_memory: 0 bytes
             last_wait: NULL
     last_wait_latency: NULL
                source: NULL
           trx_latency: NULL
             trx_state: NULL
        trx_autocommit: NULL
                   pid: 17956
          program_name: mysql
```

查询 sys.session，可以得到会话的 ID、连接用户信息、连接时间、查询影响行数、返回行数、最后执行 SQL 等信息。

4. db 相关

以表为第一维度，分析表和索引的使用情况。

查询 schema_redundant_indexes 表，可以找出冗余的索引。按照最左原则找出冗余的索引，例如表 A 有 3 个索引，index1(a,b)、index2(a)和 index3(b)。index1 最左边第一个字段包含了 index2，所以 index2 是冗余的索引，建议删掉，最后剩下 index1 和 index3 索引。例如，employees 表结构如下：

```
mysql>show create table employees\G
*************************** 1. row ***************************
       Table: employees
Create Table: CREATE TABLE `employees` (
  `emp_no` int(11) NOT NULL,
  `birth_date` date NOT NULL,
  `first_name` varchar(14) COLLATE utf8_unicode_ci NOT NULL,
  `last_name` varchar(16) COLLATE utf8_unicode_ci NOT NULL,
  `gender` enum('M','F') COLLATE utf8_unicode_ci NOT NULL,
  `hire_date` date NOT NULL,
  PRIMARY KEY (`emp_no`),
  KEY `idx_birth_date` (`birth_date`),
  KEY `idx_birth_date_gender` (`birth_date`,`gender`)
) ENGINE=InnoDB DEFAULT CHARSET=utf8 COLLATE=utf8_unicode_ci
1 row in set (0.00 sec)
```

可以看到 KEY 'idx_birth_date' ('birth_date')和 KEY 'idx_birth_date_gender' ('birth_date', 'gender')功能重复了，我们再来看看 schema_redundant_indexes 表的信息：

```
mysql>select * from schema_redundant_indexes limit 2\G
*************************** 1. row ***************************
              table_schema: employees
                table_name: employees
       redundant_index_name: idx_birth_date
    redundant_index_columns: birth_date
 redundant_index_non_unique: 1
         dominant_index_name: idx_birth_date_gender
      dominant_index_columns: birth_date,gender
   dominant_index_non_unique: 1
              subpart_exists: 0
              sql_drop_index: ALTER TABLE `employees`.`employees` DROP INDEX `idx_birth_date`
2 rows in set (0.01 sec)
```

MySQL 建议删除 idx_birth_date 索引，并在 sql_drop_index 列给出删除索引的完整命令。

查询 schema_table_statistics 表，得到实例启动以来，每个表的详细使用情况。

```
mysql>select * from schema_table_statistics order by rows_fetched + rows_inserted+ update_latency+rows_deleted desc limit 1\G
*************************** 1. row ***************************
     table_schema: employees
       table_name: salaries
    total_latency: 5.02 m
     rows_fetched: 145870818
    fetch_latency: 5.02 m
    rows_inserted: 0
   insert_latency: 0 ps
     rows_updated: 4
   update_latency: 25.30 ms
     rows_deleted: 4047
   delete_latency: 69.57 ms
  io_read_requests: 39785
          io_read: 4.10 MiB
   io_read_latency: 97.74 ms
 io_write_requests: 9352
         io_write: 145.75 MiB
  io_write_latency: 141.26 ms
  io_misc_requests: 34356
   io_misc_latency: 213.81 ms
1 row in set (0.05 sec)
```

按照表增删改查的总行数进行倒序排序，找到写操作影响行数最多的表。日常优化过程中，通过调整影响行数和执行时间的组合规则，可以找出符合规则的热点表，然后进行优化，提高数据库实例的整体性能。

5. SQL 语句相关

按照多个维度分析 SQL 执行情况。分析维护有全表扫描次数、是否经历排序、是否创建临时表等。

例如，statement_analysis 表：

```
mysql>select * from statement_analysis where rows_examined > 0 and rows_affected > 0 order by exec_count desc limit 1\G
*************************** 1. row ***************************
            query: UPDATE `percona` . `checksums` ... AND `tbl` = ? AND `chunk` = ?
               db: employees
        full_scan:
       exec_count: 2034
        err_count: 0
       warn_count: 0
    total_latency: 7.12 s
      max_latency: 448.02 ms
      avg_latency: 3.50 ms
     lock_latency: 235.85 ms
```

```
          rows_sent: 0
      rows_sent_avg: 0
      rows_examined: 2034
  rows_examined_avg: 1
      rows_affected: 2034
  rows_affected_avg: 1
         tmp_tables: 0
    tmp_disk_tables: 0
        rows_sorted: 0
  sort_merge_passes: 0
             digest: 42e6500228e8aab2ecfe255339c0fd63
         first_seen: 2018-08-07 21:29:48
          last_seen: 2018-08-20 22:53:15
1 row in set (0.02 sec)
```

该类对象比较重要的字段如下。

- query：替换 SQL 语句变量并美化。
- db：应用执行 SQL 时连接的 db 名。
- exec_count：SQL 执行总次数。
- rows_examined：SQL 执行查询的总记录数。
- rows_affected：SQL 执行影响的总记录数。

通过上面的信息，得到 SQL 的执行次数、影响行数、创建临时表次数和排序记录数。如果有异常的指标，则可以进一步分析和优化。

20.3 小结

PS 库从多个维度收集数据库实例运行的性能信息，SYS 库通过对 PS 库和 IS 库的封装，让我们能更方便查询这些性能信息。合理使用 PS 库和 SYS 库的信息，将为我们的调优之路带来很大的便利。

第 21 章 故障诊断

MySQL 的故障诊断，需要对整个系统的架构、硬件、软件都有比较深入的了解，由于数据库服务的特点，一旦出现问题，可能影响整个系统的可用性，需要在最短的时间内解决。

在这一章，笔者会结合自身的经验，对处理常见故障的一般性原则、流程和方法做一下介绍，并列举了两个实际处理故障的例子，希望能够为读者提供一些在处理故障时候的思路。当然，实际工作中遇到的故障多种多样，书中不可能完全覆盖，需要读者在解决问题的过程中，不断积累经验。

21.1 故障诊断和处理的原则

（1）沟通第一。

数据库只是整个应用系统的一个组成部分，任何故障都不会仅限于数据库层面，数据库出了问题，也可能会导致其他系统组件出现问题，同时，数据库的故障本身也可能是由系统的其他部分带来，所以在数据库出现故障时候，务必和运维、开发、产品等其他团队保持高效沟通。

沟通的内容主要包括以下几个方面。

- 及时和团队内其他 DBA 沟通情况，寻求信息和协助。其他人可能之前处理过类似故障，也可能知道故障的原因，至少能帮忙一起想办法尽快解决问题。多人协作过程中，为了提高效率，最好对人员做合理分工，主 DBA 作为核心处理者，可以给其他 DBA 分配任务，比如有的负责和开发沟通、有的负责查各类日志、有的负责查官方 BUG，重大故障还要及时通知 DBA 经理，由他来和其他团队的负责人以及上级领导沟通，以协调所需资源。

- 告知其他团队当前故障的情况、具体故障报错等，方便大家共同判断可能的原因以及可能造成的结果。例如，如果负责用户登录认证的数据库出现故障，那么及时通知其他团队，其他团队一方面可以帮忙判断当前对于数据库的访问量、访问方式是否有异常，也可以更直接地判断当前数据库故障到底会影响用户的哪些访问行为。

- 获取其他团队的信息和帮助，结合用户反馈、故障现象、应用层报错等，通常可以更加容易地判断故障原因。例如，如果应用层报出的是数据库连接错误，那么就需要重点从数据库连接的相关设置和 session 的状态上做排查。此外，完成故障定位后，在解决故障的过程中，也能从其他团队那里得到支持，例如，开发和运维团队可以帮忙暂时屏蔽部分非关键服务，降低数据库的压力；产品和客服团队可以及时通知用户可能存在的问题，引导用户的访问行为等。这些都可以在一定程度上帮助 DBA 尽快解决问题，并为 DBA 争取到处理故障的宝贵时间。

● 沟通故障处理方案和进度，确定处理方案之后要及时告知其他团队，并对可能的影响、影响的范围和时间等进行说明。有的方案需要由相关团队的人员共同确认后才可以执行。例如：需要重启数据库，那么重启的时间是否可以接受；如果可能导致短期或长期的数据丢失，是否可以接受等，这些都不能只靠 DBA 单独做出判断。此外，如果故障不能在短时间内排除，解决的过程中也要对处理进度保持持续沟通。

● 沟通故障处理结果和预防措施，故障处理完成之后，对于故障处理的结果和预防措施要做充分的沟通，例如故障是彻底解决还是临时解决，数据是否有损失等都需要知会所有相关人员。此外，为了避免此类故障所要采取的改进的措施，往往也需要其他团队配合完成，例如逻辑调整、代码优化、网络优化、集群扩容、硬件升级等。

总之，DBA 在遇到故障时，一定不要忘了沟通的重要性，即使时间紧迫，简要的沟通往往也能带来事半功倍的效果。从长远来看，也有利于培养和其他人、其他团队之间的合作和信任关系。

（2）关注人为故障。

在作者处理过的故障中，人为故障占了不小的比例。在手工运维的时代，人为故障所占的比例更高，所以在数据库出现故障时，要意识到这很有可能是由于当前操作所导致的。要通过及时沟通并查看操作系统和数据库历史记录，确认自己和其他 DBA 的操作是否有误、要和其他团队沟通是否有特殊操作、要去检查系统和数据库的参数是否符合预期。"某某系统、数据库有什么操作么"这句话永远是要第一时间问出来的。

当然，解决人为故障最好的方法还是将数据库运维自动化、标准化、规范化。随着自动化运维系统越来越广泛的应用，人为操作造成的故障已经逐渐减少，然而再自动化的系统也不可能覆盖到 100%的操作，自动化的系统也一定会存在 BUG，又由 DBA 在自动化运维系统中的操作触发。总之，人为故障永远是不可忽视的一个重要问题来源。

（3）快速解决问题、恢复服务优先。

在处理故障的时候，要明确的一个思路是要优先恢复服务，确保服务的最大可用性，其他的不一定要优先考虑。例如：是否完全确定了故障原因，解决方案是否完美、简洁，故障责任谁来承担等。

（4）三思而后行，不做破坏性操作。

有些故障处理的方式，有可能对数据库造成难以恢复的影响，例如删除数据库中的某个文件、truncate 表等，做这些操作之前，务必要谨慎，并尽量做好备份。

而有些故障处理时候的操作，一般只有在出现故障时才会去做。可能由于对这些操作本身不熟悉带来额外的问题。因此，处理故障时候，不管是通过自己分析、他人建议，还是网上搜索得到的操作命令，一定要认真考虑一下这个命令可能带来的后果，避免对系统带来二次伤害。

（5）服务分级。

平时应当对服务、应用、DB 做好分级，一旦出现大面积故障，可以按照服务的优先级来恢复核心业务。

21.2 故障处理一般流程

本节按照发现→定位→解决的顺序，对故障处理的一般流程和关键点做一下简单介绍。

21.2.1 故障发现

发现依赖于监控，对于数据库服务，一定要构建一套完整的监控系统，在出现问题的时候，能够通过邮件、短信、电话等形式将报警信息发送给 DBA 和其他相关人员，比故障更可怕的是，出现了故障还没人知道。数据库监控一般应该包括以下几个指标。

1. 操作系统层面的指标

（1）负载。

负载是衡量一个服务器整体压力最直观的指标，代表平均有多少进程在等待被 CPU 调度，可以通过 w、uptime、top 等命令来获取。

```
[root@hz_10_120_240_251 ~]# uptime
 10:58:23 up 685 days, 23:37,  1 user,  load average: 0.06, 0.12, 0.15
```

load average 后面的 3 个数值，分别代表 1min 平均负载、5min 平均负载、15min 平均负载。假设服务器的 CPU 有 n 个核，那么所有 CPU 都在满负荷运转时，负载就是 n。一般来说，健康的系统，负载应该保持在 n×0.7 以下，负载超过 n 之后就意味着系统开始拥塞。

负载的报警阈值，通常应该根据系统正常状况下的负载来制定。当系统负载显著升高时，即使还未发生拥塞，也应该及时排查原因，防患于未然。

（2）CPU 使用率。

CPU 使用率可以通过 top 或者 sar 等命令获取：

```
[root@hz_10_120_240_251 ~]# sar -u -P ALL -C 1 1
Linux 2.6.32-504.el6.x86_64 (hz_10_120_240_251)    11/12/18    _x86_64_    (8 CPU)

15:35:15        CPU     %user     %nice   %system   %iowait    %steal     %idle
15:35:16        all      5.22      0.00      3.05      0.89      0.00     90.84
15:35:16          0      6.38      0.00      4.26      6.38      0.00     82.98
15:35:16          1      5.10      0.00      5.10      0.00      0.00     89.80
15:35:16          2      8.08      0.00      2.02      0.00      0.00     89.90
15:35:16          3      8.16      0.00      2.04      0.00      0.00     89.80
15:35:16          4      4.08      0.00      2.04      0.00      0.00     93.88
15:35:16          5      5.00      0.00      2.00      0.00      0.00     93.00
15:35:16          6      4.95      0.00      1.98      0.00      0.00     93.07
15:35:16          7      3.06      0.00      2.04      0.00      0.00     94.90
```

其中 sar 后的参数含义分别如下。

- -u：统计 CPU 数据。
- -P：统计每个核的信息。
- ALL：展示 CPU 的所有指标。
- -C：显示注释信息。
- 1 1：每一秒做 1 次采集，总共采集 1 次。

结果中%idle 表示 CPU 的空闲率，100%减去%idle 就是 CPU 的使用率。如果服务器有多个 CPU，那么会分别计算每个 CPU 的使用率和所有 CPU 的平均使用率。对于 CPU 使用率的监控，一般优先考虑监控全部 CPU 的平均使用率即可。一旦发现全部 CPU 的平均使用率过高，再进一步分析到底是消耗在 user、system 还是 iowait 上。

（3）磁盘空间。

磁盘空间可以通过 df –h 命令来获取，-h 表示结果用最佳可读方式展现。磁盘空间满会直接导致数据库服务不可用，如果是根目录满的话，还会导致服务器登录困难，进一步影响故障处理速度。因此，建议对磁盘的所有分区都进行监控。监控的目标除了给 DBA 留下足够的时间来完成清理或者扩容操作之外，最好还能够对磁盘空间的异常变动进行提醒。

（4）IO 使用率。

作为 IO 密集型服务，IO 使用率是数据库监控中的重要指标，IO 使用率可以通过 iostat 命令来获取：

```
[root@hz_10_120_240_251 ~]# iostat -xd 2 2
Linux 2.6.32-504.el6.x86_64 (hz_10_120_240_251)      11/12/18      _x86_64_       (8 CPU)

Device:          rrqm/s    wrqm/s      r/s      w/s    rsec/s    wsec/s  avgrq-sz  avgqu-sz    await   r_await  w_await  svctm   %util
    sda            0.09    186.24     3.56    35.43   131.55   1773.38     48.85      0.03     0.86     5.11     0.44    0.48    1.87
    sdb            0.08     66.01    11.31   218.52  6150.59   6753.95     56.15      0.02     0.07     2.82     0.26    0.29    6.68
    sdc            0.01     72.73     0.58    53.09    22.07   1137.95     21.62      0.04     0.67     4.01     0.63    0.23    1.26

Device:          rrqm/s    wrqm/s      r/s      w/s    rsec/s    wsec/s  avgrq-sz  avgqu-sz    await   r_await  w_await  svctm   %util
    sda            0.00      4.50     0.00     3.00     0.00     60.00     20.00      0.00     0.00     0.00     0.00    0.00    0.00
    sdb            0.00     56.00     0.00     8.00     0.00    512.00     64.00      0.00     0.25     0.00     0.25    0.19    0.15
    sdc            0.00    142.50     0.00   394.00     0.00   4292.00     10.89      0.02     0.07     0.00     0.07    0.05    1.80
```

其中，-xd 表示显示扩展信息和设备的利用率信息（%util），"2 2"表示每 2s 统计一次，总共统计两次。这里需要优先关注的是最后一列%util，当这一列接近 100%时，代表磁盘的 IO 能力已经达到极限，需要 DBA 从 IO 层面关注故障的原因。

（5）SWAP 使用情况。

当系统内存不足的时候，操作系统会从 SWAP 分区中分配一部分空间来临时保存一部分原本计划保存在内存中的数据。由于 SWAP 分区实际上是保存在磁盘上，而磁盘的读写速度远远低于内存，因此，一旦发生 SWAP，往往伴随着系统整体性能的大幅下降。在 Linux 系统中，可以使用 free 或者 top 等命令来检查 SWAP 的使用情况：

```
[root@hz_10_120_240_251 ~]# free -g
                total       used       free     shared    buffers     cached
Mem:              126        124          1          0          0         27
-/+ buffers/cache:             96         29
Swap:              31          0         31
```

其中，-g 参数表示结果以 GB 为单位。由于数据库的性能对 IO 依赖比较大，理想的状况下，应该通过合理的设置内存参数，避免系统使用 SWAP 分区。对于 SWAP 的监控，也建议设置一个比较小的阈值。

2. **数据库层面的指标**

（1）数据库存活。

存活状态是数据库的基本监控。由于 MySQL 的单进程架构，相对 Oracle 等多进程架构的数据库来说，更容易因为某些异常或 BUG 导致数据库实例崩溃。对于数据库存活的监控，一般通过尝试建立数据库连接来检查即可。

（2）连接数。

如果数据库出现连接数异常增长，监控需要能够及时发现，因为一旦连接数超过数据库参数中设定的最大连接，就会出现无法建立连接，影响服务的情况。此外，由于每个连接都要消耗一定的内存，因此数据库总共的连接数也应该控制在一个合理的范围内。对于连接数的监控，一般来说可以参考正常状况下的连接数来确定。数据库连接数可以通过 information_schema.processlist 表来确定：

```
mysql> select count(1) from information_schema.processlist;
+----------+
| count(1) |
+----------+
|      414 |
+----------+
1 row in set (0.00 sec)
```

（3）慢 SQL。

慢 SQL 是导致数据库问题的最常见因素之一，因此需要及时发现数据库中执行效率偏低的 SQL。监控慢 SQL 的方法有很多种，常用的一种方法是开启慢查询日志，然后通过 pt-digest 工具来定期分析日志，这样可以获得慢 SQL 的详细信息。也可以简单地通过查询 information_schema.processlist 来监控当前是否有慢 SQL 的存在：

```
mysql> select * from information_schema.processlist where user not in ('system user', 'event_scheduler') and Command <> 'Sleep' and Time >10;
Empty set (0.00 sec)
```

（4）主从延迟。

从库作为主库的备份，如果延迟过大，一旦主库出现问题，那么从库将会无法及时顶替主库提供服务。此外，很多情况下从库会分担一部分主库的查询请求，如果延迟过大，那么发送到从库的查询请求也无法提供正确的结果。可以通过 show slave status 命令来检查主从的延迟情况。

以上提到的指标都是对于数据库最为基础的一部分监控指标，在实际工作中，需要根据实际需求，定制个性化的监控指标。

21.2.2　故障定位

DBA 收到报警之后，接下来要做的就是尽快定位故障的原因。

（1）检查当前和最近的操作。

如果 DBA、开发正在做某项操作，那么由该操作导致的故障可能性比较大，所以要首先确认当前所有相关操作。常见的导致数据库故障操作有以下几种。

- 程序发布：每次应用程序发布，都有可能引发数据库问题，例如新的业务逻辑带来的 SQL 变动、新上线的促销活动带来的访问量增加、发布的时候批量创建连接带来的连接冲击等，都是常见的造成数据库报警的原因。

- 在线表变更：造成报警的常见原因包括表变更造成的锁表、复制表过程中带来的 IO 压力等。

- 在线数据修改：造成报警的常见原因是大批量数据修改带来的压力，或者执行 SQL 不合理导致的锁问题。

- 后台任务、数据统计：后台任务和数据统计往往需要查询或者更新大量数据，导致操作系统 CPU 或 IO 使用率快速升高而报警。
- 数据库参数调整：在线的数据库参数调整也有可能造成意外的影响。
- 其他误操作：各种人为误操作也是报警的常见原因。

（2）操作系统层检查。

接下来可以对操作系统和数据库的各项指标进行检查，进一步帮助自己定位故障。在操作系统层，一般需要先从下面几个角度进行排查。

- 系统进程：通过进程占用的 CPU 和内存情况，判断当前系统压力是否主要由 MySQL 进程所致。检查是否存在其他的异常服务、是否有定时任务在运行等。
- CPU：检查当前 CPU 的使用情况，并通过 CPU 主要等待时间消耗在哪里，来判断到底是 IO 能力不足，还是运算能力不足。
- 内存、SWAP：检查内存和 SWAP 的使用情况，来判断故障是否是由内存分配不当造成的。
- IO：检查磁盘的 IO 吞吐量和 IO 使用率是否正常，可以辅助判断当前系统是否存在 IO 能力不足的问题，以及问题到底是由于 IO 吞吐量下降造成的，还是由于 IO 量过高造成的。
- 系统日志：从 dmesg、messages、secure 等系统日志中，可以辅助判断当前系统是否存在硬件故障、了解故障前系统都做过哪些操作等。

（3）数据库层检查。

在数据库层，DBA 应该重点关注下面这些指标。

- 连接：通过检查数据库当前的 session，以及活跃状态的 session，DBA 可以了解到当前系统的连接数、连接的使用状况是否正常。
- 慢查询：通过慢查询分析，可以最直观地了解到，是否有大量消耗系统资源的 SQL，从而判断问题是否由 SQL 导致，由哪些 SQL 导致。
- 锁等待：锁等待也是常见的导致连接问题或性能问题的原因，需要通过检查数据库中的锁持有和等待的情况、锁持续的时间、被阻塞的连接数量等指标来判断问题。
- QPS：通过检查数据库当前的 QPS，可以帮助判断数据库的访问量是否正常，从而判断是否由于业务量波动或者缓存失效等原因导致数据库的请求量异常。
- 错误日志：部分实例错误的信息，需要通过 MySQL 错误日志来判断，虽然当前版本中 MySQL 错误日志提供的信息还不够充分。未来随着 MySQL 版本的更新，错误日志也会越来越完善。

对于常见的故障，这里做了一个简单的图示（见图 21-1），当遇到故障时可以作为参考。

21.2.3 故障解决

通过定位，已经可以比较明确的确定导致故障的原因，针对不同的原因，需要用不同的方法来解决。下面我们来分析一些常见故障的解决方法。

1. 慢 SQL

通过分析，如果最终确定故障是由于一个或多个 SQL 执行慢导致的，那解决故障就是优化 SQL 的执行效率。常见的导致慢 SQL 的原因及解决方案有以下几种。

21.2 故障处理一般流程

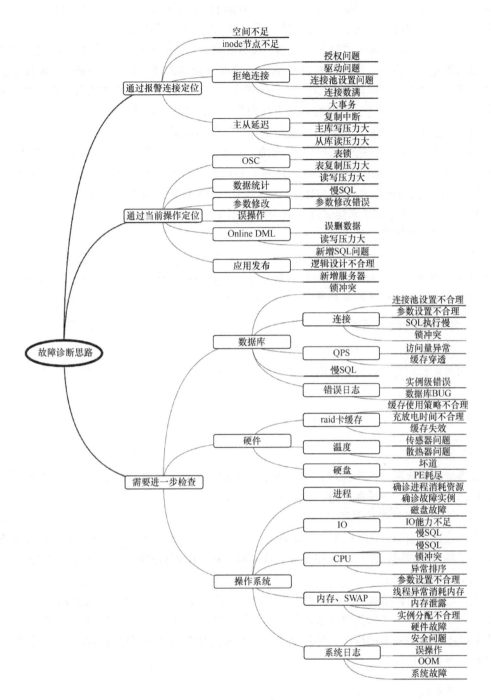

图 21-1 常见的故障图示

（1）选择条件上没有索引或者索引效率低。

如果表的数据量不太大，而 SQL 上也存在着选择度比较好的条件时，可以选择直接为 SQL 创建一个索引来解决故障。而如果表的数据量非常大，创建索引需要花的时间和资源也很大，这时可以先和其他团队沟通一下 SQL 涉及的具体业务逻辑，对于非核心业务，或许可以临时停掉这个 SQL，这样可以快速恢复核心业务，并给 DBA 留下创建索引的时间窗口。对于核心业务的 SQL，如果是由于 SQL 变更导致，可以考虑回滚应用程序版本。

此外，如果涉及的表上数据只有新增，没有修改，并且对于查询结果的准确性要求不高，可以通过创建一个空表，并和大表通过 rename 临时互换的方式，来加快服务恢复的速度。接下来，再跟产品和开发团队沟通 SQL 的具体需求，讨论是否可以通过其他高效率的方式来查询数据库，比如增加其他有索引的列作为查询条件等，从而更好地解决问题。

（2）有索引，但没有用到索引，或者选择了错误的索引。

这种情况往往是由于表的统计信息不准确，或者 SQL 过于复杂导致，当然也不排除 MySQL 的 SQL 优化器有时候也会犯错误。这时还是需要优先恢复核心服务，然后通过收集统计信息，或者使用 SQL 改写插件来强制 SQL 走正确索引。如果故障影响可控，也可以选择等待应用程序修改 SQL 写法等方式来尝试让 SQL 的执行计划符合预期。

（3）过滤条件不强，结果集太大。

这常常是一些统计分析类 SQL 造成的，从这些 SQL 本身来看可能已经无法优化。对于这种 SQL，首先应该停掉相应的统计分析任务，不让其影响线上服务，然后可以考虑将统计分析任务调整到从库或者调整到非业务高峰时段来执行。

2. SQL 执行频率高

（1）恶意攻击。

大多数数据库都是存放在内网环境，因此直接受到恶意攻击的情况不多，但是针对 Web 和应用服务器的攻击，还是会通过 SQL 请求的方式传递到数据库。应对这种情况仅靠数据库层不容易处理，可以及时发现并反馈给应用开发和运维团队来解决，例如自动封禁掉异常访问的 IP 等。

（2）缓存失效。

如果应用层缓存设计不合理，导致有过多的请求穿透缓存到达数据库，那么也需要及时和应用团队沟通，改进缓存策略、加大缓存容量。

（3）应用实现逻辑不合理。

可以通过回滚等方式临时解决。此外，作为 DBA，应该对自己负责业务线的关键业务逻辑有所了解，这样可以配合产品和开发团队一起，对应用逻辑做出合理的设计和调整。

（4）业务量突增。

这种情况相对比较棘手，可以首先考虑通过对部分重点 SQL 进行优化的方式，从一定程度上缓解压力，争取能够扛过业务峰值。之后再考虑对整个业务逻辑进行优化、数据库拆分或者对数据库集群资源进行扩容等长期方案。在紧急情况下，也可以放弃部分边缘业务，来保障核心业务的运行。

3. 锁冲突

对于锁冲突带来的连接数过高的问题，一般来说可以通过临时加大最大连接数，或者手动杀掉一部分连接的方法，避免连接数达到限制，无法创建新连接的情况发生。同时应该联系开发和运维，从应用层限制锁冲突严重的接口的并发量。待服务稳定后，再根据导致锁冲突的不同原因，调整程序逻辑。下面分别是两种导致锁冲突的常见原因的解决思路。

- 大事务：通过优化逻辑，尽可能地拆分事务，调整事务内的 SQL 顺序，减少持有锁的时间。
- 热点问题：可以通过分散热点、减小事务、串行化等方式来缓解热点带来的锁冲突。

4．硬件问题

硬件问题涉及的可能性比较广，对于数据库这种 IO 密集型服务来说，更多地需要从 IO 层面来关注可能的硬件问题。
- RAID 卡缓存问题：RAID 卡缓存设置不合理，或者电池充放电，都有可能导致 IO 性能的波动，处理方法可以参考第 18 章中相关的章节。
- 硬盘损坏：硬盘损坏也是常见的造成 IO 性能下降的原因，需要及时更换损坏的硬盘，并且在 RAID 重建期间密切关注服务器的 IO 压力。

5．参数配置不合理

如果故障是由于操作系统或数据库参数设置不合理所导致，那么在尝试调整参数来解决故障的时候，一定要很清楚自己所做的调整可能带来的影响。对于不常操作的参数，建议先查阅文档，再通过搜索引擎了解一下其他人的处理经验，避免因为错误的参数调整，带来更多的问题。

21.3 典型故障案例

接下来通过两个案例来完整地介绍一下故障诊断和处理的流程。

21.3.1 案例 1

1．场景

DBA 收到短信报警，数据库负载超过阈值，并持续上升中。

收到报警后，需要尽快登录服务器，并同时通知系统的运维和开发人员，告知数据库负载高，可能影响服务，并询问当前系统正在进行的操作。

2．诊断

（1）执行 TOP 命令，确认负载过高并持续上升中：

```
[root@hz_10_120_240_251 ~]# top
top - 11:05:57 up 702 days, 23:44, 11 users,  load average: 12.79, 5.61, 2.83
Tasks: 385 total,   2 running, 383 sleeping,   0 stopped,   0 zombie
Cpu(s): 74.6%us,  2.0%sy,  0.0%ni, 11.9%id, 10.2%wa,  0.0%hi,  0.3%si,  0.0%st
Mem:  132146300k total, 130865208k used,  1281092k free,   369816k buffers
Swap: 33554428k total,   892784k used, 32661644k free, 27139052k cached

  PID USER      PR  NI  VIRT  RES  SHR S %CPU %MEM    TIME+  COMMAND
30750 mysql330  20   0 6234m 1.7g 7296 S 673.2  1.3 447:00.39 mysqld
32054 cetus_34  20   0  240m 4788 1692 S  0.7  0.0  38:18.95 cetus
  977 root      20   0     0    0    0 S  0.3  0.0 473:06.08 jbd2/sdc1-8
 1280 root      20   0     0    0    0 S  0.3  0.0 1596:46 kondemand/0
 1283 root      20   0     0    0    0 S  0.3  0.0 1045:31 kondemand/3
```

MySQL 进程的 CPU 使用率为 673%，显著偏高；内存和 swap 正常，CPU 使用率的几个指标中，us 即用户进程消耗的时间为 74.6%，wa 即 io 等待时间为 10%，这两个指标明显高于正常值。从这些指标可以做出初步的推测，有 MySQL 进程大量消耗了 IO 资源，导致 MySQL

进程中请求积压,带来负载上升。

(2)执行 iostat -xd 3 3,观察 IO 资源的使用情况:

```
 Device:         rrqm/s    wrqm/s     r/s     w/s    rsec/s    wsec/s avgrq-sz avgqu-sz   await r_await w_await  svctm  %util
...
    sdc           0.00     43.00  1408.67  197.67  75077.33   1925.33    29.26     1.16    0.72    0.81    0.07   0.31  79.83
```

观察到%util 即 IO 使用率为 79.8%,每秒 IO 量非常大,并且绝大部分为读操作,可以确定有请求大量消耗 IO 资源,并判断应该是由读操作导致。

(3)MySQL 进程的大量读 IO,很可能是由于 SQL 导致,因此接下来可以对 MySQL 进行 SQL 分析,SQL 分析可以借助于自动化诊断工具来快速完成,有兴趣的同学可以参考第 23 章的内容,这里先假设没有自动化分析工具,那么我们应如何来定位问题 SQL 呢?

在 MySQL 中执行:

```
select * from information_schema.processlist where state <>'Sleep' \G;
......
*************************** 14. row ***************************
     ID: 6234
   USER: emp
   HOST: 192.168.1.101
     DB: employees
COMMAND: Query
   TIME: 207
  STATE: Sending data
   INFO: select * from tb_order where order_status = 0
*************************** 15. row ***************************
     ID: 6229
   USER: emp
   HOST: 192.168.1.102
     DB: employees
COMMAND: Query
   TIME: 197
  STATE: Sending data
   INFO: select * from tb_order where order_status = 0
*************************** 16. row ***************************
     ID: 6232
   USER: emp
   HOST: 192.168.1.103
     DB: employees
COMMAND: Query
   TIME: 187
  STATE: Sending data
   INFO: select * from tb_order where order_status = 0
```

可以看到,存在大量 SQL 语句 "select * from tb_order where order_status = 0",状态为 sending data,其执行时间很长,200s 左右还未执行完毕,说明这个 SQL 语句的效率很可能有问题。

(4)如果仅仅从 processlist 无法确定是哪个 SQL,那么需要进一步通过分析慢查询日志来确定问题 SQL。建议使用 percona 的 pt-query-digest 工具,安装后执行下面的命令:

```
pt-query-digest --since=15m slow.log > slow_report.log
```

分析完成后,可以通过 slow_report.log 来定位问题:

```
[root@hz_10_120_240_251 data]# more slow_report.log

# 5.6s user time, 280ms system time, 34.43M rss, 218.81M vsz
# Current date: Thu Nov 29 11:30:13 2018
# Hostname: hz_10_120_240_251
# Files: slow.log
# Overall: 363 total, 3 unique, 0.04 QPS, 46.10x concurrency _____
# Time range: 2018-11-29T11:16:11 to 2018-11-29T11:29:57
# Attribute          total     min     max     avg     95%  stddev  median
```

```
# =============    =======  =======  =======  =======  =======  =======  =======
# Exec time        38076s   1ms      125s     105s     124s     12s      102s
# Lock time        56ms     48us     32us     153us    131us    2ms      54us
# Rows sent        202      0        98       0.56     0        6.36     0
# Rows examine     60.71G   98       119.07M  182.92M  182.41M  1.67M    182.41M
# Query size       16.35k   44       69       46.12    44.60    1.58     44.60

# Profile
# Rank Query ID                            Response time        Calls  R/Call    V/M
# ==== =================================   =================    =====  ========  =====
#    1 0x2E5C1AA8E628621B6AF18E03...       38076.0023 98.2%       360  105.7667  0.45 SELECT tb_order
…
# Query 1: 0.04 QPS, 46.10x concurrency, ID 0x2E5C1AA8E628621B6AF18E035F2716DB at byte 176863585
# This item is included in the report because it matches --limit.
# Scores: V/M = 0.05
# Time range: 2018-11-29T11:16:11 to 2018-11-29T11:29:57
# Attribute    pct    total    min     max     avg     95%    stddev  median
# ==========   ===   =======  ======= ======= ======= ======= ======= =======
# Count         99    360
# Exec time     98   38076s   1031s   1259s   1067s   1248s    72s    1028s
# Lock time     41   23ms     48us    327us   63us    125us    26us   54us
# Rows sent      0    0        0       0       0       0        0      0
# Rows examine  99   60.71G   190.7M  190.7M  190.7M  190.7M    0     190.7M
# Query size    98   16.17k   46      46      46      46        0     46
# String:
# Query_time distribution
……
#   100s
#  1000s+  ##############################################################
# Tables
#    SHOW TABLE STATUS LIKE 'tb_order'\G
#    SHOW CREATE TABLE `tb_order`\G
# EXPLAIN /*!50100 PARTITIONS*/
select * from tb_order where order_status = 0\G
```

通过分析上面的报告,也能够看出"select * from tb_order where order_status = 0"这条 SQL 语句消耗了大量的系统资源,是负载升高的主要原因。

(5)确定问题 SQL 语句之后,接下来查看表结构和 SQL 执行计划,执行下面的命令:

```
mysql> show create table tb_order\G
*************************** 1. row ***************************
       Table: tb_order
Create Table: CREATE TABLE `tb_order` (
  `order_id` varchar(60) COLLATE utf8_unicode_ci NOT NULL COMMENT '订单ID',
  `order_time` timestamp(6) COLLATE utf8_unicode_ci NOT NULL COMMENT '订单时间',
  `account_id` varchar(60) COLLATE utf8_unicode_ci DEFAULT NULL COMMENT '账号ID',
  `product_id` varchar(60) COLLATE utf8_unicode_ci DEFAULT NULL COMMENT '商品ID',
  `product_amount` int(11) DEFAULT NULL COMMENT '商品数量',
  `order_amount` decimal(10,2) DEFAULT NULL COMMENT '订单金额',
  `order_status` smallint(6) DEFAULT NULL COMMENT '订单状态',
  PRIMARY KEY (`order_id`)
KEY `idx_order_time` (`order_time`)
) ENGINE=InnoDB DEFAULT CHARSET=utf8 COLLATE=utf8_unicode_ci COMMENT='订单表'
1 row in set (0.00 sec)

explain select * from tb_order where order_status = 0\G
*************************** 1. row ***************************
           id: 1
  select_type: SIMPLE
        table: tb_order
   partitions: NULL
         type: ALL
possible_keys: NULL
          key: NULL
      key_len: NULL
          ref: NULL
         rows: 214729138
     filtered: 10.00
        Extra: Using where
1 row in set, 1 warning (0.00 sec)
```

发现由于表上没有 order_status 的索引，导致 SQL 使用全表扫描，而且扫描行数非常大。

（6）将问题 SQL 和并发团队沟通，了解到这是并发团队新增的 SQL，用途是通过定时任务的方式，每 10s 扫描一次订单表中，超过 1h 仍未支付的订单，然后进行订单关闭操作。

（7）至此，问题确定，是由于新增功能使用的 SQL 产生了对大表 tb_order 的全表扫描，并且执行频率比较高，从而导致数据库服务器的负载升高。

3．故障解决

（1）应急处理。

在处理负载升高的故障时，需要注意，如果负载上升速度很快，可能导致服务器假死，同时，负载高还有可能伴随连接数过高等问题，会直接导致线上服务不可用，如果是这样的情况，应该果断要求开发团队回滚代码，DBA 也需要根据情况不断杀掉问题 SQL，直至数据库恢复稳定，同时尽可能保障服务的可用性。

在本例中，由于负载上升比较缓慢，连接数也比较稳定，因此 DBA 可以有相对充分的时间和开发团队一起沟通解决方案。这里考虑到关闭订单业务的时效性要求并不高，也不会影响核心业务流程，又是通过定时任务调用，因此首先联系开发人员，将这个 SQL 的查询频率降低到 30min 一次。通过这个调整先把系统负载降下来，避免影响到其他重要业务。

（2）接下来针对具体的这个问题，就是要降低查询超时未支付订单这个业务需求所消耗的系统资源。

通过分析这个 SQL，首先的考虑是可以通过在 status 列上创建索引，虽然 status 列本身选择度不高，但是由于存在定时关闭超时未支付订单的逻辑，实际上状态是未支付的订单并不多，status = 0 的选择度是比较好的，因此在 status 列上创建索引可以有效地优化查询效率。

（3）考虑到这个 SQL 实际上要实现的业务逻辑，还可以做进一步优化。

首先，tb_order 表本身比较大，创建索引的代价也比较大，需要由 DBA 选择业务低峰期来操作。其次，这个 SQL 每次查询会查出大量的未支付订单，其中有很多还未超过支付时限，不需要关闭的订单，这部分查询出来的数据就完全没有意义。

考虑上面的两点，可以通过增加 order_time 这个选择条件，一个作用是可以减少每次查询的结果集，另外也可以在 status 索引创建之前，先借用 order_time 上的索引来降低 SQL 的资源消耗。而考虑到关闭超时订单的任务设计为每半分钟执行一次，order_time 的时间选择距离当前时间 1～2 小时的订单就可以既满足需求，又留有一定余量。因此可以建议开发人员将 SQL 语句修改为 "select * from tb_order where order_time > date_add(now(),interval -2 hour) and order_time < date_add(now(),interval -1 hour) and order_status = 0;"。

（4）更进一步，如果 SQL 做了上述调整，并且 DBA 在 status 列上创建好了索引，此时考虑到 status 列的特殊性，列值的分布严重倾斜，SQL 优化器对 status 列的选择度判断可能不准确，再加上存在另一个 order_time 上的索引，容易使执行计划变得不稳定，出现 SQL 错误选择使用 order_time 索引的情况。此外，单独为了一个查询就在这么大一个表上新增一个索引，对于系统资源也有一定浪费。

要解决这两个问题，可以考虑由 DBA 在业务低峰期，先创建 order_time+status 的组合索引，再将 order_time 上的索引删除。

（5）因为涉及索引的调整，DBA 完成上述操作之后，需要对表上的 SQL 进行检查，确保所有 SQL 的执行计划都符合预期。

21.3.2 案例 2

1. 场景

DBA 收到短信报警,数据库连接数超过阈值,并快速上升,很快就达到数据库设置的最大连接数 2000。

收到报警后,需要尽快登录服务器,并同时通知系统的运维和开发人员,告知数据库连接数满,可能影响服务,需要运维和开发人员配合一起确认服务状况和解决问题,此外询问近期系统是否有调整。

2. 应急处理

由于数据库连接数已经达到上限,并且收到开发人员反馈,应用不断出现无法建立连接的报错,部分服务已经受到影响,因此需要首先对故障做应急处理。

首先检查数据库的 CPU、内存、IO 使用状况,发现此时 CPU、内存、IO 的使用率都不高,尤其是内存资源占用并没有显著异常,存在增加最大连接数的条件:

```
top - 17:23:35 up 700 days,  6:02,  3 users,  load average: 1.78, 1.74, 1.27
Tasks: 330 total,   1 running, 329 sleeping,   0 stopped,   0 zombie
Cpu(s):  8.6%us,  1.5%sy,  0.0%ni, 89.1%id,  0.5%wa,  0.0%hi,  0.2%si,  0.0%st
Mem:  132146300k total, 125966652k used,  6179648k free,   446772k buffers
Swap: 33554428k total,   892784k used, 32661644k free, 27002416k cached

  PID USER       PR  NI  VIRT  RES  SHR S %CPU %MEM    TIME+  COMMAND
30750 mysql330   20   0 10.9g 1.3g 7112 S 49.4  1.0 183:23.08 mysqld
  977 root       20   0     0    0    0 S  0.7  0.0 460:50.38 jbd2/sdc1-8
32051 cetus_34   20   0  240m 5116 2008 S  0.7  0.0  32:16.60 cetus
   58 root       20   0     0    0    0 S  0.3  0.0 857:41.16 kblockd/0
 1281 root       20   0     0    0    0 S  0.3  0.0 1089:13   kondemand/1
 1283 root       20   0     0    0    0 S  0.3  0.0 1041:38   kondemand/3
 1287 root       20   0     0    0    0 S  0.3  0.0 649:01.39 kondemand/7
```

因此首先调整最大连接数至 4000,并对连接数继续保持观察,有必要的话,继续增加。

检查数据库中连接的状态,发现大部分连接都是下面的状态:

```
mysql> select * from PROCESSLIST limit 10\G
……
*************************** 3. row ***************************
     ID: 4014
   USER: order_app
   HOST: hz_10_120_240_252
     DB: order
COMMAND: Query
   TIME: 94
  STATE: Sending data
   INFO: select product_price,product_stock from tb_product where product_id = ? for update
*************************** 4. row ***************************
     ID: 4117
   USER: order_app
   HOST: hz_10_120_240_252
     DB: order
COMMAND: Query
   TIME: 157
  STATE: Sending data
   INFO: select product_price,product_stock from tb_product where product_id = ? for update
*************************** 5. row ***************************
     ID: 4108
   USER: order_app
   HOST: hz_10_120_240_252
     DB: order
```

```
 COMMAND: Query
    TIME: 155
   STATE: Sending data
    INFO: select product_price,product_stock from tb_product where product_id = ? for update
……
```

此时开发反馈刚刚上线一期抢购活动，将刚才的 SQL 和开发做了确认，确定是属于新上线抢购活动的新增 SQL。因此判断连接数暴涨和抢购活动新上线的 SQL 有关，要求应用运维马上对活动接口进行限流，并发数初步限制到 100。如果不能马上限制并发数，建议临时下线抢购活动接口。

完成并发限制或下线接口之后，分批重启应用服务器，重建数据库连接。

经过上面的处置，数据库连接数下降至 500 左右，大部分服务已经恢复正常，但是抢购活动的效率仍然很低，开发根据应用日志发现抢购的效率大概在每秒 1 笔左右。

3. 进一步诊断

通过前面的处理过程，我们已经可以判断抢购活动的 SQL 设计可能存在问题，每秒钟 1 笔左右的成交效率肯定无法满足业务的需要，因此首先需要重点了解一下抢购活动的 SQL 设计，开发提供了整个抢购流程的设计如下。

（1）首先，抢购流程涉及的几个表的表结构、数据量、数据分布情况如下。

- 用户账号表：数据量在 10000 左右。

```
mysql> show create table tb_account\G
*************************** 1. row ***************************
       Table: tb_account
Create Table: CREATE TABLE `tb_account` (
  `account_id` varchar(60) COLLATE utf8_unicode_ci NOT NULL COMMENT '账号 ID',
  `user_name` varchar(200) COLLATE utf8_unicode_ci DEFAULT NULL COMMENT '用户名',
  `account_balance` decimal(10,2) DEFAULT NULL COMMENT '账户余额',
  PRIMARY KEY (`account_id`)
) ENGINE=InnoDB DEFAULT CHARSET=utf8 COLLATE=utf8_unicode_ci COMMENT='账户表'
1 row in set (0.00 sec)
```

- 活动表：记录用户报名参与活动的状态。

```
mysql> show create table tb_activity\G
*************************** 1. row ***************************
       Table: tb_activity
Create Table: CREATE TABLE `tb_activity` (
  `account_id` varchar(60) COLLATE utf8_unicode_ci NOT NULL COMMENT '账号 ID',
  `activity_id` varchar(60) COLLATE utf8_unicode_ci NOT NULL COMMENT '活动 ID',
  `status` tinyint(4) DEFAULT NULL COMMENT '活动参与状态',
  PRIMARY KEY (`account_id`,`activity_id`)
) ENGINE=InnoDB DEFAULT CHARSET=utf8 COLLATE=utf8_unicode_ci COMMENT='活动表'
1 row in set (0.00 sec)
```

- 活动商品表：只记录了参加活动的 10 种特价商品的价格、库存等信息。

```
mysql> show create table tb_product\G
*************************** 1. row ***************************
       Table: tb_product
Create Table: CREATE TABLE `tb_product` (
  `product_id` varchar(60) COLLATE utf8_unicode_ci NOT NULL COMMENT '商品 ID',
  `product_name` varchar(200) COLLATE utf8_unicode_ci DEFAULT NULL COMMENT '商品名',
  `product_price` decimal(10,2) DEFAULT NULL COMMENT '商品价格',
  `product_stock` int(11) DEFAULT NULL COMMENT '商品库存'
) ENGINE=InnoDB DEFAULT CHARSET=utf8 COLLATE=utf8_unicode_ci COMMENT='商品表'
1 row in set (0.00 sec)
```

- 订单表：记录了所有用户、所有产品的订单信息。

```
mysql> show create table tb_order\G
*************************** 1. row ***************************
```

```
        Table: tb_order
Create Table: CREATE TABLE `tb_order` (
  `order_id` varchar(60) COLLATE utf8_unicode_ci NOT NULL COMMENT '订单ID',
  `account_id` varchar(60) COLLATE utf8_unicode_ci DEFAULT NULL COMMENT '账号ID',
  `product_id` varchar(60) COLLATE utf8_unicode_ci DEFAULT NULL COMMENT '商品ID',
  `product_amount` int(11) DEFAULT NULL COMMENT '商品数量',
  `order_amount` decimal(10,2) DEFAULT NULL COMMENT '订单金额',
  `order_status` smallint(6) DEFAULT NULL COMMENT '订单状态',
  PRIMARY KEY (`order_id`)
) ENGINE=InnoDB DEFAULT CHARSET=utf8 COLLATE=utf8_unicode_ci COMMENT='订单表'
1 row in set (0.00 sec)
```

（2）用户参与抢购过程的 SQL 代码如下。

- 开启事务：

```
start transaction;
```

- 查询商品库存和价格：

```
select product_price,product_stock from tb_product where product_id = ? for update;
```

- 查询用户账户余额：

```
select account_balance from tb_account where account_id = ? for update;
```

- 查询用户是否领取了活动资格：

```
select status from tb_activity where account_id = ? and activity_id = ?;
```

- 查询用户是否有购买记录：

```
select count(1) from tb_order where product_id = ? and account_id = ?;
```

如果满足库存>=1，账户余额>=价格，用户领取了活动资格，用户购买数量没有超过限制这 4 个条件，那么接下来：

- 更新账户余额：

```
update tb_account set account_balance = ? where account_id = ?;
```

- 更新库存信息：

```
update tb_product set product_stock = ? where product_id = ?;
```

- 写入订单信息：

```
insert into tb_order values (?, ?, ?, ?, ?, ?);
```

- 提交：

```
commit;
```

（3）整个过程看起来很合理，但是其实隐藏着很多问题。

- 由于抢购商品表 tb_product 只有 10 行数据，因此在设计表结构的时候，并未给这个表创建任何索引。而实际上，由于抢购活动的特殊性，这个数据量最小的表，正是热点最为集中的表。如果表上有 product_id 字段的索引，那么每次只有一行数据被加锁，而没有给表创建索引，导致每次都会给所有记录加锁，锁竞争也增加了 10 倍。

- 上面我们已经提到过，在这个抢购业务中，记录库存的 tb_product 表，将会成为锁竞争最为严重的表，但是整个事务的 SQL 设计中，却最先给这个表加锁，导致在整个事务执行的过程中，这个锁竞争最严重的表，被加锁的时间是最长的。

- 事务中存在一些执行效率比较低的 SQL，如 "select count(1) from tb_order where product_id = ? and account_id = ?;"。如果用户在订单表中的购买记录比较多，那么这个 SQL 的执行时间也会比较长，整个事务持有锁的时间也会更长。

- 事务没有做到最小化，例如，"select status from tb_activity where account_id = ? and activity_id = ?;"。用户是否领取了活动资格,不会在事务中更新,也没有必要放在事务中查询。事务中的 SQL 越多,加锁的时间也就越长。

4.故障处理

根据上面的分析,可以通过以下几点来解决抢购活动中遇到的问题。

(1)在 tb_product 表的 product_id 列上增加主键,可以有效地降低锁粒度。如果希望进一步分散热点,可以将商品做拆分,例如,将商品 A 拆分为商品 A1 和商品 A2 两行。

(2)调整事务逻辑,将无关 SQL 从事务中拆分出去,将热点表加锁的顺序尽可能往后放,调整后的 SQL 逻辑如下:

```
select status from tb_activity where account_id = ? and activity_id = ?;
start transaction;
select account_balance from tb_account where account_id = ? for update;
select count(1) from tb_order where product_id = ? and account_id = ?;
select product_price,product_stock from tb_product where product_id = ? for update;
```

如果满足购买条件:

```
update tb_account set account_balance = ? where account_id = ?;
update tb_product set product_stock = ? where product_id = ?;
insert into tb_order values (?, ?, ?, ?, ?, ?);
commit;
```

(3)优化效率偏低的 SQL,例如这里通过对 tb_order 表做 count 的方式来判断用户购买数量,可以调整为通过对 tb_activity 表中用户状态的判断来完成。

(4)在应用层,增加对抢购每种商品的并发量的限制,使同时竞争某一个商品的锁数量不高于 3 个,对于数据库来说,也能够起到一定的保护作用。

(5)如果经过上面一系列的调整,仍然无法满足业务需求,需要进一步增加并发量,可以考虑将此处抢购的逻辑做进一步调整,不再让用户来竞争库存上的锁。在用户下单时,可以只是 insert 一条用户下单记录,再由系统中订单处理服务来做异步处理。处理完成后,将订单处理的结果反馈给用户,这样用户的下单流程就不在需要锁表,可以最大限度地保障并发性,但是相应的用户只能异步接收到订单处理的结果,对用户体验可能有一定的影响。

21.4 小结

本章总结了在故障诊断和处理过程中的一些常见思路,并用两个具体的例子做了说明,在实际工作中,可能遇到的故障多种多样,一方面需要不断积累经验来提高故障诊断和处理的能力;另一方面,也要积极尝试借助一些自动化的工具来完成。关于自动化工具的详细介绍,可以参考第 23 章中的相应内容。

第四部分
管理维护篇

第 22 章 MySQL 高级安装和升级

对于 Linux/UNIX 平台来说，用户还可以考虑采用另外两种安装包来进行安装，一种是二进制包（Binary Package），另一种是源码包（Source Package）。这两种包都可以从 MySQL 的官方网站下载（https://dev.mysql.com/downloads/mysql/），因为针对不同的硬件和操作系统安装包有所不同，所以读者在下载时请根据实际安装环境选择相应的包。

这两种安装包相对于 RPM 包的最大优点是安装配置更灵活，更适合于中高级用户，因此称为"高级"安装。本章将主要对这两种安装包的使用进行介绍。

22.1 Linux/UNIX 平台下的安装

本节主要介绍 Linux/UNIX 平台下不同安装包之间的区别，并重点介绍二进制包和源码包的安装步骤以及参数文件的设置方法。

22.1.1 安装包比较

Linux 的安装包分为 RPM 包、二进制包和源码包。表 22-1 简单描述了 3 种安装包之间的主要差异，其中文件布局指的是 MySQL 安装完毕后生成的各个目录和用途。

表 22-1　　　　　　　　　　Linux 平台下的 3 种安装包比较

	RPM	二进制	源码
优点	安装简单，适合初学者学习使用	安装简单；可以安装在任何路径下，灵活性好；一台服务器可以安装多个版本的 MySQL 软件	与平台无关，可按需定制编译，最灵活；性能最好；一台服务器可以安装多个版本的 MySQL 软件
缺点	需要单独下载客户端和服务器；安装路径不灵活，默认路径不能修改，一台服务器只能安装一个版本的 MySQL 软件	已经过编译，性能不如源码编译得好；不能灵活定制编译参数	安装过程较复杂；编译时间长
文件布局	/usr/bin（客户端程序和脚本） /usr/sbin（mysqld 服务器） /var/lib/mysql（日志文件和数据库） /usr/share/man（Linux 文档页） /usr/include/mysql（包含（头）文件） /usr/lib/mysql（库文件） /usr/share/mysql（错误消息和字符集文件）	bin（客户端程序和 mysqld 服务器） docs（文档） man（Linux 文档页） include（包含（头）文件） lib（库文件） share（错误消息和字典、安装数据库的 SQL 文件） support-files（帮助文件）	bin（客户端程序和 mysqld 服务器） docs（文档） man（Linux 文档页） include（包含（头）文件） lib（库文件） share（错误消息和字典、安装数据库的 SQL 文件） support-files（帮助文件）

22.1.2 安装二进制包

如果用户既不想安装最简单却不够灵活的 RPM 包,又不想安装复杂费时的源码包,那么,已经编译好的二进制包将是很好的选择。

具体安装步骤如下。

(1)用 root 登录操作系统,增加 mysql 用户和组,数据库将安装在此用户下:

```
shell> groupadd mysql
shell> useradd -g mysql mysql
```

(2)解压二进制安装包,假设安装文件放在/home/mysql,并对解压后的 mysql 目录加一个符号链接"mysql",这样对 mysql 目录的操作会更方便:

```
shell> cd /home/mysql
shell>tar -xzvf  /home/mysql/mysql-VERSION-OS.tar.gz
shell> ln -s mysql-VERSION-OS mysql
```

(3)在数据目录下创建系统数据库和系统表,--user 表示这些对象的操作系统 owner:

```
shell> cd mysql
shell> ./bin/mysqld --user=mysql --initialize
```

(4)设置目录权限,将 data 目录 owner 改为 mysql,其他目录和文件为 root:

```
shell> chown -R root:mysql .
shell> chown -R mysql:mysql data
```

(5)启动 MySQL:

```
shell> bin/mysqld_safe --user=mysql &
```

22.1.3 安装源码包

如果对数据库的性能要求很高,并且希望能够灵活地定制安装选项,安装源码包将是明智的选择。源码包的安装步骤与二进制包非常类似,具体如下。

(1)用 root 登录操作系统,增加 mysql 用户和组,数据库将安装在此用户下:

```
shell> groupadd mysql
shell> useradd -g mysql mysql
```

(2)解压源码安装文件 mysql-VERSION.tar.gz,并进入解压后的目录:

```
shell> tar xvfz mysql-VERSION.tar.gz
shell> cd mysql-VERSION
```

(3)用 cmake 工具来编译源码,这里可以选择很多编译参数,具体可以参考官方文档 http://dev.mysql.com/doc/refman/5.7/en/source-installation.html。这里假设 MySQL 安装在/usr/local/mysql 下:

```
shell> cmake . --DCMAKE_INSTALL_PREFIX=/usr/local/mysql
shell> make
shell> make install
```

(4)编辑配置文件。

```
shell> vi /etc/my.cnf
```

(5)在数据目录下创建系统数据库和系统表,--user 表示这些对象的操作系统 owner。

```
shell> cd /usr/local/mysql
shell> bin/mysqld --initialize --user=mysql
```

(6)设置目录权限,将 var 目录 owner 改为 mysql(源码安装,默认数据目录为 var),其他目录和文件为 root。

```
shell> chown -R root .
shell> chown -R mysql var
shell> chgrp -R mysql .
```

(7)启动 MySQL:

```
shell> bin/mysqld_safe --user=mysql &
```

当采用源码包进行安装时,可以在安装源码包的过程中,根据自己的需要进行灵活配置。下面介绍源码安装常用的两个选项,有兴趣的读者可以参考 MySQL 官方文档,以获得更详细的信息。

(1)修改默认路径。

如果不想要位于"/usr/local/var"目录下面的日志(log)文件和数据库,可以使用类似于下列 cmake 命令中的一个:

```
shell>cmake . -DCMAKE_INSTALL_PREFIX=/usr/local/mysql
shell>cmake . -DCMAKE_INSTALL_PREFIX=/usr/local -DMYSQL_DATADIR=/usr/local/mysql/data
```

第一个命令改变安装前缀以便将所有内容安装到"/usr/local/mysql"下面而非默认的"/usr/local"。第二个命令保留默认安装前缀,但是覆盖了数据库目录的默认目录(通常是"/usr/local/var")并且把它改为/usr/local/mysql/data。编译完 MySQL 后,可以通过选项文件更改这些选项。

修改 socket 的默认位置:

```
shell> cmake . -DMYSQL_UNIX_ADDR=/usr/local/mysql/tmp/mysql.sock
```

(2)只选择要使用的字符集。

MySQL 5.7 使用 latin1 和 latin1_swedish_ci 作为默认的字符集和校对规则,从 MySQL 8.0 开始使用 utf8mb4 和 utf8mb4_0900_ai_ci 作为默认的字符集和校对规则。如果想改变安装后的默认字符集和默认排序规则,则可以使用如下编译选项:

```
shell> cmake . -DDEFAULT_CHARSET=CHARSET
cmake . -DDEFAULT_COLLATION=COLLATION
```

如果不需要安装所有的字符集,那么编译时可以选择只安装用户需要的字符集。这样可以节省更多的系统资源,并且使得安装后的 MySQL 速度更快。编译选项如下:

```
shell> cmake . -with_extra_charsets=LIST
```

其中 LIST 可以是下面的任何一项:

- 以空格为间隔的一系列字符集名;
- complex,以包括不能动态装载的所有字符集;
- all,包括所有字符集。

22.1.4 参数设置方法

在 MySQL 启动过程中,首先会读取参数配置文件,如果不设置参数文件,MySQL 就按照系统中所有参数的默认值来进行启动,通过"mysqld –verbose –help"命令可以来查看参数文件中所有参数的当前设置值。

在 Windows 和 Linux 上,参数文件可以被放在多个位置,数据库启动的时候将按照不同的顺序来搜索，如果在多个位置都有参数文件，则搜索顺序靠后的文件中的参数将覆盖靠前的参数。表 22-2 和表 22-3 分别给出了在不同操作系统中数据库启动时，MySQL 搜索参数文件的顺序。

表 22-2　　　　　　　Windows 平台上 MySQL 参数文件的读取顺序

文　件　名	备　注
WINDIR\my.ini	全局选项
C:\my.cnf	全局选项
INSTALLDIR\my.ini	全局选项
defaults-extra-file	用--defaults-extra-file=path 指定的文件

表 22-3　　　　　　　Linux 平台上 MySQL 参数文件的读取顺序

文　件　名	备　注
/etc/my.cnf	全局选项
$MYSQL_HOME/my.cnf	服务器相关选项，其中$MYSQL_HOME 为环境变量中指定的 MySQL 安装目录
defaults-extra-file	用--defaults-extra-file=path 指定的文件
~/.my.cnf	用户相关选项

● WINDIR 典型名称为 C: \WINDOWS 或 C: \WINNT。用户可以使用以下命令从 WINDIR 环境变量值确定自己的确切位置：

```
C: \> echo %WINDIR%
```

● INSTALLDIR 是 MySQL 的安装目录，比如 c:\mysql。

● defaults-extra-file 是 MySQL 启动时可选择的附带选项，用此参数可以指定任何路径下的配置文件。

● "全局选项"表示如果一台服务器上安装了多个 MySQL，则每个 MySQL 服务启动的时候都会首先从此选项中读取参数。

注意：不管在 Windows 还是 Linux 平台上，为了避免混淆，建议最好只在一个位置指定配置文件。

当参数需要修改时，可以选择以下 3 种修改方式（命令行中 para_name 表示要修改的参数名，value 表示要修改的目标参数值）。

● session 级修改（只对本 session 有效），在 mysql 提示符下执行如下命令：

```
mysql>set para_name=value;
```

● 全局级修改（对所有新的连接都有效，但是对本 session 无效，数据库重启后失效），在 mysql 提示符下执行如下命令：

```
mysql>set global  para_name=value;
```

● 永久修改。将参数在 my.cnf 中增加或者修改，数据库重启后生效。

22.2　升级 MySQL

MySQL 的版本更新很快，新版本中往往包含了很多新功能，并且解决了很多旧版本中的 BUG，因此在很多情况下用户需要对数据库进行升级。

MySQL 的升级很简单，以下给出了几种不同的升级方法，每种升级方法都有一定的优缺点，用户可以按照实际需求选择合适的方法进行操作。

1. 方法 1：最简单，适合于任何存储引擎（不一定速度最快）

（1）在目标服务器上安装新版本的 MySQL。

（2）在新版本的 MySQL 上创建和老版本同名的数据库，命令如下：

```
shell> mysqladmin -h hostname -P port -u user -p passwd  create db_name
```

（3）将老版本 MySQL 上的数据库通过管道导入到新版本数据库中，命令如下：

```
shell> mysqldump --opt db_name | mysql -h  hostname -P port -u user -p passwd   db_name
```

这里的 --opt 选项表明采用优化（Optimize）方式进行导出。

注意：如果网络较慢，可以在导出选项中加上 --compress 来减少网络传输。

对于不支持管道操作符（|）的操作系统，可以先用 mysqldump 工具将旧版本的数据导出为文本文件，然后再往新版本 MySQL 中导入此文件。其实就是把上面的操作分为两步执行，具体操作如下：

```
shell> mysqldump --opt db_name > filename（旧版本 MySQL 上执行）
shell> mysql -u user -p passwd db_name < filename（新版本 MySQL 上执行）
```

（4）将旧版本 MySQL 中的 mysql 数据库目录全部 cp 过来覆盖新版本 MySQL 中的 mysql 数据库，在新版本服务器的 shell 里面执行 mysql_upgrade 命令升级权限表。重启新版本 MySQL 服务。至此，升级完毕。

2. 方法 2：适合于任何存储引擎，速度较快

（1）参照方法一中的步骤（1）安装新版本 MySQL。

（2）在旧版本 MySQL 中，创建用来保存输出文件的目录并用 mysqldump 备份数据库：

```
shell> mkdir DUMPDIR
shell>mysqldump --tab=DUMPDIR db_name
```

这里使用 --tab 选项不会生成 SQL 文本。而是在备份目录下对每个表分别生成了 .sql 和 .txt 文件，其中 .sql 保存了表的创建语句；.txt 保存了用默认分隔符生成的纯数据文本。

（3）将 DUMPDIR 目录中的文件转移到目标服务器上相应的目录中并将文件装载到新版本的 MySQL 中，具体操作如下（以下命令都在新版本服务器中执行）：

```
shell> mysqladmin create db_name                # 创建数据库
shell> cat DUMPDIR/*.sql | mysql db_name        # 创建数据库表
shell> mysqlimport db_name DUMPDIR/*.txt        # 加载数据
```

（4）参照方法一中的步骤（4）升级权限表，并重启 MySQL 服务。

3. 方法 3：适合于任何存储引擎，速度最快

（1）在目标服务器上安装新版本的 MySQL。

（2）用 mysqldump 命令导出旧版本 MySQL 数据，使用 --single-transaction 参数，备份期间旧版本 mysql 可以写入新数据。

```
shell>mysqldump --single-transaction -A> filename
```

（3）新版本 MySQL 导入 filename 中的数据，并将新版本 MySQL 配置为旧版本 MySQL 的从库。

```
shell>mysql -u user -p passwd db_name < filename（新版本 MySQL 上执行）
```

（4）使用高可用工具，例如 MHA，做在线主从切换，将新数据写入新版本 MySQL。这样就完成了 MySQL 版本升级。

上面 3 种方法中，前两种比较类似，第三种是通过搭建新版本从库，再做主从切换实现的。这里需要读者注意两点：

- 前两种升级方法都是假设升级期间旧版本 MySQL 不再进行数据更新，否则，迁移过去的数据将不能保证和原数据库一致；
- 迁移前后的数据库字符集最好能保持一致，否则可能会出现各种各样的乱码问题。

MySQL 一般很少降级使用（降到低版本）。使用 mysqldump 命令导出文本后再将其导入低版本的数据库中即可。

22.3 小结

本章主要对 Linux/UNIX 平台下 MySQL 的不同形式的安装包进行比较，并详细介绍了二进制安装包和源码安装包的安装步骤。当使用源码包进行安装时，一并介绍了一些常用的编译选项，用户可以通过这些选项灵活地进行安装定制。在本章的最后还着重介绍了 MySQL 的升级，用户可以根据实际需求进行选择。

第 23 章 MySQL 中的常用工具

在 MySQL 的日常工作和管理中，用户经常会用到 MySQL 相关的管理工具，包括官方的和第三方的。熟练使用这些工具将会大大提高工作效率。本章将介绍这些常用工具的使用。由于这些工具一般都有很多的选项参数，限于篇幅限制，本章只选择一些最常用的进行介绍。如果读者希望了解更多的选项，可以参考相应工具的帮助文档。

23.1 MySQL 官方工具

MySQL 提供了大量的客户端工具程序，用于管理和维护 MySQL 服务器。这些工具简单易用，实现的功能众多，而且不用单独安装，在日常工作中使用率非常高。下面介绍一些常用的工具。

23.1.1 mysql（客户端连接工具）

在 MySQL 提供的工具中，DBA 使用最频繁的莫过于 mysql。这里的 mysql 不是指 MySQL 服务，也不是指 mysql 数据库，而是指连接数据库的客户端工具。它类似于 Oracle 数据库里的 sqlplus，是操作者和数据库之间的纽带和桥梁。

在前面章节的例子中，曾多次使用 mysql 进行数据库的连接。大多数情况下，它的使用都非常简单，语法如下：

```
mysql [OPTIONS] [database]
```

这里的 OPTIONS 表示 mysql 的可用选项，可以一次写一个或者多个，甚至可以不写；database 表示连接的数据库，一次只能写一个或者不写，如果不写，连接成功后需要用 "use dbname" 命令来进入要操作的数据库。

下面介绍 mysql 的一些常用选项，这些选项通常有两种表达方式，一种是 "-" +选项单词的缩写字符+选项值；另一种是 "--" +选项的完整单词+ "=" +选项的实际值。例如，下面两种写法是完全等价的：

- mysql --uroot;
- mysql --user=root。

在下面的介绍中，如果有两种表达方式，都会用逗号隔开进行列出，否则将只显示一种表达方式。要了解更多的选项，可以使用 mysql --help 命令进行查看。

1. 连接选项

```
-u, --user=name            #指定用户名
-p, --password[=name]      #指定密码
-h, --host=name            #指定服务器 IP 或者域名
-P, --port=#               #指定连接端口
```

这 4 个选项经常一起配合使用。假设数据库创建一个本地用户 root@127.0.0.1，密码为 123456，数据库端口为 3307。使用 mysql 命令，指定 4 个选项的值就可以连接到数据库。

```
[root@localhost ~]$ mysql -uroot -p123456 -h127.0.0.1 -P3307
Warning: Using a password on the command line interface can be insecure.
Welcome to the MySQL monitor.  Commands end with ; or \g.
Your MySQL connection id is 48
Server version: 5.7.22-log MySQL Community Server (GPL)

Type 'help;' or '\h' for help. Type '\c' to clear the current input statement.

mysql>
```

出现上面类似的输出，证明已经连接上数据库了。细心的读者会发现，连接数据库时，mysql 客户端提示 "Warning: Using a password on the command line interface can be insecure"。这是因为在 shell 命令行中使用了明文密码，会有密码泄露的风险。安全的做法是在交互窗口中输入密码。

```
[root@localhost ~]$ mysql -uroot -p -h127.0.0.1 -P3307
Enter password:
```

在 mysql 客户端提示 "Enter password" 后，输入密码：

```
[root@localhost ~]$ mysql -uroot -p -h127.0.0.1 -P3307
Enter password:
Welcome to the MySQL monitor.  Commands end with ; or \g.
Your MySQL connection id is 51
Server version: 5.7.22-log MySQL Community Server (GPL)

Type 'help;' or '\h' for help. Type '\c' to clear the current input statement.

mysql>
```

这样，mysql 客户端就不会提示 Warning 了。

单纯输入 mysql 命令会出现什么结果呢？如果没有指定任何选项，mysql 会在 my.cnf 里面找 [client] 组内的用户名和密码，如果有，则按照此用户名和密码进行登录；如果没有，则系统会使用'root'@'localhost'进行登录。来看下面的例子，my.cnf 中有用户 emp，密码为 emp123：

```
[root@ localhost ~]# more /etc/my.cnf
[client]
user = emp
password = emp123
```

使用 mysql 命令直接登录后查看当前用户：

```
[root@localhost ~]# mysql
Welcome to the MySQL monitor.  Commands end with ; or \g.
Your MySQL connection id is 65
Server version: 5.7.22-log MySQL Community Server (GPL)

Type 'help;' or '\h' for help. Type '\c' to clear the current input statement.

mysql> select  current_user();
+----------------+
| current_user() |
+----------------+
| emp@localhost  |
+----------------+
1 row in set (0.00 sec)
```

果然，此时的默认登录用户换成了emp@localhost。这里将[client]选项中的"user = emp"注释后，重启服务器，再次用mysql命令直接登录：

```
[root@localhost ~]# mysql
Welcome to the MySQL monitor.  Commands end with ; or \g.
Your MySQL connection id is 69
Server version: 5.7.22-log MySQL Community Server (GPL)

Type 'help;' or '\h' for help. Type '\c' to clear the current input statement.

mysql> select  current_user();
+----------------+
| current_user() |
+----------------+
| root@localhost |
+----------------+
1 row in set (0.00 sec)
```

正如我们所料，此时的登录用户已经变成了'root'@'localhost'。

如果客户端和服务器位于同一台机器上，通常不需要指定-h 选项，否则要指定 MySQL 服务所在的 IP 或者主机名。如果不指定端口，默认连接到 3306 端口。以下是一个远程用户用 root 账号成功连接到服务器 192.168.7.55 上 3306 端口的例子：

```
[root@localhost ~]# mysql -h 192.168.7.55  -P 3306 -uroot -p
Enter password:
Welcome to the MySQL monitor.  Commands end with ; or \g.
Your MySQL connection id is 19
Server version: 5.7.22-log MySQL Community Server (GPL)

Type 'help;' or '\h' for help. Type '\c' to clear the current input statement.

mysql>
```

> **注意**：在正式的生产环境中，为了安全起见，一般需要创建应用账号并赋予适当权限，而不会用 root 直接操纵数据库；默认端口（3306）一般不要使用，可以改为任意操作系统未占用的端口。

2．客户端字符集选项

```
character-set-server=charset-name
```

细心的读者可能会发现，作为服务器的字符集选项，这个选项也可以配置在 my.cnf 的[mysqld]组中。同样，作为客户端字符集选项，也可以配置在 my.cnf 的[mysql]组中，这样，每次用 mysql 工具连接数据库时，就会自动使用此客户端字符集。当然，也可以在 mysql 的命令行中手工指定客户端字符集，如下所示：

```
shell>mysql -u user --default-character-set=charset
```

相当于在 mysql 客户端连接成功后执行：

```
set names charset;
```

下例描述了此选项使用前后客户端字符集的变化。

（1）正常连接到 MySQL 服务后的客户端字符集：

```
[root@localhost ~]# mysql
Welcome to the MySQL monitor.  Commands end with ; or \g.
Your MySQL connection id is 71
Server version: 5.7.22-log MySQL Community Server (GPL)

Type 'help;' or '\h' for help. Type '\c' to clear the current input statement.

mysql> show variables like 'chara%';
```

```
+--------------------------+-------------------------------------------+
| Variable_name            | Value                                     |
+--------------------------+-------------------------------------------+
| character_set_client     | utf8                                      |
| character_set_connection | utf8                                      |
| character_set_database   | utf8                                      |
| character_set_filesystem | binary                                    |
| character_set_results    | utf8                                      |
| character_set_server     | utf8                                      |
| character_set_system     | utf8                                      |
| character_sets_dir       | /home/mysql3307/mysqlhome/share/charsets/ |
+--------------------------+-------------------------------------------+
8 rows in set (0.00 sec)
```

（2）加上参数重新连接后再次观察客户端字符集（粗体显示）：

```
[root@localhost ~]# mysql --default-character-set=gbk
Welcome to the MySQL monitor.  Commands end with ; or \g.
Your MySQL connection id is 72
Server version: 5.7.22-log MySQL Community Server (GPL)

Type 'help;' or '\h' for help. Type '\c' to clear the current input statement.

mysql> show variables like 'chara%';
+--------------------------+-------------------------------------------+
| Variable_name            | Value                                     |
+--------------------------+-------------------------------------------+
| character_set_client     | gbk                                       |
| character_set_connection | gbk                                       |
| character_set_database   | utf8                                      |
| character_set_filesystem | binary                                    |
| character_set_results    | gbk                                       |
| character_set_server     | utf8                                      |
| character_set_system     | utf8                                      |
| character_sets_dir       | /home/mysql3307/mysqlhome/share/charsets/ |
+--------------------------+-------------------------------------------+
8 rows in set (0.00 sec)
```

果然，系统的客户端字符集按照之前的设置发生了变化。

3．执行选项

`-e, --execute=name`　　　　执行 SQL 语句并退出

此选项可以直接在 MySQL 客户端执行 SQL 语句，而不用连接到 MySQL 数据库后再执行，对于一些批处理脚本，这种方式尤其方便。下面的例子从客户端直接查询 mysql 数据库的 user 表中的 User 和 Host 字段：

```
[root@localhost ~]$ mysql -uroot -p -e  "SELECT User, Host FROM mysql.user"
Enter password:
+---------------+-----------+
| User          | Host      |
+---------------+-----------+
| emp           | 127.0.0.1 |
| root          | 127.0.0.1 |
| emp           | localhost |
| mysql.session | localhost |
| mysql.sys     | localhost |
| root          | localhost |
+---------------+-----------+
```

可以按这种方式连续执行多个 SQL 语句，用英文分号（;）隔开。下面的例子中连续执行了两个 SQL 语句：

```
[root@localhost ~]$ mysql -u root -p employees -e "select distinct last_name from employees where last_name like 'Gul%';select count(0) from dept_manager;"
  Enter password:
+-----------+
| last_name |
```

```
+----------+
| Gulla    |
| Gulak    |
| Gulik    |
+----------+
+----------+
| count(0) |
+----------+
|       24 |
+----------+
```

4. 格式化选项

```
-E , --vertical        将输出方式按照字段顺序竖着显示
-s , --silent          去掉 mysql 中的线条框显示
```

"-E"选项类似于 mysql 里面执行 SQL 语句后加"\G",可以将输出内容比较多的行更清晰完整地进行显示,经常和"-e"选项一起使用。下例中在 shell 命令行直接对数据库做查询,并将结果格式化输出:

```
[root@localhost ~]$ mysql -u root -p -e "SELECT User, Host FROM mysql.user"  -E
Enter password:
*************************** 1. row ***************************
User: emp
Host: 127.0.0.1
*************************** 2. row ***************************
User: root
Host: 127.0.0.1
*************************** 3. row ***************************
User: emp
Host: localhost
*************************** 4. row ***************************
User: mysql.session
Host: localhost
*************************** 5. row ***************************
User: mysql.sys
Host: localhost
*************************** 6. row ***************************
User: root
Host: localhost
```

在 mysql 的安静模式下,"-s"选项可以将输出的线条框去掉,字段之间用 tab 进行分隔,每条记录显示一行。此选项对于只显示数据的情况很有用,下例是此选项的显示结果:

```
[root@localhost ~]$ mysql -s -uroot -p employees
Enter password:
mysql> select * from departments;
dept_no  dept_name
d009     Customer Service
d005     Development
d002     Finance
d003     Human Resources
d001     Marketing
d004     Production
d006     Quality Management
d008     Research
d007     Sales
```

5. 错误处理选项

```
-f, --force            强制执行 SQL
-v, --verbose          显示更多信息
--show-warnings        显示警告信息
```

在一个批量执行的 SQL 中,如果其中一个 SQL 执行出错,正常情况下,该批处理将停止退出。加上-f 选项,则跳过出错 SQL,强制执行后面的 SQL;加上-v 选项,则显示出错的 SQL 语句;加上--show-warnings,则会显示全部错误信息。

这 3 个参数经常一起使用，在很多情况下会对用户很有帮助，比如加载数据。如果数据中有语法错误的地方，则会将出错信息记录在日志中，而不会停止使得后面的正常 SQL 无法执行；而出错的语句，也可以在日志中得以查看，进行修复。下面是一个例子。

（1）设置测试 SQL 文本 a.sql 和测试表 t2，a.sql 中记录了 3 条 insert 语句，t2 为只有一个 int 类型字段的空表。

```
[root@localhost ~]# more a.sql
insert into t2 values(1);
insert into t2 values(2a);
insert into t2 values(3);

mysql> desc t2;
+-------+---------+------+-----+---------+-------+
| Field | Type    | Null | Key | Default | Extra |
+-------+---------+------+-----+---------+-------+
| id    | int(11) | YES  |     | NULL    |       |
+-------+---------+------+-----+---------+-------+
1 row in set (0.00 sec)

mysql> select * from t2;
Empty set (0.00 sec)
```

（2）不加任何参数将数据导入表 t2。

```
[root@localhost ~]# mysql -uroot test  < a.sql
ERROR 1054 (42S22) at line 2: Unknown column '2a' in 'field list'
```

可以发现，在导入过程中出错。查看一下实际导入的记录：

```
[root@localhost ~]# mysql -uroot test -e 'select * from t2'
+------+
| id   |
+------+
| 1    |
+------+
```

可以发现，由于第二条记录出现语法错误，无法插入表 t2，系统自动退出，第一条记录成功插入，第三条记录没有插入。

（3）加参数-f 重新导入。

```
[root@localhost ~]# mysql -uroot test -f < a.sql
ERROR 1054 (42S22) at line 2: Unknown column '2a' in 'field list'
[root@localhost ~]# mysql -uroot test -e 'select * from t2'
+------+
| id   |
+------+
| 1    |
| 1    |
| 3    |
+------+
```

可以发现，虽然导入过程依旧报错，但是出错后的记录 3 却已经成功插入。但是我们只能看到部分出错信息，无法定位出错的 SQL 语句。

（4）加参数-v 后重新导入。

```
[root@localhost ~]# mysql -uroot test -f -v< a.sql
--------------
insert into t2 values(1)
--------------

--------------
insert into t2 values(2a)
--------------

ERROR 1054 (42S22) at line 2: Unknown column '2a' in 'field list'
```

```
--------------
insert into t2 values(3)
--------------

[root@localhost ~]# mysql -uroot test -e  'select * from t2'
+------+
| id   |
+------+
|    1 |
|    1 |
|    3 |
|    1 |
|    3 |
+------+
```

这时发现,出错后的 SQL 依然能正确插入,并且对出错 SQL 和错误内容都进行了提示,我们可以很容易地定位并解决问题。

(5)修改测试数据,将第二条记录中的 "2a" 改为 "2222222222222222222",显然,后者超出了 int 数据类型的范围。

```
[root@localhost ~]# more a.sql
insert into t2 values(1);
insert into t2 values(2222222222222222222);
insert into t2 values(3);
```

(6)再次将 a.sql 导入表 t2:

```
[root@localhost ~]# mysql -uroot test  < a.sql
```

这次没有出现任何错误提示,此时没有设置 SQL_MODE,是因为 SQL_MODE 设置的是非严格的数据校验,也就是第二条记录虽然可以插入表 t2,但是插入的数据是错误的。来看一下 t2 的数据:

```
mysql> select * from t2;
+------------+
| id         |
+------------+
|          1 |
| 2147483647 |
|          3 |
+------------+
3 rows in set (0.00 sec)
```

果然,插入的数据是 int 类型的最大值 "2147483647",而非 "2222222222222222222"。

但是由于只是警告,并没有停止出错 SQL 的运行,这样会导致插入错误的数据,而我们却无法得知。所以下面尝试加入 -v 参数再次导入。

(7)加 -v 参数,再次导入 t2。

```
[root@localhost ~]# mysql -uroot -v test  < a.sql
--------------
insert into t2 values(1)
--------------


--------------
insert into t2 values(2222222222222222222)
--------------


--------------
insert into t2 values(3)
```

结果显示了所有 SQL 语句的执行情况,但是对第二条出错语句并没有任何报警提示,我们仍然无法得知数据出错。最后加上 --show-warnings 选项试试。

(8)加上 --show-warnings 选项再次导入表 t2。

```
[root@localhost ~]# mysql -uroot -v --show-warnings test  < a.sql
--------------
```

```
insert into t2 values(1)
--------------

--------------
insert into t2 values(2222222222222222222)
--------------
Warning (Code 1264): Out of range value adjusted for column 'id' at row 1
--------------
insert into t2 values(3)
```

结果发现，第二条错误数据报警，提示插入的值超出了字段的范围，可以从这个日志中很容易地找到错误数据。

通过上面的测试例子，可以发现-f、-v、--show-warnings 选项在执行一些可能含有语法错误或者数据错误的批处理作业中，可以记录比较完整的日志，从而帮助用户发现并解决这些错误。

23.1.2　mysqladmin（MySQL 管理工具）

mysqladmin 是一个执行管理操作的客户端程序。我们可以用它来检查服务器的配置和当前的状态、创建并删除数据库等。它的功能和 mysql 客户端非常类似，主要区别在于它更侧重于一些管理方面的功能，比如关闭数据库。

mysqladmin 的用法如下：

```
shell> mysqladmin [options] command [command-options]
        [command [command-options]] ...
```

使用方法、常用的选项和 MySQL 非常类似，这里就不再赘述。这里将可以执行的命令行简单列举如下：

```
create databasename           Create a new database
debug                         Instruct server to write debug information to log
drop databasename             Delete a database and all its tables
extended-status               Gives an extended status message from the server
flush-hosts                   Flush all cached hosts
flush-logs                    Flush all logs
flush-status                  Clear status variables
flush-tables                  Flush all tables
flush-threads                 Flush the thread cache
flush-privileges              Reload grant tables (same as reload)
kill id,id,...                Kill mysql threads
password [new-password]       Change old password to new-password in current format
old-password [new-password]   Change old password to new-password in old format
ping                          Check if mysqld is alive
processlist                   Show list of active threads in server
reload                        Reload grant tables
refresh                       Flush all tables and close and open logfiles
shutdown                      Take server down
status                        Gives a short status message from the server
start-slave                   Start slave
stop-slave                    Stop slave
variables                     Prints variables available
version                       Get version info from server
```

这里简单给出一个关闭数据库的例子：

```
[root@localhost ~]# mysqladmin -uroot -p shutdown
Enter password:
```

23.1.3　mysqlbinlog（日志管理工具）

由于服务器生成的二进制日志文件以二进制格式保存，当我们查看这些文件时，经常需要先转换为文本格式,此时就会用到 mysqlbinlog 日志管理工具。

mysqlbinlog 的具体用法如下:

```
shell> mysqlbinlog [options] log-files1 log-files2...
```

option 有很多选项,常用的选项如下。

- -d, --database=name:指定数据库名称,只列出指定的数据库的相关操作。
- -o, --offset=#:忽略掉日志中的前 n 行命令。
- -r, --result-file=name:将输出的文本格式日志输出到指定文件。
- -s, --short-form:显示简单格式,省略掉一些信息。
- --set-charset=name:在输出为文本格式时,在文件第一行加上 SET NAMES character_set,这个选项在某些情况下装载数据时非常有用。
- --start-datetime=name --stop-datetime=name:指定日期间隔内的所有日志。
- --start-position=# --stop-position=#:指定位置间隔内的所有日志。
- -v, --verbose:重构日志格式为 row 模式下的虚拟 sql 语句,-v -v 会加上列的数据类型注释。

下面举一个例子说明这些选项的使用。

(1)创建新日志,对 mysql 和 employees 数据库做不同的 DML 操作。

```
[root@localhost ~]# mysql -uroot
Warning: Using a password on the command line interface can be insecure.
Welcome to the MySQL monitor.  Commands end with ; or \g.
Your MySQL connection id is 101
Server version: 5.7.22-log MySQL Community Server (GPL)

Type 'help;' or '\h' for help. Type '\c' to clear the current input statement.

mysql> flush logs;
Query OK, 0 rows affected (0.01 sec)

mysql> use mysql
Reading table information for completion of table and column names
You can turn off this feature to get a quicker startup with -A

Database changed

mysql> revoke ALL PRIVILEGES ON `employees`.* from  'emp'@'127.0.0.1';
Query OK, 0 rows affected (0.00 sec)

mysql> use employees
Database changed

mysql> truncate table departments;
Query OK, 0 rows affected (0.06 sec)

mysql> insert into departments (dept_no,dept_name) values ('d009','Development');
Query OK, 1 row affected (0.00 sec)
```

(2)仅加上 -v 选项,用于重构 row 模式下的 dml 语句,显示所有日志(粗体字显示上一步执行过的 SQL)。

```
[root@localhost data]# mysqlbinlog mysql-bin.000004 -v
/*!50530 SET @@SESSION.PSEUDO_SLAVE_MODE=1*/;
/*!40019 SET @@session.max_insert_delayed_threads=0*/;
/*!50003 SET @OLD_COMPLETION_TYPE=@@COMPLETION_TYPE,COMPLETION_TYPE=0*/;
DELIMITER /*!*/;
……
# at 219
#180612 16:28:25 server id 7237  end_log_pos 387 CRC32 0x63bf525a         Query    thread_id=103
exec_time=0      error_code=0
```

```
    use `employees`/*!*/;
    SET TIMESTAMP=1528792105/*!*/;
    SET @@session.pseudo_thread_id=103/*!*/;
    SET @@session.foreign_key_checks=1, @@session.sql_auto_is_null=0, @@session.unique_checks=1,
@@session.autocommit=1/*!*/;
    SET @@session.sql_mode=1411907618/*!*/;
    SET @@session.auto_increment_increment=1, @@session.auto_increment_offset=1/*!*/;
    /*!\C utf8 *//*!*/;
    SET @@session.character_set_client=33,@@session.collation_connection=33,@@session.collation_
server=192/*!*/;
    SET @@session.lc_time_names=0/*!*/;
    SET @@session.collation_database=DEFAULT/*!*/;
    revoke ALL PRIVILEGES ON `employees`.* from  'emp'@'127.0.0.1'
    /*!*/;
    # at 387
    #180612 16:29:01 server id 7237  end_log_pos 452 CRC32 0x3a733b26     GTID [commit=yes]
    SET @@SESSION.GTID_NEXT= 'ANONYMOUS'/*!*/;
    # at 452
    #180612 16:29:01 server id 7237  end_log_pos 547 CRC32 0x27ff96f5     Query    thread_id=94
     exec_time=0     error_code=0
    use `employees`/*!*/;
    SET TIMESTAMP=1528792141/*!*/;
    truncate table departments
    /*!*/;
    # at 547
    #180612 16:29:05 server id 7237  end_log_pos 612 CRC32 0x02820f2e     GTID [commit=no]
    SET @@SESSION.GTID_NEXT= 'ANONYMOUS'/*!*/;
    # at 612
    #180612 16:29:05 server id 7237  end_log_pos 686 CRC32 0xc9192bb2     Query    thread_id=94
     exec_time=0     error_code=0
    SET TIMESTAMP=1528792145/*!*/;
    BEGIN
    /*!*/;
    # at 686
    #180612 16:29:05 server id 7237  end_log_pos 747 CRC32 0xcc321e85     Table_map: `employees`
.`departments` mapped to number 224
    # at 747
    #180612 16:29:05 server id 7237  end_log_pos 801 CRC32 0x48fe0ea0     write_rows: table id
224 flags: STMT_END_F

    BINLOG '
    UYQfWxNFHAAAPQAAAOsCAAAAAOAAAAAAAAEABnRlc3RkYgALZGVwYXJ0bWVudHMAAv4PBP544C4A
    hR4yzA==
    UYQfWx5FHAAANgAAACEDAAAAAOAAAAAAAEAAgAC//wEZDAwOQsARGV2ZWxvcG1lbnSgDv5I
    '/*!*/;
    ### INSERT INTO `employees`.`departments`
    ### SET
    ###   @1='d009'
    ###   @2='Development'
    # at 801
    #180612 16:29:05 server id 7237  end_log_pos 876 CRC32 0x0cfb5409     Query    thread_id=94
     exec_time=0     error_code=0
    SET TIMESTAMP=1528792145/*!*/;
    COMMIT
    /*!*/;
    SET @@SESSION.GTID_NEXT= 'AUTOMATIC' /* added by mysqlbinlog */ /*!*/;
    # at 876
    #180612 16:29:23 server id 7237  end_log_pos 923 CRC32 0xf8e85261     Rotate to mysql-bin.
000005  pos: 4
    DELIMITER ;
    # End of log file
    ROLLBACK /* added by mysqlbinlog */;
    /*!50003 SET COMPLETION_TYPE=@OLD_COMPLETION_TYPE*/;
    /*!50530 SET @@SESSION.PSEUDO_SLAVE_MODE=0*/;
```

（3）加-v -d 选项，-d 选项将只显示对 employees 数据库的操作日志。

```
[root@localhost mysql]# mysqlbinlog mysql-bin.000004 -v -d employees
/*!50530 SET @@SESSION.PSEUDO_SLAVE_MODE=1*/;
/*!40019 SET @@session.max_insert_delayed_threads=0*/;
```

```
    /*!50003 SET @OLD_COMPLETION_TYPE=@@COMPLETION_TYPE,COMPLETION_TYPE=0*/;
DELIMITER /*!*/;
……
# at 452
#180612 16:29:01 server id 7237  end_log_pos 547 CRC32 0x27ff96f5      Query    thread_id=94
 exec_time=0     error_code=0
use `employees`/*!*/;
SET TIMESTAMP=1528792141/*!*/;
SET @@session.pseudo_thread_id=94/*!*/;
SET @@session.foreign_key_checks=1, @@session.sql_auto_is_null=0, @@session.unique_checks=1,
@@session.autocommit=1/*!*/;
SET @@session.sql_mode=1411907618/*!*/;
SET @@session.auto_increment_increment=1, @@session.auto_increment_offset=1/*!*/;
/*!\C utf8 *//*!*/;
SET @@session.character_set_client=33,@@session.collation_connection=33,@@session.collation_
server=192/*!*/;
SET @@session.lc_time_names=0/*!*/;
SET @@session.collation_database=DEFAULT/*!*/;
truncate table departments
/*!*/;
# at 547
#180612 16:29:05 server id 7237  end_log_pos 612 CRC32 0x02820f2e      GTID [commit=no]
SET @@SESSION.GTID_NEXT= 'ANONYMOUS'/*!*/;
# at 612
#180612 16:29:05 server id 7237  end_log_pos 686 CRC32 0xc9192bb2      Query    thread_id=94
 exec_time=0     error_code=0
SET TIMESTAMP=1528792145/*!*/;
BEGIN
/*!*/;
# at 686
#180612 16:29:05 server id 7237  end_log_pos 747 CRC32 0xcc321e85      Table_map: `employees`
.`departments` mapped to number 224
# at 747
#180612 16:29:05 server id 7237  end_log_pos 801 CRC32 0x48fe0ea0      Write_rows: table id
224 flags: STMT_END_F

BINLOG '
UYQfWxNFHAAAPQAAAOsCAAAAAOAAAAAAAAEABnRlc3RkYgALZGVwYXJ0bWVudHMAAv4PBP544C4A
hR4yzA==
UYQfWx5FHAAANgAAACEDAAAAAOAAAAAAAAEAAgAC//wEZDAwOQsARGV2ZWxvcG1lbnSgDv5I
'/*!*/;
### INSERT INTO `employees`.`departments`
### SET
###   @1='d009'
###   @2='Development'
# at 801
#180612 16:29:05 server id 7237  end_log_pos 876 CRC32 0x0cfb5409      Query    thread_id=94
 exec_time=0     error_code=0
SET TIMESTAMP=1528792145/*!*/;
COMMIT
/*!*/;
SET @@SESSION.GTID_NEXT= 'AUTOMATIC' /* added by mysqlbinlog */ /*!*/;
# at 876
#180612 16:29:23 server id 7237  end_log_pos 923 CRC32 0xf8e85261      Rotate to mysql-bin.
000005  pos: 4
DELIMITER ;
# End of log file
ROLLBACK /* added by mysqlbinlog */;
/*!50003 SET COMPLETION_TYPE=@OLD_COMPLETION_TYPE*/;
/*!50530 SET @@SESSION.PSEUDO_SLAVE_MODE=0*/;
```

日志中的粗体字显示，输出中仅仅包含对 employees 数据库的操作部分。

（4）加-v -o 选项，忽略掉前 6 个操作，只剩下对 departments 的 insert 操作。

```
[root@localhost data]# mysqlbinlog mysql-bin.000004 -v -o 6
/*!50530 SET @@SESSION.PSEUDO_SLAVE_MODE=1*/;
/*!40019 SET @@session.max_insert_delayed_threads=0*/;
/*!50003 SET @OLD_COMPLETION_TYPE=@@COMPLETION_TYPE,COMPLETION_TYPE=0*/;
DELIMITER /*!*/;
……
```

```
    # at 612
    #180612 16:29:05 server id 7237   end_log_pos 686  CRC32 0xc9192bb2        Query   thread_id=94
    exec_time=0       error_code=0
    SET TIMESTAMP=1528792145/*!*/;
    SET @@session.pseudo_thread_id=94/*!*/;
    SET @@session.foreign_key_checks=1, @@session.sql_auto_is_null=0, @@session.unique_checks=1,
@@session.autocommit=1/*!*/;
    SET @@session.sql_mode=1411907618/*!*/;
    SET @@session.auto_increment_increment=1, @@session.auto_increment_offset=1/*!*/;
    /*!\C utf8 *//*!*/;
    SET @@session.character_set_client=33,@@session.collation_connection=33,@@session.collation_
server=192/*!*/;
    SET @@session.lc_time_names=0/*!*/;
    SET @@session.collation_database=DEFAULT/*!*/;
    BEGIN
    /*!*/;
    # at 686
    #180612 16:29:05 server id 7237   end_log_pos 747  CRC32 0xcc321e85        Table_map: `employees`
.`departments` mapped to number 224
    # at 747
    #180612 16:29:05 server id 7237   end_log_pos 801  CRC32 0x48fe0ea0        Write_rows: table id
224 flags: STMT_END_F

    BINLOG '
    UYQfWxNFHAAAPQAAAOsCAAAAAOAAAAAAAAEABnRlc3RkYgALZGVwYXJ0bWVudHMAAv4PBP544C4A
    hR4yzA==
    UYQfWx5FHAAANgAAACEDAAAAAOAAAAAAAAEAAgAC//wEZDAwOQsARGV2ZWxvcG1lbnSgDv5I
    '/*!*/;
    ### INSERT INTO `employees`.`departments`
    ### SET
    ###    @1='d009'
    ###    @2='Development'
    # at 801
    #180612 16:29:05 server id 7237   end_log_pos 876  CRC32 0x0cfb5409        Query   thread_id=94
    exec_time=0       error_code=0
    SET TIMESTAMP=1528792145/*!*/;
    COMMIT
    /*!*/;
    SET @@SESSION.GTID_NEXT= 'AUTOMATIC' /* added by mysqlbinlog */ /*!*/;
    # at 876
    #180612 16:29:23 server id 7237   end_log_pos 923  CRC32 0xf8e85261        Rotate to mysql-bin.
000005  pos: 4
    DELIMITER ;
    # End of log file
    ROLLBACK /* added by mysqlbinlog */;
    /*!50003 SET COMPLETION_TYPE=@OLD_COMPLETION_TYPE*/;
    /*!50530 SET @@SESSION.PSEUDO_SLAVE_MODE=0*/;
```

（5）加-r 选项，将上面的结果输出到文件 resultfile 中。

```
    [root@localhost data]# mysqlbinlog mysql-bin.000004 -v -o 6 -r resultfile
    [root@localhost data]# more resultfile
    /*!50530 SET @@SESSION.PSEUDO_SLAVE_MODE=1*/;
    /*!40019 SET @@session.max_insert_delayed_threads=0*/;
    /*!50003 SET @OLD_COMPLETION_TYPE=@@COMPLETION_TYPE,COMPLETION_TYPE=0*/;
    DELIMITER /*!*/;
    ……
    # at 612
    #180612 16:29:05 server id 7237   end_log_pos 686  CRC32 0xc9192bb2        Query   thread_id=94
    exec_time=0       error_code=0
    SET TIMESTAMP=1528792145/*!*/;
    SET @@session.pseudo_thread_id=94/*!*/;
    SET @@session.foreign_key_checks=1, @@session.sql_auto_is_null=0, @@session.unique_checks=1,
@@session.autocommit=1/*!*/;
    SET @@session.sql_mode=1411907618/*!*/;
    SET @@session.auto_increment_increment=1, @@session.auto_increment_offset=1/*!*/;
    /*!\C utf8 *//*!*/;
    SET @@session.character_set_client=33,@@session.collation_connection=33,@@session.collation_
server=192/*!*/;
    SET @@session.lc_time_names=0/*!*/;
    SET @@session.collation_database=DEFAULT/*!*/;
```

```
    BEGIN
    /*!*/;
    # at 686
    #180612 16:29:05 server id 7237   end_log_pos 747  CRC32 0xcc321e85         Table_map: `employees`
.`departments` mapped to number 224
    # at 747
    #180612 16:29:05 server id 7237   end_log_pos 801  CRC32 0x48fe0ea0         Write_rows: table id
224 flags: STMT_END_F

    BINLOG '
    UYQfWxNFHAAAPQAAAOsCAAAAAOAAAAAAAAEABnRlc3RkYgALZGVwYXJ0bWVudHMAAv4PBP544C4A
    hR4yzA==
    UYQfWx5FHAAANgAAACEDAAAAAOAAAAAAAAEAAgAC//wEZDAwOQsARGV2ZWxvcG1lbnSgDv5I
    '/*!*/;
    ### INSERT INTO `employees`.`departments`
    ### SET
    ###   @1='d009'
    ###   @2='Development'
    # at 801
    #180612 16:29:05 server id 7237   end_log_pos 876  CRC32 0x0cfb5409         Query   thread_id=94
     exec_time=0    error_code=0
    SET TIMESTAMP=1528792145/*!*/;
    COMMIT
    /*!*/;
    SET @@SESSION.GTID_NEXT= 'AUTOMATIC' /* added by mysqlbinlog */ /*!*/;
    # at 876
    #180612 16:29:23 server id 7237   end_log_pos 923  CRC32 0xf8e85261         Rotate to mysql-bin.
000005  pos: 4
    DELIMITER ;
    # End of log file
    ROLLBACK /* added by mysqlbinlog */;
    /*!50003 SET COMPLETION_TYPE=@OLD_COMPLETION_TYPE*/;
    /*!50530 SET @@SESSION.PSEUDO_SLAVE_MODE=0*/;
```

（6）结果显示的内容较多，显得比较乱，加 -s 选项可将上面的内容进行简单显示。

```
    [root@localhost data]# mysqlbinlog mysql-bin.000004 -v  -s
    /*!50530 SET @@SESSION.PSEUDO_SLAVE_MODE=1*/;
    /*!40019 SET @@session.max_insert_delayed_threads=0*/;
    /*!50003 SET @OLD_COMPLETION_TYPE=@@COMPLETION_TYPE,COMPLETION_TYPE=0*/;
    DELIMITER /*!*/;
    # [empty]
    SET @@SESSION.GTID_NEXT= 'ANONYMOUS'/*!*/;
    use `employees`/*!*/;
    SET TIMESTAMP=1528792105/*!*/;
    SET @@session.pseudo_thread_id=999999999/*!*/;
    SET @@session.foreign_key_checks=1, @@session.sql_auto_is_null=0, @@session.unique_checks=1,
@@session.autocommit=1/*!*/;
    SET @@session.sql_mode=1411907618/*!*/;
    SET @@session.auto_increment_increment=1, @@session.auto_increment_offset=1/*!*/;
    /*!\C utf8 *//*!*/;
    SET @@session.character_set_client=33,@@session.collation_connection=33,@@session.collation_
server=192/*!*/;
    SET @@session.lc_time_names=0/*!*/;
    SET @@session.collation_database=DEFAULT/*!*/;
    revoke ALL PRIVILEGES ON `employees`.* from  'emp'@'127.0.0.1'
    /*!*/;
    SET @@SESSION.GTID_NEXT= 'ANONYMOUS'/*!*/;
    use `employees`/*!*/;
    SET TIMESTAMP=1528792141/*!*/;
    truncate table departments
    /*!*/;
    SET @@SESSION.GTID_NEXT= 'ANONYMOUS'/*!*/;
    SET TIMESTAMP=1528792145/*!*/;
    BEGIN
    /*!*/;
    SET TIMESTAMP=1528792145/*!*/;
    COMMIT
    /*!*/;
    SET @@SESSION.GTID_NEXT= 'AUTOMATIC' /* added by mysqlbinlog */ /*!*/;
    DELIMITER ;
```

```
# End of log file
ROLLBACK /* added by mysqlbinlog */;
/*!50003 SET COMPLETION_TYPE=@OLD_COMPLETION_TYPE*/;
/*!50530 SET @@SESSION.PSEUDO_SLAVE_MODE=0*/;
```

可以发现，内容的确比上文精简了，仅保留了 ddl 语句，dml 语句已经被过滤掉了。

（7）加 "--start-datetime --stop-datetime" 选项显示 16:27:34～16:28:26 的日志。

```
[root@localhost data]# mysqlbinlog mysql-bin.000004 --start-datetime='2018-06-12 16:27:34' --stop-datetime='2018-06-12 16:28:26' -v
/*!50530 SET @@SESSION.PSEUDO_SLAVE_MODE=1*/;
/*!40019 SET @@session.max_insert_delayed_threads=0*/;
/*!50003 SET @OLD_COMPLETION_TYPE=@@COMPLETION_TYPE,COMPLETION_TYPE=0*/;
DELIMITER /*!*/;
# at 4
#180612 16:27:34 server id 7237  end_log_pos 123 CRC32 0x32a82b82     Start: binlog v 4, server v 5.7.22-log created 180612 16:27:34
BINLOG '
9oMfWw9FHAAAdwAAAHsAAAAAAQANS43LjIyLWxvZwAAAAAAAAAAAAAAAAAAAAAAAAAAAAAAAA
AAAAAAAAAAAAAAAAAAAAAAAAEzgNAAgAEgAEBAQEEgAAXwAEGggAAAAICAgCAAAACgoKKioAEjQA
AYIrqDI=
'/*!*/;
# at 123
#180612 16:27:34 server id 7237  end_log_pos 154 CRC32 0x31567282     Previous-GTIDs
# [empty]
# at 154
#180612 16:28:25 server id 7237  end_log_pos 219 CRC32 0xa43e01c1     GTID [commit=yes]
SET @@SESSION.GTID_NEXT= 'ANONYMOUS'/*!*/;
# at 219
#180612 16:28:25 server id 7237  end_log_pos 387 CRC32 0x63bf525a     Query   thread_id=103   exec_time=0     error_code=0
use `employees`/*!*/;
SET TIMESTAMP=1528792105/*!*/;
SET @@session.pseudo_thread_id=103/*!*/;
SET @@session.foreign_key_checks=1, @@session.sql_auto_is_null=0, @@session.unique_checks=1, @@session.autocommit=1/*!*/;
SET @@session.sql_mode=1411907618/*!*/;
SET @@session.auto_increment_increment=1, @@session.auto_increment_offset=1/*!*/;
/*!\C utf8 *//*!*/;
SET @@session.character_set_client=33,@@session.collation_connection=33,@@session.collation_server=192/*!*/;
SET @@session.lc_time_names=0/*!*/;
SET @@session.collation_database=DEFAULT/*!*/;
revoke ALL PRIVILEGES ON `employees`.* from  'emp'@'127.0.0.1'
/*!*/;
SET @@SESSION.GTID_NEXT= 'AUTOMATIC' /* added by mysqlbinlog *//*!*/;
DELIMITER ;
# End of log file
ROLLBACK /* added by mysqlbinlog */;
/*!50003 SET COMPLETION_TYPE=@OLD_COMPLETION_TYPE*/;
/*!50530 SET @@SESSION.PSEUDO_SLAVE_MODE=0*/;
```

开始日期和结束日期可以只写一个。如果只写开始日期，表示范围是开始日期到日志结束；如果只写结束日期，表示日志开始到指定的结束日期。

（8）--start-position=#和--stop-position=#，与日期范围类似，不过可以更精确地表示范围。例如，将以上例子改成位置范围后：

```
[root@localhost data]# mysqlbinlog mysql-bin.000004 --start-position=219 --stop-position=387 -v
/*!50530 SET @@SESSION.PSEUDO_SLAVE_MODE=1*/;
/*!40019 SET @@session.max_insert_delayed_threads=0*/;
/*!50003 SET @OLD_COMPLETION_TYPE=@@COMPLETION_TYPE,COMPLETION_TYPE=0*/;
DELIMITER /*!*/;
# at 4
#180612 16:27:34 server id 7237  end_log_pos 123 CRC32 0x32a82b82     Start: binlog v 4, server v 5.7.22-log created 180612 16:27:34
BINLOG '
9oMfWw9FHAAAdwAAAHsAAAAAAQANS43LjIyLWxvZwAAAAAAAAAAAAAAAAAAAAAAAAAAAAAAAA
```

```
AAAAAAAAAAAAAAAAAAAAAAAAAEzgNAAgAEgAEBAQEEgAAXwAEGggAAAAICAgCAAAACgoKKioAEjQA
AYIrqDI=
'/*!*/;
# at 219
#180612 16:28:25 server id 7237  end_log_pos 387 CRC32 0x63bf525a       Query    thread_id=103
exec_time=0      error_code=0
use `employees`/*!*/;
SET TIMESTAMP=1528792105/*!*/;
SET @@session.pseudo_thread_id=103/*!*/;
SET @@session.foreign_key_checks=1, @@session.sql_auto_is_null=0, @@session.unique_checks=1,
@@session.autocommit=1/*!*/;
SET @@session.sql_mode=1411907618/*!*/;
SET @@session.auto_increment_increment=1, @@session.auto_increment_offset=1/*!*/;
/*!\C utf8 *//*!*/;
SET @@session.character_set_client=33,@@session.collation_connection=33,@@session.collation_
server=192/*!*/;
SET @@session.lc_time_names=0/*!*/;
SET @@session.collation_database=DEFAULT/*!*/;
revoke ALL PRIVILEGES ON `employees`.* from  'emp'@'127.0.0.1'
/*!*/;
DELIMITER ;
# End of log file
ROLLBACK /* added by mysqlbinlog */;
/*!50003 SET COMPLETION_TYPE=@OLD_COMPLETION_TYPE*/;
/*!50530 SET @@SESSION.PSEUDO_SLAVE_MODE=0*/;
```

23.1.4 mysqlcheck（表维护工具）

MySQL 实例开启状态下，可以用 mysqlcheck 来检查、修复、优化和分析表。维护期间，mysqlcheck 会在表上加上读锁，此时表不能执行 DML 和 DDL 操作。

mysqlcheck 的用法如下：

```
shell> mysqlcheck [options] db_name [tbl_name ...]
shell> mysqlcheck [options] --databases db_name ...
shell> mysqlcheck [options] --all-databases
```

option 有很多选项，常用的选项如下。

- --all-databases, -A: 除了 INFORMATION_SCHEMA 和 performace_schema 数据库，检查剩下所有数据库的所有表。
- --analyze, -a: 分析表，此时命令等同于 mysqlanalyze。
- --check: 检查表是否有错误，属于默认选项。
- --databases, -B: 检查指定数据库下的所有表。一般 mysqlcheck 把命令行的第一个名字识别为数据库名，后面的名字识别为表名。使用这个参数后，所有的名字都将识别为数据库名。
- --fast, -F: 仅检查表是否正确关闭，检查速度很快。
- --force, -f: 忽略发现的错误，继续执行后续操作。
- --optimize, -o: 优化表，此时命令等同于 mysqloptimize。
- --repair, -r: 修复表。出了不能修复主键有重复数据，几乎能修复其他所有问题。
- --skip-database=db_name: 反向排除数据库。
- --tables: 覆盖 --databases 或者 –B 选项，这个选项后面的所有名字都将被识别为表名。
- --write-binlog: 默认开启，开启这个选项后 mysqlcheck 生成的分析表、优化表和修复表语句将记录到二进制日志中。可以使用 --skip-write-binlog 选项，mysqlcheck 将不在二进制日志中记录操作语句，这样可以减少从库应用主库日志的延迟。

23.1 MySQL 官方工具

> **注意**：建议使用 mysqlcheck 工具前，先备份需要维护的表，因为在某些极端情况下，可能会导致数据丢失！

下面通过具体例子来说明这些选项的使用。

检查所有数据库的表：

```
[mysql3307@localhost ~]$ mysqlcheck -uroot -S/tmp/mysql_3307.sock --check -A
employees.city                                  OK
employees.country                               OK
employees.departments                           OK
employees.dept_emp                              OK
employees.dept_manager                          OK
employees.emp                                   OK
...
mysql.time_zone_transition_type                 OK
mysql.user                                      OK
```

分析 employees 数据库的 salaries 表：

```
[mysql3307@localhost ~]$ mysqlcheck -uroot -S/tmp/mysql_3307.sock -a employees salaries
employees.salaries                              OK
```

随着时间的推移，在伴随着数据被频繁写入和删除后，数据表会出现碎片化，这时可以对这些表进行优化操作。下面是一个具体例子：

查看 salaries 表的数据量，一共有 284 万条数据。

```
mysql>select count(*) as total from salaries;
+---------+
| total   |
+---------+
| 2844036 |
+---------+
```

表在磁盘上的表空间文件大小如下：

```
[mysql3307@localhost employees]$ ls -ltrh |grep salaries
-rw-r----- 1 mysql3307 mysql3307 8.5K May 23 14:52 salaries.frm
-rw-r----- 1 mysql3307 mysql3307 104M Jul 24 21:45 salaries.ibd         //104M
```

下面删除一部分数据：

```
mysql>delete from salaries where emp_no >399999;
Query OK, 946354 rows affected (9.04 sec)
```

通过 delete 删除数据，表空间文件没有缩小。

```
[mysql3307@localhost employees]$ ls -ltrh |grep salaries
-rw-r----- 1 mysql3307 mysql3307 8.5K May 23 14:52 salaries.frm
-rw-r----- 1 mysql3307 mysql3307 104M Jul 24 21:56 salaries.ibd
```

下面再对 salaries 表进行优化：

```
[mysql3307@localhost employees]$ mysqlcheck -uroot -S/tmp/mysql_3307.sock -o employees salaries
employees.salaries
note     : Table does not support optimize, doing recreate + analyze instead
status   : OK
```

再看一下表空间文件大小：

```
[mysql3307@localhost employees]$ ls -ltrh |grep salaries
-rw-r----- 1 mysql3307 mysql3307 8.5K Jul 24 21:58 salaries.frm
-rw-r----- 1 mysql3307 mysql3307  80M Jul 24 21:58 salaries.ibd
```

从这里看到，ibd 文件从 104MB 减小到了 80MB，salaries 表使用 InnoDB 引擎，mysqlcheck 对 InnoDB 引擎的表进行优化，实际上是做了 alter table 操作，对表进行重构，重新整理表空间文件。

mysqlcheck 做检查、分析和恢复操作时，默认情况下，会在 BINLOG 文件中记录下来相应语句，然而我们在某些情况下并不希望 BINLOG 文件记录这些操作，例如备份 BINLOG 文件，用于以后恢复增量数据，因为我们需要尽量减少恢复时间。mysqlcheck 提供了 --skip-write-binlog 选项来解决类似的问题。

登录 MySQL，查看当前 BINLOG 位置：

```
mysql>show master status;
*************************** 1. row ***************************
             File: mysql-bin.000035
         Position: 154
     Binlog_Do_DB:
 Binlog_Ignore_DB:
Executed_Gtid_Set:
1 row in set (0.00 sec)
```

优化 salaries 表：

```
[mysql3307@localhost ~]$ mysqlcheck -uroot -S/tmp/mysql_3307.sock -o employees salaries
employees.salaries
note     : Table does not support optimize, doing recreate + analyze instead
status   : OK
```

使用 mysqlbinlog 命令，查看 BINLOG 文件的信息：

```
[mysql3307@localhost data]$ mysqlbinlog -vvv mysql-bin.000035
# at 219
#180725 17:28:46 server id 7237  end_log_pos 328 CRC32 0x5934d815         Query     thread_id=37
 exec_time=3     error_code=0
use `employees`/*!*/;
SET TIMESTAMP=1532510926/*!*/;
SET @@session.pseudo_thread_id=37/*!*/;
SET @@session.foreign_key_checks=1, @@session.sql_auto_is_null=0, @@session.unique_checks=1, @@session.autocommit=1/*!*/;
SET @@session.sql_mode=1143472162/*!*/;
SET @@session.auto_increment_increment=1, @@session.auto_increment_offset=1/*!*/;
/*!\C utf8 *//*!*/;
SET @@session.character_set_client=33,@@session.collation_connection=33,@@session.collation_server=192/*!*/;
SET @@session.lc_time_names=0/*!*/;
SET @@session.collation_database=DEFAULT/*!*/;
OPTIMIZE TABLE `salaries`
/*!*/;
```

可以看到 BINLOG 文件记录了 optimize table 命令。

再次优化 salaries 表，此次增加 --skip-write-binlog 选项，使得 BINLOG 文件不再记录 optimize table 命令。

```
[mysql3307@localhost ~]$ mysqlcheck -uroot -S/tmp/mysql_3307.sock -o --skip-write-binlog employees salaries
employees.salaries
note     : Table does not support optimize, doing recreate + analyze instead
status   : OK
```

这时 BINLOG 文件中已经找不到 optimize table 命令，证明 --skip-write-binlog 选项已经生效了。

23.1.5 mysqldump（数据导出工具）

mysqldump 客户端工具用来备份数据库或在不同数据库之间进行数据迁移。备份内容包含创建表或装载表的 SQL 语句。

可以使用以下 3 种方式来调用 mysqldump：

```
shell> mysqldump [options] db_name [tables]              #备份单个数据库或者库中部分数据表
shell> mysqldump [options] ---database DB1 [DB2 DB3...]  #备份指定的一个或者多个数据库
shell> mysqldump [options] --all-database                #备份所有数据库
```

下面是 mysqldump 的一些常用选项，要查阅更详细的功能，可以使用 "mysqldump -help" 查看。

1. 连接选项

```
-u , --user=name         指定用户名
-p , --password[=name]   指定密码
-h , --host=name         指定服务器 IP 或者域名
-P , --port=#            指定连接端口
```

这 4 个选项经常一起配合使用，如果客户端位于服务器上，则通常不需要指定 host。如果不指定端口，那么默认连接到 3306 端口。以下是一个远程客户端连接到服务器的例子：

```
[root@localhost ~]# mysqldump -h192.168.7.55  -P3306 -uroot -p employees > employees.txt
Enter password: **********
```

2. 输出内容选项

```
--add-drop-database    每个数据库创建语句前加上 DROP DATABASE 语句
--add-drop-table       在每个表创建语句前加上 DROP TABLE 语句
```

以上这两个选项可以在导入数据库时不用先手工删除旧的数据库，而是会自动删除，提高导入效率，但是导入前一定要做好备份并且确认旧数据库的确已经可以删除，否则误操作将会造成数据的损失。在默认情况下，这两个参数会自动加上。

```
-n, --no-create-db     不包含数据库的创建语句
-t, --no-create-info   不包含数据表的创建语句
-d, --no-data          不包含数据
```

这 3 个选项分别表示备份文件中不包含数据库的创建语句、不包含数据表的创建语句、不包含数据，在不同的场合下，用户可以根据实际需求来进行选择。

下例中只导出表的创建语句，不包含任何其他信息：

```
[root@localhost ~]# mysqldump -uroot  --compact -p -d employees departments > departments.sql
[root@localhost ~]# cat departments.sql
/*!40101 SET @saved_cs_client     = @@character_set_client */;
/*!40101 SET character_set_client = utf8 */;
CREATE TABLE `departments` (
  `dept_no` char(4) COLLATE utf8_unicode_ci NOT NULL,
  `dept_name` varchar(40) COLLATE utf8_unicode_ci NOT NULL,
  PRIMARY KEY (`dept_no`),
  UNIQUE KEY `dept_name` (`dept_name`)
) ENGINE=InnoDB DEFAULT CHARSET=utf8 COLLATE=utf8_unicode_ci;
/*!40101 SET character_set_client = @saved_cs_client */;
```

3. 输出格式选项

--compact 选项使得输出结果简洁，不包括默认选项中的各种注释。下例对 employees 数据库中的表 emp 进行简洁导出：

```
[root@localhost ~]# mysqldump -uroot  --compact employees departments > departments.sql
[root@localhost ~]# cat departments.sql
/*!40101 SET @saved_cs_client     = @@character_set_client */;
/*!40101 SET character_set_client = utf8 */;
CREATE TABLE `departments` (
  `dept_no` char(4) COLLATE utf8_unicode_ci NOT NULL,
  `dept_name` varchar(40) COLLATE utf8_unicode_ci NOT NULL,
  PRIMARY KEY (`dept_no`),
  UNIQUE KEY `dept_name` (`dept_name`)
) ENGINE=InnoDB DEFAULT CHARSET=utf8 COLLATE=utf8_unicode_ci;
/*!40101 SET character_set_client = @saved_cs_client */;
INSERT INTO `departments` VALUES ('d009','Customer Service'),('d005','Development'),('d008','Research'),('d007','Sales');
```

-c 或者--complete-insert 选项使得输出文件中的 insert 语句包括字段名称，默认是不包括字段名称的。下例对 employees 数据库中的表 emp 使用此选项进行导出：

```
[root@localhost ~]# mysqldump -uroot -c --compact employees departments > departments.sql
[root@localhost ~]# cat departments.sql
/*!40101 SET @saved_cs_client     = @@character_set_client */;
/*!40101 SET character_set_client = utf8 */;
CREATE TABLE `departments` (
  `dept_no` char(4) COLLATE utf8_unicode_ci NOT NULL,
  `dept_name` varchar(40) COLLATE utf8_unicode_ci NOT NULL,
  PRIMARY KEY (`dept_no`),
  UNIQUE KEY `dept_name` (`dept_name`)
) ENGINE=InnoDB DEFAULT CHARSET=utf8 COLLATE=utf8_unicode_ci;
/*!40101 SET character_set_client = @saved_cs_client */;
INSERT INTO `departments` (`dept_no`, `dept_name`) VALUES ('d009','Customer Service'),('d005',
'Development'),('d008','Research'),('d007','Sales');
```

-T 选项将指定数据表中的数据备份为单纯的数据文本和建表 SQL 两个文件，经常和下面几个选项一起配合使用，将数据导出为指定格式显示。下面几个选项的具体使用方法将会在第 25 章中进行详细讲解。

- -T,--tab=name（备份数据和建表语句）;
- --fields-terminated-by=name（域分隔符）;
- --fields-enclosed-by=name（域引用符）;
- --fields-optionally-enclosed-by=name（域可选引用符）;
- --fields-escaped-by=name（转义字符）。

在下面的例子中，将 employees 数据库中的表 departments 导出为单纯的数据文本和建表 SQL 两个文件，并存放在当前路径下的/tmp 子目录下。

（1）将 employees 数据库下的表 departments 备份到/tmp 目录下。

```
[root@localhost ~]$ mysqldump -uroot -p employees departments -T /tmp
Enter password:
```

（2）进入/tmp 目录，发现生成了两个文件：一个是.sql；另一个是.txt。

```
[root@localhost tmp]# ls departments*
departments.sql   departments.txt
```

（3）查看两个文件的内容，.sql 文件存放了建表语句，而.txt 文件存放了实际的数据。

```
[root@localhost tmp]# cat departments.sql
--
-- Table structure for table `departments`
--

DROP TABLE IF EXISTS `departments`;
/*!40101 SET @saved_cs_client     = @@character_set_client */;
/*!40101 SET character_set_client = utf8 */;
CREATE TABLE `departments` (
  `dept_no` char(4) COLLATE utf8_unicode_ci NOT NULL,
  `dept_name` varchar(40) COLLATE utf8_unicode_ci NOT NULL,
  PRIMARY KEY (`dept_no`),
  UNIQUE KEY `dept_name` (`dept_name`)
) ENGINE=InnoDB DEFAULT CHARSET=utf8 COLLATE=utf8_unicode_ci;


[root@localhost tmp]# cat departments.txt
d009    Customer Service
d005    Development
d002    Finance
d003    Human Resources
```

4. 字符集选项

--default-character-set=name 选项可以设置导出的客户端字符集。系统默认的客户端字符集可以通过以下命令来查看：

```
[root@localhost ~]# mysqld --verbose --help|grep character-set-server|grep -v name
character-set-server                                            utf8
```

这个选项在导出数据库时非常重要，如果客户端字符集和数据库字符集不一致，数据在导出的时候就需要进行字符集转换，将数据库字符集转换为客户端字符集，经过转换后的数据很可能成为乱码或者"？"等特殊字符，使得备份文件无法恢复。

下面是一个字符集导出中文的例子。

（1）测试表 emp 字符集为 latin1，插入测试记录。

```
mysql> show create table emp\G
*************************** 1. row ***************************
       Table: emp
Create Table: CREATE TABLE `emp` (
  `id` int(11) NOT NULL DEFAULT '0',
  `name` varchar(200) DEFAULT NULL,
  `content` text,
  PRIMARY KEY (`id`)
) ENGINE=InnoDB DEFAULT CHARSET=latin1
1 row in set (0.03 sec)
mysql> set names latin1;
Query OK, 0 rows affected (0.00 sec)

mysql> INSERT INTO emp(id,name,content) VALUES (1,'z1','aa');
Query OK, 1 row affected (0.07 sec)

mysql> INSERT INTO emp(id,name,content) VALUES (2,'中国','aa');
Query OK, 1 row affected (0.06 sec)
```

（2）用默认客户端字符集导出表 emp。

```
[root@localhost ~]# mysqld --verbose --help|grep character-set-server|grep -v name
character-set-server                                            utf8
[root@localhost ~]# mysqldump -uroot --compact employees emp >emp.sql
[root@localhost ~]#  more emp.sql
/*!40101 SET @saved_cs_client     = @@character_set_client */;
/*!40101 SET character_set_client = utf8 */;
CREATE TABLE `emp` (
  `id` int(11) NOT NULL DEFAULT '0',
  `name` varchar(200) DEFAULT NULL,
  `content` text,
  PRIMARY KEY (`id`)
) ENGINE=InnoDB DEFAULT CHARSET=latin1;
/*!40101 SET character_set_client = @saved_cs_client */;
INSERT INTO `emp` VALUES (1,'z1','aa'),(2,'ä‚-å›½','aa');
```

可以发现"中国"这两个汉字已经变成了乱码。

（3）手工设置客户端字符集为 latin1，重新导出。

```
[root@localhost ~]# mysqldump -uroot --compact --default-character-set=latin1 employees emp >emp.sql
[root@localhost ~]# more emp.sql
/*!40101 SET @saved_cs_client     = @@character_set_client */;
/*!40101 SET character_set_client = utf8 */;
CREATE TABLE `emp` (
  `id` int(11) NOT NULL DEFAULT '0',
  `name` varchar(200) DEFAULT NULL,
  `content` text,
  PRIMARY KEY (`id`)
) ENGINE=InnoDB DEFAULT CHARSET=latin1;
/*!40101 SET character_set_client = @saved_cs_client */;
INSERT INTO `emp` VALUES (1,'z1','aa'),(2,'中国','aa');
```

这次，中文字符可以正确地导出。

5．其他常用选项

另外，还有两个常用的选项值得一提。

- -F --flush-logs（备份前刷新日志）：加上此选项后，备份前将关闭旧日志，生成新日志。使得进行恢复的时候直接从新日志开始进行重做，大大方便了恢复过程。
- -l --lock-tables（给所有表加读锁）：可以在备份期间使用，使得数据无法被更新，从而使备份的数据保持一致性，可以配合-F选项一起使用。

23.1.6　mysqlpump（并行的数据导出工具）

从 MySQL 5.7.8 开始，MySQL 官方提供了一个新的数据导出工具 mysqlpump。相对于 mysqldump，mysqlpump 有以下几个特点。

- 并行导出表数据，提高导出效率。
- 更好地控制数据库和表、存储过程、用户账号的导出。
- 导出用户账号通过 create user 和 grant 语句实现。
- 导出文件的同时支持在线压缩。
- 显示导出进度。
- 先导出建表语句，接着导出 insert 语句，最后再导出二级索引，提高数据导入速度。

mysqlpump 大多数选项与 mysqldump 是一致的，下面只介绍 mysqlpump 独有且常用的参数。

选项	说明
--add-drop-user	每个 create user 命令前，记录 drop user 命令。
--compress-output=algorithm	压缩结果文件，默认不开启。支持 LZ4 和 ZLIB 压缩算法。
--default-parallelism=N	默认并行线程数。默认值是 2。
--defer-table-indexes	导出文件先记录建表语句，再记录 insert 语句，最后创建二级索引。
--include-databases=db_list	导出数据库的列表，用逗号分隔
--include-tables=table_list	导出表的列表，用逗号分隔
--include-users=user_list	导出用户的列表，用逗号分隔
--parallel-schemas=[N:]db_list	创建一个队列，用于导出数据库列表的表。如果指定 N 的值，那么这个队列将开启 N 个线程；如果 N 未指定，那么将用--default- parallelism 指定的值作为默认线程数。
--users	以 create uer 和 grant 语法导出导出用户。
--watch-progress	显示导出进度。使用--skip-watch-progress 关闭显示进度。

注意：mysqlpump 默认不导出 information_schema、performance_schema、ndbinfo 和 sys 数据库。

下面来看几个例子。

导出所有数据库，默认开启一个队列，两个线程。

```
[mysql3307@localhost ~]$ mysqlpump -uroot -S/tmp/mysql_3307.sock -A > backup.sql
Dump progress: 1/2 tables, 0/0 rows
Dump progress: 10/61 tables, 1142206/2745024 rows
Dump progress: 28/61 tables, 2443005/2745024 rows
Dump completed in 2626 milliseconds
```

导出所有数据库，启用压缩功能。

```
[mysql3307@localhost ~]$ mysqlpump -uroot -S/tmp/mysql_3307.sock -A --compress-output=LZ4 > backup.lz4
Dump progress: 1/2 tables, 0/0 rows
Dump progress: 10/61 tables, 995206/2745024 rows
Dump progress: 19/61 tables, 2008086/2745024 rows
Dump completed in 3084 milliseconds
```

解压 backup.LZ4，如果本机没有 lz4 工具，则可以用 yum install lz4 来安装。

```
[mysql3307@localhost ~]$ lz4 -d backup.lz4
Decoding file backup
backup.lz4           : decoded 123398461 bytes
```

导出所有数据库。其中一个队列导出 employees 和 testdb 数据库，该队列分配 10 个线程；另一个队列导出剩下的数据库，队列分配 3 个线程。

```
mysqlpump -uroot -S/tmp/mysql_3307.sock -A --parallel-schemas=10:employees,testdb --default-parallelism=3 > backup.sql
 Dump progress: 1/2 tables, 0/0 rows
 Dump progress: 54/60 tables, 1253262/2745023 rows
 Dump progress: 62/64 tables, 2443906/2749076 rows
 Dump completed in 4492 milliseconds
```

导出数据库 db_master1 的表结构和所有用户。

```
[mysql3307@localhost ~]$ mysqlpump -uroot -S/tmp/mysql_3307.sock __skip_dump_rows --users -B db_master1 > backup.sql
 Dump completed in 1484 milliseconds

[mysql3307@localhost ~]$ more  backup.sql
-- Dump created by MySQL pump utility, version: 5.7.22, linux-glibc2.12 (x86_64)
-- Dump start time: Wed Jul 25 20:27:35 2018
-- Server version: 5.7.22

GRANT SELECT ON *.* TO 'emp'@'%';
CREATE USER 'repl'@'%' IDENTIFIED WITH 'mysql_native_password' AS '*A424E797031BF97C69A2E88CF9231C5C2038C039' REQUIRE NONE PASSWORD EXPIRE DEFAULT ACCOUNT UNLOCK;
```

23.1.7 mysqlimport（数据导入工具）

mysqlimport 是客户端数据导入工具，用来导入 mysqldump 加 -T 选项后导出的文本文件。它实际上是客户端提供了 LOAD DATA INFILEQL 语句的一个命令行接口。用法和 LOAD DATA INFILE 子句非常类似，第 25 章将对 mysqlimport 和 LOAD DATA INFILE 的用法进行详细的介绍，这里不再赘述。

mysqlimport 的基本用法如下：

```
shell> mysqlimport [options] db_name textfile1         [textfile2 ...]
```

23.1.8 mysqlshow（数据库对象查看工具）

mysqlshow 客户端对象查找工具，用于快速查找存在哪些数据库、数据库中的表、表中的列或索引。和 mysql 客户端工具很类似，不过有些特性是 mysql 客户端工具所不具备的。

mysqlshow 的使用方法如下：

```
shell> mysqlshow[option] [db_name [tbl_name [col_name]]]
```

如果不加任何选项，默认情况下会显示所有数据库。下例显示了当前 MySQL 中的所有数据库：

```
[root@localhost ~]# mysqlshow -uroot
+--------------------+
|     Databases      |
+--------------------+
| information_schema |
| employees          |
| mysql              |
| performance_schema |
| sys                |
+--------------------+
```

下面是 mysqlshow 的一些常用选项。

（1）--count（显示数据库和表的统计信息）。

如果不指定数据库，则显示每个数据库的名称、表数量、记录数量；如果指定数据库，则显示指定数据库的每个表名、字段数量，记录数量；如果指定具体数据库中的具体表，则显示表的字段信息，如下例所示。

不指定数据库：

```
[root@localhost ~]# mysqlshow -uroot --count
+--------------------+--------+------------+
|     Databases      | Tables | Total Rows |
+--------------------+--------+------------+
| information_schema |    61  |     333668 |
| employees          |    19  |    4519075 |
| mysql              |    31  |       2949 |
| performance_schema |    87  |      46532 |
| sys                |   101  |       4456 |
+--------------------+--------+------------+
6 rows in set.
```

指定数据库：

```
[root@localhost ~]# mysqlshow -uroot employees --count
Database: employees
+----------------------+---------+------------+
|        Tables        | Columns | Total Rows |
+----------------------+---------+------------+
| current_dept_emp     |    4    |    300024  |
| dept_emp             |    4    |    331603  |
| dept_emp_latest_date |    3    |    300024  |
| dept_manager         |    4    |        24  |
| employees            |    6    |    300024  |
| salaries             |    4    |   2844047  |
+----------------------+---------+------------+
6 rows in set.
```

指定数据库和表：

```
[root@localhost ~]# mysqlshow -uroot employees departments --count
Database: employees  Table: departments  Rows: 9
+-----------+-------------+-----------------+------+-----+---------+-------+------------------------------+---------+
|Field      |Type         |Collation        |Null  |Key  |Default  |Extra  |Privileges                    |Comment  |
+-----------+-------------+-----------------+------+-----+---------+-------+------------------------------+---------+
|dept_no    |char(4)      |utf8_unicode_ci  |NO    |PRI  |         |       |select,insert,update,references|        |
|dept_name  |varchar(40)  |utf8_unicode_ci  |NO    |UNI  |         |       |select,insert,update,references|        |
+-----------+-------------+-----------------+------+-----+---------+-------+------------------------------+---------+
```

（2）-k 或者--keys（显示指定表中的所有索引）。

此选项显示了两部分内容，一部分是指定表的表结构，另外一部分是指定表的当前索引信息。下例显示了 employees 库中表 departments 的表结构和当前索引信息：

```
[root@localhost ~]# mysqlshow -uroot employees departments --count -k
Database: employees  Table: departments  Rows: 9
+-----------+-------------+-----------------+------+-----+---------+-------+--------------------+
| Field     | Type        | Collation       | Null | Key | Default | Extra | Privileges         |
|           | Comment     |                 |      |     |         |       |                    |
+-----------+-------------+-----------------+------+-----+---------+-------+--------------------+
| dept_no   | char(4)     | utf8_unicode_ci | NO   | PRI |         |       | select,insert,
update,references |             |
| dept_name | varchar(40) | utf8_unicode_ci | NO   | UNI |         |       | select,insert,
update,references |             |
+-----------+-------------+-----------------+------+-----+---------+-------+--------------------+
```

```
+--------------+------------+------------+--------------+------------+-------------+----------+
| Table        | Non_unique | Key_name   | Seq_in_index | Column_name| Collation   | Cardinality | Sub_part |
| Packed | Null | Index_type | Comment    | Index_comment |
+--------------+------------+------------+--------------+------------+-------------+----------+
| departments  | 0          | PRIMARY    | 1            | dept_no    | A           | 9        |
|        |      | BTREE      |            |            |
| departments  | 0          | dept_name  | 1            | dept_name  | A           | 9        |
|        |      | BTREE      |            |            |
+--------------+------------+------------+--------------+------------+-------------+----------+
```

细心的读者可能会发现，显示的内容实际上和在 mysql 客户端执行"show full columns from departments"和"show index from departments"的结果完全一致。

（3）-i 或者--status（显示表的一些状态信息）。

下例显示了 test 数据库中 emp 表的一些状态信息：

```
[root@localhost ~]# mysqlshow -uroot employees departments -i
+-------------+--------+---------+------------+------+----------------+-------------+-----------------+
| Name        | Engine | Version | Row_format | Rows | Avg_row_length | Data_length | Max
_data_length | Index_length | Data_free | Auto_increment | Create_time | Update_time |
Check_time   | Collation  | Checksum | Create_options | Comment       |
+-------------+--------+---------+------------+------+----------------+-------------+-----------------+

| departments | InnoDB | 10      | Dynamic    | 9    | 1820           | 16384       | 0
  | 16384        | 0         |                | 2018-05-23 14:52:20 |             |
  | utf8_unicode_ci |       |                |               |
+-------------+--------+---------+------------+------+----------------+-------------+-----------------+
```

此命令和 mysql 客户端执行"show table status from employees like 'departments'"的结果完全一致。

23.1.9 perror（错误代码查看工具）

在 MySQL 的使用过程中，可能会出现各种各样的 error。这些 error 有些是由于操作系统引起的，比如文件或者目录不存在；有些则是由于存储引擎使用不当引起的。这些 error 一般都有一个代码，类似于"error: #"或者"Errcode: #"，"#"代表具体的错误号。perror 的作用就是解释这些错误代码的详细含义。

perror 的用法很简单，如下所示：

```
perror [OPTIONS] [ERRORCODE [ERRORCODE...]]
```

在下面的例子中，可以看一下错误号 30 和 60 分别指的是什么错误：

```
[root@localhost ~]# perror 30 60
OS error code  30:  Read-only file system
OS error code  60:  Device not a stream
```

23.1.10 MySQL Shell

MySQL Shell 是 MySQL Server 的高级客户端和代码编辑器。除了提供类似 MySQL 的 SQL 功能，MySQL Shell 还提供了 JavaScript 和 Python 的脚本功能，并包含用于处理 MySQL 的 API。MySQL Shell 支持批量执行代码，支持 Tabbed、Table 和 JSON 这 3 种输出格式。

MySQL Shell 的用法如下：

```
shell> mysqlsh [OPTIONS] [URI]
shell> mysqlsh [OPTIONS] [URI] -f <path> [script args...]
shell> mysqlsh [OPTIONS] [URI] --dba [command]
shell> mysqlsh [OPTIONS] [URI] -cluster
```

以下是登录 MySQL Shell 常用的选项：

```
-e, --execute=<cmd>        执行命令并退出
-f, --file=file            要在批处理模式下处理的文件
--uri=value                指定连接字符串，格式为[user[:pass]@]host[:port][/db]
-mx, --mysqlx              以 X 协议创建连接，使用 mysqlx_port 端口连接 MySQL
-mc, --mysql               以经典模式创建连接，使用 port 端口连接 MySQL
--sql                      以 SQL 模式登录，使用 port 端口连接 MySQL
--sqlx                     以 X 协议的 SQL 模式登录。使用 mysqlx_port 端口连接 MySQL
--js, --javascript         以 JS 模式登录
--py, --python             以 python 模式登录
--json[=format]            以 json 格式打印输出
--table                    以表格格式显示输出
--log-level=value          指定日志级别，可以是 1 到 8 的整数，或是[none,internal,error,warning,info,debug,
debug2,debug3]
```

登录 MySQL Shell 后，可以使用 \? 命令调出帮助文档：

```
MySQL  127.0.0.1:34880+ ssl  JS > \?
```

帮助文档主要包括 5 个方面。

● AdminAPI：包括 dba 全局 object 和 InnoDB Cluster 管理 API，用于创建和管理 InnoDB Cluster 环境。使用\? dba 命令可调出详细帮助信息。

● Shell 命令：mysqlsh 自带的 shell 命令帮助文档。

● ShellAPI：包含 shell 和 util 全局 object 的帮助信息；在 mysql 模式下，还包含 mysql 模块的帮助信息。可分别使用\? Shell、\? util、\? mysql 调出详细帮助信息。

● SQL 语法：列出指定 SQL 语句的帮助信息。使用\? sql syntax 命令调出详细帮助信息，其中 sql syntax 为具体的 sql 语法，例如使用\? select 查看 select 的语法帮助。

● X DevAPI：mysqlx 模块和 X DevAPI 的使用方法，使用\? mysqlx 调出详细帮助信息。

1．MySQL Shell 软件的安装

（1）以 8.0.12 版本为例，下载 Linux x86 二进制软件安装包：

```
[mysql3488@localhost ~]$ wget https://cdn.mysql.com//Downloads/MySQL-Shell/mysql-shell
-8.0.12-linux-glibc2.12-x86-64bit.tar.gz
```

（2）解压安装包：

```
[mysql3488@ localhost ~]$ tar xvfz mysql-shell-8.0.12-linux-glibc2.12-x86-64bit.tar.gz
[mysql3488@ localhost ~]$ cd mysql-shell-8.0.12-linux-glibc2.12-x86-64bit
```

（3）配置 MySQL Shell 环境变量：

```
[mysql3488@ localhost ~]$ export MYSQL_SHELL_HOME=/home/mysql3488/mysql-shell-8.0.12-linux-
glibc2.12-x86-64bit
[mysql3488@ localhost ~]$ export PATH=$MYSQL_SHELL_HOME/bin:$PATH
[mysql3488@ localhost ~]$ export C_INCLUDE_PATH=$MYSQL_SHELL_HOME/include:$C_INCLUDE_PATH
[mysql3488@ localhost ~]$ export LD_LIBRARY_PATH=$MYSQL_SHELL_HOME/lib:$LD_LIBRARY_PATH
[mysql3488@ localhost ~]$ which mysqlsh
~/mysql-shell-8.0.12-linux-glibc2.12-x86-64bit/bin/mysqlsh
```

（4）使用 emp 用户，以 SQL 模式登录 MySQL Shell：

```
[mysql3488@ localhost ~]$  mysqlsh 'emp':emp123@'127.0.0.1':3488/employees --sql
mysqlsh: [Warning] Using a password on the command line interface can be insecure.
Creating a session to 'emp@127.0.0.1:3488/employees'
Fetching schema names for autocompletion... Press ^C to stop.
Fetching table and column names from `employees` for auto-completion... Press ^C to stop.
Your MySQL connection id is 71
```

```
Server version: 8.0.11 MySQL Community Server - GPL
Default schema set to `employees`.
MySQL Shell 8.0.12
Type '\help' or '\?' for help; '\quit' to exit.

MySQL  127.0.0.1:3488 ssl  employees  SQL > show databases;
+--------------------+
| Database           |
+--------------------+
| employees          |
| information_schema |
+--------------------+
2 rows in set (0.0012 sec)
```

（5）从 SQL 模式切换到 JS 模式：

```
MySQL  127.0.0.1:3488 ssl  employees  SQL > \js
Switching to JavaScript mode...

MySQL  127.0.0.1:3488 ssl  employees  JS >
```

2. MySQL Shell 使用 JS 模式查询数据

（1）确认 MySQL 安装了 mysqlx 插件。

```
mysql>show plugins;
+----------------------+---------+---------+---------+---------+
| Name                 | Status  | Type    | Library | License |
+----------------------+---------+---------+---------+---------+
......
| mysqlx               | ACTIVE  | DAEMON  | NULL    | GPL     |
| mysqlx_cache_cleaner | ACTIVE  | AUDIT   | NULL    | GPL     |
......
+----------------------+---------+---------+---------+---------+
```

（2）启用 mysqlx 插件。

在 my.cnf 配置 mysqlx 参数，并重启实例。测试实例端口为 3488，mysqlx 的监听端口相应设置为 34880：

```
[mysqld]
loose_mysqlx_socket = mysqlx.sock
loose_mysqlx_port = 34880
```

确认 mysqlx 端口已启用：

```
mysql>show variables like 'mysqlx_port';
+---------------+-------+
| Variable_name | Value |
+---------------+-------+
| mysqlx_port   | 34880 |
+---------------+-------+
1 row in set (0.00 sec)
```

（3）通过 mysqlx 协议登录 MySQL Shell：

```
shell>mysqlsh 'emp':emp123@'127.0.0.1':34880 --js -mx
Creating an X protocol session to 'emp@127.0.0.1:34880/employees'
Fetching schema names for autocompletion... Press ^C to stop.
Your MySQL connection id is 13 (X protocol)
Server version: 8.0.11 MySQL Community Server - GPL
Default schema `employees` accessible through db.
MySQL Shell 8.0.12
Type '\help' or '\?' for help; '\quit' to exit.

MySQL  127.0.0.1:34880+ ssl  JS >
```

（4）在 JS 模式下查询表记录。

设置当前 db 变量：

```
MySQL    127.0.0.1:34880+ ssl  JS > var employees=session.setCurrentSchema('employees')
MySQL    127.0.0.1:34880+ ssl  JS > employees
<Schema:employees>
```

查看 db 下所有的表：

```
MySQL    127.0.0.1:34880+ ssl  JS > employees.getTables()
[
    <Table:departments>,
    <Table:dept_emp>,
    <Table:dept_manager>,
    <Table:employees>,
    <Table:sa1>,
    <Table:salaries>,
    <Table:titles>,
    <Table:titles1>
]
```

设置 departments 表变量：

```
MySQL    127.0.0.1:34880+ ssl  JS > var departments=employees.getTable('departments')
MySQL    127.0.0.1:34880+ ssl  JS > departments
<Table:departments>
```

查看 departments 表记录：

```
MySQL    127.0.0.1:34880+ ssl  JS > departments.select()
+---------+--------------------+
| dept_no | dept_name          |
+---------+--------------------+
| d009    | Customer Service   |
| d005    | Development        |
| d002    | Finance            |
| d003    | Human Resources    |
| d001    | Marketing          |
| d004    | Production         |
| d006    | Quality Management |
| d008    | Research           |
| d007    | Sales              |
+---------+--------------------+
9 rows in set (0.0012 sec)
```

通过上面的命令，读者可以了解 MySQL Shell 的一些用法，在操作上也比较简单。遇到不熟悉的操作时，多使用 MySQL Shell 提供的帮助命令。

MySQL Shell 的一个重要的功能是配置 InnoDB Cluster 环境，详细的配置方法可以参考第 31 章中的相关内容。

23.2 Percona 工具包

上面介绍了 MySQL 官方提供的一些常用工具，但在实际工作过程中，官方提供的工具并不能完全解决各种问题。MySQL 的开源社区中涌现了不少高质量的工具，Percona Toolkit（工具包）就是其中的佼佼者。

Percona Toolkit 前身源自于 Maatkit 和 Aspersa 工具，是各种高级命令行工具的集合，用于执行重复或者复杂的 MySQL 任务和系统任务。这些工具相互独立，不依赖其他安装包。该工具集涵盖了 MySQL 的开发管理、性能优化、配置管理、监控管理、复制管理、系统统计等诸多方面。下面介绍几个较为常用的工具。

23.2.1 pt-archiver（数据归档工具）

pt-archiver 是一个数据归档工具，用于迁移源表数据到另外一个表或者文件，也可以删除源表数据。pt-archiver 工具原理是根据表中索引（默认是主键索引）找到第一条记录，然后顺序找到更多需要归档的记录，从而小批量地、慢慢地归档数据，以减少对线上业务的影响。该工具主要适用于历史数据迁移或者删除过期的数据等比较耗时而重复性又比较高的场景。

pt-archive 的用法如下：

```
shell> pt-archiver [OPTIONS] --source DSN --where WHERE
```

下面是一些常用的选项：

```
--analyze              任务完成后，分析表
--ascend-first         按照索引的最左列做升序查找，避免在有多字段的索引上，根据多列字段查找数据
--charset              设置字符集
--[no]check-charset    确认连接和表的字符集是否一致，默认开启
--check-interval       如果使用--check-slave-lag 选项，指定检测从库延迟的间隔，默认 1s
--check-slave-lag      是否开启从库延迟检查，默认开启。当从库延迟超过--max-lag 指定的值后，任务将暂停，直至从库延迟的值小于--max-lag 指定的值
--columns              只迁移指定的列，列名用逗号分隔
--commit-each          归档完一批数据后，立刻提交事务。这个参数与--txn-size 互斥，与--limit 配合使用
--dest                 DSN 格式，指定数据迁移到的目的地。如果--dest 的值与--source 一致，可忽略此参数
--dry-run              仅打印归档过程运行的 sql 语句，而不真正执行
--file                 将数据归档到这个文件，导出的数据格式跟 SELECT INTO OUTFILE 一致
--host                 pt 工具连接的主机 ip
--ignore               为 insert 语句加上 ignore 参数
--limit                每次从源表取记录并归档到目的地的记录数量
--max-lag              从库最大延迟时间，默认 1s。配合--check-slave-lag 使用
--no-delete            数据归档到目的地后，不删除源表的数据。如果不指定这个参数，默认删除源表数据
--optimize             任务完成后，优化表
--password             指定密码
--pid                  运行任务的时候，生成 pid 文件，可避免重复执行归档任务
--progress             指定操作 N 条记录后，打印进度信息。进度信息包括当前时间，已消耗时间，已迁移记录数
--purge                在源表删除数据，不进行数据归档
--run-time             指定任务运行时间。超过指定时间后，即使数据未迁移完，任务也会停止
--sleep                指定每次在源表取数的间隔
--source               DSN 格式，指定数据源端的连接方式
--statistics           收集信息和打印任务总的时间统计
--txn-size             指定每个事务迁移 N 条记录。N 为 0 表示不开启事务，启用事务自动提交
--where                增加 where 条件，过滤需要归档的数据
```

下面来看几个具体例子。

删除 employees.salaries 表 from_date 小于 1987-02-25 的数据，每删除 1 万条数据打印一次进度信息，每次删除 1000 条数据。

```
[mysql3307@localhost employees]$ pt-archiver --source h=127.0.0.1,D=employees,t=salaries,u=root,P=3307 --charset=utf8 --where "from_date <'1987-02-25'" --purge --progress=10000 --limit 1000
TIME                 ELAPSED    COUNT
2018-07-28T15:44:38        0        0
2018-07-28T15:44:52       13    10000
2018-07-28T15:45:05       27    20000
2018-07-28T15:45:19       40    30000
2018-07-28T15:45:33       54    40000
2018-07-28T15:45:46       68    50000
2018-07-28T15:46:00       81    60000
2018-07-28T15:46:04       86    63250
```

将 employees.salaries 表 from_date 小于 1987-02-25 的数据，归档到 employees.salaries_history 表。每次归档 2 万条记录，归档 2 万条记录后立刻提交，每次归档间隔 2s。任务结束后，打印总的统计信息。

归档前查看 employees.salaries 符合条件的记录数。

```
mysql>select count(0) from salaries where from_date <'1987-02-25';
+----------+
| count(0) |
+----------+
|    63250 |
+----------+
1 row in set (1.20 sec)
```

查看 employees.salaries_history 表的记录数。

```
mysql>select count(0) from salaries_history;
+----------+
| count(0) |
+----------+
|        0 |
+----------+
1 row in set (0.00 sec)
```

开始把数据归档到历史表。

```
[mysql3307@localhost employees]$ pt-archiver --source h=127.0.0.1,D=employees,t=salaries,u=root,
P=3307 --dest h=127.0.0.1,D=employees,t=salaries_history,u=root,P=3307 --charset=utf8 --where "
from_date <'1987-02-25'" --limit 20000 --commit-each --sleep 2 --statistics
Started at 2018-07-28T16:02:45, ended at 2018-07-28T16:03:37
Source: A=utf8,D=employees,P=3307,h=127.0.0.1,t=salaries,u=root
Dest:   A=utf8,D=employees,P=3307,h=127.0.0.1,t=salaries_history,u=root
SELECT 63250
INSERT 63250
DELETE 63250
Action         Count       Time        Pct
deleting       63250     18.8361      36.53
inserting      63250     15.0347      29.16
sleep              4      8.0010      15.52
select             5      1.6581       3.22
commit            10      0.2996       0.58
other              0      7.7289      14.99
```

根据统计信息，可以知道每个操作类型消耗的总时间：删除数据用了 18s，插入数据用了 15s；其间 sleep 4 次，一共 8s。

看看 employees.salaries 表的记录数。

```
mysql>select count(0) from salaries where from_date <'1987-02-25';
+----------+
| count(0) |
+----------+
|        0 |
+----------+
1 row in set (1.13 sec)
```

结果显示，employees.salaries from_date 在 1987-02-25 前的 63 250 条数据已经被删掉，再看 employees.salaries_history 表的记录数：

```
mysql>select count(0) from salaries_history;
+----------+
| count(0) |
+----------+
|    63250 |
+----------+
1 row in set (0.02 sec)
```

employees.salaries_history 有 63 250 条记录，跟我们的预期是一致的。

再看看把记录归档到文件中的例子。

将 employees.salaries 表 from_date 小于 1987-02-25 的数据，归档到/tmp 目录下，归档文件名为归档时间和数据库名、表名的组合。每次归档 100 条记录，任务最多只运行 10s，不删除源表的记录，打印总的统计信息。

开始任务前，查看 employees.salaries 表符合条件的记录数：

```
mysql>select count(0) from salaries where from_date <'1987-02-25';
+----------+
| count(0) |
+----------+
|    63250 |
+----------+
1 row in set (1.20 sec)
```

开始归档数据：

```
[mysql3307@localhost tmp]$ time pt-archiver --source h=127.0.0.1,D=employees,t=salaries,u=root,
P=3307 --file '/tmp/%Y-%m-%d-%D.%t' --charset=utf8 --where "from_date <'1987-02-25'" --limit
100 --run-time 10 --no-delete --statistics
Started at 2018-07-28T16:33:36, ended at 2018-07-28T16:33:46
Source: A=utf8,D=employees,P=3307,h=127.0.0.1,t=salaries,u=root
SELECT 40000
INSERT 0
DELETE 0
Action          Count      Time       Pct
commit          39904      3.1803     31.80
select            400      1.6762     16.76
print_file      39903      0.3048      3.05
other               0      4.8389     48.39

real    0m13.256s
user    0m5.119s
sys     0m1.869s
```

注意：根据 time 命令统计的运行时间是 13.256s，pt-archiver 的 Time 统计时间累加是 10s。所以 --run-time 是按照 pt-archiver 本身的统计时间为准的。

再次查询 employees.salaries 表符合条件的记录数：

```
mysql>select count(0) from salaries where from_date <'1987-02-25';
+----------+
| count(0) |
+----------+
|    63250 |
+----------+
1 row in set (1.32 sec)
```

可以看到，employees.salaries 表的记录没有被删除。

查看归档的文件：

```
[mysql3307@localhost tmp]$ more 2018-07-28-employees.salaries
10001   60117   1986-06-26   1987-06-26
10004   40054   1986-12-01   1987-12-01
10009   60929   1985-02-18   1986-02-18
10009   64604   1986-02-18   1987-02-18
10009   64780   1987-02-18   1988-02-18
10013   40000   1985-10-20   1986-10-20
...
```

employees.salaries 表数据量很大，即使没有完全归档数据，10s 后任务自动停止了。通过限定任务的执行时间，可以减少对现有环境的影响。

23.2.2　pt-config-diff（参数对比工具）

pt-config-diff 用于找出配置文件和 MySQL 内存值不一样的参数，其使用方法：

```
shell> pt-config-diff [OPTIONS] CONFIG CONFIG [CONFIG...]
```

CONFIG 可以是配置文件名字，也可以是 DSN 连接串。pt-config-diff 只比较两个 CONFIG

都包含的变量。

以下是一些常用的选项：

```
--charset               指定连接字符集
--defaults-file         指定 mysql 的配置文件
--host                  连接的主机 ip
--[no]ignore-case       对比的参数是否区分大小写。默认为 yes
--ignore-variables      忽略比较的变量
```

比较两个配置文件，区分大小写：

```
[mysql3307@localhost mysqlhome]$ pt-config-diff --noignore-case my.cnf my1.cnf
2 config differences
Variable                    my.cnf              my1.cnf
=========================   =================   =================
init_connect                SET NAMES utf8      set NAMES utf8
port                        3307                3308
```

可以看到，命令区分大小写后，set 和 SET 被识别为两个值。

比较 MySQL 实例内存与 cnf 配置文件：

```
[mysql3307@localhost mysqlhome]$ pt-config-diff h=127.0.0.1,u=root,P=3307 my1.cnf
10 config differences
Variable                    localhost                   my1.cnf
=========================   =========================   =========================
innodb_buffer_pool_size     5368709120                  5242880000
port                        3307                        3308
secure_file_priv            /tmp/                       /tmp
slave_parallel_type         DATABASE                    LOGICAL_CLOCK
```

23.2.3 pt-duplicate-key-checker（检查冗余索引工具）

这个工具用于检查表中是否有冗余的索引和外键。工具通过 show create table 查看表的索引，如果一个索引与另外一个索引具有相同的列，并且列在索引的顺序是一样的，或者被另一个索引的前几列所覆盖，那么这个索引就是重复索引；检查冗余外键时，如果两个外键覆盖的字段一样，且字段指向的父表也一样，那么其中一个外键是冗余的。最后工具给出优化对应的索引或者外键的 SQL 语句。

pt-duplicate-key-checker 的使用方法如下：

```
shell> pt-duplicate-key-checker [OPTIONS] [DSN]
```

常用参数如下：

```
--databases,-d          只检查列出的数据库的表，每个数据库用逗号分隔
--defaults-file         从给定的配置文件读取 MySQL 参数
--ignore-databases      不检查这些数据库下的表，，每个数据库用逗号分隔
--ignore-order          忽略索引字段顺序。选项启用后，key(a,b)和 key(b,a)属于同一索引
--ignore-tables         忽略检查的表。表名前要加上 db 名，用逗号分隔
--key-types             检查的 key 类型，默认为 fk。f 为外键，k 为索引
--[no]sql               打印 drop key 的 sql 语句，默认开启
--[no]summary           打印索引统计信息，默认开启
--tables,-t             只检查指定表。表名前要加上 db 名，用逗号分隔
```

下面来看一些具体的例子。

检查重复、冗余索引，不检查外键。

```
[mysql3307@localhost ~]$ pt-duplicate-key-checker --socket /tmp/mysql_3307.sock --user root
--key-types=k
# ######################################################################
# employees.departments
# ######################################################################
```

```
# idx_dept_no is a duplicate of PRIMARY
# Key definitions:
#   KEY `idx_dept_no` (`dept_no`)
#   PRIMARY KEY (`dept_no`),
# Column types:
#         `dept_no` char(4) collate utf8_unicode_ci not null
# To remove this duplicate index, execute:
ALTER TABLE `employees`.`departments` DROP INDEX `idx_dept_no`;
```

结果显示，idx_dept_no 为冗余索引，工具给出了 drop 索引的命令。

23.2.4 pt-find（查找工具）

pt-find 工具通过 show tables 和 show table status 获得信息，查找符合特定规则的表。默认输出数据库名和表名。

pt-find 的使用方法如下：

```
shell> pt-find [OPTIONS] [DATABASES]
```

常用参数如下：

```
--case-insensitive          大小写不敏感
--charset                   设置默认连接的字符集
--or                        用 or 替代 and 连接输入的命令；or 和 and 不能混合使用
```

下面来看几个例子。

（1）查看数据库大于 100MB 的表：

```
[mysql3307@localhost ~]$ pt-find --user root -S/tmp/mysql_3307.sock --tablesize +100M
`employees`.`salaries`
```

（2）查看一天内创建的存储引擎为 innodb 的表：

```
[mysql3307@localhost ~]$ pt-find --user root -S/tmp/mysql_3307.sock --ctime -1 --engine InnoDB
`employees`.`departments`
```

23.2.5 pt-heartbeat（监控主从延迟工具）

pt-heartbeat 工具用于监控从库复制是否延迟，通过在主库更新心跳表的时间戳字段，从而对比从库该字段与主机的时间差异，来评估延迟情况。

注意：主从 MySQL 实例的主机时间要保持一致。

pt-heartbeat 工具的使用方法如下：

```
shell> pt-heartbeat [OPTIONS] [DSN] --update|--monitor|--check|--stop
```

常用参数如下：

```
--check                 只检查一次从库的延迟。如果指定 --recurse 选项，会检查所有的级联从库
--check-read-only       如果开启，工具不在 read_only=on 的实例做任何插入操作
--create-table          创建心跳表
--daemonize             把任务放到后台运行
--database              连接的数据库名
--interval              更新或检查心跳表的间隔，默认 1s
--log                   把任务放到后台运行后，可以指定输出到一个 log 文件
--master-server-id      指定当前数据库的主库的 server-id。如果未指定，工具会尝试主动获取
--monitor               监控从库延迟。每秒打印出当前从库的延迟情况
--recurse               递归检查从库延迟
--replace               在 --update 选项中，用 replace 操作代替 update 操作
--run-time              任务的最大运行时间
--slave-user            指定连接从库的用户名
--slave-password        指定连接从库的用户密码
--update                更新主库心跳表的时间戳
```

下面来看一些具体的例子,首先创建心跳表的数据库:

```
root@localhost:mysql_3307.sock   [(none)]>create database percona;
Query OK, 1 row affected (0.00 sec)

root@localhost:mysql_3307.sock   [(none)]>use percona;
Database changed

root@localhost:mysql_3307.sock   [percona]>show tables;
Empty set (0.00 sec)
```

创建心跳表:

```
[mysql3307@localhost ~]$ pt-heartbeat --user=root  --socket=/tmp/mysql_3307.sock -D percona
--create-table --master-server-id=7237  --check
1.00

root@localhost:mysql_3307.sock   [percona]>show tables;
+-------------------+
| Tables_in_percona |
+-------------------+
| heartbeat         |
+-------------------+
1 row in set (0.00 sec)

root@localhost:mysql_3307.sock   [percona]>select * from heartbeat\G
*************************** 1. row ***************************
                  ts: 2018-08-02 15:29:05
           server_id: 7237
                file: NULL
            position: NULL
  relay_master_log_file: NULL
   exec_master_log_pos: NULL
1 row in set (0.00 sec)
```

可以看到,心跳表已经创建好。如果未指定,表名默认为 heartbeat,ts 为时间戳字段。

运行后台任务,每秒更新心跳时间戳,任务只运行 30s。

```
[mysql3307@localhost ~]$ pt-heartbeat --user=root  --socket=/tmp/mysql_3307.sock -D percona
--master-server-id=7237 --interval 1 --run-time 30 --update --daemonize
```

30s 后,pt-heartbeat 进程退出,查看心跳表:

```
root@localhost:mysql_3307.sock   [percona]>select * from heartbeat\G
*************************** 1. row ***************************
                  ts: 2018-08-02T15:36:24.001270
           server_id: 7237
                file: mysql-bin.000043
            position: 40153
  relay_master_log_file: NULL
   exec_master_log_pos: NULL
1 row in set (0.00 sec)
```

时间戳 ts、file 和 position 字段已经更新了。

连接从库,监控从库的复制延迟:

```
[mysql3307@localhost ~]$ pt-heartbeat --user=root  --socket=/tmp/mysql_3308.sock -D percona
--master-server-id=7237 --interval 1 --monitor
   31.00s [  0.52s,  0.10s,  0.03s ]
   32.00s [  1.05s,  0.21s,  0.07s ]
   33.00s [  1.60s,  0.32s,  0.11s ]
   34.00s [  2.17s,  0.43s,  0.14s ]
   35.00s [  2.75s,  0.55s,  0.18s ]
   36.00s [  3.35s,  0.67s,  0.22s ]
   37.00s [  3.97s,  0.79s,  0.26s ]
   38.00s [  4.60s,  0.92s,  0.31s ]
    0.00s [  4.60s,  0.92s,  0.31s ]
    0.91s [  4.62s,  0.92s,  0.31s ]
    1.91s [  4.65s,  0.93s,  0.31s ]
    2.91s [  4.70s,  0.94s,  0.31s ]
```

```
0.00s  [  4.70s,  0.94s,  0.31s ]
0.00s  [  4.70s,  0.94s,  0.31s ]
```

pt-heartbeat 监控从库延迟输出有 4 列，第一列是当前时间跟从库时间戳 ts 字段的差值，单位为 s；后三列为 1min、5min、15min 的平均延迟。

23.2.6　pt-kill（杀死会话工具）

pt-kill 工具用于杀死找到符合条件的连接。pt-kill 可以通过执行 show processlist 得到信息，也可以通过给定文件得到信息。

pt-kill 工具的使用方法如下：

```
shell> pt-kill [OPTIONS] [DSN]
```

常用参数如下：

```
--create-log-table      创建存放 log 的表
--daemonize             把任务放到后台运行
--database              连接的数据库名
--interval              检查间隔。如果未指定--busy-time，默认为 30s；如果已指定--busy-time，默认为
                        --busy-time 指定值的一半；如果两个都指定，以--interval 的值为准
--log-dsn               在 dsn 指定的数据库，保存被杀掉的查询的信息
--query-id              打印被杀死的查询 ID。打印的 ID 格式跟 pt-query-digest 输出格式一样
--run-time              任务的最大运行时间
--slave-user            指定连接从库的用户名
--slave-password        指定连接从库的用户名
--victims               杀死匹配查询的类别。默认值为 oldest。oldest：只杀死匹配查询中，运行时间最长的；
                        all：杀掉所有匹配的查询；all-but-oldest：除了运行时间最长的查询，其他查询都杀掉
--wait-after-kill       杀掉一个查询后，等待多少时间再杀掉下一个查询。设置这个参数的目的是为了让其他被阻
                        塞的查询有机会继续执行
--wait-before-kill      杀掉一个查询前等待的时间。设置这个参数的目的是让--execute-command 指定的命令
                        有机会查看匹配的查询和收集 MySQL、系统的信息
```

下面看一些具体的例子。

（1）杀死运行时间超过 10s 的查询，打印查询 ID。

执行 sleep 语句：

```
mysql>select sleep(10);
```

运行 pt-kill 命令：

```
[mysql3307@localhost ~]$ pt-kill  --user=root  --socket=/tmp/mysql_3307.sock --busy-time 10
--kill --query-id
```

sleep 语句运行 10s 后，连接被杀掉：

```
mysql>select sleep(10);
ERROR 2013 (HY000): Lost connection to MySQL server during query
```

执行 pt-kill 命令的窗口，打印出 Query ID：

```
Query ID: 0xF9A57DD5A41825CA
```

（2）每 10s 打印 Command 为 Sleep 的所有连接，但是不杀掉连接：

```
[mysql3307@localhost ~]$ pt-kill  --user=root  --socket=/tmp/mysql_3307.sock --match-command
Sleep --print --victims all --interval 10
# 2018-08-02T17:28:58 KILL 118 (Sleep 99 sec) NULL
# 2018-08-02T17:28:58 KILL 117 (Sleep 92 sec) NULL
# 2018-08-02T17:28:58 KILL 119 (Sleep 70 sec) NULL
```

23.2.7　pt-online-schema-change（在线修改表结构工具）

该工具简称 pt-osc，支持在线修改表结构，加锁时间短，对线上业务影响小。此工具首先

根据需要修改的源表的表结构创建一个临时使用的空表，然后在源表创建触发器，接下来开始慢慢地将数据从源表复制到新表中，在复制过程中，如果源表数据发生变化，则通过触发器同步到新表中，当数据完成复制后，调用 rename table 指令，互换源表和新表名字，换下来的废弃的表默认就被删除了。另外，源表必须要有主键或者唯一索引（唯一索引不能存在空值）。

pt-online-schema-change 工具的使用方法如下：

```
shell> pt-online-schema-change [OPTIONS] DSN
```

常用参数如下：

```
--alter                用于修改表结构，不用加上 ALTER TABLE 关键字。多个修改命令用逗号分隔
--charset              设置字符集
--[no]check-alter      检查--alter 命令的正确性，默认 yes
--check-interval       检查从库延迟是否达到--max-lag 的间隔，默认为 1s
--check-slave-lag      检查从库延迟，当延迟超过--max-lag 后，停止复制数据
--chunk-size           每次复制块的记录数，默认为 1000
--chunk-size-limit     每次复制块的尺寸，默认为 4
--chunk-time           根据--chunk-time 的时间动态调节复制块的记录数，默认为 0.5s
--database             连接的数据库名
--[no]drop-new-table   如果复制源表失败，删除新表，默认为 yes。此时源表仍然保留
--[no]drop-old-table   如果成功 rename 到新表，删除源表，默认为 yes
--dry-run              仅创建和修改新表的结构，不复制数据，也不执行 rename 操作
--execute              使用 pt-osc 修改源表结构
--max-lag              如果从库延迟大于--max-lag，复制数据任务暂停，直至从库延迟小于--max-lag，默认为 1s
--new-table-name       指定新表的名字
--print                打印使用的 sql，而不真正执行，配合--dry-run 使用
--sleep                每次复制数据块的间隔，默认为 0s
```

下面看一些具体的例子。

（1）模拟数据库 employees 的 salaries_id 表添加 amount 字段，并不真正执行：

```
[mysql3307@localhost ~]$ pt-online-schema-change --user=root --socket=/tmp/mysql_3307.sock --charset=utf8 --alter "add column amount bigint(10)" D=employees,t=salaries_id --print --dry-run

Operation, tries, wait:
  analyze_table, 10, 1
  copy_rows, 10, 0.25
  create_triggers, 10, 1
  drop_triggers, 10, 1
  swap_tables, 10, 1
  update_foreign_keys, 10, 1
Starting a dry run.  `employees`.`salaries_id` will not be altered.  Specify --execute instead of --dry-run to alter the table.
Creating new table...
CREATE TABLE `employees`.`_salaries_id_new` (
  `id` int(11) NOT NULL AUTO_INCREMENT,
  `emp_no` int(11) NOT NULL,
  `salary` int(11) NOT NULL,
  `from_date` date NOT NULL,
  `to_date` date NOT NULL,
  PRIMARY KEY (`id`),
  KEY `emp_no` (`emp_no`,`from_date`)
) ENGINE=InnoDB AUTO_INCREMENT=14220236 DEFAULT CHARSET=utf8 COLLATE=utf8_unicode_ci
Created new table employees._salaries_id_new OK.
Altering new table...
ALTER TABLE `employees`.`_salaries_id_new` add column amount bigint(10)
Altered `employees`.`_salaries_id_new` OK.
Not creating triggers because this is a dry run.
Not copying rows because this is a dry run.
INSERT LOW_PRIORITY IGNORE INTO `employees`.`_salaries_id_new` (`id`, `emp_no`, `salary`, `from_date`, `to_date`) SELECT `id`, `emp_no`, `salary`, `from_date`, `to_date` FROM `employees`.`salaries_id` FORCE INDEX(`PRIMARY`) WHERE ((`id` >= ?)) AND ((`id` <= ?)) LOCK IN SHARE MODE /*pt-online-schema-change 30889 copy nibble*/
  SELECT /*!40001 SQL_NO_CACHE */ `id` FROM `employees`.`salaries_id` FORCE INDEX(`PRIMARY`) WHERE ((`id` >= ?)) ORDER BY `id` LIMIT ?, 2 /*next chunk boundary*/
```

```
Not swapping tables because this is a dry run.
Not dropping old table because this is a dry run.
Not dropping triggers because this is a dry run.
DROP TRIGGER IF EXISTS `employees`.`pt_osc_employees_salaries_id_del`
DROP TRIGGER IF EXISTS `employees`.`pt_osc_employees_salaries_id_upd`
DROP TRIGGER IF EXISTS `employees`.`pt_osc_employees_salaries_id_ins`
2018-08-07T10:33:16 Dropping new table...
DROP TABLE IF EXISTS `employees`.`_salaries_id_new`;
2018-08-07T10:33:16 Dropped new table OK.
Dry run complete. `employees`.`salaries_id` was not altered.
```

这里可以看到 pt-osc 工具的工作流程，主要的步骤是创建临时表、修改临时表、拷贝数据、切换源表和临时表，以及删除源表。

（2）将数据库 employees 下的 titles 表转换为分区表，按 emp_no 字段进行 hash 分区，分区数为 3：

```
[mysql3307@localhost ~]$ pt-online-schema-change --user=root --socket=/tmp/mysql_3307.sock --charset=utf8 --alter "PARTITION BY HASH( emp_no ) PARTITIONS 3" D=employees,t=titles --execute
Cannot connect to A=utf8,S=/tmp/mysql_3307.sock,h=127.0.0.1,u=root
No slaves found. See --recursion-method if host localhost has slaves.
Not checking slave lag because no slaves were found and --check-slave-lag was not specified.
Operation, tries, wait:
  analyze_table, 10, 1
  copy_rows, 10, 0.25
  create_triggers, 10, 1
  drop_triggers, 10, 1
  swap_tables, 10, 1
  update_foreign_keys, 10, 1
Altering `employees`.`titles`...
Creating new table...
Created new table employees._titles_new OK.
Altering new table...
Altered `employees`.`_titles_new` OK.
2018-08-07T11:06:39 Creating triggers...
2018-08-07T11:06:39 Created triggers OK.
2018-08-07T11:06:39 Copying approximately 441297 rows...
2018-08-07T11:06:49 Copied rows OK.
2018-08-07T11:06:49 Analyzing new table...
2018-08-07T11:06:49 Swapping tables...
2018-08-07T11:06:49 Swapped original and new tables OK.
2018-08-07T11:06:49 Dropping old table...
2018-08-07T11:06:49 Dropped old table `employees`.`_titles_old` OK.
2018-08-07T11:06:49 Dropping triggers...
2018-08-07T11:06:49 Dropped triggers OK.
Successfully altered `employees`.`titles`.
```

23.2.8　pt-query-digest（SQL 分析工具）

pt-query-digest 工具用于分析 MySQL 慢查询日志、全量 SQL 日志、二进制日志文件，同时也能分析 show processlist 获取的结果和 tcpdump 抓包的 MySQL 协议数据。在日常工作中，这个工具主要用于分析 MySQL 的慢日志，可以更直观、更形象地展现分析结果。

pt-query-digest 工具的使用方法如下：

```
shell> pt-query-digest [OPTIONS] [FILES] [DSN]
```

常用参数如下：

```
--limit      展示最耗时 SQL 的数量，默认 95%:20。如果同时写百分比和数量，则以先满足的条件为准
--since      指定分析 SQL 的开始时间
--type       解析的输入类型，默认为 slowlog。可选类型为 binlog、slowlogtcpdump、rawlog
--until      指定分析 SQL 的结束时间
```

下面看一些具体的例子。

分析 2018-11-05 从 16 点到 17 点的慢查询日志：

```
[mysql3307@localhost data]$ pt-query-digest --since '2018-11-05 16:00:00' --until '2018-11-05 17:00:00' --limit 100% slow3307.log
# 280ms user time, 20ms system time, 24.60M rss, 204.98M vsz
# Current date: Mon Nov  5 16:52:38 2018
# Hostname: localhost
# Files: slow3307.log
# Overall: 9 total, 1 unique, 0.16 QPS, 0.14x concurrency _____
# Time range: 2018-11-05T16:51:38 to 2018-11-05T16:52:33
# Attribute          total     min     max     avg     95%  stddev  median
# ============     ======= ======= ======= ======= ======= ======= =======
# Exec time             8s   386ms      3s   873ms      3s   920ms   412ms
# Lock time            6ms   176us     5ms   711us     5ms     1ms   185us
# Rows sent              9       1       1       1       1       0       1
# Rows examine      12.74M 924.17k   2.71M   1.42M   2.62M 739.82k 915.49k
# Query size           438      47      49   48.67   46.83       0   46.83

# Profile
# Rank Query ID                           Response time Calls R/Call V/M
# ==== ================================== ============= ===== ====== =====
#    1 0xD038FC7210F92F0689CD4E3405472409  7.8556 100.0%     9 0.8728  0.97 SELECT salaries

# Query 1: 0.16 QPS, 0.14x concurrency, ID 0xD038FC7210F92F0689CD4E3405472409 at byte 0
# This item is included in the report because it matches --limit.
# Scores: V/M = 0.97
# Time range: 2018-11-05T16:51:38 to 2018-11-05T16:52:33
# Attribute    pct   total     min     max     avg     95%  stddev  median
# ============ === ======= ======= ======= ======= ======= ======= =======
# Count        100       9
# Exec time    100      8s   386ms      3s   873ms      3s   920ms   412ms
# Lock time     99     6ms   176us     5ms   711us     5ms     1ms   185us
# Rows sent    100       9       1       1       1       1       0       1
# Rows examine 100  12.74M 924.17k   2.71M   1.42M   2.62M 739.82k 915.49k
# Query size   100     438      47      49   48.67   46.83       0   46.83
# String:
# Databases    employees
# Hosts        localhost
# Users        root
# Query_time distribution
#   1us
#  10us
# 100us
#   1ms
#  10ms
# 100ms  ################################################################
#    1s  #################
#  10s+
# Tables
#    SHOW TABLE STATUS FROM `employees` LIKE 'salaries'\G
#    SHOW CREATE TABLE `employees`.`salaries`\G
# EXPLAIN /*!50100 PARTITIONS*/
select count(0) from salaries where emp_no>1000\G
```

分析结果首先展示所有慢 SQL 总的汇总部分。第 5 行 Overall 行，显示共有 9 个慢 SQL，1 个不同的 SQL，也就是说有 1 个慢 SQL 出现了 9 次；第 6 行 Time range 行显示慢 SQL 的时间跨度；Attribute 行和接下来的 6 行展示了另外的汇总信息。Exec time 指的是慢 SQL 的执行时间，Rows sent 指的是慢 SQL 的返回行数，Rows examine 指的是慢 SQL 检索的记录行数。可以按照字面意思理解 Attribute 行的内容，例如 Exec time 行和 total 列指向的是 8s，表示所有的慢 SQL 执行的总时间是 8s；Rows sent 行和 avg 列指向的是 1，表示慢 SQL 平均的返回记录行数是 1。

接下来 Profile 以下就是每个 SQL 的汇总部分。Query ID 为 SQL 指纹的 id，同一个 SQL 如果仅是常量的不同，被识别为同一个指纹 id；Response time 为 SQL 总的执行时间和占总时间的百分比；Calls 为慢 SQL 执行次数，这里 Rank 为 1 的 SQL 执行了 9 次；R/Call 为慢 SQL 平均执行时间，例如 Rank 为 1 的 SQL，其 Response time 为 7.8556s，执行了 9 次，7.8556 除以 9 约等于 R/Call 0.8727s。

最后是每个 SQL 的明细部分，Attribute 行和接下来 7 行，跟 SQL 汇总部分含义一致。后面还列出了执行 sql 对应的数据库名、客户端的 IP 信息、用户名、慢 SQL 执行时间在每个时间段的比例。根据这些信息，可以很容易地定位慢 SQL 来自哪个应用，慢 SQL 执行时间主要分布在哪个时间区间，例子中的慢 SQL 执行时间主要分布在 100ms～1s 的区间。随后还给出慢 SQL 的一个样例，方便在数据库执行 explain 命令。

pt-query-digest 仅是一个命令行工具，缺少可视化的展示界面，当数据库规模达到一定程度时，手工操作变得很烦琐，这时可以通过 Anemometer 平台自动分析慢 SQL，具体可以参考第 24 章相关内容。

23.2.9 pt-table-checksum（数据检验工具）

pt-table-checksum 工具用于校验从库数据跟主库是否一致，校验时主库可以保持读写。工具在主库执行 SQL 统计表的数据，然后通过 BINLOG 把在主库执行的 SQL 传到从执行，所以在校验过程需要开启从库的 IO 进程和 SQL 进程，并且从库延迟要尽可能得小。

pt-table-checksum 的使用方法如下：

```
shell> pt-table-checksum [OPTIONS] [DSN]
```

常用参数如下：

参数	说明
--[no]check-binlog-format	检查所有主从实例的 binlog_format 是否一致，默认为 yes
--check-interval	检查从库延迟是否达到 --max-lag 的间隔，默认为 1s
--check-slave-lag	检查从库延迟，当延迟超过 --max-lag 后，停止检查数据
--chunk-index	指定检查表时遍历数据的索引
--chunk-index-columns	指定使用 --chunk-index 索引最左边的几个字段遍历数据
--chunk-size	每次检查块的记录数，默认为 1000
--chunk-size-limit	每次检查块的尺寸，默认为 2
--chunk-time	根据 --chunk-time 的时间动态调节检查块的记录数，默认为 0.5s
--databases,-d	只检查指定数据库下的表，多个数据库用逗号分隔
--explain	展示检查表锁使用的查询，而不真正检查表。如果指定两次 --explain，将展示工具迭代分块算法，打印每次迭代分块的上界和下界
--ignore-databases	忽略检查指定数据库下的表，多个数据库用逗号分隔
--ignore-tables	忽略检查指定的表，表名前要加上数据库名，多个表用逗号分隔
--max-lag	如果从库延迟大于 --max-lag，检查数据任务暂停，直至从库延迟小于 --max-lag，默认为 1s
--replicate	指定存放检查结果的表
--run-time	任务的最大运行时间
--tables	只检查指定的表，表名前要加上数据库名，多个表用逗号分隔
--truncate-replicate-table	每次检查任务开始前，清空结果表
--where	只检查匹配 where 语句的记录，语句里面不用写 where 关键字

下面看一些具体的例子。

校验主从数据，不检查主从实例的 BINLOG 格式，检查结果存入 percona.checksums 表，pt 工具自动找从库的信息：

```
[mysql3307@localhost ~]$ pt-table-checksum   --no-check-binlog-format --replicate=percona.checksums --host=127.0.0.1  --port 3307  -uroot
  Checking if all tables can be checksummed ...
  Starting checksum ...
            TS          ERRORS DIFFS ROWS   DIFF_ROWS CHUNKS SKIPPED TIME   TABLE
  08-20T21:42:51 0      1     9            0         1      0       0.313  employees.departments
  08-20T21:42:53 0      4     331603       0         6      0       1.652  employees.dept_emp
  08-20T21:42:53 0      1     24           0         1      0       0.31   employees.dept_manager
  08-20T21:42:55 0      0     300024       0         1      0       1.671  employees.employees
```

```
...
08-20T21:43:53   0         0       443306         0         5        0        2.938   employees.titles
```

结果显示，employees.departments 表有 9 条记录，通过 1 个 chunk 检查（chunks），diffs 的 chunk 数也为 1，代表主从的 9 条记录数据不一致。employees.dept_emp 表有 331 603 条记录，通过 6 个 chunk 检查，每个 chunk 的数据不一定均匀，其中 4 个 chunk 中的主从数据不一致。

在上面命令的基础上，每 1s 检查一次检查从库延迟，如果从库延迟大于 1s，校验数据任务将暂停，直到从库延迟小于 1s：

```
[mysql3307@localhost ~]$ pt-table-checksum   --no-check-binlog-format --replicate=percona.
checksums --host=127.0.0.1  --port 3307  -uroot  --check-interval 1 --max-lag 1
Checking if all tables can be checksummed ...
Starting checksum ...
Replica lag is 55 seconds on localhost.  Waiting.
            TS    ERRORS  DIFFS  ROWS  DIFF_ROWS  CHUNKS  SKIPPED  TIME     TABLE
08-20T21:42:51   0         1       9         0         1        0        0.313   employees.departments
08-20T21:42:53   0         4       331603    0         6        0        1.652   employees.dept_emp
08-20T21:42:53   0         1       24        0         1        0        0.31    employees.dept_manager
08-20T21:42:55   0         0       300024    0         1        0        1.671   employees.employees
....
08-20T21:43:53   0         0       443306    0         5        0        2.938   employees.titles
```

在开始任务前，从库的延迟大于 1s；校验任务处于暂停状态，当从库延迟小于 1s，校验任务正常开始。

23.2.10　pt-table-sync（数据同步工具）

pt-table-sync 工具用于从库同步主库的表数据，不同步表结构和索引，该工具主要适用于主从库少量数据不一致的场景。

pt-table-sync 的使用方法如下：

pt-table-sync [OPTIONS] DSN [DSN]

常用参数如下：

```
--charset                设置字符集
--chunk-index            指定检查表时遍历数据的索引
--chunk-index-columns    指定使用 --chunk-index 索引最左边的几个字段遍历数据
--databases,-d           只同步指定数据库下的表，多个数据库用逗号分隔
--dry-run                分析同步过程使用的算法，而不真正执行。配合 -verbose 参数可以打印模拟的结果
--execute                执行数据同步任务
--explain-hosts          打印同步任务要连接的所有主机，并且退出任务
--ignore-databases       忽略同步指定数据库下的表，多个数据库用逗号分隔
--ignore-tables          忽略检查的表。表名前要加上 db 名，用逗号分隔。
--lock                   在主库锁定正在对比的数据库。0：不锁定数据；1：每次同步的循环锁定；2：同步表数
                         据结束前，锁定表；3：锁全库所有表
--print                  不真正执行数据同步，仅打出过程使用的 sql
--replicate              同步表中记录的数据不一致的表，表中数据可以由 pt-table-checksum 校验得到
--slave-user             指定连接从库的用户名
--slave-password         指定连接从库的用户名
--sync-to-master         把命令中列的 DSN 认定为从库，根据主库同步数据
--tables,-t              只同步指定表。表名前要加上 db 名，用逗号分隔。
```

下面看一些具体的例子。

模拟同步 departments 表，但是仅打印同步过程执行的 SQL，而不是真正地执行：

```
[mysql3308@localhost ~]$ pt-table-sync  --replicate=percona.checksums  \
> h=127.0.0.1  --port 3307  -uroot  \
> --databases employees --tables departments --print
```

```
REPLACE INTO `employees`.`departments`(`dept_no`, `dept_name`) VALUES ('d001', 'Marketing')
/*percona-toolkit src_db:employees src_tbl:departments src_dsn:P=3307,h=127.0.0.1,u=root dst_db:
employees dst_tbl:departments dst_dsn:P=3308,h=127.0.0.1,u=root lock:1 transaction:1 changing_
src:percona.checksums replicate:percona.checksums bidirectional:0 pid:22485 user:mysql3308 host
:localhost*/;
```

真正同步 departments 表，:

```
[mysql3307@localhost ~]$ pt-table-sync  --replicate=percona.checksums  h=127.0.0.1  --port 330
7 -uroot  --databases employees --tables departments --execute
```

同步任务完成后，再用 pt-table-checksum 校验主从数据：

```
[mysql3307@localhost ~]$ pt-table-checksum  --replicate=percona.checksums --no-check-binlog-
format  --host=127.0.0.1  --port 3307  -uroot  --databases employees --tables departments
    Checking if all tables can be checksummed ...
    Starting checksum ...
            TS ERRORS  DIFFS     ROWS  DIFF_ROWS  CHUNKS SKIPPED   TIME TABLE
08-20T22:53:15      0      0        9          0       1       0  0.346 employees.departments
```

结果显示，从库 employees.departments 表数据已经跟主库数据一致了。

23.3 小结

本章介绍了在 MySQL 运维过程中的一些常用工具，并举例说明了它们的使用方法。熟练掌握这些工具将会给工作带来很大的便利。由于这些工具参数众多，这里不再一一列举，如果读者想了解更多工具的使用方法，读者可以参考相关的帮助文档。

第 24 章 MySQL 日志

在任何一种数据库中，都会有各种各样的日志，记录着数据库工作的方方面面，以帮助数据库管理员追踪数据库曾经发生过的各种事件。MySQL 也不例外，在 MySQL 中，有 6 种不同的日志，分别是错误日志（Error Log）、二进制日志（Binary Log）、查询日志（General Query Log）、慢查询日志（Slow Query Log）、中继日志（Relay Log）和元数据日志（DDL Log）。这些日志记录着数据库在不同方面的踪迹。

从库通过 I/O 线程拉取主库二进制日志，记录到本地，生成中继日志，中继日志跟二进制日志相差不多；元数据日志记录数据定义语句（如 alter table）的操作，用于 MySQL crash 恢复，目前用户无法配置元数据日志相关的选项和参数。由于用户很少主动使用中继日志和元数据日志，本章将详细介绍错误日志、二进制日志、查询日志和慢查询日志的作用和使用方法，希望读者能充分利用这些日志对数据库进行各种维护和调优。

24.1 错误日志

错误日志是 MySQL 中最重要的日志之一，它记录了当 mysqld 启动和停止时，以及服务器在运行过程中发生任何严重错误时的相关信息。当数据库出现任何故障导致无法正常使用时，可以首先查看此日志。

可以用 --log-error[=file_name] 选项来指定 mysqld（MySQL 服务器）保存错误日志文件的位置。如果没有给定 file_name 值，mysqld 使用错误日志名 host_name.err（host_name 为主机名）并默认在参数 DATADIR（数据目录）指定的目录中写入日志文件。

以下是 MySQL 正常启动和关闭的一段日志，不同的版本可能略有不同：

```
[root@localhost mysql]# more localhost.localdomain.err
 2018-07-21T11:01:35.588314+08:00 0 [Note] InnoDB: File '/data2/mysql3307/data/ibtmp1' size is now 12 MB.
 2018-07-21T11:01:35.589687+08:00 0 [Note] InnoDB: 96 redo rollback segment(s) found. 96 redo rollback segment(s) are active.
 2018-07-21T11:01:35.589714+08:00 0 [Note] InnoDB: 32 non-redo rollback segment(s) are active.
 2018-07-21T11:01:35.590327+08:00 0 [Note] InnoDB: Waiting for purge to start
 2018-07-21T11:01:35.649848+08:00 0 [Note] InnoDB: 5.7.22 started; log sequence number 8023211655
 2018-07-21T11:01:35.650358+08:00 0 [Note] InnoDB: Loading buffer pool(s) from /data2/mysql3307/data/ib_buffer_pool
 2018-07-21T11:01:35.650718+08:00 0 [Note] Plugin 'FEDERATED' is disabled.
 2018-07-21T11:01:36.355876+08:00 0 [Note] Server hostname (bind-address): '*'; port: 3307
 2018-07-21T11:01:36.355935+08:00 0 [Note] IPv6 is available.
 2018-07-21T11:01:36.355955+08:00 0 [Note]   - '::' resolves to '::';
 2018-07-21T11:01:36.355992+08:00 0 [Note] Server socket created on IP: '::'.
```

```
 2018-07-21T11:01:36.569714+08:00 0 [Note] Event Scheduler: Loaded 0 events
 2018-07-21T11:01:36.569926+08:00 1 [Note] Event Scheduler: scheduler thread started with id 1
 2018-07-21T11:01:36.570157+08:00 0 [Note] /home/mysql3307/mysqlhome/bin/mysqld: ready for
connections.
 Version: '5.7.22-log'  socket: '/tmp/mysql_3307.sock'  port: 3307  MySQL Community Server (GPL)
…… 上面是启动日志 ……
…… 下面是关闭日志 ……
 2018-07-21T11:01:08.827360+08:00 0 [Note] Shutting down plugin 'INNODB_LOCKS'
 2018-07-21T11:01:08.827367+08:00 0 [Note] Shutting down plugin 'INNODB_TRX'
 2018-07-21T11:01:08.827373+08:00 0 [Note] Shutting down plugin 'InnoDB'
 2018-07-21T11:01:08.903530+08:00 0 [Note] InnoDB: FTS optimize thread exiting.
 2018-07-21T11:01:08.903801+08:00 0 [Note] InnoDB: Starting shutdown...
 2018-07-21T11:01:09.008627+08:00 0 [Note] InnoDB: Dumping buffer pool(s) to /data2/mysql3307
/data/ib_buffer_pool
 2018-07-21T11:01:09.041581+08:00 0 [Note] InnoDB: Buffer pool(s) dump completed at 180721 11:01:09
 2018-07-21T11:01:10.495131+08:00 0 [Note] InnoDB: Shutdown completed; log sequence number 802
3211655
 2018-07-21T11:01:10.518026+08:00 0 [Note] InnoDB: Removed temporary tablespace data file: "ibtmp1"
 2018-07-21T11:01:10.518052+08:00 0 [Note] Shutting down plugin 'sha256_password'
 2018-07-21T11:01:10.518062+08:00 0 [Note] Shutting down plugin 'mysql_native_password'
 2018-07-21T11:01:10.518356+08:00 0 [Note] Shutting down plugin 'binlog'
 2018-07-21T11:01:10.533994+08:00 0 [Note] /home/mysql3307/mysqlhome/bin/mysqld: Shutdown complete
```

24.2 二进制日志

二进制日志（BINLOG）记录了所有的 DDL（数据定义语言）语句和 DML（数据操纵语言）语句，但是不包括数据查询语句。语句以"事件"的形式保存，它描述了数据的更改过程。此日志对于灾难时的数据恢复起着极其重要的作用。

24.2.1 日志的位置和格式

当用 --log-bin[=file_name] 选项启动时，mysqld 开始将数据变更情况写入日志文件。如果没有给出 file_name 值，默认名为主机名后面跟"-bin"。如果给出了文件名，但没有包含路径，则文件默认被写入参数 DATADIR（数据目录）指定的目录。

在 MySQL 5.7 中，二进制日志的格式分为 3 种：STATEMENT、ROW 和 MIXED，可以在启动时通过参数 --binlog_format 进行设置。

1. STATEMENT

MySQL 5.1 之前的版本都采用这种方式，顾名思义，日志中记录的都是语句（statement），每一条对数据造成修改的 SQL 语句都会记录在日志中，通过 mysqlbinlog 工具，可以清晰地看到每条语句的文本。主从复制的时候，从库（slave）会将日志解析为原文本，并在从库重新执行一次。这种格式的优点是日志记录清晰易读、日志量少，对 I/O 影响较小；缺点是在某些情况下，slave 的日志复制会出错。

2. ROW

MySQL 5.1.11 之后，出现了这种新的日志格式。现在是目前 MySQL 默认的日志格式，它将每一行的变更记录到日志中，而不是记录 SQL 语句。比如一个简单的更新 SQL，update emp set name='abc'，如果是 STATEMENT 格式，日志中会记录一行 SQL 文本；如果是 ROW，由于是对全表进行更新，也就是每一行记录都会发生变更，如果是一个 100 万行的大表，则日志中会记录 100 万条记录的变化情况，日志量大大增加。这种格式的优点是会记录每一行数据的变化细节，不会出现某些情况下无法复制的情况；缺点是日志量大，对 I/O 影响较大。

3. MIXED

该日志混合了 STATEMENT 和 ROW 两种日志。默认情况下采用 STATEMENT，但在一些特殊情况下采用 ROW 来进行记录，比如采用 NDB 存储引擎，此时对表的 DML 语句全部采用 ROW；客户端使用了临时表；客户端采用了不确定函数，比如 current_user()等，因为这种不确定函数在主从中得到的值可能不同，导致主从数据产生不一致。MIXED 格式能尽量利用两种模式的优点，而避开它们的缺点。

> 注意：可以在 global 和 session 级别对 binlog_format 进行日志格式的设置，但一定要谨慎操作，确保从库的复制能够正常进行。

24.2.2 日志的读取

由于日志以二进制方式存储，不能直接读取，需要用 mysqlbinlog 工具来查看，语法如下：

```
shell > mysqlbinlog log-file;
```

mysqlbinlog 的用法在第 23 章中已经详细介绍过，这里不再赘述。下面以 STATEMENT 格式为例演示了二进制日志的读取过程。

（1）往测试表 emp 中插入两条测试记录。

```
mysql> use employees
Reading table information for completion of table and column names
You can turn off this feature to get a quicker startup with -A

Database changed
mysql> insert into emp values(1,'z1');
Query OK, 1 row affected (0.00 sec)

mysql> insert into emp values(1,'z2');
Query OK, 1 row affected (0.00 sec)
```

（2）使用 mysqlbinlog 工具进行日志查看，粗体字显示了步骤（1）中所做的操作。

```
[root@localhost mysql]# mysqlbinlog -vv localhost-bin.000001
/*!50530 SET @@SESSION.PSEUDO_SLAVE_MODE=1*/;
/*!40019 SET @@session.max_insert_delayed_threads=0*/;
/*!50003 SET @OLD_COMPLETION_TYPE=@@COMPLETION_TYPE,COMPLETION_TYPE=0*/;
DELIMITER /*!*/;
# at 4
#181008 14:23:32 server id 1  end_log_pos 120 CRC32 0x218fea45     Start: binlog v 4, server v 5.6.34-log created 181008 14:23:32
# Warning: this binlog is either in use or was not closed properly.
BINLOG '
5Pe6ww8BAAAAdAAAAHgAAAABAAQANS42LjM0LWxvZwAAAAAAAAAAAAAAAAAAAAAAAAAAAAAAA
AAAAAAAAAAAAAAAAAAAAAAAEzgNAAgAEgAEBAQEEgAAXAAEGggAAAAICAgCAAAACgoKGRkAAUXq
jyE=
'/*!*/;
# at 120
#181008 14:23:39 server id 1  end_log_pos 192 CRC32 0xd0f6bc2a     Query     thread_id=2  exec_time=0     error_code=0
SET TIMESTAMP=1538979819/*!*/;
SET @@session.pseudo_thread_id=2/*!*/;
SET @@session.foreign_key_checks=1, @@session.sql_auto_is_null=0, @@session.unique_checks=1, @@session.autocommit=1/*!*/;
SET @@session.sql_mode=1073741824/*!*/;
SET @@session.auto_increment_increment=1, @@session.auto_increment_offset=1/*!*/;
/*!\C utf8mb4 *//*!*/;
SET @@session.character_set_client=45,@@session.collation_connection=45,@@session.collation_
```

```
server=33/*!*/;
    SET @@session.lc_time_names=0/*!*/;
    SET @@session.collation_database=DEFAULT/*!*/;
    BEGIN
    /*!*/;
    # at 192
    #181008 14:23:39 server id 1  end_log_pos 241 CRC32 0xd0ec50cf    Table_map: `employees`.`emp
mapped to number 71
    # at 241
    #181008 14:23:39 server id 1  end_log_pos 284 CRC32 0x80b83184    Write_rows: table id 71
flags: STMT_END_F

    BINLOG '
    6/e6WxMBAAAAMQAAAPEAAAAAAECAAAAAAAEABHRlc3QAA2VtcCAACAw8C/wADz1Ds0A==
    6/e6Wx4BAAAAKwAAABwBAAAAAECAAAAAAAEAAgAC//wBAAAAAAnoxhDG4gA==
    '/*!*/;
    ### INSERT INTO `employees`.`emp`
    ### SET
    ###   @1=1 /* INT meta=0 nullable=1 is_null=0 */
    ###   @2='z1' /* VARSTRING(255) meta=255 nullable=1 is_null=0 */
    # at 284
    #181008 14:23:39 server id 1  end_log_pos 315 CRC32 0x48e56802    Xid = 14
    COMMIT/*!*/;
    # at 315
    #181008 14:23:44 server id 1  end_log_pos 387 CRC32 0x8fa2d170    Query    thread_id=2 exec
_time=0     error_code=0
    SET TIMESTAMP=1538979824/*!*/;
    BEGIN
    /*!*/;
    # at 387
    #181008 14:23:44 server id 1  end_log_pos 436 CRC32 0x31df9046    Table_map: `test`.`emp`
mapped to number 71
    # at 436
    #181008 14:23:44 server id 1  end_log_pos 479 CRC32 0x840fc684    Write_rows: table id 71
flags: STMT_END_F

    BINLOG '
    8Pe6WxMBAAAAMQAAALQBAAAAAECAAAAAAAEABHRlc3QAA2VtcCAACAw8C/wADRpDfMQ==
    8Pe6Wx4BAAAAKwAAAN8BAAAAAECAAAAAAAEAAgAC//wBAAAAAAnoyhMYPhA==
    '/*!*/;
    ### INSERT INTO `employees`.`emp`
    ### SET
    ###   @1=1 /* INT meta=0 nullable=1 is_null=0 */
    ###   @2='z2' /* VARSTRING(255) meta=255 nullable=1 is_null=0 */
    # at 479
    #181008 14:23:44 server id 1  end_log_pos 510 CRC32 0xed444efe    Xid = 15
    COMMIT/*!*/;
    DELIMITER ;
    # End of log file
    ROLLBACK /* added by mysqlbinlog */;
    /*!50003 SET COMPLETION_TYPE=@OLD_COMPLETION_TYPE*/;
    /*!50530 SET @@SESSION.PSEUDO_SLAVE_MODE=0*/;
```

可以看到，二进制日志中记录了 MySQL 实例信息、会话中的参数信息、执行事件类型与事件块位置、执行语句等内容，在之后的 24.2.4 节会做简要的分析。

24.2.3 日志的删除

对于比较繁忙的 OLTP（在线事务处理）系统，由于每天生成日志量大，这些日志如果长时间不清除，将会对磁盘空间带来很大的浪费。因此，定期删除日志是 DBA 维护 MySQL 数据库的一个重要工作内容。下面介绍几种删除日志的常见方法。

1. 方法 1

执行"RESET MASTER;"命令，删除所有 BINLOG 日志，新日志编号从"000001"开始。

在 MySQL 8.0 中，支持 RESET MASTER TO + 编号的命令，可以自定义新日志的开始编号。下例中删除了当前的所有日志。

（1）查看删除前日志。

```
mysql> system ls -ltrh mysql-bin*
-rw-r----- 1 mysql3307 mysql3307 128K Sep 26 20:39 mysql-bin.000052
-rw-r----- 1 mysql3307 mysql3307  281 Sep 26 20:42 mysql-bin.000053
-rw-r----- 1 mysql3307 mysql3307  578 Sep 26 20:43 mysql-bin.000054
-rw-r----- 1 mysql3307 mysql3307  578 Sep 26 20:43 mysql-bin.000055
-rw-r----- 1 mysql3307 mysql3307  281 Sep 26 20:47 mysql-bin.000056
-rw-r----- 1 mysql3307 mysql3307  880 Sep 26 20:47 mysql-bin.000057
-rw-r----- 1 mysql3307 mysql3307  281 Sep 26 20:49 mysql-bin.000058
-rw-r----- 1 mysql3307 mysql3307  312 Sep 26 20:49 mysql-bin.index
-rw-r----- 1 mysql3307 mysql3307  828 Sep 26 20:50 mysql-bin.000059
```

结果中的"mysql-bin.index"是日志的索引文件，记录了最大的日志序号，本例中可以将此文件忽略。

（2）用 RESET MASTER 命令进行日志删除。

```
mysql> reset master;
Query OK, 0 rows affected (0.08 sec)
```

（3）查看删除后的日志。

```
mysql> system ls -ltrh mysql-bin*
-rw-r----- 1 mysql3307 mysql3307 154 Sep 26 20:54 mysql-bin.000001
-rw-r----- 1 mysql3307 mysql3307  39 Sep 26 20:54 mysql-bin.index
```

可以发现，以前的日志全部被清空。新日志重新从"000001"开始编号。

2．方法 2

执行"PURGE MASTER LOGS TO 'mysql-bin.******'"命令，删除"******"编号之前的所有日志。下例删除了"mysql-bin.000057"之前编号的所有日志。

（1）查看删除前日志。

```
mysql> system ls -ltrh mysql-bin*
-rw-r----- 1 mysql3307 mysql3307 128K Sep 26 20:39 mysql-bin.000052
-rw-r----- 1 mysql3307 mysql3307  281 Sep 26 20:42 mysql-bin.000053
-rw-r----- 1 mysql3307 mysql3307  578 Sep 26 20:43 mysql-bin.000054
-rw-r----- 1 mysql3307 mysql3307  578 Sep 26 20:43 mysql-bin.000055
-rw-r----- 1 mysql3307 mysql3307  281 Sep 26 20:47 mysql-bin.000056
-rw-r----- 1 mysql3307 mysql3307  880 Sep 26 20:47 mysql-bin.000057
-rw-r----- 1 mysql3307 mysql3307  281 Sep 26 20:49 mysql-bin.000058
-rw-r----- 1 mysql3307 mysql3307  312 Sep 26 20:49 mysql-bin.index
-rw-r----- 1 mysql3307 mysql3307  828 Sep 26 20:50 mysql-bin.000059
```

（2）用 PURGE 命令进行删除。

```
mysql> purge master logs to 'mysql-bin.000057';
Query OK, 0 rows affected (0.04 sec)
```

（3）查看删除后日志。

```
mysql> system ls -ltrh mysql-bin*
-rw-r----- 1 mysql3307 mysql3307 880 Sep 26 20:47 mysql-bin.000057
-rw-r----- 1 mysql3307 mysql3307 281 Sep 26 20:49 mysql-bin.000058
-rw-r----- 1 mysql3307 mysql3307 828 Sep 26 20:50 mysql-bin.000059
-rw-r----- 1 mysql3307 mysql3307 117 Sep 26 20:57 mysql-bin.index
```

从结果中发现，编号"000057"之前的所有日志都已经被删除。

3．方法 3

执行"PURGE MASTER LOGS BEFORE 'yyyy-mm-dd hh24:mi:ss'"命令，删除日期为

"yyyy-mm-dd hh24:mi:ss"之前产生的所有日志。下例删除了日期在"2018-09-26 20:58:00"之前的所有日志。

(1)查看删除前日志。

```
mysql> system ls -ltrh mysql-bin*
-rw-rw-r-- 1 mysql3307 mysql3307 540M Jul 25 16:22 mysql-bin.000032.log
-rw-r----- 1 mysql3307 mysql3307  880 Sep 26 20:47 mysql-bin.000057
-rw-r----- 1 mysql3307 mysql3307  281 Sep 26 20:49 mysql-bin.000058
-rw-r----- 1 mysql3307 mysql3307  875 Sep 26 20:58 mysql-bin.000059
-rw-r----- 1 mysql3307 mysql3307  281 Sep 26 20:58 mysql-bin.000060
-rw-r----- 1 mysql3307 mysql3307  281 Sep 26 20:58 mysql-bin.000061
-rw-r----- 1 mysql3307 mysql3307  281 Sep 26 20:58 mysql-bin.000062
-rw-r----- 1 mysql3307 mysql3307  273 Sep 26 20:58 mysql-bin.index
-rw-r----- 1 mysql3307 mysql3307  234 Sep 26 20:58 mysql-bin.000063
```

(2)用 PURGE 命令删除"2018-09-26 20:58:00"之前的所有日志。

```
mysql> purge master logs before '2018-09-26 20:58:00';
Query OK, 0 rows affected (0.04 sec)
```

(3)查看删除后日志,系统保留了指定删除日期后的日志。

```
mysql> system ls -ltrh mysql-bin*
-rw-r----- 1 mysql3307 mysql3307  875 Sep 26 20:58 mysql-bin.000059
-rw-r----- 1 mysql3307 mysql3307  281 Sep 26 20:58 mysql-bin.000060
-rw-r----- 1 mysql3307 mysql3307  281 Sep 26 20:58 mysql-bin.000061
-rw-r----- 1 mysql3307 mysql3307  281 Sep 26 20:58 mysql-bin.000062
-rw-r----- 1 mysql3307 mysql3307  234 Sep 26 20:58 mysql-bin.000063
-rw-r----- 1 mysql3307 mysql3307  195 Sep 26 20:59 mysql-bin.index
```

4.方法 4

设置参数--expire_logs_days=#,此参数的含义是设置日志的过期天数,过了指定的天数后日志将会被自动删除,这样将有利于减少 DBA 管理日志的工作量。下例将通过手工更改系统日期来测试此参数的使用。

(1)查看删除前日志。

```
mysql> system ls -ltr mysql-bin*
-rw-r----- 1 mysql3307 mysql3307 875 Sep 26 20:58 mysql-bin.000059
-rw-r----- 1 mysql3307 mysql3307 281 Sep 26 20:58 mysql-bin.000060
-rw-r----- 1 mysql3307 mysql3307 281 Sep 26 20:58 mysql-bin.000061
-rw-r----- 1 mysql3307 mysql3307 281 Sep 26 20:58 mysql-bin.000062
-rw-r----- 1 mysql3307 mysql3307 234 Sep 26 20:58 mysql-bin.000063
-rw-r----- 1 mysql3307 mysql3307 195 Sep 26 20:59 mysql-bin.index
```

(2)动态修改 expire_logs_days=3,如果需要持久化,记得修改 my.cnf 配置文件。

```
mysql>show variables like 'expire_logs_days';
+------------------+-------+
| Variable_name    | Value |
+------------------+-------+
| expire_logs_days | 7     |
+------------------+-------+
1 row in set (0.00 sec)

mysql>set global  expire_logs_days=3;
Query OK, 0 rows affected (0.00 sec)

mysql>show variables like 'expire_logs_days';
+------------------+-------+
| Variable_name    | Value |
+------------------+-------+
| expire_logs_days | 3     |
+------------------+-------+
1 row in set (0.00 sec)
```

(3)将系统时间改为 1 天以后。

```
[root@localhost ~]# date
Thu Sep 26 21:08:38 CST 2018
[root@localhost ~]# date -s '2018-09-27 21:10:00'
Thu Sep 27 21:10:00 CST 2018
```

（4）用"flush logs"触发日志文件更新，由于没有到 3 天，所有日志将不会被删除。

```
mysql>flush logs;
Query OK, 0 rows affected (0.00 sec)

mysql>system ls -ltr mysql-bin*
-rw-r----- 1 mysql3307 mysql3307 875 Sep 26 20:58 mysql-bin.000059
-rw-r----- 1 mysql3307 mysql3307 281 Sep 26 20:58 mysql-bin.000060
-rw-r----- 1 mysql3307 mysql3307 281 Sep 26 20:58 mysql-bin.000061
-rw-r----- 1 mysql3307 mysql3307 281 Sep 26 20:58 mysql-bin.000062
-rw-r----- 1 mysql3307 mysql3307 281 Sep 27 21:02 mysql-bin.000063
-rw-r----- 1 mysql3307 mysql3307 234 Sep 27 21:02 mysql-bin.index
-rw-r----- 1 mysql3307 mysql3307 234 Sep 27 21:02 mysql-bin.000064
```

（5）将日期改为 3 天以后，再次执行"flush logs"触发日志文件更新。

```
[root@localhost ~]# date -s '2018-09-30 21:10:00'
Sun Sep 30 21:10:00 CST 2018

mysql>flush logs;
Query OK, 0 rows affected (0.01 sec)

mysql>system ls -ltr mysql-bin*
-rw-r----- 1 mysql3307 mysql3307 281 Sep 30 21:10 mysql-bin.000064
-rw-r----- 1 mysql3307 mysql3307 234 Sep 30 21:10 mysql-bin.000065
-rw-r----- 1 mysql3307 mysql3307  78 Sep 30 21:10 mysql-bin.index
```

从结果中可以看出，3 天前日志 mysql-bin.000059-- mysql-bin.000063 都已经被删除了。

5．方法 5

MySQL 8.0 版本新增了参数 binlog_expire_logs_seconds，BINLOG 日志过期时间可以精确到秒，并将 expire_logs_days 参数定义为 deprecated，后续版本将废弃这个参数。建议使用 binlog_expire_logs_seconds 来管理 BINLOG 日志的生命周期。

（1）查看删除前参数。

```
mysql>show variables like 'expire_logs_days';
+------------------+-------+
| Variable_name    | Value |
+------------------+-------+
| expire_logs_days | 7     |
+------------------+-------+
1 row in set (0.00 sec)

mysql>show variables like 'binlog_expire_logs_seconds';
+----------------------------+-------+
| Variable_name              | Value |
+----------------------------+-------+
| binlog_expire_logs_seconds | 0     |
+----------------------------+-------+
1 row in set (0.00 sec)
```

binlog_expire_logs_seconds 参数为 0，expire_logs_days 参数为 7，实例会自动删除 7 天前的 BINLOG 日志文件。

（2）查看删除前日志。

```
mysql>select now();
+---------------------+
| now()               |
+---------------------+
| 2018-09-26 21:42:21 |
+---------------------+
```

```
1 row in set (0.01 sec)

mysql]>flush logs;
Query OK, 0 rows affected (0.00 sec)

mysql> system ls -ltrh mysql-bin*
-rw-r----- 1 mysql3488 mysql3488 242 Sep 26 21:39 mysql-bin.000013
-rw-r----- 1 mysql3488 mysql3488 242 Sep 26 21:39 mysql-bin.000014
-rw-r----- 1 mysql3488 mysql3488 195 Sep 26 21:39 mysql-bin.000015
-rw-r----- 1 mysql3488 mysql3488 117 Sep 26 21:39 mysql-bin.index
```

这里可以看到，现有的 BINLOG 文件的生成时间跟当前时间属于同一天，所以日志没有被删除。

（3）动态修改 binlog_expire_logs_seconds=300，删除 300s 前的 BINLOG 日志文件。

```
mysql>set GLOBAL  binlog_expire_logs_seconds=300;
ERROR 11079 (HY000): The option expire_logs_days cannot be used together with option binlog_expire_logs_seconds. Therefore, value of expire_logs_days is ignored.

mysql>show global variables like 'binlog_expire_logs_seconds';
+----------------------------+-------+
| Variable_name              | Value |
+----------------------------+-------+
| binlog_expire_logs_seconds | 0     |
+----------------------------+-------+
1 row in set (0.03 sec)
```

在 MySQL 8.0.11 版本中，expire_logs_days 参数和 binlog_expire_logs_seconds 参数不可同时设置。需要先将 expire_logs_days 设置为 0，再设置 binlog_expire_logs_seconds。

```
mysql>set global expire_logs_days=0;
Query OK, 0 rows affected, 1 warning (0.00 sec)

mysql >set GLOBAL  binlog_expire_logs_seconds=300;
Query OK, 0 rows affected (0.00 sec)

mysql >show global variables like 'binlog_expire_logs_seconds';
+----------------------------+-------+
| Variable_name              | Value |
+----------------------------+-------+
| binlog_expire_logs_seconds | 300   |
+----------------------------+-------+
1 row in set (0.00 sec)
```

（4）再次执行 "flush logs" 触发日志文件更新。

```
mysql>select now();
+---------------------+
| now()               |
+---------------------+
| 2018-09-26 22:00:34 |
+---------------------+
1 row in set (0.00 sec)

mysql>flush logs;
Query OK, 0 rows affected (0.00 sec)

mysql>system ls -ltrh mysql-bin*
-rw-r----- 1 mysql3488 mysql3488 242 Sep 26 21:59 mysql-bin.000016
-rw-r----- 1 mysql3488 mysql3488 195 Sep 26 21:59 mysql-bin.000017
-rw-r----- 1 mysql3488 mysql3488  78 Sep 26 21:59 mysql-bin.index
```

从结果中可以看出，300s 前日志 mysql-bin.000013-- mysql-bin.000015 都已经被删除。

24.2.4 日志的事件

在 24.2.2 节中，ROW 格式的日志可以被拆分为多个事件（event）。在解析后的文件中，每个事件的起始都是"# at + 数字"的格式，代表事件在二进制文件中起始的位置。一个二进制日志由大量的事件组成，如图 24-1 所示。

其中 FORMAT_DESCRIPTION_EVENT 与 ROTATE_EVENT 仅有一个，分别位于日志的起始和结尾，ROTATE_EVENT 在日志关闭时才会生成。

日志的中间有多个 QUERY_EVENT、TABLE_MAP_EVENT、WRITE_ROW_EVENT/UPDATE_ROW_EVENT/DELETE_ROW_EV

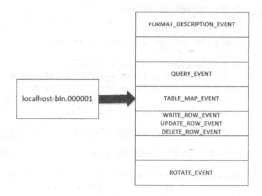

图 24-1　日志的组成

ENT，分别对应建库建表事件、DML 事务前的对应表信息和 DML 语句事件，其中 TABLE_MAP_EVENT 和 ROW_EVENT 是成对出现的。

除此之外，日志中还有 GTID_LOG_EVENT/ANONYMOUS_GTID_LOG_EVENT 和 XID_EVENT 等事件。

在这里我们拆分一下 24.2.2 节中的 DML 事务，也就是 insert into emp values(1,'z1')语句，对应 TABLE_MAP_EVENT 与 WRITE_ROW_EVENT 这两个事件，来分析一个 insert 语句在二进制日志中所生成的内容。

```
TABLE_MAP_EVENT：

# at 192    // 事件在二进制文件中的起始位置
#181008 14:23:39   // 事件发生的时间
server id 1  // MySQL 的 server_id
end_log_pos 241   // 结束位置
CRC32 0xd0ec50cf       // 循环冗余校验码
Table_map: `test`.`emp` mapped to number 71 // 表名与 id

WRITE_ROW_EVENT：

# at 241
#181008 14:23:39 server id 1  end_log_pos 284 CRC32 0x80b83184         Write_rows: table id 71 flags: STMT_END_F

BINLOG '
6/e6WxMBAAAAMQAAAPEAAAAAAECAAAAAAAEABHRlc3QAA2VtcAACAw8C/wADz1DsOA==
6/e6Wx4BAAAAKwAAABwBAAAAAECAAAAAAAEAgAC//wBAAAAAnoxhDG4gA==
'/*!*/;
### INSERT INTO `test`.`emp`
### SET
###   @1=1 /* INT meta=0 nullable=1 is_null=0 */
###   @2='z1' /* VARSTRING(255) meta=255 nullable=1 is_null=0 */    //解析后的 SQL 语句
```

可以看出，二进制日志完整地记录了一条 DML 命令的所有信息，我们可以应用这份日志做主从复制、数据恢复、闪回等功能。

24.2.5　日志闪回

日志的闪回是指当用户执行并提交了某些错误的删除或写入语句时，可以通过二进制日志将对应的事件逆向化，转为相反的写入或删除语句，并重新应用新生成的语句来抵消掉之前误操作的过程。日志闪回与数据库的正常备份恢复是不同的，因为使用日志进行闪回针对的是已知的语句，条件颗粒化到表级别、语句类型级别甚至是条件内容级别，可以在语句已经提交并同步到备库后，逆向的执行语句来恢复数据。能够做到无损地恢复数据并将错误造成的影响控制到最小。

日志闪回主要针对的是单个或少量已知误操作的紧急恢复，与数据库的备份并不相关。

下面简单介绍一下二进制日志做闪回的原理。首先我们看一个事件在二进制文件中的具体结构。

```
+=========================================+
| event    | timestamp       0 : 4        |
| header   +------------------------------+
|          | type_code       4 : 1        |
|          +------------------------------+
|          | server_id       5 : 4        |
|          +------------------------------+
|          | event_length    9 : 4        |
|          +------------------------------+
|          | next_position  13 : 4        |
|          +------------------------------+
|          | flags          17 : 2        |
|          +------------------------------+
|          | extra_headers  19 : x-19     |
+=========================================+
| event    | fixed part                   |
| data     +------------------------------+
|          | variable part                |
+=========================================+
```

所有事件的结构都如上所示，包括 event header 和 event data 两部分，其中 header 的结构是固定的，记录了该事件的时间戳、类型、事件长度、下个事件位置等内容。而 data 部分对于不同的事件而言结构都是不同的，例如建表或写入语句，会有不同的语句结构部分（fixed part）和不同的语句内容长度（variable part）。在这里我们不逐个分析所有的事件结构，仅通过 DML 语句来演示闪回语句是如何生成的。

1. insert/delete 语句

首先写入并删除一条数据。

```
mysql> insert into emp values(1,'z1');
Query OK, 1 row affected (0.00 sec)

mysql> delete from emp where a=1;
Query OK, 1 row affected (0.00 sec)
```

之后解析二进制日志并取出需要查看的事件。

```
# at 192
#181008 17:34:11 server id 1  end_log_pos 241 CRC32 0xb970ec62  Table_map: `test`.`emp` mapped to number 73
# at 241
#181008 17:34:11 server id 1  end_log_pos 284 CRC32 0x74a8cbc7  Write_rows: table id 73 flags: STMT_END_F

BINLOG '
kyS7WxMBAAAAMQAAAPEAAAAAEkAAAAAAAEABHRlc3QAA2VtcAACAw8C/wADYuxwuQ==
kyS7Wx4BAAAAKwAAABwBAAAAAEkAAAAAAAEAAgAC//wBAAAAAnoxx8uodA==
'/*!*/;
### INSERT INTO `test`.`emp`
### SET
###   @1=1 /* INT meta=0 nullable=1 is_null=0 */
###   @2='z1' /* VARSTRING(255) meta=255 nullable=1 is_null=0 */
...
# at 387
#181008 17:34:22 server id 1  end_log_pos 436 CRC32 0x1e19728f  Table_map: `test`.`emp` mapped
```

```
  to number 73
    # at 436
    #181008 17:34:22 server id 1  end_log_pos 479 CRC32 0x707bea77   Delete_rows: table id 73 flags
: STMT_END_F

BINLOG '
niS7WxMBAAAAMQAAALQBAAAAAEkAAAAAAAEABHRlc3QAA2VtcAACAw8C/wADj3IZHg==
niS7WyABAAAAKwAAAN8BAAAAAEkAAAAAAAEAAgAC//wBAAAAAnoxd+p7CA==
'/*!*/;
### DELETE FROM `test`.`emp`
### WHERE
###   @1=1 /* INT meta=0 nullable=1 is_null=0 */
###   @2='z1' /* VARSTRING(255) meta=255 nullable=1 is_null=0 */
```

可以看到日志中有一条写入和删除的事件。仔细观察可以发现,在 delete 语句中虽然只写了 a=1 这一个条件,但实际上整条数据都记录在日志中了,也就是说在二进制文件中, WRITE_ROW_EVENT 和 DELETE_ROW_EVENT 这两个事件不同之处仅在于类型的不同, 也就是 event header 中 type code 不同(WRITE_ROW_EVENT type code 为 0x1e, DELETE_ROW_EVENT v2 的 type code 为 0x20),我们可以简单地通过修改二进制文件中 type code 的内容,就能将 insert 和 delete 两种类型的语句互换,如图 24-2 所示。

图 24-2　事件类型的修改

2. update 语句

这次我们更新一条数据。

```
mysql> insert into emp values(1,'z1');
Query OK, 1 row affected (0.00 sec)

mysql> update emp set b='z2' where a=1;
Query OK, 1 row affected (0.00 sec)
Rows matched: 1  Changed: 1  Warnings: 0
```

取出对应的 TABLE_MAP_EVENT 和 UPDATE_ROW_EVENT 事件。

```
    # at 777
    #181008 18:03:23 server id 1  end_log_pos 826 CRC32 0xe7806c15     Table_map: `test`.`emp`
mapped to number 73
    # at 826
    #181008 18:03:23 server id 1  end_log_pos 878 CRC32 0x78c39704     Update_rows: table id 73
flags: STMT_END_F

BINLOG '
ayu7WxMBAAAAMQAAADoDAAAAAEkAAAAAAAEABHRlc3QAA2VtcAACAw8C/wADFWyA5w==
ayu7Wx8BAAAANAAAAG4DAAAAAEkAAAAAAAEAAgAC///8AQAAAAJ6MfwBAAAAAnoyBJfDeA==
'/*!*/;
### UPDATE `test`.`emp`
### WHERE
###   @1=1 /* INT meta=0 nullable=1 is_null=0 */
###   @2='z1' /* VARSTRING(255) meta=255 nullable=1 is_null=0 */
### SET
###   @1=1 /* INT meta=0 nullable=1 is_null=0 */
###   @2='z2' /* VARSTRING(255) meta=255 nullable=1 is_null=0 */
```

同样可以看到,日志中完整地记录了一条数据从修改前到修改后的数据内容,我们可以交换这两个数据段的位置，便可以将一个更新语句逆向化，生成一个新的闪回语句。如上两种情况中，具体的二进制文件修改功能的编码过程这里就不做介绍了。

得到逆向的语句之后，需要重新执行新语句来应用到数据库中。但是需要注意执行顺序问题，如图 24-3 所示。

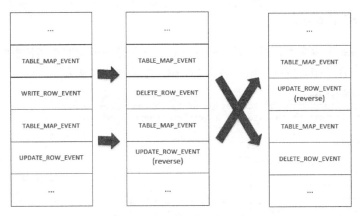

图 24-3　新事件列表反转

我们需要将整个的事件列表以（TABLE_MAP_EVENT,ROW_EVENT）为一个单位完全反转过来，然后按顺序执行图中最后一个列表的内容，才能完成一个闪回的操作。

由此可以看出，日志闪回能够有效地解决 DML 误操作所带来的问题，但当操作是 drop 或 truncate 等命令时，数据库将无法通过这种方式进行闪回。

目前 MySQL 已经有 binlog2sql、MyFlash 等可用的闪回工具可以直接使用。当然读者也可以根据上面所涉及的二进制日志闪回原理，设计并编写一套属于自己的闪回工具。

24.3　查询日志

查询日志记录了客户端的所有语句，而二进制日志不包含查询数据的语句。

24.3.1　日志的位置和格式

查询日志和慢查询日志（见 24.4 节）都可以选择保存在文件或者表中，并使用参数 --log-output[=value,...]来进行控制，value 值可以是 TABLE、FILE、NONE 的一个或者多个组合，中间用逗号进行分割，分别表示日志保存在表、文件、不保存在表和文件中，这里的表指的是 mysql 库中的 general_log（慢查询日志是 slow_log）表。其中 NONE 的优先级最高，比如--log-output =TABLE,FILE 表示日志可以同时输出到表和文件中，而 --log-output =TABLE,NONE，由于 NONE 的优先级，表示日志不保存在表和文件中。如果不显式设置此参数，则默认输出到文件。日志记录到表比记录到文件要占用更多的系统资源。如果需要更高的性能，则建议使用文件来记录日志。

如果要启用查询日志，可以通过参数--general_log[={0|1}]和--general_log_file=file_name 来进行控制。前者控制是否启用日志，后者控制日志文件的路径。--general_log 设置为 1 或者不带值都可以启用查询日志；设置为 0 则关闭查询日志，如果不指定此参数也不会启用查询

日志。如果没有指定--general_log_file=file_name 的值，且没有显式设置--log-output 参数，那么日志将写入参数 DATADIR（数据目录）指定的路径下，默认文件名是 host_name.log。这两个参数都是 global 类型，可以在系统启动时或者系统运行时进行动态修改，如果想在 session 级别控制日志是否被记录，则通过在 session 中设置参数 sql_log_off 为 on 或者 off 来进行控制。

24.3.2 日志的读取

查询日志格式是纯文本，可以直接进行读取。下面是一个读取查询日志的例子。

（1）首先在客户端对数据库做一些简单操作，包括查询和插入。

```
mysql> use employees
Reading table information for completion of table and column names
You can turn off this feature to get a quicker startup with -A

Database changed
mysql>select * from departments;
+---------+--------------------+
| dept_no | dept_name          |
+---------+--------------------+
| d009    | Customer Service   |
| d005    | Development        |
| d002    | Finance            |
+---------+--------------------+
3 rows in set (0.00 sec)

mysql>insert into departments values('d010','IT');
Query OK, 1 row affected (0.03 sec)
```

（2）然后查看查询日志中记录的客户端的所有操作，对应的内容如下：

```
[root@localhost data]# more localhost.log
...
2018-09-27T19:53:08.299817+08:00    716 Query    select * from departments
2018-09-27T19:53:51.666781+08:00    716 Query    insert into departments values('d010','IT')
2018-09-27T19:54:03.439720+08:00    716 Quit
```

注意：log 日志中记录了所有数据库的操作，对于访问频繁的系统，此日志对系统性能的影响较大，建议关闭。

24.4 慢查询日志

慢查询日志记录了所有执行时间超过参数 long_query_time（单位为 s）设置值并且扫描记录数不小于 min_examined_row_limit 的所有 SQL 语句的日志（注意，获得表锁定的时间不算作执行时间）。long_query_time 默认为 10s，最小为 0，精度可以到微秒。

在默认情况下，有两类常见语句不会记录到慢查询日志：管理语句和不使用索引进行查询的语句。这里的管理语句包括 ALTER TABLE、ANALYZE TABLE、CHECK TABLE、CREATE INDEX、DROP INDEX、OPTIMIZE TABLE 和 REPAIR TABLE。如果要监控这两类 SQL 语句，可以分别通过参数--log-slow-admin-statements 和--log-queries-not-using-indexes 进行控制。

24.4.1 文件位置和格式

慢查询日志默认是关闭的，可以用两个参数来控制开启状态和输出方式：使用--slow_query_log[={0|1}]显式指定慢查询的状态，如果不指定值或者指定值为 1 都会打开慢查

询；使用 slow_query_log_file[=file_name]来指定慢查询日志的路径，如果没有给定 file_name 的值，日志将写入参数 DATADIR（数据目录）指定的路径下，默认文件名是 host_name-slow.log。另外，如前所述，还可以使用--log-output 参数来指定日志的输出方式，默认会输出到文件，当然也可以选择输出到表。

24.4.2　日志的读取

和错误日志、查询日志一样，慢查询日志记录的格式也是纯文本，可以被直接读取。下例中演示了慢查询日志的设置和读取过程。

（1）首先查询一下 long_query_time 的值。

```
mysql> show variables like 'long%';
+-----------------+-------+
| Variable_name   | Value |
+-----------------+-------+
| long_query_time | 10    |
+-----------------+-------+
1 row in set (0.00 sec)
```

（2）为了方便测试，将修改慢查询时间为 2s。

```
mysql> set long_query_time=2;
Query OK, 0 rows affected (0.02 sec)
```

（3）依次执行下面两个查询语句。

第一个查询因为查询时间低于 2s 而不会出现在慢查询日志中：

```
mysql> select count(1) from salaries;
+----------+
| count(1) |
+----------+
|  2834339 |
+----------+
1 row in set (0.93 sec)
```

第二个查询因为查询时间大于 2s 而应该出现在慢查询日志中：

```
mysql> select count(1) from salaries t,salaries t1 where t.emp_no=t1.emp_no;
+----------+
| count(1) |
+----------+
| 33521281 |
+----------+
1 row in set (24.68 sec)
```

（4）查看慢查询日志。

```
[root@localhost data]# more localhost-slow.log
# Time: 2018-09-27T19:59:10.609135+08:00
# User@Host: root[root] @ localhost []  Id:   717
# Query_time: 24.675446  Lock_time: 0.000381 Rows_sent: 1  Rows_examined: 36355620
SET timestamp=1538049550;
select count(1) from salaries t,salaries t1 where t.emp_no=t1.emp_no;
```

从上面的日志中可以发现查询时间超过 2s 的 SQL，而小于 2s 的则没有出现在此日志中。

（5）设置微秒级慢查询。

```
mysql> set global long_query_time=0.01;
Query OK, 0 rows affected (0.00 sec)

mysql> show variables like 'long%';
+-----------------+----------+
| Variable_name   | Value    |
+-----------------+----------+
```

```
| long_query_time | 0.010000 |
+-----------------+----------+
1 row in set (0.01 sec)
```

查看日志输出方式为文件和表同时写：

```
mysql> show variables like '%output%';
+---------------+------------+
| Variable_name | Value      |
+---------------+------------+
| log_output    | FILE,TABLE |
+---------------+------------+
```

执行慢查询 SQL 如下：

```
mysql> select count(0) from dept_emp;
+----------+
| count(0) |
+----------+
|   286809 |
+----------+
1 row in set (0.09 sec)
```

查看日志文件记录，日期的确精确到微秒：

```
# Time: 2018-09-27T20:04:20.918640+08:00
# User@Host: root[root] @ localhost []  Id:    718
# Query_time: 0.090921  Lock_time: 0.000298 Rows_sent: 1  Rows_examined: 286809
SET timestamp=1538049860;
select count(0)  from dept_emp;
```

查询系统表记录，如前所述，时间无法精确到微秒：

```
mysql> select query_time,sql_text from mysql.slow_log where sql_text like '%select count(0) from dept_emp%';
+-----------------+---------------------------------+
| query_time      | sql_text                        |
+-----------------+---------------------------------+
| 00:00:00.090921 | select count(0)  from dept_emp  |
+-----------------+---------------------------------+
1 row in set (0.00 sec)
```

　　如果慢查询日志中记录内容很多，可以使用 pt-query-digest 工具进行分析，可以得到每个 SQL 的响应时间、最大执行时间、最小执行时间、执行时间对比等信息，结果非常直观。pt-query-digest 的详细用法可以参考第 23 章中的相关内容。

　　如果没有安装 pt-query-digest 工具，想简单分析慢日志信息，也可以使用 mysqldumpslow 工具（MySQL 客户端安装自带）来对慢查询日志进行分类汇总。下例中对日志文件 localhost-slow.log 进行了分类汇总，只显示汇总后的摘要结果：

```
[root@localhost mysql]# mysqldumpslow localhost-slow.log
Count: 2   Time=23.37s (46s)  Lock=0.00s (0s)  Rows=1.0 (2), root[root]@localhost
  select count(N) from salaries t,salaries t1 where t.emp_no=t1.emp_no

Count: 1   Time=0.09s (0s)  Lock=0.00s (0s)  Rows=1.0 (1), root[root]@localhost
  select count(N)  from dept_emp

Count: 3   Time=0.00s (0s)  Lock=0.00s (0s)  Rows=2.3 (7), root[root]@localhost
  select time_format(query_time,'S') time,sql_text from mysql.slow_log

Count: 1   Time=0.00s (0s)  Lock=0.00s (0s)  Rows=5.0 (5), root[root]@localhost
  select * from  mysql.slow_log
```

　　对于 SQL 文本完全一致，只是变量不同的语句，mysqldumpslow 将会自动视为同一个语句进行统计，变量值用 N 来代替。这个统计结果将大大增加用户阅读慢查询日志的效率，并迅速定位系统的 SQL 瓶颈。

注意：慢查询日志对于我们发现应用中有性能问题的 SQL 很有帮助，建议正常情况下，打开此日志并经常查看分析。

24.4.3 Anemometer 简介

慢查询日志的重要性不言而喻，在数据库优化和故障诊断时，通常会先检查慢查询日志，往往能找到一些蛛丝马迹。此时使用 pt-query-digest 工具，根据时间区间、汇总方式等需求，手工分析慢查询日志文件，得到比较直观的结果。然而随着 MySQL 实例增多，手工处理的方式已经满足不了要求；日常工作中，我们也希望给开发人员提供查看慢查询日志的接口，让开发人员自己优化 SQL。一些公司会开发分析慢查询日志的平台，减少手工操作的工作量。如果自己不希望开发类似的平台，可以使用开源工具 Anemometer。

Anemometer 是图形化展示 MySQL 慢查询日志的一个工具平台。它的好处是提供了丰富的选项和友好的显示界面，用户只需操作鼠标，就可以查看相应实例的慢查询日志；缺点是没有权限认证，数据安全性低。

Anemometer 需要以下工具和组件：
- 存放汇总慢查询日志的 MySQL 数据库；
- pt-query-digest 工具；
- 被监控实例主机上部署 crontab 任务；
- PHP 5.5 版本以上的 Web 服务。

Anemometer 的架构如图 24-4 所示。

图 24-4　Anemometer 架构示意图

在 Anemometer 的架构中，由被监控 MySQL 实例的主机运行 crontab 任务，定时用 pt-query-digest 工具分析慢查询日志文件，将分析结果推到慢查询汇总库。用户在 Web 端选择查询条件后，Anemometer 到汇总库查询，在 Web 页面显示慢查询 SQL。如果用户想了解慢查询 SQL 的执行计划，Anemometer 会到被监控数据库执行 explain 命令，并在 Web 页面显示。

Anemometer 的安装并不复杂，读者按照 GitHub 上的提示安装即可。如果安装完成 Anemometer，在浏览器中输入 http://$ip/anemometer 即可登录，其中$ip 为 Anemometer 平台所在主机，如图 24-5 所示。

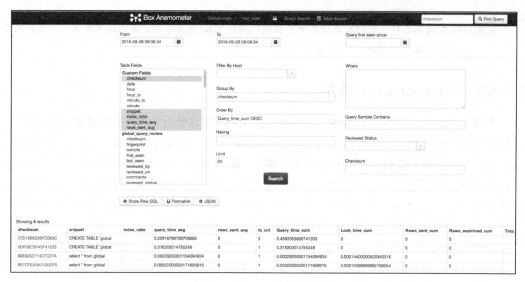

图 24-5　Web 端查询页面

从图 24-5 中能看到很多选项，在框内写入相应的限制条件，可以在汇总库找到想要的慢查询日志信息。下面介绍几个常用的选项。

- From，To：框定需要查看的时间范围。
- Query first seen since：慢 SQL 第一次出现在日志中的时间。
- Table Fields：选择显示慢 SQL 的信息字段。只有选中的字段，才会在结果中显示。
- Filter By Host：选择过滤 MySQL 服务所在的主机名称，默认为空。
- Group By：按照哪个字段进行分组，默认是 checksum 字段。
- Order By：按照哪个字段进行排序，默认是按照 Query_time_sum 字段倒序。
- Limit：只取前 N 条慢 SQL，配合上面的 Order By 使用。默认是 20，即找出 Top 20 的慢 SQL。
- Reviewed Status：根据慢 SQL 的审核状态查询，默认为空。
- Show Raw SQL：显示用户选定所有过滤条件后，Anemometer 到汇总库查询的真实 SQL。参考这个 SQL，可以知道查询逻辑跟我们期望的是否一致，方便高阶用户使用。

为了加深理解，这里做一个简单的查询，From 时间为 2018-09-02 00:00:35，To 时间为 2018-09-30 00:00:35，平台会查询这个时间区间的慢 SQL。Query first seen since 为 2018-09-05 00:00:00，在上面的查询结果中，再限定所有慢查询日志 SQL，第一次出现的时间在 2018-09-05 00:00:00 之后。Group By 为 snippet，即按照 SQL 前几个单词进行分组。Order By 为 query_time_avg DESC，即按照 query_time_avg 字段倒序排序。Limit 为 5，取结果集排序后的前 5 行。查询界面如图 24-6 所示。

如果想查看所有条件组合的原始查询 SQL，可以单击左下角的"Show Raw SQL"按钮。单击"Search"按钮，可以看到查询结果，一共查出 5 条记录，如图 24-7 所示。

单击"checksum"字段会跳转到另外一个页面，在这个页面可以对慢查询 SQL 进行分析和审核。分析功能包括查看慢 SQL 出现的频率、执行计划、表结构和表状态等。以上功能基本能覆盖用户工作中的大多数需求。如果想更详细地了解 Anemometer，可以参见 GitHub 的官方主页 *https://github.com/box/Anemometer/wiki*。

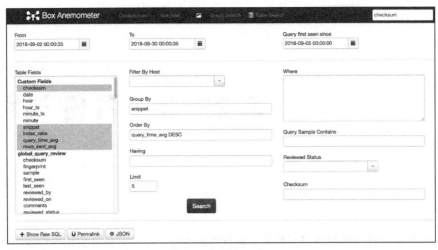

图 24-6 Web 端查询页面

Showing 5 results

checksum	snippet	index_ratio	query_time_avg	rows_sent_avg	ts_cnt	Query_time_sum	Lock_time_sum	Rows_sent_sum	Rows_examined_sum
FCFC405208ACBDFE	INSERT INTO proxy_he		3073.9261904761906	0	210	645524.5	0.1291929930448532	0	0
8AE06173AA02BA16	select count(0) from	0.00	1.5023410320281982	1	1	1.5023410320281982	0.00017899999511428177	1	0
8E3BD93378DB7BEB	CREATE TABLE sbtest1		1.0382293065388997	0	3	3.114687919616699	0	0	0
6F2F9273B7A5F89	CREATE TABLE `test_b		1.0026110410690308	0	1	1.0026110410690308	0	0	0
5066111A87FF5D08	CREATE TABLE `tb_hea		0.8293989896774292	0	1	0.8293989896774292	0	0	0

图 24-7 查询结果

Anemometer 的缺点是没有登录和权限认证。如果多个业务共用一个 MySQL 实例，所有业务的慢查询 SQL 是放在一起的，用户能查看所有业务的慢查询 SQL，在某些情况下这是不安全的。针对这种情况，一种方法是将业务拆分，每业务专用自己的 MySQL 实例；第二种方式是对 Anemometer 进行二次开发，实现登录和权限认证需求。读者可以采取适合自己的解决方法。

24.5 小结

日志是数据库中很重要的记录内容，它可以帮助我们诊断数据库出现的各种问题。本章主要介绍了 MySQL 常用的 4 种日志类型：错误日志、二进制日志、查询日志和慢查询日志。这 4 种日志各有不同的用途。

- 系统故障时，建议首先查看错误日志，以帮助用户迅速定位故障原因。
- 二进制日志记录数据的变更和 DDL 操作，是数据备份、数据复制和数据恢复等操作的基础，应默认开启此日志。
- 如果希望记录数据库发生的任何操作，包括 SELECT，则需要用--general_log 将查询日志打开，此日志默认关闭，一般情况下建议不要打开此日志，以免影响系统整体性能。
- 如果希望查看系统的性能问题，希望找到有性能问题的 SQL 语句，则需要用--slow_query_log 打开慢查询日志。对于大量的慢查询日志，建议使用 pt-query-digest 工具来进行汇总查看。

第 25 章 备份与恢复

备份和恢复在任何数据库里面都是非常重要的内容，好的备份方法和备份策略将会使得数据库备份更高效、数据更安全。与很多数据库类似，MySQL 的备份主要分为逻辑备份和物理备份。本章将重点介绍这两种方式的备份和恢复方法。

25.1 备份/恢复策略

对于一个 DBA 来说，定制合理的备份策略无疑是很重要的。以下是我们在进行备份或恢复操作时需要考虑的一些因素。

- 确定要备份的表的存储引擎是事务型还是非事务型，两种不同的存储引擎备份方式在处理数据一致性方面是不太一样的。
- 确定使用全量备份还是增量备份。全量备份的优点是备份保持最新备份，恢复的时候可以花费更少的时间；缺点是如果数据量大，将会花费很多的时间，并对系统造成较长时间的压力。增量备份则恰恰相反，只需要备份每天的增量日志，备份时间少，对负载压力也小；缺点就是恢复的时候需要全量备份加上次备份到故障前的所有日志，恢复时间会长一些。
- 可以考虑采取复制的方法来做异地备份，但是记住，复制不能代替备份，它对数据库的误操作也无能为力。
- 要定期做备份，备份的周期要充分考虑系统可以承受的恢复时间。备份要在系统负载较小的时候进行。
- 确保 MySQL 打开 log-bin 选项，有了 BINLOG，MySQL 才可以在必要的时候做完整恢复，或基于时间点的恢复，或基于位置的恢复。
- 要经常做备份恢复测试，确保备份是有效的，并且是可以恢复的。

25.2 逻辑备份和恢复

在 MySQL 中，逻辑备份的最大优点是对于各种存储引擎都可以用同样的方法来备份，而物理备份则不同，不同的存储引擎有着不同的备份方法。因此，对于不同存储引擎混合的数据库，用逻辑备份会更简单一些。本节将详细介绍逻辑备份以及相应的恢复方法。

25.2.1 备份

MySQL 中的逻辑备份是将数据库中的数据备份为文本文件，备份的文件可以被查看和编辑。在 MySQL 中，可以使用 mysqldump 工具来完成逻辑备份。可以使用以下 3 种方法来调用 mysqldump。

- 备份指定的数据库或者此数据库中的某些表。

```
shell> mysqldump [options] db_name [tables]
```

- 备份指定的一个或多个数据库。

```
shell> mysqldump [options] ---database DB1 [DB2 DB3...]
```

- 备份所有数据库。

```
shell> mysqldump [options] --all-database
```

如果没有指定数据库中的任何表，默认导出所有数据库中的所有表。以下给出一些使用 mysqldump 工具进行备份的例子。

（1）备份所有数据库：

```
[mysql3307@localhost ~]$ mysqldump -uroot -p --all-database > all.sql
Enter password:
```

（2）备份数据库 employees：

```
[mysql3307@localhost ~]$ mysqldump -uroot -p employees > employees.sql
Enter password:
```

（3）备份数据库 employees 下的表 salaries：

```
[mysql3307@localhost ~]$ mysqldump -uroot -p employees salaries > salaries.sql
Enter password:
```

（4）备份数据库 employees 下的表 dept_emp 和 salaries：

```
[mysql3307@localhost ~]$ mysqldump -uroot -p employees dept_emp salaries > dept_emp_salaries.sql
Enter password:
```

（5）备份数据库 employees 下的所有表为逗号分割的文本，备份到/tmp：

```
[mysql3307@localhost tmp]# mysqldump -uroot -T /tmp employees dept_emp --fields-terminated-by ','
[mysql3307@localhost tmp]# more dept_emp.txt
11025,d003,1998-09-07,2000-05-26
11026,d004,1991-11-08,9999-01-01
11027,d004,1986-05-07,9999-01-01
11028,d004,1986-06-18,1997-09-04
11029,d007,1992-12-27,9999-01-01
11030,d007,1988-05-09,1988-07-08
11032,d005,2000-12-15,9999-01-01
11033,d005,1991-03-14,9999-01-01
11036,d002,1988-10-24,9999-01-01
11037,d004,1988-01-28,9999-01-01
11038,d008,1985-07-15,9999-01-01
......
```

其中 mysqldump 的选项很多，具体可以使用"--help"参数查看帮助：

```
mysqldump -help
```

需要强调的是，为了保证数据备份的一致性，MyISAM 存储引擎在备份时需要加上-l 参数，表示将所有表加上读锁，在备份期间，所有表将只能读而不能进行数据更新。但是对于事务存储引擎（InnoDB）来说，可以采用--single-transaction 选项。此选项将使得 InnoDB 存储引擎得到一个快照（Snapshot），保证备份的数据能够一致。

注意：MySQL 8.0 的系统表已经废弃 MyISAM 存储引擎，默认存储引擎为 InnoDB。

日常工作中，我们也经常使用 mydumper 进行逻辑备份。它是一个多线程备份工具，通过增加线程数，极大地提高备份速度。与 mysqldump 不同，mydumper 为每个表创建一个或者多个文本文件，方便并行备份和并行恢复数据。

注意：可以在 https://github.com/maxbube/mydumper 下载安装包。

mydumper 使用方法跟 mysqldump 基本一致，对于两者相同的参数将不再赘述。下面是 mydumper 一些常用的选项，要查阅更多的选项，可以使用"--help"参数查看帮助。

```
-B, --database          需要备份的数据库，一个数据库写一条命令。如果不指定，默认备份所有数据库。
-T, --tables-list       需要备份的表
-o, --outputdir         导出文件存放目录
-s, --statement-size    按照 insert 语句插入的数据大小拆分，单位 bytes,默认值 1000000
-r, --rows              按照行数对结果文件进行拆分
-F, --chunk-filesize    按照导出文件大小进行拆分，单位为 MB
-t, --threads           线程并行数，默认 4 个线程
```

下面来看一些使用 mydumper 工具的例子。

（1）备份 employees 数据库：

```
[mysql3307@localhost ~]$ mydumper -uroot -S/tmp/mysql_3307.sock \
-o /data2/backup/mydumper -B employees -p
Enter password:

[mysql3307@localhost mydumper]$ cd /data2/backup/mydumper
[mysql3307@localhost mydumper]$ ll
total 4
drwx------ 2 mysql3307 mysql3307 4096 Oct 11 20:31 export-20181011-203134
[mysql3307@localhost mydumper]$ ll export-20181011-203134/
total 209596
-rw-rw-r-- 1 mysql3307 mysql3307       624 Oct 11 20:31 employees.dept_manager-schema.sql
-rw-rw-r-- 1 mysql3307 mysql3307      1168 Oct 11 20:31 employees.dept_manager.sql
-rw-rw-r-- 1 mysql3307 mysql3307       484 Oct 11 20:31 employees.salaries-schema.sql
-rw-rw-r-- 1 mysql3307 mysql3307 118693182 Oct 11 20:31 employees.salaries.sql
-rw-rw-r-- 1 mysql3307 mysql3307       484 Oct 11 20:31 employees.titles-schema.sql
-rw-rw-r-- 1 mysql3307 mysql3307  21708866 Oct 11 20:31 employees.titles.sql
……
-rw-rw-r-- 1 mysql3307 mysql3307  21708866 Oct 11 20:31 metadata
```

可以看到，备份结果文件一共有 3 类。第一类是 db.table.sql 文件，存放表数据；第二类是 db.table-schema.sql 文件，存放表结构；第三类是 metadata 文件，记录备份的开始时间和结束时间，以及 BINLOG 日志文件名、位置和 GTID 信息。可以根据 metadata 记录的信息进行完全恢复或基于位置的恢复。下面以 employees.dept_manager 表为例进行讲解。

查看表数据：

```
[mysql3307@localhost mydumper]$ cd export-20181011-203134
[mysql3307@localhost export-20181011-203134]$ cat employees.dept_manager.sql
/*!40101 SET NAMES binary*/;
/*!40014 SET FOREIGN_KEY_CHECKS=0*/;
/*!40103 SET TIME_ZONE='+00:00' */;
INSERT INTO `dept_manager` VALUES
(110022,"d001","1985-01-01","1991-10-01"),
(110039,"d001","1991-10-01","9999-01-01"),
(110085,"d002","1985-01-01","1989-12-17"),
(110114,"d002","1989-12-17","9999-01-01"),
(110183,"d003","1985-01-01","1992-03-21"),
……
(111784,"d009","1988-10-17","1992-09-08"),
```

```
(111877,"d009","1992-09-08","1996-01-03"),
(111939,"d009","1996-01-03","9999-01-01");
```

查看表结构：

```
[mysql3307@localhost export-20181011-203134]$ cat employees.dept_manager-schema.sql
/*!40101 SET NAMES binary*/;
/*!40014 SET FOREIGN_KEY_CHECKS=0*/;

/*!40103 SET TIME_ZONE='+00:00' */;
CREATE TABLE `dept_manager` (
  `emp_no` int(11) NOT NULL,
  `dept_no` char(4) COLLATE utf8_unicode_ci NOT NULL,
  `from_date` date NOT NULL,
  `to_date` date NOT NULL,
  PRIMARY KEY (`emp_no`,`dept_no`),
  KEY `dept_no` (`dept_no`),
  CONSTRAINT `dept_manager_ibfk_1` FOREIGN KEY (`emp_no`) REFERENCES `employees` (`emp_no`) ON DELETE CASCADE,
  CONSTRAINT `dept_manager_ibfk_2` FOREIGN KEY (`dept_no`) REFERENCES `departments` (`dept_no`) ON DELETE CASCADE
) ENGINE=InnoDB DEFAULT CHARSET=utf8 COLLATE=utf8_unicode_ci;
```

查看备份开始时间和结束时间：

```
[mysql3307@localhost export-20181011-211838]$ cat metadata
Started dump at: 2018-10-11 21:14:57
SHOW MASTER STATUS:
        Log: mysql-bin.000008
        Pos: 2750
        GTID:b509f331-cd14-11e8-866a-0024e869b4d5:1-18044

Finished dump at: 2018-10-11 21:15:00
```

（2）备份 employees 数据库的 salaries 表，开启 3 个并发线程数，每个文本文件最大 50MB。

```
[mysql3307@localhost ~]$ mydumper -u root -S /tmp/mysql_3307.sock \
-o /data2/backup/salaries -B employees -T salaries -t 3 -F 50 -p
Enter password:

[mysql3307@localhost salaries]$ cd /data2/backup/salaries
[mysql3307@localhost salaries]$ ls -ltrh
total 114M
-rw-rw-r-- 1 mysql3307 mysql3307   92 Oct 11 20:50 employees-schema-create.sql
-rw-rw-r-- 1 mysql3307 mysql3307  484 Oct 11 20:50 employees.salaries-schema.sql
-rw-rw-r-- 1 mysql3307 mysql3307  49M Oct 11 20:50 employees.salaries.00001.sql
-rw-rw-r-- 1 mysql3307 mysql3307  49M Oct 11 20:50 employees.salaries.00002.sql
-rw-rw-r-- 1 mysql3307 mysql3307  16M Oct 11 20:50 employees.salaries.00003.sql
-rw-rw-r-- 1 mysql3307 mysql3307  181 Oct 11 20:50 metadata
```

25.2.2 完全恢复

mysqldump 的恢复也很简单，将备份作为输入执行即可，具体语法如下：

```
mysql -uroot -p dbname < bakfile
```

注意，将备份恢复后数据并不完整，还需要将备份后执行的日志进行重做，语法如下：

```
mysqlbinlog binlog-file | mysql -u root -p***
```

以下是一个完整的 mysqldump 备份与恢复的例子。

（1）上午 9 点，备份数据库：

```
[mysql3307@localhost mysql]# mysqldump  -uroot -p --single-transaction -F employees >employees.dmp
Enter password:
```

其中 --single-transaction 参数表示给 InnoDB 表加生成快照，保证数据一致性，-F 表示生成一个新的日志文件，此时，employees 数据库中 departments 表的数据如下：

```
mysql> select * from employees.departments;
+---------+--------------------+
| dept_no | dept_name          |
+---------+--------------------+
| d009    | Customer Service   |
| d005    | Development        |
| d002    | Finance            |
| d003    | Human Resources    |
| d010    | IT                 |
| d001    | Marketing          |
| d004    | Production         |
| d006    | Quality Management |
| d008    | Research           |
| d007    | Sales              |
+---------+--------------------+
10 rows in set (0.00 sec)
```

（2）9点半备份完毕，然后，插入新的数据：

```
mysql> insert into departments values('d011','YEATION');
Query OK, 1 row affected (0.01 sec)

mysql> insert into departments values('d012','Koala');
Query OK, 1 row affected (0.01 sec)
```

（3）10点，数据库突然发生故障，数据无法访问，需要恢复备份：

```
[mysql3307@localhost mysql]# mysql -uroot -p employees< employees.dmp
Enter password:
```

恢复后的数据如下：

```
mysql> select * from employees.departments;
+---------+--------------------+
| dept_no | dept_name          |
+---------+--------------------+
| d009    | Customer Service   |
| d005    | Development        |
| d002    | Finance            |
| d003    | Human Resources    |
| d010    | IT                 |
| d001    | Marketing          |
| d004    | Production         |
| d006    | Quality Management |
| d008    | Research           |
| d007    | Sales              |
+---------+--------------------+
10 rows in set (0.00 sec)
```

（4）使用 mysqlbinlog 恢复自 mysqldump 备份以来的 BINLOG。

```
[mysql3307@localhost mysql]# mysqlbinlog  localhost-bin.000015 | mysql -u root -p employees
Enter password:
```

查询完全恢复的数据如下：

```
mysql> select * from employees.departments;
+---------+--------------------+
| dept_no | dept_name          |
+---------+--------------------+
| d009    | Customer Service   |
| d005    | Development        |
| d002    | Finance            |
| d003    | Human Resources    |
| d010    | IT                 |
| d012    | Koala              |
| d001    | Marketing          |
| d004    | Production         |
| d006    | Quality Management |
| d008    | Research           |
| d007    | Sales              |
| d011    | YEATION            |
```

```
+----------+-------------------+
12 rows in set (0.00 sec)
```

至此,数据库完全恢复。

25.2.3 基于时间点恢复

由于误操作,比如误删除了一张表,这时使用完全恢复是没有用的,因为日志里面还存在误操作的语句,我们需要的是恢复到误操作之前的状态,然后跳过误操作语句,再恢复后面执行的语句,完成我们的恢复。这种恢复叫作不完全恢复。在 MySQL 中,不完全恢复分为基于时间点的恢复和基于位置的恢复。

以下是基于时间点恢复的操作步骤。

(1)如果上午 10 点发生了误操作,可以用备份和 BINLOG 将数据恢复到故障前,所用语句如下:

```
shell>mysqlbinlog --stop-date="2018-10-20 9:59:59" \
/data2/mysql3307/data/mysql-bin.123456 | mysql -u root -pmypwd
```

(2)跳过故障时的时间点,继续执行后面的 BINLOG,完成恢复。

```
shell>mysqlbinlog --start-date="2018-10-20 10:01:00" \
/data2/mysql3307/data/mysql-bin.123456| mysql -u root -pmypwd
```

25.2.4 基于位置恢复

和基于时间点的恢复类似,但是更精确,因为同一个时间点可能有很多条 SQL 语句同时执行。恢复的操作步骤如下。

(1)在 shell 下执行如下命令:

```
shell>mysqlbinlog --start-date="2018-10-20 9:55:00" --stop-date="2018-10-20 10:05:00" /data2/mysql3307/data/mysql-bin.123456 > /tmp/mysql_restore.sql
```

该命令将在/tmp 目录创建小的文本文件,编辑此文件,找到出错语句前后的位置号,例如前后位置号分别是 368312 和 368315。

(2)恢复了以前的备份文件后,应从命令行中输入下面的内容:

```
shell>mysqlbinlog --stop-position="368312" /data2/mysql3307/data/mysql-bin.123456 \
    | mysql -u root -pmypwd
shell>mysqlbinlog --start-position="368315" /data2/mysql3307/data/mysql-bin.123456 \
    | mysql -u root -pmypwd
```

上面的第一行将恢复到停止位置为止的所有事务。下一行将恢复从给定的起始位置直到二进制日志结束的所有事务。因为 mysqlbinlog 的输出包括每个 SQL 语句记录之前的 SET TIMESTAMP 语句,因此恢复的数据和相关 MySQL 日志将反映事务执行的原时间。

25.2.5 并行恢复

如果使用 mydumper 进行备份,可以使用配套的 myloader 工具并行恢复数据,大大减少恢复时间。下面是 myloader 的一些常用选项,读者可以使用 "--help" 参数查看帮助,了解更多选项。

```
-d, --directory         备份文件所在目录
-o, --overwrite-tables  是否覆盖已有表
-B, --database          执行需要恢复的数据库
-e, --enable-binlog     是否记录 BINLOG。如果在主库导入数据,开启此选项,才能把数据同步到从库
-t, --threads           线程并行数,默认为 4 个线程
```

下面看一个 myloader 并行恢复的例子。

(1) 查看表数据量：

```
root@localhost:mysql_3307.sock  [employees]>select count(0) from salaries;
+----------+
| count(0) |
+----------+
|  2844047 |
+----------+
1 row in set (0.86 sec)
```

(2) 删掉部分数据：

```
root@localhost:mysql_3307.sock  [employees]>delete from salaries limit 47;
```

(3) 使用 25.2.1 节中的备份文件恢复 salaries 表。开启 3 个并发线程数，覆盖已存在的表：

```
[mysql3307@localhost ~]$ myloader -u root -S /tmp/mysql_3307.sock \
-d /data2/backup/salaries -B employees -t 3 -o -p
Enter password:
```

(4) 查看恢复后表数据量：

```
root@localhost:mysql_3307.sock  [employees]>select count(0) from salaries;
+----------+
| count(0) |
+----------+
|  2844047 |
+----------+
1 row in set (0.86 sec)
```

恢复后表记录数与备份前的记录数相同，证明表恢复成功了。

25.3 物理备份和恢复

物理备份分为冷备份和热备份两种，和逻辑备份相比，它的最大优点是备份和恢复的速度更快，因为物理备份的原理都是基于文件的 cp。本节将介绍 MySQL 中的物理备份及其恢复的方法。

25.3.1 冷备份和热备份

冷备份其实就是停掉数据库服务，cp 数据文件的方法。这种方法对 MyISAM 和 InnoDB 存储引擎都适合，但是一般很少使用，因为很多应用是不允许长时间停机的。

- 进行备份的操作如下：停掉 MySQL 服务，在操作系统级别备份 MySQL 的数据文件和日志文件到备份目录。
- 进行恢复的操作如下：首先停掉 MySQL 服务，在操作系统级别恢复 MySQL 的数据文件；然后重启 MySQL 服务，使用 mysqlbinlog 工具恢复自备份以来的所有 BINLOG。

热备份是在数据库运行的情况下，同时备份数据文件和备份期间产生的 BINLOG 的方法。即热备份是系统处于正常运转状态下的备份。

在 MySQL 中，对于不同的存储引擎热备份方法也有所不同。下面重点介绍 MyISAM 和 InnoDB 两种存储引擎的热备份方法。

25.3.2 MyISAM 存储引擎的热备份

MyISAM 存储引擎的热备份方法，本质其实就是将要备份的表加读锁，然后再 cp 数据文件到备份目录。在 5.7 版本之前，官方提供了 mysqlhotcopy 工具来完成备份功能；5.7 版本之后，由于 MyISAM 使用越来越少，此工具已从安装包中删除。如果要备份 MyISAM 表，操作步骤如下。

（1）数据库中所有表加读锁：

```
mysql> FLUSH TABLES WITH READ LOCK;
```

（2）cp 数据文件到备份目录：

```
cp source/datadir target/datadir
```

（3）释放表锁完成备份：

```
mysql>unlock tables;
```

25.3.3 InnoDB 存储引擎的热备份

InnoDB 存储引擎的物理热备份通常有两种方式，一种是使用表空间迁移技术，cp 数据文件，类似于上文 MyISAM 的备份方式，这种方式主要适用于备份指定的单表或者多表；另一种是使用专门的热备份工具，比如 MySQL 官方企业版提供的收费工具 ibbackup 或者 percona 提供的免费工具 Xtrabackup，后者使用更为广泛。下面详细介绍这两种备份方式。

1. 利用表空间迁移备份表

MySQL 5.6.6 版本开始，基于表空间迁移技术，可以快速迁移或备份 InnoDB 引擎的表。使用表空间迁移要求 innodb_file_per_table 参数设置为 ON；迁移过程，表处于只读状态，不能进行 DDL 和写操作。下面看一个具体的例子。

（1）查看表记录数：

```
root@localhost:mysql_3307.sock  [employees]>select count(0) from salaries;
+----------+
| count(0) |
+----------+
|  2843996 |
+----------+
1 row in set (0.86 sec)
```

（2）备份 salaries 表：

执行 FLUSH TABLES ... FOR EXPORT 命令，在数据库目录生成 cfg 文件，此时 salaries 处于只读状态。

```
root@localhost:mysql_3307.sock  [employees]> flush tables salaries  for export;

[mysql3307@localhost employees]$ cd /data2/mysql3308/data/employees
[mysql3307@localhost employees]$ ll salaries.*
-rw-r----- 1 mysql3308 mysql3308       849 Nov 28 17:55 salaries.cfg
-rw-r----- 1 mysql3308 mysql3308      8790 Oct 11 16:00 salaries.frm
-rw-r----- 1 mysql3308 mysql3308 239075328 Nov 19 18:22 salaries.ibd
```

备份 ibd 文件和 cfg 文件。ibd 文件保存表数据；cfg 文件保存元数据，元数据用于验证导入表空间文件时的模式。

```
[mysql3307@localhost employees]$ mkdir -p /home/mysql3307/backup
[mysql3307@localhost employees]$ cp salaries.cfg salaries.ibd /home/mysql3307/backup/
```

解锁 salaries 表，表备份完成。

```
root@localhost:mysql_3307.sock    [employees]> unlock tables;
Query OK, 0 rows affected (0.00 sec)
```

（3）模拟误删 salaries 表：

```
root@localhost:mysql_3307.sock    [employees]>drop table salaries;
```

（4）重新创建 salaries 表：

```
root@localhost:mysql_3307.sock    [employees]> CREATE TABLE `salaries` (
    ->   `emp_no` int(11) NOT NULL,
    ->   `salary` int(11) NOT NULL,
    ->   `from_date` date NOT NULL,
    ->   `to_date` date NOT NULL,
    ->   `salary_by_1k` int(11) GENERATED ALWAYS AS (round((`salary` / 1000),0)) VIRTUAL,
    ->   `dept_name` varchar(100) COLLATE utf8_unicode_ci DEFAULT NULL,
    ->   PRIMARY KEY (`emp_no`,`from_date`),
    ->   KEY `test_salaries` (`to_date`),
    ->   KEY `idx_salary_by_1k` (`salary_by_1k`),
    ->   CONSTRAINT `salaries_ibfk_1` FOREIGN KEY (`emp_no`) REFERENCES `employees` (`emp_no`) ON DELETE CASCADE
    -> ) ENGINE=InnoDB DEFAULT CHARSET=utf8 COLLATE=utf8_unicode_ci;
Query OK, 0 rows affected (0.05 sec)

root@localhost:mysql_3307.sock    [employees]>select count(0) from salaries;
+----------+
| count(0) |
+----------+
|        0 |
+----------+
1 row in set (0.00 sec)
```

（5）下线 salaries 表：

```
root@localhost:mysql_3307.sock    [employees]> alter table salaries discard tablespace;
Query OK, 0 rows affected (0.00 sec)
```

（6）拷贝步骤二备份的 ibd 和 cfg 文件到数据库目录：

```
[mysql3308@localhost employees]$ cp /home/mysql3308/backup/salaries.cfg /data2/mysql3308/data/employees/
[mysql3308@localhost employees]$ cp /home/mysql3308/backup/salaries.ibd /data2/mysql3308/data/employees/
```

（7）加载 salaries 表：

```
root@localhost:mysql_3307.sock    [employees]> alter table salaries import tablespace;
Query OK, 0 rows affected (0.00 sec)
```

（8）查看表恢复后记录数：

```
root@localhost:mysql_3307.sock    [employees]>select count(0) from salaries;
+----------+
| count(0) |
+----------+
|  2843996 |
+----------+
1 row in set (0.86 sec)
```

恢复后表记录数与备份前的记录数相同，证明表恢复成功了。

2. 使用 Xtrabackup 做全量备份

Xtrabackup 是 Percona 公司 CTO Vadim 参与开发的一款基于 InnoDB 的在线热备工具，具有开源、免费、支持在线热备、备份恢复速度快、占用磁盘空间小等特点，并且支持不同情况下的多种备份形式。Xtrabackup 的官方下载地址为 http://www.percona.com/redir/downloads/XtraBackup。

25.3 物理备份和恢复

Xtrabackup 包含两个主要的工具，即 xtrabackup 和 innobackupex，两者区别如下：

- xtrabackup 只能备份 InnoDB 和 XtraDB 两种数据表，而不能备份 MyISAM 数据表；
- innobackupex 是一个封装了 xtrabackup 的 Perl 脚本，支持同时备份 InnoDB 和 MyISAM，但在对 MyISAM 备份时需要加一个全局的读锁。

下面以 innobackupex 为例，来介绍一下此工具备份、恢复、增量备份恢复的原理。

（1）备份过程。

Innobackupex 的备份过程如图 25-1 所示。

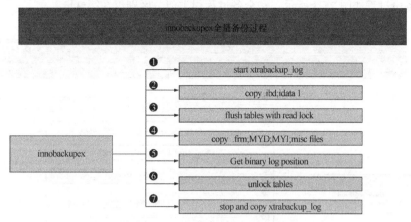

图 25-1　innobackupex 备份过程

在图 25-1 中，备份开始时首先会开启一个后台检测进程，实时检测 mysql redo 的变化，一旦发现 redo 中有新日志写入，立刻将日志记入后台日志文件 xtrabackup_log 中。之后复制 InnoDB 的数据文件和系统表空间文件 ibdata1，待复制结束后，执行 flush tables with read lock 操作，复制 .frm、.MYI、.MYD 等文件（执行 flush tables with read lock 的目的是为了防止数据表发生 DDL 操作，并且在这一时刻获得 BINLOG 的位置），最后会发出 unlock tables，把表设置为可读写状态，最终停止 xtrabackup_log。

（2）全备恢复。

这一阶段会启动 XtraBackup 内嵌的 InnoDB 实例，回放 xtrabackup 日志 xtrabackup_log，将提交的事务信息变更应用到 InnoDB 数据/表空间，同时回滚未提交的事务（这一过程类似 InnoDB 的实例恢复）。恢复过程原理如图 25-2 所示。

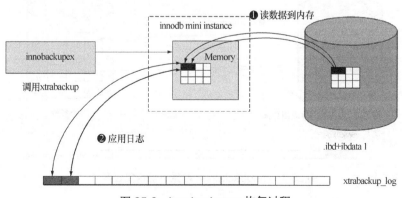

图 25-2　innobackupex 恢复过程

（3）增量备份。

innobackupex 增量备份过程中的"增量"处理，其实主要是相对 InnoDB 而言，对 MyISAM 和其他存储引擎而言，它仍然是一个全拷贝。

"增量"备份的过程主要是通过拷贝 InnoDB 中有变更的"页"（这些变更的数据页指的是"页"的 LSN 大于 xtrabackup_checkpoints 中给定的 LSN）。增量备份是基于全备的，第一次增备的数据必须要基于上一次的全备，之后的每次增备都是基于上一次的增备，最终达到一致性的增备。

增量备份过程如图 25-3 所示，和全备过程很类似，区别仅在第❷步。

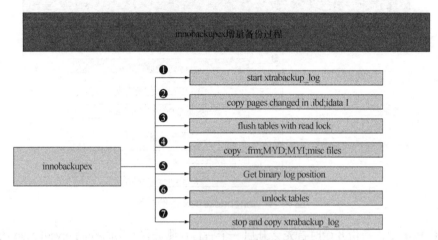

图 25-3　innobackupex 增量备份过程

（4）增量备份恢复。

和全备份恢复类似，也需要两步，一是数据文件的恢复，如图 25-4 所示，这里的数据来源由 3 部分组成：全备份、增量备份和 xtrabackup log。二是对未提交事务的回滚，如图 25-5 所示。

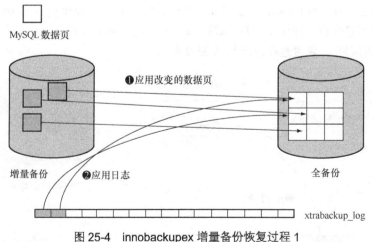

图 25-4　innobackupex 增量备份恢复过程 1

25.3 物理备份和恢复

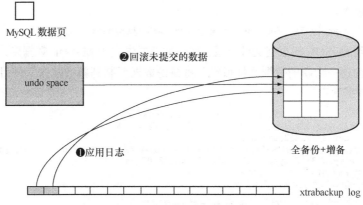

图 25-5 innobackupex 增量备份恢复过程 2

下面将通过示例向读者详细介绍 innobackupex 的使用方法。首先从 http://www.percona.com 下载和安装 Percona XtraBackup 软件：

```
[root@localhost ~]# wget -c \
https://www.percona.com/downloads/XtraBackup/Percona-XtraBackup-2.4.12/binary/redhat/6/x86_64/percona-xtrabackup-24-2.4.12-1.el6.x86_64.rpm
    [root@localhost ~]# rpm -ivh percona-xtrabackup-24-2.4.12-1.el6.x86_64.rpm
    warning: percona-xtrabackup-24-2.4.12-1.el6.x86_64.rpm: Header V4 DSA/SHA1 Signature, key ID cd2efd2a: NOKEY
    Preparing...                ########################################### [100%]
       1:percona-xtrabackup-24  ########################################### [100%]
    [root@localhost ~]# innobackupex -v
    xtrabackup: recognized server arguments: --datadir=/var/lib/mysql
    innobackupex version 2.4.12 Linux (x86_64) (revision id: 170eb8c)
```

注意：确认本机操作系统版本，并下载对应的 RPM 包。

（1）全量备份。

创建备份用户：

```
mysql> CREATE USER 'backup'@'%' IDENTIFIED BY '123456';
 mysql> GRANT RELOAD, LOCK TABLES, REPLICATION CLIENT, CREATE TABLESPACE, SUPER, PROCESS ON *.* TO 'backup'@'%';
```

规划好备份目录路径为 mkdir -p /data2/backup/hotbackup/。创建 innobackupex 的配置文件 /tmp/my.cnf，如下所示：

```
[mysqld]
datadir = "/data2/mysql3307/data"
innodb_data_home_dir = "/data2/mysql3307/data"
innodb_data_file_path = "ibdata1:10M:autoextend"
innodb_log_group_home_dir = "/data2/mysql3307/data"
innodb_log_files_in_group = 3
innodb_log_file_size = 1024M
```

创建测试表：

```
mysql>create database if not exists test;
mysql>use test;
mysql>create table test(id int auto_increment not null primary key,name varchar(20));
mysql>insert into test(name) values('test1');
mysql>insert into test(name) values('test2');
mysql>insert into test(name) values('test3');
mysql>insert into test(name) values('test4');
```

进行全量备份：

```
innobackupex --defaults-file=/tmp/my.cnf --user=backup --password=123456 --socket=/tmp/mysql_3307.sock /data2/backup/hotbackup/full  --no-timestamp
```

恢复全量备份:

```
innobackupex --apply-log --use-memory=5G /data2/backup/hotbackup/full
```

恢复备份到 MySQL 的数据文件目录，这一过程要先关闭 MySQL 数据库，重命名原数据文件目录，再创建一个新的数据文件目录，将备份数据复制到新的数据文件目录下，赋权，启动 MySQL 数据库:

```
shell>mysqladmin  -S  /tmp/mysql_3307.sock  shutdown
shell>mv /data2/mysql3307/data  /data2/mysql3307/data_bak
shell>mkdir  /data2/mysql3307/data
shell>innobackupex --defaults-file=/tmp/my.cnf --copy-back --rsync  /data2/backup/hotbackup/full/
shell>chown -R mysql3307:mysql3307  /data2/mysql3307/data
shell>cd /home/mysql3307/mysqlhome
./bin/mysqld_safe -user=mysql3307 &
```

校验恢复后数据库的一致性，查看 test 数据表:

```
mysql>use test;
mysql>select * from test;
+----+-------+
| id | name  |
+----+-------+
|  1 | test1 |
|  2 | test2 |
|  3 | test3 |
|  4 | test4 |
+----+-------+
4 rows in set (0.00 sec)
```

可以看到 test 库下 test 数据表已经成功恢复。

（2）增量备份。

在 MySQL 中进行增量备份时，首先要进行一次全量备份，第一次增量备份是基于全备的，之后的增量备份是基于上一次的增量备份。

创建基础备份 base：

```
innobackupex --defaults-file=/tmp/my.cnf --user=backup --password=123456 --socket=/tmp/mysql_3307.sock /data2/backup/hotbackup/base --no-timestamp
```

在 test 库下增加增量数据：

```
mysql> insert into test(name) values('test5');
Query OK, 1 row affected (0.00 sec)

mysql> insert into test(name) values('test6');
Query OK, 1 row affected (0.00 sec)

mysql> insert into test(name) values('test7');
Query OK, 1 row affected (0.02 sec)

mysql> insert into test(name) values('test8');
Query OK, 1 row affected (0.00 sec)

mysql> select * from test;
+----+-------+
| id | name  |
+----+-------+
|  1 | test1 |
|  2 | test2 |
|  3 | test3 |
|  4 | test4 |
|  5 | test5 |
|  6 | test6 |
|  7 | test7 |
|  8 | test8 |
+----+-------+
8 rows in set (0.00 sec)
```

创建增量备份 incremental_one:

```
innobackupex --defaults-file=/tmp/my.cnf --user=backup --password=123456 --socket=/tmp/mysql_3307
.sock --incremental /data2/backup/hotbackup/incremental_one --incremental-basedir=/data2/backup/
hotbackup/base  --no-timestamp --parallel=2
```

在 test 库下继续插入增量数据:

```
mysql> insert into test(name) values('test9');
Query OK, 1 row affected (0.00 sec)

mysql> insert into test(name) values('test10');
Query OK, 1 row affected (0.00 sec)

mysql> select * from test;
+----+--------+
| id | name   |
+----+--------+
|  1 | test1  |
|  2 | test2  |
|  3 | test3  |
|  4 | test4  |
|  5 | test5  |
|  6 | test6  |
|  7 | test7  |
|  8 | test8  |
|  9 | test9  |
| 10 | test10 |
+----+--------+
10 rows in set (0.00 sec)
```

创建增量备份 incremental_two:

```
innobackupex --defaults-file=/tmp/my.cnf --user=backup --password=123456 --socket=/tmp/mysql_3307
.sock --incremental /data2/backup/hotbackup/incremental_two --incremental-basedir=/data2/backup/
hotbackup/incremental_one  --no-timestamp --parallel=2
```

(3) 增量备份恢复。

增量备份的恢复大体分为 3 个步骤。

- 恢复基础备份（全备）。
- 恢复增量备份到基础备份（开始恢复的增量备份要添加--redo-only 参数，到最后一次增量备份去掉--redo-only 参数）。
- 对整体的基础备份进行恢复，回滚那些未提交的数据。

恢复基础备份（注意这里一定要加--redo-only 参数，该参数的意思是只应用 xtrabackup 日志中已经提交的事务数据，不回滚还未提交的数据）:

```
innobackupex --apply-log --redo-only --use-memory=5G /data2/backup/hotbackup/base
```

将增量备份 incremental_one 应用到基础备份 base:

```
innobackupex --apply-log --redo-only --use-memory=5G /data2/backup/hotbackup/base --incremental
-dir=/data2/backup/hotbackup/incremental_one/
```

将增量备份 incremental_two 应用到基础备份 base（注意恢复最后一个增量备份时需要去掉--redo-only 参数，回滚 xtrabackup 日志中那些还未提交的数据）:

```
innobackupex --apply-log --use-memory=5G /data2/backup/hotbackup/base --incremental-dir=/
data2/backup/hotbackup/incremental_two/
```

把所有合在一起的基础备份整体进行一次 apply 操作，回滚未提交的数据:

```
innobackupex --apply-log --use-memory=5G /data2/backup/hotbackup/base
```

把恢复完的备份复制到数据文件目录中，赋权，然后启动 MySQL 数据库:

```
mysqladmin  -S/tmp/mysql_3307.sock shutdown
mv  /data2/mysql3307/data   /data2/mysql3307/data_bak
mkdir   /data2/mysql3307/data
innobackupex --defaults-file=/tmp/my.cnf --copy-back --rsync  /data2/backup/hotbackup/base/
chown -R mysql3307:mysql3307   /data2/mysql3307/data
cd /home/mysql3307/mysqlhome
./bin/mysqld_safe -user=mysql3307 &
```

查看最终数据：

```
mysql> use test;
mysql> select * from test;
+----+--------+
| id | name   |
+----+--------+
|  1 | test1  |
|  2 | test2  |
|  3 | test3  |
|  4 | test4  |
|  5 | test5  |
|  6 | test6  |
|  7 | test7  |
|  8 | test8  |
|  9 | test9  |
| 10 | test10 |
+----+--------+
10 ows in set (0.00 sec)
```

（4）不完全恢复。

上文已经详细介绍了如何通过 mysqlbinlog 进行不完全恢复，mysqlbinlog 的不完全恢复的方法同样适合于 innobackup 热备的不完全恢复。

例如，在 2018 年 10 月 11 日 14:00 的时候开发人员在测试环境上进行了一次误操作，drop 掉了一张业务表，这个时候找到 DBA，由于库并不是很大，并且为测试库，并没有访问，这时可以进行基于位置和基于时间点的不完全恢复。

首先找到早上的备份：

```
cd /data2/backup/hotbackup/full/20181011
ls -l
total 3244496
-rw-r----- 1 root root        489 Oct 11 13:09 backup-my.cnf
drwxr-x--- 2 root root       4096 Oct 11 13:12 employees
-rw-r----- 1 root root     258850 Oct 11 13:09 ib_buffer_pool
-rw-r----- 1 root root 1073741824 Oct 11 13:13 ib_logfile0
-rw-r----- 1 root root 1073741824 Oct 11 13:13 ib_logfile1
-rw-r----- 1 root root 1073741824 Oct 11 13:13 ib_logfile2
-rw-r----- 1 root root   79691776 Oct 11 13:13 ibdata1
-rw-r----- 1 root root   12582912 Oct 11 13:13 ibtmp1
drwxr-x--- 2 root root       4096 Oct 11 13:12 mysql
drwxr-x--- 2 root root       4096 Oct 11 13:12 performance_schema
drwxr-x--- 2 root root      12288 Oct 11 13:12 sys
drwxr-x--- 2 root root       4096 Oct 11 13:12 test
-rw-r----- 1 root root        110 Oct 11 13:12 xtrabackup_binlog_info
-rw-r--r-- 1 root root         22 Oct 11 13:13 xtrabackup_binlog_pos_innodb
-rw-r----- 1 root root        121 Oct 11 13:13 xtrabackup_checkpoints
-rw-r----- 1 root root        791 Oct 11 13:12 xtrabackup_info
-rw-r----- 1 root root    8388608 Oct 11 13:12 xtrabackup_logfile
-rw-r--r-- 1 root root          1 Oct 11 13:13 xtrabackup_master_key_id
```

找到记录备份结束时刻的 BINLOG 的位置文件 xtrabackup_binlog_info，查看备份结束时刻 BINLOG 的名称和位置。

```
cat xtrabackup_binlog_info
mysql-bin.000001          516847
```

查看当前数据库的 BINLOG 文件和位置：

25.3 物理备份和恢复

```
mysql> show master logs;
+------------------+-----------+
| Log_name         | File_size |
+------------------+-----------+
| mysql-bin.000001 |    517340 |
| mysql-bin.000002 |       107 |
+------------------+-----------+
2 rows in set (0.00 sec)
```

从全备中恢复数据库，恢复全备，之后再从热备结束时刻 BINLOG 的位置开始，恢复到误操作时刻 14:00 之前的 BINLOG。

```
cd /data2/mysql3307/data/
mysqlbinlog --start-position="516847" --stop-datetime=" 2018-10-11 13:59:59"
mysql-bin.000001 mysql-bin.000002  | mysql -u root -pmypwd
```

跳过故障点的误操作的时间点，应用之后的所有 BINLOG 文件。

```
cd /data2/mysql3307/data/
mysqlbinlog --start-datetime="2018-10-11 14:01:00" mysql-bin.000001 mysql-bin.000002 | mysql -u
root -pmypwd
```

到此为止，一次基于位置和时间点的不完全恢复完成。

（5）克隆 slave。

在日常生活中，我们做得比较多的操作是在线添加从库，比如线上有一主一从两个数据库，由于业务的需要一台从库的读取量无法满足现在的需求，这样就需要我们在线添加从库，出于安全考虑，我们通常需要在从库上进行在线克隆 slave。

克隆 slave 时，常用参数 --slave-info 和 --safe-slave-backup。

● --slave-info 会将 master 的 Binary Log 的文件名和偏移位置保存到 xtrabackup_slave_info 文件中。

● --safe-slave-backup 则会暂停 slave 的 SQL 线程，直到没有打开的临时表的时候开始备份。待备份结束后 SQL 线程会自动启动，这样操作的目的主要是确保一致性的复制状态。

以下示例将介绍在一主一从的情况下在线搭建新的从库，环境如下：

```
Master:192.168.1.1     //主机名为master
Slave:192.168.1.2      //主机名为slave
NewSlave:192.168.1.3   //主机名为newslave
```

在上述示例中，主机名为 newslave 的主机即为新搭建的从库。在主机名为 slave 的主机上进行备份：

```
[root@slave~]# innobackupex --defaults-file=/tmp/my.cnf --user=backup --password=123456 --socket
=/tmp/mysql_3307.sock --slave-info --safe-slave-backup /data2/backup/hotbackup/cloneslave --no-
timestamp --parallel=2
```

查看目录下生成的文件：

```
[root@slave~]# ls  -lrt /data2/backup/hotbackup/cloneslave
total 78276
-rw-r----- 1 root root 79691776 Oct 11 15:14 ibdata1
drwxr-x--- 2 root root     4096 Oct 11 15:14 mysql
drwxr-x--- 2 root root     4096 Oct 11 15:14 test
drwxr-x--- 2 root root     4096 Oct 11 15:14 performance_schema
drwxr-x--- 2 root root     4096 Oct 11 15:14 testdb
drwxr-x--- 2 root root     4096 Oct 11 15:14 employees
drwxr-x--- 2 root root    12288 Oct 11 15:14 sys
-rw-r----- 1 root root       73 Oct 11 15:14 xtrabackup_slave_info
-rw-r----- 1 root root       70 Oct 11 15:14 xtrabackup_binlog_info
-rw-r----- 1 root root     2560 Oct 11 15:14 xtrabackup_logfile
-rw-r----- 1 root root      121 Oct 11 15:14 xtrabackup_checkpoints
-rw-r----- 1 root root   258850 Oct 11 15:14 ib_buffer_pool
-rw-r----- 1 root root      489 Oct 11 15:14 backup-my.cnf
-rw-r----- 1 root root      693 Oct 11 15:14 xtrabackup_info
```

查看 xtrabackup_slave_info 文件的内容，这个内容即为搭建从库时刻的 change mater to 参数：

```
[root@slave~]#cat  /data2/backup/hotbackup/cloneslave/xtrabackup_slave_info
CHANGE MASTER TO MASTER_LOG_FILE='mysql-bin.000004', MASTER_LOG_POS=553;
```

在主机名为 slave 的主机上进行还原：

```
[root@slave~]#innobackupex --apply-log --redo-only --use-memory=5G   /data2/backup/hotbackup/
cloneslave
```

将还原的文件复制到新的从库 newslave 上：

```
[root@slave~]#rsync -avprP -e ssh  /data2/backup/hotbackup/cloneslave  newslave:/home/
mysql3307/mysqlhome/data
```

在主机名为 master 的主库上添加对主机 newslave 的授权：

```
mysql> GRANT REPLICATION SLAVE ON *.* TO 'repl'@'slave2'
IDENTIFIED BY '123456';
```

在主机 newsalve 上复制主机为 slave 的 my.cnf 文件，并且修改 server-id 参数，修改完毕后，启动新的从库 newslave：

```
[root@newslave~]# scp slave:/etc/mysql/my.cnf  /etc/mysql/my.cnf
skip-slave-start
server-id=3
log-slave-updates=1
```

查找主机 slave 备份后生成的 xtrabackup_slave_info 文件，提取其中的 master_log_file 和 master_log_pos 信息，然后在新的从库 newslave 上进行 change master to 操作。

在 newslave 上：

```
mysql> CHANGE MASTER TO
MASTER_HOST='master'
MASTER_USER='repl',
MASTER_PASSWORD='123456',
MASTER_LOG_FILE='mysql-bin.000004',
MASTER_LOG_POS=553;
```

启动从库，并且检查复制是否正常：

```
mysql> START SLAVE;
          Slave_IO_Running: Yes
          Slave_SQL_Running: Yes
...
```

25.4 表的导入和导出

在数据库的日常维护中，表的导入和导出是很频繁的一类操作。本节将对 MySQL 中的这类操作进行详细的介绍，希望读者能够熟练掌握。

25.4.1 导出

在某些情况下，为了一些特定的目的，经常需要将表里的数据导出为某些符号分割的纯数据文本，而不是 SQL 语句。这些目的可能有以下一些：

- 用来作为 Excel 显示；
- 单纯为了节省备份空间；
- 为了快速地加载数据，LOAD DATA 的加载速度比普通的 SQL 加载要快 20 倍以上。

为了满足这些应用，可以使用以下两种办法来实现。

1. 方法 1

使用 SELECT ...INTO OUTFILE ...命令来导出数据，具体语法如下：

```
mysql> SELECT * FROM tablename INTO OUTFILE 'target_file' [option];
```

其中，option 参数可以是以下选项：

```
FIELDS   TERMINATED BY 'string' （字段分隔符，默认为制表符'\t'）；
FIELDS   [OPTIONALLY] ENCLOSED BY 'char'（字段引用符，如果加 OPTIONALLY 选项则只用在 char、varchar 和
         text 等字符型字段上。默认不使用引用符）；
FIELDS   ESCAPED BY 'char' （转义字符，默认为'\'）；
LINES    STARTING BY 'string' （每行前都加此字符串，默认为''）；
LINES    TERMINATED BY 'string'（行结束符，默认为'\n'）；
```

其中，char 表示此符号只能是单个字符，string 表示可以是字符串。

例如，将 dept_manager 表中数据导出为数据文本，其中，字段分隔符为"，"，字段引用符为""（双引号），记录结束符为回车符，具体实现如下：

```
mysql> select * from dept_manager into outfile '/tmp/dept_manager.txt' fields terminated by ",
" enclosed by '"' ;
Query OK, 24 rows affected (0.00 sec)

mysql>
mysql> system more /tmp/dept_manager.txt
"110022","d001","1985-01-01","1991-10-01"
"110039","d001","1991-10-01","9999-01-01"
"110085","d002","1985-01-01","1989-12-17"
"110114","d002","1989-12-17","9999-01-01"
"110183","d003","1985-01-01","1992-03-21"
"110228","d003","1992-03-21","9999-01-01"
"110303","d004","1985-01-01","1988-09-09"
"110344","d004","1988-09-09","1992-08-02"
"110386","d004","1992-08-02","1996-08-30"
......
```

发现第一列是数值型，如果不希望字段两边用引号引起，则语句改为：

```
mysql> select * from dept_manager into outfile '/tmp/dept_manager.txt'  fields terminated by
","  optionally enclosed by '"' ;
  Query OK, 24 rows affected (0.00 sec)

mysql> system more /tmp/dept_manager.txt
110022,"d001","1985-01-01","1991-10-01"
110039,"d001","1991-10-01","9999-01-01"
110085,"d002","1985-01-01","1989-12-17"
110114,"d002","1989-12-17","9999-01-01"
110183,"d003","1985-01-01","1992-03-21"
110228,"d003","1992-03-21","9999-01-01"
110303,"d004","1985-01-01","1988-09-09"
......
```

结果如我们所愿，第一列的双引号被去掉。

下面来测试一下转义字符。转义字符，顾名思义，就是由于含义模糊而需要特殊进行转换的字符，在不同的情况下，需要转义的字符是不一样的。MySQL 导出的数据中需要转义的字符主要包括以下 3 类：

- 转义字符本身；
- 字段分隔符；
- 记录分隔符。

在下面的例子中，对表 dept_manager 中的 dept_no 更新为含以上 3 类字符的字符串，然后导出：

```
mysql> update dept_manager set dept_no='\\"##!d001' where dept_no='d001';
Query OK, 2 rows affected (0.00 sec)
Rows matched: 2  Changed: 2  Warnings: 0

mysql> system rm /tmp/dept_manager.txt
mysql> select * from dept_manager into outfile '/tmp/dept_manager.txt' fields terminated by
","  optionally enclosed by '"' ;
Query OK, 24 rows affected (0.00 sec)

mysql> system more /tmp/dept_manager.txt
110022,"\\\"##!d001","1985-01-01","1991-10-01"
110039,"\\\"##!d001","1991-10-01","9999-01-01"
110085,"d002","1985-01-01","1989-12-17"
110114,"d002","1989-12-17","9999-01-01"
110183,"d003","1985-01-01","1992-03-21"
110228,"d003","1992-03-21","9999-01-01"
……
```

以上例子中，dept_no 中含有转义字符本身"\"和域引用符""""，因此，在输出的数据中我们发现这两种字符前面都加上了转义字符"\"，"\"#"变成了"\\\"#"。继续进行测试，将 dept_no 为 d002 的 dept_no 更新为含有字段分隔符","的字符串：

```
mysql> update dept_manager set dept_no='\\"#,#,!d002' where dept_no='d002';
Query OK, 2 rows affected (0.00 sec)
Rows matched: 2  Changed: 2  Warnings: 0

mysql> system rm /tmp/dept_manager.txt
mysql> select * from dept_manager into outfile '/tmp/dept_manager.txt' fields terminated by ",
" optionally enclosed by '"' ;
Query OK, 24 rows affected (0.01 sec)

mysql> system more /tmp/dept_manager.txt
110022,"d001","1985-01-01","1991-10-01"
110039,"d001","1991-10-01","9999-01-01"
110085,"\\\"#,#,!d002","1985-01-01","1989-12-17"
110114,"\\\"#,#,!d002","1989-12-17","9999-01-01"
110183,"d003","1985-01-01","1992-03-21"
110228,"d003","1992-03-21","9999-01-01"
110303,"d004","1985-01-01","1988-09-09"
……
```

注意：在 MySQL 客户端连接成功后，如果要执行操作系统的命令，那么可以用"system+操作系统命令"来进行执行。

这个时候，发现数据中的字符","并没有被转义，这是为什么呢？其实仔细想想就明白了，因为每个字符串的两边带有引用符""""（双引号），所以当 MySQL 看到数据中的","时，由于它处在前半个引用分隔符之后，后半个引用分隔符之前，所以并没有将它作为字段分隔符，而只是作为普通的一个数据字符来对待，因而不需要转义。继续做测试，将输出文件的字段引用符去掉，这个时候，我们的预期是数据中的","将成为转义字符而需要加上"\"：

```
mysql> system rm /tmp/dept_manager.txt
mysql> select * from dept_manager1 into outfile '/tmp/dept_manager.txt' fields terminated by
","  ;
Query OK, 24 rows affected (0.00 sec)

mysql> system more /tmp/dept_manager.txt
110022,d001,1985-01-01,1991-10-01
110039,d001,1991-10-01,9999-01-01
110085,\\"#\,#\,!d002,1985-01-01,1989-12-17
110114,\\"#\,#\,!d002,1989-12-17,9999-01-01
110183,d003,1985-01-01,1992-03-21
110228,d003,1992-03-21,9999-01-01
110303,d004,1985-01-01,1988-09-09
110344,d004,1988-09-09,1992-08-02
```

果然，现在的"，"前面加上了转义字符"\"。而刚才的引用符"""却没有被转义，因为它已经没有什么歧义，不需要被转义。

通过上面的测试，可以得出以下结论：

● 当导出命令中包含字段引用符时，数据中含有转义字符本身和字段引用符的字符需要被转义；

● 当导出命令中不包含字段引用符时，数据中含有转义字符本身和字段分隔符的字符需要被转义。

注意：SELECT...INTO OUTFILE...产生的输出文件如果在目标目录下有重名文件，将不会创建成功，源文件不能被自动覆盖。

2. 方法2

使用 mysqldump 导出数据为文本的具体语法如下：

```
mysqldump -u username -T target_dir dbname tablename [option]
```

其中，option 参数可以是以下选项：

● --fields-terminated-by=name（字段分隔符）；

● --fields-enclosed-by=name（字段引用符）；

● --fields-optionally-enclosed-by=name（字段引用符，只用在 char、varchar 和 text 等字符型字段上）；

● --fields-escaped-by=name（转义字符）；

● --lines-terminated-by=name（记录结束符）。

下面的例子中，采用 mysqldump 生成了指定分隔符分隔的文本：

```
[mysql3307@localhost tmp]# mysqldump -uroot -T /tmp employees dept_manager --fields-terminated
-by ',' --fields-optionally-enclosed-by '"'
[mysql3307@localhost tmp]# more /tmp/dept_manager.txt
110022,"\\\"##!d001","1985-01-01","1991-10-01"
110039,"\\\"##!d001","1991-10-01","9999-01-01"
110085,"\\\"#,#,!d002","1985-01-01","1989-12-17"
110114,"\\\"#,#,!d002","1989-12-17","9999-01-01"
110183,"d003","1985-01-01","1992-03-21"
110228,"d003","1992-03-21","9999-01-01"
110303,"d004","1985-01-01","1988-09-09"
110344,"d004","1988-09-09","1992-08-02"
```

除了生成数据文件 dept_manager.txt 之外，还生成一个 dept_manager.sql 文件，里面记录了 emp 表的创建脚本，记录的内容如下：

```
[mysql3307@localhost tmp]# more emp.sql
-- MySQL dump 10.13  Distrib 5.7.22, for linux-glibc2.12 (x86_64)
--
-- Host: localhost    Database: employees
-- ------------------------------------------------------
-- Server version       5.7.22-log

/*!40101 SET @OLD_CHARACTER_SET_CLIENT=@@CHARACTER_SET_CLIENT */;
/*!40101 SET @OLD_CHARACTER_SET_RESULTS=@@CHARACTER_SET_RESULTS */;
/*!40101 SET @OLD_COLLATION_CONNECTION=@@COLLATION_CONNECTION */;
/*!40101 SET NAMES utf8 */;
/*!40103 SET @OLD_TIME_ZONE=@@TIME_ZONE */;
/*!40103 SET TIME_ZONE='+00:00' */;
/*!40101 SET @OLD_SQL_MODE=@@SQL_MODE, SQL_MODE='' */;
/*!40111 SET @OLD_SQL_NOTES=@@SQL_NOTES, SQL_NOTES=0 */;

--
```

```
-- Table structure for table `dept_manager`
--

DROP TABLE IF EXISTS `dept_manager`;
/*!40101 SET @saved_cs_client     = @@character_set_client */;
/*!40101 SET character_set_client = utf8 */;
CREATE TABLE `dept_manager1` (
  `emp_no` int(11) NOT NULL,
  `dept_no` char(14) COLLATE utf8_unicode_ci NOT NULL,
  `from_date` date NOT NULL,
  `to_date` date NOT NULL,
  PRIMARY KEY (`emp_no`,`dept_no`),
  KEY `dept_no` (`dept_no`)
) ENGINE=InnoDB DEFAULT CHARSET=utf8 COLLATE=utf8_unicode_ci;
/*!40101 SET character_set_client = @saved_cs_client */;

/*!40103 SET TIME_ZONE=@OLD_TIME_ZONE */;

/*!40101 SET SQL_MODE=@OLD_SQL_MODE */;
/*!40101 SET CHARACTER_SET_CLIENT=@OLD_CHARACTER_SET_CLIENT */;
/*!40101 SET CHARACTER_SET_RESULTS=@OLD_CHARACTER_SET_RESULTS */;
/*!40101 SET COLLATION_CONNECTION=@OLD_COLLATION_CONNECTION */;
/*!40111 SET SQL_NOTES=@OLD_SQL_NOTES */;

-- Dump completed on 2018-10-11 17:20:28
```

可以发现，除多了一个表的创建脚本文件外，mysqldump 和 SELEC…INTO OUTFILE…的选项和语法非常类似。其实，mysqldump 实际调用的就是后者提供的接口，并在其上面添加了一些新的功能而已。

25.4.2 导入

本节只讨论用 SELECT…INTO OUTFILE 或者 mysqldump 导出的纯数据文本的导入方法。和导出类似，导入也有两种不同的方法，分别是 LOAD DATA INFILE…和 mysqlimport，它们的本质是一样的，区别只是在于一个在 MySQL 内部执行，另一个在 MySQL 外部执行。

1. 方法 1

使用"LOAD DATA INFILE…"命令，具体语法如下：

```
mysql > LOAD DATA    [LOCAL] INFILE 'filename' INTO TABLE tablename [option]
```

option 可以是以下选项：

- FIELDS　TERMINATED BY 'string'（字段分隔符，默认为制表符'\t'）；
- FIELDS　[OPTIONALLY] ENCLOSED BY 'char'（字段引用符，如果加 OPTIONALLY 选项则只用在 char、varchar 和 text 等字符型字段上。默认不使用引用符）；
- FIELDS　ESCAPED BY 'char'（转义字符，默认为'\'）；
- LINES　STARTING BY 'string'（每行前都加此字符串，默认为''）；
- LINES　TERMINATED BY 'string'（行结束符，默认为'\n'）；
- IGNORE number LINES（忽略输入文件中的前 n 行数据）；
- (col_name_or_user_var,…)（按照列出的字段顺序和字段数量加载数据）；
- SET col_name = expr,…（将列做一定的数值转换后再加载）。

其中，char 表示此符号只能是单个字符，string 表示可以是字符串。

FILELD、LINES 和前面 SELECT…INTO OUTFILE…的含义完全相同，不同的是多了几个不同的选项，下面的例子将文件"/tmp/emp.txt"中的数据加载到表 emp 中：

25.4 表的导入和导出

```
mysql> load data infile '/tmp/dept_manager.txt' into table dept_manager fields terminated by
',' enclosed by '"' ;
Query OK, 24 rows affected (0.03 sec)
Records: 24  Deleted: 0  Skipped: 0  Warnings: 0

mysql> select * from dept_manager;
+--------+---------+------------+------------+
| emp_no | dept_no | from_date  | to_date    |
+--------+---------+------------+------------+
| 110022 | d001    | 1985-01-01 | 1991-10-01 |
| 110039 | d001    | 1991-10-01 | 9999-01-01 |
| 110085 | d002    | 1985-01-01 | 1989-12-17 |
| 110114 | d002    | 1989-12-17 | 9999-01-01 |
| 110183 | d003    | 1985-01-01 | 1992-03-21 |
| 111692 | d009    | 1985-01-01 | 1988-10-17 |
……
| 111784 | d009    | 1988-10-17 | 1992-09-08 |
| 111877 | d009    | 1992-09-08 | 1996-01-03 |
| 111939 | d009    | 1996-01-03 | 9999-01-01 |
+--------+---------+------------+------------+
24 rows in set (0.00 sec)
```

如果不希望加载文件中的前两行，可以进行如下操作：

```
mysql> load data infile '/tmp/dept_manager.txt' into table dept_manager fields terminated by
',' enclosed by '"' ignore 2 lines;
Query OK, 22 rows affected (0.04 sec)
Records: 22  Deleted: 0  Skipped: 0  Warnings: 0

mysql> select * from dept_manager;
+--------+---------+------------+------------+
| emp_no | dept_no | from_date  | to_date    |
+--------+---------+------------+------------+
| 110085 | d002    | 1985-01-01 | 1989-12-17 |
| 110114 | d002    | 1989-12-17 | 9999-01-01 |
| 110183 | d003    | 1985-01-01 | 1992-03-21 |
| 110228 | d003    | 1992-03-21 | 9999-01-01 |
| 110303 | d004    | 1985-01-01 | 1988-09-09 |
| 110344 | d004    | 1988-09-09 | 1992-08-02 |
……
| 111400 | d008    | 1985-01-01 | 1991-04-08 |
| 111534 | d008    | 1991-04-08 | 9999-01-01 |
| 111692 | d009    | 1985-01-01 | 1988-10-17 |
| 111784 | d009    | 1988-10-17 | 1992-09-08 |
| 111877 | d009    | 1992-09-08 | 1996-01-03 |
| 111939 | d009    | 1996-01-03 | 9999-01-01 |
+--------+---------+------------+------------+
22 rows in set (0.00 sec)
```

此时数据只加载了 22 行。

如果发现文件中的列顺序和表中的列顺序不符，或者只想加载部分列，可以在命令行中加上列的顺序，如下所示：

```
mysql> load data infile '/tmp/dept_manager.txt' into table dept_manager fields terminated by
',' enclosed by '"' ignore 2 lines (emp_no,dept_no,to_date,from_date);
Query OK, 22 rows affected (0.05 sec)
Records: 22  Deleted: 0  Skipped: 0  Warnings: 0

mysql> select * from dept_manager;
+--------+---------+------------+------------+
| emp_no | dept_no | from_date  | to_date    |
+--------+---------+------------+------------+
| 110022 | d001    | 1991-10-01 | 1985-01-01 |
| 110039 | d001    | 9999-01-01 | 1991-10-01 |
| 110085 | d002    | 1989-12-17 | 1985-01-01 |
| 110114 | d002    | 9999-01-01 | 1989-12-17 |
| 110183 | d003    | 1992-03-21 | 1985-01-01 |
| 110228 | d003    | 9999-01-01 | 1992-03-21 |
……
```

```
|  110386 |   d004  |  1996-08-30 |  1992-08-02 |
|  111692 |   d009  |  1988-10-17 |  1985-01-01 |
|  111784 |   d009  |  1992 09 08 |  1988 10 17 |
|  111877 |   d009  |  1996-01-03 |  1992-09-08 |
|  111939 |   d009  |  9999-01-01 |  1996-01-03 |
+---------+---------+-------------+-------------+
22 rows in set (0.00 sec)
```

可以发现，文件中第三列的内容放到了 to_date 里面，第四列的内容放到了 from_date 里面。

如果只想加载第一列，字段的列表里面可以只加第一列的名称：

```
mysql> load data infile '/tmp/dept_manager.txt' into table dept_manager fields terminated by ',' enclosed by '"' ignore 2 lines (emp_no);
Query OK, 22 rows affected, 2 warnings (0.04 sec)
Records: 22  Deleted: 0  Skipped: 0  Warnings: 2

mysql> select * from dept_manager;
+---------+---------+-------------+-------------+
| emp_no  | dept_no | from_date   | to_date     |
+---------+---------+-------------+-------------+
| 110085  |  NULL   |    NULL     |    NULL     |
| 110114  |  NULL   |    NULL     |    NULL     |
| 110183  |  NULL   |    NULL     |    NULL     |
| 110228  |  NULL   |    NULL     |    NULL     |
……
| 111784  |  NULL   |    NULL     |    NULL     |
| 111877  |  NULL   |    NULL     |    NULL     |
| 111939  |  NULL   |    NULL     |    NULL     |
+---------+---------+-------------+-------------+
22 rows in set (0.00 sec)
```

如果希望将 id 列的内容加上 10 后再加载到表中，可以如下操作：

```
mysql> load data infile '/tmp/dept_manager.txt' into table dept_manager fields terminated by ',' enclosed by '"' set emp_no = emp_no +10;
Query OK, 24 rows affected (0.03 sec)
Records: 24  Deleted: 0  Skipped: 0  Warnings: 0

mysql> select * from dept_manager;
+---------+---------+-------------+-------------+
| emp_no  | dept_no | from_date   | to_date     |
+---------+---------+-------------+-------------+
| 110095  |  d002   |  1989-12-17 |  1985-01-01 |
| 110124  |  d002   |  9999-01-01 |  1989-12-17 |
| 110193  |  d003   |  1992-03-21 |  1985-01-01 |
| 110238  |  d003   |  9999-01-01 |  1992-03-21 |
……
| 111794  |  d009   |  1992-09-08 |  1988-10-17 |
| 111887  |  d009   |  1996-01-03 |  1992-09-08 |
| 111949  |  d009   |  9999-01-01 |  1996-01-03 |
+---------+---------+-------------+-------------+
24 rows in set (0.00 sec)
```

2. 方法 2

用 mysqlimport 来实现，具体命令如下。

```
Shell > mysqlimport -u root -p*** [- -LOCAL] dbname order_tab.txt [option]
```

其中 option 参数可以是以下选项：

- --fields-terminated-by=name（字段分隔符）；
- --fields-enclosed-by=name（字段引用符）；
- --fields-optionally-enclosed-by=name（字段引用符，只用在 char、varchar 和 text 等字符型字段上）；

- --fields-escaped-by=name（转义字符）；
- --lines-terminated-by=name（记录结束符）；
- -- ignore-lines=number（忽略前几行）。

这与 mysqldump 的选项几乎完全相同，这里不再详细介绍，简单来看一个例子：

```
[mysql3307@localhost tmp]# mysqlimport -uroot employees /tmp/dept_manager.txt \
 --fields-terminated-by=',' --fields-enclosed-by='"'
employees.dept_manager: Records: 24  Deleted: 0  Skipped: 0  Warnings: 0
[mysql3307@localhost tmp]#
[mysql3307@localhost tmp]# mysql -uroot employees -e 'select count(10) from dept_manager'
+-----------+
| count(10) |
+-----------+
|        24 |
+-----------+
[mysql3307@localhost tmp]# mysql -uroot employees -e 'select * from dept_manager'
+--------+---------+------------+------------+
| emp_no | dept_no | from_date  | to_date    |
+--------+---------+------------+------------+
| 110022 | d001    | 1985-01-01 | 1991-10-01 |
| 110039 | d001    | 1991-10-01 | 9999-01-01 |
| 110085 | d002    | 1985-01-01 | 1989-12-17 |
| 110114 | d002    | 1989-12-17 | 9999-01-01 |
| 110183 | d003    | 1985-01-01 | 1992-03-21 |
......
| 111877 | d009    | 1992-09-08 | 1996-01-03 |
| 111939 | d009    | 1996-01-03 | 9999-01-01 |
+--------+---------+------------+------------+
```

注意：如果导入和导出是跨平台操作的（Windows 和 Linux），那么要注意设置参数 line-terminated-by，Windows 上设置为 line-terminated-by=' \r\n'，Linux 上设置为 line- terminated-by='\n'。

25.5 小结

本章主要介绍了 MySQL 的备份和恢复方法。和其他数据库类似，MySQL 也分为逻辑备份和物理备份。两种备份方法各有优缺点，逻辑备份保存的是 SQL 文本，可以在各种条件下恢复，但是对于大数据量的系统，备份和恢复的时间都比较长；物理备份恰恰相反，由于是文件的物理 cp，备份和恢复时间都比较短，但是备份的文件在不同的平台上不一定兼容。

其中，mysqldump 是常用的逻辑备份工具，适合各种存储引擎，希望读者重点掌握。物理备份对于不同的存储引擎，备份方式有所不同。对于 MyISAM 引擎，读者了解一下手工热备份的方法；对于 InnoDB 的热备份，读者主要掌握 innobackupex 的使用方法，此工具属于开源工具，效率高且不收费的，因此普及率很高，而 MySQL 官方目前没有提供免费的 InnoDB 热备份工具。

对于数据表的导入和导出方法，应重点掌握 SELECT…INTO OUTFILE 和 LOAD DATA INFILE 的使用，mysqldump 和 mysqlimport 实际上是调用了前两种方法接口，只不过是在 MySQL 外部执行罢了。数据的导入和导出在数据库的管理与维护中使用非常频繁，而 LOAD DATA INFILE 是加载数据最快的方法，因此读者应重点掌握。

第 26 章 MySQL 权限与安全

MySQL 的权限系统主要用来对连接到数据库的用户进行权限的验证，以此来判断此用户是否属于合法的用户。如果是合法用户，则赋予相应的数据库权限。

数据库的权限和数据库的安全是息息相关的，不当的权限设置可能会导致各种各样的安全隐患，操作系统的某些设置也会对 MySQL 的安全造成影响。

本章对 MySQL 的权限系统以及相应的安全问题进行了一些探讨，希望能够帮助读者对这些方面有深入的认识。

26.1 MySQL 权限管理

本节从 MySQL 权限系统的工作原理和账号管理两个方面来进行介绍。读完本节后，希望读者能够在了解权限系统的工作原理基础上，熟练掌握账号的管理和使用方法。

26.1.1 权限系统的工作原理

MySQL 权限系统通过下面两个阶段进行认证：

- 对连接的用户进行身份认证，合法的用户通过认证，不合法的用户拒绝连接；
- 对通过认证的合法用户赋予相应的权限，用户可以在这些权限范围内对数据库做相应的操作。

对于身份的认证，MySQL 是通过 IP 地址和用户名联合进行确认的，例如 MySQL 用户 root@localhost 表示用户 root 只能从本地（localhost）进行连接才可以通过认证，此用户从其他任何主机对数据库进行的连接都将被拒绝。也就是说，同样的一个用户名，如果来自不同的 IP 地址，则 MySQL 将其视为不同的用户。

MySQL 的权限表在数据库启动的时候就载入内存，当用户通过身份认证后，就在内存中进行相应权限的存取，这样，此用户就可以在数据库中做权限范围内的各种操作了。

26.1.2 权限表的存取

在权限存取的两个过程中，系统会用到 mysql 数据库（安装 MySQL 时被创建，数据库名称叫作 mysql）中 user 和 db 这两个最重要的权限表。在 MySQL 5.7.22 版本，它们的定义如表 26-1 所示。

26.1 MySQL 权限管理

表 26-1　　　　　　　　　　mysql 数据库中的两个权限表定义

表　名	user	db
用户列	host	host
	user	user
	authentication_string	——
	——	db
权限列	select_priv	select_priv
	insert_priv	insert_priv
	update_priv	update_priv
	delete_priv	delete_priv
	create_priv	create_priv
	drop_priv	drop_priv
	reload_priv	——
	shutdown_priv	——
	process_priv	——
	file_priv	——
	grant_priv	grant_priv
	references_priv	——
	index_priv	index_priv
	alter_priv	alter_priv
	show_db_priv	——
	super_priv	——
	create_tmp_table_priv	create_tmp_table_priv
	lock_tables_priv	lock_tables_priv
	execute_priv	execute_priv
	repl_slave_priv	——
	repl_client_priv	——
	create_view_priv	create_view_priv
	show_view_priv	show_view_priv
	create_routine_priv	create_routine_priv
	alter_routine_priv	alter_routine_priv
	create_user_priv	——
	event_priv	event_priv
	trigger_priv	trigger_priv
权限列	create_tablespace_priv	——
	create_role_priv	——
	drop_role_priv	——
安全列	ssl_type	——
	ssl_cipher	——
	x509_issuer	——
	x509_subject	——
	password_expired	——
	password_last_changed	——
	password_lifetime	——
	account_locked	——
	plugin	

续表

表　名	user	db
资源控制列	max_questions	—
	max_updates	—
	max_connections	—
	max_user_connections	—

在这两个表中，最重要的表是 user 表，其次是 db 表。user 中的列主要分为 4 个部分：用户列、权限列、安全列和资源控制列。

通常用得最多的是用户列和权限列，其中权限列又分为普通权限和管理权限。普通权限主要用于数据库的操作，比如 select_priv、create_priv 等；而管理权限主要用来对数据库进行管理的操作，比如 process_priv、super_priv 等。

当用户进行连接时，权限表的存取过程有以下两个阶段。

● 先从 user 表中的 host、user 和 uthentication_string 这 3 个字段中判断连接的 IP、用户名和密码是否存在于表中，如果存在，则通过身份验证，否则拒绝连接。

● 如果通过身份验证，则按照以下权限表的顺序得到数据库权限：user→db→tables_priv→columns_priv。

在这几个权限表中，权限范围依次递减，全局权限覆盖局部权限。上面的第一阶段好理解，下面以一个例子来详细解释一下第二阶段。

（1）创建用户 emp_sel@localhost，并赋予所有数据库上的所有表的 select 权限。

```
mysql>create user emp_sel@localhost identified by 'emp_sel123';
Query OK, 0 rows affected (0.22 sec)
mysql>grant select on *.* to emp_sel@localhost;
Query OK, 0 rows affected, 1 warning (0.00 sec)
mysql>select * from user where user='emp_sel' and host='localhost'\G
*************************** 1. row ***************************
                  Host: localhost
                  User: emp_sel
           Select_priv: Y
           Insert_priv: N
           Update_priv: N
           Delete_priv: N
           Create_priv: N
             Drop_priv: N
...
 authentication_string: *FBEFDCEDE1FA35EDEA26B60FBFCD646A24D47DF6
...
```

（2）再来看 db 表：

```
mysql>select * from db where user='emp_sel' and host='localhost'\G
Empty set (0.02 sec)
```

可以发现，user 表的 select_priv 列是 Y，而 db 表中并没有记录，也就是说，对所有数据库都具有相同权限的用户记录并不需要记入 db 表，而仅仅需要将 user 表中的 select_priv 改为 Y 即可。换句话说，user 表中的每个权限都代表了对所有数据库都有的权限。

（3）将 emp_sel@localhost 上的权限改为只对 employees 数据库上所有表的 select 权限。

```
mysql>revoke select on *.* from emp_sel@localhost;
Query OK, 0 rows affected, 1 warning (0.00 sec)
mysql>grant select on employees.* to emp_sel@localhost;
Query OK, 0 rows affected, 1 warning (0.02 sec)
mysql>select * from user where user='emp_sel' and host='localhost'\G
*************************** 1. row ***************************
```

```
                     Host: localhost
                     User: emp_sel
              Select_priv: N
              Insert_priv: N
              Update_priv: N
              Delete_priv: N
              Create_priv: N
                Drop_priv: N
...
    authentication_string: *FBEFDCEDE1FA35EDEA26B60FBFCD646A24D47DF6
...
mysql>select * from db where user='emp_sel' and host='localhost'\G
*************************** 1. row ***************************
                     Host: localhost
                       Db: employees
                     User: emp_sel
              Select_priv: Y
              Insert_priv: N
              Update_priv: N
              Delete_priv: N
              Create_priv: N
                Drop_priv: N
...
1 row in set (0.00 sec)
```

这个时候发现，user 表中的 select_priv 变为 N，而 db 表中则增加了 db 为 employees 的一条记录。也就是说，当只授予部分数据库某些权限时，user 表中的相应权限列保持 N，而将具体的数据库权限写入 db 表。table 和 column 的权限机制和 db 类似，这里就不再赘述。

从上面的例子可以看出，当用户通过权限认证，进行权限分配时，将按照 user→db→tables_priv→columns_priv 的顺序进行权限分配，即先检查全局权限表 user，如果 user 中对应权限为 Y，则此用户对所有数据库的权限都为 Y，将不再检查 db、tables_priv 和 columns_priv；如果为 N，则到 db 表中检查此用户对应的具体数据库，并得到 db 中为 Y 的权限；如果 db 中相应权限为 N，则检查 tables_priv 中此数据库对应的具体表，取得表中为 Y 的权限；如果 tables_priv 中相应权限为 N，则检查 columns_priv 中此表对应的具体列，取得列中为 Y 的权限。

26.1.3 账号管理

理解了权限系统的工作原理后，本节开始介绍账号的管理。账号管理也是 DBA 日常工作中很重要的工作之一，主要包括账号的创建、权限更改和账号的删除。用户连接数据库的第一步都从账号创建开始。

1. 创建账号

有两种方法可以用来创建账号：使用 CREATE USER 配合 GRANT 的语法创建或者直接操作授权表，但更推荐使用第一种方法，因为操作简单，出错概率更少。下面详细讲述这两种方式的使用方法。

（1）CREATE USER 的常用语法如下：

```
CREATE USER [IF NOT EXISTS]
    user [auth_option] [, user [auth_option]] ...
    [REQUIRE {NONE | tls_option [[AND] tls_option] ...}]
    [WITH resource_option [resource_option] ...]
    [password_option | lock_option] ...
```

（2）GRANT 的常用语法如下：

```
GRANT
    priv_type [(column_list)]
      [, priv_type [(column_list)]] ...
    ON [object_type] priv_level
    TO user [auth_option] [, user [auth_option]] ...
    [REQUIRE {NONE | tls_option [[AND] tls_option] ...}] [WITH {GRANT OPTION | resource_
option} ...]

object_type: {
    TABLE
  | FUNCTION
  | PROCEDURE
  }
```

注意：在8.0版本以前，可以用grant命令自动创建用户；在8.0版本以后，必须先create user，再grant权限，否则会报错：ERROR 1133 (42000): Can't find any matching row in the user table.

下面来看几个例子。

例1：创建用户emp，权限为可以在所有数据库上执行所有权限，只能从本地进行连接。

```
mysql> create user  emp@localhost;
Query OK, 0 rows affected (0.00 sec)

mysql>grant all privileges on *.* to emp@localhost;
Query OK, 0 rows affected, 1 warning (0.00 sec)
mysql>select * from user where user='emp' and host='localhost' \G;
*************************** 1. row ***************************
                 Host: localhost
                 User: emp
           Select_priv: Y
           Insert_priv: Y
           Update_priv: Y
           Delete_priv: Y
           Create_priv: Y
             Drop_priv: Y
           Reload_priv: Y
         Shutdown_priv: Y
          Process_priv: Y
             File_priv: Y
            Grant_priv: N
       References_priv: Y
            Index_priv: Y
            Alter_priv: Y
          Show_db_priv: Y
            Super_priv: Y
  Create_tmp_table_priv: Y
       Lock_tables_priv: Y
          Execute_priv: Y
        Repl_slave_priv: Y
       Repl_client_priv: Y
       Create_view_priv: Y
         Show_view_priv: Y
    Create_routine_priv: Y
     Alter_routine_priv: Y
       Create_user_priv: Y
...
```

可以发现，除了Grant_priv权限外，所有权限在user表里面都是Y。

例2：在例1基础上，增加对emp的grant权限。

```
mysql>grant all privileges on *.* to emp@localhost with grant option;
Query OK, 0 rows affected (0.00 sec)

mysql>select * from user where user='emp' and host='localhost' \G
*************************** 1. row ***************************
                 Host: localhost
                 User: emp
           Select_priv: Y
```

```
            Insert_priv: Y
            Update_priv: Y
            Delete_priv: Y
            Create_priv: Y
              Drop_priv: Y
            Reload_priv: Y
          Shutdown_priv: Y
           Process_priv: Y
              File_priv: Y
             Grant_priv: Y
...
```

例 3：在例 2 基础上，设置密码为"emp123"。

```
mysql> alter user emp@localhost identified by 'emp123';
Query OK, 0 rows affected (0.00 sec)
```

从 user 表中查看修改的密码：

```
mysql> select * from user where user='emp' and host='localhost' \G
mysql>select * from user where user='emp' and host='localhost' \G
*************************** 1. row ***************************
                  Host: localhost
                  User: emp
           Select_priv: Y
           Insert_priv: Y
           Update_priv: Y
           Delete_priv: Y
           Create_priv: Y
...
   authentication_string: *31D409663B40D0EE6FCA4EE0C19F8C9576352E13
...
```

可以发现，密码变成了一堆加密后的字符串。在 MySQL 5.7 里面，密码的算法是生成一个以"*"开始的 41 位的字符串，安全性很高。

例 4：创建新用户 emp，可以从任何 IP 进行连接，权限为对 employees 数据库里的所有表进行 SELECT、UPDATE、INSERT 和 DELETE 操作，初始密码为"emp123"。

```
mysql>create user emp@'%' identified by 'emp123';
Query OK, 0 rows affected (0.10 sec)

mysql>grant select,insert,update,delete on employees.* to emp@'%';
Query OK, 0 rows affected (0.11 sec)
mysql>select * from user where user='emp' and host='%' \G
*************************** 1. row ***************************
                  Host: %
                  User: emp
           Select_priv: N
           Insert_priv: N
           Update_priv: N
           Delete_priv: N
                  ...
   authentication_string: *31D409663B40D0EE6FCA4EE0C19F8C9576352E13
                  ...
mysql>select * from db where user='emp' and host='%' \G
*************************** 1. row ***************************
                  Host: %
                    Db: employees
                  User: emp
           Select_priv: Y
           Insert_priv: Y
           Update_priv: Y
           Delete_priv: Y
...
```

如上文所述，user 表中的权限都是 N，db 表中增加的记录权限则都是 Y。一般地，我们只授予用户适当的权限，而一般不会授予过多的权限，本例中的权限适合于大多数应用账号。

本例中的 IP 限制为所有 IP 都可以连接，因此设置为"%"，mysql 数据库中是通过 user

表的 Host 字段进行控制的，Host 可以是以下类型的赋值。
- Host 值可以是主机名或 IP 号，或 localhost 指出本地主机。
- 可以在 Host 列值使用通配符字符 "%" 和 "_"。
- Host 值 "%" 匹配任何主机名，空 Host 值等价于 "%"。它们的含义与 LIKE 操作符的模式匹配操作相同。例如，"%" 的 Host 值与所有主机名匹配，而 "%.mysql.com" 匹配 mysql.com 域的所有主机。

使用 Host 值与 User 值的组合进行连接的例子如表 26-2 所示。

表 26-2　　　　　　　　　　Host 和 User 组合进行连接的例子

Host 值	User 值	被条目匹配的连接
'thomas.loc.gov'	'fred'	fred，通过 thomas.loc.gov 连接
'thomas.loc.gov'	''	任何用户，通过 thomas.loc.gov 连接
'%'	'fred'	fred，通过任何主机连接
'%'	''	任何用户，通过任何主机连接
'%.loc.gov'	'fred'	fred，通过 loc.gov 域的任何主机连接
'x.y.%'	'fred'	fred，通过 x.y.net、x.y.com、x.y.edu 等连接
'144.155.166.177'	'fred'	fred，通过 IP 地址为 144.155.166.177 的主机连接
'144.155.166.%'	'fred'	fred，从来自 144.155.166 这个 C 类子网的任何主机进行连接

可能读者会有这样的疑问，如果权限表中的 Host 既有 "thomas.loc.gov"，又有 "%"，而此时，连接从主机 thomas.loc.gov 过来。显然，user 表里面这两条记录都符合匹配条件，那系统会选择哪一个呢？

如果有多个匹配，那么服务器必须选择使用哪个条目。按照下述原则来解决：
- 服务器在启动时读入 user 表后进行排序；
- 然后当用户试图连接时，以排序的顺序浏览条目；
- 服务器使用与客户端和用户名匹配的第一行。

当服务器读取表时，它首先以最具体的 Host 值排序。主机名和 IP 号是最具体的。"%" 意味着 "任何主机" 并且是最不特定的。有相同 Host 值的条目首先以最具体的 User 值排序（空 User 值意味着 "任何用户" 并且是最不特定的）。下例是排序前和排序后的结果。

排序前：

```
+-----------+----------+-
| Host      | User     | …
+-----------+----------+-
| %         | root     | …
| %         | jeffrey  | …
| localhost | root     | …
| localhost |          | …
+-----------+----------+-
```

排序后：

```
+-----------+----------+-
| Host      | User     | …
+-----------+----------+-
| localhost | root     | …
| localhost |          | …
| %         | jeffrey  | …
| %         | root     | …
+-----------+----------+-
```

很显然，在上面的例子中应该匹配 host 为 "thomas.loc.gov" 所对应的权限。

26.1 MySQL 权限管理

注意：mysql 数据库的 user 表中 Host 的值为 "%" 或者空，表示所有外部 IP 都可以连接，但是不包括本地服务器 localhost，因此，如果要包括本地服务器，必须单独为 localhost 赋予权限。

例 5：授予 SUPER、PROCESS、FILE 权限给用户 emp@%。

```
mysql>grant super,process,file on *.* to 'emp'@'%';
Query OK, 0 rows affected (0.00 sec)
```

因为这几个权限都属于管理权限，因此不能够指定某个数据库，on 后面必须跟 "*.*"，下面的语法将提示错误：

```
mysql> grant super,process,file on employees.* to 'emp'@'%';
ERROR 1221 (HY000): Incorrect usage of DB GRANT and GLOBAL PRIVILEGES
```

例 6：只授予登录权限给 emp@localhost。

```
mysql>create user  emp@'localhost' identified by 'emp123';
Query OK, 0 rows affected (0.00 sec)
创建用户后，不用 grant，用户默认有 usage 权限。

 [mysql3307@localhost ~]$ mysql -uemp -hlocalhost -S/tmp/mysql_3307.sock  -p
Enter password:
Welcome to the MySQL monitor.  Commands end with ; or \g.
Your MySQL connection id is 217
Server version: 5.7.22-log MySQL Community Server (GPL)

Copyright (c) 2000, 2018, Oracle and/or its affiliates. All rights reserved.

Oracle is a registered trademark of Oracle Corporation and/or its
affiliates. Other names may be trademarks of their respective
owners.
Type 'help;' or '\h' for help. Type '\c' to clear the current input statement.
mysql>show databases;
+--------------------+
| Database           |
+--------------------+
| information_schema |
+--------------------+
1 row in set (0.00 sec)
```

usage 权限可以登录数据库，只能查询 information_schema 数据库下的部分表，权限非常小。

直接操作权限表也可以进行权限的创建，其实 GRANT 操作的本质就是修改权限表后进行权限的刷新，因此，GRANT 比操作权限表更简单。下面继续以上文的例 4 为例来说明一下更新权限表的用法。

创建新用户 emp1，可以从任何 IP 进行连接，权限为对 employees 数据库里的所有表进行 SELECT、UPDATE、INSERT 和 DELETE，初始密码为 emp123，具体步骤如下：

```
mysql>create user  emp1@'%' identified by 'emp123';
Query OK, 0 rows affected (0.00 sec)
mysql>grant select,insert,update,delete on employees.* to 'emp1'@'%' ;
Query OK, 0 rows affected (0.00 sec)
```

直接操作权限表如下：

```
[mysql3307@localhost ~]$ mysql -uroot  --socket=/tmp/mysql_3307.sock mysql
Reading table information for completion of table and column names
You can turn off this feature to get a quicker startup with -A

Welcome to the MySQL monitor.  Commands end with ; or \g.
Your MySQL connection id is 246
Server version: 5.7.22-log MySQL Community Server (GPL)

Copyright (c) 2000, 2018, Oracle and/or its affiliates. All rights reserved.

Oracle is a registered trademark of Oracle Corporation and/or its
affiliates. Other names may be trademarks of their respective
```

```
owners.

Type 'help;' or '\h' for help. Type '\c' to clear the current input statement.
mysql>insert into user (host,user,ssl_cipher,x509_issuer,x509_subject,authentication_string)
values ('%','emp1','','','',password('emp123'));
Query OK, 1 row affected, 1 warning (0.00 sec)

mysql>insert into db (Host,Db,User,Select_priv,Insert_priv,Update_priv,Delete_priv) values('%',
'employees','emp1','Y','Y','Y','Y');
Query OK, 1 row affected (0.00 sec)

mysql>flush privileges;
Query OK, 0 rows affected (0.00 sec)

[mysql3307@localhost ~]$ mysql -uemp1 -hlocalhost -S/tmp/mysql_3307.sock -p
Enter password:
Welcome to the MySQL monitor.  Commands end with ; or \g.
Your MySQL connection id is 247
Server version: 5.7.22-log MySQL Community Server (GPL)

Copyright (c) 2000, 2018, Oracle and/or its affiliates. All rights reserved.

Oracle is a registered trademark of Oracle Corporation and/or its
affiliates. Other names may be trademarks of their respective
owners.

Type 'help;' or '\h' for help. Type '\c' to clear the current input statement.

emp1@localhost:mysql_3307.sock   [(none)]>show databases;
+--------------------+
| Database           |
+--------------------+
| information_schema |
| employees          |
+--------------------+
2 rows in set (0.00 sec)
```

创建完账号后，时间长了可能就会忘记分配的权限而需要查看账号权限，也有可能会经过一段时间后需要更改以前的账号权限。

2．查看账号权限

账号创建好后，可以通过如下命令查看权限：

```
show grants for user@host;
```

如下例所示：

```
mysql>show grants for emp@'localhost';
+-----------------------------------------------------------------------------+
| Grants for emp@localhost                                                    |
+-----------------------------------------------------------------------------+
| GRANT USAGE ON *.* TO 'emp'@'localhost'                                     |
| GRANT SELECT, INSERT, UPDATE, DELETE ON `employees`.* TO 'emp'@'localhost'  |
+-----------------------------------------------------------------------------+
2 rows in set (0.00 sec)
```

host 可以不写，默认是 "%"，如下所示：

```
mysql> show grants for emp;
ERROR 1141 (42000): There is no such grant defined for user 'emp' on host '%'

mysql>create user  emp@'%' identified by 'emp123';
Query OK, 0 rows affected (0.00 sec)

mysql> show grants for emp;
+---------------------------------+
| Grants for emp@%                |
+---------------------------------+
| GRANT USAGE ON *.* TO 'emp'@'%' |
```

```
+-----------------------------------+
1 row in set (0.00 sec)
```

3. 更改账号权限

可以进行权限的新增和回收。和账号创建一样,权限变更也有两种方法:使用 GRANT(新增)和 REVOKE(回收)语句,或者更改权限表。

第二种方法和前面一样,直接对 user、db、tables_priv 和 columns_priv 中的权限列进行更新即可,这里重点介绍第一种方法。

和创建账号语法完全一样,GRANT 可以直接用来对账号进行增加。其实 GRANT 语句在执行的时候,如果权限表中不存在目标账号,则创建账号;如果已经存在,则执行权限的新增。下面来看一个例子。

(1) emp@localhost 目前只有登录权限。

```
mysql>show grants for emp@localhost;
+-----------------------------------------+
| Grants for emp@localhost                |
+-----------------------------------------+
| GRANT USAGE ON *.* TO 'emp'@'localhost' |
+-----------------------------------------+
1 row in set (0.00 sec)
```

(2) 赋予 emp@localhost 所有数据库上的所有表的 SELECT 权限。

```
mysql>grant select on *.* to 'emp'@'localhost';
Query OK, 0 rows affected, 1 warning (0.00 sec)
mysql>show grants for emp@localhost;
+------------------------------------------+
| Grants for emp@localhost                 |
+------------------------------------------+
| GRANT SELECT ON *.* TO 'emp'@'localhost' |
+------------------------------------------+
1 row in set (0.00 sec)
```

(3) 继续给 emp@localhost 赋予 SELECT 和 INSERT 权限,和已有的 SELECT 权限进行合并。

```
mysql>grant select,insert on *.* to 'emp'@'localhost';
Query OK, 0 rows affected, 1 warning (0.00 sec)

mysql>show grants for emp@localhost;
+--------------------------------------------------+
| Grants for emp@localhost                         |
+--------------------------------------------------+
| GRANT SELECT, INSERT ON *.* TO 'emp'@'localhost' |
+--------------------------------------------------+
1 row in set (0.00 sec)
```

REVOKE 语句可以回收已经赋予的权限,语法如下:

```
REVOKE priv_type [(column_list)] [, priv_type [(column_list)]] ...
    ON [object_type] {tbl_name | * | *.* | db_name.*}
    FROM user [, user] ...

REVOKE ALL PRIVILEGES, GRANT OPTION FROM user [, user] ...
```

对于上面的例子,这里决定要收回 emp@localhost 上的 INSERT 和 SELECT 权限:

```
mysql>revoke select,insert on *.* from emp@localhost;
Query OK, 0 rows affected, 1 warning (0.00 sec)

mysql>show grants for emp@localhost;
+-----------------------------------------+
| Grants for emp@localhost                |
+-----------------------------------------+
```

```
| GRANT USAGE ON *.* TO 'emp'@'localhost' |
+-----------------------------------------+
1 row in set (0.00 sec)
```

usage 权限不能被回收，也就是说，REVOKE 用户并不能删除用户。

```
mysql>show grants for emp@localhost;
+-----------------------------------------+
| Grants for emp@localhost                |
+-----------------------------------------+
| GRANT USAGE ON *.* TO 'emp'@'localhost' |
+-----------------------------------------+
1 row in set (0.00 sec)

mysql>revoke usage on *.* from emp@localhost;
Query OK, 0 rows affected, 1 warning (0.00 sec)

mysql>show grants for emp@localhost;
+-----------------------------------------+
| Grants for emp@localhost                |
+-----------------------------------------+
| GRANT USAGE ON *.* TO 'emp'@'localhost' |
+-----------------------------------------+
1 row in set (0.00 sec)
```

4. 修改账号密码

方法 1：可以用 mysqladmin 命令在命令行指定密码。

```
shell> mysqladmin -u user_name -h host_name password "newpwd"
```

方法 2：执行 alter user 语句。下例中将账号"jeffrey'@'%'"的密码改为"biscuit"。

```
mysql> alter user jeffrey@'%' identified by 'biscuit';
```

方法 3：还可以在全局级别使用 GRANT USAGE 语句（在"*.*"）来指定某个账户的密码而不影响账户当前的权限。

```
mysql> GRANT USAGE ON *.* TO 'jeffrey'@'%' IDENTIFIED BY 'biscuit';
```

方法 4：直接更改数据库的 user 表。

```
mysql>insert into user (host,user,ssl_cipher,x509_issuer,x509_subject,authentication_string)
values ('%','emp','','','',password('emp123'));
Query OK, 1 row affected, 1 warning (0.00 sec)

mysql>flush privileges;
Query OK, 0 rows affected (0.00 sec)
mysql>UPDATE user SET authentication_string = PASSWORD('123456') WHERE Host = '%' AND User = 'emp';
Query OK, 1 row affected, 1 warning (0.00 sec)
Rows matched: 1  Changed: 1  Warnings: 1

mysql>flush privileges;
Query OK, 0 rows affected (0.01 sec)
```

注意：MySQL 8.0 废弃了 PASSWORD 函数，所以方法 4 在 MySQL 8.0 已不适用。

方法 5：以上方法在更改密码时，用的都是明文，这样就会存在安全问题，比如修改密码的机器被入侵，那么通过命令行的历史执行记录就可以很容易地得到密码。因此，在一些重要的数据库中，可以直接使用 MD5 密码值来对密码进行更改，如以下例子：

```
GRANT USAGE ON *.* TO 'jeffrey'@'%' IDENTIFIED BY  PASSWORD '*6BB4837EB74329105EE4568DDA7DC67ED2CA2AD9';
```

或者：

```
set password = '*6BB4837EB74329105EE4568DDA7DC67ED2CA2AD9';
```

其中的 MD5 密码串可以事先用其他方式获得。

注意：MySQL 8.0 废弃了 PASSWORD 函数，所以方法 5 的 grant 命令在 MySQL 8.0 已不适用。

5. 删除账号

要彻底删除账号，同样也有两种实现方法，即 DROP USER 命令和修改权限表。
DROP USER 语法非常简单，具体如下：

```
DROP USER user [, user] ...
```

举一个简单的例子，将 emp@localhost 用户删除：

```
mysql>show grants for emp@localhost ;
+-----------------------------------------+
| Grants for emp@localhost                |
+-----------------------------------------+
| GRANT USAGE ON *.* TO 'emp'@'localhost' |
+-----------------------------------------+
1 row in set (0.00 sec)

mysql>drop user emp@localhost;
Query OK, 0 rows affected (0.00 sec)

mysql>show grants for emp@localhost ;
ERROR 1141 (42000): There is no such grant defined for user 'emp' on host 'localhost'
```

修改权限表方法只要把 user 用户中的用户记录删除即可，这里不再演示。

6. 账号资源限制

创建 MySQL 账号时，还有一类选项前面没有提及，我们称为账号资源限制（Account Resource Limit）。这类选项的作用是限制每个账号实际具有的资源限制，这里的"资源"主要包括以下内容：

- 单个账号每小时执行的查询次数；
- 单个账号每小时执行的更新次数；
- 单个账号每小时连接服务器的次数；
- 单个账号并发连接服务器的次数。

在实际应用中，可能会发生这种情景，由于程序 BUG 或者系统遭到攻击，使得某些应用短时间内发生了大量的点击，从而对数据库造成了严重的并发访问，数据库短期无法响应甚至宕机，对生产带来负面影响。为了防止这种问题的出现，我们可以通过对连接账号进行资源限制的方式来解决，比如按照日常访问量加上一定冗余设置每小时查询 1 万次，那么 1 小时内如果超过 1 万次查询，数据库就会给出资源不足的提示，而不会再分配资源进行实际查询。

设置资源限制的语法如下：

```
GRANT ...with option
```

其中，option 的选项可以是以下几个。

- MAX_QUERIES_PER_HOUR count：每小时最大查询次数。
- MAX_UPDATES_PER_HOUR count：每小时最大更新次数。
- MAX_CONNECTIONS_PER_HOUR count：每小时最大连接次数。
- MAX_USER_CONNECTIONS count：最大用户连接数。

其中，MAX_CONNECTIONS_PER_HOUR count 和 MAX_USER_CONNECTIONS count 的区别在于前者是每小时累计的最大连接次数，而后者是瞬间的并发连接数。系统还有一个全局参数 MAX_USER_CONNECTIONS，它和用户 MAX_USER_CONNECTIONS count 的区

别在于如果后者为 0,则此用户的实际值应该为全局参数值,否则就按照用户 MAX_USER_CONNECTIONS count 的值来设置。

下面举例来说明一下资源限制的使用方法。

创建用户 emp,要求具有 employees 库上的 select 权限,并且每小时查询次数小于等于 6 次,最多同时只能有 3 个此用户进行并发连接,代码如下:

```
Grant select on employees.* to emp@localhost
with MAX_QUERIES_PER_HOUR 6
MAX_USER_CONNECTIONS 3;
```

从 mysql 数据库的 user 表中可以看到相关资源的值:

```
mysql>select user,max_questions,max_updates,max_connections  from user where user='emp';
+------+---------------+-------------+-----------------+
| user | max_questions | max_updates | max_connections |
+------+---------------+-------------+-----------------+
| emp  |             6 |           0 |               0 |
+------+---------------+-------------+-----------------+
1 row in set (0.00 sec)
```

用 emp 登录后,执行下面的查询:

```
[mysql3307@localhost ~]$ mysql -uemp -S/tmp/mysql_3307.sock employees -p
Enter password:
Reading table information for completion of table and column names
You can turn off this feature to get a quicker startup with -A

Welcome to the MySQL monitor.  Commands end with ; or \g.
Your MySQL connection id is 34
Server version: 5.7.22-log MySQL Community Server (GPL)

Copyright (c) 2000, 2018, Oracle and/or its affiliates. All rights reserved.

Oracle is a registered trademark of Oracle Corporation and/or its
affiliates. Other names may be trademarks of their respective
owners.

Type 'help;' or '\h' for help. Type '\c' to clear the current input statement.

emp@localhost:mysql_3307.sock   [employees]>select count(0) from city;
+----------+
| count(0) |
+----------+
|        1 |
+----------+
1 row in set (0.00 sec)

emp@localhost:mysql_3307.sock   [employees]>select count(0) from city;
ERROR 1226 (42000): User 'emp' has exceeded the 'max_questions' resource (current value: 6)
```

可以发现,登录后执行到第 2 个查询时提示用户 emp 已经超过了最大查询的资源限制,从而提示出错。这里有些读者可能有疑问,设置为 6 应该是执行到第 7 个查询的时候出错才对,为什么第 2 个就报错了呢?其实 MySQL 里面很多非"select"语句都会归类到"查询",比如"show"语句、"desc"语句等,还有一些隐式的查询包含在内,上面的查询从日志中可以查到在登录后隐式执行了以下 SQL 语句,这才导致了上面的结果。

```
2018-07-21T14:24:41.867892+08:00          35 Query     SET NAMES utf8
2018-07-21T14:24:41.869351+08:00          35 Query     show databases
2018-07-21T14:24:41.869927+08:00          35 Query     show tables
2018-07-21T14:24:41.874695+08:00          35 Query     select @@version_comment limit 1
2018-07-21T14:24:41.877209+08:00          35 Query     select USER()
```

需要注意的是,资源限制是对某一个账号进行累计的,而不是对账号的一次连接进行累计。当资源限制到达后,账号的任何一次相关操作都会被拒绝。如果要继续操作,只能清除

26.1 MySQL 权限管理

相关的累加值。可以使用 root 执行 flush user_resources/flush privileges/mysqladmin reload 这 3 个命令中的任何一个来执行清除工作。如果数据库发生重启，则原先累计的计数值清零。

如果要对账号的资源限制进行修改或者删除，将相应参数设置为 0 即可：

```
GRANT USAGE ON *.* TO 'emp'@'localhost'
WITH MAX_CONNECTIONS_PER_HOUR 0;
```

> **注意**：账号的资源限制设置一定要非常小心，一般不建议设置。如果一定要设置，就要对系统的高峰访问情况了解清楚并加上足够的冗余后再进行设置，为了防止达到资源限制后所有功能的失效，将不同的功能分给不同的用户是一个可行的办法。

7. 用户密码管理

对于一些敏感用户，为了提高安全性，我们需要定期修改用户的密码。MySQL 提供基于时间维度定期修改密码的功能。这个功能涉及 mysql.user 表的 3 个字段。

- password_expired：密码是否过期，默认值为 N。
- password_last_changed：最后修改密码的时间。
- password_lifetime：密码有效时间，单位为天。

可以通过 alter user 命令设置 password_expired 字段值为 Y，要求用户重置账号的密码。当密码过期后，新连接需要用 alter user 命令重置密码后才能执行 SQL 语句，密码过期对旧连接没有影响。

从 mysql 数据库的 user 表中可以看到 password_expired 和 password_last_changed 的值：

```
mysql> select password_expired,password_last_changed from user where User='emp' and Host='%';
+------------------+-----------------------+
| password_expired | password_last_changed |
+------------------+-----------------------+
| N                | 2018-10-16 14:21:36   |
+------------------+-----------------------+
1 row in set (0.00 sec)
```

结果显示，password_expired 的值为 N，代表密码未过期；上次修改密码时间 password_last_changed 字段的值为 2018-10-16 14:21:36。

使用 emp 用户登录数据库的实例如下：

```
[mysql3307@localhost ~]$ mysql -uemp -pemp123 -h127.0.0.1 -P3307
Welcome to the MySQL monitor.  Commands end with ; or \g.
Your MySQL connection id is 94
Server version: 5.7.22-log MySQL Community Server (GPL)
Type 'help;' or '\h' for help. Type '\c' to clear the current input statement.

mysql>
```

修改 password_expired 字段值为 Y：

```
mysql>alter user emp@'%' password expire;
mysql> select password_expired,password_last_changed from user where User='emp' and Host='%';
+------------------+-----------------------+
| password_expired | password_last_changed |
+------------------+-----------------------+
| Y                | 2018-10-16 14:21:36   |
+------------------+-----------------------+
1 row in set (0.00 sec)
```

成功修改 password_expired 字段后，使用 emp 用户旧连接会话执行命令：

```
mysql>show databases;
+--------------------+
| Database           |
+--------------------+
```

```
| information_schema    |
| employees             |
| mysql                 |
| performance_schema    |
| sys                   |
| testdb                |
+-----------------------+
6 rows in set (0.00 sec)
```

旧的连接执行命令正常。下面用 emp 用户新建立连接,执行命令:

```
[mysql3307@localhost ~]$ mysql -uemp -pemp123 -h127.0.0.1 -P3307
Welcome to the MySQL monitor.  Commands end with ; or \g.
Your MySQL connection id is 95
Server version: 5.7.22-log MySQL Community Server (GPL)
Type 'help;' or '\h' for help. Type '\c' to clear the current input statement.

mysql>show databases;
ERROR 1820 (HY000): You must reset your password using ALTER USER statement before executing this statement.
```

新会话执行命令前做校验,发现密码过期,会报 ERROR 1820。

在以上报错的连接会话中,重置 emp 用户的密码:

```
mysql>alter user emp@'%' identified by 'emp123';
Query OK, 0 rows affected (0.00 sec)
```

再执行命令:

```
mysql>show databases;
+-----------------------+
| Database              |
+-----------------------+
| information_schema    |
| employees             |
| mysql                 |
| performance_schema    |
| sys                   |
| testdb                |
+-----------------------+
6 rows in set (0.00 sec)
mysql> select password_expired,password_last_changed from user where User='emp' and Host='%';
+------------------+-----------------------+
| password_expired | password_last_changed |
+------------------+-----------------------+
| N                | 2018-10-16 14:33:20   |
+------------------+-----------------------+
1 row in set (0.00 sec)
```

如我们所料,命令可以正常执行了,密码的上次修改时间也变成最新时间。

注意:这里重置的新密码与旧密码相同,但是为了安全,建议设置一个不同的新密码。

password_lifetime 指的是用户密码的有效天数。如果当天时间减去 password_last_changed 字段值,所得天数超过 password_lifetime 的值,password_expired 被设置为 Y。如果对所有的用户应用同一个密码有效天数,可以指定全局参数 default_password_lifetime,单位为天。设置用户密码有效时间为默认配置:

```
mysql> alter user emp@'%' password expire default;
```

8. 锁定用户

从 MySQL 5.7.6 版本开始,可以执行 alter user 命令锁定和解锁用户,用于临时禁止用户登录。

下面来看一个锁定用户的简单例子。

查看 emp 用户的锁定状态:

```
mysql>select account_locked from user where User='emp' and Host='%';
+----------------+
| account_locked |
+----------------+
| N              |
+----------------+
1 row in set (0.00 sec)
```

结果显示，emp 用户处于非锁定状态，下面将用户锁定：

```
mysql>alter user emp@'%' account lock;
Query OK, 0 rows affected (0.00 sec)
```

再次查看 emp 用户的锁定状态：

```
mysql>select account_locked from user where User='emp' and Host='%';
+----------------+
| account_locked |
+----------------+
| Y              |
+----------------+
1 row in set (0.00 sec)
```

account_locked 字段值为 Y，用户已被锁定。重新登录创建连接：

```
[mysql3307@localhost ~]$ mysql -uemp -pemp123 -h127.0.0.1 -P3307
ERROR 3118 (HY000): Access denied for user 'emp'@'127.0.0.1'. Account is locked
```

由于用户被锁定，尝试登录会报错 ERROR 3118，emp 用户已无法登录。用 root 用户解锁 emp 用户：

```
mysql>alter user emp@'%' account unlock;
Query OK, 0 rows affected (0.00 sec)
```

解锁后，尝试登录数据库：

```
[mysql3307@localhost ~]$ mysql -uemp -pemp123 -h127.0.0.1 -P3307
Welcome to the MySQL monitor.  Commands end with ; or \g.
Your MySQL connection id is 96
Server version: 5.7.22-log MySQL Community Server (GPL)
Type 'help;' or '\h' for help. Type '\c' to clear the current input statement.

mysql>
```

结果如我们所料，可以正常登录了。

26.2 MySQL 安全问题

对于任何一种数据库来说，安全问题都是非常重要的。如果数据库出现安全漏洞，轻则数据被窃取，重则数据被破坏，这些后果对于一些重要的数据库来说，都是非常严重的。本节将从操作系统和数据库两个层面对 MySQL 的安全问题进行探讨。

26.2.1 操作系统相关的安全问题

本节介绍一些常见的操作系统安全问题，这些问题主要出现在 MySQL 的安装和启动过程中，希望读者能够从安装开始就重视安全问题。

1．严格控制操作系统账号和权限

在数据库服务器上要严格控制操作系统的账号和权限，比如：

- 锁定 mysql 用户；
- 其他任何用户都采取独立的账号登录，管理员通过 mysql 专有用户管理 MySQL，或

者通过 root su 到 mysql 用户下进行管理；

- mysql 用户目录下，除了数据文件目录，其他文件和目录属主都改为 root。

2. 防止 DNS 欺骗

创建用户时，host 可以指定域名或者 IP 地址。但是，如果指定域名，就可能带来如下安全隐患：如果域名对应的 IP 地址被恶意修改，则数据库就会被恶意的 IP 地址进行访问，导致安全隐患。

在下例中，尝试改变域名对应的 IP 地址，以此来观察一下对连接的影响。

（1）创建测试用户 emp，域名指定为 test_hostname。

```
mysql>grant select on employees.* to emp@test_hostname identified by 'emp123';
Query OK, 0 rows affected (0.00 sec)

mysql> show grants for emp@test_hostname;
+-------------------------------------------------------+
| Grants for emp@test_hostname                          |
+-------------------------------------------------------+
| GRANT USAGE ON *.* TO 'emp'@'test_hostname'           |
| GRANT SELECT ON `employees`.* TO 'emp'@'test_hostname'|
+-------------------------------------------------------+
```

（2）编辑 hosts 文件，增加此域名和 IP 地址的对应关系：

```
[root@localhost ~]# vi /etc/hosts

# Do not remove the following line, or various programs
# that require network functionality will fail.
127.0.0.1               localhost.localdomain localhost
192.168.7.55            localhost.localdomain
192.168.52.24           test_hostname
```

（3）客户端尝试连接成功：

```
C:\mysql\bin>ipconfig

Windows IP Configuration


Ethernet adapter 无线网络连接:

        Media State . . . . . . . . . . . : Media disconnected

Ethernet adapter 本地连接:

        Connection-specific DNS Suffix  . : netease.internal
        IP Address. . . . . . . . . . . . : 192.168.52.24
        Subnet Mask . . . . . . . . . . . : 255.255.254.0
        Default Gateway . . . . . . . . . : 192.168.52.254

C:\mysql\bin>mysql -h192.168.7.55 -P3307 -uemp -p
Enter password:
Welcome to the MySQL monitor.  Commands end with ; or \g.
Your MySQL connection id is 6 to server version: 5.7.22-beta-log

Type 'help;' or '\h' for help. Type '\c' to clear the buffer.

mysql> exit
```

（4）修改域名 IP 地址的对应关系，将 192.168.52.24 改为 192.168.52.23：

```
[root@localhost ~]# vi /etc/hosts

# Do not remove the following line, or various programs
# that require network functionality will fail.
127.0.0.1               localhost.localdomain localhost
192.168.7.55            localhost.localdomain
192.168.52.23           test_hostname
```

(5)客户端再次尝试连接失败:

```
C:\mysql\bin>mysql -h192.168.7.55 -P3307 -uemp -p
Enter password:
ERROR 1045 (28000): Access denied for user 'emp'@'192.168.52.24' (using NO)
```

26.2.2 数据库相关的安全问题

本节介绍一些常见的数据库安全问题,这些问题大多数是由于账号的管理不当造成的。希望读者读完本节后能够认识到账号管理的重要性,同时加强对账号管理的安全意识。

1. 给 root 账号设置口令

安装 MySQL 5.7,如果使用--initialize-insecure 选项初始化数据库,则 root 的默认口令为空,需要马上修改 root 口令:

```
[mysql3307@localhost ~]$ mysql -uroot -S/tmp/mysql_3307.sock
Welcome to the MySQL monitor.  Commands end with ; or \g.
Your MySQL connection id is 4
Server version: 5.7.22-log MySQL Community Server (GPL)
Type 'help;' or '\h' for help. Type '\c' to clear the current input statement.

mysql> alter user root@'localhost' identified by 'newpassword';
Query OK, 0 rows affected (0.00 sec)
```

不写密码登录将被拒绝:

```
[mysql3307@ localhost ~]$ mysql -uroot -S/tmp/mysql_3307.sock
ERROR 1045 (28000): Access denied for user 'root'@'localhost' (using password: NO)
```

2. 设置安全密码

密码的安全体现在以下两个方面:

● 设置安全的密码,建议使用 8 位以上字母、数字、下划线和一些特殊字符组合而成的字符串;

● 使用密码期间尽量保证使用过程的安全,不会被别人窃取。

第一点就不说了,越长、越复杂、越没有规律的密码越安全。对于第二点,可以总结一下,在日常工作中,使用密码一般是采用以下几种方式。

(1)直接将密码写在命令行中:

```
[mysql3307@localhost ~]$ mysql -uroot -S/tmp/mysql_3307.sock -p123
Welcome to the MySQL monitor.  Commands end with ; or \g.
Your MySQL connection id is 8
Server version: 5.7.22-log MySQL Community Server (GPL)
```

(2)交互式方式输入密码:

```
[mysql3307@localhost ~]$ mysql -uroot -S/tmp/mysql_3307.sock -p
Enter password:
Welcome to the MySQL monitor.  Commands end with ; or \g.
Your MySQL connection id is 11
Server version: 5.7.22-log MySQL Community Server (GPL)
```

(3)将用户名和密码写在配置文件里面,连接的时候自动读取。比如应用连接数据库或者执行一些批处理脚本。对于这种方式,MySQL 供了一种方法,在 my.cnf 里面写入连接信息:

```
[client]
user=username
password=password
```

然后对配置文件进行严格的权限限制,例如:

```
chmod +600 my.cnf
```

以上是 3 种常见的密码使用方式。很显然，第 1 种最不安全，因为它将密码写成为明文；第 2 种比较安全，但是只能使用在交互式的界面下；第 3 种使用比较方便，但是需要将配置文件设置严格的存取权限，而且任何只要可以登录操作系统的用户都可以自动登录，存在一定的安全隐患。

下面举一个第 3 种方法的例子。

输入 mysql 无法登录。

```
[mysql3307@localhost ~]$ mysql
ERROR 1045 (28000): Access denied for user 'hr'@'localhost' (using password: NO)
```

在执行 mysql 客户端用户的家目录，创建 .my.cnf 文件，加入连接信息。

```
[mysql3307@localhost ~]$ cd ~
[mysql3307@localhost ~]$ vi .my.cnf
[client]
user = root
password = 123
socket=/tmp/mysql_3307.sock
[mysql3307@localhost ~]$ chmod 660  .my.cnf
```

再次输入 mysql。

```
[mysql3307@localhost ~]$ mysql
Welcome to the MySQL monitor.  Commands end with ; or \g.
Your MySQL connection id is 12
Server version: 5.7.22-log MySQL Community Server (GPL)

Copyright (c) 2000, 2018, Oracle and/or its affiliates. All rights reserved.

Oracle is a registered trademark of Oracle Corporation and/or its
affiliates. Other names may be trademarks of their respective
owners.

Type 'help;' or '\h' for help. Type '\c' to clear the current input statement.
mysql> select current_user();
+----------------+
| current_user() |
+----------------+
| root@localhost |
+----------------+
1 row in set (0.00 sec)
```

3．只授予账号必需的权限

只需要赋予普通用户必需的权限，比如：

```
Grant select,insert,update,delete on tablename to 'username'@'hostname';
```

在很多情况下，DBA 为了方便，经常赋予用户 all privileges 权限。这个 all privileges 到底具体包含哪些权限呢？来看下面的这个例子：

```
mysql>grant all privileges on employees.* to 'emp'@'localhost';
Query OK, 0 rows affected, 1 warning (0.00 sec)
mysql> select * from db where user='emp' \G
*************************** 1. row ***************************
                 Host: localhost
                   Db: employees
                 User: emp
          Select_priv: Y
          Insert_priv: Y
          Update_priv: Y
          Delete_priv: Y
          Create_priv: Y
            Drop_priv: Y
```

```
            Grant_priv: N
       References_priv: Y
            Index_priv: Y
            Alter_priv: Y
  Create_tmp_table_priv: Y
       Lock_tables_priv: Y
       Create_view_priv: Y
         Show_view_priv: Y
    Create_routine_priv: Y
     Alter_routine_priv: Y
           Execute_priv: Y
             Event_priv: Y
           Trigger_priv: Y
1 row in set (0.00 sec)
```

all privileges 里面的权限远远超过了一般应用所需要的权限。而且，有些权限如果误操作，将会产生非常严重的后果，比如 drop_priv 等。因此，赋予用户权限的时候越具体，对数据库越安全。

4．除 root 外，任何用户不应有 mysql 库 user 表的存取权限

由于 MySQL 中可以通过更改 mysql 数据库的 user 表进行权限的增加、删除、变更等操作，因此，除了 root 以外，任何用户都不应该拥有对 user 表的存取权限（SELECT、UPDATE、INSERT、DELETE 等），否则容易造成系统的安全隐患。下例中对普通用户 emp 授予了 user 表的存取权限，看看会对系统产生怎样的安全隐患。

（1）创建普通用户 emp，拥有对 mysql 数据库中 user 表的各种权限。

```
[mysql3307@localhost ~]$ mysql -uroot -p123
Welcome to the MySQL monitor.  Commands end with ; or \g.
Your MySQL connection id is 25
Server version: 5.7.22-log MySQL Community Server (GPL)

Type 'help;' or '\h' for help. Type '\c' to clear the current input statement.

mysql>grant select,update,insert,delete on mysql.user to emp@localhost;
Query OK, 0 rows affected (0.00 sec)
```

（2）用 emp 来更新 root 权限。

```
[mysql3307@localhost ~]$ mysql -uemp
Welcome to the MySQL monitor.  Commands end with ; or \g.
Your MySQL connection id is 30
Server version: 5.7.22-log MySQL Community Server (GPL)

mysql> use mysql
mysql> update user set authentication_string=password('abcd') where user='root' and host='localhost';
Rows matched: 1  Changed: 1  Warnings: 0
```

（3）当数据库重启或者 root 刷新权限表后，root 登录时密码已经被更改。

```
[mysql3307@localhost ~]$ mysql -uroot -p123
mysql: [Warning] Using a password on the command line interface can be insecure.
ERROR 1045 (28000): Access denied for user 'root'@'localhost' (using password: YES)

[mysql3307@localhost ~]$ mysql -uroot -pabcd
mysql: [Warning] Using a password on the command line interface can be insecure.
Welcome to the MySQL monitor.  Commands end with ; or \g.
Your MySQL connection id is 36
Server version: 5.7.22-log MySQL Community Server (GPL)

mysql>
```

5．不要把 FILE、PROCESS 或 SUPER 权限授予管理员以外的账号

FILE 权限主要有以下两种作用。

- 将数据库的信息通过 SELECT…INTO OUTFILE…写到服务器上有写权限的目录下，作为文本格式存放。如果导出的数据量很大，将有耗尽系统磁盘空间的风险。
- 可以将有读权限的文本文件通过 LOAD DATA INFILE…命令写入数据库表，如果这些文本文件中存放了很重要的信息，将对系统造成很大的安全隐患。

在 MySQL 5.7 版本，可以设置 secure_file_priv 参数，指定用户只能在有权限的目录中执行 SELECT…INTO OUTFILE 和 LOAD DATA INFILE 命令。

PROCESS 权限能被用来执行"show processlist"命令，查看当前所有用户执行的查询的明文文本，包括设定或改变密码的查询。在默认情况下，每个用户都可以执行"show processlist"命令，但是只能查询本用户的进程。因此，对 PROCESS 权限管理不当，有可能会使得普通用户能够看到管理员执行的命令。

下例中对普通用户赋予了 PROCESS 权限，来看看会造成什么安全隐患。

（1）将 PROCESS 权限授予普通用户：

```
[mysql3307@localhost ~]$ mysql -uroot -p
Reading table information for completion of table and column names
You can turn off this feature to get a quicker startup with -A

Welcome to the MySQL monitor.  Commands end with ; or \g.
Your MySQL connection id is 39
Server version: 5.7.22-log MySQL Community Server (GPL)

mysql>show processlist;
 mysql> grant process on *.* to 'emp'@'localhost';
Query OK, 0 rows affected (0.00 sec)
```

（2）锁定表 user，可以让进程阻塞，以方便用户看到进程内容：

```
mysql>lock table user read;
Query OK, 0 rows affected (0.00 sec)
```

（3）打开另外一个 session，用 root 执行修改密码操作，此时因为 user 表被锁定，此进程被阻塞挂起：

```
mysql>set password=password('123');
```

（4）打开第 3 个 session，用 emp 登录，执行 show processlist 语句：

```
[mysql3307@localhost ~]$ mysql -uemp -p
Enter password:
Welcome to the MySQL monitor.  Commands end with ; or \g.
Your MySQL connection id is 42
Server version: 5.7.22-log MySQL Community Server (GPL)

mysql>show processlist\G
*************************** 57. row ***************************
     Id: 40
   User: emp
   Host: localhost
     db: employees
Command: Query
   Time: 11
  State: Waiting for table metadata lock
   Info: set password=password('123')
```

可以发现，emp 显示的进程中清楚地看到了 root 的修改密码操作，并看到了明文的密码，这将对系统造成严重的安全隐患。

SUPER 权限能执行 kill 命令，终止其他用户进程。在下面的例子中，普通用户拥有了 SUPER 权限后，便可以任意 kill 任何用户的进程。

（1）emp 登录后想 kill 掉上面例子中 root 修改密码进程（进程号 40）：

26.2 MySQL 安全问题

```
mysql>kill 40;
ERROR 1095 (HY000): You are not owner of thread 40
```

（2）kill 失败后，root 将 super 权限赋予 emp：

```
mysql>grant super on *.* to emp@localhost;
Query OK, 0 rows affected (0.00 sec)
mysql>show grants for emp@'localhost';
+---------------------------------------------------------------------+
| Grants for emp@localhost                                            |
+---------------------------------------------------------------------+
| GRANT PROCESS, SUPER ON *.* TO 'emp'@'localhost'                    |
| GRANT SELECT, INSERT, UPDATE, DELETE ON `mysql`.`user` TO 'emp'@'localhost' |
+---------------------------------------------------------------------+
```

（3）emp 用户重新登录重新 kill root 的进程成功：

```
[mysql3307@localhost ~]$ mysql -uemp -p
Enter password:
Welcome to the MySQL monitor.  Commands end with ; or \g.
Your MySQL connection id is 47
Server version: 5.7.22-log MySQL Community Server (GPL)

mysql>show processlist;
*************************** 57. row ***************************
     Id: 40
   User: emp
   Host: localhost
     db: employees
Command: Query
   Time: 11
  State: Waiting for table metadata lock
   Info: set password=password('123')

mysql>kill 40;
Query OK, 0 rows affected (0.00 sec)
```

从上面的例子中，可以看到了 FILE、PROCESS、SUPER 这 3 个管理权限可能带来的安全隐患，因此，除了管理员外，不要把这些权限赋予普通用户。

6. LOAD DATA LOCAL 带来的安全问题

LOAD DATA 如果权限控制不好，会带来了以下安全问题。

- 可以任意加载本地文件到数据库。
- 在 Web 环境中，客户从 Web 服务器连接，用户可以使用 LOAD DATA LOCAL 语句来读取 Web 服务器进程有读访问权限的任何文件（假定用户可以运行 SQL 服务器的任何命令）。在这种环境中，MySQL 服务器的客户实际上是 Web 服务器，而不是连接 Web 服务器的用户运行程序。

解决方法是，通过设置 secure_file_priv 来满足实际业务的需要，secure_file_priv 的值及其含义如表 26-3 所示。

表 26-3　　　　　　　　　secure_file_priv 的值及其含义

值	含 义
空	不做任何限制
/tmp/	数据库只能导入/tmp 下有读权限的文件
NULL	不允许执行导入操作

注意：更改 secure_file_priv 的值需要重启 MySQL 实例。

7. DROP TABLE 命令并不收回以前的相关访问授权

DROP 表的时候，其他用户对此表的权限并没有被收回，这样导致重新创建同名的表时，以前其他用户对此表的权限会自动赋予,进而产生权限外流。因此,在删除表时,要同时取消其他用户在此表上的相应权限。

下面的例子说明了不收回相关访问授权的隐患。

(1)用 root 创建用户 emp,授予对 employees 下所有表的 select 权限:

```
mysql>grant select on employees.city to emp@localhost;
Query OK, 0 rows affected (0.00 sec)

mysql>show grants for emp@localhost;
+--------------------------------------------------------+
| Grants for emp@localhost                               |
+--------------------------------------------------------+
| GRANT USAGE ON *.* TO 'emp'@'localhost'                |
| GRANT SELECT ON `employees`.`city` TO 'emp'@'localhost' |
+--------------------------------------------------------+
2 rows in set (0.00 sec)
```

(2)emp 登录,测试权限:

```
[mysql3307@localhost ~]$ mysql -uemp
Welcome to the MySQL monitor.  Commands end with ; or \g.
Your MySQL connection id is 54
Server version: 5.7.22-log MySQL Community Server (GPL)

mysql>use employees
Reading table information for completion of table and column names
You can turn off this feature to get a quicker startup with -A

Database changed
emp@localhost:mysql_3307.sock  [employees]>show tables;
+---------------------+
| Tables_in_employees |
+---------------------+
| city                |
+---------------------+
1 row in set (0.00 sec)
```

(3)root 登录,删除表 city:

```
mysql>drop table city;
```

(4)emp 登录,再次测试权限:

```
[mysql3307@localhost ~]$ mysql -uemp
Welcome to the MySQL monitor.  Commands end with ; or \g.
Your MySQL connection id is 56
Server version: 5.7.22-log MySQL Community Server (GPL)

mysql>use employees
Reading table information for completion of table and column names
You can turn off this feature to get a quicker startup with -A

Database changed

emp@localhost:mysql_3307.sock  [employees]>show tables;
Empty set (0.00 sec)
```

(5)此时 city 表已经看不到了:

```
mysql>show grants for emp@localhost;
+--------------------------------------------------------+
| Grants for emp@localhost                               |
+--------------------------------------------------------+
| GRANT USAGE ON *.* TO 'emp'@'localhost'                |
| GRANT SELECT ON `employees`.`city` TO 'emp'@'localhost' |
+--------------------------------------------------------+
2 rows in set (0.00 sec)
```

emp 用户仍然有 employees 下 city 表的 SELECT 权限(安全漏洞)。

（6）root 再次登录，创建 city 表：

```
mysql>CREATE TABLE `city` (
    -> `city_id` smallint(5) unsigned NOT NULL DEFAULT '0',
    -> `city` varchar(50) CHARACTER SET utf8 NOT NULL,
    -> `country_id` smallint(5) unsigned NOT NULL,
    -> `last_update` timestamp NOT NULL DEFAULT CURRENT_TIMESTAMP ON UPDATE CURRENT_TIMESTAMP
    -> ) ENGINE=InnoDB DEFAULT CHARSET=utf8 COLLATE=utf8_unicode_ci;
Query OK, 0 rows affected (0.01 sec)
```

（7）emp 登录，对 city 权限依旧存在：

```
[mysql3307@localhost ~]$ mysql -uemp
Welcome to the MySQL monitor.  Commands end with ; or \g.
Your MySQL connection id is 58
Server version: 5.7.22-log MySQL Community Server (GPL)

mysql>use employees

emp@localhost:mysql_3307.sock  [employees]>show tables;
+---------------------+
| Tables_in_employees |
+---------------------+
| city                |
+---------------------+
1 row in set (0.00 sec)

emp@localhost:mysql_3307.sock  [employees]>select * from city;
Empty set (0.00 sec)
```

注意：对表做删除后，其他用户对此表的权限不会自动收回，一定记住要手工收回。

8. 使用 SSL

SSL（Secure Socket Layer，安全套接字层）是一种安全传输协议，最初由 Netscape 公司所开发，用以保障在 Internet 上数据传输的安全，利用数据加密（Encryption）技术，可确保数据在网络上的传输过程中不会被截取及窃听。MySQL 通过两种方式支持 SSL，一种是使用 OpenSSL，另一种是使用 YaSSL。在 MySQL 8.0 版本前，MySQL 社区版本使用的是 YaSSL；8.0 版本后，MySQL 把 OpenSSL 作为企业版本和社区版本的统一默认 TLS/SSL 库。

SSL 协议提供的服务主要有 3 种。

- 认证用户和服务器，确保数据发送到正确的客户机和服务器。
- 加密数据以防止数据中途被窃取。
- 维护数据的完整性，确保数据在传输过程中不被改变。

在 MySQL 中，要想使用 SSL 进行安全传输，需要在命令行中或选项文件中设置"--ssl"选项。

对于服务器，"--ssl"选项规定该服务器允许 SSL 连接。对于客户端程序，它允许客户使用 SSL 连接服务器。单单该选项不足以使用 SSL 连接，还必须指定--ssl-ca、--ssl-cert 和--ssl-key 选项。如果不想启用 SSL，则可以将选项指定为--skip-ssl 或--ssl=0。

注意：如果编译的服务器或客户端不支持 SSL，则使用普通的未加密的连接。

确保使用 SSL 连接的安全方式是，使用含 REQUIRE SSL 子句的 GRANT 语句在服务器上创建一个账户，然后使用该账户来连接服务器，服务器和客户端均应启用 SSL 支持。下面的例子创建了一个含 REQUIRE SSL 子句的账号：

```
mysql>grant select on *.* to emp identified by 'emp123' REQUIRE ssl;
Query OK, 0 rows affected, 1 warning (0.00 sec)
```

- --ssl-ca=file_name：含可信 SSL CA 的清单的文件的路径。
- --ssl-cert=file_name：SSL 证书文件名，用于建立安全连接。

- --ssl-key=file_name：SSL 密钥文件名，用于建立安全连接。

9. 如果可能，给所有用户加上访问 IP 限制

对数据库来说，我们希望从客户端过来的连接都是安全的，因此，就很有必要在创建用户时指定可以进行连接的服务器 IP 或者 HOSTNAME，只有符合授权的 IP 或者 HOSTNAME 才可以进行数据库访问。

10. REVOKE 命令的漏洞

当用户被多次赋予权限后，由于各种原因，需要将此用户的权限全部取消，此时，REVOKE 命令可能并不会按照我们的意愿执行，来看下面的例子。

（1）连续赋予用户两次权限，其中，第 2 次是对所有数据库的所有权限。

```
mysql>grant select,insert on employees.* to emp@localhost;
Query OK, 0 rows affected, 2 warnings (0.00 sec)

mysql>grant all privileges on *.* to emp@localhost;
Query OK, 0 rows affected, 1 warning (0.00 sec)

mysql>show grants for emp@localhost;
+-----------------------------------------------------------------+
| Grants for emp@localhost                                        |
+-----------------------------------------------------------------+
| GRANT ALL PRIVILEGES ON *.* TO 'emp'@'localhost'                |
| GRANT SELECT, INSERT ON `employees`.* TO 'emp'@'localhost'      |
+-----------------------------------------------------------------+
2 rows in set (0.00 sec)
```

（2）此时，需要取消此用户的所有权限。

```
mysql>revoke all privileges on *.* from emp@localhost;
Query OK, 0 rows affected, 1 warning (0.00 sec)
```

（3）我们很可能以为，此时用户已经没有任何权限了，而不会再去查看他的权限表。而实际上，此时的用户依然拥有 employees 上的 SELECT 和 INSERT 权限。

```
mysql>show grants for emp@localhost;
+-----------------------------------------------------------------+
| Grants for emp@localhost                                        |
+-----------------------------------------------------------------+
| GRANT USAGE ON *.* TO 'emp'@'localhost'                         |
| GRANT SELECT, INSERT ON `employees`.* TO 'emp'@'localhost'      |
+-----------------------------------------------------------------+
2 rows in set (0.00 sec)
```

这个是 MySQL 权限机制造成的隐患，在一个数据库上多次赋予权限，权限会自动合并；但是在多个数据库上多次赋予权限，每个数据库上都会认为是单独的一组权限，必须在此数据库上用 REVOKE 命令来单独进行权限收回，而 REVOKE ALL PRIVILEGES ON *.*并不会替用户自动完成这个过程。

11. 使用角色管理用户权限

MySQL 角色（role）是 8.0 版本提供的新特性。角色是一组权限的集合，角色被授予 MySQL 用户后，用户将获得相应的权限。多个用户共用一组权限，方便统一运维和管理。

下面介绍一下角色的使用方法。

（1）创建只读角色 role_sel，赋予角色拥有查询 employees 数据库的权限。

```
mysql>create role role_sel;
Query OK, 0 rows affected (0.15 sec)
mysql>grant select on employees.* to role_sel;
Query OK, 0 rows affected (0.10 sec)
```

26.3 其他安全设置选项

（2）创建只读用户 emp_sel，将 role_sel 角色赋予 emp_sel 用户。

```
mysql>create user emp_sel@'%' identified by 'emp_sel123';
Query OK, 0 rows affected (0.15 sec)
mysql>grant role_sel to emp_sel@'%';
Query OK, 0 rows affected (0.10 sec)
```

（3）查看 emp_sel 用户权限，使用 using role 选项，将列出角色所拥有的权限。

```
mysql>show grants for emp_sel@'%' using role_sel;
+------------------------------------------------+
| Grants for emp_sel@%                           |
+------------------------------------------------+
| GRANT USAGE ON *.* TO `emp_sel`@`%`            |
| GRANT SELECT ON `employees`.* TO `emp_sel`@`%` |
| GRANT `role_sel`@`%` TO `emp_sel`@`%`          |
+------------------------------------------------+
3 rows in set (0.00 sec)
```

（4）对账户在登录数据库之后要激活的 role 进行设置。

```
mysql>set default role all to emp_sel@'%';
Query OK, 0 rows affected (0.01 sec)
```

（5）用 emp_sel 用户登录数据库实例。

```
[mysql3307@localhost ~]$ mysql -uemp_sel -pemp_sel123 -h127.0.0.1 -P3488
Welcome to the MySQL monitor.  Commands end with ; or \g.
Your MySQL connection id is 47
Server version: 8.0.11 MySQL Community Server - GPL
Type 'help;' or '\h' for help. Type '\c' to clear the current input statement.
mysql> show databases;
+--------------------+
| Database           |
+--------------------+
| employees          |
| information_schema |
+--------------------+
2 rows in set (0.00 sec)

mysql>use employees
Reading table information for completion of table and column names
You can turn off this feature to get a quicker startup with -A

Database changed
mysql>select count(0) from dept_emp;
+----------+
| count(0) |
+----------+
|   331603 |
+----------+
1 row in set (0.36 sec)
```

通过命令确认 emp_sel 用户已经拥有 employees 数据库的查询权限。分别使用 REVOKE 命令、DROP ROLE 命令回收角色权限和删除角色，使用方法比较简单，读者可以自己尝试。

26.3 其他安全设置选项

除了上面介绍的那些需要注意的安全隐患外，MySQL 本身还带着一些选项，适当地使用这些选项将会使数据库更加安全。

26.3.1 密码插件

在 MySQL 8.0 版本之前，PASSWORD 函数生成的密码是 41 位；8.0 以后，MySQL 废弃了

password()函数，默认改用 caching_sha2_password 插件进行加密，使用缓存解决连接时的延时问题。

MySQL 5.7 中查看 PASSWORD 函数的结果如下：

```
mysql> SELECT PASSWORD('mypass');
+-------------------------------------------+
| PASSWORD('mypass')                        |
+-------------------------------------------+
| *6C8989366EAF75BB670AD8EA7A7FC1176A95CEF4 |
+-------------------------------------------+
```

MySQL 8.0 中执行的结果如下，可以看到 PASSWORD 函数已经被废弃。

```
mysql> SELECT PASSWORD('mypass');
ERROR 1064 (42000): You have an error in your SQL syntax; check the manual that corresponds
to your MySQL server version for the right syntax to use near '('mypass')' at line 1
```

MySQL 8.0 中查看用户默认的密码插件：

```
mysql>SELECT User, Host, plugin from mysql.user;
+------------------+-----------+-----------------------+
| User             | Host      | plugin                |
+------------------+-----------+-----------------------+
......
| emp              | 127.0.0.1 | caching_sha2_password |
| emp              | localhost | caching_sha2_password |
......
+------------------+-----------+-----------------------+
9 rows in set (0.00 sec)
```

结果显示，MySQL 8.0 默认密码插件为 caching_sha2_password，MySQL 8.0 以前的密码插件为 mysql_native_password。这样就会出现一个问题，当 8.0 以后的客户端连接 8.0 以前的服务器时，没有问题，因为新客户端可以理解新旧两种加密算法。但是反过来，当 8.0 以前的客户端需要连接 8.0 以后的服务器时，由于无法理解新的密码算法，发到服务器端的密码还是旧的算法加密后的结果，于是导致在新的服务器上出现下面无法认证的情况：

```
shell> mysql -h localhost -u root
ERROR 2059 (HY000): Authentication plugin 'caching_sha2_password' cannot be loaded: /usr/local/
mysql/lib/plugin/caching_sha2_password.so: cannot open shared object file: No such file or directory
```

对于这个问题，可以采用以下办法解决。

（1）在服务器端用 alter user 命令修改密码的加密方式，客户端可以进行正常连接：

```
mysql> alter user 'user'@'host' identified with mysql_native_password BY 'password';
```

（2）在 my.cnf 的[mysqld]中增加默认密码认证插件参数并重启服务器，这样新的数据库连接成功之后做的 grant 操作后生成的新密码全部变成旧的密码格式。

```
default_authentication_plugin=mysql_native_password
```

> **注意**：这个参数只是为了支持 8.0 版本前的客户端才进行设置，但是这将使得新建或者修改的用户密码全部变成旧的格式，降低了系统的安全性。

26.3.2 safe-user-create

此参数如果启用，用户将不能用 GRANT 语句创建新用户，除非用户有 mysql 数据库中 user 表的 INSERT 权限。如果想让用户具有授权权限来创建新用户，应给用户授予下面的权限：

```
mysql > GRANT INSERT(user) ON mysql.user TO 'user_name'@'host_name';
```

这样确保用户不能直接更改权限列，必须使用 GRANT 语句给其他用户授予该权限。以下例子描述了这个过程。

26.3 其他安全设置选项

(1) 用 root 创建用户 emp，emp 可以将权限授予其他用户：

```
[mysql3307@localhost ~]$ mysql -uroot
Welcome to the MySQL monitor.  Commands end with ; or \g.
Your MySQL connection id is 64
Server version: 5.7.22-log MySQL Community Server (GPL)

mysql>grant select,insert on employees.* to emp@localhost with grant option;
Query OK, 0 rows affected, 2 warnings (0.00 sec)
```

(2) 使用 emp 创建新用户成功：

```
mysql>grant select on employees.* to emp@localhost;
Query OK, 0 rows affected, 1 warning (0.00 sec)

mysql> exit
```

(3) 用 safe-user-create 选项重启数据库：

```
[mysql3307@localhost bin]# ./mysqld_safe --safe-user-create &
[1] 32422
[mysql3307@localhost bin]# Starting mysqld daemon with databases from /var/lib/mysql
```

(4) 重新用 emp 创建用户失败：

```
[root@localhost bin]# mysql -uemp
Welcome to the MySQL monitor.  Commands end with ; or \g.
Your MySQL connection id is 2
Server version: 5.7.22-community-log MySQL Community Edition (GPL)

Type 'help;' or '\h' for help. Type '\c' to clear the buffer.

mysql> grant select on employees.* to 'emp1'@'192.168';
ERROR 1410 (42000): You are not allowed to create a user with GRANT
mysql> exit
```

(5) 用 root 登录，给 emp 赋予 mysql 数据库中 user 表的 insert 权限：

```
[root@localhost bin]# mysql -uroot
Welcome to the MySQL monitor.  Commands end with ; or \g.
Your MySQL connection id is 5
Server version: 5.7.22-community-log MySQL Community Edition (GPL)

Type 'help;' or '\h' for help. Type '\c' to clear the buffer.

mysql> grant insert on mysql.user to emp@localhost;
Query OK, 0 rows affected (0.00 sec)

mysql> exit
Bye
```

(6) 用 emp 重新登录，授权用户成功：

```
[root@localhost bin]# mysql -uemp
Welcome to the MySQL monitor.  Commands end with ; or \g.
Your MySQL connection id is 7
Server version: 5.7.22-community-log MySQL Community Edition (GPL)

Type 'help;' or '\h' for help. Type '\c' to clear the buffer.

mysql> grant select on employees.* to 'emp1'@localhost;
Query OK, 0 rows affected (0.00 sec)
```

26.3.3 表空间加密

从 MySQL 5.7 开始，InnoDB 表空间加密功能为每个表的表空间和一般表空间（general tablespace）提供静态数据加密功能。表空间加密使用双层加密密钥体系结构，包括主加密密钥和表空间密钥。当表空间被加密时，表空间密钥被加密并存储在表空间头部。当应用程序或经过身份验证的用户想要访问加密的表空间数据时，InnoDB 使用主加密密钥来解密表空间密钥。

从 MySQL 8.0 开始，又提供了重做日志（redo log）和回滚日志（undo log）的加密。与表空间数据一样，当重做日志和回滚日志数据写入磁盘时，将数据进行加密，而从磁盘读取日志数据时再进行解密，当日志数据被载入内存以后，数据就是已经被解密的了。如果要启用重做日志和回滚日志加密，可以配置 innodb_redo_log_encrypt 和 innodb_undo_log_encrypt 参数，加密生效后，只对新的日志有效，而旧的日志数据将保持原样不变，即不再被重新加密。

开启表空间加密将大大增强数据文件和日志文件的安全性。

26.3.4　skip-grant-tables

skip-grant-tables 选项导致服务器根本不使用权限系统，从而给每个人以完全访问所有数据库的权力。通过执行 mysqladmin flush-privileges 或 mysqladmin reload 命令，或执行 flush privileges 语句，可以让一个正在运行的服务器再次开始使用授权表。

下面的例子演示了此参数的使用。

- 使用--skip-grant-tables 启动数据库。

```
[mysql3307@localhost ~]$ mysqld_safe --skip-grant-tables &
[1] 15298
[mysql3307@localhost ~]$ Starting mysqld daemon with databases from /var/lib/mysql
```

- 此时没有权限的用户可以直接登录，而不需要密码。

```
[mysql3307@localhost ~]$ mysql -uemp_test
Welcome to the MySQL monitor.  Commands end with ; or \g.
Your MySQL connection id is 9
Server version: 5.7.22-log MySQL Community Server (GPL)

emp_test@localhost:mysql_3307.sock   [(none)]>
```

- 此时执行 flush privileges 命令，重新使用权限系统。

```
mysql>flush privileges;
Query OK, 0 rows affected (0.01 sec)
```

- 退出后再次登录，将无法登录成功。

```
[mysql3307@localhost ~]$ mysql -uemp_test
ERROR 1045 (28000): Access denied for user 'emp_test'@'localhost' (using password: YES)
```

26.3.5　skip-networking

在网络上不允许 TCP/IP 连接，所有到数据库的连接必须经由命名管道（Named Pipe）、共享内存（Shared Memory）或 UNIX 套接字（SOCKET）文件进行。这个选项适合应用和数据库共用一台服务器的情况，其他客户端将无法通过网络远程访问数据库，这样大大增强了数据库的安全性，但同时也带来了管理维护上的不方便，来看下面的例子。

- 服务器上打开此选项（默认关闭）并重启 MySQL 服务。

```
[mysqld]
skip-networking
port             = 3307
…
```

- 远程客户端进行连接。

```
[mysql3307@localhost ~]$ mysql -emp -h192.168.7.55 -P3307 -p
Enter password:
ERROR 2003 (HY000): Can't connect to MySQL server on '192.168.7.55' (111)
```

- 关闭此选项后重启服务器。

```
[mysqld]
#skip-networking
port            = 3307
...
```

- 远程客户端进行连接。

```
[mysql3307@localhost ~]$ mysql -emp -h192.168.7.55 -P3307 -p
Enter password:
Welcome to the MySQL monitor.  Commands end with ; or \g.
Your MySQL connection id is 4
Server version: 5.7.22-log MySQL Community Server (GPL)
```

26.3.6 skip-show-database

使用 skip-show-database 选项，只允许有 show databases 权限的用户执行 show databases 语句，该语句显示所有数据库名。不使用该选项，允许所有用户执行 show databases，但只显示用户有 show databases 权限或部分数据库权限的数据库名。下面的例子显示了启用此选项后 show databases 的执行结果：

```
[mysql3307@localhost ~]$ mysqld_safe --skip-show-database &
[1] 15027
[mysql3307@localhost ~]$ Starting mysqld daemon with databases from /var/lib/mysql
[mysql3307@localhost ~]$ mysql -uemp -h127.0.0.1 -P3307 -p
Enter password:
Welcome to the MySQL monitor.  Commands end with ; or \g.
Your MySQL connection id is 4
Server version: 5.7.22-log MySQL Community Server (GPL)

emp@127.0.0.1:3307  [(none)]>show databases;
ERROR 1227 (42000): Access denied; you need (at least one of) the SHOW DATABASES privilege(s) for this operation
```

26.4 小结

权限和安全问题在任何数据库中都是非常重要的。本章重点介绍了 MySQL 中的权限管理以及可能存在的一些安全问题，并给出了很多例子加以详细说明。最后还讨论了 MySQL 提供的一些安全方面的参数，用户可以根据实际情况选择使用。

第 27 章 MySQL 监控

随着企业的发展壮大,在线服务器的数量越来越多,软硬件故障发生的概率也随之越来越高。这时,如果没有一个足够完善和强大的监控系统,当主机发生异常时,则可能无法及时发现和处理,从而造成业务的中断。尤其对于企业的核心业务系统,这将是不可承受之重。因此,一个好的监控系统对企业来说正变得越来越重要。

27.1 如何选择一个监控方案

在这里主要提出对两个方面的思考:第一,应该选择何种监控方式;第二,如何选择适合自己的监控工具。

27.1.1 选择何种监控方式

常见的监控方式主要有以下 3 种:
(1)自己写程序或者脚本进行监控;
(2)监控采用商业解决方案;
(3)监控开源软件方案。

第一种方案,当机器很少时,可以通过写程序或者编写脚本的方式监控线上的服务器,但是随着业务量的增大,业务上需要监控的点变得越来越细化,需要部署的脚本也越来越多,通过脚本进行监控的方法基本上无法满足业务的需求,并且脚本的后期维护成本也很大。常常会看到企业开始初期,服务器上布满了大大小小的监控脚本。

第二种方案,选择一个商业的解决方案,通过第三方为企业实现一套完整的监控系统。采用这种方案的优点是能在短时间内搭建一套监控平台,并且平台有很受欢迎的展现方式,如报表、美观的用户界面等,但是也存在一定的缺点,比如说这套平台需要花费很高的成本,并且随着业务的发展可能需要监控的粒度越来越细,当前的系统无法更好地去扩展。

第三种方案,选择一套已有的开源工具,通过开源工具对企业的生产系统进行监控,选择开源工具的优势包括完全免费、定制能力强、完全可控、集中化管理、可视性好。但是开源工具也有一定的缺点,比如说需要花费大量时间阅读相关文档。

27.1.2 如何选择适合自己的监控工具

出于成本考虑,最终确定了开源的解决方案,但是在软件选型时我们需要选择的监控系统软件具备以下几项功能。

(1)监控系统必须具有对主机的监控,包括对主机的 CPU、内存、网络、整体负载、相关进程数的监控等。

(2)监控系统必须具有对数据库的监控,主要包括对数据库的一些自身的性能指标进行监控,比如缓冲池的命中率、连接数等。

(3)监控系统要做到监控的实时性,监控系统需要具有相关触发报警的功能,当主机或者数据库发生异常时,要在第一时间进行短信报警、邮件报警等。

(4)在数据的表现形式上来讲,监控系统需要具备良好的图形展示的功能,当数据库发生异常时,数据库管理员能根据监控系统中异常时刻的趋势图迅速地定位到故障的产生原因。

(5)在协议的支持方面,监控系统客户端需要支持现有协议,如 IPMI、SNMP 等协议。

(6)在数据存储方面,要充分了解监控系统监控数据的存取,包括以文件的形式存储、用数据库的方式进行存储等,在部署监控系统时提前对数据进行规划。

(7)监控系统的部署和配置的复杂度、界面的友好性以及对中文的支持。

目前常见的开源监控系统包括 Open-Falcon、Nagios、Zabbix、Ganglia 等很多种,那么,这么多开源的监控系统,我们到底该选哪种呢?

还是要再次明确监控系统选择的目标:我们需要一套既能灵活地完成服务器各种监控信息的采集、分析、存储,又能支持快速的报警和信息发送的软件。那么上述哪些开源软件具备这些功能呢?

27.2 常用的网络监控工具

下面将分别对 Open-Falcon、Nagios、Zabbix 这几个开源的网络监控工具进行简单的介绍。

27.2.1 Open-Falcon 简介

Open-Falcon 是小米公司根据 SRE、SA、DEVS 的运维经验和反馈,结合业界众多互联网公司的监控经验而设计的一个分布式的监控系统。Open-Falcon 后端使用 Go 语言开发,可自动发现并采集安装了 Agent 的机器的各种数据和指标,主动上报至 Server 端,不需要用户在 Server 做任何配置,具有强大的可扩展性和灵活的数据采集能力。

Open-Falcon 的架构如图 27-1 所示。

在图 27-1 中,在每台被监控的机器安装 falcon-agent 代理,falcon-agent 会自动采集监控数据和指标,将数据上传至 Transfer。Transfer 接收到数据后,会检查和规整数据,再转发到多个后端系统去处理。当转发到每个后端业务系统,Transfer 根据一致性 hash 算法,进行数据分片,来达到后端业务系统的水平扩展。Transfer 将数据转发到 Graph 组件,Graph 收到数据后,以 Rrdtool 的数据归档方式来存储,同时,Transfer 也将数据转发到 Judge,数据到达 Judge 后,会触发相关策略的判定,来决定是否满足报警条件。如果满足条件,则会发送给

Alarm，Alarm 再以邮件、短信、米聊等形式通知用户。

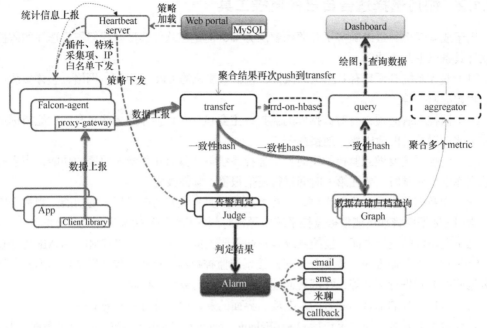

图 27-1　Open-Falco 架构图

用户可以在 Dashboard 仪表盘查看关心的数据，Query 组件通过 Graph 的 rpc 接口查询数据。由于 Rrdtool 在处理监控指标方面的效率非常高，即使查询一年的监控数据也能在秒级返回结果。

Heartbeat Server 定期加载 MySQL 中的内容，分发 Agent 要执行的插件、要监控的进程、端口；Judge 也会定期和 Heartbeat Server 保持沟通，来获取相关的报警策略。

Portal 是一个 Web 项目，无状态，支持水平扩展。用户可以在 Portal 配置报警策略，并在 MySQL 中存储报警策略。

Open-Falcon 具备以下几项监控功能。

（1）网络监控。

（2）主机系统监控。
- 网络接口流量（进出口网卡流量）。
- 监控 CPU 负载、内存使用情况等。
- 监控磁盘的空间、磁盘使用率等。
- 内核参数、NTP 偏移等。

（3）Open-Falcon 常见的检测对象。
- 服务器资源：CPU、磁盘、内存、进程、端口等。
- 服务器类型：Web、JVM、FTP、数据库、中间件。
- 操作系统：Linux、Windows。
- 网络接口：流量、转发速度、丢包率。
- 设备运行状态：风扇、电源、温度。
- 机房运行环境：电流、电压、温湿度。

27.2.2 Nagios 简介

Nagios 是一款用于系统和网络监控的应用程序。它可以在用户设定的条件下对主机和服务进行监控，当状态改变时发出相关告警信息。

Nagios 通常由 1 个主程序（Nagios）、1 个插件程序（Nagios-plugins）和 4 个可选的附件组件（NRPE、NSCA、NSClient++和 NDOUtils）组成。Nagios 的监控工作都是通过插件功能实现的，因此，Nagios 和 Nagios-plugins 是服务器端工作所必需的组件。4 个附件组件中的功能如下：

● NRPE：用来在监控的远程 Linux/UNIX 主机上执行脚本插件，以实现对这些主机资源的监控。
● NSCA：用来让被监控的远程 Linux/UNIX 主机主动将监控信息发送给 Nagios 服务器。
● NSClient++：安装在 Windows 主机上的组件，主要用来监控 Windows 主机。
● NDOUtils：用来将 Nagios 的配置信息和各 Event 产生的数据存入数据库，以实现这些数据的快速检索和处理。

这 4 个附件组件中，NRPE 和 NSClient++工作于客户端，NDOUtils 工作于服务器端，而 NSCA 则需要同时安装在服务器端和客户端。

Nagios 的工作原理如图 27-2 所示。

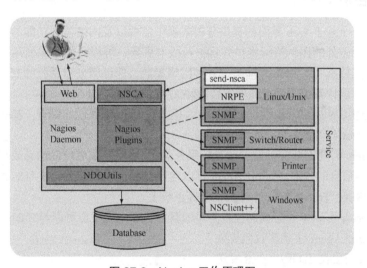

图 27-2　Nagios 工作原理图

Nagios 的功能是监控服务和主机，但是它自身并不包括这部分功能，所有的监控、检测功能都是通过各种插件来完成的。启动 Nagios 后，它会周期性地自动调用插件去检测服务器状态，同时 Nagios 会维持一个队列，所有插件返回来的状态信息都进入队列，Nagios 每次都从队首开始读取信息，并进行处理后，把状态结果通过 Web 显示出来。Nagios 提供了许多插件，利用这些插件可以方便地监控很多服务状态。安装完成后，在 Nagios 主目录下的/libexec 里放有 Nagios 自带的可以使用的所有插件，如 check_disk 是检查磁盘空间的插件，check_load 是检查 CPU 负载的，等等。每一个插件可以通过运行./check_xxx –h 来查看其使用方法和功能。Nagios 可以识别 4 种状态的返回信息，即 0（OK）表示状态正常，1（WARNING）表示出现一定的异常，2（CRITICAL）表示出现非常严重的错误，3（UNKNOWN）表示被监控的对象已经停止了。Nagios 根据插件返回来的值来判断监控对象的状态，并通过 Web 显示出来，以供管理员及时发现故障。

Nagios 支持的监控功能如下：

（1）网络监控服务（SMTP、POP3、HTTP、NNTP、PING 等）；
（2）监控主机资源（处理器负荷、磁盘利用等）；
（3）插件设计使得用户可以方便地扩展自己所需要定制的监控项；
（4）并行服务检测机制；
（5）具备定义网络分层结构的能力，用"parent"主机定义来表达网络主机间的关系，这种关系可被用来发现和明晰主机宕机或不可达状态；
（6）具有快速的消息通知功能，当服务或者主机产生问题时能及时地将告警发送给相关业务负责人（可以通过 E-mail、短信、用户定义方式），可高效地保证服务器的维护；
（7）具备定义事件句柄功能，它可以在主机或服务的事件发生时获取更多问题定位；
（8）自动的日志回滚；
（9）可以支持并实现对主机的冗余监控；
（10）友好的 Web 界面用于查看当前的网络状态、通知和故障历史、日志文件等。

27.2.3　Zabbix 简介

Zabbix 是一个基于 Web 界面的提供分布式系统监视以及网络监视功能的企业级的开源解决方案，由一个国外的团队持续维护并且进行版本更新，可以自由下载使用，运作团队靠提供收费的技术支持赢利。Zabbix 能监视各种网络参数，保证服务器系统的安全运营，并提供灵活的通知机制以让系统管理员快速定位/解决存在的各种问题。

Zabbix 通过 C/S 模式采集数据，通过 B/S 模式在 Web 端展示和配置。Zabbix 由两部分构成：Zabbix Server 与可选组件 Zabbix Agent。

Zabbix Agent 需要安装在被监视的目标服务器上，主要完成对硬件信息或与操作系统有关的内存、CPU 等信息的收集。Zabbix Agent 可以运行在 Linux、Solaris、HP-UX、AIX、Free BSD、Open BSD、OS X、Tru64/OSF1、Windows 等系统上。

Zabbix Server 可以单独监视远程服务器的服务状态，同时也可以与 Zabbix Agent 配合，可以轮询 Zabbix Agent 主动接收监视数据，同时还可被动接收 Zabbix Agent 发送的数据，通过收集 SNMP 和 Agent 发送的数据，写入 MySQL 数据库，再通过 Apache 等软件在 Web 前端显示，Zabbix Server 需运行在 LAMP（Linux+Apache+MySQL+PHP）环境下，对硬件要求低。

Zabbix 的工作原理如图 27-3 所示。

Zabbix Agent 负责数据收集操作，将定制的数据传送到 Zabbix Server，Zabbix Server 会把相关数据存入到 MySQL 数据

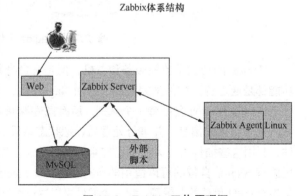

图 27-3　Zabbix 工作原理图

库中。最终用户通过 Zabbix Web 端查看数据，并且 Zabbix Web 端具备报警等功能。

Zabbix 的主要特点如下：

- 安装与配置简单，学习成本低；
- 支持多语言（包括中文）；
- 免费开源；
- 自动发现服务器与网络设备；
- 分布式监视以及 Web 集中管理功能；
- 可以无 Agent 监视；
- 用户安全认证和柔软的授权方式；
- 通过 Web 界面设置或查看监视结果；
- E-mail 等通知功能。

Zabbix 具有以下几项功能：

- 具备常见的商业监控软件所具备的功能（主机的性能监控、网络设备性能监控、数据库性能监控、FTP 等通信协议的监控、多种告警方式、详细的报表图表绘制）。
- 支持自动发现网络设备和服务器。
- 支持分布式，能集中显示、管理分布式的监控点。
- 扩展性强，Server 提供通用接口，可以自己开发完善各类监控。

27.2.4 几种常见开源软件比较

Open-Falcon 和 Nagios、Zabbix 的优缺点对比如下。

- Open-Falcon：采集数据灵活。安装 Agent 后，不需要配置 Server 端，即可自动发现并采集数据；整个系统无核心单点，可以水平扩展，支持每个周期上亿次的数据采集；单机支持每分钟 200 万 metric 的上报和存储；采用 rrdtool 的数据归档策略，秒级返回上百个 metric 一年的历史数据。它的缺点是组件太多，部署和监控有点烦琐；graph 集群缩扩容时，历史数据有损，历史数据无法迁移；系统权限控制不足，数据安全性较弱。
- Nagios：适合监视大量服务器上面的大批服务是否正常，重点并不在图形化的监控，其集成的很多功能，例如报警，在绘图以及图形塑造方面精细度比较弱。
- Zabbix：最大的优点是开源，无软件成本投入；对 Server 端的设备性能要求低，支持设备多，支持分布式集中管理，开放式接口，扩展性强，并且第三方插件 percona-zabbix-templates 专门定制了对数据库的监控。其缺点是数据量太大时，对于数据库清理不是很方便。

通过以上论述，如果监控机器数量中等，对安全性要求高，可以采用 Zabbix+percona-zabbix-templates 插件对数据库进行监控；如果监控机器数量多，监控系统放在内网，没有安全性顾虑，可以采用 Open-Falcon+mymon 对数据库进行日常监控。

出于篇幅考虑，本章将会详细介绍 Zabbix+percona-zabbix-templates 插件的使用。

27.3 Zabbix 部署

Zabbix 监控环境部署如表 27-1 所示。

表 27-1　　　　　　　　　　　　　Zabbix 监控环境部署

角　色	IP 地址	主机名字	操作系统	用途
Zabbix Server	192.168.8.81	ip81	CentOS 6	Zabbix 服务端
Zabbix Agent	192.168.7.83	ip83	CentOS 6	MySQL Server

27.3.1　Zabbix Server 软件安装

Zabbix Server 软件安装大概需要以下几个步骤。

（1）安装 LAMP 环境：

```
[root@ip81 ~]#yum install mysql-server httpd php
```

（2）安装 Zabbix Web 所需的依赖包：

```
[root@ip81 ~]#yum install gcc gcc-c++ autoconf php-mysql httpd-manual mod_ssl mod_perl mod_auth_mysql php-gd php-xml php-mbstring php-ldap php-pear php-xmlrpc php-bcmath mysql-connector-odbc mysql-devel libdbi-dbd-mysql net-snmp-devel curl-devel unixODBC-devel OpenIPMI-devel java-devel libssh2-devel.x86_64 openldap openldap-devel
```

（3）创建 Zabbix 运行的用户：

```
[root@ip81 ~]#groupadd zabbix
[root@ip81 ~]#useradd -g zabbix zabbix
```

（4）安装 Zabbix Server：

```
[root@ip81~]#wget -c https://sourceforge.net/projects/zabbix/files/ZABBIX%20Latest%20Stable/3.4.12/zabbix-3.4.12.tar.gz/download
[root@ip81 ~]#tar zxvf zabbix-3.4.12.tar.gz
[root@ip81 ~]#cd zabbix-3.4.12
[root@ip81 ~]#export  MYSQL_HOME=/home/mysql/mysqlhome
[root@ip81 ~]#export C_INCLUDE_PATH=$MYSQL_HOME/include
[root@ip81 ~]#export LD_LIBRARY_PATH=$MYSQL_HOME/lib
[root@ip81 ~] #./configure --prefix=/usr/local/zabbix --enable-server --enable-agent --enable-proxy --with-mysql=/usr/bin/mysql_config --with-net-snmp --with-libcurl --with-openipmi --with-unixodbc --with-ldap --with-ssh2 --enable-java
[root@ip81 ~]#make && make install
```

27.3.2　Zabbix Server 配置与启动

Zabbix Server 安装配置和启动大概需要以下几个步骤。

（1）创建 Zabbix 数据库和 MySQL 用户：

```
mysql> create database zabbix character set utf8;
Query OK, 1 row affected (0.05 sec)

mysql> create user 'zabbix'@'%' identified by '123456';
Query OK, 0 rows affected (0.16 sec)

mysql> grant all on zabbix.* to 'zabbix'@'%';
Query OK, 0 rows affected (0.03 sec)

mysql> flush privileges;
Query OK, 0 rows affected (0.04 sec)
```

（2）导入 Zabbix 数据库初始数据：

```
[root@ip81 ~]#cd database/mysql/
[root@ip81 ~]#mysql -S /tmp/mysql.sock zabbix < schema.sql
[root@ip81 ~]#mysql -S /tmp/mysql.sock zabbix < images.sql;
[root@ip81 ~]#mysql -S /tmp/mysql.sock zabbix < data.sql;
```

（3）配置 Zabbix 配置文件。

编辑 Zabbix Server 的配置文件/usr/local/zabbix/etc/zabbix_server.conf，修改以下参数：

```
ListenPort=10051
LogFile=/usr/local/zabbix/logs/zabbix_server.log
PidFile=/usr/local/zabbix/logs/zabbix_server.pid
DBHost=192.168.7.81
DBName=zabbix
DBUser=zabbix
DBPassword=123456
DBPort=3306
DBSocket=/tmp/mysql.sock
```

（4）配置 Zabbix 服务。

从安装目录复制 zabbix_server 脚本并编辑：

```
[root@ip81 ~]#cd zabbix-3.4.12
[root@ip81 ~]#cp misc/init.d/fedora/core/zabbix_server /etc/init.d/
[root@ip81 ~]#mkdir -p /usr/local/zabbix/logs
[root@ip81 ~]#chown -R zabbix:zabbix /usr/local/zabbix
[root@ip81 ~]#vi /etc/init.d/zabbix_server
[root@ip81 ~]#cat /etc/init.d/zabbix_server
…
BASEDIR=/usr/local/zabbix
…
}
```

（5）开启 Zabbix 安全限制。

调整防火墙规则（开放端口 10051）：

```
[root@ip81 ~]#iptables -A INPUT -p tcp -m tcp --dport 10051 -j ACCEPT
[root@ip81 ~]#service iptables save
Saving firewall rules to /etc/sysconfig/iptables: [  OK  ]
```

（6）启动 Zabbix Server：

```
[root@ip81 ~]#service zabbix_server start
Starting Zabbix Server:                                    [  OK  ]
```

（7）停止 Zabbix Server：

```
[root@ip81 ~]#service zabbix_server stop
Stopping Zabbix Server:                                    [  OK  ]
```

（8）配置 Zabbix Server 开机自动启动：

```
[root@ip81 ~]#chkconfig --add zabbix_server
[root@ip81 ~]#chkconfig --level 35 zabbix_server on
```

27.3.3　配置 Zabbix Web 服务端

配置 Zabbix Web 服务大概需要以下几个步骤。

（1）将 Zabbix Web 文件复制到 Apache Web 目录中。

将安装目录中的 frontends 复制到指定的 Web root 目录中：

```
[root@ip81 ~]#cd zabbix-3.4.12
[root@ip81 ~]#cp -ra frontends/php/* /var/www/html/
[root@ip81 ~]#chown -R apache.apache /var/www/html/
```

（2）Apache 配置：

```
[root@ip81 ~]#cat  /etc/httpd/conf/httpd.conf
…
ServerName 192.168.7.81:80
…
```

（3）PHP 配置。

/etc/php.ini 配置如下，注意这里必须这样配置，否则 Web 界面安装检查会失败。

```
…
date.timezone = Asia/Shanghai
memory_limit = 128M
post_max_size = 16M
max_execution_time = 300
max_input_time = 300
session.auto_start = 0  ;
mbstring.func_overload = 2
…
```

开启 httpd 服务：

```
[root@ip81 ~]# service httpd start
```

配置 httpd 开机自动启动：

```
[root@ip81 ~]# chkconfig httpd on
```

（4）Zabbix Web 安装。

访问 Web 界面 http://192.168.7.81，进行 Zabbix 相关的 Web 配置，配置完成后使用默认用户名为 admin（密码为 zabbix）登录即可。Zabbix 的安装界面如图 27-4 所示。

安装过程中的检查信息如图 27-5 所示，安装后的统计信息如图 27-6 所示，安装成功后的登录窗口如图 27-7 所示。

图 27-4　Zabbix 安装界面

图 27-5　Zabbix 安装前的环境检查

图 27-6　Zabbix 安装成功输出

图 27-7　Zabbix 安装成功后的登录窗口

注意这里默认密码是"zabbix"，进入系统第一件事就是修改 Zabbix 的后台默认密码，可以通过 Administration→Users→Admin→Change Password 来修改。

27.3.4 Zabbix Agent 安装和配置

安装 Zabbix Agent 大概需要以下几个步骤。

（1）下载安装 Zabbix Agent 软件：

```
[root@ip83~]#wget -c https://sourceforge.net/projects/zabbix/files/ZABBIX%20Latest%20Stable/3.4.12/zabbix-3.4.12.tar.gz
[root@ip83~]#tar zxvf zabbix-3.4.12.tar.gz
[root@ip83~]#cd zabbix-3.4.12
[root@ip83~]#./configure --prefix=/usr/local/zabbix --enable-agent
[root@ip83~]#make && make install
[root@ip83~]#cp misc/init.d/fedora/core/zabbix_agentd /etc/init.d/
```

（2）配置 zabbix_agentd：

```
[root@ip83~]#groupadd zabbix
[root@ip83~]#useradd -g zabbix zabbix
[root@ip83~]#mkdir -p /usr/local/zabbix/logs
[root@ip83~]#chown zabbix:zabbix -R /usr/local/zabbix/
[root@ip83~]#vi /usr/local/zabbix/etc/zabbix_agentd.conf
[root@ip83~]#cat /usr/local/zabbix/etc/zabbix_agentd.conf
…
PidFile=/usr/local/zabbix/logs/zabbix_agentd.pid
LogFile=/usr/local/zabbix/logs/zabbix_agentd.log
Server=192.168.7.81,127.0.0.1
ListenPort=10050
ServerActive=192.168.7.81,127.0.0.1
Hostname=zabbix_agent83
Timeout=15
Include=/usr/local/zabbix/etc/zabbix_agentd.conf.d/
…
```

为了在本机测试 zabbix_get 命令，Server 和 ServerActive 参数配置 127.0.0.1 的 ip 地址，否则会报错：

```
zabbix_get [6248]: Check access restrictions in Zabbix agent configuration
```

（3）配置 Zabbix Agent 系统服务启动脚本：

```
cat /etc/init.d/zabbix_agentd
…
# Zabbix-Directory
BASEDIR=/usr/local/zabbix
…
```

（4）防火墙设置。开启防火墙端口 10050：

```
[root@ip83~]iptables -A INPUT -p tcp -m tcp --dport 10050 -j ACCEPT
[root@ip83~]service iptables save
```

（5）启动 zabbix_agentd：

```
[root@ip83~]#/etc/init.d/zabbix_agentd start
Starting Zabbix Agent:                              [ OK ]
```

（6）配置开机自动启动：

```
[root@ip83~]#chkconfig --add zabbix_agentd
[root@ip83~]#chkconfig --level 35 zabbix_agentd on
```

（7）测试 Zabbix Agent。测试 Zabbix Agent 是否正常工作：

```
[root@ip83~]# /usr/local/zabbix/sbin/zabbix_agentd -c /usr/local/zabbix/etc/zabbix_agentd.conf -t system.uptime
system.uptime                               [u|37643579]
```

27.3.5 PMP 插件介绍和部署

PMP（Percona Monitoring Plugins for Zabbix）是一套比较完整的监控 MySQL 数据库的 Zabbix 插件。该插件中包括丰富的监控项和触发器规则，通过和 Zabbix 客户端整合，可以完美地实现对 MySQL 数据库的监控。

Zabbix 使用 PMP 插件后的监控结构，如图 27-8 所示。

图 27-8 显示了 PMP 作为 Zabbix 插件监控 MySQL 的整体架构，首先 Zabbix 客户端 zabbix_agentd 会读取

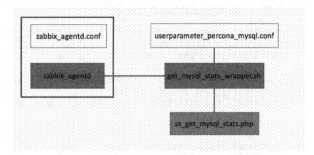

图 27-8 Zabbix 和 PMP 实现对 MySQL 监控

配置文件 zabbix_agentd.conf，zabbix_agentd.conf 则会根据设置载入 userparameter_percona_mysql.conf 文件，将 Zabbix 客户端和 PMP 插件实现挂接。PMP 的主要功能是完成对数据库的相关信息收集和上传操作。对数据库信息的收集操作，主要通过 zabbix_agentd 传入的 zabbix Key，在 userparameter_percona_mysql.conf 文件来读取相关配置信息，在收集过程中会调用插件自身的 get_mysql_stats_wrapper.sh，通过 ss_get_mysql_stats.php 收集数据，最终通过调用 zabbix_sender 实现上传。

部署 PMP 插件大概需要以下几个步骤。

（1）安装和部署 Zabbix Server 软件、Zabbix Web。
（2）下载安装 PMP 软件包及其依赖包。
（3）通过 Zabbix Web 导入 PMP 中所需要的模板文件。
（4）在 Zabbix Web 端创建 Host Group（主机组）。
（5）在 Zabbix Web 端创建 Host（注意 Web 端配置的 Hostname 必须与 zabbix Agent 配置中的 hostname 一致）。
（6）将模板关联到所创建的 Host。
（7）安装和配置 PMP Agent，将 PMP 整合到 Zabbix 中。
（8）重启 Zabbix 客户端服务。

本章的开头部分已经介绍了 Zabbix Server 和客户端的部署，后面关于这两部分的内容将略过。

1. Percona-zabbix-templates 下载及其依赖安装

安装依赖的相关 php 包：

```
[root@ip83~]#  yum install -y php.x86_64 php-mysql.x86_64  php-mysql
```

下载 rpm 包并安装：

```
[root@ip83~]# wget https://www.percona.com/downloads/percona-monitoring-plugins/percona-monitoring-plugins-1.1.8/binary/redhat/6/x86_64/percona-zabbix-templates-1.1.8-1.noarch.rpm
[root@ip83~]#ls -l percona-zabbix-templates-1.1.8-1.noarch.rpm
-rw-r--r--. 1 root root 28960 Jan 10  2018 percona-zabbix-templates-1.1.8-1.noarch.rpm
[root@ip83~]# rpm -ivh percona-zabbix-templates-1.1.8-1.noarch.rpm
warning: percona-zabbix-templates-1.1.8-1.noarch.rpm: Header V4 DSA/SHA1 Signature, key ID cd2efd2a: NOKEY
Preparing...                ########################################### [100%]
   1:percona-zabbix-template########################################### [100%]
```

```
Scripts are installed to /var/lib/zabbix/percona/scripts
Templates are installed to /var/lib/zabbix/percona/templates
```

安装 rpm 包后，在/var/lib/zabbix/percona/scripts 生成脚本，Zabbix Agent 调用这些脚本来采集数据。

```
[root@ip83~]# ll /var/lib/zabbix/percona/scripts
total 64
-rwxr-xr-x. 1 root root  1251 Jan 10  2018 get_mysql_stats_wrapper.sh
-rwxr-xr-x. 1 root root 60679 Jan 10  2018 ss_get_mysql_stats.php
```

在/var/lib/zabbix/percona/templates 生成 xml 模板，需要在 Zabbix Web 端导入这个模板。

```
[root@ip83~]# ll /var/lib/zabbix/percona/templates
total 284
-rw-r--r--. 1 root root  18866 Jan 10  2018 userparameter_percona_mysql.conf
-rw-r--r--. 1 root root 269258 Jan 10  2018 zabbix_agent_template_percona_mysql_server_ht_
2.0.9-sver1.1.8.xml
```

2. Zabbix Web 端导入 xml 模板

将 zabbix_agent_template_percona_mysql_server_ht_2.0.9-sver1.1.8.xml 文件上传到 Windows。要导入模板，可以在 Zabbix Web 中选择 Configuration→Templates，然后在"Import"栏下单击"浏览"按钮，选中所需要的 XML 文件，单击"Import"按钮即可导入相应的监控模板，如图 27-9 所示。

> 注意：percona 官方提供 xml 模板，只适用于 Zabbix 2.0，导入 Zabbix 3.0 会报错。需要先将 xml 模板先导入 Zabbix 2.0，再选择导出模板，然后把导出的模板导入 Zabbix 3.0。

在 Zabbix Web 中，选择 Configuration→Templates，可以查看刚才导入的模板，如图 27-10 所示。

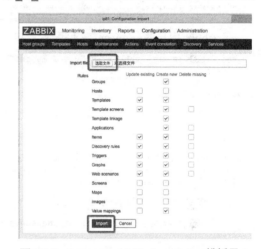

图 27-9 percona-zabbix-templates 模板导入

图 27-10 percona-zabbix-templates 模板查看

percona-zabbix-templates 中的模板带有丰富的 Item 项和 Trigger 项，基本上能覆盖 MySQL 日常运维中的常用监控项。

3. 修改 php 脚本参数

php 脚本通过 show 命令，获取 mysql 参数和状态。下面配置连接 mysql 的用户名、密码、端口和 soket 文件：

```
[root@ip83~]# cd /var/lib/zabbix/percona/scripts/
[root@ip83 scripts]# vi ss_get_mysql_stats.php
$mysql_user = 'root';
$mysql_pass = '123456';
$mysql_port = 3306;
$mysql_socket = '/tmp/mysql.sock';
```

执行命令测试 php 脚本能否采集数据：

```
[root@ip83~]# /usr/bin/php -q /var/lib/zabbix/percona/scripts/ss_get_mysql_stats.php --port 5306 --host localhost  --items mm
mm:71702
```

看到 mm:71702，表明 php 可以连接上 mysql 数据库了。

4．修改 sh 脚本参数

sh 脚本通过 show slave status 命令，获取 mysql 从库 io 进程和 sql 进程状态；通过 php 脚本获取 mysql 的其他状态。在 HOME=~zabbix mysql 之后添加连接 mysql 的用户名和 socket 文件：

```
[root@ip83~]# cd /var/lib/zabbix/percona/scripts/
[root@ip83 scripts]# vi get_mysql_stats_wrapper.sh
RES=`HOME=~zabbix mysql -uroot -S/tmp/mysql.sock -e
```

在 zabbix 用户的~/.my.cnf 文件配置连接 mysql 的密码：

```
[root@ip83~]# su - zabbix -c "vi ~/.my.cnf"
[client]
user=root
password=123456
```

sh 脚本需要在 mysql_cacti_stats.txt 之后添加"分号+mysql 的端口号"，假设被监控的 MySQL 实例端口为 3306：

```
[root@ip83~]# cd /var/lib/zabbix/percona/scripts/
[root@ip83 scripts]# vi get_mysql_stats_wrapper.sh
/tmp/$HOST-mysql_cacti_stats.txt:3306
```

执行命令测试 sh 脚本能否采集数据：

```
[root@ip83~]# /var/lib/zabbix/percona/scripts/get_mysql_stats_wrapper.sh mm
71702
```

执行命令测试 sh 脚本能否采集 io 进程和 sql 进程的状态：

```
[root@ip83~]#/var/lib/zabbix/percona/scripts/get_mysql_stats_wrapper.sh running-slave
0
```

从测试结果来看，sh 脚本能正确采集到信息。

5．拷贝 conf 文件

将/var/lib/zabbix/percona/templates 下的 conf 文件复制到 Zabbix Include 指定的目录：

```
[root@ip83~]# cat /usr/local/zabbix/etc/zabbix_agentd.conf|grep Include|grep -v "#"
Include=/usr/local/zabbix/etc/zabbix_agentd.conf.d/
[root@ip83~]# cd /var/lib/zabbix/percona/templates
[root@ip83~]# cp userparameter_percona_mysql.conf /usr/local/zabbix/etc/zabbix_agentd.conf.d/
[root@ip83~]# ll /usr/local/zabbix/etc/zabbix_agentd.conf.d/
total 20
-rw-r--r--. 1 root root 18866 Aug 27 08:35 userparameter_percona_mysql.conf
```

conf 文件记录了 Zabbix 模板的 key 和查询脚本的对应关系，例如：

```
[root@ip83~]# head -n 1 userparameter_percona_mysql.conf
UserParameter=MySQL.Sort-scan,/var/lib/zabbix/percona/scripts/get_mysql_stats_wrapper.sh kt
```

如果采集项的 key 为"MySQL.Sort-scan"，那么 agent 执行的命令是/var/lib/zabbix/percona/

scripts/get_mysql_stats_wrapper.sh kt

```
[root@ip83~]# /var/lib/zabbix/percona/scripts/get_mysql_stats_wrapper.sh kt
86
```

重启 Zabbix Agentd，载入 conf 文件。

```
# /etc/init.d/zabbix_agentd restart
```

使用 zabbix_get 命令测试：

```
[root@ip83~]# /usr/local/zabbix/bin/zabbix_get  -s127.0.0.1 -p10050 -k"MySQL.Sort-scan"
86
```

结果显示，Zabbix Agent 采集数据与直接执行 sh 脚本的结果一致。

27.3.6 Zabbix Web 端操作

在 Zabbix Web 中，要实现对被监控主机的监控大概需要以下几个步骤。

（1）通过 Zabbix Web 端创建监控数据库组。

通过 Configuration→Host groups→Create host group→输入相关组名，如图 27-11 所示，创建"mysql database"组，该组主要是为了管理所有的 MySQL 数据库。

图 27-11 创建 Zabbix 监控组

（2）在 Zabbix 监控组中创建 Zabbix 监控主机。

创建 Zabbix 监控的主机，主机名为 zabbix_agent83。在 Configuration→Hosts→Create host 中输入相关主机信息，如图 27-12 所示。

> 注意：Zabbix Web 添加的 Host name，必须与被监控主机 zabbix_agentd.conf 配置的 Hostname 一致，否则 Zabbix 无法采集数据。

（3）添加被监控主机所需的模板文件，如图 27-13 所示。

图 27-12 创建 Zabbix 被监控主机信息

图 27-13 为监控的客户端配置模板文件

（4）查看监控端主机的监控状态，如图 27-14 所示。当主机的状态为 Enabled，Availability 变为绿色时，代表该主机监控正常。

图 27-14　查看被监控主机的监控状态

（5）在 Zabbix Web 中，选择 Monitoring→Latest data，可以查看监控的具体情况，如图 27-15 所示。

图 27-15　Zabbix 数据收集状况查看

通过相关输出信息，可以看见 Agent 端的数据库监控信息已经完全发送到了 Zabbix Server 端。

（6）查看 Zabbix 绘制图形情况，这里选取连接数的绘图，单击 Connections，如图 27-16 所示。

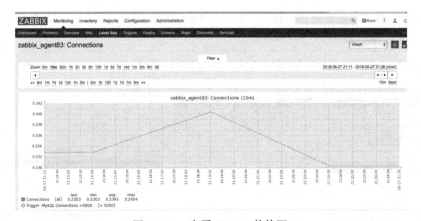

图 27-16　查看 Zabbix 趋势图

可以看到，在图 27-16 中，时间窗口可以任意拖动，窗口中显示最近一段时间的连接数变化趋势。

27.4 性能医生 orzdba

Zabbix 功能十分强大，可以监控数据库各个参数和性能指标，通过 Zabbix 可以查看较长时间的性能趋势图，但是 Zabbix 指标监控周期一般在 1~3min，线上数据库性能急剧下降或者主机负载突然升高时，通过 Zabbix 可能无法很快地诊断问题。我们需要一个监控周期更短，可以实时分析数据库的当前状态的工具，这时 orzdba 将是一个很好的选择。

orzdba 是淘宝工程师开发并开源的 mysql 监控工具，它使用 Perl 开发，用于实时收集 Linux 主机和 MySQL 的性能数据。由于要收集主机数据，需要在 MySQL 实例所在主机运行 orzdba。

orzdba 的用法如下：

```
shell> orzdba [options]...
```

option 有很多选项，常用的选项如下。

- -i,--interval: 收集信息间隔，单位为秒。
- -t,--time: 输出结果打印当前时间。
- -l,--load: 打印主机负载。
- -c,--cpu: 打印 cpu 信息。
- -s,--swap: 打印 swap 信息。
- -S,--socket: 指定 MySQL 实例的 socket 文件。
- -com: 打印每秒增删改查数量，即 TPS 和 QPS。
- -hit: 打印缓冲池命中率。
- -innodb_rows: 打印每秒增删改查的记录数。
- -lazy: 打印时间、主机负载、cpu 信息、swap 信息、TPS 和 QPS、缓冲池命中率。

27.4.1 orzdba 安装

由于 orzdba 用 Perl 语言开发，使用前需要安装 Perl 相关的包；orzdba 本身不用安装，下载即可使用。需要安装的 Perl 依赖包如下：

```
yum install perl-Test-Simple.x86_64
yum install perl-Time-HiRes
yum install perl-ExtUtils-CBuilder
yum install perl-ExtUtils-MakeMaker
yum install perl-DBD-MySQL
yum install perl-DBI
安装 Perl 的 version 模块、File:LockFfile 模块、Class-Data-Inheritable 模块和 Module-Build 模块
```

Perl 依赖包安装后，在运行 orzdba 的用户下配置数据库 root 用户的密码。例如对应的用户为 mysql，root@localhost 的密码为 123456，编辑.my.cnf 文件：

```
shell> vi /home/mysql/.my.cnf
[client]
user=root
password=123456
```

在/etc/hosts 配置本机 ip 和主机的对应关系，例如本机 ip 为 192.168.1.18，主机名为 testos，那么在/etc/hosts 文件中增加下面一行：

```
192.168.1.18    testos
```

27.4.2 orzdba 使用

完成配置后，orzdba 使用非常简单，指定收集信息的模式和间隔即可。下面以 lazy 模式，每秒收集系统和数据库的性能数据：

```
[mysql3307@localhost bin]$ perl orzdba -lazy -S /tmp/mysql_3391.sock -i 1
```

结果如图 27-17 所示，orzdba 首先打印 MySQL 实例的 DB 信息和一些重要参数，帮助用户初步了解 MySQL 的配置情况。接着收集信息，并格式化输出。

- time 列是收集信息的当前时间，间隔为 1s。
- load-avg 列下的 1m、5m、15m 分别对应操作系统 1min 内、5min 内和 15min 内的平均负载。从图 27-17 可知，15min、10min 和 1min 内负载逐渐升高，此时主机负载呈上升趋势。
- cpu-usage 列下的 usr、sys、idl、iow 分别对应操作系统的用户 cpu、系统 cpu、空闲 cpu 和 io 等待 cpu 占总 cpu 使用量的百分比。从图 27-17 可知，主要是用户使用 cpu 时间，使用率为 13%～17%。
- ins、upd、del 分别对应数据库每秒插入、更新和删除的 sql 数量。iud 为增删改的总和，即 TPS。在图 27-17 中，数据库的 TPS 为 8～12。

```
DB  : db_master1|employees|employees_hash|employees_range|ggtest|percona|performance_schema|proxy_hear
Var : binlog_format[ROW] max_binlog_cache_size[32M] max_binlog_size[1G]
      max_connect_errors[10000] max_connections[10050] max_user_connections[10000]
      open_files_limit[131072] sync_binlog[1] table_definition_cache[4096]
      table_open_cache[6144] thread_cache_size[256]

      innodb_adaptive_flushing[ON] innodb_adaptive_hash_index[ON] innodb_buffer_pool_size[5G]
      innodb_file_per_table[ON] innodb_flush_log_at_trx_commit[1] innodb_flush_method[O_DIRECT]
      innodb_io_capacity[4000] innodb_lock_wait_timeout[1] innodb_log_buffer_size[64M]
      innodb_log_file_size[1G] innodb_log_files_in_group[3] innodb_max_dirty_pages_pct[50.000000]
      innodb_open_files[6144] innodb_read_io_threads[8] innodb_thread_concurrency[0]
      innodb_write_io_threads[8]

-------- -----load-avg---- ---cpu-usage--- ---swap---                              -QPS- -TPS-         -Hit%-
  time  |  1m    5m   15m |usr sys idl iow|   si   so|  ins   upd   del    sel   iud|     lor    hit|
20:24:20| 1.19  1.05  0.67|  8   6  85   1|    0    0|    0     0     0      0     0|       0 100.00|
20:24:21| 1.19  1.05  0.67| 16   2  81   1|    0    0|    0    12     0     13    12|     221 100.00|
20:24:22| 1.19  1.05  0.67| 14   3  78   5|    0    0|    0     8     0      9     8|     147 100.00|
20:24:23| 1.18  1.05  0.67| 15   2  81   2|    0    0|    0     8     0      9     8|     381 100.00|
20:24:24| 1.18  1.05  0.67| 16   2  79   3|    0    0|    0    12     0     13    12|     206 100.00|
20:24:25| 1.18  1.05  0.67| 13   2  81   3|    0    0|    0     8     0      9     8|     151 100.00|
20:24:26| 1.18  1.05  0.67| 17   4  78   2|    0    0|    0    12     0     13    12|     205 100.00|
20:24:27| 1.18  1.05  0.67| 16   2  81   1|    0    0|    0     8     0      9     8|     151 100.00|
20:24:28| 1.16  1.05  0.67| 15   3  81   1|    0    0|    0    12     0      9    12|     444 100.00|
20:24:29| 1.16  1.05  0.67| 13   2  82   3|    0    0|    0     8     0     13     8|     156 100.00|
```

图 27-17 orzdba 实时监控

- sel 列对应数据库每秒查询的 sql 数量，即 QPS。在图 27-17 中，数据库的 QPS 为 9～13。
- lor 列对应每秒 innodb 缓冲池的读次数（逻辑读请求数）。
- hit 列对应每秒缓冲池的命中率。图 27-17 结果显示，innodb 缓冲池命中率都是 100%，表明数据库 innodb 缓冲池大小足以装下当前的热数据。

orzdba 使用简单，功能强大。可以自定义收集间隔，实时了解主机和 MySQL 的性能数据，为分析和诊断问题提供了很多便利。

27.5 小结

本章简单地介绍了常见的监控工具，并重点介绍了 Zabbix 和 orzdba 的安装和部署。由于篇幅有限，Zabbix 的一些常见功能并没有详细介绍，如自动报警、自动发现监控主机、网络拓扑显示、分布式功能等。有兴趣的读者可以仔细阅读官方文档。

第 28 章　MySQL 常见问题和应用技巧

在 MySQL 日常开发或者维护工作中，用户经常会遇到各种各样的故障或者问题，比如密码丢失、表损坏等。本章总结了一些常见的问题，希望对读者在遇到类似问题时有所帮助。

28.1　忘记 MySQL 的 root 密码

经常会有朋友或者同事问起，MySQL 的 root 密码忘了，不知道改怎么办。其实解决方法很简单，下面介绍两种方法。

方法 1：使用--skip-grant-tables 选项，以无权限认证方式启动数据库实例，再登录数据库修改密码，以下是详细操作步骤。

（1）登录到数据库所在的服务器，手工 kill 掉 MySQL 进程：

```
kill' cat /mysql-data-directory/hostname.pid'
```

其中，/mysql-data-directory/hostname.pid 指的是 MySQL 数据目录下的 .pid 文件，它记录了 MySQL 服务的进程号。

（2）使用--skip-grant-tables 选项重启 MySQL 服务：

```
[mysql3307@localhost mysqlhome]$ ./bin/mysqld_safe --defaults-file=/home/mysql3307/mysqlhome/my.cnf --user=mysql3307 --skip-grant-tables &
[1] 5647
    [mysql3307@localhost mysqlhome]$ 2018-07-22T02:36:21.163410Z mysqld_safe Logging to '/data2/mysql3307/data/error3307.log'.
    2018-07-22T02:36:21.203452Z mysqld_safe Starting mysqld daemon with databases from /data2/mysql3307/data
```

其中--skip-grant-tables 选项前面曾经介绍过,意思是启动 MySQL 服务时跳过权限表认证。启动后，连接到 MySQL 的 root 将不需要口令。

（3）用空密码的 root 用户连接到 MySQL，并且更改 root 口令：

```
[mysql3307@localhost ~]$ mysql -uroot
Welcome to the MySQL monitor.  Commands end with ; or \g.
Your MySQL connection id is 5
Server version: 5.7.22-log MySQL Community Server (GPL)

mysql> use mysql
Reading table information for completion of table and column names
You can turn off this feature to get a quicker startup with -A

Database changed
```

```
mysql> alter user root@'localhost' identified by '123';
ERROR 1290 (HY000): The MySQL server is running with the --skip-grant-tables option so it
cannot execute this statement
```

此时，由于使用了--skip-grant-tables 选项启动，直接使用"alter user"命令更改密码失败。需要执行 flush privileges 后再修改密码：

```
mysql> flush privileges;
Query OK, 0 rows affected (1.00 sec)

mysql> alter user root@'localhost' identified by '123';
Query OK, 0 rows affected (0.00 sec)
```

（4）刷新权限表，使得权限认证重新生效：

```
mysql> flush privileges;
Query OK, 0 rows affected (0.00 sec)
```

（5）重新用 root 登录时，必须输入新口令：

```
[mysql3307@localhost ~]$ mysql -uroot
ERROR 1045 (28000): Access denied for user 'root'@'localhost' (using password: NO)

[mysql3307@localhost ~]$ mysql -uroot -p123
mysql: [Warning] Using a password on the command line interface can be insecure.
Welcome to the MySQL monitor.  Commands end with ; or \g.
Your MySQL connection id is 6

Server version: 5.7.22-log MySQL Community Server (GPL)

mysql>
```

方法 2：使用--init-file 选项启动数据库实例，该参数指定数据库启动时执行包含 sql 语句的文件，以下是详细操作步骤。

（1）初始化 sql 文件，文件包含修改用户密码的 sql 语句。

```
[mysql3307@localhost ~]$ vi /tmp/alter_user.sql
alter user root@'localhost' identified by '123';
```

（2）使用--init-file 选项重启 MySQL 服务：

```
[mysql3307@localhost mysqlhome]$ ./bin/mysqld_safe --defaults-file=/home/mysql3307/mysqlhome
/my.cnf --user=mysql3307 --init-file=/tmp/alter_user.sql &
    [1] 7481
    [mysql3307@localhost mysqlhome]$ 2018-07-22T03:32:11.262410Z mysqld_safe Logging to '/data2/
mysql3307/data/error3307.log'.
    2018-07-22T03:32:11.262410Z mysqld_safe Starting mysqld daemon with databases from /data2/
mysql3307/data
```

数据库启动时会执行--init-file 选项后指定的 sql 文件。

（3）用 root 登录数据库，输入新口令：

```
[mysql3307@localhost ~]$ mysql -uroot -p123
mysql: [Warning] Using a password on the command line interface can be insecure.
Welcome to the MySQL monitor.  Commands end with ; or \g.
Your MySQL connection id is 9

Server version: 5.7.22-log MySQL Community Server (GPL)

mysql>
```

以上两种方法都可以巧妙解决忘记 MySQL root 密码的问题，读者可以选择任选其中一种。

28.2 数据目录磁盘空间不足的问题

很多系统在正式上线后，随着数据量的不断增加，会发现数据目录下的可用空间越来越小，从而给数据库造成了安全隐患。对于这类问题，MySQL 5.7 新增通用表空间功能，它允许用户自定义数据文件的存放目录，多个表可以共用同一个通用表空间。下面来看一下具体的例子。

（1）登录到数据库，查看数据目录：

```
mysql> show variables like 'datadir';
+---------------+------------------------+
| Variable_name | Value                  |
+---------------+------------------------+
| datadir       | /data2/mysql3307/data/ |
+---------------+------------------------+
```

（2）在 /data 盘创建通用表空间目录：

```
[root@localhost ~]# mkdir -p /data/mysql3307/data/
[root@localhost ~]# chown -R mysql3307:mysql3307  /data/mysql3307/data/
```

（3）创建通用表空间：

```
CREATE TABLESPACE general_ts ADD datafile '/data/mysql3307/data/general_ts.ibd' ENGINE=InnoDB;
```

（4）查看通用表空间文件：

```
[root@localhost ~]# ll /data/mysql3307/data/
total 32
-rw-r----- 1 mysql3307 mysql3307 65536 Aug 28 21:19 general_ts.ibd
```

（5）在通用表空间上，创建新表：

```
mysql> CREATE TABLE `employees` (
    -> `emp_no` int(11) NOT NULL,
    -> `birth_date` date NOT NULL,
    -> `first_name` varchar(14) COLLATE utf8_unicode_ci NOT NULL,
    -> `last_name` varchar(16) COLLATE utf8_unicode_ci NOT NULL,
    -> `gender` enum('M','F') COLLATE utf8_unicode_ci NOT NULL,
    -> `hire_date` date NOT NULL,
    -> PRIMARY KEY (`emp_no`),
    -> KEY `idx_emp_birth_date` (`birth_date`)
    -> ) /*!50100 TABLESPACE `general_ts` */ ENGINE=InnoDB DEFAULT CHARSET=utf8 COLLATE=utf8_unicode_ci
    -> ;
Query OK, 0 rows affected (0.01 sec)
```

新建的表可以指定 general_ts 通用表空间，不存储在 MySQL 数据目录。已存在的表，可以用导出、重建表、导入的方法，将表迁移到通用表空间。

28.3 mysql.sock 丢失后如何连接数据库

在 MySQL 服务器本机上连接数据库时，经常会出现 mysql.sock 不存在，导致无法连接的问题。这是因为如果指定 localhost 作为一个主机名，则 mysqladmin 默认使用 UNIX 套接字文件连接，而不是 TCP/IP。而这个套接字文件（一般命名为 mysql.sock）经常会因为各种原因而被删除。通过 --protocol= TCP | SOCKET | PIPE | MEMORY} 选项，用户可以显式地指定连接协议。下面演示了使用 UNIX 套接字失败后使用 TCP 协议连接成功的例子。

（1）UNIX 套接字连接：

```
[mysql3307@localhost ~]$ mysql -uroot --socket=/tmp/mysql_3307.sock
ERROR 2002 (HY000): Can't connect to local MySQL server through socket '/tmp/mysql_3307.sock' (2)
```

（2）TCP 连接：

```
[mysql3307@localhost ~]$ mysql --protocol=TCP -uroot -p -P3307 -hlocalhost
Enter password:
```

```
Welcome to the MySQL monitor.  Commands end with ; or \g.
Your MySQL connection id is 14
Server version: 5.7.22-log MySQL Community Server (GPL)

root@localhost:3307  [(none)]>
```

28.4 从 mysqldump 文件抽取需要恢复的表

日常工作中，通常使用 mysqldump 备份一个或则多个 db 下的所有表，所有数据备份到同一个 sql 文件中。当恢复数据时，可以直接执行备份文件。如果只恢复备份文件中一个表，而且备份文件特别大，想通过手工编辑 sql 文件的方式，找到对应的 sql 语句，效率是非常低的。此时，可以使用 sed 命令抽取需要的 sql 语句，然后执行语句来恢复这个表。下面我们模拟一下整个过程。

（1）备份 employees 数据库：

```
[mysql3307@localhost tmp]$ mysqldump -uroot -S/tmp/mysql_3307.sock employees > employees.sql
[mysql3307@localhost tmp]$ ls -ltrh employees.sql
-rw-rw-r-- 1 mysql3307 mysql3307 163M Aug 28 21:39 employees.sql
```

（2）查看已备份的表和对应的开始行号：

```
[mysql3307@localhost tmp]$ grep -n 'Table structure' employees.sql
43:-- Table structure for table `departments`
69:-- Table structure for table `dept_emp`
124:-- Table structure for table `dept_manager`
153:-- Table structure for table `employees`
198:-- Table structure for table `salaries`
340:-- Table structure for table `salaries_history`
368:-- Table structure for table `test`
391:-- Table structure for table `titles`
```

（3）结果显示，文件备份了 8 张表，如果想恢复 salaries 表，可以用 sed 命令抽取 sql 语句：

```
[mysql3307@localhost tmp]$ sed -n '198,340 p' employees.sql > salaries.sql
[mysql3307@localhost tmp]$ ls -ltrh salaries.sql
-rw-rw-r-- 1 mysql3307 mysql3307 111M Aug 28 22:07 salaries.sql
```

其中，198 代表 salaries 表起始行数，340 代表下一个表的起始行数。

（4）查看抽取的 sql 语句：

```
[mysql3307@localhost tmp]$ more salaries.sql
-- Table structure for table `salaries`
--

DROP TABLE IF EXISTS `salaries`;
/*!40101 SET @saved_cs_client     = @@character_set_client */;
/*!40101 SET character_set_client = utf8 */;
CREATE TABLE `salaries` (
  `emp_no` int(11) NOT NULL,
  `salary` int(11) NOT NULL,
........
INSERT INTO `salaries` (`emp_no`, `salary`, `from_date`, `to_date`) VALUES (10001,100,'1986-06-26','1987-06-26')
........
```

（5）执行 sql 语句，恢复 salaries 表：

```
mysql>source salaries.sql
```

到这里，salaries 表的恢复就完成了。

28.5 使用 innobackupex 备份恢复单表

在 MySQL 的数据量已经很大时，我们会考虑优先使用 innobackupex 工具来备份数据库，当然通常的做法仍然是全库备份，那么针对这种情况，如何进行单表恢复呢？下面我们模拟一下整个过程。

（1）备份前确认 dept_emp 表的数据量：

```
mysql> select count(0) from dept_emp;
+----------+
| count(0) |
+----------+
|   310309 |
+----------+
1 row in set (0.09 sec)
```

（2）开始 innobackupex 全量备份：

```
[mysql3307@localhost employees]$ innobackupex --defaults-file=/home/mysql3307/mysqlhome/my.cnf --no-timestamp --user=root --password=123 --socket=/tmp/mysql_3307.sock /home/mysql3307/tmp/fullbackup
180828 23:01:14 innobackupex: Starting the backup operation
……
xtrabackup: Transaction log of lsn (12965519373) to (12965519382) was copied.
180828 23:01:51 completed OK!
```

看到"completed OK"，证明备份成功结束了。

（3）模拟 dept_emp 被误删：

```
mysql> drop table dept_emp;
Query OK, 0 rows affected (0.03 sec)

mysql> select count(0) from dept_emp;
ERROR 1146 (42S02): Table 'employees.dept_emp' doesn't exist
```

（4）应用全备，需要加上--export 选项，用于生成 import table 所需要的.exp 文件：

```
[mysql3307@localhost employees]$ innobackupex --defaults-file=/home/mysql3307/mysqlhome/my.cnf --user=root --password=  --use-memory=256m  --redo-only  --apply-log  /home/mysql3307/tmp/fullbackup  --export
180828 23:06:26 innobackupex: Starting the apply-log operation
……
xtrabackup: export metadata of table 'employees/dept_emp' to file `./employees/dept_emp.exp` (2 indexes)
………
180828 23:06:27 completed OK!
```

看到"completed OK"，应用日志完成。查看 dept_emp 表相关文件：

```
[mysql3307@localhost employees]$ cd /home/mysql3307/tmp/fullbackup/employees
[mysql3307@localhost employees]$ ll dept_emp*
-rw-rw-r-- 1 mysql3307 mysql3307      631 Aug 28 23:06 dept_emp.cfg
-rw-r----- 1 mysql3307 mysql3307    16384 Aug 28 23:06 dept_emp.exp
-rw-r----- 1 mysql3307 mysql3307     8676 Aug 28 23:01 dept_emp.frm
-rw-r----- 1 mysql3307 mysql3307 28311552 Aug 28 23:01 dept_emp.ibd
-rw-r----- 1 mysql3307 mysql3307      887 Aug 28 23:01 dept_emp_latest_date.frm
```

（5）重新创建 dept_emp 表：

```
mysql> CREATE TABLE `dept_emp` (
    -> `emp_no` int(11) NOT NULL,
    -> `dept_no` char(4) COLLATE utf8_unicode_ci NOT NULL,
    -> `from_date` date NOT NULL,
    -> `to_date` date NOT NULL,
    -> PRIMARY KEY (`emp_no`,`dept_no`),
```

```
    -> KEY `dept_no` (`dept_no`),
    -> CONSTRAINT `dept_emp_ibfk_1` FOREIGN KEY (`emp_no`) REFERENCES `employees` (`emp_no`)
ON DELETE CASCADE,
    -> CONSTRAINT `dept_emp_ibfk_2` FOREIGN KEY (`dept_no`) REFERENCES `departments` (`dept_no`) ON DELETE CASCADE
    -> ) ENGINE=InnoDB DEFAULT CHARSET=utf8 COLLATE=utf8_unicode_ci;
Query OK, 0 rows affected (0.01 sec)
```

(6) 下线 dept_emp 表：

```
mysql>ALTER TABLE dept_emp DISCARD TABLESPACE;
Query OK, 0 rows affected (0.02 sec)
```

(7) 复制步骤（4）生成的 dept_emp.exp 和 dept_emp.ibd 文件到数据目录：

```
[mysql3307@localhost employees]$ cd /home/mysql3307/tmp/fullbackup/employees
[mysql3307@localhost employees]$ cp dept_emp.exp dept_emp.ibd /data2/mysql3307/data/employees/
```

(8) 加载 dept_emp 表：

```
mysql>ALTER TABLE dept_emp IMPORT TABLESPACE;
Query OK, 0 rows affected, 1 warning (0.32 sec)
```

(9) 查看 dept_emp 表记录数：

```
mysql>select count(0) from dept_emp;
+----------+
| count(0) |
+----------+
|   310309 |
+----------+
1 row in set (0.33 sec)
```

到这里，dept_emp 表的恢复就完成了。

注意：如果备份文件比较大，需要考虑到磁盘空间问题以及恢复所用的时间。

28.6 分析 BINLOG，找出写的热点表

日常工作中，想找出线上写（增删改）的热点表。比较简单的方法是定时收集 MySQL 内存统计值，然后计算差值。这种方法需要主动收集数据，而且时间点不够灵活。下面介绍一个解析 BINLOG 文件的方法，这样有两个好处：

- 不需要事先收集数据，统计的时间点也非常灵活；
- 分析的维度可以精确到表、字段级别。

为了简化操作，推荐使用 GitHub 上一个分析统计 BINLOG 的工具，可以大大减少手工操作的工作量。下面看看怎么安装和使用这个工具。

(1) 安装 Perl：

```
[root@localhost ~]# yum install perl
```

(2) 下载解压 pasrebinlog 脚本,：

```
[mysql3307@localhost ~]$ wget https://codeload.github.com/wubx/mysql-binlog-statistic/zip/master
[mysql3307@localhost ~]$ unzip master
[mysql3307@localhost ~]$ ll mysql-binlog-statistic-master/bin/pasrebinlog
-rwxr-xr-x 1 mysql3307 mysql3307 1776 Aug 14  2013 mysql-binlog-statistic-master/bin/pasrebinlog
```

(3) 查询 MySQL 数据库目录及 BINLOG 文件：

```
mysql>show variables like '%datadir%';
+---------------+------------------------+
| Variable_name | Value                  |
+---------------+------------------------+
```

```
| datadir         | /data2/mysql3307/data/ |
+-----------------+------------------------+
1 row in set (0.00 sec)

mysql>show master logs;
+------------------+-----------+
| Log_name         | File_size |
+------------------+-----------+
| mysql-bin.000052 |    120640 |
+------------------+-----------+
1 row in set (0.00 sec)
```

（4）修改 pasrebinlog 脚本参数：

修改 mysqlbinlog 工具目录。

```
[mysql3307@localhost ~]$ which mysqlbinlog
~/mysqlhome/bin/mysqlbinlog

[mysql3307@localhost ~]$ vi mysql-binlog-statistic-master/bin/pasrebinlog
my $binlog = "~/mysqlhome/bin/mysqlbinlog"
```

可以按照需求，增加解析 BINLOG 的限制条件，例如增加--start-datetime=name，限定解析 BINLOG 的开始时间。

```
[mysql3307@localhost ~]$ vi mysql-binlog-statistic-master/bin/pasrebinlog
my $binlog = "~/mysqlhome/bin/mysqlbinlog --start-datetime='2018-08-28 18:00:00'"
```

（5）开始解析 BINLOG，统计表的变化：

```
[mysql3307@localhost ~]$ perl mysql-binlog-statistic-master/bin/pasrebinlog  /data2/mysql3307
/data/mysql-bin.000052
File /data2/mysql3307/data/mysql-bin.000052
=================================
Table `employees`.`employees`:
Type DELETE opt:   1024
=================================

=================================
Table `employees`.`repl_filter`:
Type INSERT opt:   1
Type UPDATE opt:   1
1 col : 1
=================================

=================================
Table `employees`.`repl_normal`:
Type INSERT opt:   2
=================================

=================================
Table `employees`.`salaries`:
Type DELETE opt:   4047
=================================
```

结果显示，'employees'.'employees'表删除了 1024 条记录；'employees'.'repl_filter'表插入了 1 条记录，更新了 1 条记录，操作更新第 1 列 1 次；'employees'.'repl_normal'表插入了两条记录。表和写操作的关系一目了然，方便我们分析和优化数据库。

28.7 在线 DDL

在 MySQL 5.6 版本前，DDL 操作（例如加字段、加索引）会阻塞写操作，如果对大表进行这些操作，会导致长时间无法写入数据，对业务的影响很大，几乎是不可接受的。直到 5.6 版本，MySQL 开始提供在线 DDL（Online DDL）功能，对于某些 DDL 操作来说，不再阻塞写操作。随后 5.7 版本和 8.0 版本都对在线 DDL 进行了优化，大大减少了 DDL 操作对业务的影响。

在 MySQL 5.6 之前，操作 DDL 语句通常采取两种方式。

- copy table 方式：新建与原表相同的临时表，在临时表上做 DDL 操作，同时锁定原表，此时原表只能读不能写。复制原表数据到临时表，复制完成后，通过 rename 命令用临时表替换原表。此方式比较消耗磁盘和 CPU 资源。
- inplace 方式：直接在原表上做操作，不再复制数据，效率更高，但是操作期间原表也是不可写的。

MySQL 5.6 以后，开始支持在线 DDL。在线 DDL 操作虽然也是采用 copy table 和 inplace 方式，但当采用 inplace 方式时，执行 DDL 语句期间，原表是可写的。这样将不再阻塞业务数据写入，增加数据库的可用性。

MySQL 5.7 对在线 DDL 进一步增强，支持 tinyint、int、smallint 和 bigint 类型位数的在线增大或减小；支持 varchar 类型在[0,255]以内或者 255 字节以上增加的在线操作，例如 varchar 长度由 10 调到 200，属于[0,255]以内的调整，支持在线操作，varchar 类型长度由 10 调到 300，跨越了[0,255]和 255 以上两个段，不支持在线操作；支持在线修改索引名字。

MySQL 8.0 中支持了快速增加表字段功能；增加了在线操作分区表的功能，例如在线增加分区、删除分区和重建分区。

在线 DDL 语法如下：

```
mysql> Alter table …. , ALGORITHM [=] {DEFAULT|INPLACE|COPY}, LOCK [=] { DEFAULT| NONE| SHARED| EXCLUSIVE }
```

前面已经对 inplace 和 copy 算法进行过介绍，inplace 由于不用复制数据，所以相对来说，性能会好一些，但并不是所有的 DDL 操作都支持 inplace 方式。由于涉及内容太多，具体限制读者可以查看官方文档，这里不再赘述。

在线 DDL 一共有一下 4 个加锁方式。
- DEFAULT：由 MySQL 自动选择锁方式，优先选择锁定时间短的方式。
- NONE：不加锁，支持读写，效率最高。
- SHARED：支持读，不支持写。
- EXCLUSIVE：不支持读写，效率最低。

在线执行 DDL 语句的过程中，读者可以不指定 ALGORITHM 和 LOCK，MySQL 会根据代价和规则自动选择，即便人为指定 LOCK 模式为 NONE，实际执行过程中也会申请锁，因为在线 DDL 操作在开始时，会短暂地申请排他元数据锁。如果此时 DDL 操作挂起，将会阻塞后续对该表的操作。下面来看一个具体的例子。

第一个会话，开启事务，查询 departments 表，申请共享锁：

```
mysql>start transaction;
mysql>select * from departments;
```

第二个会话，为 departments 表增加新的字段：

```
mysql>ALTER TABLE departments ADD `dept_name` varchar(40) , ALGORITHM=INPLACE, LOCK=NONE;
```

第三个会话，查看线程运行情况：

```
mysql >show processlist\G
……
*************************** 12. row ***************************
     Id: 133
   User: root
   Host: localhost
     db: employees
Command: Query
   Time: 37
```

```
    State: Waiting for table metadata lock
     Info: ALTER TABLE departments ADD `dept_name` varchar(40) , ALGORITHM=INPLACE, LOCK=NONE
...
```

因为第一个会话持有 employees 表的共享锁,第二个会话的 DDL 操作即便使用了 inplace 算法和 none 锁,还是会挂起,State(状态)是 Waiting for table metadata lock。

第四个会话,查询 departments 表:

```
mysql>select * from departments;
```

第三个会话,查看线程运行情况:

```
*************************** 12. row ***************************
      Id: 133
    User: root
    Host: localhost
      db: employees
 Command: Query
    Time: 281
   State: Waiting for table metadata lock
    Info: ALTER TABLE departments ADD `dept_name` varchar(40) , ALGORITHM=INPLACE, LOCK=NONE
*************************** 15. row ***************************
      Id: 136
    User: root
    Host: localhost
      db: employees
 Command: Query
    Time: 18
   State: Waiting for table metadata lock
    Info: select * from departments
```

DDL 操作阻塞了第四个会话的查询请求,查询 SQL 的 State 也是 Waiting for table metadata lock。如果第一个会话的共享锁长时间不释放,后面的 DDL 操作和 departments 表的读写操作都会被阻塞,给数据库带来隐患。对于这种情况,一种办法是尽快提交第一个会话的事务;另一个办法是合理设置 lock_wait_timeout 的值,规定在 lock_wait_timeout 的时间后,DDL 操作超时退出,department 表后续的读写操作将不会被阻塞。

MySQL 5.7 版本后,可以开启 performance shema 统计,监控在线 DDL 的进度,下面来看一个示例。

开启 PS 统计:

```
mysql>use performance_schema
mysql>update setup_instruments set enabled = 'YES' where name like 'stage/innodb/alter%';
Query OK, 0 rows affected (0.01 sec)
Rows matched: 7  Changed: 0  Warnings: 0

mysql>update setup_consumers set enabled = 'YES' where name like '%stages%';
Query OK, 3 rows affected (0.00 sec)
Rows matched: 3  Changed: 3  Warnings: 0
```

修改表,添加一个新的字段:

```
mysql>ALTER TABLE salaries ADD COLUMN emp_name varchar(20) AFTER salary;
```

由于 salaries 表记录数多,增加字段会持续比较长的时间;开启第二个会话,查看增加字段的进度:

```
mysql>use performance_schema
mysql>select event_name, work_completed, work_estimated from events_stages_current;
+------------------------------------------------------+----------------+----------------+
| event_name                                           | work_completed | work_estimated |
+------------------------------------------------------+----------------+----------------+
| stage/innodb/alter table (read PK and internal sort) |           1976 |          26932 |
+------------------------------------------------------+----------------+----------------+
```

```
1 row in set (0.00 sec)

mysql>select event_name, work_completed, work_estimated from events_stages_current;
+--------------------------------+----------------+----------------+
| event_name                     | work_completed | work_estimated |
+--------------------------------+----------------+----------------+
| stage/innodb/alter table (insert) |          21869 |          27356 |
+--------------------------------+----------------+----------------+
1 row in set (0.00 sec)

mysql>select event_name, work_completed, work_estimated from events_stages_current;
+--------------------------------+----------------+----------------+
| event_name                     | work_completed | work_estimated |
+--------------------------------+----------------+----------------+
| stage/innodb/alter table (flush)  |          25882 |          27083 |
+--------------------------------+----------------+----------------+
1 row in set (0.00 sec)
```

event_name 代表增加字段当前经历的阶段，work_estimated 为 MySQL 预估总的工作量，work_completed 为已完成的工作量，当 work_completed 与 work_estimated 相等后，DDL 操作完成。通过监控 work_completed 与 work_estimated 的比值，可以估算 DDL 完成的百分比。

MySQL 8.0 版本对在线 DDL 的支持已经比较完善，但是实际使用过程中需要注意两个问题。第一是前面提到的，如果事务长时间持有表锁，再对该表进行 DDL 操作，那么 DDL 操作和后续的读写都会被阻塞；第二是在主库对大表进行在线 DDL 操作，主库需要较长时间才能完成 DDL 操作，从库一般也需要相同或者更长时间才能应用完 DDL 操作，这样将造成从库极大的延迟。希望后续版本能完善在线 DDL 存在的问题，方便用户使用。

28.8 小结

本章详细介绍了在 MySQL 中经常遇到的一些问题及其解决办法。对于实际应用来说，用户遇到的问题可能远远不止这些，希望读者在实践中能够多多总结，为管理和应用 MySQL 积累更多的经验。

第 29 章　自动化运维系统的开发

随着业务量的增长，MySQL 实例的数量不断扩增，仅凭人力去手动维护大量的数据库实例将会愈发困难。通过搭建数据库自动化运维平台将有效地提升部署与管理实例的效率，集中监控多实例的状态，能让 DBA 以最快速度响应并处理报警信息。

本章将介绍搭建一个 MySQL 自动化运维平台的基本流程，具体包括应用背景、框架选型、技术实践等，并将以实现 MySQL 自动化安装功能为例来教大家一步步搭建自己的自动化运维平台。

29.1　MySQL 自动化运维背景

在小规模的生产环境中，DBA 常通过直连客户端的方式，通过命令行来执行管理数据库的操作。但是当 MySQL 实例数量不断增长时，仅凭人工去逐台登录服务器来完成诸如安装配置 MySQL、部署监控等操作将会增加大量重复性的工作，同时也会增加配置出错的概率。在这种情况下，DBA 通过编写定制化脚本，通过批量管理工具在多台服务器上执行命令或脚本，减少了重复性工作，在一定程度上提升了运维的效率。但这种方式也会出现如数据库信息统计困难、监控时效性差、脚本多版本难以管理等问题，或者从根本上说仍然是运维人员登录服务器进行操作，并没有实质上提升运维效率。

因此我们需要一个自动化运维平台，提供清晰简洁的可视化客户端界面，有一套后台系统管理资产信息与数据库信息，后台能调度管理所有脚本，从而完成 MySQL 部署与监控的工作，并能自动地收集数据库运行时的监控信息，将异常报警信息发送给运维人员。这样在 MySQL 节点数量与压力高速增长时，运维仍能保证系统的稳定运行。下面介绍平台的基本架构。

如图 29-1 所示，一个 MySQL 自动化运维平台通常分为后台管理系统、任务调度系统、客户端三大部分。

（1）后台管理系统，也被称为 CMDB（Configuration Management Database），具体包括以下 3 个模块。

- CMDB 数据库：存储了服务器的资产信息，MySQL 部署信息与监控运行信息等系统的内部信息，使用的数据库和存储的内容依据业务内部逻辑制定。为保证数据的一致性与可靠性，系统仅通过 API 与 CMDB 数据库进行数据的交互。

- 批量管理系统：通过批量的执行相同任务，完成管理多个 MySQL 实例的系统。系统可以直接在主机上执行指令，或是传递指令给任务调度系统调度执行，从而实现在一台主机上操控多台主机的功能。本章使用 SaltStack 做批量管理工具。

● 后台 API：主要提供数据库信息的调度服务，CMDB 数据库的信息都通过 API 来读取与存储。后台 API 提供了客户端，数据库和批量管理系统间的命令调度方法。本章使用 Django REST Framework 来完成 API 的开发。

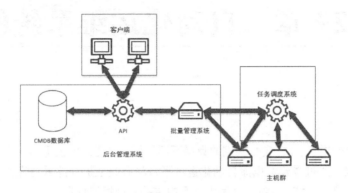

图 29-1　自动化运维平台组织结构

（2）任务调度系统，主要功能是帮助批量管理系统处理任务。由批量管理系统传递到任务调度系统的任务，由该系统进行队列分发，通过异步或定时的方式执行相关操作，完成诸如安装 MySQL 这类耗时较长的任务或 MySQL 运行状态的定期监控这类定时任务。本章选用 Celery 作为任务调度的工具。

（3）客户端，提供了查看各类信息的可视化界面，如显示数据库信息、任务的调度状态、日志信息、监控信息等内容。本章选用 Vue.js 作为客户端开发的工具。

下面来详细介绍每个部分的系统架构设计和搭建方式。

29.2　CMDB 系统搭建

CMDB 系统是整个平台的核心部分，支持系统中数据的调度与持久化，发送任务请求，管理平台上的所有服务器等功能，需要有较高的可靠性。下面我们逐个介绍组成 CMDB 系统的主要组件。

29.2.1　CMDB 数据库

CMDB 数据库是自动化运维平台的数据基础，记录了系统所有必要的数据信息，如服务器信息、MySQL 实例信息、监控信息、配置项、数据变动信息、任务调度信息等多项内容。为了保证配置库中数据与实际线上业务数据的一致性，CMDB 数据库中信息只能通过 API 调度来实现修改。为了方便理解，本章以服务器和数据库信息两个配置表为例来介绍 CMDB 数据库，而在实际的应用中还需要记录配置信息、监控信息、任务信息等多种信息。服务器和数据库信息两表中记录的主要信息如下。

● 服务器信息表：至少需要记录服务器名称、网络信息、类型、状态等内容。

● 数据库信息表：至少需要记录数据库所属服务器、启动端口、版本、实例用户、配置文件信息、状态等内容。

29.2 CMDB 系统搭建

根据以上信息，设计出如图 29-2 所示的数据库类图。

在之后会使用 Django-Model 将数据库表结构抽象化为模型，并将模型序列化从而对数据做增删改查操作，这部分将在后文的后台 API 设计章节中进行讨论。

29.2.2 批量管理系统

对于大规模的 MySQL 集群管理，靠传统的手工逐台操作是低效且不现实的，通常的解决方案是使用批量运维工具。常用的批量管理工具有 Puppet、SaltStack、Ansible 等。其中 SaltStack 是一款 Python 编写的，用于大规模批量管理服务器的工具，结合了消息队列服务 ZeroMQ，支持在主节点（Master）执行命令来远程批量配置多个从节点（Minion），支持执行特定命令、传输文件、安装服务、配置定时任务等常用操作，其基本的架构如图 29-3 所示。

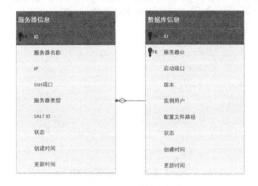

图 29-2 CMDB 数据库类图

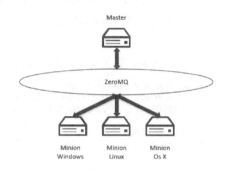

图 29-3 SaltStack 架构图

下面详细介绍 SaltStack 的安装配置过程。

1. SaltStack 安装与配置

（1）Master 节点配置。

Master 节点是集群的控制中心，salt 命令都是通过 Master 节点传到 Minion 节点上执行的。下例中选择了两台主机，Master：10.0.0.1 和 Slave：10.0.0.2（Minion1）来演示部署 SaltStack 服务的流程。

在 Master 节点安装并配置 salt-master。

```
[root@10.0.0.1 ~]# yum install -y epel-release
[root@10.0.0.1 ~]# yum install -y salt-master salt-minion
[root@10.0.0.1 ~]# vi /etc/salt/master

cachedir: /var/cache/salt
keep_jobs: 24
file_roots:
  base:
    - /srv/salt/
  dev:
    - /srv/salt/dev/services
    - /srv/salt/dev/states
  prod:
    - /srv/salt/prod/services
    - /srv/salt/prod/states
publish_port: 4505
ret_port: 4506
```

/etc/salt/master 是 salt-master 的默认配置文件，在每次修改后都需要重启该服务。在这里需要主要关注如上几个参数，并可根据需求进行修改。

● cachedir：放置 salt 命令执行的缓存信息，大量的调用会使该目录存在过多缓存，可以更换存放路径。

● keep_jobs：cachedir 中保持缓存信息的时间，默认为 24h。

● file_roots：salt 执行文件时默认的脚本所在地址，后文涉及的 sls 文件也在该目录中。

● publish_port 和 ret_port：代表 salt 的消息发布系统端口和节点间通信端口，默认为 4505 和 4506，请避免这两个端口被占用。

配置完成后启动 master 服务：

```
[root@10.0.0.1 ~]# service salt-master start
[root@10.0.0.1 ~]# service salt-master status
salt-master.service - The Salt Master Server
   Loaded: loaded (/usr/lib/systemd/system/salt-master.service; enabled; vendor preset: disabled)
   Active: active (running) since Mon 2018-07-16 19:30:59 CST; 3 weeks 6 days ago
```

可以看到服务已经正常启动。

（2）Minion 节点配置。

❶ 在 Minion 节点安装并配置 salt-minion：

```
[root@10.0.0.2 ~]# yum install -y epel-release
[root@10.0.0.2 ~]# yum install -y salt-minion
[root@10.0.0.2 ~]# vi /etc/salt/minion

master: 10.0.0.1
id: Minion1
```

/etc/salt/minion 是 salt-minion 的默认配置文件，需要关注的参数为 master 和 id。

● master：Master 主机的 IP 或 hostname。

● id：Master 向指定 Minion 发送指令时被指定 Minion 的标识 id，可以根据业务关系配置为明了易读的字符串。

❷ 配置完成后启动 minion 服务：

```
[root@10.0.0.2 ~]# service salt-minion start
[root@10.0.0.2 ~]# service salt-minion status
Redirecting to /bin/systemctl status salt-minion.service
salt-minion.service - The Salt Minion
   Loaded: loaded (/usr/lib/systemd/system/salt-minion.service; enabled; vendor preset: disabled)
   Active: active (running) since Mon 2018-07-16 19:30:55 CST; 3 weeks 6 days ago
```

启动后，为了配置 Master 和 Minion 节点间的互信关系，Minion 会生成公私钥一对，并将公钥发送给 Master。Master 需要执行命令接受该 minion-key 来确认互信关系。

❸ 配置互信关系如下：

```
[root@10.0.0.1 ~]# salt-key -L
Accepted Keys:
Denied Keys:
Unaccepted Keys:
Minion1
Rejected Keys:
[root@10.0.0.1 ~]# salt-key -A -y
The following keys are going to be accepted:
Unaccepted Keys:
Minion1
Key for minion Minion1 accepted.
```

配置完 Minion1 节点与 Master 节点的互信后，可以指定 Minion1 节点作为管理对象集，从 Master 节点发送 test.ping 指令，测试 Master 和 Minion1 节点间连接是否成功。

```
[root@10.0.0.1 ~]# salt -L Minion1 test.ping
Minion1:
    True
```

在上面的例子中，在 salt 命令后指定了-L 选项，以列表形式将 SaltStack 本次的管理对象集限定 id 为 Minion1 的节点，因此 test.ping 指令仅在 Minion1 节点上执行。除此之外，SaltStack 还支持如下的选择管理对象集的方式，可以方便地完成对多个 Minion 节点的连接测试。

```
salt '*' test.ping            //选择所有 Minion 节点
salt 'Minion*' test.ping      //根据 Minion 节点的 id 进行正则匹配
salt -L Minion1 test.ping     //根据 Minion 节点的 id 列表匹配
salt -G 'os:RedHat' test.ping //grains 类型匹配
salt -S '10.0.0.0/24' test.ping //CIDR 匹配
```

2. SaltStack 管理功能

在上文中部署并配置测试了 SaltStack 环境，下面介绍几种常用的管理功能。

（1）远程批量执行各类命令。

❶ 执行特定命令：

```
[root@10.0.0.1 ~]# salt-run manage.up    // 检查所有存活 Minion
- Minion1
[root@10.0.0.1 ~]# salt Minion1 service.get_all    // 获取 Minion1 上全部服务
...
[root@10.0.0.1 ~]# salt Minion1 cmd.run 'uptime'   // Minion1 执行 uptime 命令，并返回结果
Minion1:
    03:58:16 up 435 days, 20:56,  3 users,  load average: 0.50, 0.59, 0.48
```

❷ 复制 Master 上文件 salt://a.sh 到 Minion1 的 /tmp 下：

```
[root@10.0.0.1 ~]# salt Minion1 cp.get_file salt://a.sh /tmp/a.sh
Minion1:
    /tmp/a.sh
```

其中 salt:// 目录需要跟上文中 Master 配置的 file_roots 目录一致。要复制的文件必须指定 salt:// 参数，不能是 Master 上普通的目录，否则会报错。

❸ 远程执行 Master 上的 salt://a.sh 脚本，a.sh 的功能为打印第一个参数的内容：

```
[root@10.0.0.1 ~]# more /srv/salt/a.sh
echo $1

[root@10.0.0.1 ~]# salt Minion1 cmd.script salt://a.sh 'a'
Minion1:
    ----------
    pid:
        9368
    retcode:
        0
    stderr:

    stdout:
        a
[root@10.0.0.1 ~]# salt Minion1 cmd.script salt://a.sh 'a' runas='mysql' cwd='/home'
```

远程脚本默认在 Minion 节点的 /tmp 目录执行，可以增加 runas 和 cwd 参数修改执行用户和默认的 Minion 节点执行路径。

❹ 部署定时任务：

```
[root@10.0.0.1 ~]# salt Minion1 cron.set_job root '0' '*' '*' '*' '*' 'sh /home/a.sh' '每小时 0 分执行命令，直接写入 Minion 的对应用户的 crontab 中'
Minion1:
    new
```

通过以上介绍的几种命令，可以实现向 Minion 节点所在服务器批量地传输 MySQL 相关脚本、安装实例、部署对应的监控和定时备份脚本等基本功能。

（2）管理 Minion 节点系统数据。

这里主要介绍 SaltStack 的 grains 模块和 pillar 模块，前者用于从 Master 节点获取 Minion 节点的系统数据，后者则用于从 Master 节点配置系统数据到 Minion 节点。

- grains 模块。

通过 grains 模块提供的一些方法，Master 节点可以方便地获取到 Minion 节点的操作系统版本信息和设备型号信息等系统数据。但需要注意的是，这些系统数据是在 Minion 节点的 salt-minion 服务启动时收集到并保存在各自 /etc/salt/grains 中的静态数据，直到 salt-minion 服务下一次重启时才会再次进行更新，因此 grains 模块更适合用于资产管理。

```
[root@10.0.0.1 ~]# salt Minion1 grains.item os    //查看Minion1的os信息
Minion1:
    RedHat
[root@10.0.0.1 ~]# salt Minion1 grains.items    //查看Minion1的所有服务器属性
Minion1:
    ----------
    SSDs:
    biosreleasedate:
        04/05/2016
    biosversion:
        6.00
    cpu_flags:
...
```

通过在 Minion 节点修改配置文件，或者在 Master 节点进行指定，可以对指定的 Minion 节点新增自定义的系统数据项。同样需要重启 salt-minion 服务后才可以获取到新的系统数据项。

```
[root@10.0.0.2 ~]# vi /etc/salt/minion
grains:
  roles:
    - Database
[root@10.0.0.2 ~]# service salt-minion restart

[root@10.0.0.1 ~]# salt Minion1 grains.item roles
Minion1:
    Database
[root@10.0.0.1 ~]# salt Minion1 grains.setval environment 'dev'
Minion1:
    ----------
    environment :
        dev
[root@10.0.0.1 ~]# salt Minion1 grains.item environment
Minion1:
    ----------
    environment :
        dev
```

- pillar 模块。

如上所述，grains 模块的信息保存在各个 Minion 节点上，而 pillar 模块的信息则由 Master 节点进行配置并存储在 Master 节点的 /srv/pillar 目录中。pillar 模块的信息仅对指定的 Minion 可读，因此更适用于一些动态的、比较敏感的数据。下面的例子演示了如何使用 pillar 来保存 Minion1 节点的用户名和密码。

❶ 首先建立一个 top.sls 文件，作为 pillar 的入口文件。

```
[root@10.0.0.1 ~]# mkdir /srv/pillar
[root@10.0.0.1 ~]# vi /srv/pillar/top.sls

base:
    Minion1:
        - minion1
```

❷ 配置对应的 minion1.sls 文件，里面写入测试的用户名和密码信息。

```
[root@10.0.0.1 ~]# vi /srv/pillar/minion1.sls
db_user: username
db_passwd: passwd
```

❸ 刷新 Master 的 pillar 配置来应用新的配置内容，之后查询 pillar 信息。

```
[root@10.0.0.1 ~]# salt '*' saltutil.refresh_pillar
[root@10.0.0.1 ~]# salt Minion1 pillar.items
Minion1:
    ----------
    db_user:
        username
    db_passwd:
        passwd
```

pillar 模块的配置主要通过 SLS（Salt State）文件来完成，实际上 SLS 文件作为 SaltStack 的配置文件能够完成的工作远不止如此。下面通过安装 MySQL 实例详细介绍一下 SaltStack 中 SLS 文件的应用。

3. 利用 SLS 文件安装 MySQL 实例

SLS 是 SaltStack 的配置文件，文件存放于 Master 端，用途是配置 Minion 端的系统状态，记录的数据格式为 YAML（一种简易的标记性语言）。Minion 通过从 Master 上拉取文件来保持配置、服务的一致性，利用此特性可以很方便地为所有 Minion 安装 MySQL 实例。下面以源码安装 MySQL 为例来介绍流程。

（1）将安装过程涉及的配置文件和安装文件全部放在 file_roots 下，在本例中为默认的路径/srv/salt。安装前的目录树如下：

```
├── top.sls                        // 入口文件
├── mysql
│   ├── files
│   │   ├── my.cnf                 // 存放安装配置文件
│   │   ├── mysql-5.6.34.tar.gz    // 数据库配置文件
│   │   └── mysqld                 // 安装源码
│   ├── init.sh                    // 数据库启动文件
│   ├── init.sls                   // 配置功能文件
│   └── install.sls                // 环境配置文件
                                   // 安装文件
```

（2）top.sls 是配置管理的入口文件，从 base 标签开始解析，下一级是操作的目标 Minion 节点，可以通过正则、grains、分组名等各种方式进行匹配，再下一级是要执行的 Salt State 文件，默认扩展名为.sls。top.sls 文件内容如下：

```
[root@10.0.0.1 salt]# cat top.sls
base:
  Minion1:
    - mysql.init
    - mysql.install
```

入口文件要执行的目标 Minion 是 Minion1，Minion1 上要执行的 SLS 文件为$file_roots/mysql 目录下的 init.sls 和 install.sls。

（3）init.sls 为环境配置文件，内容如下：

```
[root@10.0.0.1 salt]# cat mysql/init.sls
pkg-init:
  pkg.installed:
    - names:
      - gcc
      - gcc-c++
      - glibc
      - make
      - cmake
```

```
        - autoconf
        - libxml2
        - libxml2-devel
        - zlib
        - zlib-devel
        - libcurl
        - libcurl-devel
        - openssl
        - openssl-devel
        - ncurses
        - ncurses-devel
        - libtool
```

其中 pkg-init 为自定义的方法名,其他模块可以直接通过方法名称来调用该方法。pkg.installed 为 SaltStack 的内部函数,检查包是否被安装以及包的版本是否正确。如果包不存在,则会被安装。names 作为可选参数,指定了所有待检查的包名。

(4) install.sls 为安装文件,指定了 Minion 端的脚本调用和命令执行流程以及相关配置文件目录,内容如下:

```
[root@10.0.0.1 salt]# cat mysql/install.sls

include:    # 包括了上文中的环境配置脚本
  - mysql.init

mysql-source-install:
  file.managed:   # 传输源文件到 Minion
    - name: /data/mysql-5.6.34.tar.gz   # Minion 端文件存放位置
    - source: salt://mysql/files/mysql-5.6.34.tar.gz   # 源文件位置
    - user: root
    - group: root
    - mode: 755
  cmd.run:    # 执行解压源码与安装服务功能
    - name: cd /data && tar xf mysql-5.6.34.tar.gz && cd mysql-5.6.34 && cmake -DCMAKE_INSTALL_PREFIX=/usr/local/mysql -DMYSQL_DATADIR=/usr/local/mysql/data -DWITH_MYISAM_STORAGE_ENGINE=1 -DWITH_INNOBASE_STORAGE_ENGINE=1 -DWITH_MEMORY_STORAGE_ENGINE=1 -DWITH_READLINE=1 -DMYSQL_UNIX_ADDR=/var/lib/mysql/mysql.sock -DMYSQL_TCP_PORT=3306 -DENABLED_LOCAL_INFILE=1 -DWITH_PARTITION_STORAGE_ENGINE=1 -DEXTRA_CHARSETS=all -DDEFAULT_CHARSET=utf8 -DDEFAULT_COLLATION=utf8_general_ci && make && make install
    - unless: test -d /usr/local/mysql    # 判断条件若为真,则不执行该模块

mysql-init:
  file.managed:   # 传输配置文件脚本
    - name: /data/init.sh
    - source: salt://mysql/init.sh
    - user: root
    - group: root
    - mode: 755
  cmd.script:    # 执行脚本
    - name: /data/init.sh
    - require:    # 执行配置文件脚本时,需要 mysql-source-install 先执行成功
      - cmd: mysql-source-install

mysql-config:
  file.managed:   # 传输 my.cnf 文件
    - name: /etc/my.cnf
    - source: salt://mysql/files/my.cnf
    - user: root
    - group: root
    - mode: 644
    - require:
      - file: mysql-init

mysql-service:
  file.managed:   # 传输 mysqld 文件
    - name: /etc/init.d/mysqld
    - source: salt://mysql/files/mysqld
    - user: root
    - group: root
```

```
      - mode: 755
    cmd.run:    # 添加启动项
      - name: chkconfig --add mysqld
      - unless: chkconfig --list |grep mysqld
      - require:
        - file: mysql-service
    service.running:    # 启动服务
      - name: mysqld
      - require:
        - cmd: mysql-service
```

（5）在 install.sls 中调用了 init.sh 配置文件，用于完成 MySQL 服务安装后的配置用户组、初始化数据等工作：

```
[root@10.0.0.1 salt]# cat mysql/init.sh

#!/bin/bash
groupadd mysql
useradd -r -g mysql mysql
chown mysql:mysql /usr/local/mysql/ -R
ln -sv /usr/local/mysql/bin/mysql /usr/bin
/usr/local/mysql/scripts/mysql_install_db --user=mysql --basedir=/usr/local/mysql --datadir=/usr/local/mysql/data
```

在上述的几个文件中，Master 节点的 SLS 文件中定义了编译安装 MySQL 的整体流程，接下来在 Master 节点执行 SlatStack 的 state.highstate 命令，使指定 Minion 节点（Minion1）的环境生效，即完成了 Minion 节点的 MySQL 服务的搭建。

```
[root@10.0.0.1 salt]# salt Minion1 state.highstate
...
```

编译安装耗时较长，需要耐心等待结果返回。安装完成后，在 Minion 节点可以查看到 MySQL 实例已经启动。

```
[root@10.0.0.2 ~]# ps -ef|grep mysql
    mysql    4981  4801  9 20:33 pts/1    00:00:00 /usr/local/mysql/bin/mysqld --basedir=/usr/local/mysql --datadir=/usr/local/mysql/data --plugin-dir=/usr/local/mysql/lib/plugin --user=mysql --log-error=/usr/local/mysql/data/Minion1.err --pid-file=/usr/local/mysql/data/Minion1.pid --socket=/tmp/mysql_3306.sock --port=3306
```

在 29.5 节中也将采用如上的流程来安装 MySQL 实例。

29.2.3 后台 API

后台 API（Application Programming Interface）是系统中最核心的部分之一，它连接了客户端、后台数据库和批量管理系统这三者，是传输数据和请求命令的唯一途径，因此后台 API 的可用性和接口的丰富性至关重要。

Django Rest Framework 是基于 Django 的扩展框架，在 Django 的基础上提供了规范的 REST（Representational State Transfer）API 模板，通过序列化的方式提供了数据的验证和渲染功能，并拥有直观的接口调试界面。这里选用 Django Rest Framework 作为 CMDB 系统的后台 API 框架。下面将介绍功能的搭建和具体实现。

1. Django Rest Framework 开发环境准备

（1）首先需要准备 Python 3 和 pip 环境，并在此基础上安装 djangorestframework 模块。

```
[root@10.0.0.1 backend]# pip install django
[root@10.0.0.1 backend]# pip install djangorestframework
```

（2）选定一个开发目录，本例中为/data/backend，新建 cmdb 项目，并在该项目目录中部署 dbms 应用。

```
[root@10.0.0.1 backend]# django-admin.py startproject cmdb
[root@10.0.0.1 backend]# cd cmdb
[root@10.0.0.1 cmdb]# django-admin.py startapp dbms
```

应用创建好之后，开发目录的目录树如下所示：

```
[root@10.0.0.1 cmdb]# tree
.
├── cmdb
│   ├── __init__.py
│   ├── settings.py
│   ├── urls.py
│   └── wsgi.py
├── dbms
│   ├── admin.py
│   ├── apps.py
│   ├── __init__.py
│   ├── migrations
│   │   └── __init__.py
│   ├── models.py
│   ├── tests.py
│   ├── views.py
│   └── serializers.py
└── manage.py
```

需要关注以下几个文件。

- settings.py：项目配置文件，包含项目的应用信息、权限管理、数据库连接串等。
- models.py：模型层文件，完成数据到代码间的 ORM（Object Relational Mapping），包含了 CMDB 系统的数据库的抽象映射关系。
- serializers.py：序列化文件，主要完成数据结构的转换与调整工作。
- views.py：视图层文件，主要完成请求处理函数的定义，并实现请求的路由和 Response。
- urls.py：路由配置文件，定义了请求 url 和视图层处理函数之间的映射关系。

2. 搭建 CMDB 系统的后台 API

在上文中，一个基本的 Django Rest Framework 开发环境已准备好，接下来我们会逐一介绍如何完善上述文件的代码，实现 CMDB 系统的后台 API 的服务搭建。下面以服务器资产信息管理为例，详细介绍如何搭建一个符合 REST 规范的接口服务。

（1）修改 cmdb/settings.py，为 cmdb 项目添加基本的配置信息。

```
[root@10.0.0.1 cmdb]# vi settings.py
# 为项目添加 dbms 应用
INSTALLED_APPS = [
    …
    'rest_framework',
    'dbms'
]

# 设置登录用户才可访问接口
REST_FRAMEWORK = {
    'DEFAULT_PERMISSION_CLASSES': [
        'rest_framework.permissions.IsAdminUser',
    ]
}
# 数据库连接串信息
DATABASES = {
    'default': {
        'ENGINE': 'django.db.backends.mysql',
        'NAME': 'xxx',
        'USER': 'xxx',
```

```
            'PASSWORD': xxx,
            'HOST': '10.0.0.2',
            'PORT': 3306,
    }
}
```

本例中只给出了其中部分配置项,感兴趣的读者可以参考官方文档获取更多配置项的细节(*https://www.django-rest-framework.org*)。

另外,由于 Django 的 MySQL 驱动识别问题,在 Python 3 环境下需要在安装了 pymysql 模块的基础上修改项目的__init__.py 文件解决兼容性问题:

```
[root@10.0.0.1 cmdb]# pip install pymysql
[root@10.0.0.1 cmdb]# vi __init__.py

import pymysql
pymysql.install_as_MySQLdb()
```

(2)修改 dbms/models.py,初步实现模型层的设计,定义一个用于记录服务器基本信息的表单结构,包括各个字段的数据类型、默认值等信息。

```
[root@10.0.0.1 dbms]# vi models.py
    from django.db import models
    from datetime import datetime

class TbServerInfo(models.Model):
    salt_id = models.CharField(max_length=20, verbose_name="服务器 SALT ID")
    ip = models.CharField(max_length=20, verbose_name="服务器 IP")
    ssh_port = models.IntegerField(verbose_name="登录端口")
    server_type = models.CharField(max_length=10, verbose_name="服务器类型")
    status = models.IntegerField(verbose_name="服务器状态")
    create_time = models.DateTimeField(default=datetime.now, verbose_name="创建时间")
    update_time = models.DateTimeField(null=True, verbose_name="更新时间")
```

(3)修改 dbms/serializers.py,对模型层定义的表单模型进行序列化。序列化是 Django REST Framework 的最大特点,能将 Python 的数据结构转换为 JSON 或 XML 等格式,并保存转换后的数据。下例中 fields 属性设置为 all,代表序列化所有字段,也可设置为包含模型中部分字段名的列表,从而序列化指定的字段。

```
[root@10.0.0.1 dbms]# vi serializers.py

from rest_framework import serializers
from .models import TbServerInfo

class ServerInfoSerializer(serializers.ModelSerializer):
    class Meta:
        model = TbServerInfo
        fields = "__all__"
```

(4)修改 dbms/views.py,借助 Django REST Framework 的 mixins 方法完成视图层的设计。

```
[root@10.0.0.1 dbms]# vi views.py

from rest_framework import viewsets, mixins
from rest_framework.permissions import IsAuthenticated
from .serializers import ServerInfoSerializer
from .models import TbServerInfo

class ServerViewSet(mixins.ListModelMixin, mixins.CreateModelMixin, mixins.UpdateModelMixin,
mixins.RetrieveModelMixin, viewsets.GenericViewSet):
    queryset = TbServerInfo.objects.all()
    serializer_class = ServerInfoSerializer
```

此处创建的 ServerViewSet 类通过继承关系,可以调用 mixins 中的 ListModelMixin、CreateModelMixin、UpdateModelMixin 等方法,简单地实现了对服务器基本信息表单的读取、插入和已有记录更新功能,极大地简化了开发流程。当然,也可以通过重写 mixins 的 create

方法来对具体的 POST 请求的响应方法进行定制，或者自定义其他 mixins 中没有的方法来实现其他请求响应方法。

（5）修改 cmdb/urls.py，完成路由层的定义。

```
[root@10.0.0.1 cmdb]# vi urls.py
from django.conf.urls import url, include
from rest_framework import routers
from dbms import views

router = routers.DefaultRouter()
router.register(r'server', views.ServerViewSet)

urlpatterns = [
    …
    url(r'^api/', include(router.urls)),
    url(r'^api-auth/', include('rest_framework.urls', namespace='rest_framework'))
]
```

以上代码表示所有访问 api/ 节点的请求都路由到 router.urls 中处理，而 router 定义了 server 路由前缀，对应的是 views.ServerViewSet 这一视图集合，即可以通过访问 api/server 来请求视图下的方法。

上述代码中完成了模型定义、序列化方法定义，以及响应方法和路由的定义，接下来通过下述命令将模型导入到数据库，并在默认的 8000 端口启动服务。

```
[root@10.0.0.1 cmdb]# python manage.py makemigrations    # 记录对 models.py 的改动
[root@10.0.0.1 cmdb]# python manage.py migrate           # 将改动同步到数据库
[root@10.0.0.1 cmdb]# python manage.py runserver
```

访问 http://127.0.0.1:8000/api-auth/login/ 来登录系统，登录成功后访问 http://127.0.0.1:8000/api/server/ 即可查看到服务器列表页，获取已录入数据库的服务器信息。

Django Rest Framework 提供了非常友好的可视化界面用于接口调试，如图 29-4 所示。通过该页面下方提供的表单录入一条服务器信息，再次刷新该接口可以看到数据已录入，如图 29-5 所示。当然除了直接在该调试界面提交数据之外，也可以直接构建 POST 请求来访问接口，例如，在后文的客户端搭建中将使用 axios 访问接口来实现各类功能。

图 29-4　服务器列表页 1

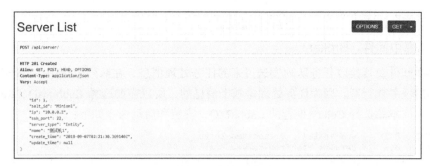

图 29-5　服务器列表页 2

29.3　任务调度系统

MySQL 自动化运维平台需要处理大量的高并发任务，其中不仅包括长耗时的实例安装任务，还包括定时类的多实例监控任务等，因此任务调度系统尤为关键。这里选用 Celery 作为任务调度系统的技术框架。

29.3.1　Celery 安装

Celery 是基于 Python 开发一个分布式任务调度模块，支持耗时较长任务的异步并行执行，并提供定时任务调度功能，因此很适合辅助 MySQL 自动化运维平台完成各类任务的队列分发和消息处理。下面将介绍基于 Celery 搭建任务调度系统的具体实现，并简介 Celery 的监控管理工具 Flower。

Celery 的调度流程包括任务队列、消息中间件、任务执行者和持久化数据四大部分，如图 29-6 所示。各个功能简介如下。

图 29-6　Celery 调度流程图

1．任务队列（Producer）

任务队列包括异步任务（Async Task）和定时任务（Celery Beat）两部分，Celery 在接收到任务后直接发送到消息中间件中等待处理，两种任务的处理流程分别如下。

（1）异步任务包括直接 API 调用，django-celery 中的请求等，可以将一段逻辑代码发送到消息中间件排队等待处理。系统中可以存在多个异步任务请求并行发送任务。

（2）定时任务以独立的进程存在，通过读取配置文件中的内容，周期性地将执行任务的请求发送到消息中间件来完成调度。系统中只能有一个定时任务调度者。

发送异步消息或定时任务请求都需要安装 Celery 模块，由于后台 API 项目基于 Django，这里同时安装上集成模块 django-celery。

```
[root@10.0.0.1 ~]# pip install celery
[root@10.0.0.1 ~]# pip install django-celery
```

2. 消息中间件（Broker）

消息中间件负责接收任务队列发送过来的任务处理消息，在队列中排序并将任务逐个发送给空闲的任务执行者。因为任务处理是基于消息的，所以需要选择 RabbitMQ、Redis 等作为消息队列。本章选择 Celery 推荐的 RabbitMQ，安装和启动命令如下：

```
[root@10.0.0.1 ~]# yum install -y rabbitmq-server
[root@10.0.0.1 ~]# service rabbitmq-server start
[root@10.0.0.1 ~]# rabbitmqctl start_app
Starting node 'rabbit@10.0.0.1' ...
...done.
```

这样 RabbitMQ 便启动了，默认监听端口为 5672。

3. 任务执行者（Worker）

任务执行者是操作系统中的一组进程，它实时监控消息中间件发来的任务处理请求，并完成任务的处理。Worker 可通过 Celery 直接启动，支持分布式部署和横向扩展，可以在多个节点增加 Worker 的数量来增加系统的高可用性。在分布式系统中，可以在不同节点上分配执行不同任务类型的 Worker 来达到模块化的目的。

4. 持久化数据（Backend）

Celery 支持将任务处理过程中的状态信息及结果保存，包括任务的执行结果、异常信息、执行参数和起止时间等内容。在生产项目中我们推荐选用 CMDB 数据库做 Backend。为了方便测试，本节我们选用 RabbitMQ 作为 Backend。

29.3.2 Celery 任务部署

本节以安装 MySQL 为例，对 Celery 的异步任务和定时任务的部署调度进行介绍。

1. 异步任务调度

（1）配置任务文件。指定消息中间件和 Celery 任务队列的进程名称。借助 app.task 装饰器，将 install_mysql 函数注册为一个可以被异步执行的任务。

```
[root@10.0.0.1 ~]# vi install.py

from celery import Celery
import time

# 配置上文中的 RabbitMQ 作为消息中间件
broker = 'amqp://guest@localhost:5672//'
# 启动一个名为 install 的 Celery 进程作为任务队列
app = Celery('install', broker=broker, backend=broker)

# 配置路由
app.conf.CELERY_ROUTES = {
    "install.install_mysql": {
        "queue": "long"
    }
}

@app.task
def install_mysql(**kwargs):
    time.sleep(10) # 模拟安装 MySQL 过程
    return 0
```

29.3 任务调度系统

（2）启动 Worker：

```
[root@10.0.0.1 ~]# celery worker -A install -Q long --loglevel=info -n query
```

其中-A 参数指定了 Celery 进程的名称。-Q 代表 Worker 接受指定名称路由的任务，在实际项目中我们需要指定多个路由名称用以区分定时任务或耗时较长的异步任务。--loglevel 则代表 worker 的日志级别。-n 指定了该 Worker 的名称。启动 Worker 后我们执行任务。

（3）执行异步任务：

```
[root@10.0.0.1 ~]# python
>>> from install import install_mysql
>>> res = install_mysql.delay(a=1)
>>> res.ready() #任务尚未完成
False
>>> res.ready() #任务已经完成
True
>>> res.get() #获取任务结果
0
```

经过测试可以看到，10s 后返回了任务执行结果，而且该任务也并未阻塞命令行中的其他功能的执行。另外，在本例中开启 Worker 时指定了 info 级别的日志，因此也可以通过 Worker 端的日志来查看任务的执行情况。

```
[tasks]
  . install.install_mysql

 [2018-10-10 18:16:03,809: INFO/MainProcess] Connected to amqp://guest:**@127.0.0.1:5672//
 [2018-10-10 18:16:03,822: INFO/MainProcess] mingle: searching for neighbors
 [2018-10-10 18:16:04,836: INFO/MainProcess] mingle: sync with 3 nodes
 [2018-10-10 18:16:04,836: INFO/MainProcess] mingle: sync complete
 [2018-10-10 18:16:04,851: WARNING/MainProcess] celery@install ready.
  [2018-10-10 18:16:18,979: INFO/MainProcess] Received task: install.install_mysql[9e891484-ac08-4d28-9c48-f44551be49b1]
  [2018-10-10 18:16:29,004: INFO/MainProcess] Task install.install_mysql[9e891484-ac08-4d28-9c48-f44551be49b1] succeeded in 10.022355429828167s: 0
```

2. 定时任务调度

Celery Beat 进程能够周期性的调度已有任务，实现定时任务部署的功能。下面详细介绍如何将上述 MySQL 安装任务配置为定时调度任务。

（1）在已有任务的任务文件中，添加以下周期调度的相关代码，指定调度的任务名称（install.install_mysql）和触发周期（一分钟一次）。

```
[root@10.0.0.1 ~]# vi install.py

from datetime import timedelta
…
app.conf.update(CELERYBEAT_SCHEDULE = {
    'check_mysql': {
        'task': 'install.install_mysql',
        'schedule': timedelta(seconds=60),
        'args': None
    },
})
…
```

（2）重启 Worker 后，再将 Celery Beat 进程启动，即完成了定时任务的配置和周期性调度。

```
[root@10.0.0.1 ~]# celery beat -A install

celery beat v3.1.26.post2 (Cipater) is starting.
__    -    ... __   -        _
Configuration ->
    . broker -> amqp://guest:**@localhost:5672//
    . loader -> celery.loaders.app.AppLoader
```

```
. scheduler -> celery.beat.PersistentScheduler
. db -> celerybeat-schedule
. logfile -> [stderr]@%INFO
. maxinterval -> now (0s)
[2018-10-10 19:13:25,889: INFO/MainProcess] beat: Starting...
[2018-10-10 19:13:25,908: INFO/MainProcess] Scheduler: Sending due task check_mysql (install
.install_mysql)
[2018-10-10 19:14:25,908: INFO/MainProcess] Scheduler: Sending due task check_mysql (install
.install_mysql)

Worker Log Info:
[2018-10-10 19:13:29,041: INFO/MainProcess] Received task: install.install_mysql[ad48db14-
bd08-4505-9d3c-d442f3f920b4]
[2018-10-10 19:13:33,212: INFO/MainProcess] Events of group {task} enabled by remote.
[2018-10-10 19:13:39,072: INFO/MainProcess] Task install.install_mysql[ad48db14-bd08-4505-
9d3c-d442f3f920b4] succeeded in 10.02820060774684s: 0
[2018-10-10 19:14:25,910: INFO/MainProcess] Received task: install.install_mysql[bbd6d7e3-
e12c-4ddb-aaf6-ff93631bfdf6]
[2018-10-10 19:14:35,943: INFO/MainProcess] Task install.install_mysql[bbd6d7e3-e12c-4ddb-aaf6
-ff93631bfdf6] succeeded in 10.030340616591275s: 0
```

29.3.3　Flower 监控

前面对 Celery 的架构和调度配置进行了介绍。本节将介绍一款基于 Web 端的 Celery 的监控和管理工具 Flower。

Flower 支持对 Celery 事件进行实时监控，包括查看任务进程与历史、显示任务的详细信息、统计事件执行频率等功能。同时 Flower 还可以远程控制 Celery，包括启动关闭 Worker、控制进程池大小、撤销或终止任务等，并提供了 API 来调度这些功能。

Flower 的安装部署非常简单，在 5555 端口启动 Flower，默认选择 RabbitMQ 作为 Broker：

```
[root@10.0.0.1 ~]# pip install flower
[root@10.0.0.1 ~]# flower --port=5555

[I 181010 19:24:31 command:139] Visit me at http://localhost:5555
[I 181010 19:24:31 command:144] Broker: amqp://guest:**@localhost:5672//
[I 181010 19:24:31 command:147] Registered tasks:
    ['celery.backend_cleanup',
     'celery.chain',
     'celery.chord',
     'celery.chord_unlock',
     'celery.chunks',
     'celery.group',
     'celery.map',
     'celery.starmap']
[I 181010 19:24:31 mixins:231] Connected to amqp://guest:**@127.0.0.1:5672//
```

服务启动后，登录 http://127.0.0.1:5555 即可访问 Flower 的 Web 界面，如图 29-7 所示。

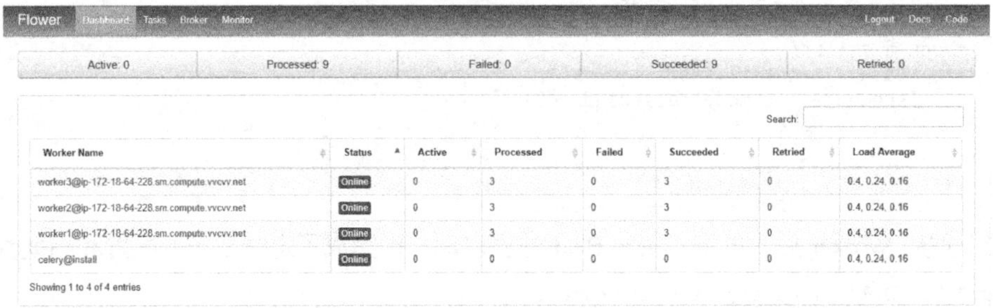

图 29-7　Flower 界面

在 Tasks 菜单中，可以查看每个任务的名称、参数、执行状态、结果、起止时间、执行时间，以及 Worker 名称等内容，可以说是记录了 Celery 调度的完整运行状态。打开 Celery Beat 进程，可以观察到 celery@install 这个 Worker 的详细信息，如图 29-8 所示。

图 29-8　Flower 任务执行详情

在 29.5 节中，将继续使用 Flower 监控任务执行情况。

29.4　客户端搭建

客户端在此主要指 Web 客户端，是在浏览器端可视化的界面。本章选择 Vue.js 作为 MySQL 自动化运维平台的客户端开发框架，实现对 CMDB 数据库信息和实例部署维护等任务调度情况的可视化展示。

29.4.1　Vue.js 简介

Vue.js 是一套用于构建用户界面的渐进式框架，可以构建复杂的单页应用。选用 Vue.js 主要由于下面两大特性。

1．响应式数据绑定

Vue.js 采用如图 29-9 所示的 MVVM 模型（Model–View–ViewModel），双向绑定 JavaScript 数据和 HTML 数据，只要一方数据更新，另一方也会随之更新，这将复杂的前端数据显示与更新自动化，有效地提高了开发效率。MVVM 模型中，View 指代页面展示视图，Model 指代模型数据，ViewModel 通过 DOM listeners 和 Data Bindings 这两个工具，分别实现 Model 和 DOM 层之间数据的双向交换。

2．组件化

Vue.js 采用组件化思想将应用界面的每个模块都抽象为一个组件，每个组件的作用域各自独立，通过定义事件完成组件之间的通信，实现复杂业务需求的拆分化简。如图 29-10 所示，一个单页应用拆分为一个由样式与功能各异的组件构成的组件树，每个组件都是一个拥有预定义选项的 Vue 实例。

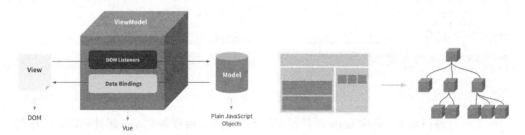

图 29-9　Vue.js 的 MVVM 模式　　　　　图 29-10　Vue 组件化框架

此外，Vue.js 也具有很强的扩展性，支持路由、AJAX、数据流等多种功能，感兴趣的读者可以前往官方网站详细了解。下面将演示使用 Vue 脚手架搭建项目的实战流程，构建请求至后台 API，完成 CMDB 数据的显示。

29.4.2　Vue 项目搭建

1. 搭建 Vue.js 开发环境

Vue-cli 是 Vue.js 的脚手架工具，可以帮助用户直接生成一个可运行的 Vue 项目，具体搭建流程如下：

（1）安装 Node.js 与 Vue-cli。

```
[root@10.0.0.1 ~]# yum install -y nodejs
[root@10.0.0.1 ~]# npm install -g vue-cli
```

（2）部署项目，初始化了一个名为 vue_tutorial 的项目。

```
[root@10.0.0.1 ~]# vue init webpack vue_tutorial

? Project name vue_tutorial
? Project description vue tutorial
? Author Mikuru
? Vue build standalone
? Install vue-router? Yes
? Use ESLint to lint your code? Yes
? Pick an ESLint preset Standard
? Set up unit tests No
? Setup e2e tests with Nightwatch? No
? Should we run `npm install` for you after the project has been created? (recommended) npm

vue-cli · Generated "vue_tutorial".

# Project initialization finished!
# ========================

To get started:

  cd vue_tutorial
  npm run dev
```

vue_tutorial 项目创建好之后，项目的目录树如下，之后开发的主要工程文件都放置在 src 目录中。

```
[root@10.0.0.1 ~]# cd vue_tutorial/
[root@10.0.0.1 vue_tutorial]# tree -aL 1
.
├── .babelrc
├── build   # 存放 webpack 信息
├── config  # 存放各种环境配置信息
├── .editorconfig
├── .eslintignore
├── .eslintrc.js
├── .gitignore
├── index.html
├── node_modules  # npm 所安装的各种依赖
├── package.json  # node_modules 中依赖的包名
├── .postcssrc.js
├── README.md
├── src  # 放置 Vue 工程文件
└── static  # 静态文件存放位置
```

src 工程目录树如下，其中 main.js 是程序的入口文件，负责各个 Vue 组件和路由的注册，App.vue 和 index.js 则分别是项目的入口文件和根路由文件。

```
[root@10.0.0.1 src]# tree
.
├── App.vue
├── assets
│   └── logo.png
├── components       # 组件库目录
│   └── HelloWorld.vue   # Vue 文件，也是主界面显示的文件内容
├── main.js
└── router
    └── index.js     # 根路由文件
```

（3）启动项目服务。

```
[root@10.0.0.1 vue_tutorial]# npm run dev

> vue_tutorial@1.0.0 dev /root/vue_tutorial
> webpack-dev-server --inline --progress --config build/webpack.dev.conf.js

 DONE  Compiled successfully in 5226ms    3:58:38 PM
Your application is running here: http://localhost:8080
```

服务默认在本地 8080 端口上，通过浏览器访问 http://localhost:8080，可以看到 Vue 应用主页如图 29-11 所示。至此，一个 Vue.js 开发环境已经搭建好。

图 29-11　Vue 应用主页

2．实现一个功能组件

项目搭建好后，下面介绍如何增加一个数据双向绑定的组件。

（1）在路由文件 index.js 添加路由 test，指向 Test.vue 文件。

```
[root@10.0.0.1 src]# vi router/index.js
import Vue from 'vue'
import Router from 'vue-router'
import HelloWorld from '@/components/HelloWorld'
import Test from '@/components/Test'

Vue.use(Router)

export default new Router({
  routes: [
    {
      path: '/',
      name: 'HelloWorld',
      component: HelloWorld
    },
    {
      path: '/test',
      name: 'Test',
      component: Test
```

（2）创建 Test.vue 文件，增加结构层（DOM）、行为逻辑层（JavaScript）和样式层（CSS）代码，实现了简易的输入框和 DOM 数据的双向绑定功能。

```
[root@10.0.0.1 src]# vi components/Test.vue
<template>
  <div id="test">
    <p>{{ message }}</p>
    <input v-model="message"/>
  </div>
</template>

<script>
export default {
  name: 'Test',
  data () {
    return {
      message: 'test'
    }
  }
}
</script>

<style>
</style>
```

Vue-cli 默认为热更新加载代码，直接访问 http://localhost:8080/#/test，即可以看到刚才添加的组件页面，如图 29-12 所示。

此时输入框中的值和 DOM 数据已经进行了双向绑定，当修改输入框的值时，DOM 中的值会同步发生变化。同样，在控制台中修改 DOM 中的值，输入框中的值也会随之同步，如图 29-13 所示。

图 29-12　Test.vue 界面　　　　图 29-13　修改内容后 DOM 发生变化

此外，Vue 在模板层还提供了大量其他功能，例如条件渲染、列表渲染、自定义事件处理、组件绑定等。在 29.5 节中会涉及这些功能，感兴趣的读者请查阅官方文档进行了解和学习。

3. Vue 构建前后端交互请求

在 CMDB 系统中，我们使用 Django Rest Framework 实现了数据调度的 API，客户端需要调用这些 API 来完成数据的读取与修改。这里我们选用官方推荐的 axios 作为后端请求的工具。axios 是一个基于 promise 的 HTTP 库，可以在浏览器中创建 XMLHTTPRequests 来请求后端 API，支持各种类型数据的转换，同时也支持拦截器和防跨域请求等。下面我们逐步介绍使用 axios 完成前后端请求的流程：

（1）安装 axios。

```
[root@10.0.0.1 vue_tutorial]# npm install axios
```

（2）在 Test.vue 文件中添加 mounted 方法，通过 axios 请求后台 API 的 server 接口，并将数据显示在文本框中。

```
[root@10.0.0.1 vue_tutorial]# vi src/components/Test.vue
```

```
<script>
import axios from 'axios'
export default {
  name: 'Test',
  data () {
    return {
      message: 'test'
    }
  },
  mounted () {
    axios
      .get('http://10.0.0.1:8000/api/server')
      .then(response => (this.message = response))
  }
}
</script>
```

完成后刷新页面,发现数据并没有被加载,查看网页日志会发现如下错误。

```
GET http://10.0.0.1:8000/api/server net::ERR_CONNECTION_TIMED_OUT
Uncaught (in promise) Error: Network Error
    at createError (createError.js?16d0:16)
    at XMLHttpRequest.handleError (xhr.js?ec6c:87)
```

发现问题是请求被拒绝,这就涉及了前后端的跨域问题。出于安全性考虑,后端的项目需要增加跨域访问许可才能接收并处理请求,因此需要修改后端 API 配置。

(3)后台 API 项目安装 django-cors-middleware,并修改配置文件。

```
[root@10.0.0.1 cmdb]# pip install django-cors-middleware
[root@10.0.0.1 cmdb]# vim cmdb/settings.py

INSTALLED_APPS = [
    …
    'corsheaders'
]

MIDDLEWARE = [
    …
    'corsheaders.middleware.CorsMiddleware',
]
CORS_ORIGIN_ALLOW_ALL = True
```

为了解决前后端的跨域问题,只有增加了跨域访问许可后,后台 API 才会接收并处理前端发过来的请求。需要注意的是,本例中开通了内网中所有服务器访问 API 的权限,出于安全方面的考虑,请不要在生产环境中这样配置。

(4)调整客户端的配置文件,此处修改的是 dev 环境的。

```
[root@10.0.0.1 vue_tutorial]# vim config/index.js

module.exports = {
  dev: {
    assetsSubDirectory: 'static',
    assetsPublicPath: '/',
    proxyTable: {
        '/api': {
            target:'http://10.0.0.1:8000/api',
            changeOrigin:true,
            pathRewrite:{
                '^/api': '/'
            }
        }
    },
```

(5)再次修改 Test.vue 文件的接口访问部分,刷新 http://localhost:8080/#/test,已经可看到如图 29-14 所示的界面。

```
[root@10.0.0.1 vue_tutorial]# vim src/components/Test.vue
  mounted () {
```

```
            axios
              .get('/api/server/')
              .then(response => (this.message = response))
        }
    }
```

```
{ "data": [ { "id": 1, "salt_id": "Minion1", "ip": "10.0.0.2", "ssh_port": 22, "server_type": "Entity", "name": "测试机1", "create_time": "2018-09-07T02:21:30.369140Z", "update_time": null } ],
  "status": 200, "statusText": "OK", "headers": { "date": "Fri, 07 Sep 2018 03:44:30 GMT", "vary": "Accept, Cookie, Accept-Encoding", "server": "WSGIServer/0.2 CPython/3.4.8", "x-frame-
  options": "SAMEORIGIN", "x-powered-by": "Express", "allow": "GET, POST, HEAD, OPTIONS", "content-type": "application/json", "connection": "keep-alive", "content-length": "166" }, "config":
  { "transformRequest": {}, "transformResponse": {}, "timeout": 0, "xsrfCookieName": "XSRF-TOKEN", "xsrfHeaderName": "X-XSRF-TOKEN", "maxContentLength": -1, "headers": { "Accept":
                                          "application/json, text/plain, */*" }, "method": "get", "url": "/api/server/", "request": {} } }
                                                                    [object Object]
```

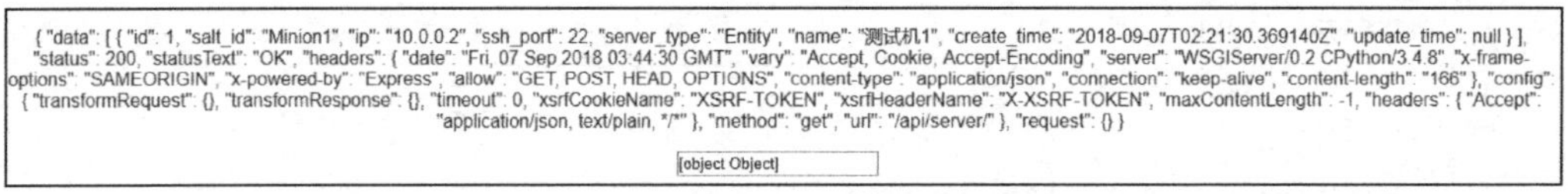

图 29-14　axios 请求 API 获取的全部信息

很明显这并不是我们想要的，我们只需要显示 data 中的信息，并且需要更清晰的数据格式。

（6）进一步修改 Test.vue 文件，添加清晰的 DOM 样式和异常处理方法。再次刷新页面，如图 29-15 所示。

```
[root@10.0.0.1 vue_tutorial]# vim src/components/Test.vue

<template>
  <div id="test">
    <h1>服务器信息</h1>

    <section v-if="error">
      <p>读取服务器信息出错</p>
    </section>

    <section v-else>
      <div v-for="i in message" :key="i">
        <div v-for="(key,item) in i" :key="key">
          {{ item }} : {{ key }}
        </div>
      </div>
    </section>

  </div>
</template>

<script>
import axios from 'axios'
export default {
  name: 'test',
  data () {
    return {
      message: null,
      error: false
    }
  },
  mounted () {
    axios
      .get('/api/server/')
      .then(response => {
        this.message = response.data
      })
      .catch(error => {
        console.log(error)
        this.error = true
      })
  }
}
</script>

<style>
</style>
```

图 29-15 中得到了一个较为清晰的显示结果，但仍与需要的样式不同，主要原因是我们

并没有修改 Vue 文件中的 style，即 CSS 配置。读者可以进一步为 Test.vue 添加 CSS 配置部分，以实现更复杂的样式。由于篇幅限制，在 29.5 节中将不会讨论样式配置相关的内容。

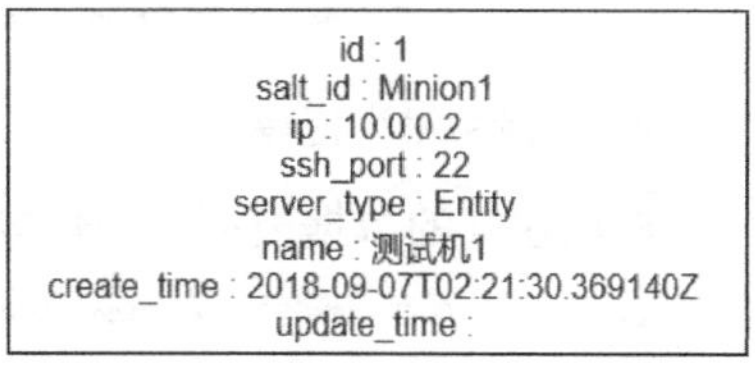

```
id : 1
salt_id : Minion1
ip : 10.0.0.2
ssh_port : 22
server_type : Entity
name : 测试机1
create_time : 2018-09-07T02:21:30.369140Z
update_time :
```

图 29-15　修改后的 axios 请求结果

29.5　自动化运维平台实战

结合上文中所提到的技术框架和技术要点，本节将带领大家从头部署一套完整的 MySQL 自动化运维平台。

29.5.1　搭建 CMDB

在前文中我们已经搭好了 SaltStack 管理环境和后台 API 项目，在此基础之上进行整个系统的开发。

1. 搭建 CMDB 数据库

修改 models.py，设置服务器和 MySQL 的主要属性，保存文件并将数据模型同步至数据库。

```
[root@10.0.0.1 cmdb]# vi dbms/models.py
from django.db import models
from datetime import datetime

# 服务器信息
class TbServerInfo(models.Model):
    SERVER_TYPE = (
        ("Entity", "实体机"),
        ("KVM", "虚拟机"),
    )
    salt_id = models.CharField(max_length=20, verbose_name="服务器 SALT ID")
    ip = models.CharField(max_length=20, verbose_name="服务器 IP")
    ssh_port = models.IntegerField(verbose_name="登录端口")
    server_type = models.CharField(max_length=10, choices= SERVER_TYPE, verbose_name="服务器类型")
    status = models.IntegerField(verbose_name="服务器状态")
    create_time = models.DateTimeField(default=datetime.now,verbose_name="创建时间")
    update_time = models.DateTimeField(null=True, verbose_name="更新时间")

    class Meta:
        verbose_name = "服务器信息"
        verbose_name_plural = verbose_name
        db_table = "tb_server_info"

# 数据库信息
class TbMysqlInfo(models.Model):
    server = models.ForeignKey(TbServerInfo, on_delete=models.CASCADE, verbose_name="所属服务器")
    port = models.IntegerField(verbose_name="数据库端口")
    version = models.CharField(max_length=50, verbose_name="版本")
    instance_user = models.CharField(max_length=20, verbose_name="实例用户")
    cnf_path = models.CharField(max_length=100, verbose_name="配置文件路径")
    status = models.IntegerField(default=0, verbose_name="数据库状态")
    create_time = models.DateTimeField(default=datetime.now, verbose_name="创建时间")
    update_time = models.DateTimeField(null=True, verbose_name="更新时间")

    class Meta:
        verbose_name = "数据库信息"
        verbose_name_plural = verbose_name
        db_table = "tb_mysql_info"

# 同步数据库
```

```
[root@10.0.0.1 cmdb]# python manage.py makemigrations
[root@10.0.0.1 cmdb]# python manage.py migrate
```

2. 后台 API 搭建

根据建好的数据模型序列化数据，完成视图层和路由的配置。

```
[root@10.0.0.1 cmdb]# vi dbms/serializers.py
from rest_framework import serializers
from .models import TbServerInfo, TbMysqlInfo

# 序列化数据库信息
class MysqlInfoSerializer(serializers.ModelSerializer):
    ip = serializers.CharField(source='server.ip')
    salt_id = serializers.CharField(source='server.salt_id')

    class Meta:
        model = TbMysqlInfo
        fields = "__all__"

# 序列化服务器信息
class ServerInfoSerializer(serializers.ModelSerializer):
    instances = MysqlInfoSerializer(many=True, read_only=True)

    class Meta:
        model = TbServerInfo
        fields = "__all__"

[root@10.0.0.1 cmdb]# vi dbms/views.py
from rest_framework import viewsets, mixins
from .serializers import ServerInfoSerializer, MysqlInfoSerializer
from .models import TbServerInfo, TbMysqlInfo

# 服务器视图
class ServerViewSet(mixins.ListModelMixin, mixins.CreateModelMixin, mixins.UpdateModelMixin,
mixins.RetrieveModelMixin, viewsets.GenericViewSet):
    queryset = TbServerInfo.objects.all()
    serializer_class = ServerInfoSerializer
    # 增加过滤字段选项
    filter_fields = ('salt_id', 'ip', 'server_type', 'status')

# 数据库视图
class MysqlViewSet(mixins.ListModelMixin, mixins.CreateModelMixin, mixins.UpdateModelMixin,
mixins.RetrieveModelMixin, viewsets.GenericViewSet):
    queryset = TbMysqlInfo.objects.all()
    serializer_class = MysqlInfoSerializer
    filter_fields = ('server', 'status')

[root@10.0.0.1 cmdb]# vi cmdb/urls.py
from django.conf.urls import url, include
from django.contrib import admin
from rest_framework import routers
from dbms import views

router = routers.DefaultRouter()
router.register(r'server', views.ServerViewSet)
router.register(r'mysql', views.MysqlViewSet)

urlpatterns = [
    url(r'^admin/', admin.site.urls),
    url(r'^api/', include(router.urls)),
    url(r'^api-auth/', include('rest_framework.urls', namespace='rest_framework')),
]
```

3. 完成分页和过滤器配置

分页的主要作用是能让 API 返回限制长度的并且排序后的数据，而过滤器则可以为视图增加过滤字段用作查询条件。下面增加一个 func.py 文件来完成相关的配置。

```
[root@10.0.0.1 cmdb]# pip install djangorestframework-filters
[root@10.0.0.1 cmdb]# vi dbms/func.py

from rest_framework.pagination import PageNumberPagination

# 分页信息配置，限制默认每页长度为 10 条数据，最多 100 条
class StandardResultsSetPagination(PageNumberPagination):
    page_size = 10
    page_size_query_param = 'limit'
    max_page_size = 100

[root@10.0.0.1 cmdb]# vi cmdb/settings.py

# 配置过滤器信息
INSTALLED_APPS = [
    …
    'django_filters'
]
REST_FRAMEWORK = {
    …
    'DEFAULT_FILTER_BACKENDS': ('django_filters.rest_framework.DjangoFilterBackend',),
    'DEFAULT_PAGINATION_CLASS': 'dbms.function.StandardResultsSetPagination',
}
```

这样数据配置相关的 API 已经完成了，之后将加入 MySQL 的安装功能。

4．安装 MySQL 功能的实现

上文中主要完成了服务器和数据库的数据调度的 API，还需要增加安装 MySQL 的接口。具体操作需要首先配置前文提到的 MySQL 安装代码，然后通过调用 SaltStack 的 Python API 来完成该逻辑。

根据需求，客户端会通过 POST 请求将创建一个 MySQL 实例的参数发送给/api/server/install/接口，接口启动一个 SaltStack 的本地客户端，通过 cmd 方法让本机的 salt-master 客户端来提交 state.highstate 请求。如果 salt 请求成功执行，则需要调用 objects.save()方法存储数据库信息在 CMDB 中。

```
[root@10.0.0.1 cmdb]# vi dbms/views.py

from .func import install_mysql
from rest_framework.response import Response
from rest_framework import status

class MysqlViewSet(…):
    …
    # 增加路由，匹配 install 接口，仅允许 POST 请求
    @action(methods=['post'], detail=False)
    def install(self, request):
        data = request.data
        # 执行安装任务
        install_mysql(infos=data)
        return Response('success', status=status.HTTP_200_OK)

[root@10.0.0.1 cmdb]# vi dbms/func.py

from .models import TbMysqlInfo
import salt.client

…
def install_mysql(**kwargs):
    # 根据客户端的传参来获取服务器的 salt id
    salt_id = TbServerInfo.objects.values_list('salt_id', flat=True).get(pk=kwargs['infos']['server_id'])
    # 启动本地客户端，通过 cmd.run 方法执行了 state.highstate 命令，来完成数据库的安装工作
    res = salt.client.LocalClient().cmd(kwargs['salt_id'][0],
                                        "state.highstate",
                                        show_timeout=True,
```

```
            )
    if res[salt_id]:
        # 安装成功,需要存储安装好的数据库信息
        return TbMysqlInfo.objects.create(**kwargs['infos'])
    else:
        raise Exception('failure')
```

以上就是数据库安装和信息保存的后台 API 的调度实现。

29.5.2 搭建任务调度平台

针对上面耗时较长的 MySQL 安装功能,可以将这个安装任务配置成异步任务。这里通过 Celery 的扩展 Django-celery 模块来配置该任务,具体步骤如下。

(1) 配置 Django-celery 并同步数据库。

```
[root@10.0.0.1 cmdb]# pip install django-celery    # 安装 django-celery
[root@10.0.0.1 cmdb]# vi cmdb/settings.py

# 配置 app
INSTALLED_APPS = [
    …
    'djcelery',
]

from celery.schedules import crontab
from datetime import timedelta

BROKER_URL = 'amqp://guest@localhost:5672//'    # Broker 配置
CELERY_RESULT_BACKEND = 'database'
# 数据库配置为 djcelery,与 Django 项目位于相同数据
CELERYBEAT_SCHEDULER = 'djcelery.schedulers.DatabaseScheduler'
CELERY_TIMEZONE = 'Asia/Shanghai'    # 配置时区
CELERY_ENABLE_UTC = False
CELERY_IMPORTS = ('dbms.views', 'dbms.func',)    # 配置 Celery 任务路径

from celery import Celery, platforms
platforms.C_FORCE_ROOT = True    # 允许 root 用户调度 Celery,在生产环境中不推荐

import djcelery
djcelery.setup_loader()    # 启动 django-celery

# 同步 django-celery 自带数据模板信息到数据库中
[root@10.0.0.1 cmdb]# python manage.py makemigrations
[root@10.0.0.1 cmdb]# python manage.py migrate
```

以上配置将 django-celery 的 Broker 和 Backend 信息配置到了 Django 中,配置完成后将表结构同步至数据库。

(2) 修改 MySQL 安装任务为异步调度任务,修改 MysqlViewSet 视图和 install_mysql 方法。

```
[root@10.0.0.1 cmdb]# vi dbms/views.py

# signature 是 celery 的自带方法,允许将任务名称和参数传递给 celery 执行,支持异步和同步执行
from celery import signature

class MysqlViewSet(…):
    …
    @action(methods=['post'], detail=False)
    def install(self, request):
        data = request.data
        # 将直接 install_mysql 的调用改为 signature 任务,并通过 delay 方法来异步执行
        signature('install_mysql', kwargs={'infos': data}).delay()
        return Response('success', status=status.HTTP_200_OK)

[root@10.0.0.1 cmdb]# vi dbms/func.py

from celery.decorators import task
```

```python
# 只要在方法前加上 task 标识即可
@task(name='install_mysql')
def install_mysql(**kwargs):
    salt_id = TbServerInfo.objects.values_list('salt_id', flat=True).get(pk=kwargs['infos']['server_id'])
    res = salt.client.LocalClient().cmd(salt_id,
                                        "state.highstate",
                                        show_timeout=True,
                                        )
    if res[salt_id]:
        return TbMysqlInfo.objects.create(**kwargs['infos'])
    else:
        raise Exception
```

这样便完成了任务调度系统的编写。

（3）启动后台服务。

上文完成了 CMDB 系统和任务调度系统的搭建，现在启动后台服务。

```
[root@10.0.0.1 backend]# python manage.py runserver
[root@10.0.0.1 backend]# python manage.py celery worker -n install
```

29.5.3 搭建客户端

对客户端而言，需要至少完成两个页面来显示系统内容，分别是 MySQL 信息的展示页和 MySQL 安装页面。同时我们也需要构建相应的请求访问 CMDB，来完成与后台的数据交互。由于篇幅限制，本节将仅演示安装页对应的 Vue 文件和 axios 跨域请求文件的部分内容，更多页面文件与样式可以参考 Vue 的相关文档。

根据上文中关于 Vue 的介绍，首先需要完成路由到 Vue 组件的配置，之后编写组件，即数据库安装页面，并在页面中加入 axios 请求方法，最后编写数据请求连通后端，完成数据库安装的异步调度任务。下面将逐条介绍安装流程。

（1）客户端路由配置。

```
[root@10.0.0.1 src]# vi router/index.js

// 引用 Vue 组件
import Install from '@/components/install'
import Mysql from '@/components/mysql'

// 定义路由
export default new Router({
  routes: [
    {
      path: '/install',
      name: 'Install',
      component: Install
    },
    {
      path: '/mysql',
      name: 'Mysql',
      component: Mysql
    }
  ]
})
```

（2）安装页文件配置。这里使用了 Element-UI 作为 UI（User Interface）视觉框架。

```
[root@10.0.0.1 src]# vi src/components/install.vue

<template>
  <div class="app-container">
    <el-form ref="mysql" :model="mysql" label-width="140px">
```

```html
          <el-form-item label="SALT ID">
            <el-select clearable class="filter-item" v-model="mysql.server_id">
              <el-option v-for="item in salt_id_list" :key="item.id" :label="item.salt_id" :value="item.id">
              </el-option>
            </el-select>
          </el-form-item>
          <el-form-item label="版本">
            <el-radio-group v-model="mysql.version">
              <el-radio label="5.6.34" value="5.6.34" />
              <el-radio label="5.7.17" value="5.7.17" />
            </el-radio-group>
          </el-form-item>
          <el-form-item label="端口">
            <el-input v-model="mysql.port"/>
          </el-form-item>
          <el-form-item label="实例用户">
            <el-input v-model="mysql.instance_user" />
          </el-form-item>
          <el-form-item label="配置文件路径">
            <el-input v-model="mysql.cnf_path" />
          </el-form-item>
          <el-form-item>
            <el-button type="primary" @click="onSubmit">创建</el-button>
            <el-button @click="onCancel">重置</el-button>
          </el-form-item>
      </el-form>
    </div>
</template>
```

```js
<script>
// 引用了安装方法
import { createMysql, serverList } from '@/api/mysql'

export default {
  name: 'Mysql',
  data() {
    return {
      mysql: {
        server_id: '',
        version: '',
        port: '',
        instance_user: '',
        cnf_path: '',
      },
      salt_id_list: undefined
    }
  },
  created() {
    // 默认请求所有服务器节点
    serverList().then(response => {
      this.salt_id_list=response.data.results
    })
  },
  methods: {
    onSubmit() {
      // 提交安装数据库表单
      createMysql(this.mysql).then(response => {
        alert('安装命令已经发出')
      })
    }
  }
}
```

（3）axios 请求页，这里仅定义了接口，在 src/utils/request 中创建 axios 请求。

```
[root@10.0.0.1 src]# vi api/mysql.js

// 将 axios 请求封装到单独的文件
import request from '@/utils/request'

// 请求全部 MySQL 信息
```

```javascript
export function mysqlList(query) {
  return request({
    url: '/api/mysql/',
    method: 'get',
    params: query
  })
}

// 请求全部服务器信息
export function serverList(query) {
  return request({
    url: '/api/server/',
    method: 'get',
    params: query
  })
}

// 数据库安装接口
export function createMysql(data) {
  return request({
    url: '/api/mysql/install/',
    method: 'post',
    data
  })
}
```

(4) 创建 axios 请求并配置拦截器。

```
[root@10.0.0.1 src]# vi utils/request.js

import axios from 'axios'
import { Message } from 'element-ui'

// 设置 axios 请求超时时间为 10000ms
const service = axios.create({
  timeout: 10000
})

// 配置拦截器，通过 JSON Web Token 传递消息
service.interceptors.request.use(
  config => {
    if (store.getters.token) {
      config.headers.Authorization = `JWT ${store.getters.token}`
    }
    return config
  },
  error => {
    Promise.reject(error)
  }
)

service.interceptors.response.use(
  response => {
    return response
  },
  error => {
    Message({
      message: error.message,
      type: 'error',
      duration: 5 * 1000
    })
    return Promise.reject(error)
  }
)

export default service
```

(5) 以上内容完成后启动服务。

```
[root@10.0.0.1 frontend]# npm run dev
```

至此项目就已经准备完成，下面来实践操作一下安装流程。

29.5.4 项目演示

在浏览器端访问 http://localhost:8080/#/install/mysql，登录客户端查看安装界面，如图 29-16 所示。

填写数据库信息，单击创建提交请求至后台 API，如图 29-17 所示。

图 29-16　安装 MySQL 客户端界面

图 29-17　提交安装 MySQL 请求

等待一段时间后，登录 http://localhost:5555，即前文提到的 Flower 系统查看任务状态，如图 29-18 所示。

图 29-18　Flower 界面显示任务状态

任务完成后登录客户端，访问 http://localhost:8080/#/mysql 查看已经安装的数据库，如图 29-19 所示。

图 29-19　客户端显示数据库情况

数据库已安装完成。至此安装功能演示完毕。

29.6　小结

本章从技术架构讲起，逐个介绍了搭建 MySQL 自动化运维平台所用到的技术，并完成了一个安装 MySQL 单实例的简易流程。

目前，搭建一套数据库的自动化运维平台来完成运维工作，已经是行业中运维技术的主要发展方向。搭建平台时可用到的技术种类繁多，读者不必拘泥于某种技术上，可根据业务的需求和本章中提到的技术思想与架构自行完成平台的搭建。

由于篇幅限制，本章仅介绍了一个自动化运维平台所用到的基本技术，详细的代码设计和技术要点还请读者自行参考相关的文档。

第五部分
架构篇

第 30 章 MySQL 复制

复制是指将主数据库的 DDL 和 DML 操作通过二进制日志传到复制服务器（也叫作从库）上，然后在从库上对这些日志重新执行（也叫作重做），从而使得从库和主库的数据保持同步。

MySQL 支持一台主库同时向多台从库进行复制，从库同时也可以作为其他服务器的主库，实现链状的复制。

MySQL 复制的优点主要包括以下 3 个方面：
- 如果主库出现问题，可以快速切换到从库提供服务；
- 可以在从库上执行查询操作，降低主库的访问压力；
- 可以在从库上执行备份，以避免备份期间影响主库的服务。

注意：如果主从库之间存在延迟，在从库上进行的查询操作需要考虑到这些数据的差异。

30.1 复制概述

MySQL 的复制原理大致如下。

（1）首先，MySQL 主库在事务提交时会把数据变更作为事件（Event）记录在二进制日志文件（BINLOG）中；MySQL 主库上的 sync_binlog 参数控制 BINLOG 日志刷新到磁盘。

（2）从库请求主库的 BINLOG 事件，并通过主库上的 BINLOG Dump 线程将需要的数据发送给从库的 I/O 线程，然后从库将接收到的数据写入中继日志 Relay Log，之后从库根据中继日志 Relay Log 重做数据变更操作，通过逻辑复制以此来达到主库和从库的数据一致。

MySQL 通过 3 个线程来完成主从库间的数据复制：其中 BINLOG Dump 线程跑在主库上，I/O 线程和 SQL 线程跑在从库上。当在从库上启动复制（Start Slave）时，首先创建 I/O 线程连接主库，主库随后创建 BINLOG Dump 线程读取数据库事件并发送给 I/O 线程，I/O 线程获取到事件数据后更新到从库的中继日志 Relay Log 中，之后从库上的 SQL 线程读取中继日志 Relay Log 中更新的数据库事件并应用，如图 30-1 所示。

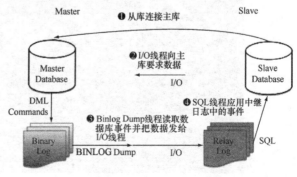

图 30-1　MySQL 复制流程

30.1 复制概述

可以通过 show processlist 命令在主库上查看 BINLOG Dump 线程，从 BINLOG Dump 线程的状态可以看到，主库将从库需要的 BINLOG 发送给从库：

```
mysql> show processlist\G;
*************************** 1. row ***************************
     Id: 20
   User: repl
   Host: 10.120.240.251:51231
     db: NULL
Command: Binlog Dump
   Time: 1984661
  State: Master has sent all binlog to slave; waiting for more updates
   Info: NULL
*************************** 2. row ***************************
     Id: 55
   User: root
   Host: localhost
     db: NULL
Command: Sleep
   Time: 531
  State:
   Info: NULL
2 rows in set (0.00 sec)
```

同样地，在从库上通过 show processlist 可以看到 Slave 的 I/O 线程和 SQL 线程，I/O 线程等待主库上的 BINLOG Dump 线程发送事件并更新到中继日志 Relay Log，SQL 线程读取中继日志 Relay Log 并应用变更到数据库：

```
mysql> show processlist\G;
*************************** 1. row ***************************
     Id: 5
   User: system user
   Host:
     db: NULL
Command: Connect
   Time: 1986092
  State: Waiting for master to send event
   Info: NULL
*************************** 2. row ***************************
     Id: 6
   User: system user
   Host:
     db: NULL
Command: Connect
   Time: 277789
  State: Slave has read all relay log; waiting for more updates
   Info: NULL
*************************** 3. row ***************************
     Id: 7
   User: system user
   Host:
     db: NULL
Command: Connect
   Time: 277789
  State: Waiting for an event from Coordinator
   Info: NULL
*************************** 4. row ***************************
     Id: 8
   User: system user
   Host:
     db: NULL
Command: Connect
   Time: 1986092
  State: Waiting for an event from Coordinator
   Info: NULL
4 rows in set (0.00 sec)

ERROR:
No query specified
```

注意：Waiting for an event from Coordinator 涉及多线程复制。

以上是 MySQL 传统的复制流程，但复制是异步的，从库上的数据和主库存在一定的延迟。

30.1.1 复制中的各类文件

从 MySQL 复制流程可以看到复制过程中涉及两类非常重要的日志文件：二进制日志文件（BINLOG）和中继日志文件（Relay Log）。

二进制日志文件（BINLOG）会把 MySQL 中的所有数据修改操作以二进制的形式记录到日志文件中，包括 Create、Drop、Insert、Update、Delete 操作等，但二进制日志文件（BINLOG）不会记录 Select 操作，因为 Select 操作并不修改数据。

可以通过 show variables 查看 BINLOG 的格式，BINLOG 支持 Statement、Row、Mixed 这 3 种格式，也对应了 MySQL 的 3 种复制技术（会在 30.1.1 节中详细讨论）：

```
mysql> show variables like 'binlog_format';
+---------------+-------+
| Variable_name | Value |
+---------------+-------+
| binlog_format | ROW   |
+---------------+-------+
1 row in set (0.00 sec)
```

中继日志文件 Relay Log 的文件格式、内容和二进制日志文件 BINLOG 一样，唯一的区别在于从库上的 SQL 线程在执行完当前中继日志文件 Relay Log 中的事件之后，SQL 线程会自动删除当前中继日志文件 Relay Log，避免从库上的中继日志文件 Relay Log 占用过多的磁盘空间（可以通过参数 relay_log_purge 改变这个行为，默认开启，在 MySQL 8.0 以后，BINLOG 文件也可以通过设置参数 binlog_expire_logs_seconds 来自动删除过期文件）。

为了保证从库 Crash 重启之后，从库的 I/O 线程和 SQL 线程仍然能够知道从哪里开始复制，从库上默认还会创建两个日志文件 master.info 和 relay-log.info 用来保存复制的进度。

注意：可以通过设置参数 --master-info-repository 和 --relay-log-info-repository 将 "FILE" 修改为 "TABLE"，创建出来的表默认是 Innodb 存储引擎，实际在线上应用中，通常设置为 "TABLE"，这样能与变更的事务保持一致性，使意外故障的恢复更高效。从 MySQL 8.0.2 以后，这两个参数的默认值为 "TABLE"。

master.info 和 relay-log.info 这两个文件在磁盘上分别记录了从库的 I/O 线程当前读取主库二进制日志 BINLOG 的进度和 SQL 线程应用中继日志 Relay Log 的进度。例如，通过 show slave status 命令能够看到当前从库复制的状态，如图 30-2 所示。

其中 master.info 记录的是 I/O 线程连接主库的一些参数，主要介绍 show slave status 显示的以下 5 项内容。

- Master Host：主库的 IP。
- Master User：主库上，主从复制使用的用户账号。
- Master Port：主库 MySQL 的端口号。
- Master_Log_File：从库的 I/O 线程当前正在读取的主库 BINLOG 的文件名。
- Read_Master_Log_Pos：从库 I/O 线程当前读取到的位置。

而 relay-log.info 记录的是 SQL 线程应用中继日志 Relay Log 的一些参数，主要介绍 show slave status 显示的以下 4 项内容。

30.1 复制概述

- Relay_Log_File：从库 SQL 线程正在读取和应用的中继日志 Relay Log 的文件名。
- Relay_Log_Pos：从库 SQL 线程正在读取并应用的中继日志 Relay Log 的位置。
- Relay_Master_Log_File：从库 SQL 线程正在读取和应用的 Relay Log 对应于主库 BINLOG 的文件名。
- Exec_Master_Log_Pos：中继日志 Relay Log 中 Relay_Log_Pos 位置对应于主库 Binlog 的位置。

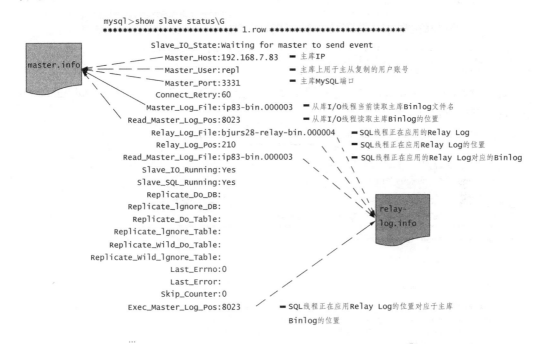

图 30-2　MySQL 从库复制中的状态值

30.1.2　3 种复制方式

二进制日志文件 BINLOG 的格式有以下 3 种。

- Statement：基于 SQL 语句级别的 BINLOG，每条修改数据的 SQL 都会保存到 BINLOG 里。
- Row：基于行级别，记录每一行数据的变化，也就是将每行数据的变化都记录到 BINLOG 里面，记录得非常详细，但是并不记录原始 SQL；在复制的时候，并不会因为存储过程或触发器导致主从库数据不一致的问题，但是记录的日志量较 Statement 格式要大得多。
- Mixed：混合 Statement 和 Row 模式，默认情况下采用 Statement 模式记录，某些情况下会切换到 Row 模式，例如 SQL 中包含与时间、用户相关的函数等。

同时也对应了 MySQL 复制的 3 种技术。

- binlog_format=Statement：基于 SQL 语句的复制，也叫作 Statement-Based Replication（SBR），MySQL 5.1.4 或之前版本仅提供基于 SQL 语句的复制。
- binlog_format=Row：基于行的复制，也叫作 Row-Based Replication（RBR）。
- binlog_format=Mixed：混合复制模式，混合了基于 SQL 语句的复制和基于行的复制。

注意：对于 MySQL 5.7.6 之前的版本，MySQL 的默认设置是基于 SQL 语句的复制，从 MySQL 5.7.7

开始之后的版本，MySQL 的默认设置变更为基于 Row 的复制。

下面通过例子来看 Statement 格式和 Row 格式的区别，这里以 Update 操作为例。在 Statement 格式下，BINLOG 记录的是原始操作 SQL：

```
mysql> show variables like 'binlog_format';
+---------------+-----------+
| Variable_name | Value     |
+---------------+-----------+
| binlog_format | STATEMENT |
+---------------+-----------+
1 row in set (0.00 sec)

mysql> update salaries set salary = salary + 10 where emp_no=10001;
Query OK, 17 rows affected (0.05 sec)
Rows matched: 17  Changed: 17  Warnings: 0
```

通过 show binlog events 命令查看到 Update 操作在 BINLOG 日志文件 mysql-bin.000013 中对应的开始位置为 308：

```
mysql> show binlog events in 'mysql-bin.000013' from 154\G;
*************************** 1. row ***************************
   Log_name: mysql-bin.000013
        Pos: 154
 Event_type: Anonymous_Gtid
  Server_id: 7237
End_log_pos: 219
       Info: SET @@SESSION.GTID_NEXT= 'ANONYMOUS'
*************************** 2. row ***************************
   Log_name: mysql-bin.000013
        Pos: 219
 Event_type: Query
  Server_id: 7237
End_log_pos: 308
       Info: BEGIN
*************************** 3. row ***************************
   Log_name: mysql-bin.000013
        Pos: 308
 Event_type: Query
  Server_id: 7237
End_log_pos: 451
       Info: use `employees`; update salaries set salary = salary + 10 where emp_no=10001
*************************** 4. row ***************************
   Log_name: mysql-bin.000013
        Pos: 451
 Event_type: Xid
  Server_id: 7237
End_log_pos: 482
       Info: COMMIT /* xid=19230 */
4 rows in set (0.00 sec)

ERROR:
No query specified
```

通过 mysqlbinlog 工具分析对应的 BINLOG 日志，会发现在 Statement 模式下，BINLOG 日志文件中（从 308 位置开始）记录了实际发生的 SQL：

```
BEGIN
/*!*/;
# at 308
#180720 10:47:23 server id 7237  end_log_pos 451 CRC32 0xddf78e5c     Query     thread_id=179
 exec_time=0    error_code=0
use `employees`/*!*/;
SET TIMESTAMP=1532054843/*!*/;
update salaries set salary = salary + 10 where emp_no=10001
/*!*/;
```

调整 BINLOG 格式为 Row，再做一个 update 操作检查一下：

```
mysql> show variables like 'binlog_format';
+---------------+-------+
| Variable_name | Value |
+---------------+-------+
| binlog_format | ROW   |
+---------------+-------+
1 row in set (0.00 sec)


mysql> update salaries set salary = salary - 10 where emp_no=10001;
Query OK, 17 rows affected (0.01 sec)
Rows matched: 17  Changed: 17  Warnings: 0
```

同样地，通过 show binlog events 命令查看到 Update 操作在 BINLOG 日志文件 mysql-bin.000013 中对应开始位置为 624：

```
mysql> show binlog events in 'mysql-bin.000013' from 547\G;
*************************** 1. row ***************************
   Log_name: mysql-bin.000013
        Pos: 547
 Event_type: Query
  Server_id: 7237
End_log_pos: 624
       Info: BEGIN
*************************** 2. row ***************************
   Log_name: mysql-bin.000013
        Pos: 624
 Event_type: Table_map
  Server_id: 7237
End_log_pos: 683
       Info: table_id: 119 (employees.salaries)
*************************** 3. row ***************************
   Log_name: mysql-bin.000013
        Pos: 683
 Event_type: Update_rows
  Server_id: 7237
End_log_pos: 1229
       Info: table_id: 119 flags: STMT_END_F
*************************** 4. row ***************************
   Log_name: mysql-bin.000013
        Pos: 1229
 Event_type: Xid
  Server_id: 7237
End_log_pos: 1260
       Info: COMMIT /* xid=19270 */
4 rows in set (0.00 sec)
```

此时再通过 mysqlbinlog 检查对应的 BINLOG 日志文件：

```
$ mysqlbinlog -vv mysql-bin.000013 --start-position=624
/*!50530 SET @@SESSION.PSEUDO_SLAVE_MODE=1*/;
/*!50003 SET @OLD_COMPLETION_TYPE=@@COMPLETION_TYPE,COMPLETION_TYPE=0*/;
DELIMITER /*!*/;
# at 4
#180720 10:35:26 server id 7237  end_log_pos 123 CRC32 0x61381108     Start: binlog v 4, server v 5.7.22-log created 180720 10:35:26
# Warning: this binlog is either in use or was not closed properly.
BINLOG '
bkpRWw9FHAAAdwAAAHsAAAABAAQANS43LjIyLWxvZwAAAAAAAAAAAAAAAAAAAAAAAAAAAAAA
AAAAAAAAAAAAAAAAAAAAAAEzgNAAgAEgAEBAQEEgAAXwAEGggAAAAICAgCAAAACgoKKioAEjQA
AQgROGE=
'/*!*/;
# at 624
#180720 15:00:40 server id 7237  end_log_pos 683 CRC32 0x02e5f4a8     Table_map: `employees`.`salaries` mapped to number 119
# at 683
#180720 15:00:40 server id 7237  end_log_pos 1229 CRC32 0x449318e0    Update_rows: table id 119 flags: STMT_END_F

BINLOG '
```

```
mIhRWxNFHAAAOwAAAKsCAAAAAHcAAAAAAAEACWVtcGxveWVlcwAIc2FsYXJpZXMMABAMDCgoAAKj0
5QI=
mIhRWx9FHAAAIgIAAM0EAAAAAHcAAAAAAAEAAgAE///wEScAAN/qAADahA/ahg/wEScAANXqAADa
hA/ahg/wEScAAKDyAADahg/ZiA/wEScAAJbyAADahg/ZiA/wEScAACQCAQDZiA/Zig/wEScAABoC
AQDZiA/Zig/wEScAAC4EAQDZig/ZjA/wEScAACQEAQDZig/ZjA/wEScAAJsFAQDZjA/Zjg/wESCA
AJEFAQDZjA/Zjg/wEScAAJAVAQDZjg/YkA/wEScAAIYVAQDZjg/YkA/wEScAAGCiAQDYkA/Ykg/w
EScAAF0iAQDYkA/Ykg/wEScAACAmAQDYkg/YlA/wEScAABymAQDYkg/YlA/wEScAAOQoAQDY1A/Y
lg/wEScAAN0oAQDYlA/Ylg/wEScAAF4sAQDYlg/XmA/wEScAAFQsAQDYlg/XmA/wEScAAJc4AQDX
mA/Xmg/wEScAAIO4AQDXmA/Xmg/wEScAAIs8AQDXmg/XnA/wEScAAIE8AQDXmg/XnA/wEScAANM8
AQDXnA/Xng/wEScAAMk8AQDXnA/Xng/wEScAAL9LAQDXng/WoA/wEScAALVLAQDXng/WoA/wESCA
AIJMAQDWoA/Wog/wEScAAHhMAQDWoA/Wog/wEScAAHNMAQDWog/WpA/wEScAAG1MAQDWog/WpA/w
EScAAIhbAQDWpA8hHk7wEScAAH5bAQDWpA8hHk7gGJNE
'/*!*/;
### UPDATE `employees`.`salaries`
### WHERE
###   @1=10001 /* INT meta=0 nullable=0 is_null=0 */
###   @2=60127 /* INT meta=0 nullable=0 is_null=0 */
###   @3='1986:06:26' /* DATE meta=0 nullable=0 is_null=0 */
###   @4='1987:06:26' /* DATE meta=0 nullable=0 is_null=0 */
### SET
###   @1=10001 /* INT meta=0 nullable=0 is_null=0 */
###   @2=60117 /* INT meta=0 nullable=0 is_null=0 */
###   @3='1986:06:26' /* DATE meta=0 nullable=0 is_null=0 */
###   @4='1987:06:26' /* DATE meta=0 nullable=0 is_null=0 */
…
```

能够清晰地看到，在 binlog_format 设置为 Row 格式时，MySQL 实际上在 BINLOG 中逐行记录数据的变更，Row 格式比 Statement 格式更能保证从库数据的一致性（复制的是记录，而不是单纯的操作 SQL）。当然，Row 格式下的 BINLOG 的日志量很可能会增大非常多，在设置时需要考虑到磁盘空间问题。

30.1.3 复制的 4 种常见架构

复制的 4 种常见架构有一主多从复制架构、多级复制架构、双主（Dual Master）复制架构和多源（Multi-Source）复制架构。

1．一主多从复制架构

在主库读取请求压力非常大的场景下，可以通过配置一主多从复制架构实现读写分离，把大量对实时性要求不是特别高的读请求通过负载均衡分布到多个从库上，降低主库的读取压力，如图 30-3 所示。

在主库出现异常宕机的情况下，可以把一个从库切换为主库继续提供服务。

2．多级复制架构

一主多从的架构能够解决大部分读请求压力特别大的场景的需求，考虑到 MySQL 的复制需要主库发送 BINLOG 日志

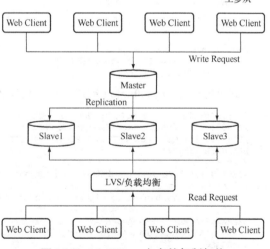

图 30-3　MySQL 一主多从复制架构

到从库的 I/O 线程，主库的 I/O 压力和网络压力会随着从库的增加而增长（每个从库都会在主库上有一个独立的 BINLOG Dump 线程来发送事件），而多级复制架构解决了一主多从场景下，主库额外的 I/O 和网络压力。MySQL 的多级复制架构如图 30-4 所示。

30.1 复制概述

对比一主多从的架构图，多级复制仅仅是在主库 Master1 复制到从库 Slave1、Slave2、Slave3 的中间增加一个二级主库 Master2，这样，主库 Master1 只需要给一个从库 Master2 发送 BINLOG 日志即可，减轻了主库 Master1 的压力。二级主库 Master2 再发送 BINLOG 日志给所有从库 Slave1、Slave2 和 Slave3 的 I/O 线程。

多级复制解决了一主多从场景下，主库的 I/O 负载和网络压力，当然也有缺点：MySQL 的传统复制是异步的，多级复制场景下主库的数据是经历两次复制才到达从库 Slave1、Slave2、Slave3 的，期间的延迟要比一主多从复制场景下只经历一次复制的还大。

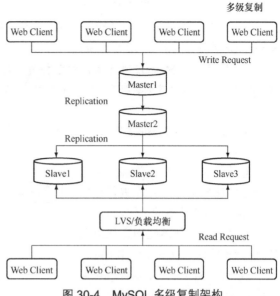

图 30-4　MySQL 多级复制架构

可以通过在二级主库 Master2 上选择表引擎为 BLACKHOLE 来降低多级复制的延迟。顾名思义，BLACKHOLE 引擎是一个"黑洞"引擎，写入 BLACKHOLE 表的数据并不会写回到磁盘上，BLACKHOLE 表永远都是一个空表，INSERT/UPDATE/DELETE 操作仅仅在 BINLOG 中记录事件。

```
mysql> CREATE TABLE blackhole(i INT, c CHAR(10)) ENGINE = BLACKHOLE;
Query OK, 0 rows affected (0.04 sec)

mysql> INSERT INTO blackhole VALUES(1,'record one'),(2,'record two');
Query OK, 2 rows affected (0.00 sec)
Records: 2  Duplicates: 0  Warnings: 0

mysql> select * from blackhole;
Empty set (0.00 sec)
```

BLACKHOLE 引擎非常适合二级主库 Master2 的场景：Master2 并不承担读写请求，仅仅负责将 BINLOG 日志尽快传送给从库。

3. 双主（Dual Master）复制架构

双主（Dual Master）复制架构适用于 DBA 做维护时需要主从切换的场景（也可以考虑使用高可用方案，具体参考第 31 章），通过双主复制架构避免了重复搭建从库的麻烦，双主复制架构如图 30-5 所示。

主库 Master1 和 Master2 互为主从，所有 Web Client 的写请求都访问主库 Master1，而读请求可以选择访问主库 Master1 或 Master2。假如，DBA 需要做日常维护操作，为了避免影响服务，需进行以下操作。

● 首先，在 Master1 库上停止 Slave 线程（STOP SLAVE），避免后续对 Master2 库的维护操作被实时复制到 Master1 库上

图 30-5　MySQL 双主复制架构

对服务造成影响。

- 其次，在 Master2 库上停止 Slave 线程（STOP SLAVE），开始日常维护操作，例如修改 varchar 字段长度从 10 增加到 300。
- 然后，在 Master2 库上完成维护操作之后，打开 Master2 库上的 Slave 线程（START SLAVE），让 Master2 库的数据和 Master1 库同步，同步完成后，把应用的读写操作切换到 Master2 库上。
- 最后，确认 Master1 库上无应用访问后，打开 Master1 库的 Slave 线程（START SLAVE）即可。

通过双主复制架构能够大大减轻一主多从架构下对主库进行维护带来的额外搭建从库的工作。

当然双主复制还能和主从复制联合起来使用：在 Master2 库下配置从库 Slave1、Slave2 等，这样既可通过从库 Slave1 等来分担读取压力，同时在 DBA 做维护的同时，避免了重建从库的额外工作，但需要注意从库的复制延迟。MySQL 的双主多级复制架构如图 30-6 所示。

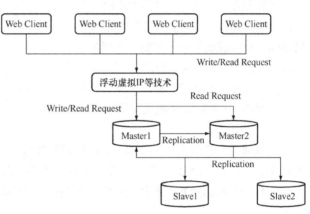

图 30-6　MySQL 双主多级复制架构

由于 MySQL 的 BINLOG 中会记录事件初始发生的 server id，以 Master1 上的 BINLOG 为例：

```
# at 282
#180720 16:16:40 server id 7237    end_log_pos 341 CRC32 0x1e63cedf    Table_map: `employees`
.`salaries` mapped to number 165
…
# at 580
#180720 16:17:27 server id 7238    end_log_pos 639 CRC32 0x02604abb    Table_map: `employees`
.`salaries` mapped to number 165
…
```

MySQL 只应用和自己 Server ID 不同的 BINLOG 日志。从上面的数据中可以看到 Server ID 有两个，分别为 7237 和 7238，其中 7237 代表是自身实例，7238 表示另一个实例。这段数据表明该实例自身执行了一些事务，又通过复制，执行了 Server ID 为 7238 所执行过的事务，MySQL 就这样通过判断 BINLOG 事件中的不同 Server ID 即可判断当前库是否是事件发生的初始发生 Server，所以双主复制加上级联复制也不会出现循环复制。

4．多源（Multi-Source）复制架构

多源（Multi-Source）复制架构适合于复杂的业务需求，既可以支撑 OLTP（联机事务处理），也可以满足 OLAP（联机分析处理）。MySQL 的多源复制架构如图 30-7 所示。

Slave1 是主库 Master1 的单独从库，Slave2 是主库 Master2 单独从库，Slave3 是主库 Master1 和 Master2 的共用从库。Web Client 客户端分业务，将不同的写请求发往主库 Master1 或者 Master2，而针对线上的读请求优先访问各自的单独从库。因主库 Master1 和 Master2 都只有线上的部分数据，不能提供全量数据；针对 OLAP 的业务需求，Slave3 能很好地提供数据，省去合并数据的烦恼。

30.2 复制搭建

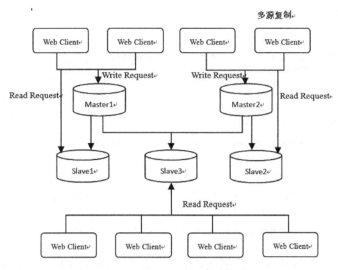

图 30-7　MySQL 多源复制架构

另外，共用从库还能作为 Slave1 和 Slave2 的备机，随时可以接替其工作，且节省购机成本（因高可用 MHA 需要一主两从，此部分内容可参见第 31 章）。

此方案的缺点也一目了然，那就是 Master1 和 Master2 如果有数据冲突，共用从库 Slave3 将会发生数据混乱。复制过程中可以显式定义通道（Channel），通过执行 show slave status 可以看到该复制线程正在使用哪个通道，后续的操作都将围绕该通道执行。

```
mysql> show slave status\G;
*************************** 1. row ***************************
            ...
            Channel_Name: channel_3305
            ...
*************************** 2. row ***************************
            ...
            Channel_Name: channel_3307
            ...
2 rows in set (0.00 sec)
            ...
```

30.2　复制搭建

30.1 节主要介绍了 MySQL 的复制原理和常见的复制架构。本节将介绍如何一步步搭建复制环境。

30.2.1　异步复制

MySQL 的复制默认是异步的，主从复制至少需要两个 MySQL 服务，这些 MySQL 服务可以分布在不同的服务器上，也可以在同一台服务器上。主从复制配置的步骤比较简单，下面进行详细介绍。

（1）确保主从库上安装了相同版本的数据库。因为主从库的角色可能会互换，同时减少问题出现的概率，所以在可能的情况下推荐安装最新的稳定版本。

（2）在主库上，设置一个复制使用的账户，并授予 REPLICATION SLAVE 权限。这里创建一个复制用户 rep1，允许 IP 为 10.120.240.251 的主机进行连接，在 MySQL 5.7 和 8.0 中，创建命令有所不同。

MySQL 5.7 可以用下面两种方式创建：

```
mysql> GRANT REPLICATION SLAVE ON *.* TO 'repl'@'10.120.240.251' IDENTIFIED BY 'repl';
Query OK, 0 rows affected (0.00 sec)
```

或者：

```
mysql> Create user 'repl'@'10.120.240.251' IDENTIFIED WITH mysql_native_password BY 'repl';
Query OK, 0 rows affected (0.00 sec)

mysql> GRANT REPLICATION SLAVE ON *.* TO 'repl'@'10.120.240.251';
Query OK, 0 rows affected (0.00 sec)
```

MySQL 8.0 中可以沿用上面的第二种方式，也可以使用下面的方式创建：

```
mysql> Create user 'repl'@'10.120.240.251' IDENTIFIED WITH BY 'repl';
Query OK, 0 rows affected (0.00 sec)

mysql> GRANT REPLICATION SLAVE ON *.* TO 'repl'@'10.120.240.251';
Query OK, 0 rows affected (0.00 sec)
```

以上代码使用了 MySQL 8.0 的默认认证方式 caching_sha2_password，执行前需要在配置文件 my.cnf 中添加如下内容，然后重启 MySQL 实例：

```
[mysqld]
default_authentication_plugin=caching_sha2_password
```

（3）修改主数据库服务器的配置文件 my.cnf，开启 BINLOG，并设置 server-id 的值。这两个参数的修改需要重新启动数据库服务才能生效。

在 my.cnf 中修改如下：

```
[mysqld]log-bin=/data2/mysql3307/data/mysql-bin
server-id=7237
```

（4）在主库上，设置读锁定有效，这个操作是为了确保没有数据库操作，以便获得一个一致性的快照：

```
mysql> flush tables with read lock;
Query OK, 0 rows affected (0.00 sec)
```

（5）然后得到主库上当前的二进制日志名和偏移量值。这个操作的目的是为了在从数据库启动以后，从这个点开始进行数据的恢复（如果已经启用了 GTID 模式，可以跳过该步骤，关于 GTID，详见下文）。

```
mysql> show master status;
+------------------+----------+--------------+------------------+-------------------+
| File             | Position | Binlog_Do_DB | Binlog_Ignore_DB | Executed_Gtid_Set |
+------------------+----------+--------------+------------------+-------------------+
| mysql-bin.000013 |     2398 |              |                  |                   |
+------------------+----------+--------------+------------------+-------------------+
1 row in set (0.00 sec)
```

（6）现在主数据库服务器已经停止了更新操作，需要生成主数据库的备份，备份的方式有很多种，可以直接在操作系统下复制全部的数据文件到从数据库服务器上，也可以通过 mysqldump 导出数据或者使用 xtrabackup 等热备份工具进行数据库的备份，这些备份操作的步骤已经在第 25 章中有详细介绍，这里就不再一一说明。如果主数据库的服务可以停止，那么直接复制数据文件应该是最快的生成快照的方法。

（7）主数据库的备份完毕后，可以恢复写操作，剩下的操作只需要在从库上执行：

```
mysql> unlock tables;
Query OK, 0 rows affected (0.00 sec)
```

（8）将主数据库的一致性备份恢复到从数据库上。如果是使用 .tar 打包的文件包，只需要

30.2 复制搭建

解开到相应的目录即可。

（9）修改从数据库的配置文件 my.cnf，增加 server-id 参数。注意 server-id 的值必须是唯一的，不能和主数据库的配置相同，如果有多个从数据库服务器，每个从数据库服务器必须有自己唯一的 server-id 值（如果已经启用了 GTID 模式，同时删除 datadir 目录下的 auto.cnf 文件，避免与主库有相同的 UUID，关于 GTID，详见下文）。

在 my.cnf 中修改如下：

```
[mysqld]
server-id=7238
```

MySQL 数据目录下的 auto.cnf 文件内容如下：

```
[mysql3489@hz_10_120_240_251 data]$ cat auto.cnf
[auto]
server-uuid=7c63d434-6edd-11e8-b378-0024e869b4d5
```

（10）在从库上，使用 --skip-slave-start 选项启动从数据库，这样不会立即启动从数据库服务上的复制进程，方便对从数据库的服务进行进一步的配置：

```
[mysql3308@hz_10_120_240_251 mysqlhome]$ ./bin/mysqld_safe --defaults-file=/home/mysql3308/
mysqlhome/my.cnf --user=mysql3308 --skip-slave-start &
```

（11）对从数据库服务器做相应设置，指定复制使用的用户，主数据库服务器的 IP、端口以及开始执行复制的日志文件和位置等，具体代码如下（如果已经启用了 GTID 模式，不需要指定复制的日志文件和位置，可以指定 MASTER_AUTO_POSITION=1，即自动发现复制起始点，关于 GTID，详见下文）：

```
mysql> CHANGE MASTER TO
    -> MASTER_HOST='master_host_name',
    -> MASTER_USER='replication_user_name',
    -> MASTER_PASSWORD='replication_password',
    -> MASTER_LOG_FILE='recorded_log_file_name',
    -> MASTER_LOG_POS=recorded_log_position;
```

举例说明如下：

```
mysql> CHANGE MASTER TO
    -> MASTER_HOST='10.120.240.251',
    -> MASTER_PORT=3307,
    -> MASTER_USER='repl',
    -> MASTER_PASSWORD='repl',
    -> MASTER_LOG_FILE='mysql-bin.000013',
    -> MASTER_LOG_POS=2398;
Query OK, 0 rows affected (0.10 sec)
```

如果想显式定义通道（Channel），详见如下：

```
mysql> CHANGE MASTER TO
    -> MASTER_HOST='10.120.240.251',
    -> MASTER_PORT=3307,
    -> MASTER_USER='repl',
    -> MASTER_PASSWORD='repl',
    -> MASTER_LOG_FILE='mysql-bin.000013',
    -> MASTER_LOG_POS=2398
    -> FOR CHANNEL 'channel_3307';
Query OK, 0 rows affected (0.10 sec)
```

（12）在从库上，启动 slave 线程：

```
mysql> start slave;
Query OK, 0 rows affected (0.00 sec)
```

注意：如果使用了 MySQL 8.0 默认的认证方式 caching_sha2_password，可以在 START SLAVE 时指定 DEFAULT_AUTH='caching_sha2_password'，详见如下：

```
mysql> start slave DEFAULT_AUTH='caching_sha2_password';
Query OK, 0 rows affected (0.00 sec)
```

（13）这时 slave 上执行 show processlist 命令将显示类似如下的进程：

```
root@localhost:mysql_3308.sock  [(none)]>show processlist\G;
*************************** 1. row ***************************
     Id: 1
   User: event_scheduler
   Host: localhost
     db: NULL
Command: Daemon
   Time: 3544378
  State: Waiting on empty queue
   Info: NULL
*************************** 2. row ***************************
     Id: 51
   User: system user
   Host: 
     db: NULL
Command: Connect
   Time: 341636
  State: Waiting for master to send event
   Info: NULL
*************************** 3. row ***************************
     Id: 52
   User: system user
   Host: 
     db: NULL
Command: Connect
   Time: 15017
  State: Slave has read all relay log; waiting for more updates
   Info: NULL
…
```

这表明 slave 已经连接上 master，并开始接受并执行日志。

（14）也可以测试复制服务的正确性，在主数据库上执行一个更新操作，观察是否同步到了从库。在主数据库上的 employees 数据库中操作一张表：

```
mysql> update salaries set salary=100 where emp_no=10001 and salary=60117;
Query OK, 1 row affected (0.01 sec)
Rows matched: 1  Changed: 1  Warnings: 0
```

（15）在从数据库上检查数据是否同步：

```
mysql> select * from salaries where emp_no=10001 and salary=100;
+--------+--------+------------+------------+
| emp_no | salary | from_date  | to_date    |
+--------+--------+------------+------------+
|  10001 |   100  | 1986-06-26 | 1987-06-26 |
+--------+--------+------------+------------+
1 row in set (0.00 sec)
```

可以看到数据可以正确同步到从数据库上，主从复制服务配置成功完成。

MySQL 主从异步复制是最常见和最简单的复制场景。数据的完整性依赖于主库 BINLOG 的不丢失，只要主库的 BINLOG 不丢失，那么就算主库宕机了，我们还可以通过 BINLOG 把丢失的部分数据通过手工同步到从库上去。

> 注意：主库宕机的情况下，DBA 可以通过 mysqlbinlog 工具手工访问主库 binlog，抽取缺失的日志并同步到从库上去；也可以通过配置高可用 MHA 架构来自动抽取缺失的数据补全从库，高可用的 MHA 架构会在其他章节介绍，或者启用 Global Transaction Identifiers（GTID）来自动抽取缺失 Binlog 到从库，此部分内容详见 30.3 节。

MySQL 在 BINLOG 中记录事务（或 SQL 语句），也就是说对于支持事务的引擎（例如

InnoDB)来说,每个事务提交时都需要写 BINLOG;对于不支持事务的引擎(例如 MyISAM)来说,每个 SQL 语句执行完成时,都需要写 BINLOG。为了保证 Binlog 的安全,MySQL 引入 sync_binlog 参数来控制 BINLOG 刷新到磁盘的频率。

```
mysql> show variables like 'sync_binlog';
+---------------+-------+
| Variable_name | Value |
+---------------+-------+
| sync_binlog   | 1     |
+---------------+-------+
1 row in set (0.10 sec)
```

- 在默认情况下,sync_binlog=1,表示事务提交之前,MySQL 都需要先把 BINLOG 刷新到磁盘,这样的话,即便出现数据库主机操作系统崩溃或者主机突然掉电的情况,系统最多损失 prepared 状态的事务;设置 sync_binlog=1,尽可能保证数据安全。
- sync_binlog=0,表示 MySQL 不控制 binlog 的刷新,由文件系统自己控制文件系统缓存的刷新。
- sync_binlog=N,如果 N 不等于 0 或者 1,刷新方式同 sync_binlog=1 类似,只不过此时会延长刷新频率至 N 次 binlog 提交组之后。

以上是传统的异步复制搭建过程,相信大家对其并不陌生,在 MySQL 5.7 的并行复制技术到来之前,为人诟病最多的还是效率问题,slave 延迟是个顽疾,虽然之前已经出现了 schema 级别的并行复制,但实际效果并不好,接下来详细介绍一下 MySQL 5.7 中的并行复制(也称为多线程复制)。

30.2.2 多线程复制

在 MySQL 5.7 中,带来了全新的多线程复制技术,解决了当 master 同一个 schema 下的数据发生了变更,从库不能并发应用的问题,同时也真正将 binlog 组提交的优势充分发挥出来,保障了从库并发应用 Relay Log 的能力,接下来还是一步步地介绍多线程复制环境的搭建过程(在 30.2.1 节的基础上继续操作)。

(1)在从库上确认以下参数的配置。

```
mysql> show variables like 'slave_parallel%';
+------------------------+---------------+
| Variable_name          | Value         |
+------------------------+---------------+
| slave_parallel_type    | LOGICAL_CLOCK |
| slave_parallel_workers | 8             |
+------------------------+---------------+
2 rows in set (0.00 sec)

mysql> show variables like '%repository';
+---------------------------+-------+
| Variable_name             | Value |
+---------------------------+-------+
| master_info_repository    | TABLE |
| relay_log_info_repository | TABLE |
+---------------------------+-------+
2 rows in set (0.01 sec)

mysql> show variables like 'relay_log_recovery';
+--------------------+-------+
| Variable_name      | Value |
+--------------------+-------+
```

```
| relay_log_recovery | ON    |
+--------------------+-------+
1 row in set (0.00 sec)

mysql> show variables like 'slave_preserve_commit_order';
+-----------------------------+-------+
| Variable_name               | Value |
+-----------------------------+-------+
| slave_preserve_commit_order | ON    |
+-----------------------------+-------+
1 row in set (0.01 sec)
```

参数 slave_parallel_type 默认是 DATABASE，兼容以前不同 schema 的并行复制，LOGICAL_CLOCK 是表示逻辑时间戳，即多线程复制依赖主库 commit 时刻的时间戳，从 BINLOG 中的 last_committed 和 sequence_number 可以来判断，相同 last_committed 作为一组事务，可以并行执行（不同的 last_committed 有时也是可以并行执行的，参见下文）。

参数 slave_parallel_workers 默认值为 0，表示禁用多线程复制；当设置该值大于 1 的时候，例如 slave_parallel_workers=8，表示有 8 个线程执行应用。如果要修改该参数值，可参考如下示例：

```
mysql> set global slave_parallel_workers=8;
Query OK, 0 rows affected (0.00 sec)

mysql> stop slave for channel 'channel_3307';
Query OK, 0 rows affected (0.00 sec)

mysql> start slave for channel 'channel_3307';
Query OK, 0 rows affected (0.02 sec)
```

参数 slave_preserve_commit_order 默认值为 OFF，即表示复制应用的过程中，不保留主库事务提交的顺序性，例如主库并发提交了事务 t1、t2、t3、t4，但从库应用过程中，提交顺序可能是 t2、t3、t1、t4，如果从库也会有业务端访问，需要考虑是否要设置为 ON，以此来保证和主库事务提交顺序的一致性。

参数 relay_log_recovery、master_info_repository、relay_log_info_repository 这样设置，则是为了考虑在多线程复制场景下，从库一旦 crash，可以保障 recovery 时数据可恢复。

（2）验证多线程复制，以下示例中 Master 启动了 3 个 session：

```
Session1:
mysql> start transaction;
Query OK, 0 rows affected (0.00 sec)

mysql> update salaries set salary=100 where emp_no=10001 and salary=60117;
Query OK, 1 row affected (0.00 sec)
Rows matched: 1  Changed: 1  Warnings: 0

Session2:
mysql> start transaction;
Query OK, 0 rows affected (0.00 sec)

mysql> update salaries set salary=200 where emp_no=10001 and salary=62102;
Query OK, 1 row affected (0.00 sec)
Rows matched: 1  Changed: 1  Warnings: 0

Session3:
mysql> start transaction;
Query OK, 0 rows affected (0.00 sec)

mysql> update salaries set salary=300 where emp_no=10001 and salary=66074;
Query OK, 1 row affected (0.00 sec)
Rows matched: 1  Changed: 1  Warnings: 0

Session3:
Commit;
```

30.2 复制搭建

```
Session2:
Commit;

Session1:
Commit;
```

用 mysqlbinlog 工具解析 BINLOG，如下所示。

```
#180730 17:06:26 server id 7237  end_log_pos 1270 CRC32 0x86e1a095    Anonymous_GTID
last_committed=4    sequence_number=5    rbr_only=yes
……
### UPDATE `employees`.`salaries`
### WHERE
###   @2=66074 /* INT meta=0 nullable=0 is_null=0 */
……
### SET
###   @2=300 /* INT meta=0 nullable=0 is_null=0 */

#180730 17:06:27 server id 7237  end_log_pos 1568 CRC32 0xf28f52d2    Anonymous_GTID
last_committed=4    sequence_number=6    rbr_only=yes
……
### UPDATE `employees`.`salaries`
### SET
###   @2=200 /* INT meta=0 nullable=0 is_null=0 */
……
#180730 17:06:27 server id 7237  end_log_pos 1866 CRC32 0xadba61fd    Anonymous_GTID
last_committed=4    sequence_number=7    rbr_only=yes
……
### UPDATE `employees`.`salaries`
### SET
###   @2=100 /* INT meta=0 nullable=0 is_null=0 */
```

以上的 last_committed 的值是一样的，都是 4，作为一组事务，表明从库应用的时候是可以并发执行的。

在 MySQL 8.0 中，多线程复制又进行了技术更新，引入了 writeset 的概念，而在之前的版本中，如果主库的同一个会话顺序执行多个不同相关对象的事务，例如，先执行了 Update A 表的数据，又执行了 Update B 表的数据，那么 BINLOG 在复制到从库后，这两个事务是不能并行执行的，writeset 的到来，突破了这个限制。下面进行详细介绍。

（1）准备好主从复制环境（以下使用了 MySQL 8.0.11 版本），此过程可参考上文描述。

（2）主库执行事务前，确认以下两个参数。

```
mysql> show variables like 'transaction_write_set_extraction';
+----------------------------------+----------+
| Variable_name                    | Value    |
+----------------------------------+----------+
| transaction_write_set_extraction | XXHASH64 |
+----------------------------------+----------+
1 row in set (0.01 sec)

mysql> show variables like 'binlog_transaction_dependency_tracking';
+----------------------------------------+--------------+
| Variable_name                          | Value        |
+----------------------------------------+--------------+
| binlog_transaction_dependency_tracking | COMMIT_ORDER |
+----------------------------------------+--------------+
```

以上第一个参数 transaction_write_set_extraction 是启用写集（writeset）的前提条件，定义了事务中写集的算法，默认是 OFF，即关闭状态，表示不启用 writeset，但从 MySQL 8.0.2 以后，该参数的默认值已经变更为 XXHASH64。

以上第二个参数 binlog_transaction_dependency_tracking，表示从库并发执行的依赖信息源，即当主库执行事务，产生 BINLOG 后，这些事务能否在开启多线程复制的从库并发执行。该参数有 3 个可选值，分别为 COMMIT_ORDER、WRITESET 和 WRITESET_SESSION，默认值是 COMMIT_ORDER。

- COMMIT_ORDER：表示依赖信息由主库提交时的时间戳产生，即 5.7 版本的默认行为。
- WRITESET：表示依赖信息由主库的 write set 产生，任何写入不同元组的事务能够并发执行，可以简单理解为，当主库的同一个会话更新不同记录时（同一个表或不同表），从库也可以并行执行。这里有一个特殊情况，如果被更新的表不包括主键或者包含了外键，即便设置了 WRITESET，也等同于参数值 COMMIT_ORDER。
- WRITESET_SESSION：与 WRITESET 类似，只不过此时又多了一个约束，即使多个操作间数据并无冲突，但如果这些操作发生在同一会话中，也等同于参数值 COMMIT_ORDER。

这些参数值初次接触可能觉得有些难以理解，下面会通过样例进行说明。

（3）在主库上新建一个会话，并顺序执行以下两个事务。

```
mysql> update employees.dept_emp set to_date='9999-02-02' limit 3;
Query OK, 3 rows affected (0.00 sec)
Rows matched: 3  Changed: 3  Warnings: 0

mysql> update employees.salaries set salary=100 limit 3 ;
Query OK, 3 rows affected (0.20 sec)
Rows matched: 3  Changed: 3  Warnings: 0
```

（4）在主库查看此时产生的 BINLOG 数据。

```
[mysql3308@hz_10_120_240_251 data]$ mysqlbinlog -vvv mysql-bin.000008 > mysql-bin.000008.log
[mysql3308@hz_10_120_240_251 data]$ cat mysql-bin.000008.log|grep "last_commit"
......
#181119 18:29:35 server id 7238  end_log_pos 1033 CRC32 0xd9819f9d        GTID      last_committed=2    sequence_number=3     rbr_only=yes
#181119 18:29:48 server id 7238  end_log_pos 1399 CRC32 0x7722ed70        GTID      last_committed=3    sequence_number=4     rbr_only=yes
```

可以看到，此时生成的 BINLOG 数据依赖的主库的时间戳，last_committed 不一致，没有在一个组内，在从库不能并发执行。

（5）修改主库的参数 binlog_transaction_dependency_tracking 变更为 WRITESET。

```
mysql> set global binlog_transaction_dependency_tracking='WRITESET';
Query OK, 0 rows affected (0.00 sec)

mysql> show variables like 'binlog_transaction_dependency_tracking';
+----------------------------------------+----------+
| Variable_name                          | Value    |
+----------------------------------------+----------+
| binlog_transaction_dependency_tracking | WRITESET |
+----------------------------------------+----------+
1 row in set (0.01 sec)
```

注意：当参数 binlog_transaction_dependency_tracking 变更为 WRITESET 后，参数 transaction_write_set_extraction 不能变更为其他值，如下所示。

```
mysql> set transaction_write_set_extraction=off;
ERROR 1221 (HY000): Incorrect usage of transaction_write_set_extraction (changed) and binlog_transaction_dependency_tracking (!= COMMIT_ORDER)
```

（6）主库上创建一个新会话，并顺序执行以下两个事务。

```
mysql> update employees.dept_emp set to_date='9999-02-01' limit 3;
Query OK, 3 rows affected (0.94 sec)
Rows matched: 3  Changed: 3  Warnings: 0

mysql> update employees.salaries set salary=200 limit 3 ;
```

```
Query OK, 3 rows affected (0.38 sec)
Rows matched: 3  Changed: 3  Warnings: 0
```

（7）在主库查看此时的 BINLOG 数据。

```
[mysql3488@hz_10_120_240_251 data]$ mysqlbinlog -vvv mysql-bin.000028 > mysql-bin.000028.log
[mysql3488@hz_10_120_240_251 data]$ cat mysql-bin.000028.log |grep "last_committed"
#181119 17:25:22 server id 7237  end_log_pos 270 CRC32 0xfa82c8fb     GTID    last_committed
=0      sequence_number=1       rbr_only=yes    original_committed_timestamp=1542619522970168
immediate_commit_timestamp=1542619522970168     transaction_length=385
#181119 17:25:35 server id 7237  end_log_pos 655 CRC32 0x111c59d1     GTID    last_committed
=1      sequence_number=2       rbr_only=yes    original_committed_timestamp=1542619535667730
immediate_commit_timestamp=1542619535667730     transaction_length=374
```

可以看到，虽然已经将参数设置正确，但并没有得到想要的效果（仍然没有在同一个组），这是为什么呢？

（8）查看会话中的事务涉及的两个表结构，具体如下：

```
mysql> show create table salaries;
……
| salaries | CREATE TABLE `salaries` (
  `emp_no` int(11) NOT NULL,
  `salary` int(11) NOT NULL,
  `from_date` date NOT NULL,
  `to_date` date NOT NULL,
  PRIMARY KEY (`emp_no`,`from_date`),
  CONSTRAINT `salaries_ibfk_1` FOREIGN KEY (`emp_no`) REFERENCES `employees` (`emp_no`) ON
DELETE CASCADE
) ENGINE=InnoDB DEFAULT CHARSET=utf8 COLLATE=utf8_unicode_ci |

mysql> show create table employees.dept_emp;
……
| dept_emp | CREATE TABLE `dept_emp` (
  `emp_no` int(11) NOT NULL,
  `dept_no` char(4) COLLATE utf8_unicode_ci NOT NULL,
  `from_date` date NOT NULL,
  `to_date` date NOT NULL,
  PRIMARY KEY (`emp_no`,`dept_no`),
  KEY `dept_no` (`dept_no`),
  CONSTRAINT `dept_emp_ibfk_1` FOREIGN KEY (`emp_no`) REFERENCES `employees` (`emp_no`) ON
DELETE CASCADE,
  CONSTRAINT `dept_emp_ibfk_2` FOREIGN KEY (`dept_no`) REFERENCES `departments` (`dept_no`)
ON DELETE CASCADE
) ENGINE=InnoDB DEFAULT CHARSET=utf8 COLLATE=utf8_unicode_ci |
```

可以看到这两个表都涉及外键，就是这个原因将导致复制不能被并发执行。接下来，删除掉这些外键约束，代码如下：

```
mysql> alter table employees.salaries DROP FOREIGN KEY `salaries_ibfk_1`;
Query OK, 0 rows affected (0.11 sec)
Records: 0  Duplicates: 0  Warnings: 0

mysql> alter table employees.dept_emp drop FOREIGN KEY dept_emp_ibfk_1;
Query OK, 0 rows affected (0.19 sec)
Records: 0  Duplicates: 0  Warnings: 0

mysql> alter table employees.dept_emp drop FOREIGN KEY dept_emp_ibfk_2;
Query OK, 0 rows affected (0.03 sec)
Records: 0  Duplicates: 0  Warnings: 0
```

（9）在主库上再次顺序执行以下两个事务。

```
mysql> update employees.dept_emp set to_date='9999-01-01' limit 3;
Query OK, 3 rows affected (0.06 sec)
Rows matched: 3  Changed: 3  Warnings: 0

mysql> update employees.salaries set salary=100 limit 3 ;
Query OK, 3 rows affected (0.01 sec)
Rows matched: 3  Changed: 3  Warnings: 0
```

（10）在主库查看此次生成的 BINLOG，具体如下：

```
[mysql3488@hz_10_120_240_251 data]$ mysqlbinlog -vvv mysql-bin.000028 > mysql-bin.000028.log
[mysql3488@hz_10_120_240_251 data]$
[mysql3488@hz_10_120_240_251 data]$ cat mysql-bin.000028.log |grep "last_committed"
......
#181119 17:26:50 server id 7237  end_log_pos 1475 CRC32 0xb55ff777      GTID    last_committed=
4       sequence_number=5       rbr_only=yes    original_committed_timestamp=1542619610265407
immediate_commit_timestamp=1542619610265407     transaction_length=385
#181119 17:26:57 server id 7237  end_log_pos 1860 CRC32 0x30ea5e65      GTID    last_committed=
4       sequence_number=6       rbr_only=yes    original_committed_timestamp=1542619617335538
immediate_commit_timestamp=1542619617335538     transaction_length=374
```

注意：实际业务场景可能比较复杂，如下所示，在经历过反复验证之后，发现在同一个会话中顺序执行上面的这两个事务（每次更新之前更改事务 1 和事务 2 的条件，使其不会更新相同记录，例如第一次，事务 1: update A where id=1，事务 2: update B where id=1；第二次，事务 1: update A where id=2，事务 2: update B where id=2），得到的 last_committed 也可能不同，类似如下情况，即 last_committed 的值是交叠相同的。事务 1 执行了两次，last_committed 的值都是 28；事务 2 执行了两次，last_committed 的值都是 29，但事务 1 与事务 2 不属于同一个组（last_committed 不相等），此时可以通过 last_committed 与 sequence_number 进行比较，如果 last_committed（28 和 29）的值<min（sequence_number）的值（32），表明数据之间不冲突，即从库仍然可以并行执行。

```
第一次，事务 1：
#181212 16:20:06 server id 7237  end_log_pos 10041 CRC32 0x1a02766e     GTID
last_committed=28       sequence_number=32      rbr_only=yes    original_committed_timestamp=
1544602806172592        immediate_commit_timestamp=1544602806172592     transaction_length=320

第一次，事务 2：
#181212 16:20:09 server id 7237  end_log_pos 10361 CRC32 0xd9147b9a     GTID
last_committed=29       sequence_number=33      rbr_only=yes    original_committed_timestamp=
1544602809135136        immediate_commit_timestamp=1544602809135136     transaction_length=320

第二次，事务 1：
#181212 16:21:02 server id 7237  end_log_pos 10681 CRC32 0xacf051bb     GTID
last_committed=28       sequence_number=34      rbr_only=yes    original_committed_timestamp=
1544602862252860        immediate_commit_timestamp=1544602862252860     transaction_length=320

第二次，事务 2：
#181212 16:21:05 server id 7237  end_log_pos 11001 CRC32 0x0fa69c25     GTID    last_committed
=29     sequence_number=35      rbr_only=yes    original_committed_timestamp=1544602865816208
immediate_commit_timestamp=1544602865816208     transaction_length=320
```

从试验结果来看，符合预期，从库可以并发应用 Relay Log 数据（这涉及多线程复制的实现机制 Commit-Parent-Based、Lock-Based，由于内容晦涩，在此不再深入讨论）。

接下来再验证一下如果将参数 binlog_transaction_dependency_tracking 的值，变更为 WRITESET_SESSION，会有什么样的结果？

（1）首先，确认主库设置以下参数值。

```
mysql> set global binlog_transaction_dependency_tracking='WRITESET_SESSION';
Query OK, 0 rows affected (0.00 sec)

mysql> show variables like 'binlog_transaction_dependency_tracking';
+----------------------------------------+------------------+
| Variable_name                          | Value            |
+----------------------------------------+------------------+
| binlog_transaction_dependency_tracking | WRITESET_SESSION |
+----------------------------------------+------------------+
1 row in set (0.00 sec)

mysql> show variables like 'transaction_write_set_extraction';
+----------------------------------+----------+
```

```
+-------------------------------+----------+
| Variable_name                 | Value    |
+-------------------------------+----------+
| transaction_write_set_extraction | XXHASH64 |
+-------------------------------+----------+
1 row in set (0.00 sec)
```

（2）主库上创建一个新会话，并顺序执行以下两个事务。

```
mysql> update employees.salaries set salary=100 limit 3 ;
Query OK, 3 rows affected (0.09 sec)
Rows matched: 3  Changed: 3  Warnings: 0

mysql> update employees.dept_emp set to_date='9999-02-01' limit 3;
Query OK, 3 rows affected (0.06 sec)
Rows matched: 3  Changed: 3  Warnings: 0
```

（3）在主库查看 BINLOG 日志。

```
[mysql3488@hz_10_120_240_251 data]$ mysqlbinlog -vv mysql-bin.000029 > mysql-bin.000029.log
[mysql3488@hz_10_120_240_251 data]$
[mysql3488@hz_10_120_240_251 data]$ cat mysql-bin.000029.log |grep "last_committed"
#181119 22:38:54 server id 7237  end_log_pos 270 CRC32 0x8d5d169d    GTID    last_committed
=0    sequence_number=1    rbr_only=yes    original_committed_timestamp=1542638334826576
immediate_commit_timestamp=1542638334826576     transaction_length=374
#181119 22:39:06 server id 7237  end_log_pos 644 CRC32 0x58131898    GTID    last_committed
=1    sequence_number=2    rbr_only=yes    original_committed_timestamp=1542638346404263
immediate_commit_timestamp=1542638346404263     transaction_length=385
```

可以看出，在将参数的值从 WRITESET 变更为 WRITESET_SESSION 之后，应用相同的试验内容，即主库在同一个会话对不同对象执行数据变更，但从库将不能并发应用。

（4）主库上再创建一个新会话，并在每一个会话分别执行一个事务。

```
session1:
mysql> update employees.salaries set salary=200 limit 3 ;
Query OK, 3 rows affected (0.09 sec)
Rows matched: 3  Changed: 3  Warnings: 0

session2:
mysql> update employees.dept_emp set to_date='9999-01-01' limit 3;
Query OK, 3 rows affected (0.06 sec)
Rows matched: 3  Changed: 3  Warnings: 0
```

（5）查看主库上的 BINLOG。

```
[mysql3488@hz_10_120_240_251 data]$ cat mysql-bin.000029.log |grep "last_committed"
......
#181119 22:47:19 server id 7237  end_log_pos 1029 CRC32 0x8aee409e   GTID    last_committed
=2    sequence_number=3    rbr_only=yes    original_committed_timestamp=1426638839795133
immediate_commit_timestamp=1542638839795133     transaction_length=374
#181119 22:47:28 server id 7237  end_log_pos 1403 CRC32 0x43e79ebc   GTID    last_committed
=2    sequence_number=4    rbr_only=yes    original_committed_timestamp=1542638848568232
immediate_commit_timestamp=1542638848568232     transaction_length=385
```

可以看出，主库产生的 BINLOG 是符合预期的。

30.2.3　增强半同步复制

前面介绍的复制是异步操作，主库和从库的数据之间难免会存在一定的延迟，这样存在一个隐患：当在主库上写入一个事务并提交成功，而从库尚未得到主库的 BINLOG 日志时，主库由于磁盘损坏、内存故障、断电等原因意外宕机，导致主库上该事务 BINLOG 丢失，此时从库就会损失这个事务，从而造成主从不一致。

为了解决这个问题，从 MySQL 5.5 开始，引入了半同步复制机制，此时的技术暂且称之为传统的半同步复制，因该技术发展到 MySQL 5.7 后，已经演变为增强半同步复制（也称为无损复制）。在异步复制时，主库执行 Commit 提交操作并写入 BINLOG 日志后即可成功返回客户端，无须等待 BINLOG 日志传送给从库，如图 30-8 所示。

而半同步复制时，为了保证主库上的每一个 BINLOG 事务都能够被可靠地复制到从库上，主库在每次事务成功提交时，并不及时反馈给前端应用用户，而是等待至少一个从库（详见参数 rpl_semi_sync_master_wait_for_slave_count）也接收到 BINLOG 事务并成功写入中继

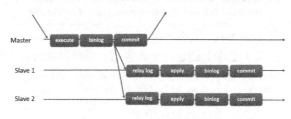

图 30-8　MySQL 的异步复制

日志后，主库才返回 Commit 操作成功给客户端（不管是传统的半同步复制，还是增强的半同步复制，目的都是一样的，只不过两种方式在一个细微地方不同，详见下文）。

半同步复制保证了事务成功提交后，至少有两份日志记录，一份在主库的 BINLOG 日志上，另一份在至少一个从库的中继日志 Relay Log 上，从而更进一步保证了数据的完整性。

在传统的半同步复制中，主库写数据到 BINLOG，且执行 Commit 操作后，会一直等待从库的 ACK，即从库写入 Relay Log 后，并将数据落盘，返回给主库消息，通知主库可以返回前端应用操作成功，这样会出现一个问题，就是实际上主库已经将该事务 Commit 到了事务引擎层，应用已经可以看到数据发生了变化，只是在等返回而已，如果此时主库宕机，有可能从库还没能写入 Relay Log，就会发生主从库数据不一致。增强半同步就是为了解决这个问题，做了微调，即主库写数据到 BINLOG 后，就开始等待从库的应答 ACK，直到至少一个从库写入 Relay Log 后，并将数据落盘，然后返回给主库消息，通知主库可以执行 Commit 操作，然后主库开始提交到事务引擎层，应用此时可以看到数据发生了变化。增强半同步复制的大致流程如图 30-9 所示。

半同步复制模式下，假如在传送 BINLOG 日志到从库时，从库宕机或者网络故障，导致 BINLOG 并没有及时地传送到从库上，此时主库上的事

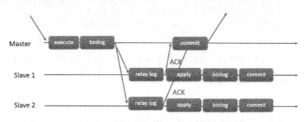

图 30-9　MySQL 增强半同步复制

务会等待一段时间（时间长短由参数 rpl_semi_sync_master_timeout 设置的毫秒数决定），如果 BINLOG 在这段时间内都无法成功发送到从库上，则 MySQL 自动调整复制为异步模式，事务正常返回提交结果给客户端。

半同步复制很大程度上取决于主从库之间的网络情况，往返时延 RTT 越小决定了从库的实时性越好。通俗地说，主从库之间网络越快，从库越实时。

注意： 往返时延 RTT（Round-Trip Time）在计算机网络中是一个重要的性能指标，它表示从发送端发送数据开始到发送端接收到接收端的确认，总共经历的时长。

半同步模式是通过一个插件来实现的，主库和从库使用不同的插件。安装比较简单，在 30.2.2 节多线程复制的环境上，安装半同步复制插件即可。

（1）首先，判断 MySQL 服务器是否支持动态增加插件：

```
mysql> select @@have_dynamic_loading;
+------------------------+
| @@have_dynamic_loading |
+------------------------+
| YES                    |
```

30.2 复制搭建

```
+-----------------------+
1 row in set (0.00 sec)
```

（2）确认支持动态增加插件后，检查 MySQL 的安装目录下是否存在插件，一般默认在 $MYSQL_HOME/lib/plugin 目录下存在主库插件 semisync_master.so 和从库插件 semisync_slave.so：

```
$MYSQL_HOME/lib/plugin/semisync_master.so
$MYSQL_HOME/lib/plugin/semisync_slave.so
```

在主库上安装 semisync_master.so 插件：

```
mysql> install plugin rpl_semi_sync_master SONAME 'semisync_master.so';
```

在从库上安装 semisync_slave.so 插件：

```
mysql> install plugin rpl_semi_sync_slave SONAME 'semisync_slave.so';
```

安装完成后，从 plugin 表中能够看到刚才安装的插件：

```
mysql> select * from mysql.plugin;
+----------------------+--------------------+
| name                 | dl                 |
+----------------------+--------------------+
| rpl_semi_sync_master | semisync_master.so |
| rpl_semi_sync_slave  | semisync_slave.so  |
+----------------------+--------------------+
2 rows in set (0.00 sec)
```

也就是说，安装完成后，MySQL 会在系统表 plugin 中记录刚才安装的插件，下次系统重启后会自动加载插件。

（3）需要分别在主库和从库上配置参数打开半同步 semi-sync，默认半同步设置是不打开的，在主库上配置全局参数：

```
mysql> set global rpl_semi_sync_master_enabled=on;
Query OK, 0 rows affected (0.02 sec)

mysql> set global rpl_semi_sync_master_timeout=1000;
Query OK, 0 rows affected (0.00 sec)

mysql> show variables like 'rpl_semi_sync_master_wait_point';
+---------------------------------+------------+
| Variable_name                   | Value      |
+---------------------------------+------------+
| rpl_semi_sync_master_wait_point | AFTER_SYNC |
+---------------------------------+------------+
1 row in set (0.01 sec)
```

在从库上配置全局参数：

```
mysql> set global rpl_semi_sync_slave_enabled=on;
Query OK, 0 rows affected (0.00 sec)
```

注意，参数 rpl_semi_sync_master_wait_point 的值默认是 AFTER_SYNC，即上文中提到的增强半同步复制；另一个可选项是 AFTER_COMMIT，即传统的半同步复制。由于之前配置的复制是异步复制，所以需要重启从库上的 I/O 线程（如果是全新配置的半同步复制，则不需要）：

```
mysql> stop slave io_thread for channel 'channel_3307';
Query OK, 0 rows affected (0.00 sec)

mysql> start slave io_thread for channel 'channel_3307';
Query OK, 0 rows affected (0.00 sec)
```

到此半同步配置完毕，下面可以进行验证。主库上通过 SHOW STATUS 命令能够看到当前半同步复制的一些状态值：

```
mysql> show status like '%semi_sync%';
+--------------------------------------------+-------+
| Variable_name                              | Value |
+--------------------------------------------+-------+
| Rpl_semi_sync_master_clients               | 1     |
| Rpl_semi_sync_master_net_avg_wait_time     | 0     |
| Rpl_semi_sync_master_net_wait_time         | 0     |
| Rpl_semi_sync_master_net_waits             | 1     |
| Rpl_semi_sync_master_no_times              | 2     |
| Rpl_semi_sync_master_no_tx                 | 2     |
| Rpl_semi_sync_master_status                | ON    |
| Rpl_semi_sync_master_timefunc_failures     | 0     |
| Rpl_semi_sync_master_tx_avg_wait_time      | 0     |
| Rpl_semi_sync_master_tx_wait_time          | 0     |
| Rpl_semi_sync_master_tx_waits              | 0     |
| Rpl_semi_sync_master_wait_pos_backtraverse | 0     |
| Rpl_semi_sync_master_wait_sessions         | 0     |
| Rpl_semi_sync_master_yes_tx                | 0     |
| Rpl_semi_sync_slave_status                 | OFF   |
+--------------------------------------------+-------+
15 rows in set (0.01 sec)
```

注意：由于试验环境不同，所以 SHOW STATUS 显示的状态值可能存在不一致，着重关注这些状态值的变化，而不是这些状态的初始值。

着重关注以下 3 个状态值。

- Rpl_semi_sync_master_status: 值为 ON，表示半同步复制目前处于打开状态。
- Rpl_semi_sync_master_yes_tx: 值为 0，表示主库当前尚未有任何一个事务是通过半同步复制到从库。
- Rpl_semi_sync_master_no_tx: 值为 2，表示当前有两个事务不是半同步模式下从库及时响应的（记住这个值，后面会有对比）。

执行一个事务，再检查一下状态：

```
mysql> update salaries set salary=66074 where emp_no=10001 and salary=300;
Query OK, 1 row affected (0.00 sec)
Rows matched: 1  Changed: 1  Warnings: 0

mysql> show status like '%semi_sync%';
+--------------------------------------------+-------+
| Variable_name                              | Value |
+--------------------------------------------+-------+
| Rpl_semi_sync_master_clients               | 1     |
| Rpl_semi_sync_master_net_avg_wait_time     | 0     |
| Rpl_semi_sync_master_net_wait_time         | 0     |
| Rpl_semi_sync_master_net_waits             | 2     |
| Rpl_semi_sync_master_no_times              | 2     |
| Rpl_semi_sync_master_no_tx                 | 2     |
| Rpl_semi_sync_master_status                | ON    |
| Rpl_semi_sync_master_timefunc_failures     | 0     |
| Rpl_semi_sync_master_tx_avg_wait_time      | 427   |
| Rpl_semi_sync_master_tx_wait_time          | 427   |
| Rpl_semi_sync_master_tx_waits              | 1     |
| Rpl_semi_sync_master_wait_pos_backtraverse | 0     |
| Rpl_semi_sync_master_wait_sessions         | 0     |
| Rpl_semi_sync_master_yes_tx                | 1     |
| Rpl_semi_sync_slave_status                 | OFF   |
+--------------------------------------------+-------+
15 rows in set (0.01 sec)
```

此时会发现 Rpl_semi_sync_master_yes_tx 的值变为 1，即刚才的 UPDATE 事务通过半同步复制到从库上了，Rpl_semi_sync_master_yes_tx 计数增加 1。到从库确认一下，数据确实被复制过去了：

```
mysql> select * from salaries where emp_no=10001 and salary=66074;
```

```
| emp_no | salary | from_date  | to_date    |
+--------+--------+------------+------------+
|  10001 |  66074 | 1988-06-25 | 1989-06-25 |
+--------+--------+------------+------------+
1 row in set (0.01 sec)
```

再尝试一下网络异常的场景下，主库在等待 rpl_semi_sync_master_timeout 毫秒超时后，自动转成异步复制的场景。

（1）首先，在主库上确认半同步复制会等待 1s 超时：

```
mysql> show variables like 'rpl_semi_sync_master_timeout';
+------------------------------+-------+
| Variable_name                | Value |
+------------------------------+-------+
| rpl_semi_sync_master_timeout | 1000  |
+------------------------------+-------+
1 row in set (0.00 sec)
```

（2）在从库上通过停掉 I/O 线程模拟复制故障（也可以使用 Iptables 等防火墙命令来模拟网络故障）：

```
mysql> stop slave io_thread for channel 'channel_3307';
Query OK, 0 rows affected (0.00 sec)
```

（3）在主库上执行一个事务并提交（默认提交即可），主库上的提交操作会被阻塞 1s：

```
mysql> update salaries set salary=62102 where emp_no=10001 and salary=200;
Query OK, 1 row affected (1.00 sec)
Rows matched: 1  Changed: 1  Warnings: 0
```

通过查看主库的错误日志，可以发现半同步复制已经关闭：

```
2018-07-31T14:57:32.648854+08:00 42 [Warning] Timeout waiting for reply of binlog (file:
mysql-bin.000043, pos: 2397), semi-sync up to file , position 0.
2018-07-31T14:57:32.648915+08:00 42 [Note] Semi-sync replication switched OFF.
```

（4）主库上再一次检查半同步复制的一些状态值：

```
mysql> show status like '%semi_sync%';
+--------------------------------------------+-------+
| Variable_name                              | Value |
+--------------------------------------------+-------+
| Rpl_semi_sync_master_clients               | 0     |
| Rpl_semi_sync_master_net_avg_wait_time     | 0     |
| Rpl_semi_sync_master_net_wait_time         | 0     |
| Rpl_semi_sync_master_net_waits             | 2     |
| Rpl_semi_sync_master_no_times              | 3     |
| Rpl_semi_sync_master_no_tx                 | 3     |
| Rpl_semi_sync_master_status                | OFF   |
| Rpl_semi_sync_master_timefunc_failures     | 0     |
| Rpl_semi_sync_master_tx_avg_wait_time      | 427   |
| Rpl_semi_sync_master_tx_wait_time          | 427   |
| Rpl_semi_sync_master_tx_waits              | 1     |
| Rpl_semi_sync_master_wait_pos_backtraverse | 0     |
| Rpl_semi_sync_master_wait_sessions         | 0     |
| Rpl_semi_sync_master_yes_tx                | 1     |
| Rpl_semi_sync_slave_status                 | OFF   |
+--------------------------------------------+-------+
15 rows in set (0.01 sec)
```

仍然看之前着重关注的 3 个状态值。

- Rpl_semi_sync_master_status：值变为 OFF 了，表示主库上半同步复制已经关闭了，目前复制模式为异步复制。

- Rpl_semi_sync_master_yes_tx：值仍然为 1，表示刚才的事务并不是通过半同步复制完成的，所以半同步成功事务仍然为 1，并不累加。

- Rpl_semi_sync_master_no_tx：值更新为 3，比原来的 2 累加了 1，表示在半同步模式下,从库没有及时响应的事务增加 1 个。

当开启从库的 I/O 线程之后,主库会从异步复制自动切换回半同步复制,此场景不再赘述。

从半同步复制的整个过程可以发现,主库和从库数据不具有强一致性,即主库已经变更的数据从库不一定发生了变更,如果想让主从库达到强一致,只能在组复制(Group Replication)实现了,关于组复制的内容可详见第 31 章。

到现在为止,文中涉及的复制均是以 file 和 postion 的方式进行的数据定位与同步,其实从 MySQL 5.6 开始就引入了 GTID(Global Transaction Identifiers),并且就多线程与组复制来说,离不开 GTID,这些技术都是直接或间接地利用了 GTID。那到底 GTID 是什么呢,接下来的章节将会进行详细介绍。

30.3 GTID(Global Transaction Identifier)

从 MySQL 5.6 开始引入了 GTID(Global Transaction Identifier)。GTID 是每个提交的事务的唯一标识,该标识不仅在 Master 端具有唯一性,在整个复制拓扑关系中,一样具有唯一性。

30.3.1 格式与存储

GTID 由 source_id 和 transaction_id 构成,表示形式如下所示:

```
GTID = source_id:transaction_id
```

其中,source_id 通常由服务端的 server_uuid 表示,而 transaction_id 通常是由连续数字表示。举例如下:

```
GTID = 3E11FA47-71CA-11E1-9E33-C80AA9429562:23
```

前半部分的"3E11FA47-71CA-11E1-9E33-C80AA9429562"表示执行该事务的 Master 的 uuid,后半部分的"23"表示这是第 23 个事务,只有主库才能生成 GTID,即能完成写事务的节点。

当 gtid_mode 为 ON 或者 ON_PERMISSIVE 时(参数解释详见下文),GTID 集合存储在表 mysql.gtid_executed 里,表中的记录样例如下所示:

```
mysql> select * from mysql.gtid_executed;
+--------------------------------------+----------------+--------------+
| source_uuid                          | interval_start | interval_end |
+--------------------------------------+----------------+--------------+
| b509f331-cd14-11e8-866a-0024e869b4d5 |              1 |        18400 |
+--------------------------------------+----------------+--------------+
1 rows in set (0.10 sec)
```

- source_uuid,表示生成该 GTID 事务的源实例的 uuid;
- interval_start,表示 GTID 的起始值;
- interval_end,表示 GTID 的结束值。

例如,"b509f331-cd14-11e8-866a-0024e869b4d5,1,18400"表示生成该 GTID 事务的源实例 uuid 为"b509f331-cd14-11e8-866a-0024e869b4d5",此实例已经执行了事务集合 b509f331-cd14-11e8- 866a-0024e869b4d5:1-18400。

另外,表 mysql.gtid_executed 的记录也可能是下面这样的形式:

```
mysql> select * from mysql.gtid_executed;
+--------------------------------------+----------------+--------------+
| source_uuid                          | interval_start | interval_end |
+--------------------------------------+----------------+--------------+
```

30.3 GTID (Global Transaction Identifier)

```
| b509f331-cd14-11e8-866a-0024e869b4d5 |               1 |             1 |
| b509f331-cd14-11e8-866a-0024e869b4d5 |               2 |             2 |
| b509f331-cd14-11e8-866a-0024e869b4d5 |               3 |             3 |
| b509f331-cd14-11e8-866a-0024e869b4d5 |               4 |             4 |
| b509f331-cd14-11e8-866a-0024e869b4d5 |               5 |             5 |
| b509f331-cd14-11e8-866a-0024e869b4d5 |               6 |             6 |
+--------------------------------------+-----------------+---------------+
6 rows in set (0.00 sec)
```

这是因为在从库上，如果关闭了 log_slave_updates 参数（即从库不写入 BINLOG），当从库在应用 Relay Log 后，会在表 mysql.gtid_executed 记录每一次执行过的事务，如果一直这样下去，这势必会造成空间的大量浪费，也会影响性能，此时可以通过参数 gtid_executed_compression_period 进行调节。

gtid_executed_compression_period 参数从 MySQL 5.7.6 版本引入，默认值为 1000，表示每当处理完 1 000 个事务后，对表 mysql.gtid_executed 进行压缩。如果参数设置为 0，表示禁用压缩特性，该参数支持动态调整。

例如，将参数 gtid_executed_compression_period 从 1 000 调整为 10：

```
mysql> set global gtid_executed_compression_period=10;
Query OK, 0 rows affected (0.00 sec)
```

当从库应用 Relay Log 数据超过 10 个事务后，开始对表 mysql.gtid_executed 进行压缩。

```
mysql> select * from mysql.gtid_executed;
+--------------------------------------+----------------+--------------+
| source_uuid                          | interval_start | interval_end |
+--------------------------------------+----------------+--------------+
| b509f331-cd14-11e8-866a-0024e869b4d5 |              1 |           11 |
+--------------------------------------+----------------+--------------+
1 row in set (0.00 sec)
```

以上内容是从库关闭了 log_slave_updates 参数的情况，如果打开这个参数会怎么样呢？接下来继续往下看。

由于参数 log_slave_updates 是静态参数，需要重启实例。当完成重启操作后，主库执行一些新事务，然后观察从库上表 mysql.gtid_executed 的内容和 slave 的执行情况：

```
mysql> select * from mysql.gtid_executed;
+--------------------------------------+----------------+--------------+
| source_uuid                          | interval_start | interval_end |
+--------------------------------------+----------------+--------------+
| b509f331-cd14-11e8-866a-0024e869b4d5 |              1 |           11 |
+--------------------------------------+----------------+--------------+
1 row in set (0.00 sec)
```

show slave status 的信息如下：

```
Retrieved_Gtid_Set: b509f331-cd14-11e8-866a-0024e869b4d5:12-55
Executed_Gtid_Set: b509f331-cd14-11e8-866a-0024e869b4d5:1-55
Auto_Position: 1
```

发现表 mysql.gtid_executed 与重启前没有变化，而通过 show slave status 命令，可以看到实际上该从库已经应用了很多 Relay Log 数据，如果此时再执行 flush logs 命令，即切换一个新的 BINLOG 文件，再看该表的数据：

```
mysql> flush logs;
Query OK, 0 rows affected (0.04 sec)

mysql> select * from mysql.gtid_executed;
+--------------------------------------+----------------+--------------+
| source_uuid                          | interval_start | interval_end |
+--------------------------------------+----------------+--------------+
| b509f331-cd14-11e8-866a-0024e869b4d5 |              1 |           55 |
+--------------------------------------+----------------+--------------+
```

此时从库已经将最新的 GTID 事务集合写入了 mysql.gtid_executed 中，其实这也表明了，一旦启用 binary logging（将事务写入 BINLOG 文件）功能，参数 gtid_executed_compression_period 也就不再起作用了。无论是在从库，还是在主库，只有当发生 BINLOG 文件切换，才会将执行过的 GTID 事务集合写入表 mysql.gtid_executed 中。

参数 gtid_executed_compression_period 的功能，是通过一个后台线程实现的。

```
mysql> SELECT * FROM performance_schema.threads WHERE NAME LIKE '%gtid%'\G
*************************** 1. row ***************************
          THREAD_ID: 45
               NAME: thread/sql/compress_gtid_table
               TYPE: FOREGROUND
     PROCESSLIST_ID: 12
   PROCESSLIST_USER: NULL
   PROCESSLIST_HOST: NULL
     PROCESSLIST_DB: NULL
PROCESSLIST_COMMAND: Daemon
   PROCESSLIST_TIME: 2406
  PROCESSLIST_STATE: Suspending
   PROCESSLIST_INFO: NULL
   PARENT_THREAD_ID: 1
               ROLE: NULL
       INSTRUMENTED: YES
            HISTORY: YES
    CONNECTION_TYPE: NULL
       THREAD_OS_ID: 3708
1 row in set (0.00 sec)
```

通过以上的演示，读者已经发现表 mysql.gtid_executed 的记录会一直增加，那怎么样才能清空该表呢？换句话说，如何才能重置执行过的 GTID 的集合，让事务序号回归初始值 1 呢？

这就需要使用 reset master 命令，一旦执行了该命令，会删除所有当下的 BINLOG 文件，并生成一个新的 BINLOG 文件，同时用新生成的 BINLOG 文件名重置 BINLOG 索引文件，也会清空系统变量 gtid_executed 和 gtid_purged 的值，以及表 mysql.gtid_executed。

例如：

```
mysql> select @@GLOBAL.gtid_executed;
+------------------------------------------+
| @@GLOBAL.gtid_executed                   |
+------------------------------------------+
| b509f331-cd14-11e8-866a-0024e869b4d5:1-55 |
+------------------------------------------+
1 row in set (0.00 sec)

mysql> select @@GLOBAL.gtid_purged;
+------------------------------------------+
| @@GLOBAL.gtid_purged                     |
+------------------------------------------+
| b509f331-cd14-11e8-866a-0024e869b4d5:1-55 |
+------------------------------------------+
1 row in set (0.00 sec)

mysql> reset master ;
Query OK, 0 rows affected (0.24 sec)

mysql> select * from mysql.gtid_executed;
Empty set (0.00 sec)

mysql> select @@GLOBAL.gtid_purged;
+----------------------+
| @@GLOBAL.gtid_purged |
+----------------------+
|                      |
+----------------------+
1 row in set (0.00 sec)
```

30.3　GTID（Global Transaction Identifier）

```
mysql> select @@GLOBAL.gtid_executed;
+------------------------+
| @@GLOBAL.gtid_executed |
+------------------------+
|                        |
+------------------------+
1 row in set (0.00 sec)
```

这里又出现了 GTID 的另一个重要变量 gtid_purged，该变量对于处理 GTID 问题有重要意义，接下来的内容将围绕该变量进行讲述。

30.3.2　gtid_purged

gtid_purged 表示该实例的二进制日志（binary log）已经被清除掉（purge）的事务集合，通俗地讲，就是这些事务的 BINLOG 已经没有了。如果该实例配置了从库，而从库在复制过程中，由于某些原因缺少这些数据，复制时就会发生错误。接下来介绍 gtid_purged 的变更会发生在什么时候，以及如何进行初始化。

gtid_purged 的变更操作分为系统自动触发和手动干预两种情况，下面先介绍系统自动触发。

当遇到 purge binary logs 操作或者 binary log 超过阈值 expire_logs_days 时（该参数表示 binary log 过期时间，前文也有过讲解，不再赘述），会触发 gtid_purged 的变更。

（1）首先，确认当前 mysql-bin.000001 文件中包含的 GTID 事务集合。

```
[mysql3307@hz_10_120_240_251 data]$ grep "b509f331-cd14-11e8-866a-0024e869b4d5" mysql-bin.000001.log
SET @@SESSION.GTID_NEXT= 'b509f331-cd14-11e8-866a-0024e869b4d5:1'/*!*/;
SET @@SESSION.GTID_NEXT= 'b509f331-cd14-11e8-866a-0024e869b4d5:2'/*!*/;
SET @@SESSION.GTID_NEXT= 'b509f331-cd14-11e8-866a-0024e869b4d5:3'/*!*/;
SET @@SESSION.GTID_NEXT= 'b509f331-cd14-11e8-866a-0024e869b4d5:4'/*!*/;
SET @@SESSION.GTID_NEXT= 'b509f331-cd14-11e8-866a-0024e869b4d5:5'/*!*/;
SET @@SESSION.GTID_NEXT= 'b509f331-cd14-11e8-866a-0024e869b4d5:6'/*!*/;
SET @@SESSION.GTID_NEXT= 'b509f331-cd14-11e8-866a-0024e869b4d5:7'/*!*/;
SET @@SESSION.GTID_NEXT= 'b509f331-cd14-11e8-866a-0024e869b4d5:8'/*!*/;
SET @@SESSION.GTID_NEXT= 'b509f331-cd14-11e8-866a-0024e869b4d5:9'/*!*/;
SET @@SESSION.GTID_NEXT= 'b509f331-cd14-11e8-866a-0024e869b4d5:10'/*!*/;
……略过中间连续序号
SET @@SESSION.GTID_NEXT= 'b509f331-cd14-11e8-866a-0024e869b4d5:53'/*!*/;
SET @@SESSION.GTID_NEXT= 'b509f331-cd14-11e8-866a-0024e869b4d5:54'/*!*/;
SET @@SESSION.GTID_NEXT= 'b509f331-cd14-11e8-866a-0024e869b4d5:55'/*!*/;
```

可以看到该文件中，包含的 GTID 集合为 b509f331-cd14-11e8-866a-0024e869b4d5:1-55。

（2）确认当前 gtid_purged 的值。

```
mysql> select @@gtid_purged;
+---------------------------------------------+
| @@gtid_purged                               |
+---------------------------------------------+
| eab6f8ea-6ed7-11e8-a897-0024e869b4d5:1-4    |
+---------------------------------------------+
1 row in set (0.00 sec)
```

由于该实例曾经执行过其他实例的 BINLOG 事务，但目前这些 BINLOG 已经不存在了，所以这里显示是有值的，而不是空值（这个不是重点，请继续往下看）。

（3）执行 purge binary logs 命令。

```
mysql>
mysql> purge binary logs to 'mysql-bin.000002';
Query OK, 0 rows affected (0.02 sec)
```

删除了 mysql-bin.000001 文件，即清除了 GTID 集合 b509f331-cd14-11e8-866a-0024e869b4d5:1-55 的 BINLOG 数据。

（4）再次查看 gtid_purged 的值。

```
mysql> select @@gtid_purged;
+------------------------------------------------------------------------------+
| @@gtid_purged                                                                |
+------------------------------------------------------------------------------+
| b509f331-cd14-11e8-866a-0024e869b4d5:1-55,eab6f8ea-6ed7-11e8-a897-0024e869b4d5:1-4|
+------------------------------------------------------------------------------+
1 row in set (0.00 sec)
```

可以看到，此时 gtid_purged 已经从 eab6f8ea-6ed7-11e8-a897-0024e869b4d5:1-4 变更为 b509f331-cd14-11e8-866a-0024e869b4d5:1-55,eab6f8ea-6ed7-11e8-a897-0024e869b4d5:1-4。

以上是通过调用系统命令 purge binary logs 执行的操作，而如果手动删除 binary log 文件，会有什么结果呢？

（1）首先，确认当下的 gtid_purged。

```
mysql> select @@gtid_purged;
+------------------------------------------------------------------------------+
| @@gtid_purged                                                                |
+------------------------------------------------------------------------------+
| b509f331-cd14-11e8-866a-0024e869b4d5:1-58,eab6f8ea-6ed7-11e8-a897-0024e869b4d5:1-8|
+------------------------------------------------------------------------------+
1 row in set (0.00 sec)
```

b509f331-cd14-11e8-866a-0024e869b4d5:1-58,eab6f8ea-6ed7-11e8-a897-0024e869b4d5:1-8 是当前该实例的 gtid_purged 的值，即已经被清除掉的 Binlog。

（2）通过 show binary logs 命令观察此时的 binary log 文件。

```
mysql> show binary logs;
+------------------+-----------+
| Log_name         | File_size |
+------------------+-----------+
| mysql-bin.000004 |      3825 |
| mysql-bin.000005 |       257 |
| mysql-bin.000006 |       234 |
+------------------+-----------+
3 rows in set (0.00 sec)
```

可以看到，当前该实例的 binary log 共包含 3 个文件，即 mysql-bin.000004、mysql-bin.000005 和 mysql-bin.000006。

mysql-bin 索引文件如下所示：

```
[mysql3307@hz_10_120_240_251 data]$ cat mysql-bin.index
/data2/mysql3307/data/mysql-bin.000004
/data2/mysql3307/data/mysql-bin.000005
/data2/mysql3307/data/mysql-bin.000006
```

通过以上观察，都可以确定 mysql-bin.000004 文件是目前存在的第一个 binary log 文件，即最旧的（oldest）文件。

（3）mysql-bin.000004 文件包含的 GTID 集合。

```
[mysql3307@hz_10_120_240_251 data]$ mysqlbinlog -vv mysql-bin.000004 > mysql-bin.000004.log
[mysql3307@hz_10_120_240_251 data]$
[mysql3307@hz_10_120_240_251 data]$ grep "GTID_NEXT" mysql-bin.000004.log
SET @@SESSION.GTID_NEXT= 'eab6f8ea-6ed7-11e8-a897-0024e869b4d5:9'  '/*!*/;
SET @@SESSION.GTID_NEXT= 'eab6f8ea-6ed7-11e8-a897-0024e869b4d5:10' '/*!*/;
SET @@SESSION.GTID_NEXT= 'eab6f8ea-6ed7-11e8-a897-0024e869b4d5:11' '/*!*/;
SET @@SESSION.GTID_NEXT= 'eab6f8ea-6ed7-11e8-a897-0024e869b4d5:12' '/*!*/;
SET @@SESSION.GTID_NEXT= 'b509f331-cd14-11e8-866a-0024e869b4d5:59' '/*!*/;
SET @@SESSION.GTID_NEXT= 'b509f331-cd14-11e8-866a-0024e869b4d5:60' '/*!*/;
SET @@SESSION.GTID_NEXT= 'b509f331-cd14-11e8-866a-0024e869b4d5:61' '/*!*/;
SET @@SESSION.GTID_NEXT= 'b509f331-cd14-11e8-866a-0024e869b4d5:62' '/*!*/;
SET @@SESSION.GTID_NEXT= 'eab6f8ea-6ed7-11e8-a897-0024e869b4d5:13' '/*!*/;
SET @@SESSION.GTID_NEXT= 'eab6f8ea-6ed7-11e8-a897-0024e869b4d5:14' '/*!*/;
```

30.3 GTID (Global Transaction Identifier)

```
SET @@SESSION.GTID_NEXT= 'eab6f8ea-6ed7-11e8-a897-0024e869b4d5:15 '/*!*/;
SET @@SESSION.GTID_NEXT= 'eab6f8ea-6ed7-11e8-a897-0024e869b4d5:16 '/*!*/;
SET @@SESSION.GTID_NEXT= 'b509f331-cd14-11e8-866a-0024e869b4d5:63 '/*!*/;
SET @@SESSION.GTID_NEXT= 'eab6f8ea-6ed7-11e8-a897-0024e869b4d5:17 '/*!*/;
SET @@SESSION.GTID_NEXT= 'AUTOMATIC' /* added by mysqlbinlog     */ /*!*/;
```

该文件中包括两个 GTID 集合，其中一个集合是 eab6f8ea-6ed7-11e8-a897-0024e869b4d5:9-17，而另一个集合是 b509f331-cd14-11e8-866a-0024e869b4d5:59-63。

（4）然后，从操作系统层面删除文件 mysql-bin.000004。

```
[mysql3307@hz_10_120_240_251 data]$ rm -rf mysql-bin.000004
```

（5）再通过 show binary logs 命令查看日志信息。

```
mysql> show binary logs;
+------------------+-----------+
| Log_name         | File_size |
+------------------+-----------+
| mysql-bin.000004 |         0 |
| mysql-bin.000005 |       257 |
| mysql-bin.000006 |       257 |
+------------------+-----------+
3 rows in set (0.02 sec)

gtid_purged:
mysql> select @@gtid_purged;
+---------------------------------------------------------------------------------+
| @@gtid_purged                                                                   |
+---------------------------------------------------------------------------------+
| b509f331-cd14-11e8-866a-0024e869b4d5:1-58,eab6f8ea-6ed7-11e8-a897-0024e869b4d5:1-8|
+---------------------------------------------------------------------------------+
1 row in set (0.00 sec)
```

虽然 show binary logs 的结果显示文件 mysql-bin.000004 的大小已经是 0，但 gtid_purged 发现和之前并没有什么差别，说明直接从操作系统层面删除 mysql-bin 文件，并不会触发 gtid_purged 的更新。

（6）重启该实例后，再次查看 gtid_purged。

```
mysql> show binary logs;
+------------------+-----------+
| Log_name         | File_size |
+------------------+-----------+
| mysql-bin.000004 |         0 |
| mysql-bin.000005 |       257 |
| mysql-bin.000006 |       257 |
| mysql-bin.000007 |       234 |
+------------------+-----------+
4 rows in set (0.02 sec)

mysql> select @@gtid_purged;
+---------------------------------------------------------------------------------+
| @@gtid_purged                                                                   |
+---------------------------------------------------------------------------------+
|b509f331-cd14-11e8-866a-0024e869b4d5:1-63,eab6f8ea-6ed7-11e8-a897-0024e869b4d5:1-17|
+---------------------------------------------------------------------------------+
1 row in set (0.00 sec)
```

重启实例后，gtid_purged 的值已经发生了变化，此时的结果是正确的，即 GTID 集合 b509f331-cd14-11e8-866a-0024e869b4d5:1-63,eab6f8ea-6ed7-11e8-a897-0024e869b4d5:1-17 相关的 binary log 已经被清除了。

以上两个场景为了说明 gtid_purged 的值是如何受外在因素的影响，从而自动触发系统检查并进行变更。在第二个示例中，由于重启了 MySQL 实例，所以这涉及 gtid_purged 的初始化过程，该过程其实是受参数 binlog_gtid_simple_recovery 影响的。

binlog_gtid_simple_recovery 参数控制着 MySQL 启动过程中，如何扫描 binary log 文件，从而获取正确的 GTID 集合，初始化 gtid_executed 和 gtid_purged。在 MySQL 5.7.6 版本以前，该参数的名字是 simplified_binlog_gtid_recovery，默认值是 FALSE，而到了 5.7.6 版本，参数更名为 binlog_gtid_simple_recovery，默认值仍为 FALSE。但从 5.7.7 版本以后，默认值调整为 TRUE，上文的两个测试场景使用的是默认值 TRUE。

- 默认值 TRUE，表示 MySQL 启动的过程中，仅需要读取最旧的（oldest）和最新的（newest）binary log 文件，查找文件中的 Previous_gtids_log_event 或者 Gtid_log_event，用于初始化 gtid_purged 和 gtid_executed，如第二个场景中，MySQL 实例重启以后，由于 mysql-bin.000004 文件已经不存在，最旧的文件是 mysql-bin.000005，此时该文件中会包含类似如下字样的信息：

```
/*!50530 SET @@SESSION.PSEUDO_SLAVE_MODE=1*/;
/*!50003 SET @OLD_COMPLETION_TYPE=@@COMPLETION_TYPE,COMPLETION_TYPE=0*/;
DELIMITER /*!*/;
# at 4
#181213 16:12:48 server id 7238  end_log_pos 123 CRC32 0x24924068     Start: binlog v 4, server v 5.7.22-log created 181213 16:12:48
# Warning: this binlog is either in use or was not closed properly.
BINLOG '
gBQSXA9GHAAAdwAAAHsAAAABAAQANS43LjIyLYWxvZwAAAAAAAAAAAAAAAAAAAAAAAAAAA
AAAAAAAAAAAAAAAAAAAAAAAAAEzgNAAgAEgAEBAQEEgAAXwAEGggAAAAICgCAAAACgoKKioAEjQA
AWhAkiQ=
'/*!*/;
# at 123
#181213 16:12:48 server id 7238  end_log_pos 250 CRC32 0x60d96f09     Previous-GTIDs
# b509f331-cd14-11e8-866a-0024e869b4d5:1-63,
# eab6f8ea-6ed7-11e8-a897-0024e869b4d5:1-17
```

gtid_purged 的值就可以确定出来了，gtid_executed 同理，扫描完成最新的 Binary Log 文件后，就可以知道该实例已经执行到哪里了。

- 可选值 FALSE，表示 MySQL 启动过程中，为了初始化 gtid_purged，系统会遍历扫描 Binary Log 文件，首先从最旧的文件开始，如果发现 Previous_gtids_log_event 事件，就停止扫描，否则就继续扫描，直到最新的文件为止，gtid_executed 也是同理，但扫描方向相反，首先扫描最新的 Binary Log 文件，如果发现 Previous_gtids_log_event 事件，就停止扫描，否则就继续扫描，直到最旧的文件为止。

> 注意：建议不要更改参数 binlog_gtid_simple_recovery 的默认值，如果 MySQL 实例是从非 GTID 升级为 GTID 模式，而且该实例有大量的 BINLOG 文件（非 GTID 模式下生成的），参数值在 FALSE 的情况下，会引发疯狂扫描 Binary Log 文件问题，导致性能急剧下降。正确做法是备份好非 GTID 模式下的 BINLOG 文件，尽快删除。

上文介绍了系统自动触发 gtid_purged 的变更，但有时仍需要手动调整，例如主从复制间出现一些错误（详见常见问题），接下来介绍如何手动设置 gtid_purged。

（1）确认当前 gtid_purged 和 gtid_executed 的值。

```
mysql> show global variables where variable_name in ('gtid_executed','gtid_purged');
+----------------+-------------------------------------------------+
| Variable_name  | Value                                           |
+----------------+-------------------------------------------------+
| gtid_executed  | eab6f8ea-6ed7-11e8-a897-0024e869b4d5:1-10       |
| gtid_purged    | eab6f8ea-6ed7-11e8-a897-0024e869b4d5:1-3        |
+----------------+-------------------------------------------------+
2 rows in set (0.01 sec)
```

如果在操作系统层面删除了 Binary Log 文件，而 MySQL 实例又不能重启，此时 gtid_purged

30.3 GTID (Global Transaction Identifier)

的值已经不正确，可能会影响到后续的主从切换，此时直接更改 gtid_purged，会收到如下错误提示：

```
mysql> set global gtid_purged='eab6f8ea-6ed7-11e8-a897-0024e869b4d5:1-5';
ERROR 1840 (HY000): @@GLOBAL.GTID_PURGED can only be set when @@GLOBAL.GTID_EXECUTED is empty.
```

该提示表明 gtid_purged 的变更，只有当 gtid_executed 为空时，才能被更改。

（2）执行 reset master。

```
mysql> reset master;
Query OK, 0 rows affected, 2 warnings (0.01 sec)
```

该命令会删除所有的 Binary Log 文件，并重置 Binary Log 索引文件，生成序号为 1 的 Binary Log 文件（在 MySQL 8.0 以后，可以指定新生成的 Binary Log 序号文件），同时清空 gtid_purged、gtid_executed 和表 mysql.gtid_executed 的记录。

（3）设置 gtid_purged。

```
mysql> set global gtid_purged='eab6f8ea-6ed7-11e8-a897-0024e869b4d5:1-5';
Query OK, 0 rows affected (0.00 sec)

mysql> show global variables where variable_name in ('gtid_executed','gtid_purged');
+---------------+------------------------------------------+
| Variable_name | Value                                    |
+---------------+------------------------------------------+
| gtid_executed | eab6f8ea-6ed7-11e8-a897-0024e869b4d5:1-5 |
| gtid_purged   | eab6f8ea-6ed7-11e8-a897-0024e869b4d5:1-5 |
+---------------+------------------------------------------+
2 rows in set (0.01 sec)
```

gtid_purged 应为 gtid_executed 的子集，当 gtid_purged 被设置完成后，gtid_executed 也同样具有了相同的值，表明该 MySQL 实例已经执行过这些事务，并且这些 binary log 已经被清除。如果该实例后续还会成为一个从库，那么其不会再复制 gtid_executed 的 GTID 集合。

从以上可见，GTID 其实理解起来也不难，只是事务的一种表示形式而已，接下来实际演练一下。如果启用了 GTID，应该如何搭建复制环境。

30.3.3 复制搭建

基于 GTID 的方式搭建复制环境，较之前异步复制、多线程复制来说，并不复杂。下面分两种情况讨论：第一种情况是在创建环境最初，就直接启用了 GTID，这个步骤最简单；第二种情况是建库之初并未启用 GTID。

1. 建库之初启用 GTID

（1）根据前文描述，安装数据库软件、初始化数据库，并创建好复制用户，修改 MySQL 配置文件，增加如下内容：

```
[mysqld]
gtid_mode=on
enforce_gtid_consistency=on
```

配置启用 GTID 模式的参数 gtid_mode 和 enforce_gtid_consistency。其中参数 enforce_gtid_consistency 表示是否启用 GTID 的一致性检查，有 3 个可选值，分别为 OFF、ON 和 WARN，默认值 OFF。

- OFF: 表示不检查 GTID 一致性，即执行的 SQL 语句可以违反 GTID 约束和限制（GTID 有很多使用上的限制，例如在同一个事务中，不能包含事务引擎（类似 Innodb）的表和非事务引擎（类似 MyISAM）表的更新操作，不能使用 CTAS（create table as select）命令等，更详细的文档可以查看官方文档）。

- ON：表示开启 GTID 一致性检查，如果需要执行的 SQL 违反该原则，操作会失败。
- WARN：介于 OFF 和 ON 之间，即允许执行违反原则的 SQL 语句，但会在错误日志中打印 Warnings 信息。

参数 gtid_mode 表示 GTID 的模式，有 4 个可选值，分别为 OFF、OFF_PERMISSIVE、ON_PERMISSIVE 和 ON，默认值 OFF，每个值含义如下。

- OFF：表示不开启 GTID 模式，即执行的事务都属于 ANONYMOUS 事务（该事务的表示方式，可参考下文）。
- OFF_PERMISSIVE：表示新事务是 ANONYMOUS，但是当从库复制时，允许应用 ANONYMOUS 事务和 GTID 事务。
- ON_PERMISSIVE：表示新事务是 GTID 事务，当从库复制时，允许应用 ANONYMOUS 事务和 GTID 事务。所以参数值 OFF_PERMISSIVE 和 ON_PERMISSIVE 的区别仅仅针对新产生的事务，表示方式不同而已，也是为了将事务逐渐进行过渡，全部转向 GTID，并且不影响从库的复制。
- ON：表示新事务是 GTID 事务，从库应用时也只接受 GTID 事务。

另外，如果 gtid_mode 设置为 ON，则 enforce_gtid_consistency 必须同时设置为 ON。

以下是 ANONYMOUS 事务示例。

```
# at 24092480
#181111  4:10:40 server id 703513  end_log_pos 24092545 CRC32 0xbb5393d3    Anonymous_GTID
last_committed=61934    sequence_number=61935
SET @@SESSION.GTID_NEXT= 'ANONYMOUS'/*!*/;
# at 24092545
#181111  4:10:40 server id 703513  end_log_pos 24092620 CRC32 0x2698b539    Query    thread
_id=37    exec_time=0    error_code=0
SET TIMESTAMP=1541880640/*!*/;
BEGIN
/*!*/;
```

（2）重启主从数据库，如下所示：

```
mysqladmin -uroot -S /tmp/mysql_3307.sock shutdown
mysqld_safe --defaults-file=/home/mysql3307/mysqlhome/my.cnf --user=mysql3307 &

mysqladmin -uroot -S /tmp/mysql_3308.sock shutdown
mysqld_safe --defaults-file=/home/mysql3308/mysqlhome/my.cnf --user=mysql3308 &
```

（3）配置从库的复制进程，并启动。

```
mysql> CHANGE MASTER TO
    -> MASTER_HOST='10.120.240.251',
    -> MASTER_PORT=3307,
    -> MASTER_USER='repl',
    -> MASTER_PASSWORD='repl',
    -> MASTER_AUTO_POSITION=1;
Query OK, 0 rows affected (0.10 sec)

mysql> start slave;
Query OK, 0 rows affected (0.03 sec)
```

注意：配置 GTID 模式下的主从复制关系时，不再需要指定 MASTER_LOG_FILE 和 MASTER_LOG_POS，而是用 MASTER_AUTO_POSITION=1 代替。

当从库配置 MASTER_AUTO_POSITION=1 并启动 slave 后，从库会将当前接收到（Retrieved_Gtid_Set）的和已经提交（Executed_Gtid_Set）的 GTID 集合做一个并集发给主库，主

30.3 GTID（Global Transaction Identifier）

库收到这些信息后，会与自身的 Executed_Gtid_Set 做对比，将从库缺少的 GTID 事务发送过去。

例如：

```
从库 show slave status:
        Retrieved_Gtid_Set: eab6f8ea-6ed7-11e8-a897-0024e869b4d5:4-5
        Executed_Gtid_Set: b509f331-cd14-11e8-866a-0024e869b4d5:1-10,
eab6f8ea-6ed7-11e8-a897-0024e869b4d5:1-5

主库：
mysql> show master status\G;
*************************** 1. row ***************************
             File: mysql-bin.000010
         Position: 764
     Binlog_Do_DB:
 Binlog_Ignore_DB:
Executed_Gtid_Set: b509f331-cd14-11e8-866a-0024e869b4d5:1-10,
eab6f8ea-6ed7-11e8-a897-0024e869b4d5:1-6
1 row in set (0.00 sec)
```

当从库执行 start slave 后，此时主库需要将 eab6f8ea-6ed7-11e8-a897-0024e869b4d5:6 的事务发给从库。

show slave status 中的 Retrieved_Gtid_Set 和 Executed_Gtid_Set 解释如下。

- Retrieved_Gtid_Set：表示 slave 的 I/O Thread 已经接收到的 GTID 事务集合，这些 GTID 集合可能仍然在 Relay log 文件里，也可能已经被清除掉。另外，Retrieved_Gtid_Set 受多种因素影响，值会被清空，例如，从库执行了 reset slave 或者 change master to 等，但如果启用了 relay_log_purge，即自动清理 Relay log 功能；当 Relay log 被自动删除后，该值并不会被清空。Retrieved_Gtid_Set 也可以通过 Performance Schema 中的 replication_connection_status 查看，查询命令如下所示：

```
SELECT RECEIVED_TRANSACTION_SET FROM PERFORMANCE_SCHEMA.replication_connection_status
```

- Executed_Gtid_Set：表示 slave 的 SQL Thread 已经执行的 GTID 事务集合，该值等同于变量 gtid_executed，也与 show master status 结果中的 Executed_Gtid_Set 一致。

这样就完成了 GTID 模式的主从复制搭建。

2. 建库之初并未启用 GTID

在实际工作中，可能在建库之初并未开启 GTID，针对这种情况，可以采取如下步骤（需要主从的 MySQL 为 5.7.6 或更高的版本，本例中传统异步复制环境已搭建好，3307 为主库，3308 为从库）。

（1）如果主库可以申请短时间的停机维护时间，首先可以设置主从库为 Read Only 模式，反之，则跳过该步骤。

```
root@localhost:mysql_3307.sock  [(none)]>show variables like 'read_only';
+---------------+-------+
| Variable_name | Value |
+---------------+-------+
| read_only     | ON    |
+---------------+-------+
1 row in set (0.00 sec)


root@localhost:mysql_3308.sock  [(none)]>show variables like 'read_only';
+---------------+-------+
| Variable_name | Value |
+---------------+-------+
| read_only     | ON    |
+---------------+-------+
1 row in set (0.06 sec)
```

（2）如果主从库设置了 Read Only 模式，执行该步骤，确认从库是否已经完全追上了主库后停止复制进程。

```
root@localhost:mysql_3307.sock    [(none)]>show master status;
+------------------+----------+--------------+------------------+-------------------+
| File             | Position | Binlog_Do_DB | Binlog_Ignore_DB | Executed_Gtid_Set |
+------------------+----------+--------------+------------------+-------------------+
| mysql-bin.000047 | 3366313  |              |                  |                   |
+------------------+----------+--------------+------------------+-------------------+
1 row in set (0.03 sec)

root@localhost:mysql_3308.sock    [(none)]>show slave status\G;
*************************** 1. row ***************************
             Slave_IO_State: Waiting for master to send event
                Master_Host: 10.120.240.251
                Master_User: repl
                Master_Port: 3307
              Connect_Retry: 60
            Master_Log_File: mysql-bin.000047
        Read_Master_Log_Pos: 3366313
       Relay_Master_Log_File: mysql-bin.000047
           Slave_IO_Running: Yes
          Slave_SQL_Running: Yes
        Exec_Master_Log_Pos: 3366313

root@localhost:mysql_3308.sock    [(none)]>stop slave;
Query OK, 0 rows affected (0.16 sec)
```

以上两个步骤如果条件不允许，可以从步骤（3）开始执行，如果步骤（1）、（2）有条件执行，那么在完成以后，可以继续以下操作。

接下来在主从库上开启 GTID 模式。

（3）分别在主从库执行以下命令，并持续关注错误日志中可能出现的任何 Warnings 和 Errors 信息，如果发现异常，需要先着手解决，才能继续操作。

```
root@localhost:mysql_3307.sock    [(none)]>set global enforce_gtid_consistency=WARN;
Query OK, 0 rows affected (0.00 sec)

root@localhost:mysql_3308.sock    [(none)]>set global enforce_gtid_consistency=WARN;
Query OK, 0 rows affected (0.00 sec)
```

该步骤是非常重要的一步，一定要确保错误日志中不再出现任何 Warnings 信息。

（4）继续在主从库执行以下命令（参数解释详见前文）。

```
root@localhost:mysql_3307.sock    [(none)]>set global enforce_gtid_consistency=on;
Query OK, 0 rows affected (0.00 sec)

root@localhost:mysql_3308.sock    [(none)]>set global enforce_gtid_consistency=ON;
Query OK, 0 rows affected (0.00 sec)
```

（5）在主从库执行以下命令，调整 gtid_mode 为 OFF_PERMISSIVE。

```
root@localhost:mysql_3307.sock    [(none)]>set global gtid_mode=OFF_PERMISSIVE;
Query OK, 0 rows affected (0.20 sec)

root@localhost:mysql_3308.sock    [(none)]>set global gtid_mode=OFF_PERMISSIVE;
Query OK, 0 rows affected (0.17 sec)
```

（6）在主从库升级 gtid_mode 从 OFF_PERMISSIVE 到 ON_PERMISSIVE。

```
root@localhost:mysql_3307.sock    [(none)]>set global gtid_mode=ON_PERMISSIVE;
Query OK, 0 rows affected (0.04 sec)

root@localhost:mysql_3308.sock    [(none)]>set global gtid_mode=ON_PERMISSIVE;
Query OK, 0 rows affected (0.00 sec)
```

30.3 GTID (Global Transaction Identifier)

（7）分别在主从库上查看状态变量 ONGOING_ANONYMOUS_TRANSACTION_COUNT 的值，需要等待直到该值为 0。

```
mysql> show status like 'ONGOING_ANONYMOUS_TRANSACTION_COUNT';
+-------------------------------------+-------+
| Variable_name                       | Value |
+-------------------------------------+-------+
| Ongoing_anonymous_transaction_count | 0     |
+-------------------------------------+-------+
1 row in set (0.43 sec)
```

此处要注意查看从库的应用情况，确保从库已经接收了所有 ANONYMOUS 事务，然后才能继续操作。另外，还有一个办法可以检查从库是否已经应用完所需日志，具体步骤如下。

首先，在主库执行 show master status 命令。

```
mysql> show master status\G;
*************************** 1. row ***************************
             File: mysql-bin.000033
         Position: 5134
     Binlog_Do_DB:
 Binlog_Ignore_DB:
Executed_Gtid_Set: 6fb578bb-572a-11e8-8aa5-0024e869b4d5:1-8247296
1 row in set (0.00 sec)
```

该命令会得到当前时刻，主库执行到的 BINLOG 位置，找到 File 和 Position 信息。

然后，在从库执行 SELECT MASTER_POS_WAIT(file, position)命令，其中 file 和 position 就是刚刚在主库查询出来的信息，替换参数后，执行命令如下：

```
mysql> SELECT MASTER_POS_WAIT('mysql-bin.000033', 5134);
+-------------------------------------------+
| MASTER_POS_WAIT('mysql-bin.000033', 5134) |
+-------------------------------------------+
|                                         0 |
+-------------------------------------------+
1 row in set (0.06 sec)
```

直到等待命令结束，会返回结果值 0，表示从库已经应用完成所需要的 BINLOG 日志。如果从库延迟非常大，可以隔段时间后再执行。

以下示例是因为从库延迟很大，执行该命令后的情况。

```
mysql> SELECT MASTER_POS_WAIT('mysql-bin.000034', 5134);
等待……

show processlist 的结果：
*************************** 12. row ***************************
     Id: 42352
   User: root
   Host: 10.120.240.251:22423
     db: NULL
Command: Query
   Time: 24
  State: Waiting for the slave SQL thread to advance position
   Info: SELECT MASTER_POS_WAIT('mysql-bin.000034', 5134)
```

（8）当步骤（7）执行完成，并确认所有的从库应用无误之后，开启 GTID 模式。

```
root@localhost:mysql_3307.sock  [(none)]>set global gtid_mode=ON;
Query OK, 0 rows affected (0.03 sec)

root@localhost:mysql_3308.sock  [(none)]>set global gtid_mode=ON;
Query OK, 0 rows affected (0.00 sec)
```

成功执行上述命令后，主从库距离完成 GTID 模式的切换都还差一点点，继续如下的操作。

（9）在主从库确认此时的状态。

```
mysql> show variables like '%gtid%';
+----------------------------------+-----------+
| Variable_name                    | Value     |
+----------------------------------+-----------+
| binlog_gtid_simple_recovery      | ON        |
| enforce_gtid_consistency         | ON        |
| gtid_executed_compression_period | 1000      |
| gtid_mode                        | ON        |
| gtid_next                        | AUTOMATIC |
| gtid_owned                       |           |
| gtid_purged                      |           |
| session_track_gtids              | OFF       |
+----------------------------------+-----------+
8 rows in set (0.00 sec)
```

（10）在参数文件中持久化变更的参数。

```
[mysql3307@hz_10_120_240_251 mysqlhome]$ cat my.cnf |grep "gtid"
gtid_mode=on
enforce_gtid_consistency=on

[mysql3308@hz_10_120_240_251 mysqlhome]$ cat my.cnf |grep "gtid"
gtid_mode=on
enforce_gtid_consistency=on
```

（11）改变从库复制为 MASTER_AUTO_POSITION=1。

```
root@localhost:mysql_3308.sock  [(none)]>stop slave;
Query OK, 0 rows affected (0.03 sec)

root@localhost:mysql_3308.sock  [(none)]>change master to MASTER_AUTO_POSITION=1;
Query OK, 0 rows affected (0.04 sec)

root@localhost:mysql_3308.sock  [(none)]>start slave;
Query OK, 0 rows affected (0.03 sec)

root@localhost:mysql_3308.sock  [(none)]>show slave status\G;
*************************** 1. row ***************************
              Slave_IO_State: Waiting for master to send event
                 Master_Host: 10.120.240.251
                 Master_User: repl
                 Master_Port: 3307
               Connect_Retry: 60
             Master_Log_File: mysql-bin.000050
         Read_Master_Log_Pos: 154
              Relay_Log_File: hz_10_120_240_251-relay-bin.000002
               Relay_Log_Pos: 367
       Relay_Master_Log_File: mysql-bin.000050
            Slave_IO_Running: Yes
           Slave_SQL_Running: Yes
            Master_Server_Id: 7237
                 Master_UUID: 400f59b6-5e55-11e8-bd15-0024e869b4d5
            Master_Info_File: mysql.slave_master_info
               Auto_Position: 1
```

执行完成后，查看从库的复制状态等各项信息，是否正确。

（12）该步骤可选，如果在开始阶段，主从库打开了 Read Only 模式，此时需要关闭 Read Only 模式（建议从库打开 Read Only，避免一些误操作）。

```
root@localhost:mysql_3307.sock  [employees]>set global read_only=off;
Query OK, 0 rows affected (0.00 sec)

root@localhost:mysql_3307.sock  [employees]>show variables like 'read_only';
+---------------+-------+
| Variable_name | Value |
+---------------+-------+
| read_only     | OFF   |
+---------------+-------+
1 row in set (0.00 sec)
```

30.3 GTID (Global Transaction Identifier)

至此，完成了在线升级为 GTID 模式，接下来是一些验证工作（如果是线上环境，此时应该已经得到了检验）。

（13）在主库执行一个数据变更操作。

```
root@localhost:mysql_3307.sock  [employees]>update employees.salaries set salary=62101 where salary=100 and emp_no=10001;
Query OK, 1 row affected (0.15 sec)
Rows matched: 1  Changed: 1  Warnings: 0
```

（14）从库验证复制是否成功。

```
root@localhost:mysql_3308.sock  [employees]>select * from employees.salaries where salary=62101 and emp_no=10001;
+--------+--------+------------+------------+--------------+
| emp_no | salary | from_date  | to_date    | salary_by_1k |
+--------+--------+------------+------------+--------------+
|  10001 |  62101 | 1986-06-26 | 1987-06-26 |           62 |
+--------+--------+------------+------------+--------------+
1 row in set (0.00 sec)

root@localhost:mysql_3308.sock  [employees]>show slave status\G;
*************************** 1. row ***************************
               Slave_IO_State: Waiting for master to send event
                  Master_Host: 10.120.240.251
                  Master_User: repl
                  Master_Port: 3307
                Connect_Retry: 60
              Master_Log_File: mysql-bin.000050
          Read_Master_Log_Pos: 461
               Relay_Log_File: hz_10_120_240_251-relay-bin.000002
                Relay_Log_Pos: 674
        Relay_Master_Log_File: mysql-bin.000050
             Slave_IO_Running: Yes
            Slave_SQL_Running: Yes
             Master_Server_Id: 7237
                  Master_UUID: 400f59b6-5e55-11e8-bd15-0024e869b4d5
           Retrieved_Gtid_Set: 400f59b6-5e55-11e8-bd15-0024e869b4d5:1
            Executed_Gtid_Set: 400f59b6-5e55-11e8-bd15-0024e869b4d5:1
                Auto_Position: 1
```

可以看到数据已经在从库得到正确应用，主从复制服务验证完成。

从以上的搭建过程，可以发现在使用 GTID 方式以后，复制进程的配置不再需要指定 MASTER_LOG_FILE 和 MASTER_LOG_POS，从此省去了查找起始点的烦恼，只需指定 MASTER_AUTO_POSITION=1 即可，其余就交给系统去处理就好了。

这一点也极大地方便了日常运维，一旦主从出现切换，无须费时费力，需要从什么地点开始继续复制，系统自己会处理好，如果在一主多从的环境中，被提升为新主的节点，缺少部分数据，还可以先从具有全部数据的节点复制过来缺少的内容，然后开始对外提供服务，这在 GTID 方式出现之前，是比较麻烦的。

既然切换变得如此容易，接下来就实际演练一下。

30.3.4 主从切换

关于主从切换，通常是自动触发，可以借助 MHA（可以参考第 31 章）或者 MySQL 官方工具，这里不做过多介绍，接下来的内容是如何手动处理主从切换。

（1）首先将主从库设置为 Read Only 模式。

（2）主库执行 flush logs 并确认从库是否已经完全追上了主库后（可以参考 30.3.3 节的方

法），停止复制进程，并执行 reset slave all，即将作为新主库。

```
mysql> stop slave;
Query OK, 0 rows affected (0.12 sec)

mysql> reset slave all;
Query OK, 0 rows affected (0.00 sec)
```

reset slave all 执行的操作清除了 slave 复制位置相关信息，以及连接 master 的信息（清空 slave_master_info 表和 slave_relay_log_info 表），并且删除所有 relay log 和 index 文件，同时生成一个初始编号的的 relay log 和 index 文件。

注意：reset slave 与 reset slave all 命令稍有不同，执行 reset salve all 意味着完全清除了 slave 上的复制相关信息，该从库有计划提升为新主库的可能，而 reset slave 所做的工作虽然与 reset slave all 几乎相同，但不同的是 reset slave 命令执行后，在 MySQL 内存里依然会记录该从库连接主库的一些信息，比如 Master_Host、Master_User、Master_Port 等。如果此时从库重启实例，这些连接信息会消失（在 MySQL 5.7.24/8.0.13 之后的版本中，reset slave 命令的作用又发生了变化，即当从库执行 reset slave 时，表 mysql.slave_master_info 的记录不会被清除，从而可以快速开启复制）。

（3）在原主库（新从库）创建复制进程，并启动 slave。

```
mysql> CHANGE MASTER TO
    ->     MASTER_HOST = '10.120.240.251',
    ->     MASTER_PORT = 3308,
    ->     MASTER_USER = 'repl',
    ->     MASTER_PASSWORD = 'repl',
    ->     MASTER_AUTO_POSITION = 1;
Query OK, 0 rows affected, 2 warnings (0.08 sec)

mysql> start slave;
Query OK, 0 rows affected (0.02 sec)
```

现在主从已经完成了切换，原主库现在已经变成了从库，而原从库现在已经切换成了新主库，接下来就是主从复制验证环节。

（4）在新主库 flush logs，确认从库复制是否正常。

```
mysql> flush logs;
Query OK, 0 rows affected (0.00 sec)
```

（5）关闭新主库的 Read Only 模式。

```
mysql> set global read_only=off;
Query OK, 0 rows affected (0.00 sec)
```

以上介绍了主从切换的整个过程，并不复杂，相信大家都能轻松学会。但在从库应用 Relay Log 的过程中，难免会出现错误。接下来继续讨论 GTID 模式下的一些常见问题以及解决办法。

30.3.5 常见问题

由于 GTID 的引入，之前的方法（没有启用 GTID）有些已经不再适用，例如从库复制时使用 skip 方法跳过这些错误，也有些工具悄然发生了改变，例如 mysqldump，稍不注意，就会进入误区，接下来将分别进行介绍。

1. 复制出现错误，如何忽略该事务？

（1）在主库模拟一个导致从库应用冲突的场景。

30.3 GTID (Global Transaction Identifier)

```
mysql> create table ignore_tran(id bigint auto_increment,primary key (id));
Query OK, 0 rows affected (0.04 sec)
```

（2）从库查看复制情况。

```
mysql> select LAST_SEEN_TRANSACTION,LAST_ERROR_NUMBER,LAST_ERROR_MESSAGE from performance_
schema.replication_applier_status_by_worker where LAST_ERROR_NUMBER<>0\G;
*************************** 1. row ***************************
LAST_SEEN_TRANSACTION: eab6f8ea-6ed7-11e8-a897-0024e869b4d5:1
    LAST_ERROR_NUMBER: 1049
   LAST_ERROR_MESSAGE: Worker 8 failed executing transaction 'eab6f8ea-6ed7-11e8-a897-
0024e869b4d5:1' at master log mysql-bin.000002, end_log_pos 372; Error 'Unknown database 'db_
master2'' on query. Default database: 'db_master2'. Query: 'create table ignore_tran(id bigint
auto_increment,primary key (id))'
1 row in set (0.00 sec)
```

复制的执行情况，可以通过 performance_schema.replication_applier_status_by_worker 进行查看，该对象包括执行的线程号、服务状态、错误信息等。从以上信息，可以发现从库没有找到 db_mater2 这个 database，导致应用 Relay Log 时报错，接下来采取下面这个办法忽略这个错误。

（3）从库忽略这个事务，继续进行复制。

```
mysql> SET GTID_NEXT='eab6f8ea-6ed7-11e8-a897-0024e869b4d5:1';
Query OK, 0 rows affected (0.00 sec)

mysql> BEGIN;COMMIT;
Query OK, 0 rows affected (0.00 sec)

mysql> SET GTID_NEXT='AUTOMATIC';
Query OK, 0 rows affected (0.00 sec)

mysql> start slave;
Query OK, 0 rows affected (0.02 sec)
```

首先找到报错时候的 GTID 信息，以上为"eab6f8ea-6ed7-11e8-a897-0024e869b4d5:1"，然后执行的 BEGIN；COMMIT;表示这是一个空事务，接下来重置 GTID_NEXT 为 AUTOMATIC，完成这些之后，启动 slave。

2. 主从切换后，新从库不能获取缺少的 GTID 集合

```
Last_IO_Errno: 1236
             Last_IO_Error: Got fatal error 1236 from master when reading data from binary
log: 'The slave is connecting using CHANGE MASTER TO MASTER_AUTO_POSITION = 1, but the master has
purged binary logs containing GTIDs that the slave requires.'
```

这个问题的出现，是因为主从切换前，从库已经清除了一部分 BINLOG 数据（从库在做维护时，没有开启 read_only 或者 super_read_only，导致写入部分数据，并且对应的 binary log 文件已经被清除掉），但随后该从库成为了新的主库，而原来的主库成为了新的从库，当新从库启动复制进程，开始复制数据的时候，发现此时主库的 gtid_executed 还包含另一个 GTID 集合，即此前写入的部分数据，但 BINLOG 数据已经不在。

如果遇到以上问题，解决方法如下。

（1）查看当前主库的 GTID 信息。

```
mysql> show master status\G;
*************************** 1. row ***************************
             File: mysql-bin.000052
         Position: 234
     Binlog_Do_DB:
 Binlog_Ignore_DB:
Executed_Gtid_Set: 400f59b6-5e55-11e8-bd15-0024e869b4d5:1,
eab6f8ea-6ed7-11e8-a897-0024e869b4d5:1-4
1 row in set (0.00 sec)
```

（2）查看从库的复制情况。

```
                Last_IO_Errno: 1236
                Last_IO_Error: Got fatal error 1236 from master when reading data from binary
log: 'The slave is connecting using CHANGE MASTER TO MASTER_AUTO_POSITION = 1, but the master has
purged binary logs containing GTIDs that the slave requires.'
                  Master_UUID: 400f59b6-5e55-11e8-bd15-0024e869b4d5
           Executed_Gtid_Set: eab6f8ea-6ed7-11e8-a897-0024e869b4d5:1-4
                Auto_Position: 1
```

可以看到，从库对比主库而言，缺少 GTID 集合 "400f59b6-5e55-11e8-bd15-0024e869b4d5:1"。

（3）从库设置 gtid_purged。

```
mysql> reset master;
Query OK, 0 rows affected (0.09 sec)

mysql> set global gtid_purged='400f59b6-5e55-11e8-bd15-0024e869b4d5:1,eab6f8ea-6ed7-11e8-a897-
0024e869b4d5:1-4';
Query OK, 0 rows affected (0.00 sec)
```

设置从库的 gtid_purged，从而放弃主库清除掉的部分 BINLOG 数据，当然这也意味着主从的数据不是完全一致的（如果主库清除掉的 BINLOG 数据仅仅是用户权限等相关内容，还是比较容易补录的）。另外关于参数 gtid_purged 在前文也有过详细介绍，不再赘述。

（4）查看从库 gtid 信息，并启动复制进程。

```
mysql> select * from mysql.gtid_executed;
+--------------------------------------+----------------+--------------+
| source_uuid                          | interval_start | interval_end |
+--------------------------------------+----------------+--------------+
| 400f59b6-5e55-11e8-bd15-0024e869b4d5 |              1 |            1 |
| eab6f8ea-6ed7-11e8-a897-0024e869b4d5 |              1 |            4 |
+--------------------------------------+----------------+--------------+
2 rows in set (0.00 sec)

mysql> start slave;
Query OK, 0 rows affected (0.03 sec)

mysql> show slave status\G;
*************************** 1. row ***************************
               Slave_IO_State: Waiting for master to send event
                  Master_Host: 10.120.240.251
                  Master_User: repl
                  Master_Port: 3307
                Connect_Retry: 60
              Master_Log_File: mysql-bin.000052
          Read_Master_Log_Pos: 234
               Relay_Log_File: hz_10_120_240_251-relay-bin.000002
                Relay_Log_Pos: 367
        Relay_Master_Log_File: mysql-bin.000052
             Slave_IO_Running: Yes
            Slave_SQL_Running: Yes
           Retrieved_Gtid_Set:
            Executed_Gtid_Set: 400f59b6-5e55-11e8-bd15-0024e869b4d5:1,
eab6f8ea-6ed7-11e8-a897-0024e869b4d5:1-4
                Auto_Position: 1
```

可以看到，此时从库已经可以正常复制。

3. 主库导入 mysqldump 导出后的数据，从库并没有同步

（1）通过 mysqldump 备份数据。

```
[mysql3308@hz_10_120_240_251 ~]$ mysqldump -uroot -S /tmp/mysql_3308.sock -d employees >
employees.sql
Warning: A partial dump from a server that has GTIDs will by default include the GTIDs of
all transactions, even those that changed suppressed parts of the database. If you don't
want to restore GTIDs, pass --set-gtid-purged=OFF. To make a complete dump, pass --all-
databases --triggers --routines --events.
```

通过输出的 warning 信息，可以看到关于 gtid_purged 的提示--set-gtid-purged=OFF，这个

30.3 GTID (Global Transaction Identifier)

有什么作用呢？请继续往下阅读。

（2）查看备份数据的前一小段内容。

```
[mysql3308@hz_10_120_240_251 ~]$ head -30 employees.sql
-- MySQL dump 10.13  Distrib 5.7.22, for linux-glibc2.12 (x86_64)
--
-- Host: localhost    Database: employees
-- ------------------------------------------------------
-- Server version    5.7.22-log

/*!40101 SET @OLD_CHARACTER_SET_CLIENT=@@CHARACTER_SET_CLIENT */;
/*!40101 SET @OLD_CHARACTER_SET_RESULTS=@@CHARACTER_SET_RESULTS */;
/*!40101 SET @OLD_COLLATION_CONNECTION=@@COLLATION_CONNECTION */;
/*!40101 SET NAMES utf8 */;
/*!40103 SET @OLD_TIME_ZONE=@@TIME_ZONE */;
/*!40103 SET TIME_ZONE='+00:00' */;
/*!40014 SET @OLD_UNIQUE_CHECKS=@@UNIQUE_CHECKS, UNIQUE_CHECKS=0 */;
/*!40014 SET @OLD_FOREIGN_KEY_CHECKS=@@FOREIGN_KEY_CHECKS, FOREIGN_KEY_CHECKS=0 */;
/*!40101 SET @OLD_SQL_MODE=@@SQL_MODE, SQL_MODE='NO_AUTO_VALUE_ON_ZERO' */;
/*!40111 SET @OLD_SQL_NOTES=@@SQL_NOTES, SQL_NOTES=0 */;
SET @MYSQLDUMP_TEMP_LOG_BIN = @@SESSION.SQL_LOG_BIN;
SET @@SESSION.SQL_LOG_BIN= 0;

--
-- GTID state at the beginning of the backup
--

SET @@GLOBAL.GTID_PURGED='eab6f8ea-6ed7-11e8-a897-0024e869b4d5:1-8';
```

可以看到在备份文件的开头部分，标注了 SQL_LOG_BIN=0，即后续的命令均不写入 binary log 文件，并提示这些数据在导出时 gtid_executed 的集合为 eab6f8ea-6ed7-11e8-a897-0024e869b4d5:1-8。

（3）指定-set-gtid-purged=OFF 导出数据。

```
[mysql3308@hz_10_120_240_251 ~]$ mysqldump -uroot -S /tmp/mysql_3308.sock -d employees --set-gtid-purged=OFF > employees_gtid_off.sql

[mysql3308@hz_10_120_240_251 ~]$ head -30 employees_gtid_off.sql
-- MySQL dump 10.13  Distrib 5.7.22, for linux-glibc2.12 (x86_64)
--
-- Host: localhost    Database: employees
-- ------------------------------------------------------
-- Server version    5.7.22-log

/*!40101 SET @OLD_CHARACTER_SET_CLIENT=@@CHARACTER_SET_CLIENT */;
/*!40101 SET @OLD_CHARACTER_SET_RESULTS=@@CHARACTER_SET_RESULTS */;
/*!40101 SET @OLD_COLLATION_CONNECTION=@@COLLATION_CONNECTION */;
/*!40101 SET NAMES utf8 */;
/*!40103 SET @OLD_TIME_ZONE=@@TIME_ZONE */;
/*!40103 SET TIME_ZONE='+00:00' */;
/*!40014 SET @OLD_UNIQUE_CHECKS=@@UNIQUE_CHECKS, UNIQUE_CHECKS=0 */;
/*!40014 SET @OLD_FOREIGN_KEY_CHECKS=@@FOREIGN_KEY_CHECKS, FOREIGN_KEY_CHECKS=0 */;
/*!40101 SET @OLD_SQL_MODE=@@SQL_MODE, SQL_MODE='NO_AUTO_VALUE_ON_ZERO' */;
/*!40111 SET @OLD_SQL_NOTES=@@SQL_NOTES, SQL_NOTES=0 */;

--
-- Table structure for table `checksums`
--
```

与第一次的备份文件 employees.sql 对比，可以看到没有出现 SQL_LOG_BIN=0 的字样，也没有相关 gtid 集合信息。

如果使用第一次导出的文件进行数据恢复，从库就不可能会同步数据了。SQL_LOG_BIN=0 多用于维护操作，例如主从复制已经配置完成，开始同步数据，但发现从库并没有提前创建复制账号（防止主从故障切换，从库一旦被提升为主库，此时新主库如果没有相应的账号，新从库将无法连接到新主库，复制就会终止），此时可以首先设置 SQL_LOG_BIN=0，然后创建账号。从这可以知晓，为防止 GTID 模式开启后，维护期间产生无用的 GTID 集合，可以通过此方法解决。

当然实际工作中还可能会遇到其他问题，只要对 GTID 理解清楚了，解决起来就没什么困难了。

30.4 主要复制启动选项

下面介绍几个常用的启动选项，如 log-slave-updates、read-only、slave-skip-errors 等。

30.4.1 log-slave-updates

log-slave-updates 参数用来配置从库上的更新操作是否写二进制日志，默认是 OFF，即不写入 BINLOG（在 MySQL 8.0.3 版本以后，默认值为 ON）。如果这个从库同时也要作为其他服务器的主库，搭建一个链式的复制，或者高可用环境下（比如开启 GTID 模式下的 MHA），那么就需要打开这个选项，这样它的从库将获得它的二进制日志以进行同步操作。该参数需要和 --log-bin 结合使用，即如果没有开启 --log-bin，即便将该参数设置为 ON，也不会写二进制日志。

注意：如果是启用了 GTID 模式的三节点以上的 MHA 环境，当主库故障的情况下，候选主库节点在人为干预（指定某节点为候选 master）影响之下可能不包含最新数据，此时需要从其他从库复制 BINLOG 数据来补足缺少的数据，这个场景下 log-slave-updates 就需要开启。

30.4.2 read-only/super_read_only

read-only 参数用来设置数据库只能接受超级用户的更新操作，从而限制应用程序错误地对数据库进行更新操作。该参数默认值是 OFF，即不限制对数据库的更新操作。

super-read-only 参数用来设置数据库不接受任何用户的更新操作，包括超级用户，对 read-only 参数起到补充的作用。

关于这两个参数的关系，详见如下示例：

（1）read-only 打开，super-read-only 关闭。

```
mysql> show variables like '%read_only';
+------------------------+-------+
| Variable_name          | Value |
+------------------------+-------+
| innodb_read_only       | OFF   |
| read_only              | ON    |
| super_read_only        | OFF   |
| transaction_read_only  | OFF   |
| tx_read_only           | OFF   |
+------------------------+-------+
5 rows in set (0.00 sec)
```

（2）read-only 关闭，super-read-only 打开。

```
mysql> set global read_only=off;
Query OK, 0 rows affected (0.00 sec)

mysql> set global super_read_only=on;
Query OK, 0 rows affected (0.02 sec)


mysql> show variables like '%read_only';
```

```
+----------------------+-------+
| Variable_name        | Value |
+----------------------+-------+
| innodb_read_only     | OFF   |
| read_only            | ON    |
| super_read_only      | ON    |
| transaction_read_only| OFF   |
| tx_read_only         | OFF   |
+----------------------+-------+
5 rows in set (0.00 sec)
```

从以上可以发现，如果 super-read-only 一旦打开，read-only 就会默认打开。

（3）super-read-only 关闭，read-only 不受影响。

（4）read-only 关闭，super-read-only 一起关闭。

以上的逻辑也比较好理解，read-only 仅限制普通用户，super-read-only 限制所有用户。如果 super-read-only 打开了，那么 read-only 必然一起打开；如果 read-only 已经关闭，super-read-only 就没必要再打开了。

30.4.3 指定复制的数据库或者表

可以使用 replicate-do-db、replicate-do-table、replicate-ignore-db、replicate-ignore-table、replicate-wild-do-table 和 replicate_wild_ignore_table 来指定从主数据库复制到从数据库的数据库或者表。有时用户只需要将关键表备份到从库上，或者只需要将提供查询操作的表复制到从库上，这样就可以通过配置这几个参数来筛选进行同步的数据库和表。

下面演示的例子是设置 replicate-do-table 的情况，首先在主数据库的 employees 数据库中创建两个表 repl_filter 和 repl_normal，然后在从数据库的复制进程中，设置 replicate_ignore_table=employees.repl_filter，即忽略表 employees.repl_filter。最后在主数据库中更新两个表，检查从数据库中数据复制的情况。

（1）首先检查主从数据库上两个表的记录，都是空表。

```
mysql> select * from repl_filter;
Empty set (0.01 sec)

mysql> select * from repl_normal;
Empty set (0.00 sec)
```

（2）在从数据库，改变复制进程中的 replicate_ignore_table。

```
mysql> stop slave;
Query OK, 0 rows affected (0.00 sec)

mysql> change replication filter REPLICATE_IGNORE_TABLE=(employees.repl_filter);
Query OK, 0 rows affected (0.00 sec)

mysql> start slave;
Query OK, 0 rows affected (0.02 sec)
```

（3）下面更新主数据库上的两个表 repl_filter 和 repl_normal。

```
mysql> insert into repl_normal(id) values(1);
Query OK, 1 row affected (0.04 sec)

mysql> insert into repl_filter(id) values(1);
Query OK, 1 row affected (0.01 sec)
```

（4）再检查从数据库的复制情况：

```
mysql> select * from repl_filter;
Empty set (0.00 sec)
```

```
mysql> select * from repl_normal;
+----+
| id |
+----+
|  1 |
+----+
1 row in set (0.03 sec)
```

从测试的结果可以看到，只有 repl_normal 表的记录被复制到从库上，而 repl_filter 表的记录被过滤掉。

这里有一点需要提醒，被过滤的内容也会同样复制到从库的，只不过在应用的过程中做了过滤而已。下面就对比一下 Relay Log 和 BINLOG 的内容。

```
[mysql3308@hz_10_120_240_251 data]$ mysqlbinlog -vv hz_10_120_240_251-relay-bin.000003
……
# at 868
#180829 17:08:32 server id 7237  end_log_pos 120033 CRC32 0x5c2ecd57    Table_map: `employees
`.`repl_filter` mapped to number 404
# at 927
#180829 17:08:32 server id 7237  end_log_pos 120073 CRC32 0x7a31ae01    Write_rows: table
id 404 flags: STMT_END_F

BINLOG '
kGKGWxNFHAAAOwAAAOHUAQAAAJQBAAAAAAEACWVtcGxveWVlcwALcmVwbF9maWx0ZXIAAQMAAQMAAFfN
Llw=
kGKGWx5FHAAAKAAAAAnVAQAAAJQBAAAAAAEAgAB//4BAAAAAa4xeg==
'/*!*/;
### INSERT INTO `employees`.`repl_filter`
### SET
###   @1=1 /* INT meta=0 nullable=0 is_null=0 */
# at 967
#180829 17:08:32 server id 7237  end_log_pos 120104 CRC32 0x6747f7f7    xid = 129526
COMMIT/*!*/;
SET @@SESSION.GTID_NEXT= 'AUTOMATIC' /* added by mysqlbinlog */ /*!*/;
DELIMITER ;
# End of log file


[mysql3308@hz_10_120_240_251 data]$ mysqlbinlog -vv mysql-bin.000001
……
# at 838
#180829 17:08:32 server id 7237  end_log_pos 903 CRC32 0x49b3d57d       GTID     last_committed=3
 sequence_number=4       rbr_only=no
SET @@SESSION.GTID_NEXT= '400f59b6-5e55-11e8-bd15-0024e869b4d5:52'/*!*/;
# at 903
#180829 17:08:32 server id 7237  end_log_pos 966 CRC32 0x7c88d9f3       Query    thread_id=637
 exec_time=0    error_code=0
SET TIMESTAMP=1535533712/*!*/;
SET @@session.sql_mode=1143472162/*!*/;
BEGIN
/*!*/;
# at 966
#180829 17:08:32 server id 7237  end_log_pos 1030 CRC32 0x5bb73c67      Query    thread_id=637
 exec_time=0    error_code=0
SET TIMESTAMP=1535533712/*!*/;
COMMIT
/*!*/;
SET @@SESSION.GTID_NEXT= 'AUTOMATIC' /* added by mysqlbinlog */ /*!*/;
DELIMITER ;
# End of log file
```

从以上对比的内容，可以发现复制进程在过滤后，对该事务执行了一个空事务（因为开启了 GTID），这点正式使用了 GTID 模式下的跳过事务的方法。不过，如果启用了过滤复制功能，需要考虑该从库是否有可能会作为主库，如果切换为主库，会缺失很多数据。

30.4.4 slave-skip-errors

在复制过程中，由于各种原因，从库可能会遇到执行 BINLOG 中的 SQL 出错的情况（比如主键冲突），默认情况下，从库将会停止复制进程，不再进行同步，等待用户介入处理。这种问题如果不能及时发现，将会对应用或者备份产生影响。此参数的作用就是用来定义复制过程中从库可以自动跳过的错误号，这样当复制过程中遇到定义中的错误号时，便可以自动跳过，直接执行后面的 SQL 语句，以此来最大限度地减少人工干预。此参数可以定义多个错误号，或者通过定义成 all 跳过全部的错误，具体语法如下：

```
--slave-skip-errors=[err_code1,err_code2,... | all]
```

如果从库主要是作为主库的备份，那么就不应该使用这个启动参数，设置不当，很可能造成主从库的数据不同步。但是，如果从库仅仅是为了分担主库的查询压力，且对数据的完整性要求不是很严格，那么这个选项的确可以减轻数据库管理员维护从库的工作量。

30.5 日常管理维护

复制环境配置完成后，数据库管理员需要经常进行一些日常监控和管理维护工作，以便能及时发现复制中的一些问题，并尽快解决，以此来保持复制能够正常工作。本节将向读者介绍一些常用的监控和管理维护方法。

30.5.1 查看从库复制状态和进度

为了防止复制过程中出现故障，从而导致复制进程停止，需要经常检查从库的复制状态。一般使用 show slave status 命令来检查。

在查询结果的各种指标中，主要关心 "Slave_IO_Running" 和 "Slave_SQL_Running" 这两个进程状态是否是 "yes"，这两个进程的含义分别如下。

- Slave_IO_Running：此进程负责从库（Slave）从主库（Master）上读取 BINLOG 日志，并写入从库上的中继日志中。
- Slave_SQL_Running：此进程负责读取并且执行中继日志中的 BINLOG 日志。

只要其中有一个进程的状态是 no，则表示复制进程停止，错误原因可以从 "Last_Errno" 字段的值中看到。

除了查看上面的信息，用户还可以通过这个命令了解从库的配置情况以及当前和主库的同步情况，包括指向哪个主库、主库的端口、复制使用的用户、当前日志恢复到的位置等。这些信息都是记录在从库这一端的，主库上并没有相应的信息。

由于从库复制可能会出现延迟，影响线上业务，所以通常需要监控从库的复制进度指标。这个值可以通过 show slave status 中的 Seconds_Behind_Master 获得，单位是秒。由于该值是通过从库的 I/O 进程与 SQL 进程间的差距计算出来的，预估的结果，并不是特别准确。

可以通过如下方法获得比较准确的延迟。

（1）通过心跳策略，即主库频繁插入当前的系统时间戳，然后从库去检查该值，并与系统时间做对比，可以借助 pt-heartbeat 工具，此处不再赘述。

（2）MySQL 8.0 版本以后，在每个事务中又增加了两个属性 Original Commit Timestamp（OCT）和 Immediate Commit Timestamp（ICT）。前者表示原始 master 执行该事务时 Commit 的时间戳，ICT 表示执行该事务的节点（从节点）执行时 Commit 的时间戳，这两个的差值，即可以理解为延迟时间。

查看 BINLOG 文件，观察相关事务的延迟情况。

```
# at 124
#181113 16:09:59 server id 7238   end_log_pos 155 CRC32 0x2ac55a1b    Previous-GTIDs
# [empty]
# at 155
#181113 16:07:45 server id 7237   end_log_pos 237 CRC32 0x9c9e50de    GTID    last_committed
=0    sequence_number=1    rbr_only=yes    original_committed_timestamp=1542096465712566   immediate_commit_
timestamp=1542096964014971    transaction_length=467
/*!50718 SET TRANSACTION ISOLATION LEVEL READ COMMITTED*//*!*/;
# original_commit_timestamp=1542096465712566 (2018-11-13 16:07:45.712566 CST)
# immediate_commit_timestamp=1542096964014971 (2018-11-13 16:16:04.014971 CST)
/*!80001 SET @@session.original_commit_timestamp=1542096465712566*//*!*/;
SET @@SESSION.GTID_NEXT= '6fb578bb-572a-11e8-8aa5-0024e869b4d5:6608562'/*!*/;
```

从上面这个示例中可以知道，该事务在原主库执行的时间是 2018-11-13 16:07:45.712566，而从库执行的时间是 2018-11-13 16:16:04.014971。

另外，也可以查看性能视图 performance_schema.replication_applier_status_by_worker、performance_schema.replication_applier_status_by_coordinator 或者 performance_schema.replication_connection_status 中的相关记录，如下示例：

```
mysql> select * from performance_schema.replication_applier_status_by_worker\G;
*************************** 1. row ***************************
                                         CHANNEL_NAME : slave8
                                            WORKER_ID : 1
                                            THREAD_ID : 132
                                        SERVICE_STATE : ON
                                    LAST_ERROR_NUMBER : 0
                                   LAST_ERROR_MESSAGE :
                                 LAST_ERROR_TIMESTAMP : 0000-00-00 00:00:00.000000
                             LAST_APPLIED_TRANSACTION : 6fb578bb-572a-11e8-8aa5-0024e869b4d5:2
   LAST_APPLIED_TRANSACTION_ORIGINAL_COMMIT_TIMESTAMP : 2018-08-31 12:01:22.380463
  LAST_APPLIED_TRANSACTION_IMMEDIATE_COMMIT_TIMESTAMP : 2018-08-31 12:01:22.380463
       LAST_APPLIED_TRANSACTION_START_APPLY_TIMESTAMP : 2018-08-31 12:01:22.381192
         LAST_APPLIED_TRANSACTION_END_APPLY_TIMESTAMP : 2018-08-31 12:01:22.388115
                                 APPLYING_TRANSACTION :
       APPLYING_TRANSACTION_ORIGINAL_COMMIT_TIMESTAMP : 0000-00-00 00:00:00.000000
      APPLYING_TRANSACTION_IMMEDIATE_COMMIT_TIMESTAMP : 0000-00-00 00:00:00.000000
           APPLYING_TRANSACTION_START_APPLY_TIMESTAMP : 0000-00-00 00:00:00.000000
```

30.5.2 主从复制问题集锦

在主从复制关系中，常见的问题就是主从库的数据不一致，原因多种多样，接下来总结一些常见原因以及解决办法，便于快速定位问题，处理故障。

- 当主从切换时，有可能因为一些原因（例如原主库保持业务端的连接未断开），导致切换期间在原主库发生了数据写入，建议使用成熟的高可用方案（例如 MHA 等工具）解决主从切换问题，或者在切换期间，原主库开启 read only/super read only 模式。另一个办法就是启用主从的半同步复制（after_sync 模式），但这种方案在复制中断、网络延迟大的情况下，也存在丢失数据的风险。
- 在主从复制拓扑关系中，每个实例（主库或者从库）都需要有不一样的 server_id。如果出现相同的值，会引起复制混乱，所以要在完成复制配置前确认每个实例具有不同的值。

30.5 日常管理维护

- 主从复制中如果使用了不同的 MySQL 版本，首先要确认兼容性问题，因为复制技术演进很快，如果数据库版本相差较大，会出现这个问题。

- 在从库配置 slave 时可以指定复制哪些库表，也可以忽略一些库表，对于一些不怎么重要的数据（例如日志类数据），有可能会被过滤掉，这样做虽然能缓解从库的压力，但是如果该实例一旦被提升为主库，会缺少很多数据，所以在配置过滤规则时，需要慎重。

- 主从库配置了不同的 SQL mode（与 SQL mode 相关内容在前文有过介绍，不再赘述），导致主库写入数据后，从库得到了不同结果，建议配置主从复制环境初期，检查 SQL mode 参数，保持一致。

- BINLOG 配置了非 Row 格式，例如 Statement，当 SQL 语句中含有函数时，从库可能生成不同于主库的值，建议主从库的 BINLOG 格式统一配置为 Row 格式。

以上列举了一些常见的故障情况，那么如果发生了主从库数据不一致，如何修复呢？除了利用前文介绍过的工具 pt-table-sync 进行数据修复，还有一个办法就是忽略错误（建议操作前先确认是否可以进行此操作），尽快恢复从库。如果拖延时间太久，会造成主从复制被长时间中断，积压大量的 BINLOG 数据，面临大量丢失数据的风险。

下面介绍一下忽略错误的方法。

在从库跳过来自主库的语句的命令为 SET GLOBAL SQL_SLAVE_SKIP_COUNTER = n（如果启用了 GTID 复制，此方法不能采用，请详见 30.3.5 节），其中 n 表示 n 个事件（event），而实际上 binary log 是由事件组（event group）按顺序组织起来的，每个事件组包括一系列的事件。如何理解事件组呢？

当执行一个 insert/update/delete 时，BINLOG 中记录的是一组连续 event，如下所示：

```
前提：BINLOG 格式=Row

mysql> update employees.departments set dept_name='a' limit 1;
Query OK, 1 row affected (0.01 sec)
Rows matched: 1  Changed: 1  Warnings: 0

解析 BINLOG 数据：
# at 813
#181221 14:25:59 server id 7238  end_log_pos 878 CRC32 0x7ea2013a   Anonymous_GTID    last
_committed=2    sequence_number=3     rbr_only=yes
/*!50718 SET TRANSACTION ISOLATION LEVEL READ COMMITTED*//*!*/;
SET @@SESSION.GTID_NEXT= 'ANONYMOUS'/*!*/;
# at 878
#181221 14:25:59 server id 7238  end_log_pos 958 CRC32 0x9663bb8a   Query     thread_id=3
exec_time=0     error_code=0
SET TIMESTAMP=1545373559/*!*/;
BEGIN
/*!*/;
# at 958
#181221 14:25:59 server id 7238  end_log_pos 1022 CRC32 0x1b15c8fa    Table_map: `employees
`.`departments` mapped to number 112
# at 1022
#181221 14:25:59 server id 7238  end_log_pos 1074 CRC32 0x7bb7db4e    Update_rows: table id
112 flags: STMT_END_F

BINLOG '
d4ccXBNGHAAAQAAAAP4DAAAAAHAAAAAAAAMACWVtcGxveeWVlcwALZGVwYXJ0bWVudHMAAv4PBP4M
eAAA+sgVGw==
d4ccXB9GHAAANAAAADIEAAAAAHAAAAAAAEAAgAC///8BGQwMDEBY/wEZDAwMQFhTtu3ew==
'/*!*/;
### UPDATE `employees`.`departments`
### WHERE
###   @1='d001' /* STRING(12) meta=65036 nullable=0 is_null=0 */
###   @2='c' /* VARSTRING(120) meta=120 nullable=0 is_null=0 */
```

```
### SET
###   @1='d001' /* STRING(12) meta=65036 nullable=0 is_null=0 */
###   @2='a'    /* VARSTRING(120) meta=120 nullable=0 is_null=0 */
# at 1074
#181221 14:25:59 server id 7238  end_log_pos 1105 CRC32 0x1d4362a0    Xid = 14
COMMIT/*!*/;
```

一次 update 操作，在 BINLOG 中主要分为 begin、update、commit 这 3 个阶段，多个 event 事件，这就是一个事件组。对事务表而言（类似 Innodb 引擎的表），此事件组可以理解为一个事务；但对非事务表来说（类似 MyISAM 引擎的表），此事件组可以理解为单一 SQL 语句（因非事务表不支持事务，当执行一个数据变更操作时，不会等待 commit，直接变更数据）。如果需要跳过的事件刚好在一个事件组中间，会直接跳至该事件组的结尾，即开始执行下一个事务。

举例如下：

（1）首先，在从库端先停止复制进程，并设置 sql_slave_skip_counter=2。

```
mysql> select * from skip_test;
Empty set (0.00 sec)

mysql> stop slave;
Query OK, 0 rows affected (0.00 sec)

mysql> set global sql_slave_skip_counter=2;
Query OK, 0 rows affected (0.00 sec)
```

（2）然后在主库写入两条记录。

```
mysql> insert into skip_test values();
Query OK, 1 row affected (0.00 sec)

mysql> insert into skip_test values();
Query OK, 1 row affected (0.00 sec)

mysql> select * from skip_test;
+----+
| id |
+----+
|  1 |
|  2 |
+----+
2 rows in set (0.00 sec)
```

（3）从库开启复制进程。

```
mysql> start slave;
Query OK, 0 rows affected (0.02 sec)

mysql> select * from skip_test;
+----+
| id |
+----+
|  2 |
+----+
1 row in set (0.00 sec)
```

可以看到，此时从库只跳过了一条记录（一个事务），即第一次 insert 的数据，如果仍然是这个场景，只是将 sql_slave_skip_counter=2 改为 sql_slave_skip_counter=1 或者 3，其结果是一样的。

通过以上这些介绍，相信读者处理起主从复制故障，会变得容易得多。

30.5.3 多主复制时的自增长变量冲突问题

在大多数情况下，一般只使用单主复制（一台主库对一台或者多台从库），但是在某些情况下，可能会需要使用多主复制（多台主库对一台从库）。这时，如果主库的表采用自动增长变量，那么复制到从库的同一张表后很可能会引起主键冲突，因为系统参数 auto_increment_increment 和 auto_increment_offset 的默认值为1,这样多台主库的自增变量列迟早会发生冲突。

在单主复制时，可以采用默认设置，不会有主键冲突发生。但是使用多主复制时，就需要定制 auto_increment_increment 和 auto_increment_offset 的设置,保证多主之间复制到从数据库不会有重复冲突。比如，两个 master 的情况可以按照以下设置。

- Master1 上: auto_increment_increment = 2, auto_increment_offset = 1; (1,3,5,7...序列)。
- Master2 上: auto_increment_increment = 2, auto_increment_offset = 0; (0,2,4,6...序列)。

下面的例子在 employees 库中创建了测试表 repl_increment，只有一个自增字段 id，我们开始演示修改这两个参数的效果。首先在参数是默认值时，往表 repl_increment 中插入记录,可以看到自动增长列的值是连续的。

```
Mysql> CREATE TABLE  repl_increment (
    ->         id bigint(20) NOT NULL AUTO_INCREMENT,
    ->         PRIMARY KEY (id)
    ->        );
Query OK, 0 rows affected (0.12 sec)

mysql> show variables like '%auto_incr%';
+-------------------------+-------+
| Variable_name           | Value |
+-------------------------+-------+
| auto_increment_increment | 1     |
| auto_increment_offset    | 1     |
+-------------------------+-------+
2 rows in set (0.00 sec)

mysql> insert into repl_increment values(null),(null),(null);
Query OK, 3 rows affected (0.07 sec)
Records: 3  Duplicates: 0  Warnings: 0
```

然后把参数 auto_increment_increment 的值修改成 10，再插入记录:

```
mysql> SET @@auto_increment_increment=10;
Query OK, 0 rows affected (0.02 sec)

mysql> show variables like '%auto_incr%';
+-------------------------+-------+
| Variable_name           | Value |
+-------------------------+-------+
| auto_increment_increment | 10    |
| auto_increment_offset    | 1     |
+-------------------------+-------+
2 rows in set (0.00 sec)

mysql> insert into repl_increment values(null),(null),(null);
Query OK, 3 rows affected (0.00 sec)
Records: 3  Duplicates: 0  Warnings: 0

mysql> select * from repl_increment;
+----+
| id |
+----+
|  1 |
|  2 |
|  3 |
| 11 |
| 21 |
| 31 |
+----+
6 rows in set (0.00 sec)
```

从测试的结果上看，新插入的记录不再连续了，每次增加 10。接着再修改 auto_increment_

offset 参数，了解插入记录的效果：

```
mysql> SET @@auto_increment_offset=5;
Query OK, 0 rows affected (0.00 sec)

mysql> show variables like '%auto_incr%';
+--------------------------+-------+
| Variable_name            | Value |
+--------------------------+-------+
| auto_increment_increment | 10    |
| auto_increment_offset    | 5     |
+--------------------------+-------+
2 rows in set (0.00 sec)

mysql> insert into repl_increment values(null),(null),(null);
Query OK, 3 rows affected (0.00 sec)
Records: 3  Duplicates: 0  Warnings: 0

mysql> select * from repl_increment;
+----+
| id |
+----+
|  1 |
|  2 |
|  3 |
| 11 |
| 21 |
| 31 |
| 45 |
| 55 |
| 65 |
+----+
9 rows in set (0.00 sec)
```

从插入记录的结果上可以了解，auto_increment_offset 参数设置的是每次增加后的偏移量，也就是每次按照 10 累加之后，还需要增加 5 个偏移量。

通过这两个参数可以方便地设置不同的主库上的自动增长列的值的范围，这样在这些数据复制到从库上时可以有效地避免主键的重复。

30.5.4 如何提高复制的性能

在某些业务繁忙的系统上，由于主库更新频繁，导致从库复制延迟，当然出现延迟的情况有很多种，比如主从数据库服务器硬件差异，从库硬件资源不足以支撑业务的发展规模，出现这种情况，除了提升硬件资源、优化业务系统的设计、重点解决高频 SQL 或热点数据这些办法（关于这方面内容，请参考优化篇），还可以考虑从复制环境上下手，比如调整一些参数或者调整系统架构。

1. 几个重要参数

关于参数，重点讨论以下几个。

- sync_binlog 和 innodb_flush_log_at_trx_commit：关于这两个参数的解释在前文详细介绍过，在此不再赘述。为了加快从库的数据写入，可以在评估之后，调整该参数的值为非 1，当然这也需要考虑带来的风险，例如该从库将来还可能切换为主库。

- binlog_group_commit_sync_delay：该参数表示在 binlog 组提交下，BINLOG 数据刷到磁盘的等待时间，默认值为 0，即不等待，最大值为 1 000 000（单位为毫秒）。如果该参数的值大于 0（建议为 10 的倍数，否则容易引起 BUG），则可以增加每次组提交的事务数量，使得更多的事务在一次刷盘便能完成操作，对于 I/O 负荷较大的系统，会有一些帮助，因为适当的等待时间能缓解高 IOPS 需求。另外，也会对事务并发复制有益处，因同一单位时间内能有更多的事务被处理，有效增加主从库并发执行效率。但也有负面影响，比如响应时延增加，如下所示。

参数 binlog_group_commit_sync_delay 值为 0：

```
mysql> show variables like 'binlog_group_commit_sync_delay';
+--------------------------------+-------+
| Variable_name                  | Value |
+--------------------------------+-------+
| binlog_group_commit_sync_delay | 0     |
+--------------------------------+-------+
1 row in set (0.00 sec)

mysql> update employees.dept_emp set to_date='9999-02-02' limit 3;
Query OK, 3 rows affected (0.04 sec)
Rows matched: 3  Changed: 3  Warnings: 0
```

参数 binlog_group_commit_sync_delay 的值为 1 000 000：

```
mysql> set global binlog_group_commit_sync_delay=10000000;
Query OK, 0 rows affected, 1 warning (0.03 sec)

mysql>
mysql> show variables like 'binlog_group_commit_sync_delay';
+--------------------------------+---------+
| Variable_name                  | Value   |
+--------------------------------+---------+
| binlog_group_commit_sync_delay | 1000000 |
+--------------------------------+---------+
1 row in set (0.00 sec)

mysql> update employees.salaries set salary=200 limit 3 ;
Query OK, 3 rows affected (1.04 sec)
Rows matched: 3  Changed: 3  Warnings: 0
```

从以上示例可以看到，默认情况下，执行例子中的 SQL 需要 0.04s；增加参数的值为 1 000 000 之后，该 SQL 需要执行 1.04s。

所以在修改该参数时，需要考虑对前端应用的影响。

- binlog_group_commit_sync_no_delay_count：该参数表示 binlog 组提交时需要等待的事务数量，需要与上一个参数 binlog_group_commit_sync_delay 结合使用。如果参数 binlog_group_commit_sync_delay 的值为 0，则该参数即便设置大于 0，也不会起作用。该参数默认值为 0，最大值为 1 000 000。由于该参数的含义同上类似，所以不再赘述。

- sync_master_info：该参数表示 slave 的 I/O 线程更新 master_info_repository 的频率，默认值为 10 000，如何进行更新也与参数 master_info_repository 的值有关系。

master_info_repository=FILE，当参数 sync_master_info=n（其中 n 大于 0，例如默认值为 10 000）的情况下，slave 会在每经过 n 次事件（event）后，更新文件 master.info，该文件记录了 slave 的 I/O 线程与主库的 BINLOG 位置的映射关系；当参数 sync_master_info=0 的时候，此时更新文件 master.info 工作将由 MySQL 转交给操作系统，即操作系统去控制何时刷新文件。由于 master_info_repository 参数当配置为 FILE 选项时，会引起很多问题，例如非原子性操作、效率低下等，建议使用 TABLE 选项。

master_info_repository=TABLE，如果 sync_master_info 的值大于 0，slave 会在每经过 n 次事件（event）后（而非事务，事件、事件组、事务的区别在前文有介绍，不再赘述），更新表 mysql.slave_master_info。该表的内容如下所示：

```
mysql> select * from mysql.slave_master_info\G;
*************************** 1. row ***************************
       Number_of_lines: 25
       Master_log_name: mysql-bin.000001
        Master_log_pos: 5401
                  Host: 10.120.240.251
……
1 row in set (0.00 sec)
```

接下来简要介绍当参数 master_info_repository=TABLE 时，参数 sync_master_info 在不同值的环境下，如何影响表 mysql.slave_master_info 的更新。

（1）从库停止 slave 的 I/O 进程。

```
mysql> show variables like 'sync_master_info';
+------------------+-------+
| Variable_name    | Value |
+------------------+-------+
| sync_master_info | 10000 |
+------------------+-------+
1 row in set (0.01 sec)

mysql> stop slave io_thread;
Query OK, 0 rows affected (0.00 sec)

mysql> select * from mysql.slave_master_info\G;
*************************** 1. row ***************************
       Number_of_lines: 25
       Master_log_name: mysql-bin.000001
        Master_log_pos: 5401
                  Host: 10.120.240.251
……
```

此时 sync_master_info 为默认值 10 000，mysql.slave_master_info 对应主库的位置是文件 mysql-bin.000001，position：5401。

（2）调整从库参数 sync_master_info 从 10 000 到 10。

```
mysql> set global sync_master_info=10;
Query OK, 0 rows affected (0.00 sec)

mysql> show variables like 'sync_master_info';
+------------------+-------+
| Variable_name    | Value |
+------------------+-------+
| sync_master_info | 10    |
+------------------+-------+
1 row in set (0.00 sec)
```

（3）主库执行两次 insert 操作。

```
mysql> insert into sync_test values();
Query OK, 1 row affected (0.00 sec)

mysql> insert into sync_test values();
Query OK, 1 row affected (0.00 sec)
```

（4）从库开启 slave 的 I/O 线程。

```
mysql> start slave io_thread;
Query OK, 0 rows affected (0.00 sec)

mysql> select * from mysql.slave_master_info\G;
*************************** 1. row ***************************
       Number_of_lines: 25
       Master_log_name: mysql-bin.000001
        Master_log_pos: 5914
                  Host: 10.120.240.251
……
```

可以看到 mysql.slave_master_info 表已经更新为 mysql-bin.000001，position：5914。

如果将参数 sync_master_info 调整为 0，mysql.slave_master_info 将不再更新。所以如果从库复制性能较差，磁盘 I/O 很高，可以考虑调整该参数。

2．直接利用多从架构来解决

通过拆分减少一个从库上需要数据同步的表来解决。首先考虑配置一主多从的架构，然后在不同的从库上，通过设置不同的 replicate-do-db、replicate-do-table、replicate-ignore-db、replicate-ignore-table、replicate-wild-do-table 和 replicate_wild_ignore_table 参数，使得不同的从库复制不同的库/表，减少每个从库上需要写入的数据。

例如，假设主库为 M1，从库为 S1、S2、S3，其中设置从库 S1 仅需要复制 databaseA，而从库 S2 仅需要复制 databaseB，从库 S3 仅需要复制 databaseC，那么每个从库只需要执行自己需要复制的库/表相关的 SQL 就可以了，如图 30-10 所示。

这时，由于主库 M1 需要给 S1、S2、S3 这 3 个从库（或者更多从库）都发送完整的 BINLOG 日志，I/O 和网络压力较大，

图 30-10　MySQL 拆分 database 复制

再改进一下架构：配置 MySQL 多级主从架构减轻主库压力，如图 30-11 所示。

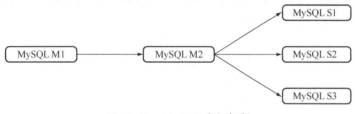

图 30-11　MySQL 多级复制

（1）主库 M1 首先给二级主库 M2 发送完整的 BINLOG。

（2）二级主库 M2 打开 log-slave-updates 配置，保证主库 M1 传送过来的 BINLOG 能够被记录在二级主库 M2 的 Relay Log 和 BINLOG 中；二级主库 M2 选择 BLACKHOLE 引擎作为表引擎，降低二级主库上 I/O 的压力。

（3）为二级主库 M2 配置 3 个从库 S1、S2、S3，这 3 个从库通过配置不同的 replicate-do-db 等参数让 S1、S2、S3 复制不同的库/表。

通过多级主从的方式提高从库的复制性能，同时尽量降低对主库的影响。

> 注意：BLACKHOLE 引擎就是一个"黑洞"引擎，在创建表的时候，选择 BLACKHOLE 引擎，那么写入表的数据不会真实地写入磁盘，仅仅记录 Binlog 日志，极大地降低了磁盘的 I/O。

该方案的优点在于能够自由拆分从库，方便地把热点数据分散开来；缺点在于维护起来不够简洁，并且由于从库 S1、S2、S3 上都没有主库完整的数据，在主库 M1 出现意外宕机的情况，应用处理较为麻烦。需要提前和应用沟通好异常的处理解决方案。

3．升级 MySQL 版本

可以利用 MySQL 5.7 的多线程复制技术，较之前的版本性能提高了很多，当然就是主库的并发度要比较高，这样才更明显。如果主库的更新操作以串行为主，那该技术带来的性能

提升就大打折扣了，甚至没有提升。不过针对这种情况，我们还可以选择直接升级至 MySQL8.0 的 GA 版本，使用复制中的 writeset 技术，这样即便主库以串行为主，从库依然可以并行执行，正好解决了 MySQL 5.7 多线程复制遗露的问题。

4．使用 MySQL 中间件

MySQL 中间件产品有很多可供选择，比如开源的或者商业化的，大多都提供丰富的垂直扩展或者水平扩展功能，也相应的提供了连接池功能，从而减少反复创建连接、销毁连接的资源开销，提高 MySQL 的性能。另外，中间件也有负载均衡、读写分离功能，可以分担线上业务的部分压力，亦可以成为高可用架构中的一员，对整体架构起着举足轻重的作用。关于该部分内容，可以参考第 32 章，此处不再详述。

30.6 小结

复制是 MySQL 数据库中经常使用的一个功能，也是 MySQL 技术发展的重要关注点，它可以有效地保证主数据库的数据安全，并减轻主数据库的备份压力，以及分担主数据库的一部分查询压力。

MySQL 复制环境的搭建非常简单，在实际使用中，建议给重要的数据库配置复制，如果没有足够的服务器可以使用，那么也可以在一个主数据库上启动另外一个 MySQL 服务，以作为另外一个 MySQL 服务的从数据库。

本章重点介绍了复制环境的搭建、GTID、日常管理和主要启动选项等，希望对读者在搭建和日常使用复制功能时有所帮助。由于 MySQL 的复制技术在不断迭代，功能越来越强大，读者也可以参考 MySQL 最新的官方文档。

第 31 章 高可用架构

对于一个企业来讲，设计一个高可用的架构非常重要，包括前端的高可用和后端数据库的高可用。企业业务上每暂停一分钟，可能会造成大量的金钱流失，因此只有在整个架构的设计上足够的高可用，才可以保证应用程序对外提供不间断的服务，进而把因软件、硬件、人为造成的故障对业务的影响降低到最小程度，把损失降低到最低。本章将重点介绍数据库高可用方面的 3 个主流架构：MHA、MGR 和 Innodb Cluster。

31.1 MHA 架构

MHA（Master High Availability）目前在 MySQL 高可用方面是一个相对成熟的解决方案，它由日本籍工程师 youshimaton 开发，是一套在 MySQL 高可用性环境下进行故障切换和主从提升的优秀软件。在 MySQL 故障切换过程中，MHA 能做到在 0～30s 之内自动完成数据库的故障切换操作，并且在故障切换过程中，最大程度地保证数据的完整性和一致性，以达到真正意义上的高可用。

MHA 由两部分组成：MHA Manager（管理节点）和 MHA Node（数据节点）。MHA Manager 建议单独部署在一台独立的机器上，可以管理多个 master-slave 集群。MHA Node 运行在每台 MySQL 服务器上，MHA Manager 会定时探测集群中的 master 节点，当 master 出现故障时，它可以自动将最新数据的 slave 提升为新的 master，然后将所有其他的 slave 重新指向新的 master。整个故障转移过程对应用程序是完全透明的。

在 MHA 自动故障切换过程中，MHA 试图从宕掉的主服务器上保存二进制日志，最大程度地保证数据不丢失，但这并不总是可行的。例如，如果主服务器硬件故障或无法通过 SSH 访问，MHA 没法保存二进制日志，只进行故障转移而丢失了最新数据，又或者 MySQL 环境启用了 GTID 复制模式，即便主服务器可以通过 SSH 访问，MHA 也不会考虑复制二进制日志应用到新主库。为了最大程度地避免主库日志的丢失，建议 MHA 与 5.7 的增强半同步复制结合起来。即使只有一个 slave 已经收到了最新的二进制日志，MHA 也可以将最新的二进制日志应用于其他所有的 slave 服务器上，来保持彼此的一致性。

目前 MHA 主要支持一主多从的架构，当然如果非要使用一主一从，技术上也是没问题的，不过为了安全考虑，一个复制集群中建议最少有 3 台数据库服务器，一主二从，即一台充当 master，一台 slave 充当切换后的备用 master，另一台 slave 充当从库，因为需要 3 台服务器。出于机器成本考虑，可以考虑选用其中一台服务器（备用 master 或者从库）作为共用备机，即可以将两套或者多套环境的备库都搭建在该机器上。

图 31-1 所示为如何通过 MHA Manager 管理多组主从复制。我们可以将 MHA 工作原理总结为以下几条。

传统复制模式（非 GTID 复制模式）如下：

（1）从宕机崩溃的 master 保存二进制日志事件（BINLOG event）；

（2）识别含有最新更新的 slave；

（3）应用差异的中继日志（Relay Log）到其他 slave；

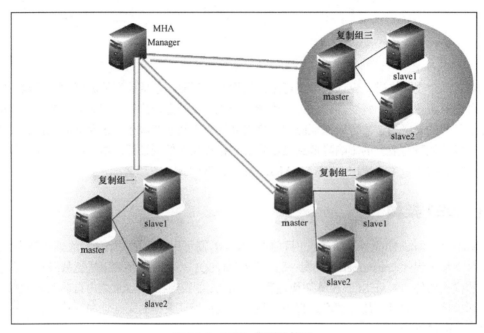

图 31-1　MHA 集群监控图

（4）应用从 master 保存的二进制日志事件（BINLOG event）；

（5）提升一个 slave 为新 master；

（6）使其他的 slave 连接新的 master 进行复制。

GTID 复制模式如下：

（1）识别含有最新更新的 slave；

（2）复制差异的 BINLOG 数据到其他 slave；

（3）提升一个 slave 为新 master；

（4）使其他的 slave 连接新的 master 进行复制。

从以上可以看到，传统复制模式和 GTID 复制模式，处理方式上有一些不同，也就是说如果启用了 GTID，需要打开从库的 log-slave-updates（该参数在第 30 章有过介绍），以减少数据丢失的风险。

MHA 软件由两部分组成，即 Manager 工具包和 Node 工具包。

Manager 工具包主要包括以下几个工具。

- masterha_check_ssh：检查 MHA 的 SSH 配置状况。
- masterha_check_repl：检查 MySQL 复制状况。

- masterha_manager：启动 MHA。
- masterha_check_status：检测当前 MHA 运行状态。
- masterha_master_monitor：监测 master 是否宕机。
- masterha_master_switch：控制故障转移（自动或手动）。
- masterha_conf_host：添加或删除配置的 server 信息。

Node 工具包（这些工具通常由 MHA Manager 的脚本触发，无须人手操作）主要包括以下几个工具。

- save_binary_logs：保存和复制 master 的二进制日志。
- apply_diff_relay_logs：识别差异的中继日志事件并将其差异的事件应用于其他 slave。
- purge_relay_logs：清除中继日志（不会阻塞 SQL 线程）。

注意：为了尽可能地减少因为主库硬件损坏宕机造成的数据丢失，因此在配置 MHA 的同时建议配置增强半同步复制。

31.1.1 安装部署 MHA

接下来开始安装部署 MHA，具体搭建环境如表 31-1 所示。

表 31-1　　　　　　　　　　搭建环境信息

角色	IP 地址	主机名	Server ID	类型
master	192.168.7.81	ip81	1	写入
candicate master	192.168.7.83	ip83	2	读
slave	192.168.7.185	ip185	3	读
monitor host	192.168.7.186	ip186	—	监控集群组

其中 master 对外提供写服务，candicate master 提供读服务，slave 也提供相关的读服务，一旦 master 宕机，将会把 candicate master 提升为新的 master，slave 指向新的 master。

1．安装 MHA Node

（1）在所有的 MySQL 服务器上安装 MHA Node 所需的 Perl 模块（DBD::mysql）。

本环境为 CentOS 部署 MHA，需要在所有 MHA Node 上安装 Perl 模块（DBD::mysql）。安装脚本（install.sh）如下：

```
install.sh
#!/bin/bash
wget http://xrl.us/cpanm --no-check-certificate
mv cpanm /usr/bin/
chmod 755 /usr/bin/cpanm
cat >/root/list<<EOF
install DBD::mysql
EOF
for package in `cat /root/list`
do
cpanm $package
done
```

（2）在所有的节点上安装 MHA Node：

二进制安装包 git 地址 https://github.com/cuichunhua/MHA.git
下载文件 mha4mysql-node-0.57.tar.gz

tar xvzf mha4mysql-node-0.57.tar.gz

```
cd mha4mysql-node-0.57
perl Makefile.PL
make
make install
```

安装后会在/usr/bin/下生成以下脚本文件：

```
/usr/bin/save_binary_logs
/usr/bin/apply_diff_relay_logs
/usr/bin/filter_mysqlbinlog
/usr/bin/purge_relay_logs
```

关于上述脚本的功能 MHA 介绍已经在上文中介绍，这里不再赘述。

2. 安装 MHA Manager

MHA Manager 中主要包括了几个管理员的命令行工具，例如 masterha_manager、masterha_master_switch 等。MHA Manager 也是依赖于一些 Perl 模块的，具体如下。

（1）安装 MHA Node 软件包。注意在 MHA Manager 的主机上也要安装 MHA Node。

```
install.sh
#!/bin/bash
wget http://xrl.us/cpanm --no-check-certificate
mv cpanm /usr/bin/
chmod 755 /usr/bin/cpanm
cat >/root/list<<EOF
install DBD::mysql
EOF
for package in `cat /root/list`
do
cpanm $package
done
```

安装 MHA Node 软件包：

```
二进制安装包 git 地址 https://github.com/cuichunhua/MHA.git
下载文件 mha4mysql-node-0.57.tar.gz

tar xvzf mha4mysql-node-0.57.tar.gz
cd mha4mysql-node-0.57
perl Makefile.PL
make
make install
```

（2）安装 MHA Manager 软件。

安装 MHA Manager 所需要的 Perl 模块：

```
#!/bin/bash
wget http://xrl.us/cpanm --no-check-certificate
mv cpanm /usr/bin/
chmod 755 /usr/bin/cpanm
cat >/root/list<<EOF
install DBD::mysql
install Config::Tiny
install Log::Dispatch
install Parallel::ForkManager
install Time::HiRes
EOF
for package in `cat /root/list`
do
cpanm $package
done
```

安装 MHA Manager 软件包：

```
二进制安装包 git 地址 https://github.com/cuichunhua/MHA.git
下载文件 mha4mysql-manager-0.57.tar.gz

tar -zxf mha4mysql-manager-0.57.tar.gz
cd mha4mysql-manager-0.57
```

```
perl Makefile.PL
make
make install
```

安装后会在/usr/bin/下生成以下脚本文件:

```
/usr/bin/masterha_check_repl
/usr/bin/masterha_check_ssh
/usr/bin/masterha_check_status
/usr/bin/masterha_conf_host
/usr/bin/masterha_manager
/usr/bin/masterha_master_monitor
/usr/bin/masterha_master_switch
/usr/bin/masterha_secondary_check
/usr/bin/masterha_stop
```

3. 配置 SSH 登录无密码验证

(1) 在 manager 上配置到所有 Node 节点的无密码验证:

```
ssh-keygen -t rsa
ssh-copy-id -i /root/.ssh/id_rsa.pub root@192.168.7.185
ssh-copy-id -i /root/.ssh/id_rsa.pub root@192.168.7.83
ssh-copy-id -i /root/.ssh/id_rsa.pub root@192.168.7.81
```

(2) 在 MHA Node ip81 上:

```
ssh-keygen -t rsa
ssh-copy-id -i /root/.ssh/id_rsa.pub root@192.168.7.83
ssh-copy-id -i /root/.ssh/id_rsa.pub root@192.168.7.185
```

(3) 在 MHA Node ip83 上:

```
ssh-keygen -t rsa
ssh-copy-id -i /root/.ssh/id_rsa.pub root@192.168.7.81
ssh-copy-id -i /root/.ssh/id_rsa.pub root@192.168.7.185
```

(4) 在 MHA Node ip185 上:

```
ssh-keygen -t rsa
ssh-copy-id -i /root/.ssh/id_rsa.pub root@192.168.7.81
ssh-copy-id -i /root/.ssh/id_rsa.pub root@192.168.7.83
```

4. 搭建主从复制环境

(1) 在 ip81 上执行备份:

```
[mysql@ip81~]$mysqldump --master-data=2 --single-transaction --default-character-set= utf8 -R --triggres -A >all.sql
```

其中, --master-data=2 代表备份时刻记录 master 的 BINLOG 位置和 Position; --single-transaction 表示获取一致性快照; -R 表示备份相关的存储过程; --triggres 表示备份触发器相关信息; -A 表示备份所有 schema。

(2) 在 ip81 上创建复制用户:

```
mysql>grant replication slave on *.* to 'repl'@'192.168.7.%' identified by '123456';
```

(3) 查看主库上备份时刻 BINLOG 的名称和位置, MASTER_LOG_FILE 和 MASTER_LOG_POS:

```
head -n 30 backup.sql |grep -i "CHANGE MASTER TO"
--CHANGE MASTER TO MASTER_LOG_FILE='mysql-bin.000043', MASTER_LOG_POS=178;
```

(4) 将备份复制到 ip83 和 ip185:

```
[mysql@ip81 ~]$scp backup.sql ip83:/home/mysql/
[mysql@ip81 ~]$scp backup.sql ip185:/home/mysql/
```

(5) 在 ip83 上搭建备库:

```
[mysql@ip83 ~]mysql -f -default-character-set=utf8 <all.sql
[mysql@ip83 ~]mysql -S /tmp/mysql_3307.sock
```

```
Mysql>change master to master_host='192.168.7.81',master_user='repl',master_password= '123456
',master_Port=3307,MASTER_LOG_FILE=' mysql-bin.000043', MASTER_LOG_POS=178;
start slave;
```

查看复制状态:

```
show slave status\G;
Master_Host: 192.168.7.81
Master_User: repl
Master_Port: 3307
Slave_IO_Running: Yes
Slave_SQL_Running: Yes
```

可以看到复制成功。

(6) 在 ip185 上搭建备库:

```
[mysql@ip185~]mysql -f -default-character-set=utf8 <all.sql
[mysql@ip185~]mysql -S /tmp/mysql_3307.sock
mysql>change master to master_host='192.168.7.81',master_user='repl',master_password= '123456',
master_Port=3307,MASTER_LOG_FILE=' mysql-bin.000043', MASTER_LOG_POS=178;
mysql>start slave;
```

查看复制状态:

```
show slave status\G;
Master_Host: 192.168.7.81
Master_User: repl
Master_Port: 3307
Slave_IO_Running: Yes
Slave_SQL_Running: Yes
```

可以看到复制成功。

(7) slave 服务器设置 read only。

将每个 slave 设置为 read only:

```
Mysql>mysql -e "set global read_only=1;"
```

从库对外提供读操作,这里将 read_only 设置为 1。

(8) 创建监控用户。

整个复制集群已经搭建完毕,这时还需要创建监控所需的用户,在 ip81 上执行:

```
mysql>grant all privileges on *.* to 'root'@'192.168.7.%' identified by '123456';
```

至此, MHA 软件已经基本安装完毕。接下来就开始配置 MHA 软件。

5. 配置 MHA

配置 MHA 的大体步骤如下。

(1) 创建 MHA 工作目录,并且创建相关配置文件:

```
mkdir -p /etc/masterha/
vi /etc/masterha/app1.cnf
[server default]
manager_log=/masterha/app1/app1.log              //设置 manager 的日志
manager_workdir=/masterha/app1                   //设置 manager 的工作日志
master_ip_failover_script=/usr/local/bin/master_ip_failover       //设置自动 failover 时候的
                                                                  //切换脚本
master_ip_online_change_script=/usr/local/bin/master_ip_online_change  //设置手动切换时候的切换脚本
user=root                                        //设置 mysql 中具有 super 权限的用户名
password=123456                                  //设置 user 对应的密码
ping_interval=1                                  //设置监控主库,发送检测命令的时间间隔,默认的是每隔 3s,尝试 3 次没有回应
                                                 //的时候进行自动 failover
ping_type=connect                                //设置检测方式,也可以选择 select、insert 方式
remote_workdir=/tmp                              //设置远端 mysql 在发生切换时保存 binlog 的具体位置
repl_password=123456                             //设置复制用户的密码
repl_user=repl                                   //设置复制环境中的复制用户名
report_script=/usr/local/bin/send_report         //设置发生切换后发送报警的脚本
```

```
secondary_check_script=/usr/bin/masterha_secondary_check -s ip83 -s ip81 --user=root --master_
host=ip81 --master_ip=192.168.7.81   --master_port=3306   // 一旦 MHA 到 ip81 的监控之间网络出现问题，
MHA Manager 将会尝试从 ip83 登录到 ip81
shutdown_script=""                    //设置故障发生后关闭故障主机脚本（该脚本主要作用是关闭主机防止发生脑裂）
ssh_user=root                         //设置 ssh 的登录用户名
ssh_port=22                           //设置 ssh 使用的端口

[server1]
hostname=192.168.7.81
master_binlog_dir=/home/binlog        //设置 MySQL 实例的 binlog 存储目录
port=3307

[server2]
hostname=192.168.7.83
master_binlog_dir=/home/binlog        //设置 MySQL 实例的 binlog 存储目录
port=3307
candidate_master=1                    //设置为候选 master,如果设置该参数后,发生主从切换后将会将此从库提升为
                                      //主,即使这个库不是集群中最新的 slave
check_repl_delay=0                    //默认情况下如果一个 slave 落后 master 100M 的 relay logs 的话,MHA
                                      //将不会选择该 slave 作为一个新的 master,因为对于这个 slave 的恢复需要
                                      //花费很长时间,通过设置 check_repl_delay=0,MHA 触发切换在选择一个
                                      //新的 master 的时候将会忽略复制延时,这个参数对于设置 candidate_master=1
                                      //的主机非常有用,因为它保证了这个候选主在切换过程中一定是新的 master

[server3]
hostname=192.168.7.185
master_binlog_dir=/home/binlog        //设置 MySQL 实例的 binlog 存储目录
port=3307
```

（2）设置 Relay Log 清除方式（在每个 slave 上）：

```
Mysql>mysql -e  "set global relay_log_purge=0; "
```

MHA 在发生切换过程中，从库的恢复过程中可能会需要 Relay Log 的相关信息，所以这里我们要将 Relay Log 的自动清除设置为 OFF，采用手动清除 Relay Log 的方式。

> 注意：如果 MySQL 主从复制环境使用了 GTID 模式，可以将参数 relay_log_purge 设置为 1，即采用自动清除 Relay Log 的方式。

MHA 提供了清除 Relay Log 的工具 pure_relay_logs，该工具参数如下所示：

```
--user mysql 用户名
--password mysql 密码
--host mysql 服务器地址
--port 端口号
--workdir 指定创建 relay log 的硬链接的位置，默认的是/var/tmp,由于系统不通分区创建硬链接文件会失败,故需要
          执行硬链接具体位置,成功执行脚本后,硬链接的中继日志文件将被删除
--disable_relay_log_purge 默认情况下,如果 relay_log_purge=1 的情况下,脚本会什么都不处理,自动退出。通过设定这
个参数,当 relay_log_purge=1 的情况下将会将 relay_log_purge 设置为 0.清理 relay log,清理之后,最后将参数设置为 OFF
```

（3）设置定期清理 relay 脚本。

使用如下命令设置 crontab 来定期清理 Relay Log：

```
vi /etc/cron.d/purge_relay_logs

0 4 * * * /usr/bin/purge_relay_logs --user=root --password=123456 -disable_relay_log_purge
--port=3307 --workdir=/home/mysql_3307/mysqlhome/ -disable_relay_log_purge>>/usr/local/
masterha/log/purge_relay_logs.log 2>&1
```

purge_relay_logs 脚本删除中继日志不会阻塞 SQL 线程。因此在每台从服务器上设置计划任务定期清除中继日志。最好在每台从服务器上的不同时间点执行计划任务。

下面列出了脚本清理过程：

```
[root@ip83 ~]#/usr/bin/purge_relay_logs --user=root  -disable_relay_log_purge --port=3307 --
workdir=/home/mysql_3307/mysqlhome/
2018-07-23 19:42:42: purge_relay_logs script started.
Found relay_log.info: /home/mysql_3307/mysqlhome/data/relay-log.info
Removing hard linked relay log files ip185-relay-bin* under /home/mysql_3307/mysqlhome.. done.
```

```
Current relay log file: /home/mysql_3307/mysqlhome/data/ip185-relay-bin.000005
Archiving unused relay log files (up to
/home/mysql_3307/mysqlhome/data/ip185-relay-bin.000004) ...
Creating hard link for /home/mysql_3307/mysqlhome/data/ip185-relay-bin.000004 under /home/
mysql_3307/mysqlhome/ip185-relay-bin.000004 .. ok.
Creating hard links for unused relay log files completed.
Executing SET GLOBAL relay_log_purge=1; FLUSH LOGS; sleeping a few seconds so that SQL thread
can delete older relay log files (if it keeps up); SET GLOBAL relay_log_purge=0; .. ok.
Removing hard linked relay log files ip185-relay-bin* under /home/mysql_3307/mysqlhome.. done.
2018-07-23 19:42:45: All relay log purging operations succeeded.
```

（4）设置 mysqlbinlog（在每个 slave 上），编辑~/.bashr 或者/etc/bashrc 文件，在文件的末尾处添加以下内容：

```
PATH="$PATH:/home/mysql/mysqlhome/bin"
export PATH
```

MHA 在切换过程中会直接调用 mysqlbinlog 命令，故需要在环境变量中指定 mysqlbinlog 的具体路径。

6. 检查 SSH 的配置

检查 MHA Manager 到所有 MHA Node 的 SSH 连接状态：

```
[root@ip186 home]# masterha_check_ssh --conf=/etc/masterha/app1.cnf
  Fri Jul 19 18:21:09 2018 - [warning] Global configuration file /etc/masterha_default.cnf not
found. Skipping.
  Fri Jul 19 18:21:09 2018 - [info] Reading application default configurations from /etc/masterha/
app1.cnf..
  Fri Jul 19 18:21:09 2018 - [info] Reading server configurations from /etc/masterha/app1.cnf..
  Fri Jul 19 18:21:09 2018 - [info] Starting SSH connection tests..
  Fri Jul 19 18:21:12 2018 - [debug]
  Fri Jul 19 18:21:09 2018 - [debug]  Connecting via SSH from root@192.168.7.83(192.168.7.83:
22) to root@192.168.7.81(192.168.7.81:22)..
  Fri Jul 19 18:21:12 2018 - [debug]   ok.
  Fri Jul 19 18:21:12 2018 - [debug]  Connecting via SSH from root@192.168.7.83(192.168.7.83:
22) to root@192.168.7.185(192.168.7.185:22)..
  Fri Jul 19 18:21:12 2018 - [debug]   ok.
  Fri Jul 19 18:21:12 2018 - [debug]
  Fri Jul 19 18:21:10 2018 - [debug]  Connecting via SSH from root@192.168.7.185(192.168.7.
185:22) to root@192.168.7.81(192.168.7.81:22)..
  Fri Jul 19 18:21:12 2018 - [debug]   ok.
  Fri Jul 19 18:21:12 2018 - [debug]  Connecting via SSH from root@192.168.7.185(192.168.7.
185:22) to root@192.168.7.83(192.168.7.83:22)..
  Fri Jul 19 18:21:12 2018 - [debug]   ok.
  Fri Jul 19 18:21:13 2018 - [debug]
  Fri Jul 19 18:21:09 2018 - [debug]  Connecting via SSH from root@192.168.7.81(192.168.7.
81:22) to root@192.168.7.83(192.168.7.83:22)..
  Fri Jul 19 18:21:11 2018 - [debug]   ok.
  Fri Jul 19 18:21:11 2018 - [debug]  Connecting via SSH from root@192.168.7.81(192.168.7.
81:22) to root@192.168.7.185(192.168.7.185:22)..
  Fri Jul 19 18:21:13 2018 - [debug]   ok.
  Fri Jul 19 18:21:13 2018 - [info] All SSH connection tests passed successfully.
```

从输出可以看出，ip83 到 ip81 和 ip85 SSH ok，ip85 到 ip81 和 ip83 SSH ok，ip81 到 ip83 和 ip185 SSH ok。

7. 检查整个复制环境状况

通过 masterha_check_repl 脚本查看整个集群的状态：

```
masterha_check_repl --conf=/etc/masterha/app1.cnf
192.168.7.81 (current master)
 +--192.168.7.83
 +--192.168.7.185
Fri Nov 23 09:45:03 2018 - [info] Checking replication health on 192.168.7.83..
Fri Nov 23 09:45:03 2018 - [info]  ok.
Fri Nov 23 09:45:03 2018 - [info] Checking replication health on 192.168.7.185..
```

```
Fri Nov 23 09:45:03 2018 - [info]  ok.
……
MySQL Replication Health is OK.
```

8. 检查 MHA Manager 的状态

通过 masterha_check_status 脚本查看 Manager 的状态：

```
[root@ip186 home]# masterha_check_status --conf=/etc/masterha/app1.cnf
app1 is stopped(2:NOT_RUNNING).
```

注意：如果正常，会显示"PING_OK"，否则会显示"NOT_RUNNING"，这代表 MHA 监控没有开启。

9. 开启 MHA Manager 监控

```
[root@ip186 home]# nohup masterha_manager --conf=/etc/masterha/app1.cnf -- remove_dead_master_
conf --ignore_last_failover< /dev/null >/masterha/app1/manager.log 2>&1 &
```

对启动中的参数说明如下。

● --remove_dead_master_conf：该参数代表当发生主从切换后，老的主库的 IP 将会从配置文件中删除。

● --ignore_last_failover：在默认情况下，如果 MHA 检测到连续发生宕机，且两次宕机时间间隔不足 8h 的话，则不会进行 Failover，之所以这样限制是为了避免 ping-pong 效应。该参数代表忽略上次 MHA 触发切换产生的文件，默认情况下，MHA 发生切换后将会在 /masterha/app1 下产生 app1.failover.complete 文件，下次再次切换的时候如果发现目录下存在该文件将不允许触发切换，除非在第一次切换后手动 rm -f /masterha/app1/app1.failover.complete，出于方便考虑，我们每次在启动 MHA 时会添加--ignore_last_failover 参数。

查看 MHA Manager 监控是否正常：

```
[root@ip186 home]# masterha_check_status --conf=/etc/masterha/app1.cnf
app1 (pid:2960) is running(0:PING_OK), master:192.168.7.81
```

在默认情况下，10s 内状态会为 10:INITIALIZING_MONITOR，当状态转变为 0:PING_OK 后表明已经开启了到 master 端的监控，master 主机为 192.168.7.81。

10. 查看启动日志

通过 tail 命令查看启动过程中的日志输出信息：

```
[root@ip186 home]# tail -f  /masterha/app1/app1.log
……
192.168.7.81(192.168.7.81:3307) (current master)
 +--192.168.7.83(192.168.7.83:3307)
 +--192.168.7.185(192.168.7.185:3307)

Wed Aug 29 11:59:33 2018 - [warning] master_ip_failover_script is not defined.
Wed Aug 29 11:59:33 2018 - [warning] shutdown_script is not defined.
Wed Aug 29 11:59:33 2018 - [info] Set master ping interval 3 seconds.
Wed Aug 29 11:59:33 2018 - [info] Set secondary check script: /usr/bin/masterha_secondary_
check --user=root --master_ip=192.168.7.81 --master_port=3307 -s 192.168.7.185 -s 192.168.7.83
Wed Aug 29 11:59:33 2018 - [info] Starting ping health check on 192.168.7.81 (192.168.7.81:3307)..
Wed Aug 29 11:59:33 2018 - [info] Ping(SELECT) succeeded, waiting until MySQL doesn't respond..
```

其中"Ping(SELECT) succeeded"输出说明整个系统监控已经开始了。

11. 关闭 MHA Manager 监控

关闭 MHA 命令如下：

```
masterha_stop --conf=/etc/masterha/app1.cnf
Stopped app1 successfully.
```

31.1.2 应用连接配置

实际应用中，如果 MHA 成功完成了 MySQL 主从的切换，但由于切换前后主从 IP 发生了变更，需要修改连接信息来适配新环境，这个过程在线业务会受到影响。有 3 种办法来解决这个问题，第一种方式是通过 keepalived 来管理 VIP，即通过对浮动 IP 的管理来解决 IP 的改变；第二种方式是通过自定义脚本方式，自动迁移 VIP，原理与第一个办法类似；第三种方式是采用 MySQL 中间件，即应用与后端 MySQL 环境之间，增加中间件，从而通过中间件来"感知"后端环境的变化。

接下来对这 3 种方式进行详细介绍。

1．keepalived 方式

keepalived 配置步骤如下。

（1）下载集群心跳软件 keepalived，并进行安装：

```
wget www.keepalived.org/software/keepalived-1.4.2.tar.gz
tar -xvzf keepalived-1.4.2.tar.gz
cd keepalived-1.4.2.tar.gz
./configure --prefix=/usr/local/keepalived
make && make install
cp /usr/local/keepalived/etc/sysconfig/keepalived   /etc/sysconfig/
mkdir /etc/keepalived
cp /usr/local/keepalived/etc/keepalived/keepalived.conf  /etc/keepalived/
cp /usr/local/keepalived/sbin/keepalived /usr/sbin/
```

（2）配置 keepalived。

在 ip81 上设置：

```
[root@ip81 tools]# cat /etc/keepalived/keepalived.conf
! Configuration File for keepalived
        global_defs {
            notification_email {
              onduty@corp.netease.com
            }
            notification_email_from dba@corp.netease.com
            smtp_server 127.0.0.1
            smtp_connect_timeout 30
            router_id MySQL-ha
            }

        vrrp_instance VI_1 {
            state BACKUP
            interface eth0
            virtual_router_id 51
            priority 150
            advert_int 1
            nopreempt

            authentication {
            auth_type PASS
            auth_pass 1111
            }
            virtual_ipaddress {
            192.168.7.201/23
            }
            }
```

注意黑体部分的代码，其中 router_id MySQL-ha 表示设定 keepalived 组的名称，将 192.168.7.201/23 这个 IP 绑定到 ip81 主机 interface eth0 上，并且设置了状态为 backup 模式，将 keepalived 的模式设置为非抢占模式（nopreempt），priority 150 优先级别设置为 150。

31.1 MHA 架构

在候选主 ip83 上设置：

```
[root@ip83 ~]# cat /etc/keepalived/keepalived.conf
onfiguration File for keepalived
      global_defs {
           notification_email {
            onduty@corp.netease.com
           }
           notification_email_from dba@corp.neteash.com
           smtp_server 127.0.0.1
           smtp_connect_timeout 30
           router_id MySQL-ha
           }
      vrrp_instance VI_1 {
           state BACKUP
           interface eth0
           virtual_router_id 51
           priority 120
            advert_int 1
            nopreempt
           authentication {
           auth_type PASS
           auth_pass 1111
           }
           virtual_ipaddress {
           192.168.7.201/23
           }
           }
```

注意黑体部分的代码，其中 router_id MySQL-ha 表示设定 keepalived 组的名称，并且设置了状态为 backup 模式，将 keepalived 的模式设置为非抢占模式（nopreempt），priority 120 优先级别设置为 120。

这里 ip81 和 ip83 都要设置为 BACKUP 模式。在 keepalived 中有两种模式，分别是 master→backup 模式和 backup→backup 模式，这两种模式分别有什么区别呢？

在 master→backup 模式下，一旦主库宕掉，虚拟 IP 会自动漂移到从库，当主库修复好，keepalived 启动后，还会把虚拟 IP 抢过来，即使设置了 nopreempt（不抢占）的方式抢占 IP 的动作也会发生。在 backup→backup 模式下，当主库宕掉后虚拟 IP 会自动漂移到从库上，当原主恢复之后重启 keepalived 服务，并不会抢占新主的虚拟 IP，即使是优先级高于从库的优先级别，也不会抢占 IP。为了减少 IP 的漂移次数，生产中通常是把修复好的主库当作新主库的备库。

（3）启动 keepalived 服务。

在 ip81 上启动 keepalived 服务：

```
[root@ip81 tools]# /usr/sbin/keepalived -D
[root@ip81 tools]# tail -f /var/log/messages
Nov 23 16:07:15 ip81 Keepalived[25214]: Starting Keepalived v1.4.2 (02/24,2018), git commit v1.4.1-41-g6a2987e+
Nov 23 16:07:15 ip81 Keepalived[25214]: Running on Linux 2.6.32-642.1.1.el6.x86_64 #1 SMP Fri May 6 14:54:05 EDT 2016 (built for Linux 2.6.32)
Nov 23 16:07:15 ip81 Keepalived[25214]: Opening file '/etc/keepalived/keepalived.conf'.
Nov 23 16:07:15 ip81 Keepalived_healthcheckers[25216]: Opening file '/etc/keepalived/keepalived.conf'.
Nov 23 16:07:15 ip81 Keepalived[25215]: Starting Healthcheck child process, pid=25216
Nov 23 16:07:15 ip81 Keepalived[25215]: Starting VRRP child process, pid=25217
Nov 23 16:07:15 ip81 Keepalived_vrrp[25217]: Registering Kernel netlink reflector
Nov 23 16:07:15 ip81 Keepalived_vrrp[25217]: Registering Kernel netlink command channel
Nov 23 16:07:15 ip81 Keepalived_vrrp[25217]: Registering gratuitous ARP shared channel
Nov 23 16:07:15 ip81 Keepalived_vrrp[25217]: Opening file '/etc/keepalived/keepalived.conf'.
Nov 23 16:07:15 ip81 Keepalived_vrrp[25217]: VRRP_Instance(VI_1) removing protocol VIPs.
Nov 23 16:07:15 ip81 Keepalived_vrrp[25217]: Using LinkWatch kernel netlink reflector...
Nov 23 16:07:15 ip81 Keepalived_vrrp[25217]: VRRP_Instance(VI_1) Entering BACKUP STATE
Nov 23 16:07:15 ip81 Keepalived_vrrp[25217]: VRRP sockpool: [ifindex(2), proto(112), unicast
```

```
(0), fd(10,11)]
    Nov 23 16:07:18 ip81 Keepalived_vrrp[25217]: VRRP_Instance(VI_1) Transition to MASTER STATE
    Nov 23 16:07:19 ip81 Keepalived_vrrp[25217]: VRRP_Instance(VI_1) Entering MASTER STATE
    Nov 23 16:07:19 ip81 Keepalived_vrrp[25217]: VRRP_Instance(VI_1) setting protocol VIPs.
    Nov 23 16:07:19 ip81 Keepalived_vrrp[25217]: Sending gratuitous ARP on eth0 for 192.168.7.201
    ……
    Nov 23 16:07:24 ip81 Keepalived_vrrp[25217]: VRRP_Instance(VI_1) Sending/queueing gratuitous
ARPs on eth0 for 192.168.7.201
    Nov 23 16:07:24 ip81 Keepalived_vrrp[25217]: Sending gratuitous ARP on eth0 for 192.168.7.201
    Nov 23 16:07:24 ip81 Keepalived_vrrp[25217]: Sending gratuitous ARP on eth0 for 192.168.7.201
    Nov 23 16:07:24 ip81 Keepalived_vrrp[25217]: Sending gratuitous ARP on eth0 for 192.168.7.201
    Nov 23 16:07:24 ip81 Keepalived_vrrp[25217]: Sending gratuitous ARP on eth0 for 192.168.7.201
```

通过输出可以看到虚拟 IP（192.168.7.201）已经绑定到 ip81 的 eth0 网卡上了。

（4）查看绑定情况。

在 ip81 上查看 IP 地址绑定情况：

```
[root@ip81 mysqlhome]# ip addr
2: eth0: <BROADCAST,MULTICAST,UP,LOWER_UP> mtu 1500 qdisc pfifo_fast qlen 1000
    link/ether 00:11:43:de:78:78 brd ff:ff:ff:ff:ff:ff
    inet 192.168.7.81/23 brd 192.168.7.255 scope global eth0
    inet 192.168.7.201/23 scope global secondary eth0
    inet6 fe80::211:43ff:fede:7878/64 scope link
       valid_lft forever preferred_lft forever
```

从 keepalived 的输出信息和系统的网卡信息可以看到，虚拟 IP（192.168.7.201）已经添加成功。在 ip83 上：

```
[root@ip83 ~]# /usr/sbin/keepalived -D
[root@ip83 ~]# tail -f /var/log/messages
    Nov 23 03:28:08 ip83 Keepalived[26040]: Starting Keepalived v1.4.2 (02/24,2018), git commit
v1.4.1-41-g6a2987e+
    Nov 23 03:28:08 ip83 Keepalived[26040]: Running on Linux 3.10.0-327.el7.x86_64 #1 SMP Thu
Oct 29 17:29:29 EDT 2015 (built for Linux 3.10.0)
    Nov 23 03:28:08 ip83 Keepalived[26040]: Opening file '/etc/keepalived/keepalived.conf'.
    Nov 23 03:28:08 ip83 Keepalived[26041]: Starting Healthcheck child process, pid=26042
    Nov 23 03:28:08 ip83 Keepalived[26041]: Starting VRRP child process, pid=26043
    Nov 23 03:28:08 ip83 Keepalived_healthcheckers[26042]: Opening file '/etc/keepalived/
keepalived.conf'.
    Nov 23 03:28:08 ip83 Keepalived_vrrp[26043]: Registering Kernel netlink reflector
    Nov 23 03:28:08 ip83 Keepalived_vrrp[26043]: Registering Kernel netlink command channel
    Nov 23 03:28:08 ip83 Keepalived_vrrp[26043]: Registering gratuitous ARP shared channel
    Nov 23 03:28:08 ip83 Keepalived_vrrp[26043]: Opening file '/etc/keepalived/keepalived.conf'.
    Nov 23 03:28:08 ip83 Keepalived_vrrp[26043]: VRRP_Instance(VI_1) removing protocol VIPs.
    Nov 23 03:28:08 ip83 Keepalived_vrrp[26043]: Using LinkWatch kernel netlink reflector...
    Nov 23 03:28:08 ip83 Keepalived_vrrp[26043]: VRRP_Instance(VI_1) Entering BACKUP STATE
    Nov 23 03:28:08 ip83 Keepalived_vrrp[26043]: VRRP sockpool: [ifindex(2), proto(112), unicast
(0), fd(10,11)]
```

从上面的输出可以看到 keepalived 已经配置成功。

需要注意的是，keepalived 存在一种脑裂状况，当主从间网络出现问题，这时主库会持有虚拟 IP 不变，从库失去和主库的联系后，从库也会抢夺 IP（即便采用 backup→backup 非抢占模式），这样造成的后果是主从数据库都持有虚拟 IP。于是造成 IP 冲突，业务也会受到影响，因此在网络不是很好的状况下，不建议采用 keepavlived 服务。

（5）MHA 引入 keepalived。

那么如何才能把 keepalived 服务引入 MHA 呢？很简单，只需修改切换时触发的脚本文件 master_ip_failover 即可，在该脚本中添加在 master 发生宕机时对 keepalived 的处理。

例如，编辑 MHA 切换时调用的配置文件 master_ip_failover，如下所示：

```
vi /usr/local/bin/master_ip_failover

#!/usr/bin/env perl
```

```perl
#  Copyright (C) 2011 DeNA Co.,Ltd.
#
#  This program is free software; you can redistribute it and/or modify
#  it under the terms of the GNU General Public License as published by
#  the Free Software Foundation; either version 2 of the License, or
#  (at your option) any later version.
#
#  This program is distributed in the hope that it will be useful,
#  but WITHOUT ANY WARRANTY; without even the implied warranty of
#  MERCHANTABILITY or FITNESS FOR A PARTICULAR PURPOSE.  See the
#  GNU General Public License for more details.
#
#  You should have received a copy of the GNU General Public License
#   along with this program; if not, write to the Free Software
#  Foundation, Inc.,
#  51 Franklin Street, Fifth Floor, Boston, MA  02110-1301  USA

## Note: This is a sample script and is not complete. Modify the script based on your environment.
use strict;
use warnings FATAL => 'all';

use Getopt::Long;
use MHA::DBHelper;

my (
  $command,          $ssh_user,         $orig_master_host, $orig_master_ip,
  $orig_master_port, $new_master_host,  $new_master_ip,    $new_master_port,
  $new_master_user,            $new_master_password
);
GetOptions(
  'command=s'              => \$command,
  'ssh_user=s'             => \$ssh_user,
  'orig_master_host=s'     => \$orig_master_host,
  'orig_master_ip=s'       => \$orig_master_ip,
  'orig_master_port=i'     => \$orig_master_port,
  'new_master_host=s'      => \$new_master_host,
  'new_master_ip=s'        => \$new_master_ip,
  'new_master_port=i'      => \$new_master_port,
  ' new_master_user =s'                 => \$ new_master_user,
  ' new_master_password =s'             => \$ new_master_password,
);

exit &main();

sub main {
  if ( $command eq "stop" || $command eq "stopssh" ) {

    # $orig_master_host, $orig_master_ip, $orig_master_port are passed.
    # If you manage master ip address at global catalog database,
    # invalidate orig_master_ip here.
    my $exit_code = 1;
    eval {
      # updating global catalog, etc
      $exit_code = 0;
    };
    if ($@) {
      warn "Got Error: $@\n";
      exit $exit_code;
    }
    exit $exit_code;
  }
  elsif ( $command eq "start" ) {

    # all arguments are passed.
    # If you manage master ip address at global catalog database,
    # activate new_master_ip here.
    # You can also grant write access (create user, set read_only=0, etc) here.
    my $exit_code = 10;
    eval {
      my $new_master_handler = new MHA::DBHelper();
```

```perl
        # args: hostname, port, user, password, raise_error_or_not
        $new_master_handler->connect( $new_master_ip, $new_master_port,
          $new_master_user, $new_master_password, 1 );

        ## Set read_only=0 on the new master
        #$new_master_handler->disable_log_bin_local();
        print "Set read_only=0 on the new master.\n";
        $new_master_handler->disable_read_only();

        ## Creating an app user on the new master
        #print "Creating app user on the new master..\n";
        #FIXME_xxx_create_user( $new_master_handler->{dbh} );
        #$new_master_handler->enable_log_bin_local();
        $new_master_handler->disconnect();

        ## Update master ip on the catalog database, etc
        #FIXME_xxx;
        $cmd = 'ssh '.$ssh_user.'@'.$orig_master_ip.' \'./lvs-admin stop\'';
        system($cmd);

        $exit_code = 0;
    };
    if ($@) {
      warn $@;

      # If you want to continue failover, exit 10.
      exit $exit_code;
    }
    exit $exit_code;
  }
  elsif ( $command eq "status" ) {

    # do nothing
    exit 0;
  }
  else {
    &usage();
    exit 1;
  }
}


sub usage {
  print
"Usage: master_ip_failover --command=start|stop|stopssh|status --orig_master_host=host --orig_master_ip=ip --orig_master_port=port --new_master_host=host --new_master_ip=ip --new_master_port=port\n";
}
```

上述黑体加粗部分即为添加的代码。在主库 ip81 上的根目录下编辑 lvs-admin 脚本：

```bash
case "$1" in
     "stop")
          echo;
          echo "stop keepalived.....";
          service keepalived stop
          ;;
     "start")
               echo
          echo "start keepalive....."
          service keepalived start
          ;;
     *)
          echo "please input you select!"
          ;;
esac
```

/usr/local/bin/master_ip_failover 添加内容的大体意思是，当主库数据库发生故障时刻，会触发 MHA 切换，MHA Manager 会执行停掉主库上 keepalived 服务，触发虚拟 IP 漂移到备选从库，从而完成切换。

2. 通过脚本的方式

修改触发故障切换脚本/usr/local/bin/master_ip_failover，修改内容如下：

```perl
#!/usr/bin/env perl
use strict;
use warnings FATAL => 'all';

use Getopt::Long;

my (
    $command,          $ssh_user,         $orig_master_host, $orig_master_ip,
    $orig_master_port, $new_master_host,  $new_master_ip,    $new_master_port
);

my $vip = '192.168.7.201/23';  # Virtual IP
my $key = "2";
my $ssh_start_vip = "/sbin/ifconfig eth0:$key $vip";
my $ssh_stop_vip = "/sbin/ifconfig eth0:$key down";

GetOptions(
    'command=s'          => \$command,
    'ssh_user=s'         => \$ssh_user,
    'orig_master_host=s' => \$orig_master_host,
    'orig_master_ip=s'   => \$orig_master_ip,
    'orig_master_port=i' => \$orig_master_port,
    'new_master_host=s'  => \$new_master_host,
    'new_master_ip=s'    => \$new_master_ip,
    'new_master_port=i'  => \$new_master_port,
);

exit &main();

sub main {

    if ( $command eq "stop" || $command eq "stopssh" ) {

        # $orig_master_host, $orig_master_ip, $orig_master_port are passed.
        # If you manage master ip address at global catalog database,
        # invalidate orig_master_ip here.
        my $exit_code = 1;
        eval {
            print "Disabling the VIP on old master: $orig_master_host \n";
            &stop_vip();
            $exit_code = 0;
        };
        if ($@) {
            warn "Got Error: $@\n";
            exit $exit_code;
        }
        exit $exit_code;
    }
    elsif ( $command eq "start" ) {

        # all arguments are passed.
        # If you manage master ip address at global catalog database,
        # activate new_master_ip here.
        # You can also grant write access (create user, set read_only=0, etc) here.
        my $exit_code = 10;
        eval {
            print "Enabling the VIP - $vip on the new master - $new_master_host \n";
            &start_vip();
            $exit_code = 0;
        };
        if ($@) {
            warn $@;
            exit $exit_code;
        }
        exit $exit_code;
    }
```

```perl
    elsif ( $command eq "status" ) {
        print "Checking the Status of the script.. OK \n";
            # do nothing
            exit 0;
    }
    else {
        &usage();
        exit 1;
    }
}

# A simple system call that enable the VIP on the new master
sub start_vip() {
    `ssh $ssh_user\@$new_master_host \" $ssh_start_vip \"`;
}
# A simple system call that disable the VIP on the old master
sub stop_vip() {
    `ssh $ssh_user\@$orig_master_host \" $ssh_stop_vip \"`;
}

sub usage {
    print
    "Usage: master_ip_failover --command=start|stop|stopssh|status --orig_master_ host= host
 --orig_master_ip=ip --orig_master_port=port --new_master_host=host --new_master_ip=ip --new_
master_port=port\n";
    }
```

该方式与 keepalived 方式比较类似，都是通过管理浮动 IP，使其随着主从的切换从而漂移到正确的机器上，并继续提供服务。

3．通过 MySQL 中间件的方式

MySQL 中间件可以选择的有很多，在此仅介绍网易公司自研的 Cetus（参考第 32 章），即前端应用连接 Cetus 组件，Cetus 负责连接后端的 MySQL。

注意：网易对 MHA 源码进行了功能扩充，增加了新的功能模块，并调整了若干切换逻辑，具体内容不再详细介绍，以下仅进行简单描述，感兴趣的读者可以参考 GitHub 上的文档。

（1）部署新的 MHA 代码，可以参考 GitHub 上的文档 https://github.com/Lede-Inc/cetus/blob/master/doc/cetus-mha.md。

调整后的功能模块加入切换前后 MHA 与 Cetus 间的关系协调，即切换前通知 Cetus，故障主库已经不可用，即将进行切换，以及切换完成之后，通知 Cetus 新主库和新从库的角色。

（2）MHA 增加了 Cetus 的配置文件 cetus.cnf，示例如下。

```
middle_ipport=192.168.7.40:34061
middle_user=xxxxxx
middle_pass=xxxxxx
```

此配置文件，供 MHA 切换期间调用，以此连接并管理 Cetus。

以上 3 种方式，其实都是为了提供一个可靠的地址，供前端进行访问，为了防止 keepalived 脑裂情况的发生，同时也为了能高效地提供连接服务，以及更严谨的处理方式，这里将采用第三种方式。

到此为止，基本上 MHA 集群环境已经配置完毕，接下来我们将通过生产中的一些案例来看一下 MHA 到底是如何进行工作的。下面将从 MHA 自动 failover、手动 failover、在线切换 3 种方式来介绍 MHA 的工作情况。

31.1.3 自动 failover

先来看一下自动 failover，本节测试环境如表 31-2 所示。

表 31-2　　　　　　　　　　测试环境信息

角　色	主　机　名
mater	ip81
slave	ip83
slave	ip185
slave	ip186
cetus	ip40

其中，cetus 连接的后端信息如表 31-2 中的内容所示，主库所在机器为 ip81，3 个从库为 ip83、ip185 和 ip186。

自动 failover 模拟测试的操作步骤如下。

（1）sysbench 生成测试数据。

在 ip40 上进行 sysbench（版本 1.1.0）数据生成，在 sbtest 库下生成 sbtest 表，共 100 万条数据。

```
[root@ip40 tools]# sysbench --db-driver=mysql /home/sysbench/src/lua/oltp_read_write.lua --mysql-host=192.168.7.40 --mysql-port=3307 --mysql-user=sbtest --mysql-password=sbtest --mysql-db=sbtest --tables=1 --table-size=1000000 --threads=16 --time=300 prepare
```

（2）依次停掉两个从库的 slave 线程，模拟主从延迟。

在 ip186 上停掉 slave 线程。

```
mysql> stop slave ;
Query OK, 0 rows affected (0.00 sec)
```

等待 30s 后，在 ip83 上停掉 slave 线程。

```
mysql> stop slave ;
Query OK, 0 rows affected (0.00 sec)
```

ip185 继续传输日志。

（3）模拟 sysbench 压力测试。

在 ip40 上进行 sysbench 测试，运行 5min，产生大量的 BINLOG。

```
[root@ip40 tools]# sysbench --db-driver=mysql /home/sysbench/src/lua/oltp_read_write.lua --mysql-host=192.168.7.40 --mysql-port=3307 --mysql-user=sbtest --mysql-password=sbtest --mysql-db=sbtest --tables=1 --table-size=1000000 --threads=16 --time=300 run

sysbench 1.1.0 (using bundled LuaJIT 2.1.0-beta3)

Running the test with following options:
Number of threads: 16
Initializing random number generator from current time


Initializing worker threads...

Threads started!

SQL statistics:
queries performed:
    read:                            2568356
    write:                           733816
    other:                           366908
    total:                           3669080
```

```
transactions: 183454 (611.45 per sec.)
queries: 3669080 (12229.08 per sec.)
ignored errors: 0 (0.00 per sec.)
reconnects: 0 (0.00 per sec.)

Throughput:
events/s (eps): 611.4539
time elapsed: 300.0292s
total number of events: 183454

Latency (ms):
     min:                                    8.98
     avg:                                   26.16
     max:                                  950.34
     95th percentile:                       47.47
     sum:                              4798524.56

Threads fairness:
    events (avg/stddev):           11465.8750/128.38
    execution time (avg/stddev):   299.9078/0.01
```

（4）杀掉主库 mysql 进程，模拟主库发生故障，进行自动 failover 操作，在 ip81 数据库上：

```
[root@ip81 ~]# ps axu|grep mysql_3307|awk '{print $2}'|xargs kill -9
```

（5）查看 MHA 切换日志，了解整个切换过程。

因主从复制环境是否启用 GTID，将对应不同的切换逻辑，下面将分别进行介绍。

1. 非 GTID 模式

```
  Mon Aug 15 10:58:06 2016 - [warning] Got error on MySQL select ping: 2006 (MySQL server has
gone away)
  Mon Aug 15 10:58:06 2016 - [info] Executing secondary network check script: /usr/bin/masterha_
secondary_check --user=mysql --port=22 --master_ip=192.168.7.81 --master_port=3307  -s 192.168.
7.185 -s 192.168.7.83 -s 192.168.7.186 --user=mysql   --master_host=192.168.7.81  --master_ip=192.
168.7.81  --master_port=3307 --master_user=root --master_password=123456 --ping_type=SELECT
  Mon Aug 15 10:58:06 2016 - [info] Executing SSH check script: save_binary_logs --command=test
--start_pos=4 --binlog_dir=/home/mysql/mhamysqlhome/data --output_file=/data/save_binary_logs_test
--manager_version=0.56 --binlog_prefix=mysqlbin
  Mon Aug 15 10:58:07 2016 - [info] HealthCheck: SSH to 192.168.7.81 is reachable.
  Monitoring server 192.168.7.185 is reachable, Master is not reachable from 192.168.7.185. OK.
  Monitoring server 192.168.7.186 is reachable, Master is not reachable from 192.168.7.186. OK.
  Monitoring server 192.168.7.83 is reachable, Master is not reachable from 192.168.7.83. OK.
  Mon Aug 15 10:58:07 2016 - [info] Master is not reachable from all other monitoring servers.
Failover should start.
  Mon Aug 15 10:58:07 2016 - [warning] Got error on MySQL connect: 2013 (Lost connection to
MySQL server at 'reading initial communication packet', system error: 111)
  Mon Aug 15 10:58:07 2016 - [warning] Connection failed 2 time(s)..
  Mon Aug 15 10:58:08 2016 - [warning] Got error on MySQL connect: 2013 (Lost connection to
MySQL server at 'reading initial communication packet', system error: 111)
  Mon Aug 15 10:58:08 2016 - [warning] Connection failed 3 time(s)..
  Mon Aug 15 10:58:09 2016 - [warning] Got error on MySQL connect: 2013 (Lost connection to
MySQL server at 'reading initial communication packet', system error: 111)
  Mon Aug 15 10:58:09 2016 - [warning] Connection failed 4 time(s)..
  Mon Aug 15 10:58:09 2016 - [warning] Master is not reachable from health checker!
  Mon Aug 15 10:58:09 2016 - [warning] Master 192.168.7.81(192.168.7.81:3307) is not reachable!
  Mon Aug 15 10:58:09 2016 - [warning] SSH is reachable.
  Mon Aug 15 10:58:09 2016 - [info] Connecting to a master server failed. Reading configuration
file /etc/masterha_default.cnf and /masterha/mhatest/test.cnf again, and trying to connect to all
servers to check server status..
  Mon Aug 15 10:58:09 2016 - [warning] Global configuration file /etc/masterha_default.cnf not
found. Skipping.
  Mon Aug 15 10:58:09 2016 - [info] Reading application default configuration from /masterha/
mhatest/test.cnf..
  Mon Aug 15 10:58:09 2016 - [info] Reading server configuration from /masterha/mhatest/test.cnf..
  Mon Aug 15 10:58:09 2016 - [warning] SQL Thread is stopped(no error) on 192.168.7.186(192.
168.7.186:3307)
  Mon Aug 15 10:58:09 2016 - [warning] SQL Thread is stopped(no error) on 192.168.7.83(192.168
.7.83:3307)
  Mon Aug 15 10:58:09 2016 - [info] GTID failover mode = 0
  Mon Aug 15 10:58:09 2016 - [info] Dead Servers:
```

```
    Mon Aug 15 10:58:09 2016 - [info]     192.168.7.81(192.168.7.81:3307)
    Mon Aug 15 10:58:09 2016 - [info] Alive Servers:
    Mon Aug 15 10:58:09 2016 - [info]     192.168.7.185(192.168.7.185:3307)
    Mon Aug 15 10:58:09 2016 - [info]     192.168.7.186(192.168.7.186:3307)
    Mon Aug 15 10:58:09 2016 - [info]     192.168.7.83(192.168.7.83:3307)
    Mon Aug 15 10:58:09 2016 - [info] Alive Slaves:
    Mon Aug 15 10:58:09 2016 - [info]     192.168.7.185(192.168.7.185:3307)  Version=5.6.25-log
(oldest major version between slaves) log-bin:enabled
    Mon Aug 15 10:58:09 2016 - [info]       Replicating from 192.168.7.81(192.168.7.81:3307)
    Mon Aug 15 10:58:09 2016 - [info]       Not candidate for the new Master (no_master is set)
    Mon Aug 15 10:58:09 2016 - [info]     192.168.7.186(192.168.7.186:3307)  Version=5.6.25-log
(oldest major version between slaves) log-bin:enabled
    Mon Aug 15 10:58:09 2016 - [info]       Replicating from 192.168.7.81(192.168.7.81:3307)
    Mon Aug 15 10:58:09 2016 - [info]       Not candidate for the new Master (no_master is set)
    Mon Aug 15 10:58:09 2016 - [info]     192.168.7.83(192.168.7.83:3307)  Version=5.6.25-log
(oldest major version between slaves) log-bin:enabled
    Mon Aug 15 10:58:09 2016 - [info]       Replicating from 192.168.7.81(192.168.7.81:3307)
    Mon Aug 15 10:58:09 2016 - [info] Checking slave configurations..
    Mon Aug 15 10:58:09 2016 - [info]  read_only=1 is not set on slave 192.168.7.185(192.168.7.
185:3307).
    Mon Aug 15 10:58:09 2016 - [warning]  relay_log_purge=0 is not set on slave 192.168.7.185
(192.168.7.185:3307).
    Mon Aug 15 10:58:09 2016 - [info]  read_only=1 is not set on slave 192.168.7.83(192.168.7.
83:3307).
    Mon Aug 15 10:58:09 2016 - [warning]  relay_log_purge=0 is not set on slave 192.168.7.83(192
.168.7.83:3307).
    Mon Aug 15 10:58:09 2016 - [info] Checking replication filtering settings..
    Mon Aug 15 10:58:09 2016 - [info]  Replication filtering check ok.
    Mon Aug 15 10:58:09 2016 - [info] Master is down!
    Mon Aug 15 10:58:09 2016 - [info] Terminating monitoring script.
    Mon Aug 15 10:58:09 2016 - [info] Got exit code 20 (Master dead).
    Mon Aug 15 10:58:10 2016 - [info] MHA::MasterFailover version 0.56.
    Mon Aug 15 10:58:10 2016 - [info] Reading server configuration from /masterha/mhatest/test.cnf..
    Mon Aug 15 10:58:10 2016 - [info] [ProxyManager::setproxy] proxy_conf: /masterha/mhatest/
proxy1.cnf
    Mon Aug 15 10:58:10 2016 - [info] type    : failover
    Mon Aug 15 10:58:10 2016 - [info] status  : maintaining
    Mon Aug 15 10:58:10 2016 - [info] addr    : 192.168.7.81:3307
    Mon Aug 15 10:58:10 2016 - [info] dbtype  : ro
    Mon Aug 15 10:58:10 2016 - [info] ** 192.168.7.40 started, pid: 1071
    Mon Aug 15 10:58:10 2016 - [info] exec command: /usr/bin/mysql -h192.168.7.40 -uadmin -P13307
 -pxxxxxx  -e " update backends set state='maintaining' , type='ro' where address='192.168.7.81
:3307';"
    Mon Aug 15 10:58:10 2016 - [info] Starting master failover.
    Mon Aug 15 10:58:10 2016 - [info]
    Mon Aug 15 10:58:10 2016 - [info] * Phase 1: Configuration Check Phase..
    Mon Aug 15 10:58:10 2016 - [info]
    Mon Aug 15 10:58:10 2016 - [warning] SQL Thread is stopped(no error) on 192.168.7.186(192.
168.7.186:3307)
    Mon Aug 15 10:58:10 2016 - [warning] SQL Thread is stopped(no error) on 192.168.7.83(192.168
.7.83:3307)
    Mon Aug 15 10:58:10 2016 - [info] GTID failover mode = 0
    Mon Aug 15 10:58:10 2016 - [info] Dead Servers:
    Mon Aug 15 10:58:10 2016 - [info]   192.168.7.81(192.168.7.81:3307)
    Mon Aug 15 10:58:10 2016 - [info] Checking master reachability via MySQL(double check)...
    Mon Aug 15 10:58:10 2016 - [info]   ok.
    Mon Aug 15 10:58:10 2016 - [info] Alive Servers:
    Mon Aug 15 10:58:10 2016 - [info]   192.168.7.185(192.168.7.185:3307)
    Mon Aug 15 10:58:10 2016 - [info]   192.168.7.186(192.168.7.186:3307)
    Mon Aug 15 10:58:10 2016 - [info]   192.168.7.83(192.168.7.83:3307)
    Mon Aug 15 10:58:10 2016 - [info] Alive Slaves:
    Mon Aug 15 10:58:10 2016 - [info]     192.168.7.185(192.168.7.185:3307)  Version=5.6.25-log
(oldest major version between slaves) log-bin:enabled
    Mon Aug 15 10:58:10 2016 - [info]       Replicating from 192.168.7.81(192.168.7.81:3307)
    Mon Aug 15 10:58:10 2016 - [info]       Not candidate for the new Master (no_master is set)
    Mon Aug 15 10:58:10 2016 - [info]     192.168.7.186(192.168.7.186:3307)  Version=5.6.25-log
(oldest major version between slaves) log-bin:enabled
    Mon Aug 15 10:58:10 2016 - [info]       Replicating from 192.168.7.81(192.168.7.81:3307)
    Mon Aug 15 10:58:10 2016 - [info]       Not candidate for the new Master (no_master is set)
    Mon Aug 15 10:58:10 2016 - [info]     192.168.7.83(192.168.7.83:3307)  Version=5.6.25-log
(oldest major version between slaves) log-bin:enabled
```

```
Mon Aug 15 10:58:10 2016 - [info]       Replicating from 192.168.7.81(192.168.7.81:3307)
Mon Aug 15 10:58:10 2016 - [info]    Starting SQL thread on 192.168.7.186(192.168.7.186:3307) ..
Mon Aug 15 10:58:10 2016 - [info]   done.
Mon Aug 15 10:58:10 2016 - [info]    Starting SQL thread on 192.168.7.83(192.168.7.83:3307) ..
Mon Aug 15 10:58:10 2016 - [info]   done.
Mon Aug 15 10:58:10 2016 - [info] Starting Non-GTID based failover.
Mon Aug 15 10:58:10 2016 - [info]
Mon Aug 15 10:58:10 2016 - [info] ** Phase 1: Configuration Check Phase completed.
Mon Aug 15 10:58:10 2016 - [info]
Mon Aug 15 10:58:10 2016 - [info] * Phase 2: Dead Master Shutdown Phase..
Mon Aug 15 10:58:10 2016 - [info]
Mon Aug 15 10:58:10 2016 - [info] Forcing shutdown so that applications never connect to the current master..
Mon Aug 15 10:58:10 2016 - [warning] master_ip_failover_script is not set. Skipping invalidating dead master IP address.
Mon Aug 15 10:58:10 2016 - [warning] shutdown_script is not set. Skipping explicit shutting down of the dead master.
Mon Aug 15 10:58:10 2016 - [info] * Phase 2: Dead Master Shutdown Phase completed.
Mon Aug 15 10:58:10 2016 - [info]
Mon Aug 15 10:58:10 2016 - [info] * Phase 3: Master Recovery Phase..
Mon Aug 15 10:58:10 2016 - [info]
Mon Aug 15 10:58:10 2016 - [info] * Phase 3.1: Getting Latest Slaves Phase..
Mon Aug 15 10:58:10 2016 - [info]
Mon Aug 15 10:58:10 2016 - [info] The latest binary log file/position on all slaves is mysqlbin.000008:516
Mon Aug 15 10:58:10 2016 - [info] Latest slaves (Slaves that received relay log files to the latest):
Mon Aug 15 10:58:10 2016 - [info]     192.168.7.185(192.168.7.185:3307)  Version=5.6.25-log (oldest major version between slaves) log-bin:enabled
Mon Aug 15 10:58:10 2016 - [info]       Replicating from 192.168.7.81(192.168.7.81:3307)
Mon Aug 15 10:58:10 2016 - [info]       Not candidate for the new Master (no_master is set)
Mon Aug 15 10:58:10 2016 - [info] The oldest binary log file/position on all slaves is mysqlbin.000008:120
Mon Aug 15 10:58:10 2016 - [info] Oldest slaves:
Mon Aug 15 10:58:10 2016 - [info]     192.168.7.186(192.168.7.186:3307)  Version=5.6.25-log (oldest major version between slaves) log-bin:enabled
Mon Aug 15 10:58:10 2016 - [info]       Replicating from 192.168.7.81(192.168.7.81:3307)
Mon Aug 15 10:58:10 2016 - [info]       Not candidate for the new Master (no_master is set)
Mon Aug 15 10:58:10 2016 - [info]
Mon Aug 15 10:58:10 2016 - [info] * Phase 3.2: Saving Dead Master's Binlog Phase..
Mon Aug 15 10:58:10 2016 - [info]
Mon Aug 15 10:58:10 2016 - [info] Fetching dead master's binary logs..
Mon Aug 15 10:58:10 2016 - [info] Executing command on the dead master 192.168.7.81(192.168.7.81:3307): save_binary_logs --command=save --start_file=mysqlbin.000008  --start_pos=516 --binlog_dir=/home/mysql/mhamysqlhome/data --output_file=/data/saved_master_binlog_from_192.168.7.81_3307_20160815105810.binlog --handle_raw_binlog=1 --disable_log_bin=0 --manager_version=0.56
  Creating /data if not exists..    ok.
  Concat binary/relay logs from mysqlbin.000008 pos 516 to mysqlbin.000008 EOF into /data/saved_master_binlog_from_192.168.7.81_3307_20160815105810.binlog ..
 Binlog Checksum enabled
  Dumping binlog format description event, from position 0 to 120.. ok.
  No need to dump effective binlog data from /home/mysql/mhamysqlhome/data/mysqlbin.000008 (pos starts 516, filesize 516). Skipping.
 Binlog Checksum enabled
  /data/saved_master_binlog_from_192.168.7.81_3307_20160815105810.binlog has no effective data events.
 Event not exists.
Mon Aug 15 10:58:10 2016 - [info] Additional events were not found from the orig master. No need to save.
Mon Aug 15 10:58:10 2016 - [info]
Mon Aug 15 10:58:10 2016 - [info] * Phase 3.3: Determining New Master Phase..
Mon Aug 15 10:58:10 2016 - [info]
Mon Aug 15 10:58:10 2016 - [info] Finding the latest slave that has all relay logs for recovering other slaves..
Mon Aug 15 10:58:11 2016 - [info] HealthCheck: SSH to 192.168.7.185 is reachable.
Mon Aug 15 10:58:11 2016 - [info] Checking whether 192.168.7.185 has relay logs from the oldest position..
Mon Aug 15 10:58:11 2016 - [info] Executing command: apply_diff_relay_logs --command=find --latest_mlf=mysqlbin.000008 --latest_rmlp=516 --target_mlf=mysqlbin.000008 --target_rmlp=120 --server_id=109 --workdir=/data --timestamp=20160815105810 --manager_version=0.56 --relay_dir=/data/mysqlhome/data --current_relay_log=192.168.7.185-relay-bin.000006  :
  Relay log found at /data/mysqlhome/data, up to 192.168.7.185-relay-bin.000006
 Fast relay log position search succeeded.
```

31.1 MHA 架构

```
    Target relay log file/position found. start_file:192.168.7.185-relay-bin.000006, start_pos:282.
    Target relay log FOUND!
  Mon Aug 15 10:58:11 2016 - [info] OK. 192.168.7.185 has all relay logs.
  Mon Aug 15 10:58:11 2016 - [info] HealthCheck: SSH to 192.168.7.186 is reachable.
  Mon Aug 15 10:58:12 2016 - [info] HealthCheck: SSH to 192.168.7.83 is reachable.
  Mon Aug 15 10:58:12 2016 - [info] Searching new master from slaves..
  Mon Aug 15 10:58:12 2016 - [info]  Candidate masters from the configuration file:
  Mon Aug 15 10:58:12 2016 - [info]  Non-candidate masters:
  Mon Aug 15 10:58:12 2016 - [info]   192.168.7.185(192.168.7.185:3307)  Version=5.6.25-log
(oldest major version between slaves) log-bin:enabled
  Mon Aug 15 10:58:12 2016 - [info]     Replicating from 192.168.7.81(192.168.7.81:3307)
  Mon Aug 15 10:58:12 2016 - [info]     Not candidate for the new Master (no_master is set)
  Mon Aug 15 10:58:12 2016 - [info]   192.168.7.186(192.168.7.186:3307)  Version=5.6.25-log
(oldest major version between slaves) log-bin:enabled
  Mon Aug 15 10:58:12 2016 - [info]     Replicating from 192.168.7.81(192.168.7.81:3307)
  Mon Aug 15 10:58:12 2016 - [info]     Not candidate for the new Master (no_master is set)
  Mon Aug 15 10:58:12 2016 - [info]  Searching from all slaves which have received the latest
relay log events..
  Mon Aug 15 10:58:12 2016 - [info]   Not found.
  Mon Aug 15 10:58:12 2016 - [info]  Searching from all slaves..
  Mon Aug 15 10:58:12 2016 - [info] New master is 192.168.7.83(192.168.7.83:3307)
  Mon Aug 15 10:58:12 2016 - [info] Starting master failover..
  Mon Aug 15 10:58:12 2016 - [info]
  From:
  192.168.7.81(192.168.7.81:3307) (current master)
   +--192.168.7.185(192.168.7.185:3307)
   +--192.168.7.186(192.168.7.186:3307)
   +--192.168.7.83(192.168.7.83:3307)

  To:
  192.168.7.83(192.168.7.83:3307) (new master)
   +--192.168.7.185(192.168.7.185:3307)
   +--192.168.7.186(192.168.7.186:3307)
  Mon Aug 15 10:58:12 2016 - [info]
  Mon Aug 15 10:58:12 2016 - [info] * Phase 3.3: New Master Diff Log Generation Phase..
  Mon Aug 15 10:58:12 2016 - [info]
  Mon Aug 15 10:58:12 2016 - [info] Server 192.168.7.83 received relay logs up to: mysqlbin.
000008:318
  Mon Aug 15 10:58:12 2016 - [info] Need to get diffs from the latest slave(192.168.7.185) up
to: mysqlbin.000008:516 (using the latest slave's relay logs)
  Mon Aug 15 10:58:12 2016 - [info] Connecting to the latest slave host 192.168.7.185, generating
diff relay log files..
  Mon Aug 15 10:58:12 2016 - [info] Executing command: apply_diff_relay_logs --command=generate
_and_send --scp_user=mysql --scp_host=192.168.7.83 --latest_mlf=mysqlbin.000008 --latest_rmlp=
516 --target_mlf=mysqlbin.000008 --target_rmlp=318 --server_id=109 --diff_file_readtolatest=/
data/relay_from_read_to_latest_192.168.7.83_3307_20160815105810.binlog --workdir=/data --timestamp
=20160815105810 --handle_raw_binlog=1 --disable_log_bin=0 --manager_version=0.56 --scp_port=22
--relay_dir=/data/mysqlhome/data --current_relay_log=192.168.7.185-relay-bin.000006
  Mon Aug 15 10:58:13 2016 - [info]
    Relay log found at /data/mysqlhome/data, up to 192.168.7.185-relay-bin.000006
   Fast relay log position search succeeded.
   Target relay log file/position found. start_file:192.168.7.185-relay-bin.000006, start_pos:480.
   Concat binary/relay logs from 192.168.7.185-relay-bin.000006 pos 480 to 192.168.7.185-relay
-bin.000006 EOF into /data/relay_from_read_to_latest_192.168.7.83_3307_20160815105810.binlog ..
   Binlog Checksum enabled
   Binlog Checksum enabled
    Dumping binlog format description event, from position 0 to 282.. ok.
    Dumping effective binlog data from /data/mysqlhome/data/192.168.7.185-relay-bin.000006
position 480 to tail(678).. ok.
   Binlog Checksum enabled
   Binlog Checksum enabled
   Concat succeeded.
   Generating diff relay log succeeded. Saved at /data/relay_from_read_to_latest_192.168.7.83_
3307_20160815105810.binlog .
    scp 192.168.7.185:/data/relay_from_read_to_latest_192.168.7.83_3307_20160815105810.binlog to
mysql@192.168.7.83(22) succeeded.
  Mon Aug 15 10:58:13 2016 - [info]  Generating diff files succeeded.
  Mon Aug 15 10:58:13 2016 - [info]
  Mon Aug 15 10:58:13 2016 - [info] * Phase 3.4: Master Log Apply Phase..
  Mon Aug 15 10:58:13 2016 - [info]
  Mon Aug 15 10:58:13 2016 - [info] *NOTICE: If any error happens from this phase, manual
```

```
recovery is needed.
    Mon Aug 15 10:58:13 2016 - [info] Starting recovery on 192.168.7.83(192.168.7.83:3307)..
    Mon Aug 15 10:58:13 2016 - [info]  Generating diffs succeeded.
    Mon Aug 15 10:58:13 2016 - [info] Waiting until all relay logs are applied.
    Mon Aug 15 10:58:13 2016 - [info]  done.
    Mon Aug 15 10:58:13 2016 - [info] Getting slave status..
    Mon Aug 15 10:58:13 2016 - [info] This slave(192.168.7.83)'s Exec_Master_Log_Pos equals to
Read_Master_Log_Pos(mysqlbin.000008:318). No need to recover from Exec_Master_Log_Pos.
    Mon Aug 15 10:58:13 2016 - [info] Connecting to the target slave host 192.168.7.83, running
recover script..
    Mon Aug 15 10:58:13 2016 - [info] Executing command: apply_diff_relay_logs --command=apply -
-slave_user='root' --slave_host=192.168.7.83 --slave_ip=192.168.7.83  --slave_port=3307 --apply
_files=/data/relay_from_read_to_latest_192.168.7.83_3307_20160815105810.binlog --workdir=/data
--target_version=5.6.25-log --timestamp=20160815105810 --handle_raw_binlog=1 --disable_log_bin=
0 --manager_version=0.56 --slave_pass=xxx
    Mon Aug 15 10:58:14 2016 - [info]
MySQL client version is 5.6.25. Using --binary-mode.
    Applying differential binary/relay log files /data/relay_from_read_to_latest_192.168.7.83_330
7_20160815105810.binlog on 192.168.7.83:3307. This may take long time...
    Applying log files succeeded.
    Mon Aug 15 10:58:14 2016 - [info] All relay logs were successfully applied.
    Mon Aug 15 10:58:14 2016 - [info] Getting new master's binlog name and position..
    Mon Aug 15 10:58:14 2016 - [info]  mysqlbin.000006:314
    Mon Aug 15 10:58:14 2016 - [info] All other slaves should start replication from here. Statement
should be: CHANGE MASTER TO MASTER_HOST='192.168.7.83', MASTER_PORT=3307, MASTER_LOG_FILE=
'mysqlbin.000006', MASTER_LOG_POS=314, MASTER_USER='repl', MASTER_PASSWORD='xxx';
    Mon Aug 15 10:58:14 2016 - [warning] master_ip_failover_script is not set. Skipping taking
over new master IP address.
    Mon Aug 15 10:58:14 2016 - [info] ** Finished master recovery successfully.
    Mon Aug 15 10:58:14 2016 - [info] [ProxyManager::setproxy] proxy_conf: /masterha/mhatest/
proxy1.cnf
    Mon Aug 15 10:58:14 2016 - [info] type   : setmaster
    Mon Aug 15 10:58:14 2016 - [info] status : unknown
    Mon Aug 15 10:58:14 2016 - [info] addr   : 192.168.7.83:3307
    Mon Aug 15 10:58:14 2016 - [info] dbtype : rw
    Mon Aug 15 10:58:14 2016 - [info] ** 192.168.7.40 started, pid: 1110
    Mon Aug 15 10:58:14 2016 - [info] exec command: /usr/bin/mysql -h192.168.7.40 -uadmin -P13307 -
pxxxxxx  -e " update backends set state='unknown' , type='rw' where address='192.168.7.83:3307';"
    Mon Aug 15 10:58:14 2016 - [info] * Phase 3: Master Recovery Phase completed.
    Mon Aug 15 10:58:14 2016 - [info]
    Mon Aug 15 10:58:14 2016 - [info] * Phase 4: Slaves Recovery Phase..
    Mon Aug 15 10:58:14 2016 - [info]
    Mon Aug 15 10:58:14 2016 - [info] * Phase 4.1: Starting Parallel Slave Diff Log Generation Phase..
    Mon Aug 15 10:58:14 2016 - [info]
    Mon Aug 15 10:58:14 2016 - [info] -- Slave diff file generation on host 192.168.7.185(192.
168.7.185:3307) started, pid: 1113. Check tmp log /masterha/mhatest/192.168.7.185_3307_
20160815105810.log if it takes time..
    Mon Aug 15 10:58:14 2016 - [info] -- Slave diff file generation on host 192.168.7.186(192.
168.7.186:3307) started, pid: 1114. Check tmp log /masterha/mhatest/192.168.7.186_3307_
20160815105810.log if it takes time..
    Mon Aug 15 10:58:14 2016 - [info]
    Mon Aug 15 10:58:14 2016 - [info] Log messages from 192.168.7.185 ...
    Mon Aug 15 10:58:14 2016 - [info]
    Mon Aug 15 10:58:14 2016 - [info]  This server has all relay logs. No need to generate diff
files from the latest slave.
    Mon Aug 15 10:58:14 2016 - [info] End of log messages from 192.168.7.185.
    Mon Aug 15 10:58:14 2016 - [info] -- 192.168.7.185(192.168.7.185:3307) has the latest relay
log events.
    Mon Aug 15 10:58:15 2016 - [info]
    Mon Aug 15 10:58:15 2016 - [info] Log messages from 192.168.7.186 ...
    Mon Aug 15 10:58:15 2016 - [info]
    Mon Aug 15 10:58:14 2016 - [info] Server 192.168.7.186 received relay logs up to: mysqlbin.
000008:120
    Mon Aug 15 10:58:14 2016 - [info] Need to get diffs from the latest slave(192.168.7.185) up
to: mysqlbin.000008:516 (using the latest slave's relay logs)
    Mon Aug 15 10:58:14 2016 - [info] Connecting to the latest slave host 192.168.7.185, generating
diff relay log files..
    Mon Aug 15 10:58:14 2016 - [info] Executing command: apply_diff_relay_logs --command=generate_and
_send --scp_user=mysql --scp_host=192.168.7.186 --latest_mlf=mysqlbin.000008 --latest_rmlp=516 -
-target_mlf=mysqlbin.000008 --target_rmlp=120 --server_id=109 --diff_file_readtolatest=/data/
```

31.1 MHA 架构

```
relay_from_read_to_latest_192.168.7.186_3307_20160815105810.binlog --workdir=/data --timestamp=
20160815105810 --handle_raw_binlog=1 --disable_log_bin=0 --manager_version=0.56 --scp_port=22 --
relay_dir=/data/mysqlhome/data --current_relay_log=192.168.7.185-relay-bin.000006
    Mon Aug 15 10:58:15 2016 - [info]
        Relay log found at /data/mysqlhome/data, up to 192.168.7.185-relay-bin.000006
    Fast relay log position search succeeded.
    Target relay log file/position found. start_file:192.168.7.185-relay-bin.000006, start_pos:282.
    Concat binary/relay logs from 192.168.7.185-relay-bin.000006 pos 282 to 192.168.7.185-relay
-bin.000006 EOF into /data/relay_from_read_to_latest_192.168.7.186_3307_20160815105810.binlog ..
    Binlog Checksum enabled
    Binlog Checksum enabled
     Dumping binlog format description event, from position 0 to 282.. ok.
     Dumping effective binlog data from /data/mysqlhome/data/192.168.7.185-relay-bin.000006
position 282 to tail(678).. ok.
    Binlog Checksum enabled
    Binlog Checksum enabled
    Concat succeeded.
    Generating diff relay log succeeded. Saved at /data/relay_from_read_to_latest_192.168.7.186_
3307_20160815105810.binlog .
    scp 192.168.7.185:/data/relay_from_read_to_latest_192.168.7.186_3307_20160815105810.binlog to
mysql@192.168.7.186(22) succeeded.
    Mon Aug 15 10:58:15 2016 - [info]  Generating diff files succeeded.
    Mon Aug 15 10:58:15 2016 - [info] End of log messages from 192.168.7.186.
    Mon Aug 15 10:58:15 2016 - [info] -- Slave diff log generation on host 192.168.7.186(192.168
.7.186:3307) succeeded.
    Mon Aug 15 10:58:15 2016 - [info] Generating relay diff files from the latest slave succeeded.
    Mon Aug 15 10:58:15 2016 - [info]
    Mon Aug 15 10:58:15 2016 - [info] * Phase 4.2: Starting Parallel Slave Log Apply Phase..
    Mon Aug 15 10:58:15 2016 - [info]
    Mon Aug 15 10:58:15 2016 - [info] -- Slave recovery on host 192.168.7.185(192.168.7.185:3307)
started, pid: 1127. Check tmp log /masterha/mhatest/192.168.7.185_3307_20160815105810.log if it
 takes time..
    Mon Aug 15 10:58:15 2016 - [info] -- Slave recovery on host 192.168.7.186(192.168.7.186:3307)
started, pid: 1128. Check tmp log /masterha/mhatest/192.168.7.186_3307_20160815105810.log if it
 takes time..
    Mon Aug 15 10:58:15 2016 - [info]
    Mon Aug 15 10:58:15 2016 - [info] Log messages from 192.168.7.185 ...
    Mon Aug 15 10:58:15 2016 - [info]
    Mon Aug 15 10:58:15 2016 - [info] Starting recovery on 192.168.7.185(192.168.7.185:3307)..
    Mon Aug 15 10:58:15 2016 - [info]  This server has all relay logs. Waiting all logs to be applied..
    Mon Aug 15 10:58:15 2016 - [info]   done.
    Mon Aug 15 10:58:15 2016 - [info]  All relay logs were successfully applied.
    Mon Aug 15 10:58:15 2016 - [info]  Resetting slave 192.168.7.185(192.168.7.185:3307) and
starting replication from the new master 192.168.7.83(192.168.7.83:3307)..
    Mon Aug 15 10:58:15 2016 - [info]  Executed CHANGE MASTER.
    Mon Aug 15 10:58:15 2016 - [info]  Slave started.
    Mon Aug 15 10:58:15 2016 - [info] End of log messages from 192.168.7.185.
    Mon Aug 15 10:58:15 2016 - [info] -- Slave recovery on host 192.168.7.185(192.168.7.185:3307)
succeeded.
    Mon Aug 15 10:58:15 2016 - [info] [ProxyManager::setproxy] proxy_conf: /masterha/mhatest/proxy1.cnf
    Mon Aug 15 10:58:15 2016 - [info] type   : setslave
    Mon Aug 15 10:58:15 2016 - [info] status : unknown
    Mon Aug 15 10:58:15 2016 - [info] addr   : 192.168.7.185:3307
    Mon Aug 15 10:58:15 2016 - [info] dbtype : ro
    Mon Aug 15 10:58:15 2016 - [info] ** 192.168.7.40 started, pid: 1132
    Mon Aug 15 10:58:15 2016 - [info] exec command: /usr/bin/mysql -h192.168.7.40 -uadmin -P13307 -
pxxxxxx -e " update backends set state='unknown' , type='ro' where address='192.168.7.185:3307';"
    Mon Aug 15 10:58:15 2016 - [info]
    Mon Aug 15 10:58:15 2016 - [info] Log messages from 192.168.7.186 ...
    Mon Aug 15 10:58:15 2016 - [info]
    Mon Aug 15 10:58:15 2016 - [info] Starting recovery on 192.168.7.186(192.168.7.186:3307)..
    Mon Aug 15 10:58:15 2016 - [info]  Generating diffs succeeded.
    Mon Aug 15 10:58:15 2016 - [info] Waiting until all relay logs are applied.
    Mon Aug 15 10:58:15 2016 - [info]  done.
    Mon Aug 15 10:58:15 2016 - [info] Getting slave status..
    Mon Aug 15 10:58:15 2016 - [info] This slave(192.168.7.186)'s Exec_Master_Log_Pos equals to
Read_Master_Log_Pos(mysqlbin.000008:120). No need to recover from Exec_Master_Log_Pos.
    Mon Aug 15 10:58:15 2016 - [info] Connecting to the target slave host 192.168.7.186, running
 recover script..
    Mon Aug 15 10:58:15 2016 - [info] Executing command: apply_diff_relay_logs --command=apply -
-slave_user='root' --slave_host=192.168.7.186 --slave_ip=192.168.7.186  --slave_port=3307 --apply_
```

```
files=/data/relay_from_read_to_latest_192.168.7.186_3307_20160815105810.binlog --workdir=/data -
-target_version=5.6.25-log --timestamp=20160815105810 --handle_raw_binlog=1 --disable_log_bin=0
--manager_version=0.56 --slave_pass=xxx
    Mon Aug 15 10:58:15 2016 - [info]
    MySQL client version is 5.6.25. Using --binary-mode.
    Applying differential binary/relay log files /data/relay_from_read_to_latest_192.168.7.186_
3307_20160815105810.binlog on 192.168.7.186:3307. This may take long time...
    Applying log files succeeded.
    Mon Aug 15 10:58:15 2016 - [info] All relay logs were successfully applied.
    Mon Aug 15 10:58:15 2016 - [info] Resetting slave 192.168.7.186(192.168.7.186:3307) and
starting replication from the new master 192.168.7.83(192.168.7.83:3307)..
    Mon Aug 15 10:58:15 2016 - [info] Executed CHANGE MASTER.
    Mon Aug 15 10:58:15 2016 - [info] Slave started.
    Mon Aug 15 10:58:15 2016 - [info] End of log messages from 192.168.7.186.
    Mon Aug 15 10:58:15 2016 - [info] -- Slave recovery on host 192.168.7.186(192.168.7.186:3307)
succeeded.
    Mon Aug 15 10:58:15 2016 - [info] [ProxyManager::setproxy] proxy_conf: /masterha/mhatest/
proxy1.cnf
    Mon Aug 15 10:58:15 2016 - [info] type    : setslave
    Mon Aug 15 10:58:15 2016 - [info] status  : unknown
    Mon Aug 15 10:58:15 2016 - [info] addr    : 192.168.7.186:3307
    Mon Aug 15 10:58:15 2016 - [info] dbtype  : ro
    Mon Aug 15 10:58:15 2016 - [info] ** 192.168.7.40 started, pid: 1142
    Mon Aug 15 10:58:15 2016 - [info] exec command: /usr/bin/mysql -h192.168.7.40 -uadmin -P13307 -
pxxxxxx -e " update backends set state='unknown' , type='ro' where address='192.168.7.186:3307';"
    Mon Aug 15 10:58:15 2016 - [info] All new slave servers recovered successfully.
    Mon Aug 15 10:58:15 2016 - [info]
    Mon Aug 15 10:58:15 2016 - [info] * Phase 5: New master cleanup phase..
    Mon Aug 15 10:58:15 2016 - [info]
    Mon Aug 15 10:58:15 2016 - [info] Resetting slave info on the new master..
    Mon Aug 15 10:58:15 2016 - [info]  192.168.7.83: Resetting slave info succeeded.
    Mon Aug 15 10:58:15 2016 - [info] Master failover to 192.168.7.83(192.168.7.83:3307) completed
successfully.
    Mon Aug 15 10:58:15 2016 - [info]

----- Failover Report -----

test: MySQL Master failover 192.168.7.81(192.168.7.81:3307) to 192.168.7.83(192.168.7.83:3307)
succeeded

Master 192.168.7.81(192.168.7.81:3307) is down.

Check MHA Manager logs at ip233:/masterha/mhatest/app1.log for details.

Started automated(non-interactive) failover.
The latest slave 192.168.7.185(192.168.7.185:3307) has all relay logs for recovery.
Selected 192.168.7.83(192.168.7.83:3307) as a new master.
192.168.7.83(192.168.7.83:3307): OK: Applying all logs succeeded.
192.168.7.185(192.168.7.185:3307): This host has the latest relay log events.
192.168.7.186(192.168.7.186:3307): Generating differential relay logs up to 192.168.7.185
(192.168.7.185:3307)succeeded.
Generating relay diff files from the latest slave succeeded.
192.168.7.185(192.168.7.185:3307): OK: Applying all logs succeeded. Slave started, replicating
from 192.168.7.83(192.168.7.83:3307)
192.168.7.186(192.168.7.186:3307): OK: Applying all logs succeeded. Slave started, replicating
from 192.168.7.83(192.168.7.83:3307)
192.168.7.83(192.168.7.83:3307): Resetting slave info succeeded.
Master failover to 192.168.7.83(192.168.7.83:3307) completed successfully.
```

从上面的输出可以看出非 GTID 模式下整个 MHA 的切换过程，共包括以下几个步骤：

- 故障 master 的一系列检测，确认是否需要进行切换；
- 通知 cetus 故障 master 信息，做好切换前的准备工作；
- 切换前的复制环境检查，这个阶段将会检查以及完成主从切换前的准备工作；
- 宕机的 master 处理，这个阶段可以调用外部脚本，例如虚拟 IP 摘除操作、主机关机等操作；
- 复制 dead master 和最新 slave 相差的 Relay Log，并保存到 MHA Manager 指定目录下；

- 识别含有最新更新的 slave;
- 应用差异的中继日志（relay log）到候选 master 的 slave;
- 应用从 master 保存的二进制日志事件（BINLOG event）到候选 master 的 slave;
- new master 恢复完成，并输出其余 slave 应该执行的 change master to 命令;
- 通知 cetus 新 master 已经产生，并可以对外提供服务;
- 恢复其他 slave，并在恢复完成后将信息发送至 Cetus;
- 输出切换报告，概述了此次切换涉及的重要过程。

2. GTID 模式

由于日志量比较多，此次仅列出切换过程中变动最大的内容。

```
……
Thu Aug 25 17:35:38 2016 - [info] New master is 192.168.7.83(192.168.7.83:3307)
Thu Aug 25 17:35:38 2016 - [info] Starting master failover..
Thu Aug 25 17:35:38 2016 - [info]
From:
192.168.7.81(192.168.7.81:3307) (current master)
 +--192.168.7.185(192.168.7.185:3307)
 +--192.168.7.83(192.168.7.83:3307)

To:
192.168.7.83(192.168.7.83:3307) (new master)
 +--192.168.7.185(192.168.7.185:3307)
Thu Aug 25 17:35:38 2016 - [info]
Thu Aug 25 17:35:38 2016 - [info] * Phase 3.3: New Master Recovery Phase..
Thu Aug 25 17:35:38 2016 - [info]
Thu Aug 25 17:35:38 2016 - [info]  Waiting all logs to be applied..
Thu Aug 25 17:35:38 2016 - [info]   done.
Thu Aug 25 17:35:38 2016 - [info]  Replicating from the latest slave 192.168.7.185(192.168.7.185:3307) and waiting to apply..
Thu Aug 25 17:35:38 2016 - [info]  Waiting all logs to be applied on the latest slave..
Thu Aug 25 17:35:38 2016 - [info]  Resetting slave 192.168.7.83(192.168.7.83:3307) and starting replication from the new master 192.168.7.185(192.168.7.185:3307)..
Thu Aug 25 17:35:38 2016 - [info]  Executed CHANGE MASTER.
Thu Aug 25 17:35:39 2016 - [info]  Slave started.
Thu Aug 25 17:35:39 2016 - [info]  Waiting to execute all relay logs on 192.168.7.83(192.168.7.83:3307)..
Thu Aug 25 17:35:44 2016 - [info]  master_pos_wait(mysqlbin.000020:21080053) completed on 192.168.7.83(192.168.7.83:3307). Executed 36 events.
Thu Aug 25 17:35:44 2016 - [info]   done.
Thu Aug 25 17:35:44 2016 - [info]   done.
Thu Aug 25 17:35:44 2016 - [info] Getting new master's binlog name and position..
Thu Aug 25 17:35:44 2016 - [info] mysqlbin.000020:25035445
Thu Aug 25 17:35:44 2016 - [info]  All other slaves should start replication from here. Statement should be: CHANGE MASTER TO MASTER_HOST='192.168.7.83', MASTER_PORT=3307, MASTER_AUTO_POSITION=1, MASTER_USER='repl', MASTER_PASSWORD='xxx';
Thu Aug 25 17:35:44 2016 - [info] Master Recovery succeeded. File:Pos:Exec_Gtid_Set: mysqlbin.000020, 25035445, 2537c70b-441e-11e6-8f0b-0015c5efe1f1:1-8,
b03aaa9c-3d14-11e6-a127-52540035c143:1-82,
fc87d64d-471b-11e6-a28c-525400f33597:1-23
Thu Aug 25 17:35:44 2016 - [warning] master_ip_failover_script is not set. Skipping taking over new master IP address.
Thu Aug 25 17:35:44 2016 - [info] Setting read_only=0 on 192.168.7.83(192.168.7.83:3307)..
Thu Aug 25 17:35:44 2016 - [info]  ok.
Thu Aug 25 17:35:44 2016 - [info] ** Finished master recovery successfully.
……
```

从上面的输出可用看出在 GTID 模式下，候选 master 如果不是含有最新数据的节点，将作为包含最新数据节点的 slave，从而补足所缺少的数据。完成之后再提升为 master，并输出 change master to 命令。

31.1.4 网络问题触发的 failover 操作

本节将模拟网络中断情况下，MHA Manager 无法连接到主库，候选主也无法连接到主库，此时 MHA 的切换情况。

在 ip81 上设置防火墙，drop 掉来自 ip186 和 ip83 的包，以模拟网络问题。

```
[root@ip81 ~]# iptables -A INPUT -s 192.168.7.186 -j DROP;iptables -A INPUT -s 192.168.7.83 -j DROP
```

在 ip186 上：

```
Wed Nov 28 11:15:04 2018 - [warning] Got timeout on MySQL Ping(SELECT) child process and killed it! at /usr/share/perl5/vendor_perl/MHA/HealthCheck.pm line 492.
Wed Nov 28 11:15:04 2018 - [info] Executing secondary network check script: /usr/bin/masterha_secondary_check --user=mysql57 --port=22 --master_ip=192.168.7.81 --master_port=3307 -s 192.168.7.83  -s 192.168.7.185  --user=mysql57  --master_host=192.168.7.81  --master_ip=192.168.7.81 --master_port=3307 --master_user=root --master_password=xxxxxx --ping_type=SELECT
Wed Nov 28 11:15:04 2018 - [info] Executing SSH check script: save_binary_logs --command=test --start_pos=4 --binlog_dir=/home/mysql57/mysql-5.7.14/data --output_file=/data/save_binary_logs_test --manager_version=0.56 --binlog_prefix=mysql-bin
Wed Nov 28 11:15:07 2018 - [warning] Got error on MySQL connect: 2003 (Can't connect to MySQL server on '192.168.7.81' (4))
Wed Nov 28 11:15:07 2018 - [warning] Connection failed 2 time(s)..
Wed Nov 28 11:15:09 2018 - [warning] HealthCheck: Got timeout on checking SSH connection to 192.168.7.81! at /usr/share/perl5/vendor_perl/MHA/HealthCheck.pm line 403.
Monitoring server 192.168.7.83 is reachable, Master is not reachable from 192.168.7.83. OK.
Wed Nov 28 11:15:10 2018 - [warning] Got error on MySQL connect: 2003 (Can't connect to MySQL server on '192.168.7.81' (4))
Wed Nov 28 11:15:10 2018 - [warning] Connection failed 3 time(s)..
Wed Nov 28 11:15:13 2018 - [warning] Got error on MySQL connect: 2003 (Can't connect to MySQL server on '192.168.7.81' (4))
Wed Nov 28 11:15:13 2018 - [warning] Connection failed 4 time(s)..
Monitoring server 192.168.7.83 is reachable, Master is not reachable from 192.168.7.83. OK.
Master is reachable from 192.168.7.185!
Wed Nov 28 11:15:15 2018 - [warning] Master is reachable from at least one of other monitoring servers. Failover should not happen.
Wed Nov 28 11:15:15 2018 - [warning] Secondary network check script returned errors. Failover should not start so checking server status again. Check network settings for details.
```

从上述日志输出可以看到，即便是主库 ip81 禁止所有来自 ip186 和 ip83 的包，切换也不会发生，只是反复地进行重试，提示 master 可以通过 ip185 进行连接，即还存在正常的网络链路，并且提示检查网络状况。因此在这种网络环境下，MHA 并不会进行误切换。当然这里还有一点需要提醒，如果在多机房环境下部署 MHA 集群，尽量在不同机房设置多个网络观测点，以免网络分区带来的隐患。

查看 monitor 状态：

```
[root@ip186 home]# masterha_check_status --conf=/etc/masterha/app1.cnf
App1 master maybe down(20:PING_FAILING). master:192.168.7.81
```

可以看到，这个时候 monitor 的状况是 master maybe down(20:PING_FAILING)状态，但是 MHA 没有进行故障切换操作，需要 DBA 介入，检查是否是网络问题造成，是否需要手动进行数据库切换操作。

31.1.5 手动 failover

手动 failover 这种场景可能发生在原主从环境并没有配置或启用 MHA 自动切换功能，而当主库服务器故障时，手动调用 MHA 脚本来进行故障切换操作，具体命令如下：

```
masterha_master_switch --master_state=dead --conf=/masterha/app1/app1.cnf --dead_master_host=192.168.7.81 --dead_master_ip=192.168.7.81 --dead_master_port=3307 --ignore_last_failover
```

31.1 MHA 架构

切换过程中的部分输出信息如下：

```
Fri Aug 28 17:55:39 2015 - [info] * Phase 3.3: Determining New Master Phase..
Fri Aug 28 17:55:39 2015 - [info]
Fri Aug 28 17:55:39 2015 - [info] Finding the latest slave that has all relay logs for recovering other slaves..
Fri Aug 28 17:55:39 2015 - [info] All slaves received relay logs to the same position. No need to resync each other.
Fri Aug 28 17:55:39 2015 - [info] Searching new master from slaves..
Fri Aug 28 17:55:39 2015 - [info]  Candidate masters from the configuration file:
Fri Aug 28 17:55:39 2015 - [info]   192.168.7.83(192.168.7.83:3307)  Version=5.6.25-log (oldest major version between slaves) log-bin:enabled
Fri Aug 28 17:55:39 2015 - [info]     Replicating from 192.168.7.81(192.168.7.81:3307)
Fri Aug 28 17:55:39 2015 - [info]     Primary candidate for the new Master (candidate_master is set)
Fri Aug 28 17:55:39 2015 - [info]  Non-candidate masters:
Fri Aug 28 17:55:39 2015 - [info]  Searching from candidate_master slaves which have received the latest relay log events..
Fri Aug 28 17:55:39 2015 - [info] New master is 192.168.7.83(192.168.7.83:3307)
Fri Aug 28 17:55:39 2015 - [info] Starting master failover..
Fri Aug 28 17:55:39 2015 - [info]
From:
192.168.7.81(192.168.7.81:3307) (current master)
 +--192.168.7.83(192.168.7.83:3307)
 +--192.168.7.185(192.168.7.185:3307)

To:
192.168.7.83(192.168.7.83:3307) (new master)
 +--192.168.7.185(192.168.7.185:3307)

Starting master switch from 192.168.7.81(192.168.7.81:3307) to 192.168.7.83(192.168.7.83:3307)
? (yes/NO): yes
Fri Aug 28 17:55:42 2015 - [info] New master decided manually is 192.168.7.83(192.168.7.83:3307)
```

上述模拟了 ip81 宕机的情况下手动切换主从的操作。

31.1.6 在线进行切换

在日常工作下，难免会遇到主库服务器需要进行维护操作，需要暂时停止提供读写服务，例如数据库升级或者硬件维修等，这时需要用到在线切换功能。MHA 提供快速切换功能，这个切换过程通常只需要 0.5~2s 的时间，这段时间内数据是无法写入的。在很多情况下，0.5~2s 的阻塞写入是可以接受的，但建议操作之前与相关人员协调一致，确认没有问题的情况下，再实施切换操作。

MHA 在线切换过程如下：
（1）检测主从复制环境设置；
（2）确定 new master，准备切换；
（3）阻塞写入到当前 master；
（4）new master 完成切换，并提供服务；
（5）恢复其余 slave 节点；
（6）配置原 master 作为新 slave 节点（可选）。

注意，在线切换的时候应用架构需要考虑以下两个问题：
- 自动识别 master 和 slave 的问题，即前端应用采用哪种连接配置方式；
- 负载均衡的问题（我们可以定义大概的读写比例，每台机器可承担的负载比例，当有机器离开集群时，需要考虑这个问题）。

为了保证数据完全一致，在最快的时间内完成切换，MHA 的在线切换必须在满足以下条件下才会切换成功,否则会切换失败。

- 所有 slave 的 IO 线程都在运行。
- 所有 slave 的 SQL 线程都在运行。
- 所有 slave 的 show slave status\G ;的输出中 Seconds_Behind_Master 参数小于或者等于 running_updates_limit 秒,如果在切换过程中不指定 running_updates_limit,那么默认情况下 running_updates_limit 为 1s。
- 在 master 端,通过 show processlist 输出,没有一个更新需要消耗的时间大于 running_updates_limit 秒。

在线切换的一般操作步骤如下。

首先,停掉 MHA 监控:

```
masterha_stop --conf=/masterha/app1/app1.cnf
```

其次,进行在线切换操作(模拟在线切换主库操作,原主库 ip81 变为 slave,ip83 提升为新的主库):

```
masterha_master_switch --master_state=alive --conf=/masterha/app1/app1.cnf --new_master_host=192.168.7.83 --new_master_port=3307 --orig_master_is_new_slave --running_updates_limit=100
```

其中--orig_master_is_new_slave 的意思是将原 master 变换成 slave,默认情况下,MHA 下不做该操作。另外,在切换时,当指定--running_updates_limit 属性后,将增加主从切换成功的概率,因放宽了限制条件,尽量保证 MHA 切换成功。

最后查看切换过程,执行过程中的部分输出信息如下:

```
Fri Aug 28 17:32:35 2015 - [info] Alive Slaves:
    Fri Aug 28 17:32:35 2015 - [info]   192.168.7.83(192.168.7.83:3307)  Version=5.6.25-log (oldest major version between slaves) log-bin:enabled
    Fri Aug 28 17:32:35 2015 - [info]     Replicating from 192.168.7.81(192.168.7.81:3307)
    Fri Aug 28 17:32:35 2015 - [info]     Primary candidate for the new Master (candidate_master is set)
    Fri Aug 28 17:32:35 2015 - [info]   192.168.7.185(192.168.7.185:3307)  Version=5.6.25-log (oldest major version between slaves) log-bin:enabled
    Fri Aug 28 17:32:35 2015 - [info]     Replicating from 192.168.7.81(192.168.7.81:3307)

It is better to execute FLUSH NO_WRITE_TO_BINLOG TABLES on the master before switching. Is it ok to execute on 192.168.7.81(192.168.7.81:3307)? (YES/no): yes
    Fri Aug 28 17:32:38 2015 - [info] Executing FLUSH NO_WRITE_TO_BINLOG TABLES. This may take long time..
    Fri Aug 28 17:32:38 2015 - [info]  ok.
    Fri Aug 28 17:32:38 2015 - [info] Checking MHA is not monitoring or doing failover..
    Fri Aug 28 17:32:38 2015 - [info] Checking replication health on 192.168.7.83..
    Fri Aug 28 17:32:38 2015 - [info]  ok.
    Fri Aug 28 17:32:38 2015 - [info] Checking replication health on 192.168.7.185..
    Fri Aug 28 17:32:38 2015 - [info]  ok.
    Fri Aug 28 17:32:38 2015 - [info] 192.168.7.83 can be new master.
    Fri Aug 28 17:32:38 2015 - [info]
From:
192.168.7.81(192.168.7.81:3307) (current master)
 +--192.168.7.83(192.168.7.83:3307)
 +--192.168.7.185(192.168.7.185:3307)

To:
192.168.7.83(192.168.7.83:3307) (new master)
 +--192.168.7.185(192.168.7.185:3307)
 +--192.168.7.81(192.168.7.81:3307)

Starting master switch from 192.168.7.81(192.168.7.81:3307) to 192.168.7.83(192.168.7.83:3307)? (yes/NO): yes
    Fri Aug 28 17:33:36 2015 - [info] Checking whether 192.168.7.83(192.168.7.83:3307) is ok for the new master..
    Fri Aug 28 17:33:36 2015 - [info]  ok.
    Fri Aug 28 17:33:36 2015 - [info] 192.168.7.81(192.168.7.81:3307): SHOW SLAVE STATUS returned empty result. To check replication filtering rules, temporarily executing CHANGE MASTER to a
```

```
dummy host.
    Fri Aug 28 17:33:36 2015 - [info] 192.168.7.81(192.168.7.81:3307): Resetting slave pointing
to the dummy host.
……
```

31.1.7 修复宕掉的 Master

通常情况下，如果发生了自动切换，原 master 因故障可能会进行修复，当修复完成后，也可能希望继续使用该服务器，例如将该服务器作为一个新的 slave 节点，这时可以借助当时自动切换时产生的 MHA 日志，来完成这一操作，示例如下：

```
grep -i "All other slaves should start replication from" /masterha/app1/app1.log
```

非 GTID：

```
Mon Aug 15 10:58:14 2016 - [info] All other slaves should start replication from here.
Statement should be: CHANGE MASTER TO MASTER_HOST='192.168.7.83', MASTER_PORT=3307, MASTER_LOG_
FILE='mysqlbin.000006', MASTER_LOG_POS=314, MASTER_USER='repl', MASTER_PASSWORD='xxx';
```

GTID：

```
Thu Aug 25 17:35:44 2016 - [info] All other slaves should start replication from here.
Statement should be: CHANGE MASTER TO MASTER_HOST='192.168.7.83', MASTER_PORT=3307, MASTER_AUTO
_POSITION=1, MASTER_USER='repl', MASTER_PASSWORD='xxx';
```

获取上述信息后，就可以直接在修复后的节点上执行 change master to 操作了。

31.2 MGR 架构

MGR（MySQL Group Replication）又称为组复制，是 MySQL 官方推出的一个高可用解决方案。从 MySQL 5.7.17 开始引入，也是支撑 Innodb Cluster 的重要技术基石。

GR 技术从 2014 年开始，作为实验室项目，到后来支持 corosync 2.x 协议，但由于 MGR 对高可用的高要求，从 0.6.0 版本开始支持 paxos 协议，最终实现了现在的事务强一致性要求，即集群中的事务，必须经过半数以上节点同意才能提交，这是与前文介绍的复制技术相对比，区别最大的地方。图 31-2 说明了传统的异步复制。

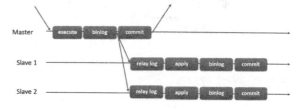

图 31-2 传统的异步复制

从图 31-2 可知，主库事务的提交与从库没有什么依赖性。

半同步复制如图 31-3 所示。根据上文可知，半同步复制可以配置 AFTER_COMMIT 或者 AFTER_SYNC，但如果等待时间超过阈值，依然会演变成传统的异步复制。

组复制技术如图 31-4 所示。组复制架构下的事务，如要提交必须先经过集群中所有节点的验证，在得到超半数节点的成功反馈后，才能真正提交。

MGR 也实现了故障自动切换功能，所以建议配置 MGR 环境时，集群中的节点要达到 3 个节点以上。图 31-5 说明了集群中对故障节点的容忍度。

由图 31-5 可以得知，集群中如果只包含 1 个或者 2 个节点，那么只要有 1 个节点故障，集群将不可用；如果有 3 个或者 4 个节点，可以容忍 1 个节点故障。

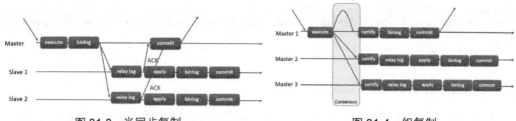

图 31-3　半同步复制　　　　　　　图 31-4　组复制

组复制的实现方式依然采用插件形式，即仅在需要加入组复制集群的节点上安装并启用一个插件 group_replication 即可，非常简单。插件架构如图 31-6 所示。

- Capture 组件：负责与正在执行的事务时刻保持联系。
- Applier 组件：负责执行其他节点执行的事务。
- Recovery 组件：负责从集群中选择一个节点获取需要的数据，并负责进行恢复以及加入集群环境。
- Replication Protocol 模块：负责处理事务冲突检测，并在集群中接收和发送事务。
- Group Communication System（GCS）：负责构建复制状态机。
- Group Communication Engine（GCE）：负责集群成员的消息通信。

Group Size	Majority	Instant Failures Tolerated
1	1	0
2	2	0
3	2	1
4	3	1
5	3	2
6	4	2
7	4	3

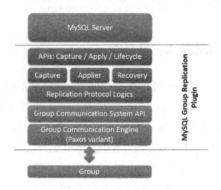

图 31-5　组复制集群节点个数故障容忍度　　　图 31-6　组复制插件架构

MGR 支持单主（Single-Primary）模式和多主（Multi-Primary）模式，接下来对这两种模式分别进行实战演练。

31.2.1　安装部署 MGR

在安装部署之前，需要先介绍一些必要的参数。

```
第一部分：
binlog_checksum=NONE
transaction_isolation=READ-COMMITTED

第二部分：
gtid_mode=on
enforce_gtid_consistency=on

第三部分：
log_slave_updates=ON
binlog_format=ROW
master_info_repository=TABLE
relay_log_info_repository=TABLE
```

```
第四部分:
transaction_write_set_extraction=XXHASH64
loose-group_replication_group_name='aaaaaaaa-aaaa-aaaa-aaaa-aaaaaaaaaaaa'
loose-group_replication_start_on_boot=off
loose-group_replication_local_address='127.0.0.1:24901'
loose-group_replication_group_seeds='127.0.0.1:24901,127.0.0.1:24902,127.0.0.1:24903'
loose-group_replication_bootstrap_group=off
```

其中,参数的第一部分是因为限制(组复制需要关闭 binlog_checksum 功能)或建议配置,官方推荐将参数 transaction_isolation(该参数指定事务隔离级别,可参考第 16 章)配置为 READ-COMMITTED,而并不是默认的 REPEATABLE-READ,当事务隔离级别设置为 REPEATABLE-READ 时,有 Next-Key 锁问题(也称为 Gap Lock,详见第 16 章),而组复制的认证过程中无法处理 Gap Lock,第二部分是启用 GTID 模式,第三部分在 MySQL 8.0.3 以后,已经是默认值,如果使用的是这个版本之前的 MySQL,需要按照以上参数值进行设置,第四部分是组复制相关参数配置,详情如下。

- transaction_write_set_extraction 表示针对每个事务的 writeset 进行 XXHASH64 算法编码。
- loose-group_replication_group_name 代表这个集群的组名,也会作为集群中 GTID 的前半部分 UUID 的值,例如 aaaaaaaa-aaaa-aaaa-aaaa-aaaaaaaaaaaa。
- loose-group_replication_start_on_boot 表示组复制是否随实例启动而自动启动。
- loose-group_replication_local_address 表示该节点启用组复制后,使用的通信地址,格式是 IP(也可以是 hostname):Port。
- loose-group_replication_group_seeds 表示当集群有新成员加入时,可以从哪些节点(IP:Port)获取需要的数据进行 recovery。
- loose-group_replication_bootstrap_group 只用于集群初始化的时候。

以上是这些参数的简要说明,也是读者在进行环境配置时需要了解的,另外还有一些组复制相关的参数,会在下文中出现的时候再做解释,接下来就开始实战演练,配置组复制下的单主模式。

1. 单主(single-primary)模式

下面要搭建的环境均在同一台物理机上进行,不同的实例使用不同的端口进行区分,集群包括 3 个 MySQL 节点、1 个 primary 节点、2 个 secondary 节点。

(1)初始化 3 个 MySQL 实例,确保使用了相同的版本(也可以先行配置一个节点,后两个进行复制,在此不再赘述)。

(2)在每一个 MySQL 节点创建复制使用的账号,并授予 REPLICATION SLAVE 权限。这里创建一个复制用户 rep1,允许 IP 为 10.120.240.251 的主机进行连接,并且执行 change master 命令,为后续该节点在加入集群时,进行 recovery 操作提供支持。

```
mysql> set sql_log_bin=0;

mysql> GRANT REPLICATION SLAVE ON *.* TO 'repl'@'10.120.240.251' IDENTIFIED BY 'repl';
Query OK, 0 rows affected (0.00 sec)

mysql> CHANGE MASTER TO MASTER_USER='repl', MASTER_PASSWORD='repl' FOR CHANNEL 'group_replication_recovery';
Query OK, 0 rows affected, 2 warnings (0.01 sec)

mysql> set sql_log_bin=1;
```

(3)修改每一个 MySQL 节点的配置文件 my.cnf,增加以下参数,并重启数据库实例:

```
[mysqld]
master-info-repository=TABLE
relay-log-info-repository=TABLE
gtid_mode=on
enforce_gtid_consistency=on
slave_preserve_commit_order=on
binlog_checksum=NONE
transaction_write_set_extraction=XXHASH64
```

（4）选择一个 MySQL 节点作为主节点，安装组复制插件。

```
mysql> INSTALL PLUGIN group_replication SONAME 'group_replication.so';
Query OK, 0 rows affected (0.11 sec)
```

（5）在主节点配置组复制参数，并将以下参数写入 my.cnf 文件。

```
mysql> set global group_replication_group_name='aaaaaaaa-aaaa-aaaa-aaaa-aaaaaaaaaaaa';
Query OK, 0 rows affected (0.00 sec)

mysql> set global group_replication_local_address= '10.120.240.251:24903';
Query OK, 0 rows affected (0.00 sec)

mysql> set global group_replication_group_seeds= "10.120.240.251:24901,10.120.240.251:24902,10.120.240.251:24903";
Query OK, 0 rows affected (0.00 sec)

mysql> set global group_replication_start_on_boot=off;
Query OK, 0 rows affected (0.00 sec)

mysql> set global group_replication_bootstrap_group=on;
Query OK, 0 rows affected (0.00 sec)

mysql> set global group_replication_ip_whitelist='10.120.240.0/24';
Query OK, 0 rows affected (0.00 sec)
```

注意：写入 my.cnf 文件时，group_replication 开头的参数应该增加 loose-开头，例如 loose-group_replication_group_name。

（6）在主节点启动组复制，并查看日志。

```
mysql> start group_replication;
Query OK, 0 rows affected (2.18 sec)

 2018-09-10T19:16:11.429754+08:00 23 [Note] 'CHANGE MASTER TO FOR CHANNEL 'group_replication_applier' executed'. Previous state master_host='<NULL>', master_port= 0, master_log_file='', master_log_pos= 293, master_bind=''. New state master_host='<NULL>', master_port= 0, master_log_file='', master_log_pos= 4, master_bind=''.
 2018-09-10T19:16:11.463556+08:00 2 [Note] Plugin group_replication reported: 'Group Replication applier module successfully initialized!'
 2018-09-10T19:16:11.463618+08:00 2 [Note] Plugin group_replication reported: 'auto_increment_increment is set to 7'
 2018-09-10T19:16:11.463633+08:00 2 [Note] Plugin group_replication reported: 'auto_increment_offset is set to 7237'
 2018-09-10T19:16:11.463618+08:00 2 [Note] Plugin group_replication reported: 'auto_increment_increment is set to 7'
 2018-09-10T19:16:11.463633+08:00 2 [Note] Plugin group_replication reported: 'auto_increment_offset is set to 7237'
 2018-09-10T19:16:11.463585+08:00 26 [Note] Slave SQL thread for channel 'group_replication_applier' initialized, starting replication in log 'FIRST' at position 0, relay log './hz_10_120_240_251-relay-bin-group_replication_applier.000002' position: 4
 2018-09-10T19:16:11.466884+08:00 0 [Note] Plugin group_replication reported: 'XCom protocol version: 3'
 2018-09-10T19:16:11.466969+08:00 0 [Note] Plugin group_replication reported: 'XCom initialized and ready to accept incoming connections on port 24903'
 2018-09-10T19:16:12.472096+08:00 37 [Note] Plugin group_replication reported: 'Only one server alive. Declaring this server as online within the replication group'
 2018-09-10T19:16:12.472254+08:00 0 [Note] Plugin group_replication reported: 'Group membership changed to 10.120.240.251:3325 on view 15365781724712422:1.'
 2018-09-10T19:16:12.774037+08:00 0 [Note] Plugin group_replication reported: 'This server was
```

```
declared online within the replication group'
  2018-09-10T19:16:12.774209+08:00 0 [Note] Plugin group_replication reported: 'A new primary
with address 10.120.240.251:3325 was elected, enabling conflict detection until the new primary
applies all relay logs.'
  2018-09-10T19:16:12.774328+08:00 39 [Note] Plugin group_replication reported: 'This server is
working as primary member.'
```

以上是启动组复制后，初始化的过程，可以看到首先会启动 applier 线程，负责集群中不同的节点事务的执行，由于组复制技术自身是可以支持多主的，这样自增量的值就不能再是 1 了，以免数据冲突，默认组复制开启后，自增量的 increment 是 7，即可以满足 7 节点的多主环境，再往后就是消息通信，采用了 XCom 协议，用以实现 Paxos 算法的思想，然后就是构建集群的初始时间点 view 15365781724712422:1，后续如果集群有任务变化（新增或者退出节点），都会重新构造一个新的 view，用以区分新旧集群的变化时刻，一切就绪后，该节点就正式作为 primary 成员，可以正常提供服务了。

（7）在主节点也可以通过以下视图查看现在的集群状态。

```
mysql> SELECT * FROM performance_schema.replication_group_members\G;
*************************** 1. row ***************************
CHANNEL_NAME: group_replication_applier
   MEMBER_ID: ba4fcb42-b4d5-11e8-9550-0024e869b4d5
 MEMBER_HOST: 10.120.240.251
 MEMBER_PORT: 3325
MEMBER_STATE: ONLINE
1 row in set (0.00 sec)

mysql> SELECT VARIABLE_VALUE FROM performance_schema.global_status WHERE VARIABLE_NAME= 'group_
replication_primary_member'\G;
*************************** 1. row ***************************
VARIABLE_VALUE: ba4fcb42-b4d5-11e8-9550-0024e869b4d5
1 row in set (0.06 sec)
```

其中，performance_schema.replication_group_members 视图显示集群中的成员情况，由于刚初始化一个主节点，所以此处就只有一条信息，另一个视图 performance_schema.global_status 显示实例的统计数据，与 show global status 类似，由于启用了组复制，所以其中增加了一条 VARIABLE_NAME 名字是 group_replication_primary_member 的信息，该变量的 value 即表示现在集群中 Primary 成员的 UUID，可以根据 performance_schema.replication_group_members 视图，确定当前哪一个节点是集群的 primary 成员。

（8）在主节点关闭 group_replication_bootstrap_group。

```
mysql> set global group_replication_bootstrap_group=off;
Query OK, 0 rows affected (0.00 sec)
```

下面的过程是在集群中加入另外两个节点，组成 3 节点集群，即 1 个 primary 和 2 个 secondary。

（9）在需要加入集群的两个节点分别安装组复制插件。

```
mysql> INSTALL PLUGIN group_replication SONAME 'group_replication.so';
Query OK, 0 rows affected (0.11 sec)
```

（10）两个节点分别配置组复制参数，并写入 my.cnf 文件。

```
节点1
mysql> set global group_replication_group_name='aaaaaaaa-aaaa-aaaa-aaaa-aaaaaaaaaaaa';
Query OK, 0 rows affected (0.00 sec)

mysql> set global group_replication_local_address='10.120.240.251:24901';
Query OK, 0 rows affected (0.00 sec)

mysql> set global group_replication_group_seeds= "10.120.240.251:24901,10.120.240.251:24902,
10.120.240.251:24903";
```

```
Query OK, 0 rows affected (0.00 sec)

mysql> set global group_replication_start_on_boot=off;
Query OK, 0 rows affected (0.00 sec)

mysql> set global group_replication_ip_whitelist='10.120.240.0/24';
Query OK, 0 rows affected (0.00 sec)
```

节点2
```
mysql> set global group_replication_group_name='aaaaaaaa-aaaa-aaaa-aaaa-aaaaaaaaaaaa';
Query OK, 0 rows affected (0.00 sec)

mysql> set global group_replication_local_address='10.120.240.251:24902';
Query OK, 0 rows affected (0.00 sec)

mysql> set global group_replication_group_seeds= "10.120.240.251:24901,10.120.240.251:24902,
10.120.240.251:24903";
Query OK, 0 rows affected (0.00 sec)

mysql> set global group_replication_start_on_boot=off;
Query OK, 0 rows affected (0.00 sec)

mysql> set global group_replication_ip_whitelist='10.120.240.0/24';
Query OK, 0 rows affected (0.00 sec)
```

注意：不同的节点，参数 group_replication_local_address 不同，其余相同。

（11）依次在两个节点启动组复制，加入集群环境，可以通过观察日志或者视图 performance_schema.replication_group_members 确认是否加入成功。

```
mysql> start group_replication;
Query OK, 0 rows affected (5.88 sec)
```

日志内容
```
2018-09-11T15:05:38.441446+08:00 9 [Note] Plugin group_replication reported: 'This server is
working as secondary member with primary member address 10.120.240.251:3325.'
……
2018-09-11T15:05:38.707533+08:00 0 [Note] Plugin group_replication reported: 'This server was
declared online within the replication group'
```

注意：当加入集群后，在 BINLOG 同级别目录下还会生成两种不同文件，relay-bin-group_replication_
　　　recovery 和 relay-bin-group_replication_applier，分别对应 recovery 组件和 applier 组件，格式
　　　与 BINLOG 相同。

（12）完成以上所有的步骤后，查看最新的集群成员，确认已经完成了集群的搭建。

```
mysql> SELECT * FROM performance_schema.replication_group_members\G;
*************************** 1. row ***************************
  CHANNEL_NAME: group_replication_applier
     MEMBER_ID: 22559fb9-8e60-11e8-85d2-0024e869b4d5
   MEMBER_HOST: 10.120.240.251
   MEMBER_PORT: 3305
  MEMBER_STATE: ONLINE
*************************** 2. row ***************************
  CHANNEL_NAME: group_replication_applier
     MEMBER_ID: 6066c689-b4d4-11e8-886f-0024e869b4d5
   MEMBER_HOST: 10.120.240.251
   MEMBER_PORT: 3315
  MEMBER_STATE: ONLINE
*************************** 3. row ***************************
  CHANNEL_NAME: group_replication_applier
     MEMBER_ID: ba4fcb42-b4d5-11e8-9550-0024e869b4d5
   MEMBER_HOST: 10.120.240.251
   MEMBER_PORT: 3325
  MEMBER_STATE: ONLINE
3 rows in set (0.00 sec)
```

至此 3 节点的组复制集群搭建完成，接下来对该集群进行验证。

（13）集群搭建成功之后，可以先通过上面介绍的方法找出 Primary 成员，因为 MGR 默认是单主模式，所以只有这个节点才能写入数据，secondary 节点会自动开启 read only 模式。

```
mysql> show variables like '%read_only';
+-----------------------+-------+
| Variable_name         | Value |
+-----------------------+-------+
| innodb_read_only      | OFF   |
| read_only             | OFF   |
| super_read_only       | OFF   |
| transaction_read_only | OFF   |
| tx_read_only          | OFF   |
+-----------------------+-------+
5 rows in set (0.00 sec)

mysql> show variables like '%read_only';
+-----------------------+-------+
| Variable_name         | Value |
+-----------------------+-------+
| innodb_read_only      | OFF   |
| read_only             | ON    |
| super_read_only       | ON    |
| transaction_read_only | OFF   |
| tx_read_only          | OFF   |
+-----------------------+-------+
5 rows in set (0.00 sec)
```

以上信息来自 Primary 成员节点和其中一个 secondary 成员节点，从中可以得知即便"找错了" primary 节点，数据也不会成功写入，这样就可以避免由于一些错误从而导致数据不一致。

（14）primary 节点进行数据变更。

```
mysql> update salaries set salary=100 where salary=60117 and emp_no=10001;
Query OK, 1 row affected (0.08 sec)
Rows matched: 1  Changed: 1  Warnings: 0
```

（15）secondary 节点确认数据。

```
mysql> select * from employees.salaries where emp_no=10001 and salary=100;
+--------+--------+------------+------------+
| emp_no | salary | from_date  | to_date    |
+--------+--------+------------+------------+
|  10001 |    100 | 1986-06-26 | 1987-06-26 |
+--------+--------+------------+------------+
1 row in set (0.00 sec)
```

验证完成，至此 3 节点集群的工作已经全部完成。

2. 多主（multi-primary）模式

上文中对单主集群模式进行了实战演练，接下来将继续为读者一步步搭建多主下的集群环境，此次仍部署 3 节点集群。

（1）初始化 3 个 MySQL 实例，并正确进行复制账号的配置，以及在 3 个实例正确安装组复制插件，并按照单主集群的参数进行正确配置（这些步骤可以详见上文，此处略过）。

（2）根据上文单主模式参数配置的基础上，再对每一个 MySQL 节点配置如下参数，并写入 my.cnf 配置文件。

```
mysql> set global group_replication_single_primary_mode=off;
Query OK, 0 rows affected (0.00 sec)

mysql> set global group_replication_enforce_update_everywhere_checks=on;
Query OK, 0 rows affected (0.00 sec)
```

（3）在其中一个节点，打开 group_replication_bootstrap_group 为 on，并开启组复制，成

功后，关闭 group_replication_bootstrap_group。

```
mysql> set global group_replication_bootstrap_group=on;
Query OK, 0 rows affected (0.00 sec)

mysql> start group_replication;
Query OK, 0 rows affected (2.10 sec)

mysql> set global group_replication_bootstrap_group=off;
Query OK, 0 rows affected (0.00 sec)
```

（4）另外两个节点逐一开启组复制，加入集群环境。

（5）通过以下视图查看现在的集群状态。

```
mysql> select * from performance_schema.replication_group_members\G;
*************************** 1. row ***************************
  CHANNEL_NAME: group_replication_applier
     MEMBER_ID: 22559fb9-8e60-11e8-85d2-0024e869b4d5
   MEMBER_HOST: 10.120.240.251
   MEMBER_PORT: 3305
  MEMBER_STATE: ONLINE
*************************** 2. row ***************************
  CHANNEL_NAME: group_replication_applier
     MEMBER_ID: 6066c689-b4d4-11e8-886f-0024e869b4d5
   MEMBER_HOST: 10.120.240.251
   MEMBER_PORT: 3315
  MEMBER_STATE: ONLINE
*************************** 3. row ***************************
  CHANNEL_NAME: group_replication_applier
     MEMBER_ID: ba4fcb42-b4d5-11e8-9550-0024e869b4d5
   MEMBER_HOST: 10.120.240.251
   MEMBER_PORT: 3325
  MEMBER_STATE: ONLINE
3 rows in set (0.00 sec)
```

通过以上视图，可以得知多主 3 节点集群环境已经搭建完成，接下来对该集群进行验证，由于是多主环境，所以我们通过 3 个场景来进行验证。

场景 1：在每一个节点对相同表同时写入不同数据，通过端口来区分写入的数据来自于哪个节点。

以下是数据的写入情况：

```
mysql> select instance_port,count(*) from demo_multi_primary group by instance_port;
+---------------+----------+
| instance_port | count(*) |
+---------------+----------+
|          3305 |     1625 |
|          3315 |     1513 |
|          3325 |     1454 |
+---------------+----------+
3 rows in set (0.00 sec)
```

在每个不同实例节点，都运行一个数据写入脚本，对相同表进行同时写入操作，通过验证，集群工作正常。

场景 2：在不同节点同时修改相同的数据。

（1）实例 3305 和实例 3325 修改相同数据行，修改前如下。

```
mysql> select * from demo_multi_primary where id=458;
+-----+---------------+------------+---------------------+
| id  | instance_port | account_id | mtime               |
+-----+---------------+------------+---------------------+
| 458 |          3315 |      73161 | 2018-09-18 16:52:29 |
+-----+---------------+------------+---------------------+
1 row in set (0.00 sec)
```

（2）实例 3305 修改该行数据的 instance_port 列为 3305，暂时不提交。

```
mysql>start transaction;
Query OK, 0 rows affected (0.00 sec)

mysql>update employees.demo_multi_primary set instance_port=3305 where id=458;
Query OK, 1 row affected (0.01 sec)
Rows matched: 1  Changed: 1  Warnings: 0
```

（3）实例 3325 修改该行数据的 instance_port 列为 3325，并提交。

```
mysql> start transaction;
Query OK, 0 rows affected (0.00 sec)

mysql> update employees.demo_multi_primary set instance_port=3325 where id=458;
Query OK, 1 row affected (0.00 sec)
Rows matched: 1  Changed: 1  Warnings: 0

mysql> commit;
Query OK, 0 rows affected (0.00 sec)
```

（4）实例 3305 提交修改。

```
mysql>commit;
ERROR 1180 (HY000): Got error 149 during COMMIT
```

因为不同节点修改了同一行数据，最先提交的会成功，后提交的会失败，自动回滚，验证成功。

场景 3：在不同节点同时执行不同对象的 DML 和 DDL 操作。

（1）在实例 3325 节点持续对表 employees.demo_multi_primary 进行数据变更。

（2）在实例 3315 节点对表 employees.country 执行 alter 操作。

```
mysql> alter table employees.country add extra int;
Query OK, 0 rows affected (0.10 sec)
Records: 0  Duplicates: 0  Warnings: 0

mysql> select * from employees.country;
+------------+---------+---------------------+-------+
| country_id | country | last_update         | extra |
+------------+---------+---------------------+-------+
|      10000 | China   | 2018-05-28 14:36:20 |  NULL |
+------------+---------+---------------------+-------+
1 row in set (0.00 sec)

mysql> select count(*) from demo_multi_primary;
+----------+
| count(*) |
+----------+
|    11432 |
+----------+
1 row in set (0.01 sec)
```

通过以上验证，验证成功，即多主模式下，集群环境中不同节点对不同对象同时进行 DML 和 DDL，可以正常执行。

以上通过 3 个不同场景，分别对多主模式进行了验证，证实是可靠的。不过要提醒一下，组复制在多主模式下，如果想要在线上使用，首先需要注意一些限制，比如不可以在不同节点同时对相同对象执行 DML 和 DDL 操作，更多的限制读者可以详细阅读官方文档，以便更加合理的用好多主模式。

31.2.2 监控

当组复制集群投入使用后，日常监控是少不了的，接下来介绍一下适合组复制架构的日常监控内容。

- 成员信息 performance_schema.replication_group_members，用来显示集群内成员信息，包括各节点的 UUID、HOST、PORT 以及成员状态。
- 成员事务统计 performance_schema.replication_group_member_stats，用来统计节点队列中的事务、冲突检测的事务数量、冲突检测数据的大小等。在 8.0.11 版本中又增加了 4 个统计指标，如本地事务数量、已经执行的远程事务数量、本地回滚事务数量和正在队列中准备执行的远程事务，可用于更细粒度的监控，为后续参数的调整提供数据参考。
- 复制配置信息 performance_schema.replication_connection_configuration，用来查看一些复制的连接配置。
- 连接状态信息 performance_schema.replication_connection_status，用来查看连接状态信息，包括集群组名字、组件服务状态等。
- 组件服务状态 performance_schema.replication_applier_status，用来查看复制组件状态、配置的复制延迟以及应用事务时的重试次数。
- 组件服务状态 performance_schema.replication_applier_status_by_coordinator，其用途与上类似，这个统计的来源是协调线程。
- 组件服务状态 performance_schema.replication_applier_status_by_worker，其用途与上类似，这个统计的来源是 worker 线程。

除了以上这些视图外，还有一个办法可以查看 recovery 阶段的执行情况，因集群有新成员加入时，需要经历两个阶段，首先是阶段一，进行 recovery，而这个阶段的复制仍然采用传统的异步复制模式（并不是组复制模式），直到追上 view 时间点后，开始进入阶段二，进行 applier 的工作，所以如果在 recovery 阶段时，复制数据出现错误，可以使用我们很熟悉的命令 show slave status for channel 'group_replication_recovery'来定位问题，并进行故障处理。

31.2.3 primary 成员切换

当组复制集群相关工作完成后，就可以投入使用了，但在实际工作中，服务器不可能 100%可靠，数据库实例也是一样，难免会有故障发生，此时组复制功能就显示出了便捷性，因为不需要借助外部工具或者做什么特殊的高可用配置，组复制技术本身就保证了这一功能，只要集群内有超过半数以上的节点保持可用，集群就能够继续提供服务，如果是 primary 成员发生了故障，集群会自动进行选主，并启用新的 primary 成员，而如果是 secondary 成员发生了故障，只需要踢出该节点即可，这些都是自动完成的。

另一个读者可能比较关心的问题是，一旦发生选主，那么该选择哪个节点作为新主呢？关于这个问题，其实是有一个相关的参数进行控制的，接下来就实战演练一下这个场景，测试环境如表 31-3 所示。

表 31-3　　　　　　　　　　测试环境信息

角　　色	本　地　地　址
primary	10.120.240.251: 3325
secondary	10.120.240.251: 3315
secondary	10.120.240.251: 3305

（1）生成测试数据。

（2）模拟压力测试，在运行 5min 后，杀掉主库 mysql 进程，模拟主库发生故障。

(3) 新集群的成员信息。

```
mysql> select * from performance_schema.replication_group_members\G;
*************************** 1. row ***************************
CHANNEL_NAME: group_replication_applier
  MEMBER_ID: 22559fb9-8e60-11e8-85d2-0024e869b4d5
MEMBER_HOST: 10.120.240.251
MEMBER_PORT: 3305
MEMBER_STATE: ONLINE
*************************** 2. row ***************************
CHANNEL_NAME: group_replication_applier
  MEMBER_ID: 6066c689-b4d4-11e8-886f-0024e869b4d5
MEMBER_HOST: 10.120.240.251
MEMBER_PORT: 3315
MEMBER_STATE: ONLINE
2 rows in set (0.00 sec)

mysql> select VARIABLE_VALUE from performance_schema.global_status WHERE VARIABLE_NAME= 'group_
replication_primary_member'\G;
*************************** 1. row ***************************
VARIABLE_VALUE: 6066c689-b4d4-11e8-886f-0024e869b4d5
1 row in set (0.01 sec)
```

从以上可以得知，新的 primary 成员已经变更为 10.120.240.251:3315 节点。

(4) 以下内容是切换日志。

```
2018-09-11T23:21:24.803040+08:00 0 [Warning] Plugin group_replication reported: 'Member with
address 10.120.240.251:3325 has become unreachable.'
2018-09-11T23:21:24.895508+08:00 0 [Note] Plugin group_replication reported: '[GCS] Removing
members that have failed while processing new view.'
2018-09-11T23:21:25.815816+08:00 0 [Warning] Plugin group_replication reported: 'Members
removed from the group: 10.120.240.251:3325'
2018-09-11T23:21:25.815864+08:00 0 [Note] Plugin group_replication reported: 'Primary server
with address 10.120.240.251:3325 left the group. Electing new Primary.'
2018-09-11T23:21:25.816000+08:00 0 [Note] Plugin group_replication reported: 'A new primary
with address 10.120.240.251:3315 was elected, enabling conflict detection until the new primary
applies all relay logs.'
2018-09-11T23:21:25.816082+08:00 81 [Note] Plugin group_replication reported: 'This server is
working as primary member.'
2018-09-11T23:21:25.816082+08:00 81 [Note] Plugin group_replication reported: 'This server is
working as primary member.'
2018-09-11T23:21:25.816140+08:00 0 [Note] Plugin group_replication reported: 'Group membership
changed to 10.120.240.251:3305, 10.120.240.251:3315 on view 15366391443682293:7.'
```

从上面的输出，可以看到整个切换过程，大致流程如下：

- 集群环境探测到原 primary 成员 unreachable。
- 集群环境变更，踢出失败节点。
- 开始进行新的投票选举。
- 选举出新的 primary 成员，等待该节点应用完所有的 relay log。
- 新的 primary 成员可用，产生新的集群 view 点。

由于这次的选主过程受到了人为干预，即哪个节点将会成为新的 primary 成员，提前已经做好了配置，这个参数就是 group_replication_member_weight，即组复制成员权重，默认值为 50，最大值为 100，如果在一个集群里，该参数设置了不同的值，那么具有最大权重的节点将会在切换时成为新的 primary 成员。如果有多个节点设置的值是相同的，而且在整个集群中权重较大，那么这多个节点就会再次根据节点的 UUID 值进行从小到大排序，排在前边的优先作为新的 primary 成员。

另外，经过此次切换演练，我们发现了另一个问题，那就是成员的角色虽然顺利进行了切换，但前端应用是感知不到的，也就是说前端应用如果没有外部干预的情况下，仍会不停地向故障节点发送读、写请求，此时就需要能有一个解决方案来避免这个问题的发生，能够做到如果成员间角色一旦发生改变，能及时地捕捉到这个信息。针对这个情况，可以选择使用 MySQL 数据库中间件，例如官方的 MySQL Router 或者网易公司自行研发的 Cetus，两者均能够及时地捕捉到组复制环境成员间角色的变化，并将前端应用的新连接请求路由到正确的 MySQL 节点。关于这部分可以参考第 32 章。

31.2.4 重要特性

组复制技术作为一个重量级"嘉宾"隆重登场，是与其带来的价值相符的，同时也附带了很多新的特性，下面将重点介绍两个方面：压缩和流控。

1. 压缩

组复制技术的特性之一就是压缩，事务在提交之前需要在集群所有节点都验证冲突，并分发 BINLOG event，如果事务比较大，压缩能节省网络传输的压力，这是通过参数 group_replication_compression_threshold 来设置的，默认 1 000 000 bytes，即如果事务产生的复制消息大小超过这个阈值，将会被压缩。如果是 OLTP（联机事务处理）环境，小事务会比较多，可以适当调小该参数阈值，而如果"大"事务（这里说的大事务不是指参数 group_replication_transaction_size_limit，这个参数讨论见下文）比较多，可以适当调大该参数阈值。

压缩动作发生在 Primary 成员端，因为只有在执行事务的节点，才需要将事务（也可以称之为 event）分发给集群其他成员，而作为接收事务的成员，会根据收到的数据包来判断是否需要解压，压缩使用了 LZ4 算法。压缩特性示意图如图 31-7 所示。

从图 31-7 可见，压缩和解压缩特性由 Group Communication System API 提供技术支撑。因集群中数据的交互会占用网络资源，所以控制好网络通信量可以极大地缓解"交通"压力，提高通信效率，降低网络时延。

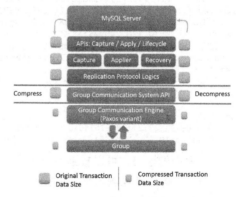

图 31-7　组复制压缩特性

2. 流控

流控这个概念估计读者并不陌生，可以简单理解为流量控制，类似水阀开关，太快了就调小一些，待管网恢复后，再恢复之前的水平。

组复制中的多个成员构成了一个集群，那么每个成员的资源不一定都是完全相同的，或者说即便拥有相同的资源，上层业务访问不一定是完全一样的，这就必然会导致有的成员任务重，有的任务轻，如果没有流控技术做保证，可能成员之间的差距会越拉越大，最终导致集群不可用，所以流控技术也是为了保证组复制的高可用性，起到一个很重要的作用。同时为了保证在流控阶段，不会导致流速大起大落，组复制引入了渐进策略，即逐渐减少队列中事务数量（涉及参数 group_replication_flow_control_hold_percent，该参数在 MySQL 8.0 版本中已经可供用户调整），待问题得以缓解后，再逐渐进行恢复。

流控技术也提供了几个相关参数，其中包括以下几项。

- group_replication_flow_control_mode：默认值为 QUOTA，另一个可选项是 DISABLED。

该参数控制流控模式是否被启用，即 Primary 成员节点在执行事务的过程中，是否需要考虑并回应发送给其余节点的 event 执行速度，如果选择 DISABLED，就表示 Primay 成员根据自身的能力执行事务即可，无须关心其他集群成员，反之，如果配置 QUOTA，Primary 成员就需要根据以下两个参数来适当调整自身的事务执行速度。

- group_replication_flow_control_applier_threshold：默认值为 25 000，最大值为 2 147 483 647，该参数表示在流控过程中，applier 队列中的等待事务数量。如果实际队列中的值超过该值，Primary 成员将在与之通信过程中，得知有节点压力比较大，需要减缓自身的事务执行速度；如果设置为 0，将忽略 applier 队列中等待的事务数量，也可以理解为关闭 applier 流控。

- group_replication_flow_control_certifier_threshold：默认值为 25 000，最大值 2 147 483 647，该参数的意思与上类似，只不过发生在事务认证阶段，如果设置为 0，将忽略认证队列中等待的事务数量。

从 MySQL 8.0.2 版本开始，又引入了大量的流控控制参数，quota 类的参数默认值都是 0，即不会对事务执行速度进行限制，percent 类的参数只是将 MySQL 5.7 中未开放出来的阈值通过参数形式，给了一个接口，用户可以进行干预，比如 group_replication_flow_control_hold_percent 默认值为 10，在 MySQL 5.7 中是在代码层面直接使用的，并未开放。在集群正常运行过程中，组复制的不同成员间会周期性地发送和接收很多统计数据，例如各个队列中的事务数量，以及上一次执行的情况，然后成员间会根据这些数据，进行一个流量调节。

总体来说，从 MySQL 8.0 以后，官方在持续改善各方面的性能，也开放了更多的接口，用户可以根据实际业务需求，进行调试，从而达到一个平衡。

新增参数列表如下（每一个参数的含义可以参考官方文档）：

- group_replication_flow_control_hold_percent；
- group_replication_flow_control_max_quota 和 group_replication_flow_control_min_quota；
- group_replication_flow_control_min_recovery_quota；
- group_replication_flow_control_member_quota_percent；
- group_replication_flow_control_period；
- group_replication_flow_control_release_percent。

31.2.5 常见问题

由于组复制"诞生"还不是很久，在生产环境投入使用之前，相信大家都会进行严格的验证和测试，以便对其尽可能多的熟悉，了解其"秉性"，接下来将从 3 个方面介绍需要重点关注的点，大事务、集群成员变化和网络质量。

1. 大事务

大事务顾名思义，就是指在一个事务中发生了过多的数据变更，例如对一个表的全表更新或者删除（这里指的是使用 Delete 命令，而不是 Truncate），如果该表包含很多数据，此时该事务可以称为一个大事务。

为什么这里专门把大事务拿出来说，是因为组复制对成员间的通信质量和时延有非常高的依赖性，这一点在下文有讲解，这里不再展开。如果一个事务在 5s 的时间窗口期内没有被成功复制到集群其他成员，组复制通信会失败，从而造成严重影响。如果此时业务量非常大，有可能集群被挂起，从而不可用。同其他阈值控制一样，依然有一个参数 group_replication_transaction_ size_limit 对事务大小进行控制,group_replication_transaction_size_limit 的默认值为 0,即事务大小不受限制,但从 MySQL 8.0.2 版本以后该参数的默认值变更为 150 000 000,单位为字节。

下面继续进行实战演练。

(1)在 primary 成员节点放开 max_binlog_cache_size 限制,构造试验环境,新建表 employees_bigtran,插入 90 万条记录。

```
mysql> set global max_binlog_cache_size=104857600000;
Query OK, 0 rows affected (0.00 sec)

mysql> show variables like 'max_binlog_cache_size';
+-----------------------+--------------+
| Variable_name         | Value        |
+-----------------------+--------------+
| max_binlog_cache_size | 104857600000 |
+-----------------------+--------------+
1 row in set (0.01 sec)

mysql> use employees;
Reading table information for completion of table and column names
You can turn off this feature to get a quicker startup with -A
Database changed

mysql> CREATE TABLE `employees_bigtran` (
    ->   `emp_no` int(11) NOT NULL,
    ->   `birth_date` date NOT NULL,
    ->   `first_name` varchar(14) COLLATE utf8_unicode_ci NOT NULL,
    ->   `last_name` varchar(16) COLLATE utf8_unicode_ci NOT NULL,
    ->   `gender` enum('M','F') COLLATE utf8_unicode_ci NOT NULL,
    ->   `hire_date` date NOT NULL,
    ->   PRIMARY KEY (`emp_no`)
    -> ) ENGINE=InnoDB DEFAULT CHARSET=utf8 COLLATE=utf8_unicode_ci ;
Query OK, 0 rows affected (0.28 sec)

mysql> insert into employees_bigtran select * from employees;
Query OK, 300024 rows affected (19.53 sec)
Records: 300024  Duplicates: 0  Warnings: 0

mysql> insert into employees_bigtran select emp_no+499999 as emp_no,birth_date,first_name,last_name,gender,hire_date from employees;
Query OK, 300024 rows affected (13.12 sec)
Records: 300024  Duplicates: 0  Warnings: 0

mysql> insert into employees_bigtran select emp_no+999998 as emp_no,birth_date,first_name,last_name,gender,hire_date from employees;
Query OK, 300024 rows affected (11.85 sec)
Records: 300024  Duplicates: 0  Warnings: 0

mysql>select count(*) from employees_bigtran;
+----------+
| count(*) |
+----------+
|   900072 |
+----------+
1 row in set (0.33 sec)
```

(2)在 primary 成员节点执行大事务。

```
mysql> update employees_bigtran set hire_date=now() ;
ERROR 3101 (HY000): Plugin instructed the server to rollback the current transaction.
```

该事务执行失败,被回滚,以下是该节点的日志信息。

```
2018-09-15T08:00:23.948346+08:00 0 [Warning] Plugin group_replication reported: 'Member with address 10.120.240.251:3305 has become unreachable.'
2018-09-15T08:00:23.948436+08:00 0 [Warning] Plugin group_replication reported: 'Member with address 10.120.240.251:3325 has become unreachable.'
2018-09-15T08:00:23.948449+08:00 0 [ERROR] Plugin group_replication reported: 'This server is not able to reach a majority of members in the group. This server will now block all updates.
```

31.2 MGR 架构

```
The server will remain blocked until contact with the majority is restored. It is possible to
use group_replication_force_members to force a new group membership.'
    2018-09-15T08:00:24.939308+08:00 0 [Warning] Plugin group_replication reported: 'Member with
address 10.120.240.251:3305 is reachable again.'
    ......
    2018-09-15T08:00:31.380988+08:00 0 [Warning] Plugin group_replication reported: 'Member with
address 10.120.240.251:3325 has become unreachable.'
    ......
    2018-09-15T08:00:39.380827+08:00 0 [ERROR] Plugin group_replication reported: 'Member was
expelled from the group due to network failures, changing member status to ERROR.'
    2018-09-15T08:00:39.381006+08:00 0 [Warning] Plugin group_replication reported: 'Due to a
plugin error, some transactions can't be certified and will now rollback.'
```

从日志内容可以得知，在 primary 节点执行大事务之后，该节点与其他节点失去了联系，在经历几次尝试之后，最终还是认为发生了网络故障，导致 primary 节点被踢出集群，执行的事务被回滚。

（3）查看 secondary 成员节点日志内容，以下内容是其中一个节点的日志信息。

```
    2018-09-15T08:00:23.300654+08:00 0 [Warning] Plugin group_replication reported: 'Member with
address 10.120.240.251:3315 has become unreachable.'
    2018-09-15T08:00:23.301392+08:00 0 [Warning] Plugin group_replication reported: 'Member with
address 10.120.240.251:3325 has become unreachable.'
    2018-09-15T08:00:23.301418+08:00 0 [ERROR] Plugin group_replication reported: 'This server is
not able to reach a majority of members in the group. This server will now block all updates.
The server will remain blocked until contact with the majority is restored. It is possible to
use group_replication_force_members to force a new group membership.'
    2018-09-15T08:00:25.335584+08:00 0 [Warning] Plugin group_replication reported: 'Member with
address 10.120.240.251:3315 is reachable again.'
    ......
    2018-09-15T08:00:32.051206+08:00 0 [Warning] Plugin group_replication reported: 'Member with
address 10.120.240.251:3315 has become unreachable.'
    ......
    2018-09-15T08:00:39.029867+08:00 0 [Note] Plugin group_replication reported: '[GCS] Removing
members that have failed while processing new view.'
    2018-09-15T08:00:39.031337+08:00 0 [Note] Plugin group_replication reported: '[GCS] Removing
members that have failed while processing new view.'
    2018-09-15T08:00:41.108324+08:00 0 [Warning] Plugin group_replication reported: 'Members
removed from the group: 10.120.240.251:3315'
    2018-09-15T08:00:41.108373+08:00 0 [Note] Plugin group_replication reported: 'Primary server
with address 10.120.240.251:3315 left the group. Electing new Primary.'
    2018-09-15T08:00:41.108654+08:00 0 [Note] Plugin group_replication reported: 'A new primary
with address 10.120.240.251:3305 was elected, enabling conflict detection until the new primary
applies all relay logs.'
    2018-09-15T08:00:41.134958+08:00 78 [Note] Plugin group_replication reported: 'This server is
working as primary member.'
    2018-09-15T08:00:41.135197+08:00 0 [Note] Plugin group_replication reported: 'Group membership
changed to 10.120.240.251:3305, 10.120.240.251:3325 on view 15366391443682293:9.'
    2018-09-15T08:00:49.442056+08:00 41 [Note] Multi-threaded slave statistics for channel 'group_
replication_applier': seconds elapsed = 32896; events assigned = 3073; worker queues filled over
overrun level = 0; waited due a Worker queue full = 0; waited due the total size = 0; waited at
clock conflicts = 0 waited (count) when Workers occupied = 0 waited when Workers occupied = 0
    2018-09-15T08:00:49.559564+08:00 41 [Note] Multi-threaded slave: Coordinator has waited 1
times hitting slave_pending_jobs_size_max; current event size = 8196.
    ......
    2018-09-15T08:01:05.249165+08:00 41 [Note] Multi-threaded slave: Coordinator has waited 3991
times hitting slave_pending_jobs_size_max; current event size = 8180.
    2018-09-15T08:01:07.603277+08:00 42 [ERROR] Slave SQL for channel 'group_replication_applier
': Worker 1 failed executing transaction 'aaaaaaaa-aaaa-aaaa-aaaa-aaaaaaaaaaaa:40' at master log ,
 end_log_pos 37766736; Could not execute Update_rows event on table employees.employees_bigtran;
 Multi-statement transaction required more than 'max_binlog_cache_size' bytes of storage;
increase this mysqld variable and try again, Error_code: 1197; handler error HA_ERR_RBR_LOGGING_
FAILED; the event's master log FIRST, end_log_pos 37766736, Error_code: 1197
    2018-09-15T08:01:07.603442+08:00 41 [Warning] Slave SQL for channel 'group_replication_applier
': ... The slave coordinator and worker threads are stopped, possibly leaving data in inconsistent
state. A restart should restore consistency automatically, although using non-transactional
storage for data or info tables or DDL queries could lead to problems. In such cases you have
to examine your data (see documentation for details). Error_code: 1756
    2018-09-15T08:01:07.603442+08:00 41 [Warning] Slave SQL for channel 'group_replication_applier
': ... The slave coordinator and worker threads are stopped, possibly leaving data in inconsistent
```

```
state. A restart should restore consistency automatically, although using non-transactional
storage for data or info tables or DDL queries could lead to problems. In such cases you have
to examine your data (see documentation for details). Error_code: 1756
  2018-09-15T08:01:07.603480+08:00 41 [Note] Error reading relay log event for channel 'group_
replication_applier': slave SQL thread was killed
  2018-09-15T08:01:07.603519+08:00 41 [ERROR] Plugin group_replication reported: 'The applier thread
execution was aborted. Unable to process more transactions, this member will now leave the group.'
  2018-09-15T08:01:07.603645+08:00 38 [ERROR] Plugin group_replication reported: 'Fatal error
during execution on the Applier process of Group Replication. The server will now leave the group.'
  2018-09-15T08:01:07.616556+08:00 38 [ERROR] Plugin group_replication reported: 'The server
was automatically set into read only mode after an error was detected.'
  2018-09-15T08:01:11.420849+08:00 0 [Note] Plugin group_replication reported: 'Group membership
changed: This member has left the group.'
  2018-09-15T08:01:13.401119+08:00 38 [Note] Plugin group_replication reported: 'The group
replication applier thread was killed'
```

日志信息已经反映出这期间发生的问题以及处理过程，首先该 secondary 节点发现与 primary 节点失去了联系，经过几次尝试之后，踢出了"失败"的 primary 节点，并选择出了新的 primary 节点，与另一个 secondary 节点构造了一个新的集群环境，变更了 view 点，表示新集群由此开始。

接下来新 primary 节点开始处理已经接收到的事务（就是原 primary 节点执行的 Update 大事务，GTID 序列 aaaaaaaa-aaaa-aaaa-aaaa-aaaaaaaaaaaa:40，虽然原 primary 执行的时候最终失败回滚了，但事务在写入 BINLOG 之前，还是会先通过 applier 组件将 BINLOG event 发往其他集群成员），由于此节点的 max_binlog_cache_size 设置大小为 16MB（当然这个不是试验重点，因为即便可以设置更大的阈值，还可能会遇到比阈值更大的事务），所以日志显示 Slave SQL for channel 'group_replication_applier' 执行失败，并给出了认为发生该问题可能的原因，最终 applier thread 被终止工作，该节点离开集群。

另一个 secondary 节点，经历过程与以上节点相似，同样因 applier 组件执行大事务，而最终停止工作，离开集群。

（4）通过视图查看成员的组复制情况。

```
mysql> select * from performance_schema.replication_group_members\G;
*************************** 1. row ***************************
CHANNEL_NAME: group_replication_applier
   MEMBER_ID: 22559fb9-8e60-11e8-85d2-0024e869b4d5
 MEMBER_HOST: 10.120.240.251
 MEMBER_PORT: 3305
MEMBER_STATE: ERROR
1 row in set (0.00 sec)

mysql> select * from performance_schema.replication_group_members\G;
*************************** 1. row ***************************
CHANNEL_NAME: group_replication_applier
   MEMBER_ID: 6066c689-b4d4-11e8-886f-0024e869b4d5
 MEMBER_HOST: 10.120.240.251
 MEMBER_PORT: 3315
MEMBER_STATE: ERROR
1 row in set (0.00 sec)

mysql> select * from performance_schema.replication_group_members\G;
*************************** 1. row ***************************
CHANNEL_NAME: group_replication_applier
   MEMBER_ID: ba4fcb42-b4d5-11e8-9550-0024e869b4d5
 MEMBER_HOST: 10.120.240.251
 MEMBER_PORT: 3325
MEMBER_STATE: ERROR
1 row in set (0.00 sec)
```

至此所有节点已全部离开集群。

以上这个场景只是想说明线上环境一定要做好事务的拆分，并合理配置参数 group_

replication_transaction_size_limit 和 max_binlog_cache_size，这两个参数相互配合使用，可以对大事务起到限制的作用，尽可能使得集群环境不被破坏。

2. 集群成员 crash

组复制集群中的半数以内成员如果出现故障，会发生以下情况：

- 如果 primary 成员不可用，会进行切换选新主，该节点的读写业务会受影响（单主环境）；
- 如果 secondary 成员不可用，不影响前端的写业务，如果该节点同时负责读业务，那这部分业务会受到影响（单主环境）；
- 如果 primary 成员不可用，会影响该节点的读写业务（多主环境）。

以上情况的发生比较容易理解，因组复制的技术特点，要求半数以上节点同时可用，该集群才能提供服务，但如果集群中仅有半数以内节点存活，又会怎么样呢？下面进行实战演练。

首先是单主环境下（一个 primary 节点和两个 secondary 节点），两个 secondary 成员接连故障。

（1） secondary 和 secondary。

❶ 为了更接近真实环境，操作之前先在 primary 成员节点写入一段时间的数据，并持续写入，此时杀掉两个 secondary 节点的 MySQL 进程（两个 Secondary 节点同时故障与接连故障，不影响试验最终结果）。

```
[mysql3305@hz_10_120_240_251 ~]$ ps -ef|grep 33[0-1]5
   539    9179     1  0 Sep11 ?        00:00:00 /bin/sh ./bin/mysqld_safe --defaults-file=
/home/mysql3305/mysqlhome/my.cnf --user=mysql3305
   540   10018     1  0 Sep11 ?        00:00:00 /bin/sh ./bin/mysqld_safe --defaults-file=
/home/mysql3315/mysqlhome/my.cnf --user=mysql3315
   539   10734  9179  0 Sep11 ?        00:37:14 /home/mysql3305/mysqlhome/bin/mysqld --defaults-
file=/home/mysql3305/mysqlhome/my.cnf --basedir=/home/mysql3305/mysqlhome --datadir=/data2/
mysql3305/data --plugin-dir=/home/mysql3305/mysqlhome/lib/plugin --log-error=/data2/mysql3305/
data/error3305.log --pid-file=hz_10_120_240_251.pid --socket=/tmp/mysql_3305.sock --port=3305
   540   11692 10018  0 Sep11 ?        00:37:54 /home/mysql3315/mysqlhome/bin/mysqld --defaults-
file=/home/mysql3315/mysqlhome/my.cnf --basedir=/home/mysql3315/mysqlhome --datadir=/data2/
mysql3315/data --plugin-dir=/home/mysql3315/mysqlhome/lib/plugin --log-error=/data2/mysql3315/
data/error3315.log --pid-file=hz_10_120_240_251.pid --socket=/tmp/mysql_3315.sock --port=3315

[root@hz_10_120_240_251 ~]# kill -9 9179 10018 10734 11692
```

❷ 此时查看 primary 节点的集群成员信息以及日志情况。

```
mysql> select * from performance_schema.replication_group_members\G;
*************************** 1. row ***************************
  CHANNEL_NAME: group_replication_applier
     MEMBER_ID: 22559fb9-8e60-11e8-85d2-0024e869b4d5
   MEMBER_HOST: 10.120.240.251
   MEMBER_PORT: 3305
  MEMBER_STATE: UNREACHABLE
*************************** 2. row ***************************
  CHANNEL_NAME: group_replication_applier
     MEMBER_ID: 6066c689-b4d4-11e8-886f-0024e869b4d5
   MEMBER_HOST: 10.120.240.251
   MEMBER_PORT: 3315
  MEMBER_STATE: UNREACHABLE
*************************** 3. row ***************************
  CHANNEL_NAME: group_replication_applier
     MEMBER_ID: ba4fcb42-b4d5-11e8-9550-0024e869b4d5
   MEMBER_HOST: 10.120.240.251
   MEMBER_PORT: 3325
  MEMBER_STATE: ONLINE
3 rows in set (0.00 sec)

2018-09-17T11:00:01.093598+08:00 0 [Warning] Plugin group_replication reported: 'Member with
```

```
address 10.120.240.251:3305 has become unreachable.'
  2018-09-17T11:00:01.093662+08:00 0 [Warning] Plugin group_replication reported: 'Member with
address 10.120.240.251:3315 has become unreachable.'
  2018-09-17T11:00:01.093678+08:00 0 [ERROR] Plugin group_replication reported: 'This server
is not able to reach a majority of members in the group. This server will now block all updates.
The server will remain blocked until contact with the majority is restored. It is possible to
use group_replication_force_members to force a new group membership.'
```

在 primary 节点查看集群成员状态，两个 secondary 节点的状态已经变成 UNREACHABLE，primary 节点的写入操作已经被阻塞或者超时离开集群（这里涉及参数 group_replication_unreachable_majority_timeout，默认值为 0，表示无限等待）。

❸ 接下来启动这两个 secondary 节点，重新加入集群环境。

```
  2018-09-17T11:06:13.254236+08:00 0 [ERROR] Plugin group_replication reported: '[GCS] The
member was unable to join the group. Local port: 24902'
  2018-09-17T11:07:13.205234+08:00 3 [ERROR] Plugin group_replication reported: 'Timeout on
wait for view after joining group'
  2018-09-17T11:07:13.205322+08:00 3 [Note] Plugin group_replication reported: 'Requesting to
leave the group despite of not being a member'
  2018-09-17T11:07:13.205359+08:00 3 [ERROR] Plugin group_replication reported: '[GCS] The
member is leaving a group without being on one.'
```

从日志内容可以发现，重新加入集群（当前仅有 primary 节点）是失败的，即不能正常加入集群，需要重新对该集群进行初始化。

下一步是在多主环境中（3 个 primary 节点），其中两个 primary 成员接连故障。

（2）primary 和 primary。

❶ 在每个节点全部开启数据加载，并在写入一段时间后，逐一杀掉其中两个 primary 成员节点。

❷ 此时查看存活的 primary 节点的集群成员信息以及日志情况。

```
mysql> select * from performance_schema.replication_group_members\G;
*************************** 1. row ***************************
  CHANNEL_NAME: group_replication_applier
     MEMBER_ID: 6066c689-b4d4-11e8-886f-0024e869b4d5
   MEMBER_HOST: 10.120.240.251
   MEMBER_PORT: 3315
  MEMBER_STATE: UNREACHABLE
*************************** 2. row ***************************
  CHANNEL_NAME: group_replication_applier
     MEMBER_ID: ba4fcb42-b4d5-11e8-9550-0024e869b4d5
   MEMBER_HOST: 10.120.240.251
   MEMBER_PORT: 3325
  MEMBER_STATE: ONLINE
2 rows in set (0.00 sec)


  2018-09-18T18:03:44.116848+08:00 0 [Warning] Plugin group_replication reported: 'Member with
address 10.120.240.251:3305 has become unreachable.'
  2018-09-18T18:03:44.117390+08:00 0 [Note] Plugin group_replication reported: '[GCS] Removing
members that have failed while processing new view.'
  2018-09-18T18:03:45.117824+08:00 0 [Warning] Plugin group_replication reported: 'Members
removed from the group: 10.120.240.251:3305'
  2018-09-18T18:03:45.117971+08:00 0 [Note] Plugin group_replication reported: 'Group membership
changed to 10.120.240.251:3315, 10.120.240.251:3325 on view 15372647062341114:6.'
  2018-09-18T18:04:15.103536+08:00 0 [Warning] Plugin group_replication reported: 'Member with
address 10.120.240.251:3315 has become unreachable.'
  2018-09-18T18:04:15.103622+08:00 0 [ERROR] Plugin group_replication reported: 'This server is
not able to reach a majority of members in the group. This server will now block all updates.
The server will remain blocked for the next 60 seconds. Unless contact with the majority is restored,
 after this time the member will error out and leave the group. It is possible to use group_
replication_force_members to force a new group membership.'
  2018-09-18T18:05:15.107156+08:00 0 [ERROR] Plugin group_replication reported: 'This member
could not reach a majority of the members for more than 60 seconds. The member will now leave
the group as instructed by the group_replication_unreachable_majority_timeout option.'
```

31.2 MGR 架构

```
2018-09-18T18:05:15.107315+08:00 0 [ERROR] Plugin group_replication reported: 'The server was
automatically set into read only mode after an error was detected.'
2018-09-18T18:05:15.107342+08:00 0 [Warning] Plugin group_replication reported: 'Due to a
plugin error, some transactions can't be certified and will now rollback.'
2018-09-18T18:05:15.107342+08:00 0 [Warning] Plugin group_replication reported: 'Due to a
plugin error, some transactions can't be certified and will now rollback.'
2018-09-18T18:05:45.107810+08:00 0 [ERROR] Plugin group_replication reported: '[GCS] Timeout
while waiting for the group communication engine to exit!'
2018-09-18T18:05:45.107923+08:00 0 [ERROR] Plugin group_replication reported: '[GCS] The member
has failed to gracefully leave the group.'
2018-09-18T18:05:45.108444+08:00 0 [Warning] Plugin group_replication reported: 'read failed'
2018-09-18T18:05:45.155813+08:00 0 [Note] Plugin group_replication reported: 'Group membership
changed: This member has left the group.'
```

在仅剩下的一个 primary 节点查看集群成员状态，只有两个节点信息，其中一个是故障节点，此时状态是 UNREACHABLE，另一个是存活的 primary 节点，此时状态是 ONLINE，由于是逐一杀掉的，所以在第一个节点失败的时候，集群是可用的，于是从集群环境踢出了失败的节点，所以视图中已经看不到该节点的状态信息，在杀掉第二个节点后，仅剩下的存活的 primary 节点执行的事务被阻塞，由于设置了超时参数 group_replication_unreachable_majority_timeout 为 60，该 primary 节点将在 60s 后超时离开集群，并回滚未提交的事务，此时集群环境已经被破坏。

接下来就需要考虑重构多主环境，这里有一个注意事项，由于需要重新启动并修复失败实例，如果此时有前端业务一直在反复尝试连接该节点，并准备写入数据，那么首先需要停掉前端业务对该节点的访问请求，否则有可能新数据被写入该节点，从而导致数据在不同节点间发生不一致，最终导致集群重构失败，且数据被污染。

3. 网络质量

网络质量对组复制影响很大，从前文的一些实战演练，相信读者已经对这方面有了一些认识和了解。组复制技术需要集群中大多数节点投票一致后才认为该操作是有效的，可以继续，如果达不到这个要求，就会阻塞操作甚至集群不可用。从这一点来看，脑裂理论上不会发生。当然如果为了某种原因，有意而为之（通过参数可以进行强制激活少数节点组成新集群），这个就是另一个话题了。

显然为了实现高可用、不脑裂，还是需要付出一定代价的，比如网络抖动、间断对复制效率影响有多大呢？这有一组数据（该数据仅用于验证网络质量带来的影响，不代表实际性能），在三节点的单主集群环境下，如果其中一个 secondary 节点网络发生了隔离，即形成了{S1,S2}{S3}这种情况，那么性能会有多大的损失呢？下面来看一下如图 31-8 所示的数据。

从图 31-8 中可以发现，其中有几组数据 tps 和 reads、writes 都比较高，而有两个时间段的数据非常低，tps 降到了 50+，响应时间增加到了 1300+ms，这正是因为期间发生了两次 secondary 节点的网络隔离，虽然集群可以正常提供读写服务，但效率已经降得很低了，由此可见，想要在组复制中持续提供高效服务，高质量的网络必不可少。

网络质量对组复制的影响不容小觑，官方也已经意识到这个问题的严重性，并正在逐渐引入新特性以降低网络间断带来的负面影响，例如在 MySQL 8.0.13 版本中，官方又引入了新参数 group_replication_member_expel_timeout。该参数的出现将避免组复制集群成员间一旦出现短时网络故障，将导致集群成员被迅速剔除出局，大大增加了集群的可靠性，更便于系统管理员进行在线维护操作，例如给集群成员做 snapshot，或者在虚拟机间进行迁移，相信组复制技术会越来越成熟。

图 31-8　网络抖动压测效果

31.3　InnoDB Cluster

随着 MGR 技术的推出，又出现了一个新的 Cluster 架构，即 InnoDB Cluster。相信有很多读者对 MySQL Cluster 比较熟悉。这两个 Cluster 有什么区别呢？其实区别很简单，InnoDB Cluster 使用的存储引擎是 InnoDB，而后者使用的是 NDB，InnoDB Cluster 是对 MGR 的封装，本质上还是 MGR。

31.2 节介绍的是 MGR，而 InnoDB Cluster 正是采用了组复制技术，并结合一些工具，从而构成了一个 InnoDB Cluster 架构，如图 31-9 所示。

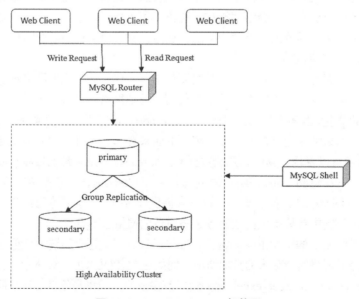

图 31-9　InnoDB Cluster 架构图

图 31-9 中，虚线框起来的部分是一个三节点单主集群（也可以是多主集群），从而构造了一个 High Availability Cluster，虚线外围涉及两个工具，其中一个是 MySQL Shell，另一个

是 MySQL Router，前者用于创建和集群维护，后者用来接收前端的访问请求并路由到 MGR 中的写节点或者读节点。

注意：关于 MySQL Shell 如何安装和使用在前文中已经进行过讲解，此处仅讲述与 InnoDB Cluster 有关的内容，而 MySQL Router 具体的安装配置也会在后续内容中找到。

读者在阅读完 MGR 内容之后，可能会发现一个问题：在组复制集群搭建完成之后，前端应该访问哪个节点？通常需要先查询数据字典，确认是主节点后才能进行写操作，而读操作可以选择任何一个节点；如果要进行读写分离，还需要知道哪些节点是只读节点；如果发生了角色切换，需要再次查询数据字典，以便找到新的正确节点。每一次连接变更都需要修改前端的配置，对线上业务来说，这种纯手工操作非常不友好。而 InnoDB Cluster 的推出真正的解决了这些烦恼，连接配置非常简单，MGR 集群中主从信息的变更对前端将变得透明。下面通过实战演练来教大家一步步搭建 InnoDB Cluster 集群。

31.3.1 安装部署

InnoDB Cluster 需要借助 MySQL Shell 进行管理，所以首先要安装 MySQL Shell 工具，然后才能进行集群的创建和维护工作。在 MySQL Shell（版本 8.0.12，后续内容也将采用该版本）安装完成以后，可以通过帮助功能了解到与 InnoDB Cluster 有关的模块和对象，如下所示。

```
MySQL    JS > \? dba
NAME
      dba - Global variable for InnoDB cluster management.

DESCRIPTION
      The global variable dba is used to access the AdminAPI functionality and
      perform DBA operations. It is used for managing MySQL InnoDB clusters.

PROPERTIES
      verbose
            Enables verbose mode on the dba operations.

FUNCTIONS
      checkInstanceConfiguration(instance[, options])
            Validates an instance for MySQL InnoDB Cluster usage.

      configureInstance([instance][, options])
            Validates and configures an instance for MySQL InnoDB Cluster
            usage.

      configureLocalInstance(instance[, options])
            Validates and configures a local instance for MySQL InnoDB Cluster
            usage.

      createCluster(name[, options])
            Creates a MySQL InnoDB cluster.

      deleteSandboxInstance(port[, options])
            Deletes an existing MySQL Server instance on localhost.

      deploySandboxInstance(port[, options])
            Creates a new MySQL Server instance on localhost.

      dropMetadataSchema(options)
            Drops the Metadata Schema.

      getCluster([name][, options])
            Retrieves a cluster from the Metadata Store.

      help([member])
            Provides help about this object and it's members
```

```
killSandboxInstance(port[, options])
      Kills a running MySQL Server instance on localhost.

rebootClusterFromCompleteOutage([clusterName][, options])
      Brings a cluster back ONLINE when all members are OFFLINE.

startSandboxInstance(port[, options])
      Starts an existing MySQL Server instance on localhost.

stopSandboxInstance(port[, options])
      Stops a running MySQL Server instance on localhost.

MySQL  JS > \? cluster
NAME
      Cluster - Represents an InnoDB cluster.

DESCRIPTION
      The cluster object is the entry point to manage and monitor a MySQL
      InnoDB cluster.

      A cluster is a set of MySQLd Instances which holds the user's data.

      It provides high-availability and scalability for the user's data.

PROPERTIES
      name
            Retrieves the name of the cluster.

FUNCTIONS
      addInstance(instance[, options])
            Adds an Instance to the cluster.

      checkInstanceState(instance[, password])
            Verifies the instance gtid state in relation with the cluster.

      describe()
            Describe the structure of the cluster.

      disconnect()
            Disconnects all internal sessions used by the cluster object.

      dissolve([options])
            Dissolves the cluster.

      forceQuorumUsingPartitionOf(instance[, password])
            Restores the cluster from quorum loss.

      getName()
            Retrieves the name of the cluster.

      help([member])
            Provides help about this class and it's members

      rejoinInstance(instance[, options])
            Rejoins an Instance to the cluster.

      removeInstance(instance[, options])
            Removes an Instance from the cluster.

      rescan()
            Rescans the cluster.

      status()
            Describe the status of the cluster.
```

MySQL Shell 的 dba 和 cluster 模块一共提供了以上功能，可以用于创建实例、配置实例和创建集群等日常操作，接下来将会围绕这些功能展开介绍。

关于集群的创建分两种情况进行演示，第一种情况是全部使用 MySQL Shell 进行实例的创建、配置以及集群的搭建；另一种情况是在已配置好的 MGR 基础上进行集群的初始化，

31.3 InnoDB Cluster

这两个场景也是我们经常遇到的。

1. 创建新集群

下面我们构建一个 sandbox 模式的集群，即只在同一个服务器上操作，这种环境仅用来演示和测试使用，如果要使用 MySQL Shell 创建一个线上使用的生产环境，应该采用 production 模式（该部分内容请参考官方文档）。

注意：在本节中 MySQL 使用 8.0.11 版本，该版本的 MySQL 已经支持参数的持久化保存。

（1）创建 3 个实例，端口分别为 3505、3515 和 3525，用于集群环境。

```
MySQL  JS > dba.deploySandboxInstance(3505)
A new MySQL sandbox instance will be created on this host in
/home/mysqlshell/mysql-sandboxes/3505

Warning: Sandbox instances are only suitable for deploying and
running on your local machine for testing purposes and are not
accessible from external networks.

Please enter a MySQL root password for the new instance: ****************
Deploying new MySQL instance...

Instance localhost:3505 successfully deployed and started.
Use shell.connect('root@localhost:3505'); to connect to the instance.
```

同理，创建出另外两个实例 3515 和 3525，创建完成后，可以查看一下现在的目录结构，如下所示。

```
[mysqlshell@hz_10_120_240_251 ~]$ ll mysql-sandboxes
total 24
drwxrwxr-x 4 mysqlshell mysqlshell 4096 Oct 17 17:28 3505
drwxrwxr-x 4 mysqlshell mysqlshell 4096 Oct 17 17:45 3515
drwxrwxr-x 4 mysqlshell mysqlshell 4096 Oct 17 17:47 3525

[mysqlshell@hz_10_120_240_251 ~]$ ll mysql-sandboxes/3505
total 676548
-rw-r----- 1 mysqlshell mysqlshell         6 Oct 17 17:28 3505.pid
-rw------- 1 mysqlshell mysqlshell       737 Oct 17 17:28 my.cnf
-rwxr-xr-x 1 mysqlshell mysqlshell 692755976 Oct 17 17:28 mysqld
drwxrwxr-x 2 mysqlshell mysqlshell      4096 Oct 17 17:28 mysql-files
drwxr-x--- 5 mysqlshell mysqlshell      4096 Oct 17 17:28 sandboxdata
-rwx------ 1 mysqlshell mysqlshell       150 Oct 17 17:28 start.sh
-rwx------ 1 mysqlshell mysqlshell       191 Oct 17 17:28 stop.sh
[mysqlshell@hz_10_120_240_251 ~]$
```

（2）初始化集群 sandbox_cluster。

```
MySQL  JS > shell.connect('root@localhost:3505')
Creating a session to 'root@localhost:3505'
Please provide the password for 'root@localhost:3505': ****************
Save password for 'root@localhost:3505'? [Y]es/[N]o/Ne[v]er (default No): yes
Fetching schema names for autocompletion... Press ^C to stop.
Your MySQL connection id is 12
Server version: 8.0.11 MySQL Community Server - GPL
No default schema selected; type \use <schema> to set one.
<ClassicSession:root@localhost:3505>

 MySQL  localhost:3505 ssl  JS > dba.createCluster('sandbox_cluster')
A new InnoDB cluster will be created on instance 'root@localhost:3505'.

Validating instance at localhost:3505...
Instance detected as a sandbox.
Please note that sandbox instances are only suitable for deploying test clusters for use within the same host.

This instance reports its own address as hz_10_120_240_251

Instance configuration is suitable.
Creating InnoDB cluster 'sandbox_cluster' on 'root@localhost:3505'...
Adding Seed Instance...
```

```
Cluster successfully created. Use Cluster.addInstance() to add MySQL instances.
At least 3 instances are needed for the cluster to be able to withstand up to
one server failure.

<Cluster:sandbox_cluster>
```

连接其中一个实例，然后初始化集群，集群名称为 sandbox_cluster。此过程类似于执行 MGR 中的 set group_replication_bootstrap_group=on，并 start group_replication，只是多了一些集群元数据的增加以及专属用户（该部分内容详见下文）。

（3）添加集群成员 3515 和 3525。

```
MySQL  localhost:3505 ssl  JS > var cluster=dba.getCluster()

MySQL  localhost:3505 ssl  JS > cluster.addInstance('root@localhost:3515')
A new instance will be added to the InnoDB cluster. Depending on the amount of
data on the cluster this might take from a few seconds to several hours.

Adding instance to the cluster ...

Please provide the password for 'root@localhost:3515': ****************
Save password for 'root@localhost:3515'? [Y]es/[N]o/Ne[v]er (default No): yes
Validating instance at localhost:3515...
Instance detected as a sandbox.
Please note that sandbox instances are only suitable for deploying test clusters for use within
the same host.

This instance reports its own address as hz_10_120_240_251

Instance configuration is suitable.
The instance 'root@localhost:3515' was successfully added to the cluster.
```

以上增加了端口为 3515 的实例，同理，可以增加端口为 3535 的实例。

（4）查看集群状态。

```
MySQL  localhost:3505 ssl  JS > cluster.status()
{
    "clusterName": "sandbox_cluster",
    "defaultReplicaSet": {
        "name": "default",
        "primary": "localhost:3505",
        "ssl": "REQUIRED",
        "status": "OK",
        "statusText": "Cluster is ONLINE and can tolerate up to ONE failure.",
        "topology": {
            "localhost:3505": {
                "address": "localhost:3505",
                "mode": "R/W",
                "readReplicas": {},
                "role": "HA",
                "status": "ONLINE"
            },
            "localhost:3515": {
                "address": "localhost:3515",
                "mode": "R/O",
                "readReplicas": {},
                "role": "HA",
                "status": "ONLINE"
            },
            "localhost:3525": {
                "address": "localhost:3525",
                "mode": "R/O",
                "readReplicas": {},
                "role": "HA",
                "status": "ONLINE"
            }
        }
    },
    "groupInformationSourceMember": "mysql://root@localhost:3505"
}
```

集群搭建完成之后，可以通过 cluster.status() 查看集群各节点信息，例如集群名称、复制集、主节点地址、只读节点地址等。

至此，集群搭建完成。

注意：连接验证部分需要借助 MySQL Router 中间件才能完成，该部分操作详见本章后续内容。

2. 利用 MGR 集群创建新集群

上一小节借助 MySQL Shell 完整的搭建了一个 InnoDB Cluster，下面将演示如何在一个现有的 MGR 集群基础上，创建一个 InnoDB Cluster。

前提：存在一个可用的 MGR 集群环境，端口分别为 3305、3315 和 3325。

利用 MySQL Shell 连接集群中的任一节点，创建集群。

```
MySQL  10.120.240.251:3305  JS > var cluster = dba.createCluster('mysql7Cluster', {adoptFromGR: true});
A new InnoDB cluster will be created based on the existing replication group on instance
'root@10.120.240.251:3305'.

Creating InnoDB cluster 'mysql7Cluster' on 'root@10.120.240.251:3305'...
Adding Seed Instance...
Adding Instance '10.120.240.251:3315'...
Adding Instance '10.120.240.251:3325'...

Cluster successfully created based on existing replication group.
```

以上连接的是 primary 节点，并且指定了 adoptFromGR: true，表明新创建的 InnoDB Cluster 是在 MGR 的基础上产生的。

如果连接的是 secondary 节点，并且没有指定 adoptFromGR: true，会怎么样呢？示例如下。

```
MySQL  10.120.240.251:3325  JS > var cluster=dba.createCluster('m7cluster')
You are connected to an instance that belongs to an unmanaged replication group.
Do you want to setup an InnoDB cluster based on this replication group? [Y/n]: y
A new InnoDB cluster will be created based on the existing replication group on instance
'root@10.120.240.251:3325'.

The MySQL instance at '10.120.240.251:3325' currently has the super_read_only
system variable set to protect it from inadvertent updates from applications.
You must first unset it to be able to perform any changes to this instance.
For more information see: https://dev.mysql.com/doc/refman/en/server-system-variables.
html#sysvar_super_read_only.

Note: there are open sessions to '10.120.240.251:3325'.
You may want to kill these sessions to prevent them from performing unexpected updates:

1 open session(s) of 'root@10.120.240.251'.

1 open session(s) of 'root@localhost'.

Do you want to disable super_read_only and continue? [y/N]: y

Creating InnoDB cluster 'm7cluster' on 'root@10.120.240.251:3325'...
Adding Seed Instance...
Adding Instance '10.120.240.251:3315'...
Adding Instance '10.120.240.251:3325'...

Cluster successfully created based on existing replication group.
```

当连接的是 secondary 节点时，在创建集群过程，首先会遇到第一个交互，即是否根据现在的 MGR 集群创建 InnoDB Cluster，这次交互相当于设置了 adoptFromGR: true 或者 false。再往下可以看到第二次交互，会询问是否关闭 super_read_only 属性，因为在创建 InnoDB Cluster 时，需要创建相关的元数据（具体内容详见下文），而 MGR 配置成功以后，secondary 节点的 read_only 和 super_read_only 都是开启的，此次只有同意关闭 super_read_only 属性，即设置为 off，才能创建成功，但需要注意的是，该节点的 super_read_only 不会还原回以前的 on 状态，如果不手动设置 off 的话，一旦出现误操作，会破坏集群一致性。

至此,该集群搭建完成。

31.3.2 初始化 MySQL Router

MySQL Router 属于 InnoDB Cluster 的一部分,是一个轻量级的中间件,负责对前端的访问进行路由,可以与前端应用放在一起,也可以分开部署。MySQL Router 从之前的 2.1 版本,一跃到了 8.0 版本,从而兼容 MySQL 5.7 和 MySQL 8.0。具体安装配置请参考第 32 章,下面直接用安装好的 MySQL Router(版本 8.0.12)来演示如何正确连接集群(端口分别为 3505、3515 和 3525):

(1)连接任一个集群成员,进行初始化配置。

```
[mysqlrouter@hz_10_120_240_251 ~]$ mysqlrouter --bootstrap localhost:3515 --directory /home/mysqlrouter/sandbox_cluster
Please enter MySQL password for root:

Bootstrapping MySQL Router instance at '/home/mysqlrouter/sandbox_cluster'...
Executing statements failed with: 'Error executing MySQL query: The MySQL server is running with the --super-read-only option so it cannot execute this statement (1290)' (1290), trying to connect to another node
Fetching Group Replication Members
disconnecting from mysql-server
trying to connecting to mysql-server at hz_10_120_240_251:3505
Checking for old Router accounts
Creating account mysql_router2_0l6bc1915h2r@'%'
MySQL Router  has now been configured for the InnoDB cluster 'sandbox_cluster'.

The following connection information can be used to connect to the cluster.

Classic MySQL protocol connections to cluster 'sandbox_cluster':
- Read/Write Connections: localhost:6446
- Read/Only Connections: localhost:6447
X protocol connections to cluster 'sandbox_cluster':
- Read/Write Connections: localhost:64460
- Read/Only Connections: localhost:64470
```

由于初始化时选了集群的一个 secondary 节点,所以在执行过程中,遇到只读实例不能创建初始化信息,于是 MySQL Router 自动切换到了 primary 节点,最后成功进行了初始化。

在配置过程中,如果端口没有手动绑定(可以参考选项--conf-base-port),Classic MySQL 协议 Read/Write 默认端口是 6446,Read/Only 默认端口是 6447,X 协议 Read/Write 默认端口是 64460,Read/Only 默认端口是 64470。

(2)启动 MySQL Router 进程。

```
[mysqlrouter@hz_10_120_240_251 ~]$ ll sandbox_cluster
total 32
drwx------ 2 mysqlrouter mysqlrouter 4096 Oct 18 16:51 data
drwx------ 2 mysqlrouter mysqlrouter 4096 Oct 18 16:51 log
-rw------- 1 mysqlrouter mysqlrouter 1374 Oct 18 16:51 mysqlrouter.conf
-rw------- 1 mysqlrouter mysqlrouter  104 Oct 18 16:51 mysqlrouter.key
-rw-rw-r-- 1 mysqlrouter mysqlrouter    6 Oct 18 17:33 mysqlrouter.pid
drwx------ 2 mysqlrouter mysqlrouter 4096 Oct 18 16:51 run
-rwx------ 1 mysqlrouter mysqlrouter  211 Oct 18 16:51 start.sh
-rwx------ 1 mysqlrouter mysqlrouter  209 Oct 18 16:51 stop.sh

[mysqlrouter@hz_10_120_240_251 ~]$ sh sandbox_cluster/start.sh
```

在上一步初始化完成后,找到初始化时 directory 指定的目录,可以看到已经产生了很多文件,其中包括配置文件 mysqlrouter.conf,启动和停止脚本 start.sh/stop.sh。

(3)验证 InnoDB Cluster 的正确性。

```
[mysqlrouter@hz_10_120_240_251 ~]$ mysql -uroot -h10.120.240.251 -P6446 -p
Enter password:
Welcome to the MySQL monitor.  Commands end with ; or \g.
Your MySQL connection id is 21542
Server version: 8.0.11 MySQL Community Server - GPL
```

```
Copyright (c) 2000, 2018, Oracle and/or its affiliates. All rights reserved.

Oracle is a registered trademark of Oracle Corporation and/or its
affiliates. Other names may be trademarks of their respective
owners.

Type 'help;' or '\h' for help. Type '\c' to clear the current input statement.

mysql> SELECT member_id, member_host, member_port, member_state FROM performance_schema.replication_
group_members WHERE channel_name = 'group_replication_applier';
+--------------------------------------+-------------------+-------------+--------------+
| member_id                            | member_host       | member_port | member_state |
+--------------------------------------+-------------------+-------------+--------------+
| 01340b8a-d1ef-11e8-8f57-0024e869b4d5 | hz_10_120_240_251 |        3505 | ONLINE       |
| 67f6d70f-d1f1-11e8-ac5a-0024e869b4d5 | hz_10_120_240_251 |        3515 | ONLINE       |
| 9ab15f6a-d1f1-11e8-aa41-0024e869b4d5 | hz_10_120_240_251 |        3525 | ONLINE       |
+--------------------------------------+-------------------+-------------+--------------+
3 rows in set (0.00 sec)

mysql> show status like 'group_replication_primary_member';
+----------------------------------+--------------------------------------+
| Variable_name                    | Value                                |
+----------------------------------+--------------------------------------+
| group_replication_primary_member | 01340b8a-d1ef-11e8-8f57-0024e869b4d5 |
+----------------------------------+--------------------------------------+
1 row in set (0.00 sec)
```

以上是通过 MySQL Router 对访问请求进行了路由，可以看到通过 6446 端口，连接到的节点是集群的 primary 成员，同理，如果连接 6447 端口，连接到的则是集群的 secondary 节点，读者可以自行试验。

通过 MySQL Router 访问集群的好处显而易见，前端可以不需要关心哪个地址是 primary 节点，哪些是 secondary 节点，一旦发生角色转换，MySQL Router 会自动感知到，对后续的连接会进行正确的路由，对前端来说是透明的，即前端无须调整相应的配置，这点对线上环境来说，至关重要。

31.3.3 集群 Metadata

前文一直在介绍集群的搭建过程，那么为什么 InnoDB Cluster 需要通过 MySQL Shell 来进行管理呢？这其中有什么联系？MySQL Router 是一个中间件，它也属于集群的一部分，那么它又产生了哪些元数据？接下来我们对 InnoDB Cluster 的元数据进行介绍，一起揭开 InnoDB Cluster 的神秘面纱。

首先来查看一下集群节点的 database，如下所示。

```
mysql> show databases;
+-----------------------------+
| Database                    |
+-----------------------------+
| information_schema          |
| mysql                       |
| mysql_innodb_cluster_metadata |
| performance_schema          |
| sys                         |
+-----------------------------+
5 rows in set (0.03 sec)
```

可以看到在集群成员实例上，都多了一个 database，名字为 mysql_innodb_cluster_metadata，其中包含的对象列表如下所示：

```
mysql> show tables in mysql_innodb_cluster_metadata;
+---------------------------------------+
| Tables_in_mysql_innodb_cluster_metadata |
+---------------------------------------+
| clusters                              |
```

```
| hosts          |
| instances      |
| replicasets    |
| routers        |
| schema_version |
+----------------+
6 rows in set (0.02 sec)
```

可以看到在 mysql_innodb_cluster_metadata 这个 database 中，包括 6 个对象，从名字就能大概了解其中内容，共涉及集群信息、实例信息、复制集信息和路由信息等，而路由信息比较特殊，需要由 MySQL Router 来进行初始化，即当 MySQL Router 配置完成后，会写入相关路由数据。

同时，在数据库用户方面，也相应产生了几个新的 user，如下所示。

```
mysql> select host,user from mysql.user;
+-----------+-----------------------------------+
| host      | user                              |
+-----------+-----------------------------------+
| %         | mysql_innodb_cluster_r6152329938  |
| %         | mysql_innodb_cluster_r6152342181  |
| %         | mysql_innodb_cluster_r6152349294  |
| %         | mysql_router3_o91xsw2bx4qh        |
| %         | root                              |
| localhost | mysql.infoschema                  |
| localhost | mysql.session                     |
| localhost | mysql.sys                         |
| localhost | mysql_innodb_cluster_r6152329938  |
| localhost | mysql_innodb_cluster_r6152342181  |
| localhost | mysql_innodb_cluster_r6152349294  |
| localhost | root                              |
+-----------+-----------------------------------+
12 rows in set (0.00 sec)

mysql> show grants for mysql_innodb_cluster_r6152329938@'%';
+--------------------------------------------------------------------------------+
| Grants for mysql_innodb_cluster_r6152329938@%                                  |
+--------------------------------------------------------------------------------+
| GRANT REPLICATION SLAVE ON *.* TO `mysql_innodb_cluster_r6152329938`@`%`       |
+--------------------------------------------------------------------------------+
1 row in set (0.00 sec)
```

其中，集群会针对%和 localhost 分别创建 n 个新用户（n 等于集群成员个数），其中前缀为 mysql_innodb_cluster_r 的用户是在 MySQL Shell 创建集群时生成，用于集群成员间的复制（FOR CHANNEL 'group_replication_recovery'），每一个集群成员用一个独立的用户名，例如：端口为 3505 的实例，使用 mysql_innodb_cluster_r6152329938，而端口为 3515 的实例，使用 mysql_innodb_cluster_r6152342181。另外，还可以发现有一个 router 相关的用户名 mysql_router3_o91xsw2bx4qh，此用户名由 MySQL Router 生成，用于连接集群 primary 节点，监控集群状态。

> **注意**：以上示例使用 sandbox 方式创建的 Innodb Cluster，默认会创建%和 localhost 下的多个用户，但如果是基于 MGR 创建的集群，只产生一个 InnoDB Cluster 相关用户和一个 MySQL Router 需要的用户，因此不需要 MySQL Shell 去配置 MGR 的环境。InnoDB Cluster 相关用户命名规则为：前缀 mysql_innodb_cluster_r+10 个数字。

通过以上内容，可以发现配置 InnoDB Cluster 后，新产生的元数据并不多，也不复杂，更加便于使用 MGR。

31.3.4 集群成员角色切换

在 InnoDB Cluster 完成配置之后，可以通过 MySQL Router 进行路由，连接集群节点，一旦集群节点出现故障，前端不需要改变配置，即可快速完成切换。

31.3 InnoDB Cluster

集群初始环境如下所示。

```
MySQL    localhost:3505 ssl    JS > cluster.status()
{
    "clusterName": "m8cluster",
    "defaultReplicaSet": {
        "name": "default",
        "primary": "localhost:3505",
        "ssl": "REQUIRED",
        "status": "OK",
        "statusText": "Cluster is ONLINE and can tolerate up to ONE failure.",
        "topology": {
            "localhost:3505": {
                "address": "localhost:3505",
                "mode": "R/W",
                "readReplicas": {},
                "role": "HA",
                "status": "ONLINE"
            },
            "localhost:3515": {
                "address": "localhost:3515",
                "mode": "R/O",
                "readReplicas": {},
                "role": "HA",
                "status": "ONLINE"
            },
            "localhost:3525": {
                "address": "localhost:3525",
                "mode": "R/O",
                "readReplicas": {},
                "role": "HA",
                "status": "ONLINE"
            }
        }
    },
    "groupInformationSourceMember": "mysql://root@localhost:3505"
}
```

localhost:3505 是 primary 节点，接收读写请求，localhost:3515 和 localhost:3525 是 secondary 节点，只接收读请求。下面将对角色切换场景进行验证。

（1）前端通过 MySQL Router 连接到 6446 端口（默认 6446 是读写端口，6447 是只读端口），并持续写入数据。

（2）通过 MySQL Shell 杀掉 primary 节点，并观察相关日志信息以及数据的写入情况。

```
MySQL    localhost:3505 ssl    JS > dba.killSandboxInstance(3505)
The MySQL sandbox instance on this host in
/home/mysqlshell/mysql-sandboxes/3505 will be killed

Killing MySQL instance...

Instance localhost:3505 successfully killed.
```

MySQL Router 日志如下。

```
2018-10-23 15:43:02 routing WARNING [7ff9ebfff700] Timeout reached trying to connect to MySQL Server localhost:3505: Invalid argument
2018-10-23 15:43:02 metadata_cache WARNING [7ff9ebfff700] Instance 'localhost:3505' [3d6f6e93-d3a8-11e8-8e9f-0024e869b4d5] of replicaset 'default' is unreachable. Increasing metadata cache refresh frequency.
2018-10-23 15:43:02 metadata_cache WARNING [7ff9f06c9700] Failed connecting with Metadata Server 127.0.0.1:3505: Can't connect to MySQL server on '127.0.0.1' (111) (2003)
2018-10-23 15:43:02 metadata_cache ERROR [7ff9f06c9700] Failed to connect to metadata server
```

从 MySQL Router 的日志可以发现，primary 节点被杀掉以后，Router 会立即发现连接失败。

数据的写入日志和 MySQL 的日志，具体如下。

```
mysql> select * from demo_innodb_cluster where mtime > '2018-10-23 15:43:00' and mtime < '2018-10-23 15:44:00';
```

```
+------+---------------+------------+---------------------+
| id   | instance_port | account_id | mtime               |
+------+---------------+------------+---------------------+
| 2047 |          3505 |      61311 | 2018-10-23 15:43:01 |
| 2054 |          3505 |      26159 | 2018-10-23 15:43:01 |
| 2061 |          3505 |      25475 | 2018-10-23 15:43:01 |
| 2068 |          3505 |      52467 | 2018-10-23 15:43:01 |
| 2075 |          3505 |       2887 | 2018-10-23 15:43:01 |
| 2082 |          3505 |      94181 | 2018-10-23 15:43:01 |
| 2089 |          3505 |       4930 | 2018-10-23 15:43:02 |
| 2096 |          3505 |       6282 | 2018-10-23 15:43:02 |
| 2103 |          3505 |      79539 | 2018-10-23 15:43:02 |
| 2110 |          3505 |      31664 | 2018-10-23 15:43:02 |
| 2111 |          3515 |      69523 | 2018-10-23 15:43:10 |
| 2112 |          3515 |      94601 | 2018-10-23 15:43:11 |
| 2113 |          3515 |      62128 | 2018-10-23 15:43:11 |
| 2114 |          3515 |       2574 | 2018-10-23 15:43:11 |
......

2018-10-23T07:43:08.187128Z 0 [Warning] [MY-011493] [Repl] Plugin group_replication reported
: 'Member with address hz_10_120_240_251:3505 has become unreachable.'
2018-10-23T07:43:10.192874Z 0 [Warning] [MY-011499] [Repl] Plugin group_replication reported
: 'Members removed from the group: hz_10_120_240_251:3505'
```

从以上信息可以得知，当 primary 节点失败时，数据停止写入，直至集群重构完成，期间经历了 8s，此值仅供参考，读者可以自行验证切换需要经历的时间。从数据加载过程来看，切换期间对前端是透明的，重构完成后，前端不需要进行任何操作，会自动重新连接新的 primary 节点。

（3）新集群的状态。

```
MySQL  localhost:3515 ssl  JS > cluster.status()
{
    "clusterName": "m8cluster",
    "defaultReplicaSet": {
        "name": "default",
        "primary": "localhost:3515",
        "ssl": "REQUIRED",
        "status": "OK_NO_TOLERANCE",
        "statusText": "Cluster is NOT tolerant to any failures. 1 member is not active",
        "topology": {
            "localhost:3505": {
                "address": "localhost:3505",
                "mode": "R/O",
                "readReplicas": {},
                "role": "HA",
                "status": "(MISSING)"
            },
            "localhost:3515": {
                "address": "localhost:3515",
                "mode": "R/W",
                "readReplicas": {},
                "role": "HA",
                "status": "ONLINE"
            },
            "localhost:3525": {
                "address": "localhost:3525",
                "mode": "R/O",
                "readReplicas": {},
                "role": "HA",
                "status": "ONLINE"
            }
        }
    },
    "groupInformationSourceMember": "mysql://root@localhost:3515"
}
```

原 primary 节点的状态已经变成了 MISSING，R/W 角色已经切换到了 3515 端口的节点。

31.3.5 集群删除/增加节点

在日常运维过程中，集群难免会因为一些原因增加或删除节点，本节将介绍具体操作过

31.3 InnoDB Cluster

程，以及需要注意的问题。

利用 MySQL Shell 删除一个节点的命令如下：

```
MySQL localhost:3505 ssl JS > cluster.removeInstance('localhost:3515')
The instance will be removed from the InnoDB cluster. Depending on the instance
being the Seed or not, the Metadata session might become invalid. If so, please
start a new session to the Metadata Storage R/W instance.

Attempting to leave from the Group Replication group...

The instance 'localhost:3515' was successfully removed from the cluster.
```

如上所示，一旦从 Cluster 环境中删除该节点后，将同时删除 InnoDB Cluster 创建的用户以及集群元数据中与该节点相关的记录，但是 mysql_innodb_cluster_metadata 数据库依然被保留，也可以再将删除的节点重新加入集群，如下所示。

```
MySQL 10.120.240.251:3305 JS > cluster.addInstance('root@10.120.240.251:3315')
A new instance will be added to the InnoDB cluster. Depending on the amount of
data on the cluster this might take from a few seconds to several hours.

Adding instance to the cluster ...

Validating instance at 10.120.240.251:3315...

This instance reports its own address as 10.120.240.251
The instance 'root@10.120.240.251:3315' was successfully added to the cluster.
```

但是在经历如上的一系列删除、添加之后，此时集群环境中的各个节点间的数据不再是"完全一致"。新加入的节点由于之前经历过删除操作，已经删除了部分用户，重新加入集群后，又会产生一个新的用户，而这个用户在集群中其余节点也都会同步创建，所以在集群环境中，会出现用户信息在不同节点间并不相同，如下所示。

```
10.120.240.251:3315
mysql> select host,user from mysql.user;
+-----------------+-------------------------------------+
| host            | user                                |
+-----------------+-------------------------------------+
| %               | emp                                 |
| %               | mysql_innodb_cluster_r6185220430    |
| %               | mysql_router3_b3caybczsmbp          |
| %               | root                                |
| 10.120.240.251  | root                                |
| 127.0.0.1       | root                                |
| localhost       | emp                                 |
| localhost       | mysql.session                       |
| localhost       | mysql.sys                           |
| localhost       | mysql_innodb_cluster_r6185220430    |
| localhost       | root                                |
+-----------------+-------------------------------------+
11 rows in set (0.00 sec)

10.120.240.251:3305
mysql> select host,user from mysql.user;
+-----------------+-------------------------------------+
| host            | user                                |
+-----------------+-------------------------------------+
| %               | emp                                 |
| %               | mysql_innodb_cluster_r6149324649    |
| %               | mysql_innodb_cluster_r6185220430    |
| %               | mysql_router3_b3caybczsmbp          |
| %               | root                                |
| 10.120.240.251  | root                                |
| 127.0.0.1       | root                                |
| localhost       | emp                                 |
| localhost       | mysql.session                       |
| localhost       | mysql.sys                           |
| localhost       | mysql_innodb_cluster_r6149324649    |
| localhost       | mysql_innodb_cluster_r6185220430    |
| localhost       | root                                |
```

```
+----------------+--------------------------------+
13 rows in set (0.00 sec)
```

可以看到,原集群节点比新加入的节点在%和 localhost 方面,多了用户 mysql_innodb_cluster_r6149324649,如果后期涉及用户的操作需要注意,以免集群节点复制错误导致挂起。

上面演示了三节点集群环境中只删除一个节点的情况,如果要销毁整个集群,可以使用下面的命令:

```
MySQL  localhost:3505 ssl  JS > cluster.dissolve({force: true})
The cluster was successfully dissolved.
Replication was disabled but user data was left intact.
```

一旦执行了上述命令,就表示该集群已经消亡了,想再次使用该集群,就只能重新构建。但集群销毁后只是删除了数据字典信息,各节点资源并不会释放,要想完全释放资源,可以通过 MySQL Shell 执行停止节点实例、删除实例两个命令来完成操作:

```
MySQL    JS > dba.stopSandboxInstance(3515)
The MySQL sandbox instance on this host in
/home/mysqlshell/mysql-sandboxes/3515 will be stopped

Please enter the MySQL root password for the instance 'localhost:3515': *****************

Stopping MySQL instance...

Instance localhost:3515 successfully stopped.


MySQL    JS > dba.deleteSandboxInstance(3515)
The MySQL sandbox instance on this host in
/home/mysqlshell/mysql-sandboxes/3515 will be deleted

Deleting MySQL instance...

Instance localhost:3515 successfully deleted.
```

31.3.6 重新加入节点

在日常使用过程中,也难免会遇到集群中的节点因异常原因而出现故障,此时查看集群的状态,可能会是以下这个情况。

```
MySQL  localhost:3505 ssl  JS > cluster.status()
{
    "clusterName": "m8cluster",
    "defaultReplicaSet": {
        "name": "default",
        "primary": "hz_10_120_240_251:3505",
        "ssl": "REQUIRED",
        "status": "OK_NO_TOLERANCE",
        "statusText": "Cluster is NOT tolerant to any failures. 1 member is not active",
        "topology": {
            "hz_10_120_240_251:3505": {
                "address": "hz_10_120_240_251:3505",
                "mode": "R/W",
                "readReplicas": {},
                "role": "HA",
                "status": "ONLINE"
            },
            "hz_10_120_240_251:3515": {
                "address": "hz_10_120_240_251:3515",
                "mode": "R/O",
                "readReplicas": {},
                "role": "HA",
                "status": "(MISSING)"
            },
            "hz_10_120_240_251:3525": {
                "address": "hz_10_120_240_251:3525",
                "mode": "R/O",
```

```
                "readReplicas": {},
                "role": "HA",
                "status": "ONLINE"
            }
        },
        "groupInformationSourceMember": "mysql://root@localhost:3505"
}
```

故障节点的状态此时是 MISSING，当执行 rescan 后，会得到关于该节点的更多的提示信息，如下所示。

```
MySQL  localhost:3505 ssl  JS > cluster.rescan()
Rescanning the cluster...

Result of the rescanning operation:
{
    "defaultReplicaSet": {
        "name": "default",
        "newlyDiscoveredInstances": [],
        "unavailableInstances": [
            {
                "host": "hz_10_120_240_251:3515",
                "label": "hz_10_120_240_251:3515",
                "member_id": "b88ce45c-cd3d-11e8-a1c9-0024e869b4d5"
            }
        ]
    }
}

The instance 'hz_10_120_240_251:3515' is no longer part of the HA setup. It is either offline
or left the HA group.
You can try to add it to the cluster again with the cluster.rejoinInstance('hz_10_120_240_25
1:3515') command or you can remove it from the cluster configuration.
Would you like to remove it from the cluster metadata? [Y/n]: n
```

当然，根据交互信息，可以选择是否从集群删除该节点，如果不想删除，可以处理完故障后，重新加入集群，如下所示。

```
MySQL  10.120.240.251:3505  JS > cluster.rejoinInstance('root@10.120.240.251:3515')
Rejoining the instance to the InnoDB cluster. Depending on the original
problem that made the instance unavailable, the rejoin operation might not be
successful and further manual steps will be needed to fix the underlying
problem.

Please monitor the output of the rejoin operation and take necessary action if
the instance cannot rejoin.

Rejoining instance to the cluster ...

The instance '10.120.240.251:3515' was successfully rejoined on the cluster.
```

如果选择直接删除该节点，如下所示。

```
Would you like to remove it from the cluster metadata? [Y/n]: y
Removing instance from the cluster metadata...

The instance 'hz_10_120_240_251:3515' was successfully removed from the cluster metadata.
```

节点被删除的过程在前文也已经有所讲解，此处不再赘述。

31.4 小结

目前有多种高可用方案可以在一定程度上实现数据库的高可用，比如本章介绍的 MHA、MGR 和 InnoDB Cluster。除此之外，还有一些本章未提到的其他方案，比如说 Percona 的 Galera Cluster 和 MySQL Cluster 等。这些高可用软件都各有优势和缺点，在进行高可用性方案选择时，主要看业务场景和对数据的一致性方面要求。最后，出于应用成熟度的考虑，MHA 更胜一筹。但如果从长远考虑，MGR 会是主流方向，最终 InnoDB Cluster 将会应用更为广泛。

第 32 章 MySQL 中间件

MySQL 中间件产品是一个庞大的家族，有官方的、自行研发的、商业化的，也有开源的，还有一些已经退出了历史舞台，那么为什么会出现这么多款不同产品？究其原因，还是因为我们对 MySQL 又爱又恨。爱的是 MySQL 简单易用，但在数据严谨性方面一点不输 Oracle；恨的是单实例性能有限，扩展性一直被人诟病。

为了应对高并发、高吞吐、高响应的应用需求，大家不得不在 MySQL 分库分表上下一番工夫，而数据一旦被拆分，势必带来主动路由的需求，因为不可能把分库配置固化到前端，配置是会随着节点变化而发生变更的，当然这只是其中一个方面，实际上 MySQL 中间件能做的事情很多，这也是为什么 MySQL 中间件市场如此复杂的原因。由于"江湖豪杰"众多，在此不妄自评论各家长短，这里重点推荐两款优秀的 MySQL 中间件，一个是来自官方的 MySQL Router，另一个是网易公司自行开发的 Cetus。

32.1 MySQL Router

MySQL Router 是官方出品，是一个轻量级的中间件，主要功能就是提供读写分离，但自从 MGR 集群的推出，目前已经被当作是 InnoDB Cluster 的一个组成部分，架构如图 32-1 所示。

从图 32-1 中可以看出，MySQL Router 负责接收前端应用的读写请求，并路由到正确的集群节点。如果后端的 High Availability Cluster 环境（此处仅指使用 MGR 集群）发生变化，比如集群成员角色切换、增加或减少成员等，MySQL Router 可以及时感知到集群中各节点的新角色，从而将新的读写请求发送到正确节点。这点还是比较友好的，对前端应用透明，尽管用就好。

在 MySQL Router 8.0 之前，Router 的最高版本是 2.1.6，此时的后端通常使用其他高可用方案来保证节点可用性，比如 MHA，或者 Percona 的 XtrDB Cluster 等。

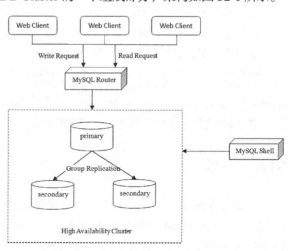

图 32-1　InnoDB Cluster 架构图

下面具体演示如何安装并进行正确配置。

32.1.1 MySQL Router 的安装

MySQL Router 的 GA 版本并不多，起初只有 2.0 和 2.1 版本。随着 MySQL 8.0 的推出，MySQL Router 也随之出现了 8.0 的 GA 版本，当前最新的版本是 8.0.13。为了更好地支持 MySQL 5.7 和 MySQL 8.0，建议安装最新的 GA 版本。

MySQL Router 安装方式有好多种，比如可以直接用 YUM 安装，官网有 YUM 源，这方面可以参考前文内容 MySQL 的安装，另外也可以直接使用 RPM 安装包，这个也非常简单，相信读者对安装 RPM 都不陌生，另一种安装方式是直接使用二进制安装包，还有就是直接使用源码进行编译安装。本文将介绍后两种安装方式，尤其是最后的源码安装，如果希望能自定义安装配置，那么选择编译安装会更合适。

1. 二进制包安装

选择二进制包进行安装时，可以在 MySQL 官网的 DOWNLOAD 页面下单击左侧的 MySQL Router，下载最新版本的安装包，然后解压后即可使用，如下所示。

```
[mysqlrouter@hz_10_120_240_251 ~]$ tar -xvJf mysql-router-8.0.13-linux-glibc2.12-x86_64.tar.xz

[mysqlrouter@hz_10_120_240_251 ~]$ ll mysql-router-8.0.13-linux-glibc2.12-x86_64
total 116
drwxrwxr-x 2 mysqlrouter mysqlrouter   4096 Oct 26 17:20 bin
drwxrwxr-x 3 mysqlrouter mysqlrouter   4096 Oct 26 17:20 lib
-rw-r--r-- 1 mysqlrouter mysqlrouter 101807 Oct  7 16:44 LICENSE.router
-rw-r--r-- 1 mysqlrouter mysqlrouter    700 Oct  7 16:44 README.router
drwxrwxr-x 3 mysqlrouter mysqlrouter   4096 Oct 26 17:20 share
```

2. 源码编译安装

选择源码编译安装时，当前最新的版本 MySQL Router 8.0.13 在 MySQL 官网 DOWNLOAD 页面的 MySQL Router 下已经不存在源码安装包了。如需要源码安装，可以到 DOWNLOAD 页面的 MySQL Community Server 下去下载 MySQL Community Server 8.0.13 的源码安装包，此安装包中已经包括了 MySQL Router 的源码，即通过编译安装 MySQL 来安装 MySQL Router 和相关工具。除源码安装包外，其他形式的安装包都是有的，如果需要之前版本的源码包，可以到 MySQL 官网的 DOWNLOAD 页面下单击 Archievs，然后再单击左侧的 MySQL Router，从其页面中下载。也可以通过 *https://github.com/mysql/mysql-router* 下载。

以下将通过编译 MySQL 方式来安装 MySQL Router。

（1）下载 MySQL 源码安装包 mysql-boost-8.0.13.tar.gz，解压后，使用 cmake 编译。

```
[mysqlrouter@oratest ~]$ tar -xvzf mysql-boost-8.0.13.tar.gz
[mysqlrouter@oratest ~]$ cd mysql-8.0.13
[root@oratest mysql-8.0.13]# mkdir build
[root@oratest mysql-8.0.13]# cd build
[root@oratest mysql-8.0.13]# cmake .. -DCMAKE_INSTALL_PREFIX=/home/mysqlrouter/mysqlhome -DWITH_BOOST=/home/mysqlrouter/mysql-8.0.13/boost
-- Running cmake version 2.8.12.2
-- Found Git: /bin/git (found version "1.8.3.1")
-- MySQL 8.0.13
-- CMAKE_GENERATOR: Unix Makefiles
-- SIZEOF_VOIDP 8
-- Packaging as: mysql-8.0.13-Linux-x86_64
-- Local boost dir /home/mysqlrouter/mysql-8.0.13/boost/boost_1_67_0
-- Found /home/mysqlrouter/mysql-8.0.13/boost/boost_1_67_0/boost/version.hpp
-- BOOST_VERSION_NUMBER is #define BOOST_VERSION 106700
```

```
......
-- Build files have been written to: /home/mysqlrouter/mysql-8.0.13/build
......
```

由于检查内容比较多，如果缺少系统包会退出并给出错误提示，以上内容是检查通过后的结果，也可以提前安装好编译时依赖的系统包。

（2）开始安装。

```
[root@oratest build]# make
......
[100%] Built target mysqlrouter
Scanning dependencies of target routing
[100%] Building CXX object router/src/routing/CMakeFiles/routing.dir/src/routing_plugin.cc.o
[100%] Building CXX object router/src/routing/CMakeFiles/routing.dir/src/plugin_config.cc.o
[100%] Building CXX object router/src/routing/CMakeFiles/routing.dir/src/mysql_routing.cc.o
[100%] Building CXX object router/src/routing/CMakeFiles/routing.dir/src/utils.cc.o
[100%] Building CXX object router/src/routing/CMakeFiles/routing.dir/src/destination.cc.o
[100%] Building CXX object router/src/routing/CMakeFiles/routing.dir/src/dest_metadata_cache.cc.o
[100%] Building CXX object router/src/routing/CMakeFiles/routing.dir/src/dest_first_available.cc.o
[100%] Building CXX object router/src/routing/CMakeFiles/routing.dir/src/dest_next_available.cc.o
[100%] Building CXX object router/src/routing/CMakeFiles/routing.dir/src/dest_round_robin.cc.o
[100%] Building CXX object router/src/routing/CMakeFiles/routing.dir/src/routing.cc.o
[100%] Building CXX object router/src/routing/CMakeFiles/routing.dir/src/protocol/classic_protocol.cc.o
[100%] Building CXX object router/src/routing/CMakeFiles/routing.dir/src/connection.cc.o
[100%] Building CXX object router/src/routing/CMakeFiles/routing.dir/src/context.cc.o
[100%] Building CXX object router/src/routing/CMakeFiles/routing.dir/src/mysql_routing_common.cc.o
[100%] Building CXX object router/src/routing/CMakeFiles/routing.dir/src/connection_container.cc.o
[100%] Building CXX object router/src/routing/CMakeFiles/routing.dir/src/protocol/x_protocol.cc.o
Linking CXX shared library ../../../plugin_output_directory/routing.so
[100%] Built target routing
Scanning dependencies of target syslog
[100%] Building CXX object router/src/syslog/CMakeFiles/syslog.dir/src/syslog.cc.o
Linking CXX shared library ../../../plugin_output_directory/syslog.so
[100%] Built target syslog


[mysqlrouter@oratest build]$ make install
[  0%] Built target INFO_BIN
[  0%] Built target INFO_SRC
[  0%] Built target abi_check
[  0%] Built target zlib
[  2%] Built target edit

[100%] Built target mysqlrouter_plugin_info
[100%] Built target mysqlrouter
[100%] Built target routing
[100%] Built target syslog
Install the project...
......
-- Installing: /home/mysqlrouter/mysqlhome/lib/mysqlrouter/routing.so
-- Set runtime path of "/home/mysqlrouter/mysqlhome/lib/mysqlrouter/routing.so" to "$ORIGIN/../:$ORIGIN/:$ORIGIN/../lib/"
-- Installing: /home/mysqlrouter/mysqlhome/lib/mysqlrouter/syslog.so
-- Set runtime path of "/home/mysqlrouter/mysqlhome/lib/mysqlrouter/syslog.so" to "$ORIGIN/../:$ORIGIN/:$ORIGIN/../lib/"
```

如果看到类似以上这些信息的输出，证明已经编译安装成功。以上编译安装过程大概耗时 2h，仅供读者参考。另外从以上编译的信息，也可以简单了解到 Router 涉及的内容，比如我们可以看到 Router 支持 round-robin 策略，支持两种协议，一个是 classic_protocol，另一个是 x_protocol。

MySQL Router 安装完毕后，下面将详细介绍 Router 的初始化和使用方法。

32.1.2 MySQL Router 的初始化

MySQL Router 目前已经被当作 Innodb Cluster 的一部分，首先通过 MGR 集群来初始化一个 Router 环境，然后也会介绍一下非 MGR 集群环境下，Router 的不同配置。

1. MGR 集群环境下的 Router

首先，利用 MySQL Shell 构造一个 Innodb Cluster 的环境，关于这部分可以参考前文的内容。然后，执行以下命令。

```
[mysqlrouter@oratest ~]$ mysqlrouter --bootstrap root@10.120.240.251:3315 --directory=/home/mysqlrouter/m7cluster --force=true
Please enter MySQL password for root:
WARNING: The MySQL server does not have SSL configured and metadata used by the router may be transmitted unencrypted.

Bootstrapping MySQL Router instance at '/home/mysqlrouter/m7cluster'...
Executing statements failed with: 'Error executing MySQL query: The MySQL server is running with the --super-read-only option so it cannot execute this statement (1290)' (1290), trying to connect to another node
Fetching Group Replication Members
disconnecting from mysql-server
trying to connecting to mysql-server at 10.120.240.251:3305
Checking for old Router accounts
Creating account mysql_router19_g8gupsilgfca@'%'
MySQL Router  has now been configured for the InnoDB cluster 'm7cluster'.

The following connection information can be used to connect to the cluster after MySQL Router has been started with generated configuration..

Classic MySQL protocol connections to cluster 'm7cluster':
- Read/Write Connections: localhost:6446
- Read/Only Connections: localhost:6447
X protocol connections to cluster 'm7cluster':
- Read/Write Connections: localhost:64460
- Read/Only Connections: localhost:64470
```

在初始化时，如果指定使用 --directory 选项，会将生成的文件全部放入该目录，便于管理，否则会生成在默认目录下。如果有多个 MySQL Router 配置，将会给管理带来麻烦。

另外，初始化过程需要连接集群其中一个节点，Primary 或者 Secondary 节点都可以，如果连接的是 Secondary 节点，会自动连接相应的 Primary 节点，因为初始化过程需要写入一些数据到数据库后端，而 force 的选项适用于之前该 Innodb Cluster 已经创建过相关 router 信息，此次如果需要重新初始化，需要指定 force=true 进行强行覆盖。

Router 在初始化过程中，需要创建一个"%"相关的账号，用户名命名规则是 mysql_router+< router_id >+12 位字符，例如 mysql_router19_g8gupsilgfca。另外，也需要在集群中注册 Router 的相关信息，如下所示。

```
mysql> select * from mysql_innodb_cluster_metadata.routers;
+-----------+-------------+---------+------------+
| router_id | router_name | host_id | attributes |
+-----------+-------------+---------+------------+
|        19 |             |      12 | NULL       |
+-----------+-------------+---------+------------+
1 rows in set (0.00 sec)
```

端口相关信息，从初始化日志中可以了解到，如果使用 Classic 协议连接 Innodb Cluster，Read/Write 连接的端口是 6446，Read/Only 连接的端口是 6447，而如果改用 X 协议连接，Read/Write 连接的端口是 64460，Read/Only 连接的端口是 64470。

初始化完成后，生成的文件目录结构如下所示。

```
[mysqlrouter@oratest ~]$ ll m7cluster/
total 28
drwx------ 2 mysqlrouter mysqlrouter 4096 Oct 30 14:27 data
drwx------ 2 mysqlrouter mysqlrouter 4096 Oct 30 14:27 log
-rw------- 1 mysqlrouter mysqlrouter 1301 Oct 30 14:27 mysqlrouter.conf
-rw------- 1 mysqlrouter mysqlrouter   98 Oct 30 14:27 mysqlrouter.key
drwx------ 2 mysqlrouter mysqlrouter 4096 Oct 30 14:27 run
-rwx------ 1 mysqlrouter mysqlrouter  169 Oct 30 14:27 start.sh
-rwx------ 1 mysqlrouter mysqlrouter  191 Oct 30 14:27 stop.sh
```

其中有一个配置文件 mysqlrouter.conf，该文件正是 MySQL Router 初始化后的配置信息，包括连接信息和连接策略等，文件内容如下所示。

```
[mysqlrouter@oratest ~]$ cat m7cluster/mysqlrouter.conf
# File automatically generated during MySQL Router bootstrap
[DEFAULT]
logging_folder=/home/mysqlrouter/m7cluster/log
runtime_folder=/home/mysqlrouter/m7cluster/run
data_folder=/home/mysqlrouter/m7cluster/data
keyring_path=/home/mysqlrouter/m7cluster/data/keyring
master_key_path=/home/mysqlrouter/m7cluster/mysqlrouter.key
connect_timeout=30
read_timeout=30

[logger]
level = INFO

[metadata_cache:m7cluster]
router_id=19
bootstrap_server_addresses=mysql://10.120.240.251:3305,mysql://10.120.240.251:3325,mysql://10.120.240.251:3315
user=mysql_router19_g8gupsilgfca
metadata_cluster=m7cluster
ttl=0.5

[routing:m7cluster_default_rw]
bind_address=0.0.0.0
bind_port=6446
destinations=metadata-cache://m7cluster/default?role=PRIMARY
routing_strategy=round-robin
protocol=classic

[routing:m7cluster_default_ro]
bind_address=0.0.0.0
bind_port=6447
destinations=metadata-cache://m7cluster/default?role=SECONDARY
routing_strategy=round-robin
protocol=classic

[routing:m7cluster_default_x_rw]
bind_address=0.0.0.0
bind_port=64460
destinations=metadata-cache://m7cluster/default?role=PRIMARY
routing_strategy=round-robin
protocol=x

[routing:m7cluster_default_x_ro]
bind_address=0.0.0.0
bind_port=64470
destinations=metadata-cache://m7cluster/default?role=SECONDARY
routing_strategy=round-robin
protocol=x
```

配置文件中的参数，读者是否觉得眼熟呢？其实这些内容在编译安装时有过提示，如果读者想深入研究其含义，可以查看官方文档或者直接读源码。下文也会挑选几个常用参数进行说明。

2. MHA 环境下的 Router

首先创建一个三节点的 MHA 环境，关于这部分可以参考前文内容，此处忽略，然后创建 MySQL Router 需要的目录和文件。

```
[mysqlrouter@hz_10_120_240_251 ~]$ mkdir -p /home/mysqlrouter/m8cluster/data
[mysqlrouter@hz_10_120_240_251 ~]$ mkdir -p /home/mysqlrouter/m8cluster/log
[mysqlrouter@hz_10_120_240_251 ~]$ mkdir -p /home/mysqlrouter/m8cluster/run
[mysqlrouter@hz_10_120_240_251 ~]$ touch /home/mysqlrouter/m8cluster/mysqlrouter.conf
[mysqlrouter@hz_10_120_240_251 ~]$ touch /home/mysqlrouter/m8cluster/start.sh
[mysqlrouter@hz_10_120_240_251 ~]$ touch /home/mysqlrouter/m8cluster/stop.sh
```

创建完成之后，开始配置 mysqlrouter.conf 文件，如下所示。

```
[mysqlrouter@hz_10_120_240_251 m8cluster]$ cat mysqlrouter.conf
[DEFAULT]
logging_folder=/home/mysqlrouter/m8cluster/log
runtime_folder=/home/mysqlrouter/m8cluster/run
data_folder=/home/mysqlrouter/m8cluster/data
connect_timeout=30
read_timeout=30

[logger]
level = INFO

[routing:default_rw]
bind_address=0.0.0.0
bind_port=6446
destinations=10.120.240.251:3308,10.120.240.251:3309
routing_strategy=first-available
protocol=classic

[routing:default_ro]
bind_address=0.0.0.0
bind_port=6447
destinations=10.120.240.251:3307,10.120.240.251:3308,10.120.240.251:3309
routing_strategy=round-robin
protocol=classic
```

并配置好启动和停止脚本。

```
[mysqlrouter@hz_10_120_240_251 m8cluster]$ cat start.sh
#!/bin/bash
basedir=/home/mysqlrouter/m8cluster
ROUTER_PID=$basedir/mysqlrouter.pid /home/mysqlrouter/mysql-router-8.0.13-linux-glibc2.12-x86_64/bin/mysqlrouter -c $basedir/mysqlrouter.conf &
disown %-

[mysqlrouter@hz_10_120_240_251 m8cluster]$ cat stop.sh
#!/bin/bash
if [ -f /home/mysqlrouter/m8cluster/mysqlrouter.pid ]; then
    kill -TERM `cat /home/mysqlrouter/m8cluster/mysqlrouter.pid` && rm -f /home/mysqlrouter/m8cluster/mysqlrouter.pid
fi
```

以上两部分分别介绍了在 MGR 环境和 MHA 环境下的 MySQL Router 的各自配置，下面针对几个关键参数进行介绍。

- destinations：路由的目标地址，如果有多个地址，中间用","进行分割；如果后端是 InnoDB Cluster 环境，初始化后会自动生成 metadata-cache 的相关链接地址。

- routing_strategy：路由策略，从版本 8.0.4 之后有 4 个值可以使用，分别为 first-available、next-available、round-robin 和 round-robin-with-fallback，默认是 round-robin。关于这些路由策略的区别，详见下文。

- bind_address：Router 提供服务的地址，如果该地址未指定端口信息，则需要配置

bind_port 参数。

- bind_port：Router 提供服务的端口地址，该参数可选。

32.1.3　MySQL Router 策略验证

MySQL Router 配置完成后，就可以投入使用了。接下来将针对不同的路由策略，进行几个场景下的验证。

首先，启动 MySQL Router。

```
[mysqlrouter@hz_10_120_240_251 m8cluster]$ sh start.sh
```

如果是在 InnoDB Cluster 的集群环境中，那么一旦启动了 MySQL Router，就会有短连接持续不断地从 Router 到 InnoDB Cluster 的 primary 节点，频繁执行监测命令，用于确认当前 MySQL Router 存储的 InnoDB Cluster 角色是正确的，执行的命令如下：

```
SELECT R.replicaset_name, I.mysql_server_uuid, I.role, I.weight, I.version_token, H.location, I.addresses->>'$.mysqlClassic', I.addresses->>'$.mysqlX' FROM mysql_innodb_cluster_metadata.clusters AS F JOIN mysql_innodb_cluster_metadata.replicasets AS R ON F.cluster_id = R.cluster_id JOIN mysql_innodb_cluster_metadata.instances AS I ON R.replicaset_id = I.replicaset_id JOIN mysql_innodb_cluster_metadata.hosts AS H ON I.host_id = H.host_id WHERE F.cluster_name = 'm7cluster';

show status like 'group_replication_primary_member';

SELECT member_id, member_host, member_port, member_state, @@group_replication_single_primary_mode FROM performance_schema.replication_group_members WHERE channel_name = 'group_replication_applier';
```

可见，Router 为了保证连接的正确性，在 InnoDB Cluster 环境下，需要通过以上这些命令的实时监测才行，一旦发现后端节点故障，及时更新可用节点信息，从而在创建新连接的过程中放弃选择该节点。但如果是在非 InnoDB Cluster 环境中，比如 MHA 架构下，就不存在这些实时监测了。如果主从发生切换，要提前在配置文件中设计好路由和策略，否则可能会将数据写入 slave 节点。

1. round-robin 策略

这个路由策略是默认的，也比较容易理解，轮询方式访问每一个 destinations 中的地址，以下是演示过程，不再赘述。

```
[mysqlrouter@hz_10_120_240_251 m8cluster]$ mysql -uroot -h10.120.240.251 -P6447
mysql> select @@port;
+--------+
| @@port |
+--------+
|   3308 |
+--------+
1 row in set (0.00 sec)

[mysqlrouter@hz_10_120_240_251 m8cluster]$ mysql -uroot -h10.120.240.251 -P6447
mysql> select @@port;
+--------+
| @@port |
+--------+
|   3307 |
+--------+
1 row in set (0.00 sec)
```

2. first-available 策略

当路由策略是 first-available 时，如果 destinations 列表中第一个地址即可访问，则后续的连接请求全部路由至该节点，如果该地址访问失败，则会开始访问下一个地址。

```
[mysqlrouter@hz_10_120_240_251 m8cluster]$ mysql -uroot -h10.120.240.251 -P6447
mysql> select @@port;
+--------+
| @@port |
+--------+
| 3307   |
+--------+
1 row in set (0.00 sec)
……
```

此时，关闭 3307 节点的实例，然后再次连接原地址。

```
[mysqlrouter@hz_10_120_240_251 m8cluster]$ mysql -uroot -h10.120.240.251 -P6447
mysql> select @@port;
+--------+
| @@port |
+--------+
| 3308   |
+--------+
1 row in set (0.00 sec)
……
```

发现以后的连接将持续路由 3308 实例的地址，只有当该地址的实例不可访问或者重启 Router 后，才会路由至其他地址，并且之前不可访问的地址，一旦可以继续进行访问，又将会重新进入候选队列。

3. next-available 策略

当路由策略是 next-available 时，如果 destinations 列表中第一个地址即可访问，则后续的连接请求全部路由至该节点，如果该地址访问失败，则会开始访问下一个地址，这个规则与 first-available 策略是一样的。与 first-available 唯一不同的是，如果失败的地址即便恢复可用，也不会被重新加入候选队列，只有通过重启 router 重置连接列表，才能解决这个问题。

接下来验证一下这个策略。

开始连接 3307 节点的实例，然后关闭该实例，再次请求连接后，路由到 3308 节点的实例，依次类推，当 destinations 列表中的实例全部被列为不可用地址后，该路由将不再可用，此时只能重启 MySQL Router 才能解决。

```
[mysqlrouter@hz_10_120_240_251 m8cluster]$ mysql -uroot -h10.120.240.251 -P6447
mysql> select @@port;
+--------+
| @@port |
+--------+
| 3308   |
+--------+
1 row in set (0.00 sec)

[mysqlrouter@hz_10_120_240_251 m8cluster]$ mysql -uroot -h10.120.240.251 -P6447
ERROR 2003 (HY000): Can't connect to remote MySQL server for client connected to '0.0.0.0:6447'
```

4. round-robin-with-fallback 策略

该路由策略需要在 InnoDB Cluster 环境中使用，如果后端使用的是 MHA 环境，将会得到以下错误。

```
2018-10-31 16:49:51 main ERROR [7ffd3b237720] Configuration error: option routing_strategy in [routing:default_ro] is invalid; valid are first-available, next-available, and round-robin (was 'round-robin-with-fallback')
```

在 InnoDB Cluster 环境中，路由策略的选择还与 role 有关。如果 role 是 PRIMARY，可以选择 round-robin 和 first-available；如果 role 是 SECONDARY，可以选择 round-robin， first-available 和 round-robin-with-fallback；如果 role 是 PRIMARY_AND_SECONDARY，则可以选择 round-robin 和 first-available。

当 InnoDB Cluster 集群成员全部可以正常访问时，如果配置了该路由策略，Router 是根据 round-robin 方式进行路由的，如下所示。

```
[mysqlrouter@hz_10_120_240_251 m8cluster]$ mysql -uroot -h10.120.240.252 -P6447
mysql> select @@port;
+--------+
| @@port |
+--------+
|   3315 |
+--------+
1 row in set (0.00 sec)

[mysqlrouter@hz_10_120_240_251 m8cluster]$ mysql -uroot -h10.120.240.252 -P6447
mysql> select @@port;
+--------+
| @@port |
+--------+
|   3325 |
+--------+
1 row in set (0.00 sec)
```

而如果集群的 secondary 成员全部不可访问时，此时再访问 Router 的 6447 端口（Secondary 成员），会发现其实路由到的节点是 primary，所以如果此时访问 Router 的 6446 端口，连接的是同一个后端。

将 InnoDB Cluster（单主）环境下的 secondary 节点实例全部关闭，然后通过 Router 进行连接，如下所示。

```
[mysqlrouter@hz_10_120_240_251 m8cluster]$ mysql -uroot -h10.120.240.252 -P6447
mysql> select @@port;
+--------+
| @@port |
+--------+
|   3305 |
+--------+
1 row in set (0.00 sec)

[mysqlrouter@hz_10_120_240_251 m8cluster]$ mysql -uroot -h10.120.240.252 -P6446
mysql> select @@port;
+--------+
| @@port |
+--------+
|   3305 |
+--------+
1 row in set (0.00 sec)
```

所以该策略与 round-robin 的区别就是，当 secondary 成员全部不可访问后，如果再次请求 Router 进行路由，将会路由到 primary 节点，以提供服务。

32.2　Cetus 架构

Cetus 是由 C 语言开发的数据库中间件，提供全面的数据库访问代理功能，包括连接池、读写分离、数据分片（Sharding）、透明数据路由及分布式事务处理、故障自动切换等特性。Cetus 连接方式与 MySQL 完全兼容，应用程序不用修改即可通过 Cetus 访问数据库，实现数据库层水平扩展和高可用。Cetus 的架构如图 32-2 所示。

从 Cetus 架构中可以看出，为了避免 Cetus 单点，前端可以使用 LVS 等负载均衡服务，当前端请求发送到 Cetus 端后，Cetus 负载解析前端的 SQL，并路由到后端节点。后端既可以是 MGR 集群，也可以是普通的主从环境。另外，Cetus 提供了调用接口，支持 MySQLdb 和 pymysql，如果后端环境发生变化，可以主动调用这些接口，通知 Cetus 集群的环境变化。

Cetus 对读写服务自动分离（读写分离版本和 Sharding 版本均如此），前端不论是写请求还是读请求，只配置同一个地址即可。

截至目前，Cetus 提供两个主要分支，master 分支为多进程版本，single-process 分支提供的是单进程版本，这两个分支在使用上无差异，仅在管理功能方面有些许不同，这点在下文也会介绍。另外，从功能上来区分，也可以分为两个版本，其中一个是读写分离版本，在该版本中，后端的数据库会被看成都是全量数据，Cetus 此时负责解析 SQL，用来区分哪些是写请求，哪些是读请求，然后分别路由至指定节点；另一个版本是 Sharding 版本（分片版本），此时后端的数据库会被当作一个一个的分片节点，Cetus 负责解析 SQL，并将数据按照设置好的分片规则，比如 hash，分别路由至目标节点。

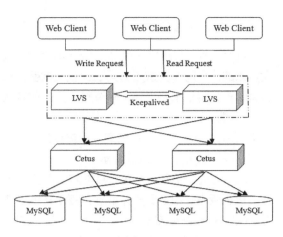

图 32-2　Cetus 架构图

Cetus 有很多优秀的特性，这使得 Cetus 在 MySQL 数据库中间件这一系列产品中，能够脱颖而出。例如分布式事务，Cetus 使用 MySQL 原生的 XA 协议，能够自动识别并高效地执行分布式事务；支持 Tcp Stream 流式处理，无须在等待后端结果集的过程中浪费大量时间，并且可以降低内存消耗，避免内存炸裂，性能显著提升；支持结果集压缩，从而减少网络质量带来的影响，有效缓解网络延迟。多进程的 Cetus 也可以从容应对大量高并发的访问请求，性能方面表现优异，经若干次多维度性能验证，Cetus 在各 MySQL 中间件产品中堪称一流。其余特性，不再一一列举，读者可以参考 GitHub 相关文档 https://github.com/Lede-Inc/cetus。在后端数据库方面，Cetus 也紧跟 MySQL 的脚步。目前支持 MySQL 数据库的多个不同版本，可以是 MySQL 5.6、5.7、8.0（需要明确数据库连接用户使用 mysql_native_password 认证方式，关于支持 MySQL 8.0 默认的 caching_sha2_password 认证方式，目前也在排期中，相信不久就会实现）。

下面将用实际案例，演示 Cetus 的安装配置与功能介绍。

32.2.1　Cetus 的安装配置

Cetus 代码已经开源，可以从 GitHub 自行下载。目前没有提供编译好的 RPM 安装包，但已经提供了自行编译 RPM 包的代码，读者可以参考 https://github.com/Lede-Inc/cetus/blob/master/doc/cetus-rpm.md 文档。

由于 Cetus 在读写分离版本和 Sharding 版本上采用了不同的解析器，所以在编译安装时需要通过参数的不同值加以区分（编译参数 DSIMPLE_PARSER），接下来将分别演示不同版本的安装过程。

1. 读写分离版本

读写分离其实很容易理解，Cetus 不会对前端发过来的读写请求做 SQL 改写，而且仅判断请求类型，然后将请求路由到正确节点（主库或者从库）；后端可以配置一主一从或多从环境；如果是多个从库，Cetus 默认会以轮询方式进行访问（下文有详细说明），如图 32-3 所示。

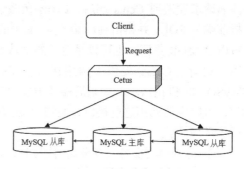

图 32-3　读写分离示意图

（1）首先，下载安装包，目前默认的主分支是多进程版本。如果要选择单进程版本，可以选择 single-process 分支。

```
[cetus_33060@hz_10_120_240_251 ~]$ git clone https://github.com/Lede-Inc/cetus.git
Cloning into 'cetus'...
......
```

（2）开始编译安装（读写分离版本的编译参数 DSIMPLE_PARSER=ON）。

```
[cetus_33060@hz_10_120_240_251 ~]$ cd cetus;mkdir build; cd build
[cetus_33060@hz_10_120_240_251 ~]$ cmake ../ -DCMAKE_BUILD_TYPE=Debug -DCMAKE_INSTALL_PREFIX=
/home/cetus_33060/cetus_20181026_9e349f5_multi -DSIMPLE_PARSER=ON
-- The C compiler identification is GNU 4.4.7
-- Check for working C compiler: /usr/bin/cc
-- Check for working C compiler: /usr/bin/cc -- works
-- Detecting C compiler ABI info
-- Detecting C compiler ABI info - done
......
-- Found FLEX: /usr/bin/flex (found version "2.5.35")
** Using simple parser [-DSIMPLE_PARSER=ON]
-- Configuring done
-- Generating done
-- Build files have been written to: /home/cetus_33060/cetus/build

[cetus_33060@hz_10_120_240_251 build]$ make
Scanning dependencies of target ev-cetus
[  1%] Building C object libev/CMakeFiles/ev-cetus.dir/ev.c.o
[  2%] Building C object libev/CMakeFiles/ev-cetus.dir/event.c.o
Linking C shared library libev-cetus.so
[  2%] Built target ev-cetus
......
[ 98%] Built target admin
Scanning dependencies of target proxy
[100%] Building C object plugins/proxy/CMakeFiles/proxy.dir/proxy-plugin.c.o
Linking C shared library libproxy.so
[100%] Built target proxy

[cetus_33060@hz_10_120_240_251 build]$ make install
[  2%] Built target ev-cetus
[  4%] Built target mysql-chassis-glibext
[  5%] Built target mysql-chassis-timing
......
-- Installing: /home/cetus_33060/cetus_20181026_9e349f5_multi/lib/cetus/plugins/libproxy.so
-- Removed runtime path from "/home/cetus_33060/cetus_20181026_9e349f5_multi/lib/cetus/plugins/
libproxy.so"
-- Installing: /home/cetus_33060/cetus_20181026_9e349f5_multi/lib/libsqlparser.so
```

编译完成之后，会在指定的 DCMAKE_INSTALL_PREFIX 目录下生成需要的文件。目前可以使用 MySQL 5.6 或者 MySQL 5.7 的版本进行编译，MySQL 8.0 的版本正在支持中。

以下是编译完成后，Cetus 的文件目录结构。

```
[cetus_33060@hz_10_120_240_251 cetus_20181026_9e349f5_multi]$ ll
total 20
```

```
drwxrwxr-x 2 cetus_33060 cetus_33060 4096 Nov  1 16:59 bin
drwxrwxr-x 2 cetus_33060 cetus_33060 4096 Nov  1 16:59 conf
drwxrwxr-x 4 cetus_33060 cetus_33060 4096 Nov  1 16:59 lib
drwxrwxr-x 2 cetus_33060 cetus_33060 4096 Nov  1 16:59 libexec
drwxrwxr-x 2 cetus_33060 cetus_33060 4096 Nov  1 16:59 logs
```

在 conf 目录下，有配置文件样例，读者可以参考这些样例进行配置，读写分离版本的配置仅需要 proxy.conf、users.json 和 variables.json，其中 proxy.conf 是 Cetus 的主配置文件，后两个是 json 格式文件，Cetus 使用的用户和密码信息就配置在文件 users.json 中，而 variables.json 文件中存放了 Cetus 支持的会话级别环境变量的设置，例如 sql_mode，读者可以根据需要修改为适合业务系统的值。

下面对配置文件中的重要参数进行简单说明，更详细的解释请参考文档 https://github.com/Lede-Inc/cetus/blob/master/doc/cetus-rw-profile.md，有些参数是支持动态修改的，但有一些则必须重启 Cetus 实例才能生效。关于该部分的描述可以参见 32.2.4 节。

（3）主配置文件 proxy.conf，如下所示：

```
[cetus_33060@hz_10_120_240_251 conf]$ cat proxy.conf
[cetus]
# For mode-switch
daemon = true

# Loaded Plugins
plugins=proxy,admin

# Defines the number of worker processes.
worker-processes=4

# Proxy Configuration, For example: MySQL master and salve host ip are both 192.0.0.1
proxy-address=0.0.0.0:33060
proxy-backend-addresses=10.120.240.251:3325
proxy-read-only-backend-addresses=10.120.240.251:3315,10.120.240.251:3305

# Admin Configuration
admin-address=0.0.0.0:33062
admin-username=admin
admin-password=admin

# Backend Configuration, use test db and username created
default-db=employees
default-username=emp
default-pool-size=100
max-resp-size=10485760
long-query-time=100

# File and Log Configuration, put log in /data and marked by proxy port, /data/cetus needs to be created manually and has rw authority for cetus os user
max-open-files = 65536
pid-file = cetus33060.pid
plugin-dir=lib/cetus/plugins
log-file=/home/cetus_33060/cetus_20181105_3fac75c_multi/logs/cetus_33060.log
log-level=debug

# Check salve delay
disable-threads=false
check-slave-delay=true
slave-delay-down=3
slave-delay-recover=1

# For trouble
keepalive=true
verbose-shutdown=true
log-backtrace-on-crash=true

group-replication-mode=1
```

其中：
- daemon 表示进程启动时是否作为后台进程，如果值是 False，则表示该 Cetus 属于前台进程，此进程不能退出，一旦退出，Cetus 也就关闭了；
- plugins 表示该 Cetus 启用的插件，该示例由于是读写分离版本，所以启用 proxy 和 admin 即可；
- worker-processes 表示 worker 进程个数，如果选择 GitHub 上的主分支，即多进程版本，该示例中配置了 4 个 worker 进程；
- proxy-address 表示该 Cetus 实例对外提供的服务地址；
- proxy-backend-addresses/proxy-read-only-backend-addresses，前者表示后端 MySQL 实例的主库地址，后者表示从库地址，即只读从库；
- default-db/default-username 表示 Cetus 创建连接时所使用的默认 db 和用户名，该用户名必须出现在 users.json 文件中，且在每个 MySQL 后端有相应权限，Cetus 的监控线程也会使用该用户进行后端连接；
- default-pool-size 表示 Cetus 每一个 worker 进程初始创建到每一个 MySQL 后端的连接数量，该示例中配置了 4 个 worker 进程，default-pool-size 为 100，即当 Cetus 启动后需要在每一个后端创建 400 个连接；
- long-query-time，该参数类似 MySQL 的慢查询阈值，单位为 ms，该示例配置了 100，表示如果一个 SQL 执行超过该阈值，将进入 Cetus 慢日志文件；
- check-slave-delay/slave-delay-down/slave-delay-recover，这 3 个参数为一组，用于 Cetus 对只读从库的复制延迟监控，如果 check-slave-delay 打开，一旦从库复制延迟超过阈值 slave-delay-down，会通知 Cetus 该后端将不可用；如果从库复制进度恢复正常，延迟已经小于 slave-delay-recover 的阈值，将恢复该后端，调整为可用；
- group-replication-mode 是否开启 MySQL 组复制监控，如果后端 MySQL 配置了 MGR，需要将其配置为 1，表示开启监控，当 MGR 集群中的成员发生切换或者有节点不可用，都会及时通知 Cetus 调整相应后端状态。

接下来再对另外两个配置文件 users.json 和 variables.json 进行简单说明。

（4）users.json 和 variables.json 配置文件如下：

```
[cetus_33060@hz_10_120_240_251 conf]$ cat users.json
{
    "users":    [{
            "user":        "emp",
            "client_pwd":    "emp123",
            "server_pwd":    "emp123"
    }, {
            "user":        "emp_sel",
            "client_pwd":    "emp123",
            "server_pwd":    "emp123"
    }]
}

[cetus_33060@hz_10_120_240_251 conf]$ cat variables.json
{
    "variables": [
        {
            "name": "sql_mode",
            "type": "string-csv",
            "allowed_values":   ["STRICT_TRANS_TABLES",
                            "NO_AUTO_CREATE_USER",
```

```
                    "NO_ENGINE_SUBSTITUTION"]
        },
        {
            "name": "connect_timeout",
            "type": "string",
            "allowed_values": ["*"],
            "silent_values": ["10", "100"]
        }
    ]
}
```

其中：
- client_pwd 是前端应用连接 Cetus 所使用的密码；
- server_pwd 是 Cetus 连接后端 MySQL 所使用的密码，该密码可以与 client_pwd 一致，也可以不同。

而在 variables.json 文件中，示例中配置了 sql_mode，也可以配置其他的 session 变量。

（5）配置完成后，启动 Cetus。

```
[cetus_33060@hz_10_120_240_251 cetus_20181105_3fac75c_multi]$ cat start.sh
#!/bin/bash
/home/cetus_33060/cetus_20181105_3fac75c_multi/bin/cetus --defaults-file=/home/cetus_33060/cetus_20181105_3fac75c_multi/conf/proxy.conf

[cetus_33060@hz_10_120_240_251 cetus_20181105_3fac75c_multi]$ sh start.sh
```

建议安装完成 Cetus 后，便配置好启动和关闭脚本，方便后期的管理维护。

Cetus 启动后，可以查看日志文件：

```
(pid=29951) 2018-11-29 16:17:56: (message) starting cetus 2.0.0
(pid=29951) 2018-11-29 16:17:56: (message) build revision: v1.1.0-180-g3fac75c
(pid=29951) 2018-11-29 16:17:56: (message) glib version: 2.28.8
(pid=29951) 2018-11-29 16:17:56: (message) libevent version: 4.24
(pid=29951) 2018-11-29 16:17:56: (message) config dir: /home/cetus_33060/cetus_20181105_3fac75c_multi/conf
(pid=29951) 2018-11-29 16:17:56: (message) read 2 users
(pid=29951) 2018-11-29 16:17:56: (message) ssl is false
(pid=29951) 2018-11-29 16:17:56: (message) plugin proxy v1.1.0-180-g3fac75c started
(pid=29951) 2018-11-29 16:17:56: (message) plugin admin v1.1.0-180-g3fac75c started
(pid=29951) 2018-11-29 16:17:56: (message) src/mysql-proxy-cli.c:652:SO_REUSEPORT is defined
(pid=29951) 2018-11-29 16:17:56: (message) set worker processes:4
(pid=29951) 2018-11-29 16:17:56: (message) set default pool size:100
(pid=29951) 2018-11-29 16:17:56: (message) set max pool size:200
(pid=29951) 2018-11-29 16:17:56: (message) set max resp len:10485760
(pid=29951) 2018-11-29 16:17:56: (message) set max alive time:7200
(pid=29951) 2018-11-29 16:17:56: (message) src/mysql-proxy-cli.c:693:set merged output size:8192
(pid=29951) 2018-11-29 16:17:56: (message) src/mysql-proxy-cli.c:696:set max header size:65536
(pid=29951) 2018-11-29 16:17:56: (message) set client_found_rows false
……
```

在启动过程中，需要观察日志中有无 critical 的信息出现，例如：

```
(pid=29960) 2018-11-29 16:17:59: (message) src/network-mysqld.c:4605: self conn timeout, state:0, con:0xf6c3b0, server:0xf6c420
(pid=29960) 2018-11-29 16:17:59: (critical) src/network-mysqld.c:4614: get conn timeout from master
(pid=29960) 2018-11-29 16:17:59: (warning) src/network-mysqld.c:4737: con:0xf6c3b0 failed for server:0xf6c420
```

在解决完所有的问题后，Cetus 就可以正常启动了。

2. Sharding 版本

Sharding 版本（分片版本）解决了 MySQL 数据库水平扩展的问题，此时后端的数据库会被当作一个一个的分片节点，Cetus 将前端发送过来的请求进行解析，并对部分 SQL 进行改写，例如前端采用 Insert 多 Values 方式批量插入数据，Cetus 会将其拆分并重新打包后路由至相应节点。图 32-4 是该版本的一个示意图。

在图 32-4 中,后端配置了 3 组主从环境,分别对应 3 套主从实例(分表功能正在紧张开发中,即支持单一实例上对表进行水平拆分,相信很快就会有正式版本发布),另外,在每一组主从环境中,同时也是支持读写分离这一功能的。

Sharding 版本较读写分离版本复杂一些,并引入了几个新的概念。由于分片环境的配置与业务关联度很高,所以在接下来的演示过程中,将分为安装、数据库设计和配置 3 个部分介绍。

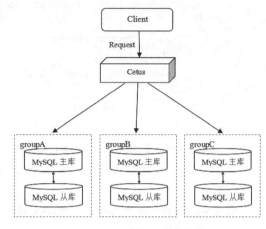

图 32-4　Sharding 环境示意图

(1)安装。

❶ 首先下载安装包,可以参考前面的内容,此处略过。

❷ 开始编译安装(Sharding 版本的编译参数 DSIMPLE_PARSER=OFF)。

```
[cetus_34060@hz_10_120_240_251 ~]$ cd cetus;mkdir build; cd build
[cetus_34060@hz_10_120_240_251 build]$ cmake ../ -DCMAKE_BUILD_TYPE=Debug -DCMAKE_INSTALL_PREFIX=/home/cetus_34060/cetus_20181026_9e349f5_multi -DSIMPLE_PARSER=OFF
[cetus_34060@hz_10_120_240_251 build]$ make
[cetus_34060@hz_10_120_240_251 build]$ make install
[  2%] Built target ev-cetus
[  4%] Built target mysql-chassis-glibext
……
-- Installing: /home/cetus_34060/cetus_20181026_9e349f5_multi/lib/cetus/plugins/libshard.so
-- Removed runtime path from "/home/cetus_34060/cetus_20181026_9e349f5_multi/lib/cetus/plugins/libshard.so"
-- Installing: /home/cetus_34060/cetus_20181026_9e349f5_multi/lib/libsqlparser.so
```

到此为止,Sharding 版本的 Cetus 已经编译完成。

(2)分片设计。

在进行下一步的配置环节之前,先介绍 Cetus 在 Sharding 版本中的一些设计思想和概念,有助于理解下文的内容(见图 32-5)。

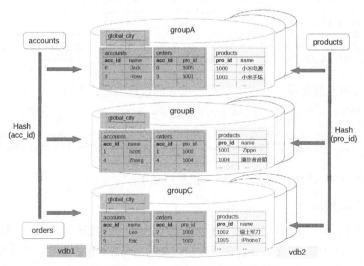

图 32-5　Cetus 数据分片设计示意图

32.2 Cetus 架构

在该示意图中，可以看到包含两个 VDB（Virtual DataBase）——vdb1 和 vdb2，即逻辑 DB，主要对应分片表，表现在业务数据层，代表此 VDB 内的数据有相同属性值，可以根据此属性值，进行数据的进一步拆分，从而构成分片表（Sharding 表），例如订单数据 orders 和用户数据 accounts 属于同一个 VDB，都可以根据用户 ID（acc_id）进行分片，从而进行关联查询，同理，产品数据 products 由于经常需要通过产品 ID（prod_id）进行访问，或者是关联查询，所以放入另一个 VDB 内。需要注意的是，不同 VDB 之间的数据不能进行关联查询，只有在同一个 VDB 内才支持。

当 VDB 设计完成之后，也就确定好了分片键（Sharding Key）和分片表。在图 32-5 中，在 vdb1 中，分片键为 acc_id，分片表为 orders 和 accounts；在 vdb2 中，分片键为 prod_id，分片表为 products。

继续观察图 32-5，当分片键和分片表都确定下来后，应该选用何种分片办法呢？即数据到底如何落到每一个后端？这时就需要考虑分片规则，应该选用 hash 还是 range，当前 hash 方法支持数字类型、字符串类型，range 方法支持数字类型、字符串类型和时间类型（date/datetime）。

沿着数据流继续走，接下来便到了底层存储。图 32-5 中包括 3 个逻辑数据组 groupA、groupB 和 groupC。通过示意图中的数据记录，可以发现之前定义的 3 个分片表 orders、accounts 和 products 分别根据分片规则 hash 方法进行数据拆分，最终落入不同的逻辑数据组内，而每一个逻辑数据组最终将映射到一个物理 DB，例如 groupA 存储到 10.120.240.251:3308（详见下文 Sharding 的配置），至此，数据流工作完成。

从以上这些内容可以得知，在设计 VDB 时，要尽早确定分片表、分片键以及分片规则，规划好 VDB，这些工作最好是在业务之初，因后续的底层数据存储以及上层数据操作都会与此息息相关。

此外，还有一种表不同于分片表，该类型的表称为全局表，支持关联查询，如图 32-5 中的 global_city，该类型的表在不同的底层物理节点中，数据是相同的，即都是全量数据，当然这也需要是在同一个 VDB 内。如果是不同的 VDB，全局表也会被认为是不同的，例如配置表 conf，vdb1 和 vdb2 都需要，此时就需要前端应用分别将变更数据写入 vdb1 和 vdb2 中。可见，针对全局表数据的操作每次都涉及多个后端，需要用到 XA 事务。如果是一个更新特别频繁的表，将会产生大量 XA 事务，从而影响性能。

为了缓解这种现象，减少 XA 事务，Cetus 又引入了另一种类型的全局表：单点全局表（Single 表），此种类型的表数据仅会存于某一个后端节点，其余节点是没有数据的，例如仅将单点全局表配置在逻辑数据组 data2 中（详见下文 Sharding 的配置），需要注意的是，单点全局表仅是为了解决全局表的上述问题，但该类型的表是不支持关联查询的，这个不同于传统意义的全局表，可以说单点全局表是全局表的一个特殊情况。

在介绍完这些内容后，开始 Cetus 的配置工作。

（3）配置。

该版本的 Cetus 在目录结构方面与读写分离版本相同，配置文件需要 shard.conf、sharding.json、users.json 和 variables.json，其中 shard.conf 是 Cetus 的主配置文件，后 3 个 json 文件分别对应分片配置、用户密码配置和会话环境变量配置。

接下来就对配置文件中的重要参数进行说明，由于在读写分离版本的安装配置时，已经介绍了一些参数,这里将挑选不同的参数或者有不同含义的部分进行介绍,更详细的内容可参考文档 *https://github.com/Lede-Inc/cetus/blob/master/doc/cetus-shard-profile.md*。

❶ 主配置文件 shard.conf,如下所示:

```
[cetus]
verbose-shutdown=true
daemon=true
basedir=/home/cetus_34060/cetus_20181105_3fac75c_multi
conf-dir=/home/cetus_34060/cetus_20181105_3fac75c_multi/conf
pid-file=/home/cetus_34060/cetus_20181105_3fac75c_multi/cetus34060.pid
plugin-dir=lib/cetus/plugins
plugins=shard,admin
log-level=debug
log-file=/home/cetus_34060/cetus_20181105_3fac75c_multi/logs/cetus_34060.log
log-xa-file=/home/cetus_34060/cetus_20181105_3fac75c_multi/logs/xa.log
max-open-files=65536
default-charset=utf8
default-username=emp
default-db=employees
max-pool-size=400
worker-processes=4
worker-id=2307
ssl=false
slave-delay-down=3.000000
slave-delay-recover=1.000000
long-query-time=100
sql-log-bufsize=10485760
sql-log-switch=ON
sql-log-prefix=cetus
sql-log-path=/home/cetus_34060/cetus_20181105_3fac75c_multi/logs
sql-log-maxsize=1024
sql-log-mode=BACKEND
proxy-address=0.0.0.0:34060
proxy-backend-addresses=10.120.240.251:3308@data1,10.120.240.251:3488@data2
proxy-read-only-backend-addresses=10.120.240.251:3307@data1,10.120.240.251:3489@data2
allow-nested-subquery=false
admin-address=0.0.0.0:34061
admin-username=admin
admin-password=admin
```

其中:

● plugins 表示该 Cetus 启用的插件,该示例是 Sharding 版本,所以启用 shard 和 admin 即可;

● log-xa-file 在读写分离版本中是不需要的,该参数表示 xa 日志文件的路径;

● sql-log 开头的这一组参数表示当启动 Cetus 时,同时打开全量日志功能,当然也可以启动 Cetus 后再动态开启,该功能在读写分离版本中也是有的,一样的用途和配置方式;

● proxy-backend-addresse 和 proxy-read-only-backend-addresses 这一组参数在读写分离版本中同样进行过说明,但是当其出现在 Sharding 版本后,参数值需要做出一些调整,如示例内容,在后端地址中多了@data1 和@data2 的信息,这表示逻辑数据组,且这些信息需要明确配置在配置文件 sharding.json(关于该文件的介绍和配置,详见下文);

● allow-nested-subquery 该参数表示是否允许执行多层嵌套子查询,因子查询写法多种多样,形式灵活自由,这会给 Cetus 的解析工作带来很大压力。另外,由于数据已经被打散在不同的后端,多层嵌套子查询可能涉及不同后端的不同组合操作,Cetus 本身对关联查询还是有一定限制的(关于 Cetus 的使用限制,可以参考文档 https://github.com/Lede-Inc/cetus/blob/master/ doc/cetus-constraint.md),所以为了保证结果的准确性,默认是不支持该操作的。如果使用者在确认打开该功能后,返回结果集是正确的,就可以启用该功能。

32.2 Cetus 架构

❷ 配置文件 sharding.json。

由于该文件的配置与业务息息相关，所以在配置规则期间，请根据自身业务逻辑定义分片规则（如何进行分片设计，详见上文介绍），样例文件如下：

```
[cetus_34060@hz_10_120_240_251 conf]$ cat sharding.json
{
  "vdb": [
    {
      "id": 1,
      "type": "int",
      "method": "hash",
      "num": 8,
      "partitions": {"data1": [0,1,2,3], "data2": [4,5,6,7]}
    },
    {
      "id": 2,
      "type": "int",
      "method": "range",
      "num": 0,
      "partitions": {"data1": 124999, "data2": 249999}
    }
  ],
  "table": [
    {"vdb": 1, "db": "employees_hash", "table": "demo_emp", "pkey": "emp_no"},
    {"vdb": 1, "db": "employees_hash", "table": "demo_titles", "pkey": "emp_no"},
    {"vdb": 2, "db": "employees_range", "table": "demo_salaries", "pkey": "emp_no"}
  ],
  "single_tables": [
    {"table": "demo_single_table", "db": "employees_range", "group": "data2"}
  ]
}
```

在示例中，需要在每个 MySQL 后端创建两个 DB，一个是 employees_hash，另一个是 employees_range，即对应两个 VDB（id=1 和 id=2）；前者包含两个分片表（Sharding 表）demo_emp、demo_titles，分片键（Sharding Key）为 emp_no，后一个 DB 包括一个 Sharding 表 demo_salaries（Sharding Key 为 emp_no）和一个单点全局表（Single 表）demo_single_table，表结构要在每一个后端进行创建。

存储部分共分成了两个逻辑数据组（group），即 data1 和 data2，对应 shard.conf 配置文件中的 proxy-backend-addresse 和 proxy-read-only-backend-addresses 这两个参数，分片键 emp_no 将按照 hash、range 规则进行分组，从而存储到不同的物理节点，例如，data1 数据组的数据写入后端地址 10.120.240.251:3308，而 data2 数据组的数据写入到 10.120.240.251:3488。

❸ 配置文件 users.json 和 variables.json 配置文件与读写分离版一致，不再赘述。

❹ 启动 Cetus。

日志中也可以看到 Sharding 相关的信息。

```
  (pid=32051) 2018-11-21 18:20:02: (message) added read/write backend: 10.120.240.251:3308@data1, state: unknown
  (pid=32050) 2018-11-21 18:20:02: (message) src/cetus-process-cycle.c:277: pass channel s:2 pid:32053 fd:11 to s:1 pid:32052 fd:9, ev base:0x1148670, ev:0x7ffbeb2b3a00
  (pid=32050) 2018-11-21 18:20:02: (message) src/cetus-process.c:177: after call fork, pid:0
  (pid=32050) 2018-11-21 18:20:02: (message) src/cetus-process.c:85: create channel 13:14
  (pid=32051) 2018-11-21 18:20:02: (message) added read/write backend: 10.120.240.251:3488@data2, state: unknown
  (pid=32053) 2018-11-21 18:20:02: (message) src/cetus-process.c:191: after call fork, channel:2, pid:32053
  (pid=32050) 2018-11-21 18:20:02: (message) src/cetus-process.c:174: before call fork, channel:3, pid:0
  (pid=32053) 2018-11-21 18:20:02: (message) src/cetus-process-cycle.c:528: call cetus_worker_process_cycle
  ……
```

至此，Sharding 版的 Cetus 配置完成。

3．远程配置库

前文介绍 Cetus 启动的方式，都是直接读取本地的配置文件，以此来加载需要的参数。其实 Cetus 还支持另一种启动方式，即读取远程配置库，这些配置参数需要提前写入配置库。这样做的好处是如果 Cetus 同时启用了多个实例，而这些实例都服务于同一个业务，当后端环境发生变化，需要考虑 Cetus 实例间参数一致性的问题，避免服务于同一业务的 Cetus 因配置不同，引起不必要的麻烦（例如，不同的实例配置了不同的用户，在轮询路由时有的连接正常，有的连接失败）。

下面将示范 Cetus 如何利用远程配置库启动，并变更配置。

（1）首先，在配置库中需要创建 3 张表，分别为 settings、objects 和 services。创建脚本的方法可参考文档 https://github.com/Lede-Inc/cetus/blob/master/doc/cetus-configuration.md。

（2）启动 Cetus。

```
bash-4.2# /home/cetus/bin/cetus --remote-conf-url=mysql:// catalog:catalog@127.0.01:3306/catalog
root      2567     1  0 06:23 ?        00:00:00 /home/cetus/libexec/cetus --remote-conf-url=mysql://catalog:catalog@127.0.01:3306/catalog
root      2568  2567  3 06:23 ?        00:00:00 /home/cetus/libexec/cetus --remote-conf-url=mysql://catalog:catalog@127.0.01:3306/catalog
root      2569  2567  5 06:23 ?        00:00:01 /home/cetus/libexec/cetus --remote-conf-url=mysql://catalog:catalog@127.0.01:3306/catalog
root      2570  2567  5 06:23 ?        00:00:01 /home/cetus/libexec/cetus --remote-conf-url=mysql://catalog:catalog@127.0.01:3306/catalog
root      2649  2567  5 06:23 ?        00:00:01 /home/cetus/libexec/cetus --remote-conf-url=mysql://catalog:catalog@127.0.01:3306/catalog
```

如上所示，Cetus 启动时指定参数--remote-conf-url 连接串，即远端配置库地址，例如示例中的信息 mysql://catalog:catalog@127.0.0.1:3306/catalog。

- mysql 表示远端配置库类型，目前支持 MySQL 和 SQLite 方式；
- catalog:catalog@127.0.0.1:3306/catalog 分别表示连接 MySQL/SQLite 数据库的连接信息，即用户名:密码@服务器 ip:端口/database。

由于该示例启用了 4 个 worker 进程，所以看到共有 1 个父进程和 4 个子进程。

（3）变更 Cetus 用户配置信息。

在使用了远程配置库方式启动 Cetus 后，如果需要修改用户配置信息，可以直接通过 Cetus 管理端用命令进行更改，Cetus 会负责自动同步至远程配置库，或者先在配置库进行更改，然后调用 Cetus 提供的 reload 命令，如下所示：

```
mysql> config reload user;
Query OK, 0 rows affected (0.03 sec)
```

另外，除了以上命令，还有 config reload variables 和 config reload 命令，具体说明可以参考文档 https://github.com/Lede-Inc/cetus/blob/master/doc/cetus-configuration.md。

4．Docker 环境

近来，Docker 环境已经被越来越多的人所接受并投入使用，因其部署简便快捷，使用者只需要拉取已经打包好的镜像，并启动该容器，就可以直接使用该环境，尤其适用于体验一些新软件、新特性。接下来，为了读者可以更方便地感受 Cetus 的使用效果，研发团队也专门定制了 Cetus 的镜像。虽然目前仅有读写分离版本，但相信不久之后，也会提供 Sharding 版本的镜像。

32.2 Cetus 架构

下面介绍一下如何使用该镜像。

（1）拉取镜像文件。

```
[root@localhost docker]# docker pull docker.io/ledetech/cetus
Using default tag: latest
Trying to pull repository docker.io/ledetech/cetus ...
latest: Pulling from docker.io/ledetech/cetus
e64f6e679e1a: Pull complete
799d60100a25: Pull complete
85ce9d0534d0: Pull complete
d3565df0a804: Pull complete
ace49781913f: Pull complete
7ae33d256be8: Pull complete
eca020eb19bf: Pull complete
8cb681e3355b: Pull complete
4ba691b11ae0: Pull complete
d920d873626b: Pull complete
7b2754e72b14: Pull complete
6f3b62f2f427: Pull complete
92648656de4e: Pull complete
7804ccf39eda: Pull complete
f4e1106e7694: Pull complete
33f99b44c2d0: Pull complete
Digest: sha256:e66e14a93a915728d4438c1e887948015a47ae25518543d809c47af0e56abfa5
Status: Downloaded newer image for docker.io/ledetech/cetus:latest

[root@localhost docker]# docker images
REPOSITORY                TAG       IMAGE ID       CREATED      SIZE
docker.io/ledetech/cetus  latest    1f53efcfa105   7 days ago   296.9 MB
```

可以看到，该镜像已经成功获取。

（2）创建并启动 Docker 容器。

```
[root@localhost docker]# docker run -d -P -it -p 3306:3306 -p 3307:3307 -p 3308:3308 -p 3309
:3309 -p 4306:4306 -p 5306:5306 1f53efcfa105
74e08d9cecadac1d9076315fe1758e620666546416219f2e5c943eddc3145966


[root@localhost docker]# docker ps
CONTAINER ID     IMAGE              COMMAND                  CREATED          STATUS
    PORTS                                                                                NAMES
74e08d9cecad     1f53efcfa105       "/bin/sh -c 'bash /en"   5 seconds ago    Up 3
seconds     0.0.0.0:3306-3309->3306-3309/tcp, 0.0.0.0:4306->4306/tcp, 0.0.0.0:5306->5306/tcp,
 0.0.0.0:8192->33060/tcp    angry_goldstine
```

启动容器时，指定了端口映射信息，服务器的 3306 端口映射为容器内的 3306，以此类推，分别映射 3307、3308、3309、4306 和 5306 端口，其中 3306 端口对应后端 MySQL 主库，3307、3308 端口对应后端 MySQL 从库，即一主两从架构，4306 是 Cetus 服务端口，5306 是 Cetus 的管理端口。

（3）验证 Cetus 服务。

```
[root@localhost ~]# mysql -ucetus_app -p123456 -h127.0.0.1 -P4306
mysql> show databases;
+--------------------+
| Database           |
+--------------------+
| cetus              |
| information_schema |
| proxy_heart_beat   |
+--------------------+
3 rows in set (0.02 sec)


[root@localhost ~]# mysql -uadmin -padmin -h127.0.0.1 -P5306
mysql> select * from backends\G;
*************************** 1. row ***************************
          PID: 388
  backend_ndx: 1
```

```
            address: 127.0.0.1:3306
              state: up
               type: rw
    slave delay(ms): NULL
         idle_conns: 20
         used_conns: 0
        total_conns: 20
*************************** 2. row ***************************
                PID: 388
        backend_ndx: 2
            address: 127.0.0.1:3307
              state: up
               type: ro
    slave delay(ms): 11
         idle_conns: 20
         used_conns: 0
        total_conns: 20
......
```

分别通过 4306 和 5306 端口对 Cetus 进行验证，从以上可以看到，Cetus 服务已经正常启动，可以开始进行一系列的体验之旅。关于这方面的更详细的介绍，可以查看 GitHub 文档 https://hub.docker.com/r/ledtech/cetus/。

32.2.2　Cetus 的使用

前面已经介绍了 Cetus 的安装和配置，那么接下来就在实际应用过程中体验一下。

1. 读写分离版本

此次部署后端采用 MGR 集群模式，部署完成以后，可以通过管理端口查看此时后端在 Cetus 里的连接状态（关于 Cetus 后端管理的内容，下文中有详细介绍）。

```
mysql> select * from backends\G;
*************************** 1. row ***************************
            PID: 15522
    backend_ndx: 1
        address: 10.120.240.251:3305
          state: up
           type: rw
    slave delay: NULL
     idle_conns: 100
     used_conns: 0
    total_conns: 100
*************************** 2. row ***************************
            PID: 15522
    backend_ndx: 2
        address: 10.120.240.251:3315
          state: up
           type: ro
    slave delay: 51
     idle_conns: 100
     used_conns: 0
    total_conns: 100
......
```

如上所示，PID 表示 Cetus 的进程号，即哪一个进程在提供服务；state 状态为 up，表示此后端正常，可以提供服务；type 为 rw，表示该后端为可写节点，如果 type 的值是 ro，表示该后端为只读节点。

下面继续做一下读写请求验证。

```
[cetus_33060@hz_10_120_240_251 ~]$ mysql -uemp -h10.120.240.251 -P33060 -p
mysql> update employees.salaries set salary=62101 where emp_no=10001 and salary=100;
Query OK, 1 row affected (0.00 sec)
Rows matched: 1  Changed: 1  Warnings: 0
```

```
mysql> select * from employees.salaries where emp_no=10001 and salary=62101;
+--------+--------+------------+------------+
| emp_no | salary | from_date  | to_date    |
+--------+--------+------------+------------+
|  10001 |  62101 | 1986-06-26 | 1987-06-26 |
+--------+--------+------------+------------+
1 row in set (0.02 sec)
```

另外，也可以添加提示，强制指定事务走读写后端还是只读后端，代码如下所示。

```
[cetus_33060@hz_10_120_240_251 ~]$ mysql -uemp -h10.120.240.251 -P33060 -p --comments
mysql> update /*# mode=readonly */ employees.salaries set salary=62101 where emp_no=10001 and salary=100;
ERROR 1105 (07000): Force write on read-only slave
mysql>
mysql> update /*# mode=readwrite */ employees.salaries set salary=62101 where emp_no=10001 and salary=100;
Query OK, 0 rows affected (0.00 sec)
Rows matched: 0  Changed: 0  Warnings: 0
```

如果想让提示功能生效，需要打开--comments 选项，从以上测试可以看到，当强制 update 语句在只读节点执行时，会给出错误提示信息，以避免路由至错误节点，导致数据不一致，而这些提示功能，服务于一些特殊需求，具体可以参考 GitHub 相关文档 https://github.com/Lede-Inc/cetus/blob/master/doc/cetus-routing-strategy.md。

由于后端使用了 MySQL 的 Group Replication 集群，所以验证一下如果发生 MGR 集群成员角色切换，Cetus 会作何反应。

首先，连接 Cetus 服务，开启数据持续加载脚本（串行加载），初始时 3315 实例为 Primary 成员，然后 kill 掉该实例，后 MGR 集群发生切换，新 Primary 成员为 3305 实例。

```
mysql> select * from demo_single_primary where id > 780 limit 5;
+-----+---------------+------------+---------------------+
| id  | instance_port | account_id | mtime               |
+-----+---------------+------------+---------------------+
| 781 |          3315 |      10000 | 2018-11-02 17:15:08 |
| 782 |          3315 |      10000 | 2018-11-02 17:15:08 |
| 783 |          3305 |      93542 | 2018-11-02 17:15:14 |
| 784 |          3305 |      36765 | 2018-11-02 17:15:14 |
| 785 |          3305 |       6184 | 2018-11-02 17:15:14 |
+-----+---------------+------------+---------------------+
5 rows in set (0.00 sec)
```

从以上可以得知，在后端 MGR 集群发生角色切换以后，大约经过了 6s，业务恢复正常，当然这个也与测试环境有关，此数据仅供参考。

2. Sharding 版本

以下演示操作延用上文的配置文件，如下所示：

```
{
  "vdb": [
    {
      "id": 1,
      "type": "int",
      "method": "hash",
      "num": 8,
      "partitions": {"data1": [0,1,2,3], "data2": [4,5,6,7]}
    },
    {
      "id": 2,
      "type": "int",
      "method": "range",
      "num": 0,
      "partitions": {"data1": 124999, "data2": 249999}
    }
```

```
    ],
    "table": [
      {"vdb": 1, "db": "employees_hash", "table": "demo_emp", "pkey": "emp_no"},
      {"vdb": 1, "db": "employees_hash", "table": "demo_titles", "pkey": "emp_no"},
      {"vdb": 2, "db": "employees_range", "table": "demo_salaries", "pkey": "emp_no"}
    ],
    "single_tables": [
      {"table": "demo_single_table", "db": "employees_range", "group": "data2"}
    ]
}
```

关于这些信息，从 Cetus 的管理端也可以进行查询，如下所示（关于 Cetus 后端管理的内容，下文中有详细介绍）。

```
mysql> select * from groups;
+-------+----------------------+----------------------+
| group | master               | slaves               |
+-------+----------------------+----------------------+
| data1 | 10.120.240.251:3308  | 10.120.240.251:3307  |
| data2 | 10.120.240.251:3488  | 10.120.240.251:3489  |
+-------+----------------------+----------------------+
2 rows in set (0.00 sec)

mysql> select * from vdb;
+--------+-----------+----------------------------------------------------------------+
| VDB id | Method    | Partitions                                                     |
+--------+-----------+----------------------------------------------------------------+
| 1      | hash(INT) | [0,1,2,3]->data1; [4,5,6,7]->data2                             |
| 2      | range(INT)| (-2147483648, 124999]->data1; (124999, 249999]->data2          |
+--------+-----------+----------------------------------------------------------------+
2 rows in set (0.00 sec)

mysql> select sharded table;
+-------------------------------+--------+--------+
| Table                         | VDB id | Key    |
+-------------------------------+--------+--------+
| employees_hash.demo_emp       | 1      | emp_no |
| employees_hash.demo_titles    | 1      | emp_no |
| employees_range.demo_salaries | 2      | emp_no |
+-------------------------------+--------+--------+
3 rows in set (0.00 sec)

mysql> select single table;
+----------------------------------+-------+
| Table                            | Group |
+----------------------------------+-------+
| employees_range.demo_single_table| data2 |
+----------------------------------+-------+
1 row in set (0.00 sec)
```

接下来开始操作 Sharding 表，针对 hash 类型的表插入一些数据。

```
[cetus_34060@hz_10_120_240_251 ~]$ mysql -uemp -h10.120.240.251 -P34060 --comments -p
mysql> use employees_hash;
Database changed

mysql> insert into employees_hash.demo_emp(emp_no,dept_no,from_date,to_date) values(10001,'d005',
'1986-06-26','9999-01-01'),(10002,'d007','1996-08-03','9999-01-01'),(10003,'d004','1995-12-03',
'9999-01-01');
Query OK, 3 rows affected (0.01 sec)
Records: 3  Duplicates: 0  Warnings: 0

mysql> insert into employees_hash.demo_emp(emp_no,dept_no,from_date,to_date) values(10005,'d003',
'1989-09-12','9999-01-01');
Query OK, 1 row affected (0.04 sec)

mysql> select /*# group=data1 */ count(*) from demo_emp;
+----------+
| count(*) |
+----------+
|        3 |
```

```
+----------+
1 row in set (0.01 sec)

mysql> select /*# group=data2 */ count(*) from demo_emp;
+----------+
| count(*) |
+----------+
|        1 |
+----------+
1 row in set (0.00 sec)
```

在 Sharing 表 employees_hash.demo_emp 中,插入了两批数据,第一批有 3 条记录,第二批有 1 条记录。利用提示语法,可以看到在 data1 组中包括 3 条记录,在 data2 组中包括 1 条记录。

针对 range 类型的表插入一些数据,演示过程同上,读者可以自行操作。下面演示如何操作全局表。

- 单点全局表:

```
mysql> insert into employees_range.demo_single_table(emp_no,birth_date,first_name,last_name,
gender,hire_date) values(10003,'1959-12-03','Parto','Bamford','M','1986-08-28');
Query OK, 1 row affected (0.01 sec)
……
mysql>
mysql> select /*# group=data1 */ count(*) from employees_range.demo_single_table;
+----------+
| count(*) |
+----------+
|        0 |
+----------+
1 row in set (0.00 sec)

mysql> select /*# group=data2 */ count(*) from employees_range.demo_single_table;
+----------+
| count(*) |
+----------+
|        3 |
+----------+
1 row in set (0.00 sec)
```

在向单点全局表写入数据时,该数据只会写在预先定义的存储节点中,而全局表则不一样,具体如下。

- 全局表:

```
mysql> insert into employees_hash.departments(dept_no,dept_name) values('d009','Customer Service');
Query OK, 1 row affected (0.01 sec)

mysql> select /*# group=data2 */ * from employees_hash.departments;
+---------+------------------+
| dept_no | dept_name        |
+---------+------------------+
| d009    | Customer Service |
+---------+------------------+
1 row in set (0.00 sec)

mysql> select /*# group=data1 */ * from employees_hash.departments;
+---------+------------------+
| dept_no | dept_name        |
+---------+------------------+
| d009    | Customer Service |
+---------+------------------+
1 row in set (0.00 sec)
```

在向全局表写入数据时,该数据会被同时写入后端所有存储节点,原来普通的单机事务会被改写为分布式事务。

这样设计出来的数据库,可以进行如下的关联查询。

（1）Sharding 表之间的关联查询。

```
mysql> select demo_emp.emp_no,demo_emp.dept_no,demo_titles.title from demo_emp, demo_titles
where demo_emp.emp_no=demo_titles.emp_no;
+--------+---------+-----------------+
| emp_no | dept_no | title           |
+--------+---------+-----------------+
|  10001 | d005    | Senior Engineer |
|  10002 | d007    | Staff           |
|  10005 | d003    | Senior Staff    |
+--------+---------+-----------------+
3 rows in set (0.03 sec)
```

当 Sharding 表和 Sharding 表相互关联的时候，需要通过 Sharding Key 作为条件，因同一个 VDB 内的 Sharding 表是有相同业务含义的，所以可以通过 where 限制条件，将不同表的数据在相同节点内关联查询。

（2）Sharding 表与全局表的关联查询。

```
mysql> select demo_emp.emp_no,demo_emp.dept_no,departments.dept_name from demo_emp left join
departments on demo_emp.dept_no=departments.dept_no;
+--------+---------+-----------+
| emp_no | dept_no | dept_name |
+--------+---------+-----------+
|  10001 | d005    | NULL      |
|  10002 | d007    | NULL      |
|  10003 | d004    | NULL      |
|  10005 | d003    | NULL      |
+--------+---------+-----------+
4 rows in set (0.00 sec)
```

当 Sharding 表与全局表相互关联的时候，由于每个节点都有相同的数据，所以 Sharding 表的不同存储节点将分别与全局表关联，最后将所有数据再进行归并。

以上是对 Cetus 的简单操作，如果想要了解更详细的操作说明，可以查看 github 文档 https://github.com/Lede-Inc/cetus/blob/master/doc/cetus-sharding.md。

32.2.3　Cetus 日志文件

Cetus 包括多种日志，这对于诊断问题来说，很有帮助。当前 Cetus 日志文件共有 4 类，文件默认位置在安装目录的 logs 下，如下所示。

```
[cetus_34060@hz_10_120_240_251 logs]$ ll
total 8724
-rw-rw-rw- 1 cetus_34060 cetus_34060       0 Nov  6 16:40 cetus-21524.clg
-rw-rw-rw- 1 cetus_34060 cetus_34060       0 Nov  6 16:40 cetus-21525.clg
-rw-rw-rw- 1 cetus_34060 cetus_34060       0 Nov  6 16:40 cetus-21526.clg
-rw-rw-rw- 1 cetus_34060 cetus_34060       0 Nov  6 16:40 cetus-21527.clg
-rw-rw---- 1 cetus_34060 cetus_34060 8914615 Nov  6 16:51 cetus_34060.log
-rw-rw-rw- 1 cetus_34060 cetus_34060       0 Nov  5 17:30 cetus_34060.log.slowquery.log
-rw-r--r-- 1 cetus_34060 cetus_34060       5 Nov  5 14:43 README_XA
-rw-r--r-- 1 cetus_34060 cetus_34060     148 Nov  6 11:56 xa.log_11
-rw-r--r-- 1 cetus_34060 cetus_34060     296 Nov  6 12:00 xa.log_12
-rw-r--r-- 1 cetus_34060 cetus_34060     148 Nov  6 15:05 xa.log_15
-rw-r--r-- 1 cetus_34060 cetus_34060       0 Nov  5 17:30 xa.log_17
```

其中，文件 cetus_xxxx.log 是主日志，即 Cetus 的相关信息，如果需要诊断问题，该文件是第一个需要查看的，文件 xa.log_xx 是分布式事务的日志，如果执行了分布式事务，都会在该文件中记载，文件 cetus_xxxx.log.slowquery.log 是慢查询日志，文件格式与 MySQL 慢日志格式一致，最后一类文件 cetus-xxxx.clg 是全量日志文件，若要细粒度诊断问题，可以从全量日志入手，信息非常全面，但使用之前需要确认该功能已经打开。

下面一起来看下这几个日志文件的使用。

（1）主日志文件，包括进程相关的重要信息，相信读者对这个应该不会陌生，不再赘述。

（2）分布式事务日志文件，该文件记录了 Cetus 执行的所有分布式事务，如下所示。

```
[cetus_34060@hz_10_120_240_251 logs]$ cat xa.log_12
 2018/11/06 12:00:18 +993 [info] XA COMMIT 'clt-00:24:e8:69:b4:d5-0.0.0.0:34061-21525_11_
17251416' 10.120.240.251:3308@9176,10.120.240.251:3488@4505
 2018/11/06 12:00:29 +475 [info] XA COMMIT 'clt-00:24:e8:69:b4:d5-0.0.0.0:34061-21525_12_
17251417' 10.120.240.251:3308@9176,10.120.240.251:3488@4505
```

分布式事务 XID 为 clt-00:24:e8:69:b4:d5-0.0.0.0:34061-21525_11_17251416 的事务涉及两个后端，并且执行了 Commit 操作。

（3）慢日志，文件中记录了超过参数 long-query-time 的 SQL 信息，如下所示。

```
# Time: 2018-11-07T15:16:19.397217+08:00
# User@Host: emp @ 10.120.240.251:8245 Id: 0
# Query_time: 0.001000 Lock_time: 0.000000 Rows_sent: 0 Rows_examined: 0
SET timestamp=1541574979;
insert into employees_hash.departments(dept_no,dept_name) values("d012","Koala");
```

如何解读这个日志文件，可以参考上文中关于 MySQL 慢日志的内容。

（4）全量日志，该文件记录的内容更为详细，具体如下。

```
  2018-11-07 15:16:19.394: #backend-sharding# C_ip:10.120.240.251:8245 C_db:employees_hash C_
usr:emp C_tx:false C_id:50331650 trans(in_xa:false xa_state:UNKNOWN) attr_adj_state:2
  2018-11-07 15:16:19.396: #backend-sharding# C_ip:10.120.240.251:8245 C_db:employees_hash C_
usr:emp C_tx:true C_id:50331650 trans(in_xa:true xa_state:NEXT_ST_XA_START) attr_adj_state:8
  2018-11-07 15:16:19.396: #backend-sharding# C_ip:10.120.240.251:8245 C_db:employees_hash C_
usr:emp C_tx:true C_id:50331650 trans(in_xa:true xa_state:NEXT_ST_XA_START) attr_adj_state:8
  2018-11-07 15:16:19.397: #backend-sharding# C_ip:10.120.240.251:8245 C_db:employees_hash C_
usr:emp C_tx:true C_id:50331650 S_ip:10.120.240.251:3308 S_db:employees_hash S_usr:emp S_id:408
trans
(in_xa:true xa_state:NEXT_ST_XA_QUERY) latency:1.550(ms) OK type:Query insert into employees_
hash.departments(dept_no,dept_name) values("d012","Koala")
  2018-11-07 15:16:19.407: #backend-sharding# C_ip:10.120.240.251:8245 C_db:employees_hash C_
usr:emp C_tx:true C_id:50331650 S_ip:10.120.240.251:3488 S_db:employees_hash S_usr:emp S_id:10043
trans(in_xa:true xa_state:NEXT_ST_XA_CANDIDATE_OVER) latency:11.650(ms) OK type:Query commit
  2018-11-07 15:16:19.407: #backend-sharding# C_ip:10.120.240.251:8245 C_db:employees_hash C_
usr:emp C_tx:true C_id:50331650 S_ip:10.120.240.251:3308 S_db:employees_hash S_usr:emp S_id:408 trans
(in_xa:true xa_state:NEXT_ST_XA_CANDIDATE_OVER) latency:12.175(ms) OK type:Query commit
```

以上内容对应慢日志中的 insert 全过程，具体如何解读，可以参考 GitHub 文档 https://github.com/Lede-Inc/cetus/blob/master/doc/cetus-sqllog-usage.md。

32.2.4　Cetus 的后端管理

Cetus 除了上述介绍，还有强大的后端管理功能，方便日常运维操作，提供的命令接口也是尽量效仿 SQL 写法（支持的命令参考 select help 的输出），易于使用和掌握。另外，Web 版的控制台也已经开源，在 GitHub 上的地址为 https://github.com/Lede-Inc/cetus-GUI，读者可以自行下载使用。接下来将通过示例介绍后端管理的一些常用指令。

连接后端管理控制台：

```
[cetus_33060@hz_10_120_240_251 ~]$ mysql -uadmin -padmin -h127.0.0.1 -P33062
mysql> select help;
……
```

这里使用 MySQL 客户端工具 mysql 进行连接，连接串需要的信息在主配置文件有相应的定义，例如用户名对应配置文件中的 admin-username，密码对应 admin-password，地址信息来自于 admin-address。当连接上 Cetus 后端管理之后，可以执行 select help 查看所支持的各种命令。由于内容较多，在此没有打印这些输出信息。

接下来介绍两个常用操作：增加新用户和在线修改参数。

1. 增加新用户

日常运维过程中，由于业务的变动可能需要增加新用户或修改连接密码，在 Cetus 端记录了前端连接 Cetus 的用户密码，以及 Cetus 连接 MySQL 后端的用户密码，登录后台管理，可以通过 select * from app_user_pwd 和 select * from user_pwd 命令查看：

```
mysql> select * from app_user_pwd;
+------+------------------------------------------+
| user | password(sha1)                           |
+------+------------------------------------------+
| emp  | B89863E61D75FA82BFF16FB596575A30EDBEBD84 |
| root | FB298E44C7AC885B36FCC2D4BF15F748C9E3209E |
+------+------------------------------------------+
2 rows in set (0.00 sec)

mysql> select * from user_pwd;
+------+------------------------------------------+
| user | password(sha1)                           |
+------+------------------------------------------+
| emp  | B89863E61D75FA82BFF16FB596575A30EDBEBD84 |
| root | FB298E44C7AC885B36FCC2D4BF15F748C9E3209E |
+------+------------------------------------------+
2 rows in set (0.00 sec)
```

其中，app_user_pwd 用于前端应用到 Cetus 的连接，user_pwd 用于 Cetus 到 MySQL 后端的连接。

如果需要增加新的用户，可以通过如下命令操作：

```
mysql> update user_pwd set password='emp123' where user='emp_sel';
Query OK, 4 rows affected (0.02 sec)

mysql> update app_user_pwd set password='emp123' where user='emp_sel';
Query OK, 0 rows affected (0.00 sec)
```

分别在 app_user_pwd 和 user_pwd 增加相应的配置。

然后，查看增加后的结果。

```
mysql> select * from app_user_pwd;
+---------+------------------------------------------+
| user    | password(sha1)                           |
+---------+------------------------------------------+
| emp_sel | B89863E61D75FA82BFF16FB596575A30EDBEBD84 |
| emp     | B89863E61D75FA82BFF16FB596575A30EDBEBD84 |
| root    | FB298E44C7AC885B36FCC2D4BF15F748C9E3209E |
+---------+------------------------------------------+
3 rows in set (0.00 sec)

mysql> select * from user_pwd;
+---------+------------------------------------------+
| user    | password(sha1)                           |
+---------+------------------------------------------+
| emp_sel | B89863E61D75FA82BFF16FB596575A30EDBEBD84 |
| emp     | B89863E61D75FA82BFF16FB596575A30EDBEBD84 |
| root    | FB298E44C7AC885B36FCC2D4BF15F748C9E3209E |
+---------+------------------------------------------+
3 rows in set (0.00 sec)
```

新用户增加完成，前端应用可以使用新用户进行连接。

注意：如果老用户在 MySQL 后端修改了权限，旧连接（连接池中在用的连接）不会生效，仅对新连接有效。如果想使所有连接立即生效，建议重启 Cetus 实例。

2. 在线修改参数

Cetus 支持两类参数，一类是动态参数，另一类是静态参数。如何知道哪些参数可以动态调整呢？

可以通过命令 show variables 查看，例如查看从库延迟阈值。

```
mysql> show variables like 'slave-delay%';
+---------------------+---------------+----------+
| Variable_name       | Value         | Property |
+---------------------+---------------+----------+
| slave-delay-down    | 3.000000 (s)  | Dynamic  |
| slave-delay-recover | 1.000000 (s)  | Dynamic  |
+---------------------+---------------+----------+
2 rows in set (0.00 sec)
```

从以上可以发现，该参数的属性值（Property）显示为 Dynamic，即表示该参数支持动态调整，即调整后立即生效。如果显示为 Static，则表示该参数需要在配置文件中进行更改，并重启 Cetus 使其生效。

调整参数 slave-delay-down 为 5s：

```
mysql> config set slave-delay-down=5;
Query OK, 4 rows affected (0.08 sec)

mysql> show variables like 'slave-delay%';
+---------------------+---------------+----------+
| Variable_name       | Value         | Property |
+---------------------+---------------+----------+
| slave-delay-down    | 5.000000 (s)  | Dynamic  |
| slave-delay-recover | 1.000000 (s)  | Dynamic  |
+---------------------+---------------+----------+
2 rows in set (0.01 sec)
```

停止从库的 sql_thread：

```
mysql> stop slave sql_thread;
Query OK, 0 rows affected (0.01 sec)
```

等待 5s，观察 Cetus 日志，可以发现以下内容：

```
......
  (pid=32051) 2018-12-04 11:34:35: (message) change backend: 10.120.240.251:3307 from type:
readonly, state: down to type: readonly, state: down
  (pid=32051) 2018-12-04 11:34:35: (critical) Slave delay 5.118 seconds. Set slave to DOWN.
  (pid=32054) 2018-12-04 11:34:35: (message) change backend: 10.120.240.251:3307 from type:
readonly, state: down to type: readonly, state: down
  (pid=32054) 2018-12-04 11:34:35: (critical) Slave delay 5.117 seconds. Set slave to DOWN.
  (pid=32054) 2018-12-04 11:34:35: (critical) Slave delay 5.117 seconds. Set slave to DOWN.
  (pid=32053) 2018-12-04 11:34:35: (message) change backend: 10.120.240.251:3307 from type:
readonly, state: down to type: readonly, state: down
  (pid=32052) 2018-12-04 11:34:35: (message) change backend: 10.120.240.251:3307 from type:
readonly, state: down to type: readonly, state: down
  (pid=32053) 2018-12-04 11:34:35: (critical) Slave delay 5.117 seconds. Set slave to DOWN.
  (pid=32052) 2018-12-04 11:34:35: (critical) Slave delay 5.117 seconds. Set slave to DOWN.
......
```

此时 Cetus 的监控线程已经检测到其中一个从库（端口 3307）发生了复制延迟，时间超过设定阈值，该从库已经不可用。

恢复从库的 sql_thread 后，观察 Cetus 日志：

```
......
  (pid=32054) 2018-12-04 11:35:00: (message) change backend: 10.120.240.251:3307 from type:
readonly, state: online to type: readonly, state: online
  (pid=32052) 2018-12-04 11:35:00: (message) change backend: 10.120.240.251:3307 from type:
readonly, state: online to type: readonly, state: online
  (pid=32052) 2018-12-04 11:35:00: (message) Slave delay 0.112 seconds. Recovered. Set slave to UP.
  (pid=32054) 2018-12-04 11:35:00: (message) Slave delay 0.114 seconds. Recovered. Set slave to UP.
```

```
  (pid=32051) 2018-12-04 11:35:00: (message) change backend: 10.120.240.251:3307 from type:
readonly, state: online to type: readonly, state: online
  (pid=32053) 2018-12-04 11:35:00: (message) change backend: 10.120.240.251:3307 from type:
readonly, state: online to type: readonly, state: online
  (pid=32051) 2018-12-04 11:35:00: (message) Slave delay 0.113 seconds. Recovered. Set slave to UP.
  (pid=32053) 2018-12-04 11:35:00: (message) Slave delay 0.114 seconds. Recovered. Set slave to UP.
  (pid=32051) 2018-12-04 11:35:00: (message) Slave delay 0.113 seconds. Recovered. Set slave to UP.
  (pid=32053) 2018-12-04 11:35:00: (message) Slave delay 0.114 seconds. Recovered. Set slave to UP.
  ……
```

Cetus 的每一个 worker 进程都检测到从库已恢复复制进度，该从库已经可用，从而标记为 UP 状态。

以上仅是两个小例子，通过命令 select help/show variables 读者可以发现 Cetus 其实提供了很多功能，包括服务器黑/白名单（allow_ip/deny_ip）、释放连接（reduce_conns）、增加/减少后端配置（add master/slave）等，具体使用方法和功能描述可以通过 select help 帮助信息获取，也可以翻阅 GitHub 文档 https://github.com/Lede-Inc/cetus/blob/master/doc/cetus-configuration.md。

32.2.5　Cetus 的路由策略

Cetus 提供了读写分离功能（无论是读写分离版本还是 Sharding 版本），通过一个指定的端口，路由前端应用过来的读写请求，当然要做到这点，需要对 SQL 进行解析，用以判断哪些请求是需要发送到主库的，而哪些请求是路由到从库的，这点与 MySQL Router 不同，后者定义了两个不同的端口，用于提供写服务和读服务，用户需要自行选择。

Cetus 对 SQL 解析的同时，即完成了 SQL 的分类。通常情况下，写请求是发送到主库的，而读请求则会根据不同的策略有选择的路由到从库或者主库。

接下来将从两方面讨论 Cetus 对读请求的处理。

1. 从库间的读请求

默认情况下，读请求都会路由到从库（特殊情况详见下文），并以 Round-Robin 方式在不同从库间分配任务。但在分配任务过程中，有些 MySQL 数据库中间件是基于 SQL 的维度做负载均衡，即不会考虑多个 SQL 请求是来自于同一个连接还是不同的连接，中间件依次将接收到的多个 SQL 请求按照既定策略发往后端，如图 32-6 所示。

在图 32-6 中可以看到，后端 MySQL 集群包括 1 个主节点和 3 个从节点，前端应用发过来的 3 个 SQL 请求，其中两个来自于同一个连接，即连续发送了两次请求，另一个连接发送了一个 SQL 请求，当其经过中间件后，被分发到了 3 个从库，即每个后端一个 SQL 请求。这个策略如果遇到长连接的场景，会造成大量的连接切换，从而造成 session 级变量的频繁调整，影响 SQL 执行效率。

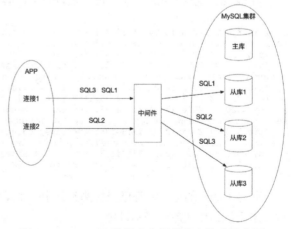

图 32-6　MySQL 数据库中间件路由策略示意图 1

Cetus 在遇到此种场景时，并非完全按照 SQL 的维度做负载均衡，而是考虑了同一个连接连续发送 SQL 请求的情况，即不会立即将当前使用完的连接放回连接池复用，而是持有短暂时间（256ms），以期后续仍有 SQL 执行，从而避免了 session 级变量的频繁调整，大大增

32.2 Cetus 架构

加了 SQL 执行的效率，如图 32-7 所示。

下面将演示一下前端读请求的路由策略（验证环境中，3306 端口为主库，3307 端口和 3308 端口为两个从库）。

首先，使用 mysql 客户端连接 Cetus，进行查询。

```
[root@localhost ~]# mysql -ucetus_app -p123456 -h127.0.0.1 -P4306
mysql> select * from read_demo;
+----+------+
| id | n1   |
+----+------+
|  1 |    1 |
|  2 |    2 |
+----+------+
2 rows in set (0.01 sec)
```

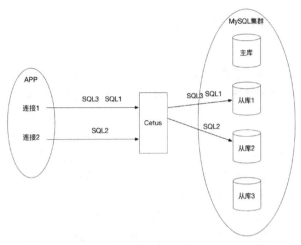

图 32-7 MySQL 数据库中间件路由策略示意图 2

从 Cetus 的全量日志中可以看到，读请求以 Round-Robin 方式，负载均衡的路由到了不同从库，示例中的两个从库分别为 3307 端口和 3308 端口。

```
 2018-12-04 09:25:44.915: #backend-rw# C_ip:172.17.0.1:59313 C_db:cetus C_usr:cetus_app C_tx:
false C_id:16777218 S_ip:127.0.0.1:3308 S_db:cetus S_usr:cetus_app S_id:2555 inj(type:3 bytes:16
rows:2) latency:0.484(ms) OK type:Query select * from read_demo
 2018-12-04 09:25:45.240: #backend-rw# C_ip:172.17.0.1:59313 C_db:cetus C_usr:cetus_app C_tx:
false C_id:16777218 S_ip:127.0.0.1:3307 S_db:cetus S_usr:cetus_app S_id:2558 inj(type:3 bytes:16
rows:2) latency:0.478(ms) OK type:Query select * from read_demo
```

接下来，再通过一段 Python 测试代码，来验证来自相同连接的多个 SQL 请求，是否可以复用同一个后端连接。

Python 验证代码如下：

```
[root@localhost ~]# cat r_cetus.py
#!/usr/bin/python
# -*- coding: utf-8 -*-
import MySQLdb as connector

conn = connector.connect(host="127.0.0.1", user="cetus_app", passwd="123456", port=4306, autocommit = True)
cursor = conn.cursor()
cursor.execute("select * from cetus.read_demo")
data = cursor.fetchone()
cursor.close()
cursor = conn.cursor()
cursor.execute("select * from cetus.read_demo")
data = cursor.fetchone()
```

在该段测试代码中，共包含两次读请求，且这两次都使用同一个 connection，但 cursor 还是单独创建的。

下面是部分全量日志信息：

```
 2018-12-04 10:21:56.937: #backend-rw# C_ip:172.17.0.1:60127 C_db: C_usr:cetus_app C_tx:false
C_id:33554436 S_ip:127.0.0.1:3308 S_db:cetus S_usr:cetus_app S_id:2575 inj(type:8 bytes:0 rows:0)
latency:0.603(ms) OK type:Query SET NAMES latin1
 2018-12-04 10:21:56.937: #backend-rw# C_ip:172.17.0.1:60127 C_db: C_usr:cetus_app C_tx:false C_
id:33554436 S_ip:127.0.0.1:3308 S_db:cetus S_usr:cetus_app S_id:2575 inj(type:3 bytes:16 rows:2)
latency:1.444(ms) OK type:Query select * from cetus.read_demo
 2018-12-04 10:21:56.939: #backend-rw# C_ip:172.17.0.1:60127 C_db: C_usr:cetus_app C_tx:false
C_id:33554436 S_ip:127.0.0.1:3308 S_db:cetus S_usr:cetus_app S_id:2575 inj(type:3 bytes:16 rows
:2) latency:0.796(ms) OK type:Query select * from cetus.read_demo
```

从全量日志中观察，可以发现这两次的读请求，都发到了 3308 端口的从库，并没有路由到其他从库，说明连接被复用。

2. 主从库间的读请求

Cetus 的读请求并不总是路由到从库的，有以下几种情况也会路由到主库。

- 事务中的读请求，即如果将查询语句嵌入到 start transaction 等控制事务开启的语句逻辑代码段中，该请求将被当作"写"请求，路由到主库。
- 开启了路由控制参数，例如 read-master-percentage 和 master-preferred，第一个参数控制主从库读请求分配比例，例如 read-master-percentage=50，表示主库和从库（所有从库）以 Round-Robin 方式分配路由；第二个参数如果开启，连接将全部路由至主库。
- 如果使用了注释功能，强制路由主库，例如在读请求中增加/*# mode=READWRITE*/，则该请求同样会发送到主库。

以上列出了一些特例，其中注释功能在前文也有过介绍，不再赘述，接下来分别就事务中的读请求和主库优先原则进行简单验证。

首先，开启参数 master-preferred，该参数为静态参数，修改后需要重启 Cetus。

```
mysql> show variables like 'master-preferred';
+------------------+-------+----------+
| Variable_name    | Value | Property |
+------------------+-------+----------+
| master-preferred | true  | Static   |
+------------------+-------+----------+
1 row in set (0.01 sec)
```

然后执行一些简单查询，Cetus 全量日志信息如下：

```
 2018-12-06 02:19:38.506: #backend-rw# C_ip:172.17.0.1:17698 C_db: C_usr:cetus_app C_tx:false
C_id:10 S_ip:127.0.0.1:3306 S_db:cetus S_usr:cetus_app S_id:328625 inj(type:3 bytes:33 rows:1)
latency:0.249(ms) OK type:Query select @@version_comment limit 1
 2018-12-06 02:19:38.506: #backend-rw# C_ip:172.17.0.1:17698 C_db: C_usr:cetus_app C_tx:false
C_id:10 S_ip:127.0.0.1:3306 S_db:cetus S_usr:cetus_app S_id:328625 inj(type:3 bytes:16 rows:2)
latency:0.301(ms) OK type:Query select * from read_demo
……
```

可以发现，所有的读请求全部路由到主库执行，这就是参数 master-preferred 的作用。

接下来，再来验证一下事务中的读请求。

```
mysql> select * from cetus.read_demo;
+----+----+
| id | n1 |
+----+----+
|  1 |  1 |
|  2 |  2 |
+----+----+
2 rows in set (0.00 sec)

mysql> start transaction;
Query OK, 0 rows affected (0.00 sec)

mysql> select * from cetus.read_demo;
+----+----+
| id | n1 |
+----+----+
|  1 |  1 |
|  2 |  2 |
+----+----+
2 rows in set (0.00 sec)
```

以下是 Cetus 的部分全量日志信息：

```
 2018-12-06 02:30:23.172: #backend-rw# C_ip:172.17.0.1:17783 C_db: C_usr:cetus_app C_tx:false
C_id:1 S_ip:127.0.0.1:3307 S_db:cetus S_usr:cetus_app S_id:3188 inj(type:3 bytes:16 rows:2) latency
:0.382(ms) OK type:Query select * from cetus.read_demo
 2018-12-06 02:30:36.529: #backend-rw# C_ip:172.17.0.1:17783 C_db: C_usr:cetus_app C_tx:false C_
id:1 S_ip:127.0.0.1:3306 S_db:cetus S_usr:cetus_app S_id:328633 inj(type:8 bytes:0 rows:0) latency
:1.133(ms) OK type:Query SET NAMES utf8
```

```
2018-12-06 02:30:36.529: #backend-rw# C_ip:172.17.0.1:17783 C_db: C_usr:cetus_app C_tx:false
C_id:1 S_ip:127.0.0.1:3306 S_db:cetus S_usr:cetus_app S_id:328633 inj(type:3 bytes:0 rows:0)
latency:1.475(ms) OK type:Query start transaction
2018-12-06 02:31:09.197: #backend-rw# C_ip:172.17.0.1:17783 C_db: C_usr:cetus_app C_tx:true
C_id:1 S_ip:127.0.0.1:3306 S_db:cetus S_usr:cetus_app S_id:328633 inj(type:3 bytes:16 rows:2)
latency:0.617(ms) OK type:Query select * from cetus.read_demo
```

可以看到，如果不开启事务，执行一个简单的查询，Cetus 会路由至其中一个从库；而如果处于事务中的查询，则会路由至主库。

32.2.6 常见问题

在日常使用 Cetus 的过程中，可能会遇到一些问题，接下来进行简单总结。虽然都是小问题，但如果稍不留意，就可能遗漏某些细节，导致问题的发生。

1. 从库延迟检测异常

该问题常见于初始化配置 Cetus，或者 Cetus 运行期间，后端 MySQL 环境发生了改变。

首先，回收监控 db（proxy_heart_beat）的授权，如下所示。

```
bash-4.2# mysql -uroot -h127.0.0.1 -P3306
mysql> revoke all on proxy_heart_beat.* from cetus_app@'%';
Query OK, 0 rows affected (0.02 sec)
```

可以看到在 Cetus 的日志中，开始报出 critical 类型的错误，从错误提示信息，得知 Cetus 在操作延迟监控表 tb_heartbeat 时，执行被拒绝，导致监控任务失败。

```
  (pid=2498) 2018-12-06 03:44:57: (message) Update heartbeat error: 1142, text: INSERT, UPDATE
command denied to user 'cetus_app'@'localhost' for table 'tb_heartbeat', backend: 127.0.0.1:3306
  (pid=2498) 2018-12-06 03:44:57: (message) monitor thread connected to backend: 127.0.0.1:3307,
cached 2 conns
  (pid=2498) 2018-12-06 03:44:57: (critical) Select heartbeat error: 1142, text: SELECT command
denied to user 'cetus_app'@'localhost' for table 'tb_heartbeat', backend: 127.0.0.1:3307
  (pid=2498) 2018-12-06 03:44:57: (message) monitor thread connected to backend: 127.0.0.1:3308,
cached 3 conns
  (pid=2498) 2018-12-06 03:44:57: (critical) Select heartbeat error: 1142, text: SELECT command
denied to user 'cetus_app'@'localhost' for table 'tb_heartbeat', backend: 127.0.0.1:3308
```

接下来，将其中一个从库的 slave 进程停止。

```
[root@localhost~]# mysql -uroot -h127.0.0.1 -P3307
mysql> stop slave;
Query OK, 0 rows affected (0.06 sec)
```

通过 Cetus 的后端管理，查看 MySQL 后端状态，可以发现从库的状态是 unknown（或者up），而不是 down（开启 slave 延迟检测功能后，如果 slave 延迟超过阈值，会将其状态调整为 down），此时执行数据查询请求，仍然会路由到 3307 端口的从库。

```
[root@localhost ~]# mysql -uadmin -padmin -h127.0.0.1 -P5306
mysql> select * from backends\G;
*************************** 1. row ***************************
            PID: 2498
    backend_ndx: 1
        address: 127.0.0.1:3306
          state: up
           type: rw
 slave delay(ms): NULL
      idle_conns: 20
      used_conns: 0
     total_conns: 20
*************************** 2. row ***************************
            PID: 2498
    backend_ndx: 2
        address: 127.0.0.1:3307
          state: unknown
           type: ro
 slave delay(ms): 0
```

```
        idle_conns: 20
        used_conns: 0
       total_conns: 20
*************************** 3. row ***************************
               PID: 2498
       backend_ndx: 3
           address: 127.0.0.1:3308
             state: unknown
              type: ro
    slave delay(ms): 0
        idle_conns: 20
        used_conns: 0
       total_conns: 20
3 rows in set (0.00 sec)
```

通过以上的故障模拟，可以得知，如果不能正常检测从库的延迟状况，而又路由到该节点去执行一些查询任务，此时返回的结果集是错误的。解决方法相信读者已经完全清楚了，这里只模拟了没有权限，实际工作中，还有可能会缺少监控 db/监控表或者没有在 users.json 配置文件中配置参数 default-username 的用户连接信息等，总而言之，解决思路是一致的，一定要确认好相关权限是否进行了正确配置。

2．无可用连接

当前端收到 service unavailable 错误时，表明 Cetus 服务不可用，当然导致该问题的发生，最可能的原因就是连接池中的可用连接已经不能够满足当前连接需求，此时可能会在 Cetus 日志中，发现以下信息：

```
......
 (pid=3053) 2018-12-06 06:46:20: (message) src/network-mysqld.c:4973: create cetus_app connection for backend ndx:0, ptr:0x1828440
 (pid=3053) 2018-12-06 06:46:20: (message) src/network-mysqld.c:4998: connected_clients add, backend ndx:0, for server:0x1866fe0, faked con:0x18672a0
 (pid=3053) 2018-12-06 06:46:20: (warning) src/network-mysqld.c:4721:READ_AUTH_RESULT:-1, server:cetus_app, errinfo:Too many connections
 (pid=3053) 2018-12-06 06:46:20: (warning) src/network-mysqld.c:4774: con:0x18672a0 failed for server:0x1866fe0
 (pid=3053) 2018-12-06 06:46:20: (critical) src/network-mysqld.c:2508: wait failed and no server backend for user:cetus_app, ret:4
 (pid=3053) 2018-12-06 06:46:20: (message) src/network-mysqld.c:2348: service unavailable for con:0x1867640
......
```

从日志中可以看到一些 too many connections 的信息，由于 Cetus 配置了多进程，例如进程数等于 4，那么每个 MySQL 后端的默认连接数就是 4* default-pool-size（假设 100），即需要初始 400 个连接，如果仍然不能满足前端应用消费，Cetus 会继续创建新连接，一旦超出后端 MySQL 的 max_connections，则会出现上述问题。

所以在配置 Cetus 连接数量时，应考虑后端 MySQL 最大连接数，以及 Cetus 的进程个数和前端应用所需的连接数量，应符合如下约束：

前端应用需要的连接数量 ＜ Cetus 进程数量*max-pool-size ＜ MySQL 最大连接数

32.3　小结

本章介绍了 MySQL 中间件 MySQL Router 以及 Cetus 的概念和架构，并通过几个实例详细讨论了各自的安装配置和使用方法等内容。由于这两个中间件也在不断地更迭和完善，有些内容文中并没有提及，感兴趣的读者可以到官网或者 GitHub 去发现更多的功能，期待能有更大的收获！